VOLUME SIXTY TWO

ADVANCES IN ECOLOGICAL RESEARCH

Tropical Ecosystems in the 21st Century

ADVANCES IN ECOLOGICAL RESEARCH

Series Editors

DAVID A. BOHAN
Directeur de Recherche
UMR Agroécologie, AgroSup Dijon
INRA, University of Bourgogne Franche-Comté
Dijon, France

ALEX J. DUMBRELL
School of Life Sciences
University of Essex
Wivenhoe Park, Colchester
Essex, United Kingdom

VOLUME SIXTY TWO

ADVANCES IN ECOLOGICAL RESEARCH

Tropical Ecosystems in the 21st Century

Edited by

ALEX J. DUMBRELL
School of Life Sciences
University of Essex
Wivenhoe Park, Colchester
Essex, United Kingdom

EDGAR C. TURNER
Department of Zoology, University of Cambridge
Cambridge, United Kingdom

TOM M. FAYLE
Institute of Entomology, Biology Centre, Czech Academy of Sciences, Ceske Budejovice, Czech Republic; Institute of Tropical Biology and Conservation, Universiti Malaysia Sabah, Kota Kinabalu, Sabah, Malaysia

ELSEVIER

ACADEMIC PRESS
An imprint of Elsevier

Academic Press is an imprint of Elsevier
125 London Wall, London, EC2Y 5AS, United Kingdom
The Boulevard, Langford Lane, Kidlington, Oxford OX5 1GB, United Kingdom
525 B Street, Suite 1650, San Diego, CA 92101, United States
50 Hampshire Street, 5th Floor, Cambridge, MA 02139, United States

First edition 2020

Notices
Knowledge and best practice in this field are constantly changing. As new research and experience broaden our understanding, changes in research methods, professional practices, or medical treatment may become necessary.

Practitioners and researchers must always rely on their own experience and knowledge in evaluating and using any information, methods, compounds, or experiments described herein. In using such information or methods they should be mindful of their own safety and the safety of others, including parties for whom they have a professional responsibility.

To the fullest extent of the law, neither the Publisher nor the authors, contributors, or editors, assume any liability for any injury and/or damage to persons or property as a matter of products liability, negligence or otherwise, or from any use or operation of any methods, products, instructions, or ideas contained in the material herein.

ISBN: 978-0-12-821134-2
ISSN: 0065-2504

For information on all Academic Press publications
visit our website at https://www.elsevier.com/books-and-journals

Publisher: Zoe Kruze
Acquisitions Editor: Jason Mitchell
Editorial Project Manager: Joanna Collett
Production Project Manager: Abdulla Sait
Cover Designer: Alan Studholme

Typeset by SPi Global, India

Contents

Contributors

Oscar J. Abelleira Martínez
Department of Agroenvironmental Sciences, University of Puerto Rico, Mayagüez, Puerto Rico

Aura M. Alonso-Rodríguez
U.S.D.A. Forest Service, International Institute of Tropical Forestry, Río Piedras, PR, United States

Oliver S. Ashford
Integrative Oceanography Division, Scripps Institution of Oceanography, University of California, San Diego, CA, United States

Gerardo Aymard
UNELLEZ-Guanare, Programa de Ciencias del Agro del Mar, Herbario Universitario (PORT), Mesa de Cavacas, estado Portuguesa, Venezuela

Alessia Bani
School of Life Sciences, University of Essex, Essex, United Kingdom

Yves Basset
Institute of Entomology, Biology Centre, Czech Academy of Sciences; Faculty of Sciences, University of South Bohemia, Ceske Budejovice, Czech Republic; ForestGEO, Smithsonian Tropical Research Institute, Balboa, Ancon; Maestria de Entomologia, Universidad de Panama, Panama, Panama

Maria Beger
School of Biology, Faculty of Biological Sciences, University of Leeds, Leeds, United Kingdom; Centre for Biodiversity and Conservation Science, School of Biological Sciences, University of Queensland, Brisbane, QLD, Australia

Megan E. Berberich
U.S.D.A. Forest Service, International Institute of Tropical Forestry, Río Piedras, PR, United States

James D. Bever
Department of Ecology and Evolutionary Biology and Kansas Biological Survey, University of Kansas, Lawrence, KS, United States

Christopher G. Bousfield
Department of Animal and Plant Science, University of Sheffield, Sheffield, United Kingdom

Laëtitia M. Bréchet
Department of Biology, Centre of Excellence PLECO (Plant and Ecosystems), University of Antwerp, Wilrijk, Belgium; INRA, UMR Ecology of Guiana Forests, AgroParisTech, Cirad, CNRS, Université des Antilles, Université de Guyane, Kourou, French Guiana

Sébastien Brosse
Evolution et Diversité Biologique (EDB), Université Toulouse 3 Paul Sabatier, CNRS, IRD, UMR 5174, Toulouse, France

Biancolini Castro
Smithsonian Tropical Research Institute, Panama, Republic of Panama

Molly A. Cavaleri
College of Forest Resources and Environmental Science, Michigan Technological University, Houghton, MI, United States

Gianluca R. Cerullo
Department of Zoology, University of Cambridge, Cambridge, United Kingdom

Bin Chen
Department of Land, Air and Water Resources, University of California, Davis, CA, United States

Simon Creer
School of Natural Sciences, Bangor University, Bangor, Gwynedd, United Kingdom

Maarten De Brauwer
School of Biology, Faculty of Biological Sciences, University of Leeds, Leeds, United Kingdom

Julian Donald
Evolution et Diversité Biologique (EDB), Université Toulouse 3 Paul Sabatier, CNRS, IRD, UMR 5174, Toulouse, France

Alex J. Dumbrell
School of Life Sciences, University of Essex, Essex, United Kingdom

David P. Edwards
Department of Animal and Plant Science, University of Sheffield, Sheffield, United Kingdom

Tom M. Fayle
Institute of Entomology, Biology Centre, Czech Academy of Sciences, Ceske Budejovice, Czech Republic; Institute of Tropical Biology and Conservation, Universiti Malaysia Sabah, Kota Kinabalu, Sabah, Malaysia

Shyamolina Ghosh
Department of Ecology and Evolutionary Biology and Kansas Biological Survey, University of Kansas, Lawrence, KS, United States

Mailyn Adriana Gonzalez
Programa Ciencias de la Biodiversidad, Instituto de Investigación de Recursos Biológicos Alexander von Humboldt. Bogotá, Colombia

Tamara Heartsill Scalley
International Institute of Tropical Forestry, USDA Forest Service, Río Piedras, Puerto Rico

Andrew Hector
Department of Plant Sciences, University of Oxford, Oxford, United Kingdom

Mark T. Holder
Department of Ecology and Evolutionary Biology and Biodiversity Institute, University of Kansas, Lawrence, KS, United States

Amaia Iribar
Evolution et Diversité Biologique (EDB), Université Toulouse 3 Paul Sabatier, CNRS, IRD, UMR 5174, Toulouse, France

Robert B. Jackson
Department of Earth System Science; Woods Institute for the Environment; Precourt Institute for Energy, Stanford University, Stanford, CA, United States

Jamaluddin Jompa
Graduate School, Hasanuddin University, Makassar, Indonesia

Deirdre Kerdraon-Byrne
Lancaster Environment Centre, Lancaster University, Lancaster, United Kingdom

Benita C. Laird-Hopkins
Institute of Entomology, Biology Centre, Czech Academy of Sciences; Faculty of Sciences, University of South Bohemia, Ceske Budejovice, Czech Republic

Greg P.A. Lamarre
Institute of Entomology, Biology Centre, Czech Academy of Sciences; Faculty of Sciences, University of South Bohemia, Ceske Budejovice, Czech Republic; ForestGEO, Smithsonian Tropical Research Institute, Balboa, Ancon, Panama

Inara R. Leal
Departamento de Botânica, Universidade Federal de Pernambuco, Recife, Pernambuco, Brazil

Céline Leroy
AMAP, Univ Montpellier, CIRAD, CNRS, INRAE, IRD, Montpellier; INRAE, CNRS, AgroParisTech, CIRAD, Université de Guyane, Université des Antilles, UMR Ecologie des Forêts de Guyane (EcoFoG), Campus agronomique, Kourou, France

Shunlin Liang
Department of Geographical Sciences, University of Maryland College Park, College Park, MD, United States

Gino Limmon
Maritime and Marine Science Center of Excellence, Pattimura University, Ambon, Indonesia

Terrance D. Loecke
Environmental Studies Program and Kansas Biological Survey, University of Kansas, Lawrence, KS, United States

Carla Martins Lopes
Departamento de Zoologia, Instituto de Biociências, Universidade Estadual Paulista (UNESP), Rio Claro, SP, Brazil

Luis Lopez-Sangil
Teagasc, Johnstown Castle, Wexford, Ireland

Ariel E. Lugo
International Institute of Tropical Forestry, USDA Forest Service, Río Piedras, Puerto Rico

Mike R. Massam
Department of Animal and Plant Science, University of Sheffield, Sheffield, United Kingdom

Ernesto Medina
International Institute of Tropical Forestry, USDA Forest Service, Río Piedras, Puerto Rico; Centro de Ecología, Instituto Venezolano de Investigaciones Científicas, Caracas, Venezuela

Jérôme Murienne
Evolution et Diversité Biologique (EDB), Université Toulouse 3 Paul Sabatier, CNRS, IRD, UMR 5174, Toulouse, France

Akihiro Nakamura
CAS Key Laboratory of Tropical Forest Ecology, Chinese Academy of Sciences, Yunnan, China

Michael J. O'Brien
Southeast Asia Rainforest Research Partnership (SEARRP), Kota Kinabalu, Malaysia; Área de Biodiversidad y Conservación, Universidad Rey Juan Carlos, Móstoles, Spain

Jérôme Orivel
INRAE, CNRS, AgroParisTech, CIRAD, Université de Guyane, Université des Antilles, UMR Ecologie des Forêts de Guyane (EcoFoG), Campus agronomique, Kourou, France

Adam F.A. Pellegrini
Department of Earth System Science, Stanford University, Stanford, CA, United States; Department of Plant Sciences, University of Cambridge, Cambridge, United Kingdom

Carlos A. Peres
School of Environmental Sciences, University of East Anglia, Norwich Research Park, Norwich, United Kingdom; Departamento de Sistemática e Ecologia, Universidade Federal da Paraíba, João Pessoa, Paraíba, Brazil

Bruno X. Pinho
Departamento de Botânica, Universidade Federal de Pernambuco, Recife, Pernambuco, Brazil

Sasha C. Reed
U.S. Geological Survey, Southwest Biological Science Center, Moab, UT, United States

Robin Reibold
U.S. Geological Survey, Southwest Biological Science Center, Moab, UT, United States

Philip C. Reid
School of Biological & Marine Sciences, University of Plymouth, Drake Circus; Continuous Plankton Recorder Survey, The Marine Biological Association, The Laboratory, Citadel Hill, Plymouth, United Kingdom

Daniel C. Reuman
Department of Ecology and Evolutionary Biology and Kansas Biological Survey, University of Kansas, Lawrence, KS; Laboratory of Populations, Rockefeller University, New York, NY, United States

Glen Reynolds
Southeast Asia Rainforest Research Partnership (SEARRP), Kota Kinabalu, Malaysia

Chadtip Rodtassana
Department of Botany, Faculty of Science, Chulalongkorn University, Bangkok, Thailand

Emma J. Sayer
Lancaster Environment Centre, Lancaster University, Lancaster, United Kingdom; Smithsonian Tropical Research Institute, Panama, Republic of Panama

Heidy Schimann
INRAE, CNRS, AgroParisTech, CIRAD, Université de Guyane, Université des Antilles, UMR Ecologie des Forêts de Guyane (EcoFoG), Campus agronomique, Kourou, France

Simon T. Segar
Department of Crop and Environment Sciences, Harper Adams University, Newport, United Kingdom

Merlin Sheldrake
Smithsonian Tropical Research Institute, Panama, Republic of Panama; Department of Plant Sciences, University of Cambridge, Cambridge, United Kingdom

Lawrence W. Sheppard
Department of Ecology and Evolutionary Biology and Kansas Biological Survey, University of Kansas, Lawrence, KS, United States

Daniel Souto-Vilarós
Institute of Entomology, Biology Centre, Czech Academy of Sciences, Ceske Budejovice; Department of Ecology, Faculty of Science, Charles University Prague, Staré Město, Czech Republic

Marcelo Tabarelli
Departamento de Botânica, Universidade Federal de Pernambuco, Recife, Pernambuco, Brazil

Pierre Taberlet
Laboratoire d'Ecologie Alpine (LECA), CNRS, Université Grenoble Alpes, Grenoble, France; UiT—The Arctic University of Norway, Tromsø Museum, Tromsø, Norway

Edmund V.J. Tanner
Department of Plant Sciences, University of Cambridge, Cambridge, United Kingdom

Benjamin L. Turner
Smithsonian Tropical Research Institute, Panama, Republic of Panama

Sophie von der Heyden
Evolutionary Genomics Group, Department of Botany and Zoology, University of Stellenbosch, Stellenbosch, South Africa

Shuntaro Watanabe
Field Science Education and Research Center, Kyoto University, Kyoto; Faculty of Sciences, Kagoshima University, Kagoshima, Japan

Tana E. Wood
U.S.D.A. Forest Service, International Institute of Tropical Forestry, Río Piedras, PR, United States

S. Joseph Wright
Smithsonian Tropical Research Institute, Panama, Republic of Panama

Jinhui Wu
State Key Laboratory of Remote Sensing Science, Faculty of Geographical Science, Beijing Normal University, Beijing, China; Department of Plant Sciences, University of Oxford, Oxford, United Kingdom

Jun Xie
School of Environmental Sciences, University of Hull, Hull, United Kingdom

Lucie Zinger
Ecole Normale Supérieure, PSL Research University, CNRS, Inserm, Institut de Biologie de l'Ecole Normale Supérieure (IBENS), Paris, France

Preface

1. Introduction

Tropical ecosystems are home to a disproportionate number of the world's species (Gaston, 2000), but are under increasing pressure from anthropogenic impacts (Pimm et al., 2014). These include overhunting in both terrestrial and marine ecosystems (McCauley et al., 2015; Ripple et al., 2019), habitat degradation, fragmentation and conversion to agriculture (Gibbs et al., 2010; Phalan et al., 2013), transmission of invasive species worldwide (Seebens et al., 2018), and ongoing climate change (Deutsch et al., 2008). These species losses have negative impacts on ecosystem functioning and provision of ecosystem services. Most importantly, many of these factors are taking place concurrently and interact, resulting in synergistic impacts on biodiversity and ecosystem processes (Brook et al., 2008). In this context, it is not surprising that levels of biodiversity loss, although uncertain, are far higher than background rates and are accelerating (Ceballos et al., 2015; Pimm et al., 2014). With an increasing global population and rising per-capita caloric and protein demands (Tilman et al., 2011a), there has never been more pressure on tropical ecosystems. At the same time, there is an increasing body of information demonstrating the important role that biodiversity can play in supporting ecosystem services, including food production (e.g. Garibaldi et al., 2011; Tscharntke et al., 2005). Reconciling these pressures with biodiversity conservation and the maintenance of functioning productive tropical landscapes, represents perhaps the biggest challenge facing the world today.

The last year has seen worrying reports of continued losses and change, with deforestation and associated fire increasing in the Brazilian Amazon (Escobar, 2019), and long-term changes being reported in even small taxa, such as insects (Wagner, 2020), which because of their short generation times and generally large population sizes, might be expected to be more resilient than other groups. On the other-hand, there has been a recent increase in public interest in human impacts on the natural world, particularly among younger age groups (Gardner and Wordley, 2019). From a scientific perspective, recent advances in technology, particularly remote sensing (Turner et al., 2003) and molecular techniques (Anderson and Cairney, 2004), are enabling human impacts on the natural world to be studied over larger areas and at finer scales than ever before, bringing a heightened level of understanding that has not been possible in the past.

It is clear that the natural world and tropical ecosystems in particular, have never been more at risk. However, the opportunities for understanding changes and hence informing conservation and more-sustainable management of tropical landscapes are equally unprecedented. Now is an opportune time to explore the current state of knowledge in tropical ecosystems at a global scale and to predict future change. With this aim, this special edition brings together 11 publications that investigate drivers behind changes in the tropics, both from a whole-ecosystem perspective (e.g. Lugo et al., 2020; Pellegrini and Jackson, 2020) and with regard to particular focal taxa (e.g. Lamarre et al., 2020; Pinho et al., 2020). These publications explore how different drivers can interact, the differences between one-off and longer-term perturbations, and the impacts of interactions between components of tropical systems that influence long-term effects. We include several papers discussing large-scale experiments, where biodiversity (Wu et al., 2020), litter dynamics (Sayer et al., 2020) and climate (Reed et al., 2020) have been manipulated, to understand the impacts of changes on whole ecosystems and to forecast future change. We also include papers that highlight the importance of long-term and large-scale comparative studies (Lamarre et al., 2020; Pellegrini and Jackson, 2020) for understanding species decline and the role of modern molecular methods (Bani et al., 2020; Zinger et al., 2020) and statistical techniques (Ghosh et al., 2020) in cataloguing change and informing conservation action.

2. Papers in this edition

Commonly the first stage in habitat alteration of tropical forests involves selective logging and timber removal, with logged and disturbed forests now representing the commonest forest type in parts of the tropics (e.g. Edwards et al., 2011). Increasingly, it is also this habitat that is converted to agriculture (Gibbs et al., 2010), making an assessment of the biodiversity and ecosystem service value of logged forests particularly timely. In Chapter 1, Bousfield et al. (2020) review the known environmental and socioeconomic value of logged forests. The authors describe how logging alters forest structure, and hence community composition for a wide range of animals and plants. However, they find that logging has more limited effects on species richness, especially compared with conversion to agriculture.

This value of logged forests is also reflected in the maintenance of important ecosystem functions, such as carbon storage, climate regulation and, to some extent, watershed maintenance. Logging can reduce the socioeconomic value of forests for local communities, but, if conducted in a socially responsible manner, can also result in benefits. In particular, strategies such as reduced impact logging and better spatial management of harvesting and road networks can reduce negative impacts on biodiversity and ecosystems services for local communities, while tailored management can also improve recovery of forests.

In some cases, disturbed forests can also develop novel communities of plants, with their own characteristics and functions. Such novel ecosystems are a key feature of the Anthropocene and can represent valuable habitats in their own right (Lindenmayer et al., 2008). In Chapter 2, Lugo et al. (2020) take a long-term perspective and review information on novel tropical forest ecosystems. They introduce case studies from an island (Puerto Rico), a montane continental region (the Andes), and the lowland continental tropics (Venezuela). The forests in all three areas show novel species composition, with a high dominance of certain species and differences in functional traits from the forests historically found in the area. These novel ecosystems also differ in functioning, with generally higher growth rates, and more-rapid turnover of wood and leaf litter nutrients.

Ultimately the functioning of communities can be heavily influenced by interactions between species, as well as the diversity of the system itself (e.g. Tilman et al., 2011b). In order to understand such effects in tropical forests, large-scale experiments can be particularly important, as they enable driving factors to be identified at a scale that is appropriate for high-complexity systems (Fayle et al., 2015). In Chapter 3, Wu et al. (2020) present findings from the large-scale Sabah Biodiversity Experiment and investigate the effects of species diversity on forest structure. This experiment was established in 2001, and comprises 124 four-hectare plots, each containing different tree diversity treatments (unplanted, monocultures, 4 and 16 species mixtures). Using high spatial resolution satellite imagery and advanced remote sensing, Wu et al. (2020) find higher vegetation cover in mixed-species treatments than monocultures and greater vegetation cover in enrichment planted plots compared to nonenrichment plots. This is one of the first experimental demonstrations of biodiversity effects (i.e. overyielding) in a tropical forest system and backs up classic and contemporary studies from grassland ecosystems (Eisenhauer et al., 2019; Tilman et al., 2011b). Such a result is particularly critical, given current levels of biodiversity loss.

Fire is still a common method used to clear forest vegetation for agriculture in large parts of the tropics, despite the negative impacts that such practices can have on biodiversity, soil condition and carbon storage (Cochrane, 2001). Negative impacts can also be exacerbated when fire is employed in low-rainfall conditions, which may become more frequent in some parts of the tropics as a result of climate change (Timmermann et al., 2002). The majority of studies on fire have considered single-burn events, while human action and positive feedback effects through increased tree mortality and standing deadwood, can mean that these single-burns are followed by long-term differences in the fire regime (Cochrane et al., 1999). In Chapter 4, Pellegrini and Jackson (2020) combine a literature review and metaanalysis to investigate the impact of single-fire events (pulse) vs sustained changes (press) in a fire regime. They find that levels and availability of different nutrients respond differently to single or repeated burns, depending on how their availability is regulated in the system. Similarly, they show that microbial biomass and decomposition rates are likely to shift with the fire regime, particularly as a result of heat-induced mortality and declines in organic material. Taken together, these findings have important implications within the tropics, where shifts in fire regimes, often associated with weather patterns and land-use change, are accelerating (Cochrane et al., 1999). The study also highlights how sustained change can have differing impacts on an ecosystem than a single perturbation event.

Within tropical ecosystems, there can also be complex interactions between biotic and abiotic components of the system and between aboveground and belowground parts of the system (Richardson et al., 2010), making it difficult to determine which factors are responsible for supporting functions such as primary productivity. Large-scale experiments can again be important in teasing apart these effects. In Chapter 5, Sayer et al. (2020) present findings from a large-scale, long-term litter manipulation project—the Gigante Litter Manipulation Project. Based in Panama, this project has manipulated leaf litter levels (in 15 large litter removal, addition and control plots) over the last 15 years, to experimentally test the long-held assumption that nutrient recycling from high litterfall in tropical forests is essential for maintaining primary productivity. The study demonstrates clear effects of changing litter levels on some soil nutrients (particularly nitrogen, calcium and magnesium) but more variable effects on others. The study also demonstrates a variety of knock-on effects to the rest of the ecosystem. For example, decreased availability of nitrogen in litter removal soils is likely to have stimulated microbial degradation of soil organic matter, resulting in reduced soil organic carbon concentrations and microbial biomass. This work again demonstrates the value of experiments to tease apart complex links within structurally and biologically complex tropical systems (Fayle et al., 2015).

Processes operating in tropical regions can be heavily influenced by changing weather patterns and climate change (e.g. Eycott et al., 2019). Once again, manipulative experiments can be key to understanding the complex nature of these effects. In Chapter 6, Reed et al. (2020) present findings from a recently established experiment based in Puerto Rico, the Tropical Responses to Altered Climate Experiment, that is experimentally warming a tropical forest for the first time. The experiment consists of six 4-m diameter plots, half of which are heated with infrared heaters. The study also takes advantage of two hurricanes that occurred during the experimental warming to investigate the synergistic effects of large-scale disturbance and warming. The study finds that experimental warming affected carbon, nitrogen, and phosphorus cycling, and increased microbial biomass, that hurricane damage affected a wide range of soil biogeochemical measurements, and that there was an interaction between the effects of warming and hurricane damage. These results indicate that biogeochemical cycling and, by association, many other processes within tropical systems will likely change in the future with global warming and changes in the frequency of extreme weather events.

It is, therefore, clear that the biodiversity of tropical systems and complex interactions between components of these ecosystems can have marked effects on ecosystem functioning. In Chapter 7, Pinho et al. (2020) review the importance of a specific aspect of tropical systems—very large tropical forest trees (mega-trees), to assess their value for supporting biodiversity and ecosystem services. In particular, they find that tropical mega-trees represent a key habitat for a diverse suite of organisms from all trophic levels. They provide nonredundant niche-space, buffer microclimate, have a disproportionate role in carbon sequestration and aboveground biomass, and provide important ecosystem services for people. However, because of their size and relative rarity, such trees are disproportionately affected by habitat change and conversion, threatening biodiversity and associated ecosystem services. Indeed, it is the large and relatively long-lived trees that Lugo et al. (2020) highlight as being missing from novel forest ecosystems. Such an apparently important role of a small subset of the total community that is particularly vulnerable to anthropogenic impacts is reminiscent of the loss of ecologically important megafauna globally (Malhi et al., 2016), and highlights the importance of targeted conservation action.

While the vulnerability of large, relatively scarce and long-lived organisms is fairly well established (e.g. Ripple et al., 2019), in Chapter 8 Lamarre et al. (2020) investigate the current state of knowledge of declines in a numerous and small-bodied group: the insects. The paper highlights the recent increase in studies demonstrating declines in insects worldwide, with links to multiple interrelated drivers, including habitat and climate change. They also highlight the current lack of knowledge of the state of insect communities in the tropics, despite the fact that these areas contain a far higher diversity of species than better-studied temperate regions (Gaston, 2000). Intriguingly, the long-term tropical studies that exist seem to indicate that patterns of insect change in the tropics are less-clear than in temperate regions. Generally, studies show large fluctuations in insect abundance over short time-scales linked to stochastic events, but no consistent long-term declines. The large fluctuations in numbers shown by the insects, demonstrates the clear need for more long-term monitoring in the tropics. The authors go on to introduce the ForestGEO arthropods initiative, which is based within seven of the Forest Global Earth Observatory (ForestGEO) plots across the tropics, and is already gathering data on long-term tropical insect change. A key issue with studying tropical insects is their extreme diversity and lack of information on their ecology, making it hard to draw general conclusions about the ecosystem impacts of any change. To address this, the paper highlights the importance of focussing on aspects of the community such as functional and physiological traits, to allow inferences to be drawn about the effects of change on the wider ecosystem. The study also highlights the importance of molecular approaches, such as DNA metabarcoding, and machine learning to increase the speed of data-gathering and processing, and the scope of data available.

Indeed, it is clear that modern molecular and computational techniques are leading to a phase-shift in understanding of tropical ecosystems, providing information that is likely to be key to contemporary conservation efforts. In Chapters 9 and 10, Zinger et al. (2020) and Bani et al. (2020) discuss the application of environmental DNA (eDNA) based approaches in different tropical ecosystems and how these may be used to inform conservation research. Zinger et al. (2020) focus their study on tropical rainforests (particularly neotropical forests) to explore the current application of different eDNA methods in addressing ecologically relevant research questions, as well as challenges associated with both using and improving these methods. They also discuss potential applications of eDNA-based approaches to future tropical forest research. Bani et al. (2020) provide a complementary examination of eDNA-based methods, but targeted on specific conservation actions associated with spatial planning decisions within marine ecosystems. They highlight the important role that eDNA-based approaches are likely to have in the future in informing spatial planning for conservation and global biomonitoring initiatives.

Given the wealth of large datasets, increased monitoring efforts in the tropics, and modern techniques generating unprecedented amounts of data, statistical techniques to study tropical ecosystems also need to become more sophisticated to take full advantage of this level of information and reliably quantify linkages between different components of an ecosystem. In Chapter 11, Ghosh et al. (2020) provide a new statistical toolbox for 21st century ecology. The authors propose ecologists move beyond standard correlation and regression-based approaches when analysing links between biological and environmental variables, to the application of 'copula'. Like regressions, this approach describes the association between two or more variables, but separates information on marginal distributions and remaining information related to the association. Copulas are regularly applied within other quantitative disciplines (e.g. finance), but have yet to be applied in ecology, although they represent a promising avenue for developing a more-sophisticated understanding of associations between ecological variables.

3. Conclusion

The chapters included in this volume of *Advances in Ecological Research* provide an overview of the state of knowledge of tropical ecosystems, current threats, and likely future change. It is clear that tropical ecosystems are experiencing a range of interacting threats, but that current understanding of these systems is still lacking, especially when compared to temperate regions. Several papers in this edition emphasize that current knowledge of human impacts on the environment is disproportionately informed by temperate studies (e.g. Lamarre et al., 2020; Pellegrini and Jackson, 2020), where ecosystem processes may differ from the tropics, making it difficult to draw reliable inferences on the long-term impacts of environmental change in the tropics. However, comparative studies drawing information from across regions, and long-term whole-ecosystem manipulative studies are yielding important novel information to fill this gap. Added to this, modern molecular approaches are increasing the scope of options for monitoring high-diversity tropical ecosystems on a scale that has not been possible before.

The next big challenge in tropical ecology is to bring this information together to inform and test conservation action, reverse tropical biodiversity loss, and support healthy tropical ecosystems into the future.

Alex J. Dumbrell
School of Life Sciences University of Essex, Wivenhoe Park,
Colchester Essex, United Kingdom
Edgar C. Turner
Department of Zoology, University of Cambridge,
Cambridge, United Kingdom
Tom M. Fayle
Institute of Entomology, Biology Centre, Czech Academy of Sciences,
Ceske Budejovice, Czech Republic; Institute of Tropical
Biology and Conservation, Universiti Malaysia Sabah,
Kota Kinabalu, Sabah, Malaysia

References

Anderson, I.C., Cairney, J.W.G., 2004. Diversity and ecology of soil fungal communities: increased understanding through the application of molecular techniques. Environ. Microbiol. 6 (8), 769–779. https://doi.org/10.1111/j.1462-2920.2004.00675.x.

Bani, A., De Brauwer, M., Creer, S., Dumbrell, A.J., Limmon, G., Jompa, J., … Beger, M., 2020. Informing marine spatial planning decisions with environmental DNA. In: Dumbrell, A.J., Turner, E.C., Fayle, T.M. (Eds.), Advances in Ecological Research. vol. 62. Elsevier, pp. 375–407.

Bousfield, C.G., Cerullo, G.R., Massam, M.R., Edwards, D.P., 2020. Protecting environmental and socio-economic values of selectively logged tropical forests in the Anthropocene. In: Dumbrell, A.J., Turner, E.C., Fayle, T.M. (Eds.), Advances in Ecological Research. vol. 62. Elsevier, pp. 1–52.

Brook, B.W., Sodhi, N.S., Bradshaw, C.J.a., 2008. Synergies among extinction drivers under global change. Trends Ecol. Evol. 23 (8), 453–460. https://doi.org/10.1016/j.tree.2008.03.011.

Ceballos, G., Ehrlich, P.R., Barnosky, A.D., García, A., Pringle, R.M., Palmer, T.M., 2015. Accelerated modern human-induced species losses: entering the sixth mass extinction. Sci. Adv. 1 (5), 9–13. https://doi.org/10.1126/sciadv.1400253.

Cochrane, M.A., 2001. In the line of fire understanding the impacts of tropical forest fires. Environ. Sci. Policy Sustain. Dev. 43 (8), 28–38. https://doi.org/10.1080/00139150109604505.

Cochrane, M., Alencar, A., Schulze, M., Souza, C., Nepstad, D., Lefebvre, V., Davidson, E., 1999. Positive feedbacks in the fire dynamic of closed canopy tropical forests. Science 284 (5421), 1832–1835. https://doi.org/10.1126/science.284.5421.1832.

Deutsch, C.A., Tewksbury, J.J., Huey, R.B., Sheldon, K.S., Ghalambor, C.K., Haak, D.C., Martin, P.R., 2008. Impacts of climate warming on terrestrial ectotherms across latitude. Proc. Natl. Acad. Sci. U. S. A. 105 (18), 6668–6672.

Edwards, D.P., Larsen, T.H., Docherty, T.D.S., Ansell, F.A., Hsu, W.W., Derhé, M.A., … Wilcove, D.S., 2011. Degraded lands worth protecting: the biological importance of Southeast Asia's repeatedly logged forests. Proc. Royal Soc. B 278 (1702), 82–90. https://doi.org/10.1098/rspb.2010.1062.

Eisenhauer, N., Bohan, D.A., Dumbrell, A.J., 2019. Preface: Mechanistic links between biodiversity and ecosystem functioning. In: Advances in Ecological Research. vol. 61. Elsevier, pp. xix–xxviii.

Escobar, H., 2019. Amazon fires clearly linked to deforestation, scientists say. Science 365 (6456), 853. https://doi.org/10.1126/science.365.6456.853.

Eycott, A.E., Advento, A.D., Waters, H.S., Luke, S.H., Aryawan, A.A.K., Hood, A.S., … Turner, E.C., 2019. Resilience of ecological functions to drought in an oil palm agroecosystem. Environ. Res. Commun. 1 (10), 101004. https://doi.org/10.1088/2515-7620/ab48da.

Fayle, T.M., Turner, E.C., Basset, Y., Ewers, R.M., Reynolds, G., Novotny, V., 2015. Whole-ecosystem experimental manipulations of tropical forests. Trends Ecol. Evol. 30 (6), 1–13. https://doi.org/10.1016/j.tree.2015.03.010.

Gardner, C.J., Wordley, C.F.R., 2019. Scientists must act on our own warnings to humanity. Nat. Ecol. Evol. 3 (9), 1271–1272. https://doi.org/10.1038/s41559-019-0979-y.

Garibaldi, L.A., Steffan-dewenter, I., Kremen, C., Morales, J.M., Bommarco, R., Cunningham, S.A., … Klein, A.M., 2011. Stability of pollination services decreases with isolation from natural areas despite honey bee visits. Ecol. Lett. 14, 1062–1072. https://doi.org/10.1111/j.1461-0248.2011.01669.x.

Gaston, K.J., 2000. Global patterns in biodiversity. Nature 405 (6783), 220–227. https://doi.org/10.1038/35012228.

Ghosh, S., Sheppard, L.W., Holder, M.T., Loecke, T.D., Reid, P.C., Bever, J.D., Reuman, D.C., 2020. Copulas and their potential for ecology. In: Dumbrell, A.J., Turner, E.C., Fayle, T.M. (Eds.), Advances in Ecological Research. vol. 62. Elsevier, pp. 409–468.

Gibbs, H.K., Ruesch, A.S., Achard, F., Clayton, M.K., Holmgren, P., Ramankutty, N., Foley, J.A., 2010. Tropical forests were the primary sources of new agricultural land in the 1980s and 1990s. Proc. Natl. Acad. Sci. U. S. A. 107 (38), 16732–16737. https://doi.org/10.1073/pnas.0910275107.

Lamarre, G.P.A., Fayle, T.M., Segar, S.T., Laird-Hopkins, B.C., Nakamura, A., Souto-Vilarós, D., … Basset, Y., 2020. Monitoring tropical insects in the 21st century. In: Dumbrell, A.J., Turner, E.C., Fayle, T.M. (Eds.), Advances in Ecological Research. vol. 62. Elsevier, pp. 295–330.

Lindenmayer, D.B., Fischer, J., Felton, A., Crane, M., Michael, D., Macgregor, C., … Hobbs, R.J., 2008. Novel ecosystems resulting from landscape transformation create dilemmas for modern conservation practice. Conserv. Lett. 1, 129–135. https://doi.org/10.1111/j.1755-263X.2008.00021.x.

Lugo, A.E., Abelleira Martínez, O.J., Medina, E., Aymard, G., Heartsill Scalley, T., 2020. Novelty in the tropical forests of the 21st century. In: Dumbrell, A.J., Turner, E.C., Fayle, T.M. (Eds.), Advances in Ecological Research. vol. 62. Elsevier, pp. 53–116.

Malhi, Y., Doughty, C.E., Galetti, M., Smith, F.A., Svenning, J.C., Terborgh, J.W., 2016. Megafauna and ecosystem function from the Pleistocene to the Anthropocene. Proc. Natl. Acad. Sci. U. S. A. 113 (4), 838–846. https://doi.org/10.1073/pnas.1502540113.

McCauley, D.J., Pinsky, M.L., Palumbi, S.R., Estes, J.A., Joyce, F.H., Warner, R.R., 2015. Marine defaunation: animal loss in the global ocean. Science 347 (6219), 247. https://doi.org/10.1126/science.1255641.

Pellegrini, A.F.A., Jackson, R.B., 2020. The long and short of it: a review of the timescales of how fire affects soils using the pulse-press framework. In: Dumbrell, A.J., Turner, E.C., Fayle, T.M. (Eds.), Advances in Ecological Research. vol. 62. Elsevier, pp. 147–171.

Phalan, B., Bertzky, M., Butchart, S.H.M., Donald, P.F., Stattersfield, A.J., Balmford, A., 2013. Crop expansion and conservation priorities in tropical countries. PLoS One 8 (1), e51759. https://doi.org/10.1371/journal.pone.0051759.

Pimm, S.L., Jenkins, C.N., Abell, R., Brooks, T.M., Gittleman, J.L., Joppa, L., … Sexton, J., 2014. The biodiversity of species and their rates of extinction, distribution, and protection. Science 344 (6187), 987–998. https://doi.org/10.1126/science.1246752.

Pinho, B.X., Peres, C.A., Leal, I.R., Tabarelli, M., 2020. Critical role and collapse of tropical mega-trees: a key global resource. In: Dumbrell, A.J., Turner, E.C., Fayle, T.M. (Eds.), Advances in Ecological Research. vol. 62. Elsevier, pp. 253–294.

Reed, S.C., Reibold, R., Cavaleri, M.A., Alonso-Rodríguez, A.M., Berberich, M.E., Wood, T.E., 2020. Soil biogeochemical responses of a tropical forest to warming and hurricane disturbance. In: Dumbrell, A.J., Turner, E.C., Fayle, T.M. (Eds.), Advances in Ecological Research. vol. 62. Elsevier, pp. 225–252.

Richardson, B.A., Richardson, M.J., González, G., Shiels, A.B., Srivastava, D.S., 2010. A canopy trimming experiment in Puerto Rico: the response of litter invertebrate communities to canopy loss and debris deposition in a tropical forest subject to hurricanes. Ecosystems 13 (2), 286–301. https://doi.org/10.1007/s10021-010-9317-6.

Ripple, W.J., Wolf, C., Newsome, T.M., Betts, M.G., Ceballos, G., Courchamp, F., … Worm, B., 2019. Are we eating the world's megafauna to extinction? Conserv. Lett. 12 e12627 1–10https://doi.org/10.1111/conl.12627.

Sayer, E.J., Rodtassana, C., Sheldrake, M., Bréchet, L.M., Ashford, O.S., Lopez-Sangil, L., … Tanner, E.V.J., 2020. Revisiting nutrient cycling by litterfall—insights from 15 years of litter manipulation in old-growth lowland tropical forest. In: Dumbrell, A.J., Turner, E.C., Fayle, T.M. (Eds.), Advances in Ecological Research. vol. 62. Elsevier, pp. 173–223.

Seebens, H., Blackburn, T.M., Dyer, E.E., Genovesi, P., Hulme, P.E., Jeschke, J.M., … Essl, F., 2018. Global rise in emerging alien species results from increased accessibility of new source pools. Proc. Natl. Acad. Sci. U. S. A. 115 (10), E2264–E2273. https://doi.org/10.1073/pnas.1719429115.

Tilman, D., Balzer, C., Hill, J., Befort, B.L., 2011a. Global food demand and the sustainable intensi fi cation of agriculture. Proc. Natl. Acad. Sci. 108 (50). https://doi.org/10.1073/pnas.1116437108.

Tilman, D., Reich, P.B., Knops, J., Wedin, D., Mielke, T., Lehman, C., 2011b. Diversity and productivity in a long-term grassland experiment. Science 2001, 843–845. https://doi.org/10.1126/science.1060391.

Timmermann, A., Oberhuber, J., Bacher, A., Esch, M., Latif, M., Roeckner, E., 2002. Increased El Niño frequency in a climate model forced by future greenhouse warming. Nature 398 (6729), 694–697. https://doi.org/10.1038/19505.

Tscharntke, T., Klein, A.-M., Kruess, A., Steffan-Dewenter, I., Thies, C., 2005. Landscape perspectives on agricultural intensification and biodiversity—ecosystem service management. Ecol. Lett. 8, 857–874.

Turner, W., Spector, S., Gardiner, N., Fladeland, M., Sterling, E., Steininger, M., 2003. Remote sensing for biodiversity science and conservation. Trends Ecol. Evol. 18 (6), 306–314. https://doi.org/10.1016/S0169-5347(03)00070-3.

Wagner, D.L., 2020. Insect declines in the Anthropocene. Annu. Rev. Entomol. 65, 457–480.

Wu, J., Chen, B., Reynolds, G., Xie, J., Liang, S., O'Brien, M.J., Hector, A., 2020. Monitoring tropical forest degradation and restoration with satellite remote sensing: a test using Sabah Biodiversity Experiment. In: Dumbrell, A.J., Turner, E.C., Fayle, T.M. (Eds.), Advances in Ecological Research. vol. 62. Elsevier, pp. 117–146.

Zinger, L., Donald, J., Brosse, S., Gonzalez, M.A., Iribar, A., Leroy, C., … Martins Lopes, C., 2020. Advances and prospects of environmental DNA in neotropical rainforests. In: Dumbrell, A.J., Turner, E.C., Fayle, T.M. (Eds.), Advances in Ecological Research. vol. 62. Elsevier, pp. 331–373.

CHAPTER ONE

Protecting environmental and socio-economic values of selectively logged tropical forests in the Anthropocene

Christopher G. Bousfield[a], Gianluca R. Cerullo[b], Mike R. Massam[a], David P. Edwards[a,*]
[a]Department of Animal and Plant Science, University of Sheffield, Sheffield, United Kingdom
[b]Department of Zoology, University of Cambridge, Cambridge, United Kingdom
*Corresponding author: e-mail address: david.edwards@sheffield.ac.uk

Contents

Advances in Ecological Research, Volume 62
ISSN 0065-2504
https://doi.org/10.1016/bs.aecr.2020.01.006

Abstract

Global timber demands have penetrated vast tracts of tropical rainforest, with selective logging the most powerful driver of human disturbance within these ecosystems. We review the consequences of timber harvesting on rainforest ecosystems, from individual species through to biological communities. We also investigate the benefits that forest-dwelling communities and wider society can derive from well-managed production forests. Selectively logged forests maintain a wide array of ecosystem services, and orders of magnitude higher levels of biodiversity and carbon storage than the agricultural lands that often imperil them. To unlock the full potential of logged forests in conserving biodiversity, tackling climate change and delivering ecosystem services into the Anthropocene will require several major transitions. These include a pantropical shift towards more environmentally sensitive logging practises, the fostering of new norms in tropical forestry based around long-term forest management, as well as coordinated efforts to limit the negative effects of logging roads, illegal logging and climate change. We discuss potential incentive and enforcement-based mechanisms for precipitating such transitions and highlight key avenues for future enquiry. Research frontiers include determining how best to spatially design logging operations to limit environmental impacts, quantifying the costs and benefits of post-logging activities aimed at recovering logged forest values, and investigating the role of concession licence structures in spawning sustainable forest management. Given their spatial dominance and high conservation value, lofty planetary goals to limit climate change and global biodiversity declines cannot be achieved unless selectively logged tropical forests take a central role in emerging conservation and emissions-reducing agendas.

Tropical forests comprise ~11% of the world's terrestrial surface (Keenan et al., 2015) yet are estimated to harbour more than 80% the world's species (FAO, 2012). Their global significance is recognised in the 2015 Paris Climate Agreement (United Nations, 2015) and Sustainable Development Goals (United Nations, 2018), both of which emphasise the importance of halting deforestation and degradation.

Beyond their importance as the Earth's biodiversity hub, tropical forests provide humankind with a diverse range of economic, environmental, social and aesthetic benefits making them critical ecosystems at the local and global scale (Sudarshana et al., 2012). Perhaps most important are the range of environmental services that tropical forests provide. Globally, tropical forests contain 27% of the world's terrestrial carbon (UNEP-WCMC, 2009) and absorb 31% of annual global emissions (Pan et al., 2011). Therefore, tropical rainforests have a considerable role to play in mitigating anthropogenic climate change. At the mesoscale, rainforests are a critical component of the water cycle by facilitating increased evapotranspiration (Makarieva and Gorshkov, 2007;

Salati and Vose, 1984). At the local scale, tropical forests also play a key role in watershed dynamics by modifying water supply, quality and sedimentation with potential implications for downstream populations and agriculture (Dessie and Bredemeier, 2013).

Selective logging, whereby large, commercially viable adult trees are harvested, represents the most powerful force of tropical forest degradation (Hosonuma et al., 2012; Houghton, 2012; Pearson et al., 2017). Currently responsible for 6% of tropical greenhouse gas emissions (Ellis et al., 2019), the extent of logging will likely expand to match population growth and increasing urbanisation (DeFries et al., 2010). At least 403 million hectares of tropical forest are presently committed to selective logging regimes (Blaser et al., 2011), not including the substantial forest area that is logged illegally (Pacheco et al., 2016). So extensive is this disturbance that, with exception of the Amazon, the area of selectively logged forests now exceeds that of primary forests in the tropics (Laurance et al., 2014).

This review will highlight the potential of selectively logged forests in conserving global biodiversity and ecosystem services, and it will examine the latest thinking on how improved logging management and revenue generation for environmentally sensitive practises can maintain the importance of selectively logged forests as we enter the Anthropocene.

1. Impacts of logging on forest structure

Tropical wet forests have tall canopies punctured by even taller emergent trees. Their understorey is humid and dark. The felling of large trees and their extraction along skid trails and roads damages neighbouring unharvested trees, opens gaps in the forest canopy and simplifies forest age structure. The fragmented canopy allows sunlight to reach sub-canopy trees that were previously overshadowed and to reach the forest floor. The sudden high light conditions promote forest regeneration, either via growth of the remaining subcanopy trees, understory saplings and pioneer light-demanding species that occupy logging gaps or, especially where heavy machinery has compacted soils, by vigorous growth of non-tree species, including climbing vines and bamboos, that impede forest regeneration (Putz et al., 2008) and long-term productivity (Hawthorne et al., 2012). What remains after large-scale mechanised logging is thus a disturbed tropical forest.

These logged forests are also more vulnerable to forest fires (Broadbent et al., 2008; Cochrane, 2003; Matricardi et al., 2013), either by allowing the desiccation of forests through fragmentation of the canopy and/or via increases in ignition sources. This vulnerability can persist for decades following cessation of logging (Siegert et al., 2001), and forests are particularly at risk under drought conditions, which can be extremely intense during El Niño events (Cochrane, 2003). Forest fires represent a major additional degrading force that can further reduce forest integrity post-logging. In the Amazon, fires increase tree mortality by 226–464%, reducing canopy cover by 23–31% and above-ground biodiversity by 12–20% (Brando et al., 2014). Moreover, areas that have suffered recent fires are more likely to burn again, risking the loss of tree diversity and a deflected successional community composed of savanna and shrub vegetation (Lewis et al., 2015) with an accompanying loss of species and ecosystem functions (Barlow and Peres, 2008).

Ultimately, the substantially reduced timber value of selectively logged forests combined with increased accessibility from logging roads make them more susceptible to reclassification for agriculture or plantations, which are more profitable than continued logging cycles (Fisher et al., 2011a). In the Amazon, ~16% of all logged forests were cleared within one year of logging and the overall probability of deforestation is four times higher than in primary forest (Asner et al., 2006). In Indonesia, the moratorium on clearance of primary forests and deep peatlands failed to protect 46.7 million hectares of selectively logged forest (Murdiyarso et al., 2011), leaving huge areas open for conversion to plantations. Clearance massively simplifies habitat structure resulting in large losses in the species richness and abundance of a range of taxa (Foster et al., 2011).

2. Selective logging: Two words for many things

Since selective logging is the most widespread force of human disturbance, its impacts can vary substantially, both geographically, and according to how it proceeds on the ground. Logging intensity, for instance, can vary by more than two orders of magnitude (<1 to >100 m^3/ha volume of timber removed) according to region, largely driven by high variation in the natural occurrence of high-valued timber species. In many areas of South America and Africa, logging intensity can be extremely low, with only 1–2 high value trees being harvested per hectare due to the patchy distribution of merchantable trees (Edwards et al., 2014a). Conversely, across Southeast Asia, logging intensities reach extremely high levels (120 m^3/ha; Edwards et al., 2010a).

Logging can also vary significantly according to the government oversight it receives, the corruption it attracts and the care taken to reduce its structural and spatial footprint—not to mention how a selectively logged forest is managed after the chainsaws have gone silent: from protection to degazettement, and from passive to active management. The spatial extent of selective logging and the variability in these factors means that it would be miraculous if logging had identical impacts on biodiversity and ecosystem services across the tropics. However, for the purposes of this review, we focus on general patterns revealed through the literature on the effects of logging, while highlighting key findings where logging's impact has been shown to vary substantially in space or time.

3. Biodiversity impacts at the community level

The impact of selective logging on biodiversity has been studied extensively and has been implicated in driving local extinctions and declines in a wide range of species groups (Fimbel et al., 2001). Such changes to community composition can have major impacts on ecosystem functioning (Dobson et al., 2006). We review the impacts of logging on species richness, community composition, and functional diversity, and then consider how fire or conversion could further degrade the biodiversity values of logged forests.

Meta-analyses spanning the tropics found that selective logging has a limited effect on biodiversity levels (Gibson et al., 2011), with 85–100% of the species richness of mammals, birds, invertebrates and plants in primary forest retained after logging (Putz et al., 2012; Fig. 1). Much of the research on the biodiversity impact of logging is focused on a few well-studied and easily sampled taxa, such as birds, mammals, and dung beetles. Important hyper-diverse taxa are often ignored, particularly arthropods, which represent the majority of rainforest fauna (2.5–3.7 million tropical species; Hamilton et al., 2010). Edwards et al. (2014b) used DNA metabarcoding (Ji et al., 2013) to assess the impacts of logging in Malaysia on over 2300 invertebrate species (operational taxonomic units). They found once-logged and twice-logged forests to retain 72% and 69.4% of species, respectively, and identified six indicator groups as consistently strong predictors of overall response to logging (birds, leaf-litter ants, beetles, aerial hymenopterans, flies, and true bugs). Focusing on these six groups will improve the accuracy and validity of future biodiversity studies.

Biodiversity impacts of selectively logging are, however, heavily impacted by harvest intensity. Intensities of 38 and 63 m^3/ha are enough

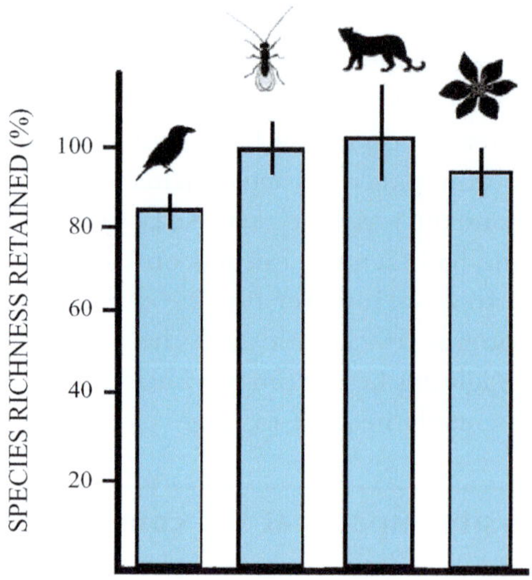

Fig. 1 Species richness of birds, invertebrates, mammals, and plants in selectively logged forests compared to undisturbed old-growth forests, based on a meta-analysis of 109 studies. Bars represent standard errors. *Adapted from Putz, F.E., Zuidema, P.A., Synnott, T., Peña-Claros, M., Pinard, M.A., Sheil, D., Vanclay, J.K., Sist, P., Gourlet-Fleury, S., Griscom, B., Palmer, J., 2012. Sustaining conservation values in selectively logged tropical forests: the attained and the attainable. Conserv. Lett. 5, 296–303.*

to cause a 50% decline in mammal and amphibian species richness, respectively (Burivalova et al., 2014), although these relationships did not include much high-intensity logging in Asia, where many mammal and amphibian species still remain (e.g. Foster et al., 2011). Given variation in the biodiversity impacts of selective logging across the tropics, multiple literature reviews have found the number of studies reporting biodiversity increases to be similar to the number reporting decreases (Burivalova et al., 2014; Hill and Hamer, 2004; Ramage et al., 2013). Variation in logging intensities and practises across the tropics therefore reflects the variation of logging impacts reported in the literature, suggesting that we need more nuanced understanding of how these interact across space to impact biodiversity.

Despite the potential for species richness to be retained, species composition can exhibit significant change as primary forest specialists disappear while disturbance-tolerant generalists prosper (Hamer et al., 2003). For example, bird species from single-source feeding guilds are more vulnerable to declines after selective logging than species using two or more sources of food (Hamer et al., 2015), with dietary specialists more vulnerable to decline after logging (Edwards et al., 2013a). Meta-analysis of logging impacts on Bornean birds and mammals demonstrated variation in species-level responses depending on specific characteristics (Costantini et al., 2016). In both birds and mammals, larger species were more susceptible to selective logging and, for birds in particular, species classified as IUCN vulnerable or near-threatened declined in abundance after logging compared to species classified as least-concern (Costantini et al., 2016).

Selective logging often facilitates increased access to remote areas for illegal hunters (Poulsen et al., 2011; Wilkie et al., 2000), which disproportionately impacts on large-bodied vertebrates (Cardillo et al., 2005). Even moderate hunting pressure can alter mammal communities (Laurance et al., 2006a), and overhunting in logging concessions is particularly detrimental in the Congo. Increased access for hunters and migration into logging towns results in a 64% increase in bush meat supplies and significant increases in the use of wire snares and diversification of species hunted (Poulsen et al., 2009).

Many functional processes, including predation, seed dispersal, pollination and decomposition, are maintained and regulated by a complex network of interactions between species. Modifications to the arrangement of these interactions (Ewers et al., 2015) can have marked implications for forest structure (Cazzolla Gatti et al., 2015; Terborgh et al., 2008), the capacity to regenerate (Bonnell et al., 2011; Leck et al., 2008) and, consequently, the ecosystem services we derive from them (Dobson et al., 2006).

Edwards et al. (2013a,b) found a strong effect of logging on bird species composition in Borneo, yet no significant effect on the functional diversity of species. The ecosystem functions performed by birds were retained after logging, despite species turnover and extremely high levels of timber extraction ($120\,m^3$/ha). Similar results were found when considering functional diversity of dung beetles in the same area (Edwards et al., 2014c). Despite a decline in functional importance of key invertebrate groups, functional processes are resilient and absorbed via changes in the abundance of other taxa in logged forests (Ewers et al., 2015; see chapter "Novelty in the tropical forests of the 21st century" by Lugo et al., this issue), with evidence suggesting even complex mutualisms (such as those between ferns and ant species, whereby ferns gain protection from herbivores while ants gain housing in fern roots) are also maintained (Fayle et al., 2015). Animal species increasing in abundance or colonising the forest after the first logging cycle thus appear to provide similar resource-use traits to the species replaced, fulfilling ecological roles vital to the forest ecosystem.

Tree species richness is also affected by selective logging, but a true consensus on the nature of this effect is lacking. Logged sites in Madagascar are incapable of recovering primary forest tree species diversity even 150 years post logging (Brown and Gurevitch, 2004) because invasion of non-native species prevents the recolonization of native species. Similarly, in the Solomon Islands, pioneer species restrict floristic recovery (Katovai et al., 2016). Conversely, in Hainan Island, China, and Borneo species richness recovers faster than species composition and structure, and can match the richness of primary forests within 20–40 years of cutting (Magrach et al., 2016; Xu et al., 2015). Beyond trees, logging intensity had no significant effect on the species richness and community composition of understory herbs in the Amazon (Costa and Magnusson, 2002), whereas in Borneo, selective logging decreased herb species richness (Cleary, 2017) but increased species richness and altered species composition of lianas (Cleary, 2017; Magrach et al., 2016).

More generally, meta-analysis of palaeoecological records reveals that forests in Central America and Africa recover faster from past disturbances than those in South America and Asia (Cole et al., 2014). Moreover, recovery rates are elevated over time for forests that experience increased frequency of disturbances, suggesting resilience among forest types to disturbance (Cole et al., 2014). The debate concerning the relationship between disturbance and diversity thus continues, with further research required to ascertain general patterns of how floral diversity and composition may be impacted far into the future by industrialised logging.

While it is clear that primary forests are irreplaceable for sustaining tropical biodiversity (Gibson et al., 2011) and that selective logging has some negative impacts on biodiversity, the impacts of rival land-uses are comparatively far worse (Edwards et al., 2014a). Of greatest concern is the vulnerability of selectively logged forests to clearance for agriculture. For example, when logged forest is replaced by oil palm in Borneo, there is a 200-fold loss of imperilled bird species (Edwards et al., 2010c), and dramatically lower functional diversity of birds and dung beetles, including the loss of all bark-foraging birds and dung-rolling beetle species (Edwards et al., 2013a, 2014c). The preservation of logged forests from clearing prevents habitat fragmentation, retaining connectivity between sub-populations, which is often necessary for meta-population viability (Elkin and Possingham, 2008) and to maintain climate connectivity (Senior et al., 2019).

The potential for selectively logged forest to increase vulnerability to fire (Broadbent et al., 2008; Matricardi et al., 2010) represents a further threat to biodiversity. Forest fires in the Amazon cause up to 50% tree mortality (Peres et al., 2010), which can create wholesale shifts in forest structure and tree species composition (Barlow and Peres, 2008), with significant negative impacts on biodiversity. Less common and more specialised species of birds are threatened by forest fires (Barlow et al., 2002), with insectivorous and frugivorous birds particularly sensitive to forest structure changes that accompany forest fires (Barlow and Peres, 2004; Haugaasen et al., 2003). Furthermore, previously burned forests struggle to recover and are increasingly susceptible to repeat burning (Cochrane et al., 1999). Thus, preventing conversion and fire, including under climate change (see below), are critical components of protecting the long-term conservation values of selectively logged tropical forests (see chapter "The long and short of it: A review of the timescales of how fire affects soils using the pulse-press framework" by Pellegrini and Jackson, this issue).

4. Biodiversity impacts at the individual level

Beyond community or population-level impacts, selective logging can result in sub-lethal effects on individuals that precede population-level impacts (Dantzer et al., 2014). Functional traits are an important aspect for determining a species' response to logging. In Malaysian Borneo, the trophic status of bird species increased after selective logging, perhaps due to changes in the abundance of higher trophic level prey (Hamer et al., 2015), and trophic niche widths declined (Edwards et al., 2013b). Higher trophic level, larger body size and lower foraging height were all strong predictors of negative species-level responses (Hamer et al., 2015). Consequently, logging particularly impacts dietary specialists that already feed on a restricted range of food items, perhaps reflecting a serious consequence that remains hidden when only considering species richness or functional diversity.

Logging also drives behavioural responses in Bornean birds, with species belonging to smaller populations after logging producing fewer songs per individual, and species known to prefer old-growth forests responding most negatively in terms of singing behaviour (Pillay et al., 2016; Fig. 2A i). Logging also affects the behaviour of Bornean orangutan, which move more and rest less in logged forests (Hardus et al., 2012). Food sources can also differ for orangutans, most likely due to altered abundances of different sources. Orangutans spend less time feeding on figs in logged forest but the same amount of time feeding on tree bark and fruit (Hardus et al., 2012).

Despite improvements in our understanding of functional and morphological traits in determining the response of a given species to habitat degradation, there is still a lack of understanding of the physiological mechanisms that mediate this response (Messina et al., 2018). A pan-tropical

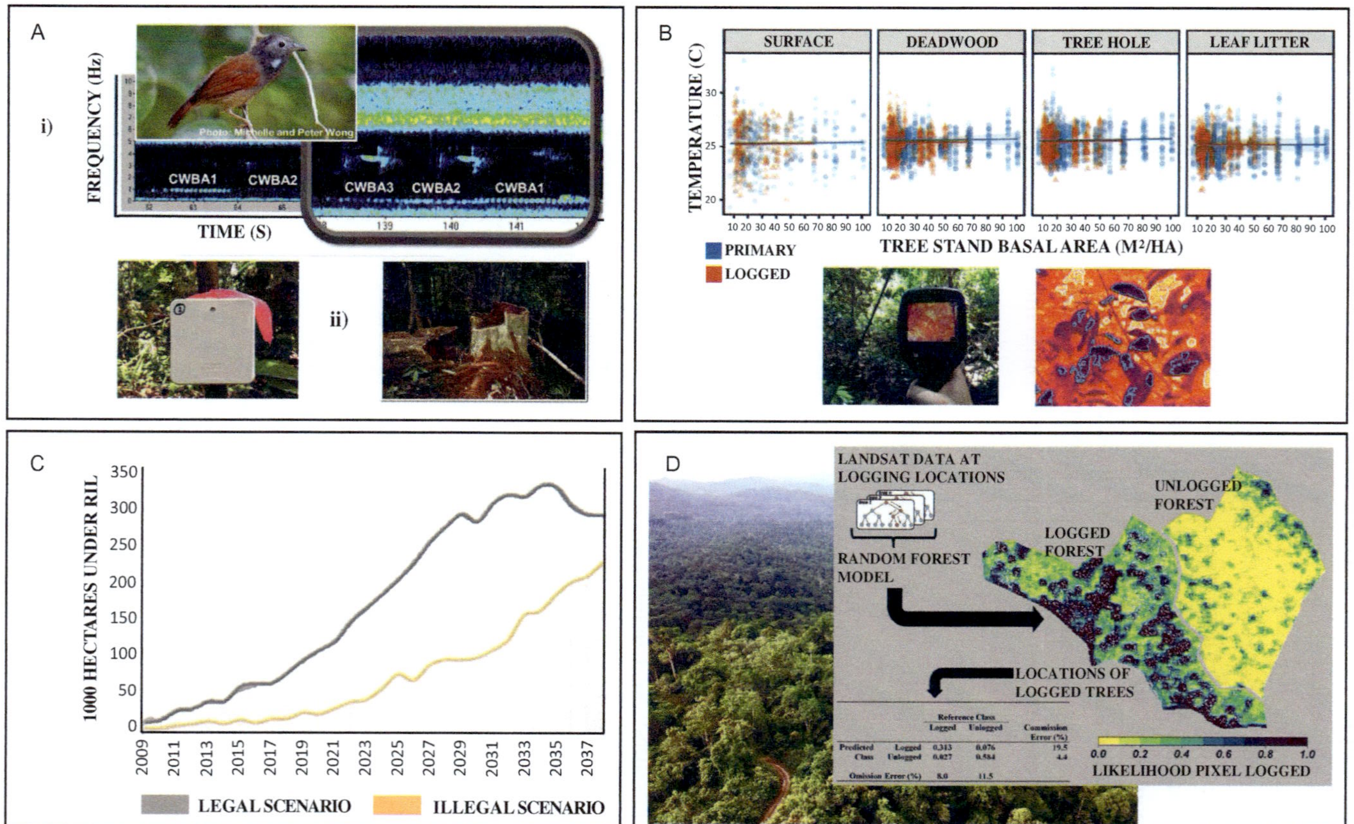

Fig. 2 See legend on opposite page.

meta-analysis by Messina et al. (2018) comparing the physiological and immunological endpoints of birds and mammals in logged and undisturbed forests found stress hormones and immunity markers show consistent changes in response to habitat degradation. Importantly, IUCN threatened birds exhibited significantly greater effect sizes than did mammals and least-concern bird species, highlighting the potential for sub-lethal responses to help predict species conservation responses.

Studies focusing on other taxonomic groups are extremely limited, in particular for invertebrates despite their importance within tropical ecosystems. In the sole study we are aware of, the individual-level effects of degradation on body fat storage of dung beetles increased as a stress response to selective logging (França et al., 2016). Increases in body fat are thought to represent a strategy to conserve energy necessary for survival processes

Fig. 2 Frontiers in detecting logging and its effects. (A) Bioacoustic monitoring in forests can help uncover species responses to logging and detect illegal activities: (i) In Borneo, breeding songbirds alter their song rates with selective logging, with individuals from bird populations that decline after logging producing fewer songs. The diagram shows multiple calls from chestnut-winged bablers; (ii) across various tropical forests, bioacoustic monitoring is being piloted as a method for detecting hunting and illegal logging by alerting rangers of chainsaw and gunshot noises in real time. (B) Thermal imagery offers insights into the climate-buffering role of tropical forests. In Malaysian Borneo, heavily logged forests show indistinguishable microclimatic temperatures across varying levels of forest quality, implying that logged forests will continue to provide habitat for temperature-sensitive species into the future. (C) Illegal logging can compete with legal reduced-impact logging within Brazilian forest concessions. In a spatially-explicit simulation of the Amazonian timber industry where logging is allowed to take place outside logging concessions (Illegal scenario), this leads to a 59% decrease in the area harvested within legal concessions (legal scenarios) across a 30-year cycle. Illegal logging can thus undercut the establishment of a sustainable forestry sector. (D) Technological advances are enabling better monitoring of even low-intensity logging activities. A machine-learning algorithm that used Landsat imagery has been applied in the Brazilian Amazon and detected 40% of low-intensity logging activity. *Figures from Panel (A) Pillay, R., Fletcher, R.J., Sieving, K.E., Udell, B.J., Bernard, H., 2019. Bioacoustic monitoring reveals shifts in breeding songbird populations and singing behaviour with selective logging in tropical forests. J. Appl. Ecol. Panel (B) from Senior, R.A., Hill, J.K., Benedick, S., Edwards, D.P. (2018) Tropical forests are thermally buffered despite intensive selective logging. Glob. Chang. Biol. 24, 1267–1278. Adapted from Panel (C) Santos De Lima, L.S., Merry, F., Soares-Filho, B., Rodrigues, H.O., dos Santos Damaceno, C., Bauch, M.A., 2018. Illegal logging as a disincentive to the establishment of a sustainable forest sector in the Amazon. PloS One. 13, p. e0207855 and Panel (D) Hethcoat, M.G., Edwards, D.P., Carreiras, J.M., Bryant, R.G., França, F.M., Quegan, S., 2019. A machine learning approach to map tropical selective logging. Remote Sens. Environ. 221, 569–582. Photos by Gianluca Cerullo and rosewood stump from Wikimedia Commons.*

(Schneider et al., 2014), and body fat increases mirrored changes in the relative abundance of dung beetle species (França et al., 2016).

Considering sub-lethal effects is important when assessing the biodiversity impacts of selective logging, since they can function as early warning mechanisms allowing the prediction of future population-level responses to logging and the identification of conservation priorities (Dantzer et al., 2014). However, our poor understanding of how these sub-lethal responses are distributed across taxa represents a barrier to effective conservation planning, and future research should focus on improved clarification on the role of physiology and immunity in species responses to logging-induced tropical forest degradation.

5. Impacts on ecosystem services

While important ecosystem services provided by primary forests, such as carbon storage, temperature regulation and watershed services, are negatively impacted by selective logging (particularly at higher intensities), they are often conserved, especially following a period of forest recovery (Edwards et al., 2014a). Thus, in combination with their vast extent and retention of high biodiversity and species richness levels (Edwards et al., 2014a; Putz et al., 2012), logged forests are now often viewed through their capacity to contribute towards global conservation and ecosystem service protection targets.

On average, 76% of above-ground organic carbon is retained in once-logged forests relative to primary forest (Putz et al., 2012), though this varies substantially according to logging intensity, forest conditions and logging techniques. During the initial phase of post-logging recovery, logged forests accumulate carbon at five times the rate of natural primary forest. However, recovering forest in Malaysian Borneo still holds 36% and 28% less carbon than primary forest after 18 and 22 years, respectively (Berry et al., 2010; Saner et al., 2012). Indeed, research suggests the carbon deficit in logged forest remains for at least 60 years (Berry et al., 2010), even as market demand can lead to re-logging within just 20 years of the initial harvest (Van Gardingen et al., 2003). Methods such as reduced-impact logging can improve carbon recovery, allowing 100% of above-ground biomass within a primary forest to be restored within 16 years of harvest in the Southern Amazon (West et al., 2014). In turn, this suggests that both efforts to increase the proportion of reduced-impact, climate-smart logging in the tropics, and attempts to avoid logging of certain old-growth or high-carbon (e.g. peaty) forests are needed to improve forest carbon storage.

There is mounting evidence that tree cover plays a major role in influencing local temperature and rainfall (Makarieva et al., 2014). Local and regional climates are largely driven by cycles of rainfall, evaporation, and cloud formation within rainforest biomes. It has been suggested that contiguous forest helps to generate winds that carry rainfall far into continental interiors and stabilise rainfall (Makarieva et al., 2014). In Sumatra, surface temperature derived from LandSat was similar in logged and primary forest (Ramdani et al., 2014), while in Borneo and the south-west Amazon, understorey air and surface temperatures (derived from iButtons and thermal imaging cameras, respectively) were initially hotter within treefall gaps immediately after logging, but by five years of recovery temperatures were similar (Mollinari et al., 2019; Senior et al., 2018). By maintaining a near-contiguous forest canopy a few years after logging, it appears likely that tracts of selectively logged forests could function like primary forests in sustaining regional rainfall. This is, however, a controversial suggestion and more studies are required.

Old-growth tropical forests provide important watershed services, such as maintaining stream flows during dry periods, preventing flash floods, improving water quality, and conserving soils (Lele, 2009). However, selective logging increases water runoff (Bruijnzeel, 2004), which mainly stems from higher runoff along skid trails and roads (Hartanto et al., 2003). For example, in Indonesian Borneo, harvest and control plots within two catchments differed in runoff only marginally, but skid trails and roads had 10-times more runoff (Hartanto et al., 2003). In Southeast Asia, the additional runoff after logging was insufficient to produce detectable flooding downstream (Chan and Parker, 1996). Forest soils are also prone to erosion after logging, largely as a result of higher water runoff, causing sedimentation of rivers and reduced water quality (Lele, 2009). Soil erosion is most severe on skid trails and roads, often in association with small-scale landslide events (Walsh et al., 2011). In Borneo, this resulted in 100- to 3000-times higher soil loss in recently logged plots compared to unlogged plots (Hartanto et al., 2003). Despite the initial pulse of erosion and sediment runoff, by several years after logging, total soil runoff (including skid trails) was similar to that of primary forest (Douglas, 1999).

Despite the (often short-term) negative impacts of selective logging on ecosystem services, the impacts of forest conversion and associated fragmentation are much greater (Edwards et al., 2014a). Cleared areas of forest have substantially reduced carbon stocks, with stocks also reduced within remaining forests at their edges (Laurance et al., 2006b). Deforested areas also have higher temperatures (Senior et al., 2017), lower evapotranspiration rates (Wang and Dickinson, 2012), higher water runoff (Bruijnzeel, 2004) and higher soil erosion (Douglas, 1999) than forested areas. As forest cover declines, the cycle of rainfall, evaporation, and cloud formation can be disrupted. For example, in the Brazilian Atlantic forest, declining forest cover caused the number of rain days to decline and inter-annual variability in rainfall to increase, especially where forests were increasingly fragmented (Webb et al., 2005). In the Amazon, large-scale deforestation causes higher temperatures and lower rates of evapotranspiration (Mahmood et al., 2014; Wang and Dickinson, 2012), resulting in lower precipitation (Spracklen et al., 2012) and potentially longer dry seasons (Mahmood et al., 2014; Wang and Dickinson, 2012). On Sumatra, oil palm has higher air temperatures than logged or primary forest (Ramdani et al., 2014), with rural communities on Borneo considering rising temperatures the most harmful environmental impact of deforestation (Meijaard et al., 2013).

Forest conversion results in 100–800% increases in annual water flow (Bruijnzeel, 2004), because of enhanced runoff in rainstorms and greatly reduced evapotranspiration (Wang and Dickinson, 2012). In turn, this results in a massive pulse of soil erosion that reduces water quality. For example, in Southeast Asia, soil loss increased from ~20 t km^{-2} $year^{-1}$ to between 1100 and 8940 t km^{-2} $year^{-1}$, with sediment loads in streams rising from ~28 to 125 t km^{-2} $year^{-1}$ (Douglas, 1999). The replacement of tree cover by savanna and shrub vegetation following multiple forest fires within logged forest would presumably degrade ecosystem services in a similar manner as clearance, although potentially not as severely. Quantifying this potential risk to logged forests is an area for urgent research.

6. Impacts on local communities

Beyond their value in providing a range of ecosystem services, tropical forests are important sources of revenue and livelihoods for over 1.35 billion forest-dependent peoples (FAO, 2014). Selective logging has the potential to restrict land access for local people, cause conflict, and restrict the availability of livelihood resources. For example, logging targets species that provide non-timber forest products for local people (Iponga et al., 2018), including fruits and oil from Moabi *Baillonella toxisperma* and edible caterpillars from Sapelli *Entandrophragma cylindricum* and Tali *Erythrophleum suaveolens*, which are frequently harvested trees in the Congo (Noutcheu et al., 2016). Logging can also create a hostile environment for the recovery of non-timber forest products, with one review finding that the availability of forest products was negatively affected by timber harvesting in 82% of cases (Rist et al., 2012).

The most pernicious and poorly tracked social impacts of logging often stem from road creation or from predatory logging within indigenous areas. In remote forest regions, logging roads can secondarily become invasion points for illegal miners, loggers, land prospectors or drug smugglers, with accompanying human abuses (Alamgir et al., 2017). Roads can also foment social unrest and disease outbreaks via the influx of migrants or colonists into forested areas. Logging roads, for example, are associated with increased incidence of diarrheal disease among communities in the Choco rainforest of coastal Ecuador, with villages closer to roads suffering higher infection rates (Eisenberg et al., 2006).

Corruption, ineffective management of land tenure rights, and the injection of huge sums of foreign capital by timber companies into poorly governed forest frontiers continue to drive social issues to varying extents across timber exporting tropical nations. Papua New Guinea's programme of granting Special Agricultural Business Leases, for instance, was intended to jumpstart agricultural development in rural areas. Instead, between 2003 and 2013, it was widely used as a guise by timber industries to smuggle unsustainable logging operations onto huge areas of customary lands, without fair consultation or compensation of landowners (Nelson et al., 2014).

Forest loss associated with selective logging can further threaten local communities. Negative social impacts range from worsening of localised climatic conditions (Meijaard et al., 2013) to growing malnutrition following the establishment of oil palm plantations in previously intact forest, an outcome likely driven by subsistence-based communities losing access to forest foods and medicines (Santika et al., 2019). Fires can also exert a heavy socioeconomic toll at local scales and beyond. In the Amazon, fires are associated with a significant drop in the availability of key game and fruit species (Peres et al., 2003), whereas In Indonesia, infants raised under the air pollution of 1997s forest fires appear to have stunted growth and earning-ability as adults (Tan-Soo and Pattanayak, 2019).

Where legal logging is conducted in a socially responsible manner and avoids future forest clearance it can, however, bring extra sources of revenue into local communities, reduce inequality, and empower those who live in or near the forest. Under forest certification schemes, most outcomes provide improvements or no change to livelihoods and community well-being of local people (Burivalova et al., 2019a). Combining certification with reduced-impact logging (RIL) can provide positive economic outcomes, but these often fall outside local communities and benefit logging companies. Community forest management is a way of handing the benefits of sustainable logging back to local people, where joint forest management is practised within a community to enhance community welfare and ecological sustainability (Burivalova et al., 2017). Systematic review of the socio-economic impacts of community forest management reveals mixed results (Burivalova et al., 2019a). For example, implementation of CFM reduced inequality in Mexico, had no effect in Bolivia and Kenya, but increased inequality in Uganda (Coleman and Fleischman, 2012). Furthermore, in Malawi, the social impacts of CFM were either positive or negative depending on the individual reserve in question, with more benefits being captured by male and richer participants (Jumbe and Angelsen, 2006). Further exploration of the benefit-sharing mechanisms, and enabling conditions and policies that drive socially favourable logging outcomes will allow up-scaling of successful interventions.

7. Delivering more sustainable outcomes via improved logging management

Given their potential to retain substantial biodiversity and ecosystem service values at or near primary forest levels, combined with the substantial losses upon fire or conversion to farmland, the value of preserving 'intact' selectively logged forests as a conservation measure has garnered increasing attention and recognition in the last decade (Berry et al., 2010; Chazdon et al., 2009; Edwards et al., 2010c; Ghazoul and Sheil, 2010; Gibson et al., 2011; Putz et al., 2012; Sheil and Meijaard, 2010; Wilson et al., 2010). While it is important to continue to conserve primary forests (Gibson et al., 2011), developing the role of well-managed logging in improving the retention and recovery of biodiversity and ecosystem services within logged forest is key. We have an excellent opportunity to maximise the conservation value of much of the 403 million hectares of tropical forest designated for timber production through improved logging techniques and strategies.

The concept of sustainable logging has been recognised as far back as the 18th century (Lowood, 1990). While these past motives related to financial sensibility, a push towards sustainable logging with conservation in mind has emerged in the last ~30 years. This proposed framework, with the goal of protecting forest biodiversity and ecosystem services, is often referred to as sustainable forest management (SFM). SFM is strongly endorsed by the UN and World Bank, yet adherence is optional (UNFF, 2007). While they may not often enforce them, the majority of nations implement regulations supporting SFM, meaning 99% of the world's forests are protected by it (FAO, 2015).

SFM calls for regulations in harvest intensity, minimum felling diameters, minimum cutting cycle lengths and seed-tree retention rates, applied using reduced-impact logging (RIL) techniques (Gardner, 2010). The basis of these regulations is generally well founded. High logging intensity inhibits regeneration (Anitha et al., 2010; Struhsaker et al., 1997; Van Gardingen et al., 2003) through increased canopy openness, which favours pioneer species, causing recruitment failure (Peña-Claros et al., 2008), increased seed tree loss (Plumptre, 1995) and induces forest fires (Matricardi et al., 2010). Small felling diameters and short cutting cycles essentially manifest as increased logging intensity, but beyond that, as more frequent disturbance that exacerbates other aspects of selective logging, such as physical disturbance, further preventing recovery. Despite broad on-paper adherence to certain SFM practises, many tropical governments lack the enabling regulatory conditions and monitoring capacities needed to enforce prescribed sustainable techniques. As of 2015, only 37% of low-income countries report forest inventories detailing how forests are managed forests, and 90% of all forest certification schemes are limited to temperate and boreal environments (MacDicken et al., 2015).

7.1 Reduced-impact logging (RIL)

RIL encompasses a suite of harvest practises designed to minimise the impact of selective logging within the forest. Specialist techniques result in careful harvest planning and implementation to minimise forest and soil damage. First described by Putz and Pinard (1993), RIL creates a solid set of guidelines for responsible logging. Methods employed in RIL include: (1) detailed pre-harvest tree mapping and subsequent planning of roads, skid trail and log landings allowing for minimal soil and waterway disturbance; (2) directional felling, increased worker safety practises and optimum stem crosscutting to minimise waste and avoid incidental damage; and (3) post-harvest assessments allowing for evaluation of the application of RIL guidelines. RIL guidelines can vary extensively between regions, and regional specific RIL practises are implemented to reflect the high variation in forest structure that exists across the tropics (Putz et al., 2008).

RIL incurs less incidental tree damage and mortality during the felling process (Bertault and Sist, 1997), as well as reducing the extent of skid trails required for timber transportation (Pinard et al., 2000). For example, in Malaysian Borneo, the total area within concessions disturbed by roads and skid trails fell from 17% under conventional logging to 6% under RIL, with less than half the subsoil disturbance (Pinard et al., 2000). The lower disturbance to the forest associated with RIL better maintains forest structure, and results in considerably faster biomass regeneration (West et al., 2014), which improves future harvestable yield (Pinard and Putz, 1996). For instance, Amazonian forest plots subjected to RIL recovered 100% of above-ground biomass within 16 years, compared to a recovery of 77% under conventional logging in the same time period (West et al., 2014). In turn, a pan-tropical meta-analysis of the biodiversity impacts of RIL demonstrates a consistent reduction of negative logging impacts on birds, mammals and arthropods, even after correcting for the reductions in harvest intensity associated with RIL (Bicknell et al., 2014). Considering the available literature, RIL had superior environmental outcomes to conventional logging in 76% of comparisons (Burivalova et al., 2017).

Given the reduction in damage and improved forest recovery rates, RIL can play a critical role in limiting the contribution of tropical forest emissions to climate change. Since a number of current RIL practises were not designed with this aim in mind, 'RIL-C' has been promoted as a subset of RIL practises that focus specifically on achieving emission reductions (Griscom et al., 2014). According to the largest field-based study of its kind to date, the pan-tropical adoption of RIL-C practises has the potential to effectively halve (by 468 million tonnes) yearly logging emissions versus conventional logging practises (Fig. 3).

Despite its environmental benefits, concern over the financial burdens imposed by RIL is often expressed as the major constraint to its widespread adoption (Putz et al., 2000). Frequent investment in new harvesting and safety equipment is required, as well as regular (re)training of crews and the hiring of technically qualified supervisors, has the potential to put RIL at an economic disadvantage (Putz et al., 2008). There are, however, economic benefits to RIL's implementation making it more profitable than conventional logging, due to improved productivity and reduced waste (Bacha and Rodriguez, 2007; Holmes et al., 2002). With conventional logging, skid crews often work in dense forest without adequate maps, leaving behind up to 20% of marketable logs (Richter, 2001).

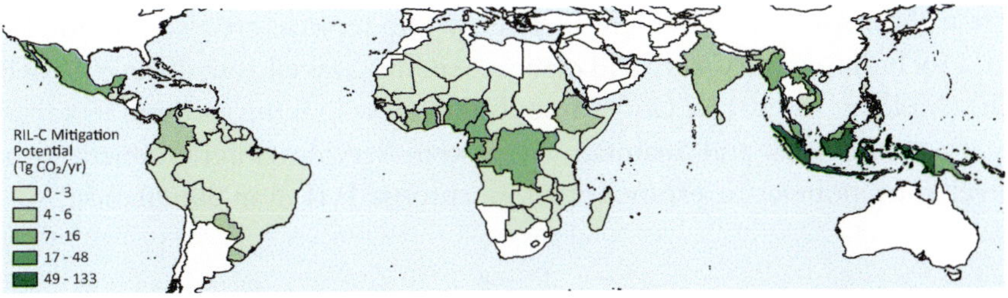

Fig. 3 Potential for reduced-impact logging for climate (RIL-C) to deliver emissions reductions relative to conventional logging across the tropics. Field studies were conducted in 7 countries and extrapolated across 83 tropical nations. RIL-C activities have the potential to nearly halve tropical logging emissions, preventing the emission of an estimated 468 million tons of CO_2 per year (Tg CO_2/year). *Adapted from Ellis, P.W., Gopalakrishna, T., Goodman, R.C., Putz, F.E., Roopsind, A., Umunay, P.M., Zalman, J., Ellis, E.A., Mo, K., Gregoire, T.G., Griscom, B.W., 2019. Reduced-impact logging for climate change mitigation (RIL-C) can halve selective logging emissions from tropical forests. For. Ecol. Manag. 438, 255–266.*

Furthermore, effective use of RIL allows logging firms to apply for premium certification schemes providing access to lucrative timber markets (Kollert and Lagan, 2007). Overall, the economic benefits of employing RIL vary between sites and are inconsistent. Despite improved management plans, lower skidding costs and market premiums, RIL was unable to compensate for the lower pre-harvest costs of conventional logging in more than half of the outcomes assessed by Burivalova et al. (2019a,b). The financial benefits or drawbacks tied to RIL thus show high levels of variation, meaning that it can be seen as a potentially risky investment.

Beyond financial implications, RIL offers socio-economic benefits to a range of stakeholders, which cannot be directly included in economic analysis. Adoption of RIL involves the training of specialist crews (Pokorny et al., 2005), providing opportunities for the development of skilled labourers. RIL safety guidelines provide better protection for workers (Pereira et al., 2002), also benefitting the institutional reputations of both employers and local governments. Additionally, RIL requires respecting the land rights of indigenous people (Burivalova et al., 2017) and can protect the livelihoods of local communities by reducing the impact of logging on non-timber forest products (Guariguata et al., 2010; Ndoye and Tieguhong, 2004). For example, the biomass of locally important epiphytes in the montane forests of Costa Rica were equal in unlogged plots and those where RIL had been implemented (Romero, 1999), and RIL forests in the Amazon hold similar yields of game and tree fruit compared to unlogged forests (Menton et al., 2009). Wider ecosystem services are also provided by RIL, including biodiversity and carbon benefits, as well as hydrological benefits (Miller et al., 2011). Governments that view logging through its capacity to create jobs and maintain ecosystem services might benefit from developing policies to promote and incentivise RIL implementation.

7.2 Spatial management of harvests

Logging intensity is an important factor in managing logged forests, as harvest intensity directly impacts the responses of biodiversity (Burivalova et al., 2014) and carbon (Griscom et al., 2017). As previously mentioned, logging intensity varies by more than two orders of magnitude (<1 to $>100\,m^3$/ha) depending on the region, and temporal variation also exists depending on local timber shortages and market willingness to accept lesser-known species (Putz et al., 2001).

Reducing logging intensity could be a fruitful option for minimising harm. However, reduced intensity might generate logging across a wider spatial scale to meet market demand. Given this, logging could be considered in the context of land sparing versus land sharing to improve spatial management of concessions. Land-sparing logging entails high-intensity cutting throughout a section of the concession, while another section is left unlogged as protected primary forest. Alternatively, land sharing entails a harvesting at lower intensity throughout the entirety of the concession, such that no large area of land remains unlogged (Edwards et al., 2014d). In many scenarios, however, simply adopting a purely land sparing or land sharing method may not accommodate for the structural complexity and variation within and between logging landscapes, therefore adopting a blended approach that incorporates the two might prove more suitable.

In Malaysian Borneo, more species of birds, dung beetles, and ants had higher abundance under land-sparing than land-sharing logging, resulting in higher overall species richness in land-sparing scenarios compared to land sharing (Edwards et al., 2014d). This disparity was extended further when considering species richness of only primary forest species, as land sparing maintained vital patches of undisturbed primary forest to sustain these specialists. However, more recent studies suggest that whether land-sharing or land-sparing logging is more beneficial for biodiversity retention depends upon deforestation pressure and land tenure. Griscom et al. (2017) found land sparing to be the least harmful only in the absence of deforestation pressure and secure land tenure. On the other hand, when land tenure is insecure and deforestation pressures exist, land sharing (with certified RIL) provides more beneficial conservation outcomes (Griscom et al., 2017). These divergent findings are driven by the increased ability of the landowner to resist deforestation due to their commercial interest in maintaining future harvests.

Focusing on butterflies in the Amazon, Montejo-Kovacevich et al. (2018) found more species benefit from land-sharing logging than land sparing, but that the species with the highest abundance increase between strategies tended to be those that benefitted from land sparing. The authors thus proposed an intermediate method where some primary forest patches are retained within a mosaic of lower-intensity logged forest. The study sampled species abundance responses across a range of local logging intensities (as opposed to only two broad-scale intensity levels in Edwards et al. (2014d)), allowing for a more detailed understanding of the fine-scale spatial impacts of logging intensity. Abundance-yield curves were produced for individual species across a range of logging intensities (Montejo-Kovacevich et al., 2018) and this method should be extended to a wider range of taxa and repeated across the tropics. For instance, focusing on the six indicator taxa identified by Edwards et al. (2014b) would deepen our understanding of the broader biodiversity impacts across a spectrum of logging intensities.

Intermediate strategies, such as those suggested by Montejo-Kovacevich et al. (2018), dispels the dichotomic nature of the land-sparing and land-sharing debate which has been previously criticised for being too rigid, ignoring optimal strategies that integrate both forms of land management (Butsic and Kuemmerle, 2015; Ekroos et al., 2016; Kremen, 2015; Legras et al., 2018). Much success has been had in North America using a 'triad' management system, which maintains an extensive shared area, and smaller areas of intensively managed land and protected areas. However, these are extensive plots of land securely owned by few tenants (Tittler et al., 2016) and so the implementation of 'triad' systems may become more complex in tropical regions with insecure land tenure and unclear governance (Griscom and Goodman, 2015). Abandoning the dichotomy in favour of intermediate strategies may create mosaics of logged and unlogged habitat that sustains yields and biodiversity, protecting primary forest specialists (Ramage et al., 2013) and maintaining heterogeneity in forest structure (Chaudhary et al., 2016).

Some authors question the benefits of either land sparing or land sharing and argue the benefits of improving management techniques outweigh those from altering harvest intensities across landscapes (Runting et al., 2019). Using spatial optimisation models across East Kalimantan, Runting et al. found improved management techniques (e.g. RIL) resulted in better conservation outcomes than altering the balance between land sparing and land sharing. However, in their scenarios, Runting et al. used plantation forestry to represent complete land sparing, but plantations may not always represent an economically viable alternative to the harvesting of natural timber. While Southeast Asian forests are dominated by dipterocarps used to make low-value products that could otherwise be produced from plantation wood, logging in Africa and the Neotropics mainly targets rarer species used for high-value furniture and flooring (Fisher et al., 2014).

Another possibility is to better manage the spatial arrangement of logging across concessions. Harvest plans can be designed to minimise species extinctions by maintaining a matrix of different-aged patches in proximity or by creating habitat blocks of similar successional stage. In simulation models of trees in a concession that is to be entirely logged, harvest plans with large contiguous harvest units yield high extinction probabilities for dispersal-limited species with clustered pre-harvest distributions (Ramage et al., 2013). These results suggest that small, randomly located harvest units can reduce extinction rates in tropical production forests, although they do not consider the potential negative impacts of a larger network of roads and skid trails. The key question is how protecting primary features (blocks, riparian strips, etc.) impacts upon these predictions and the optimal size and arrangement of such primary patches. For instance, across 10,000 hectares of logged forest sampled in Africa (Gabon, the Republic and Democratic Republic of Congo), Suriname, Mexico and Indonesia, an average of 54% (range 22–97%) of the total logging areas sampled were not directly affected by timber harvests. Better spatial planning to target these intact zones towards ecologically sensitive areas, including riparian buffers and slopes, could yield conservation benefits even without any changes in timber harvesting intensity (Putz et al., 2019).

Ultimately, the optimal spatial strategy for logging will differ substantially depending on the region under consideration, as well as the key outcome priorities (i.e. maximising profit, carbon retention, or biodiversity protection). Developing a comprehensive model capable of comparing a consistent set of conservation priorities spanning land-sparing, land-sharing, and intermediate strategies across a range of spatial arrangements to set site-specific management strategies should be a priority.

7.3 Roads

The creation of roads and skid trails to extract timber is one of the most ecologically destructive elements of selective logging (Hendrison, 1990).

Roads tend to be used for an entire logging season or across multiple seasons and can be (semi-)permanent in construction. Road width can average 20 m in logging concessions (Kleinschroth et al., 2016a), in part to allow the sun to reach and dry the surface, though in better-managed concessions this is often reduced to 6–8 m in width with a more connected canopy above. On the other hand, skid trails created by tractors are typically 4–5 m wide (Nabe-Nielsen et al., 2007; Pinard et al., 2000; Swaine and Agyeman, 2008), but may easily increase in width with repeated passes or excessive manoeuvring. In contrast to roads, many skid trails only receive single passes (Buckley et al., 2003) and are thus made for the single-use purpose of extracting trees. Yet skid trails constitute up to 40% of the area of a concession in poorly managed operations (Pinard et al., 1996; Wei et al., 2015), and the tractors or bulldozers responsible for skid creation can cause considerable damage to the stand (Pinard et al., 1996).

Soil disturbance and increased compaction from mechanised vehicles and dragged timber can persist for over 20 years (Hattori et al., 2013). Furthermore, networks of skid trails can fragment the forest resulting in edge effects, leading to decreased tree survival and seed dispersal (Broadbent et al., 2008) and providing corridors for further disturbances such as fire (Putz et al., 2001). In selectively logged forests in the Brazilian Amazon, a network of skid trails under selective logging can result in six times the area of forest edges than clear-cut fragments (Broadbent et al., 2008). Other potentially deleterious effects to the forest environment from skid trail creation include reductions in soil moisture, increased surface runoff, soil erosion (Putz et al., 2001), decreased vegetation cover, and increased canopy openness (Brais, 2001; Nabe-Nielsen et al., 2007).

Recruitment, growth and survival of the forest plant community can be severely impacted by skid trails (Pinard et al., 1996). For example, in Malaysian Borneo, the nutrient depleted and compacted soil on skid trails resulted in 80% loss of artificially sown seeds within just three rainstorms (Pinard et al., 1996). Such unfavourable conditions limit the establishment of trees on skid trails and delay recovery of pre-logging forest structure. Furthermore, logging roads negatively impact dung beetle communities, with edge effects discernable up to 170 m into the forest (Edwards et al., 2017).

Beyond the damage caused by wasteful skid trails, the construction of logging roads is responsible for 50% of carbon emissions from logging activities in the Congo (Umunay et al., 2019). Road construction therefore represents one of the most harmful aspects of selective logging, and the expansion of logging operations globally has led to annual expansion of road networks by 40 km per 10,000 km^2 of forest (Brandão and Souza, 2006). In the Congo Basin, road lengths within logging concessions have doubled since 2003, compared to just 40% growth outside of concessions. Fortunately, the impact of expanding road networks within concessions is partly ameliorated by the higher abandonment rates of logging roads (44%) compared to roads outside concessions (12%; Kleinschroth et al., 2019a; Fig. 4). Road construction leads to a loss of 0.6–8% forest cover within concessions, and roads that take on permanency facilitate increased human migration, forest clearance and hunting after cessation of logging (Kleinschroth and Healey, 2017). In the Brazilian Amazon, 95% of deforestation occurs less than 5 km from roads (Barber et al., 2014).

Tropical governments looking to expand infrastructure to stimulate economic growth will need to be convinced of the benefits of restricting road development. For example, recent plans for a Pan-Borneo Highway would drastically restrict the integration of protected areas and facilitate increased forest exploitation. Planned roads would separate two major clusters of protected area and attempts to mitigate damage through forest corridors and highway underpasses are unlikely to achieve meaningful ecological connectivity (Sloan et al., 2019).

In Borneo as in other tropical regions, tackling the issue of expanding road networks will likely require convincing regional governments of the benefits of undertaking large-scale, coordinated efforts to implement widespread decommissioning of logging roads. Looking forward, the expansion of permanent roads should be restricted (these are more likely to cause fragmentation and forest conversion; Kleinschroth et al., 2019a), with incentives for strategic, conservative road network expansion paired with rapid post-logging road closure and reopening of old roads during reharvest (Bicknell et al., 2015; Kleinschroth et al., 2016b).

8. Natural and Enhanced Recovery of Logged Forest Values

8.1 Natural Recovery

Natural recovery of logged forests and timber stocks, even under RIL, are often insufficient to maintain economically attractive cutting cycles. Post-logging accumulation of carbon is five times greater in logged forest than natural forest (1.4 and 0.28 MgC/ha/year, respectively; Berry et al., 2010), and recovery of above-ground biomass can be twice as high (4.82 ± 1.22 and 2.58 ± 1.73/ Mg/ha/year; Gourlet-Fleury et al., 2013), albeit this does not represent a net gain of carbon due to carbon losses

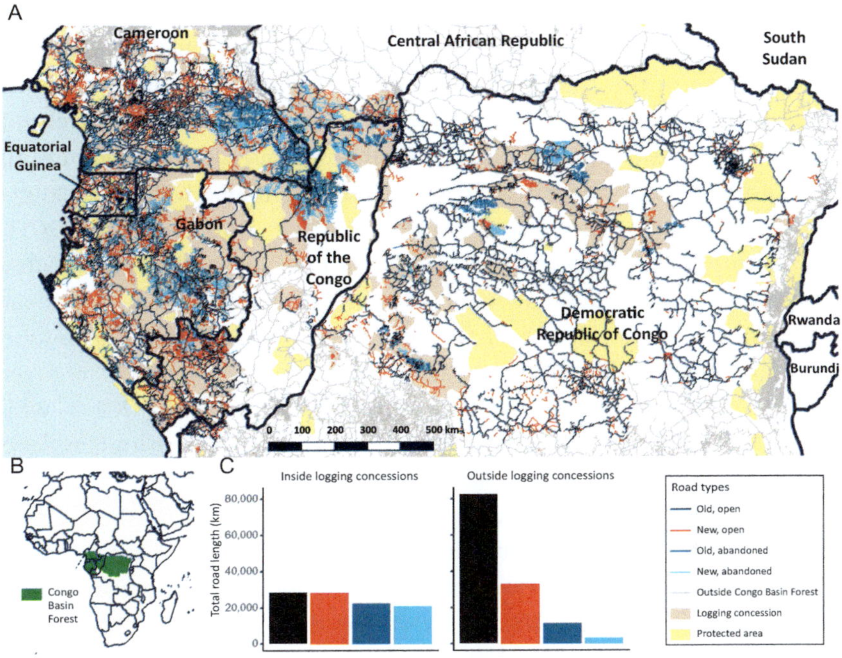

Fig. 4 The proliferation of logging roads across the Congo basin. Road lengths within logging concessions have doubled since 2003, compared to just 40% growth outside of concessions. The impact of expanding road networks within concessions is partly ameliorated by the higher abandonment rates of logging roads (44%) compared to roads outside concessions (12%). *Adapted from Kleinschroth, F., Laporte, N., Laurance, W.F., Goetz, S.J., Ghazoul, J., 2019a. Road expansion and persistence in forests of the Congo Basin. Nat. Sustain. 2, 628–634; Bélanger, L., Mertens, B., 2011. Atlas Forestier Interactif de la République Démocratique du Congo Version 1.0. World Resources Institute; Boisrobert, L., Bégoto, G., 2010. Atlas Forestier Interactif de la République Centrafricaine Version 1.0. World Resources Institute; Mertens, B., Makak, J.S., 2009. Interactive Forest Atlas for Gabon—Atlas Forestier Interactif du Gabon (Version pilote). World Resources Institute; Mertens, B., Shu, G.N., Steil, M., Tessa, B., Minnemeyer, S., Douard, P., Blang, J.M., Leach, A., Mbouna, D., Mboua, P., Fuwzing, A., Methot, P., Nglilambi, H., 2012. Interactive Forest Atlas of Cameroon Version 3.0. World Resources Institute; Tessa, B., Ibara, M., Methot, P., Minnemeyer, S., Douard, P., Steil, M., Mertens, B., 2012. Atlas Forestier Interactif du Congo Version 3.0. World Resources Institute; Tessa, B., Mboro, F.E., Steil, M., Douard, P., Ondo, J.A., Martins, A., 2013. Atlas Forestal Interactivo de la República de Guinea Ecuatorial Version 1.0: Documento de Síntesis. World Resources Institute, 2013. Panels (B) and (C).*

during logging. Despite impressive recovery rates, typical harvest cycles are short and range from 20–35 years, where even modest harvest intensities (e.g. $20\,m^3/ha$ in the Brazilian Amazon) under RIL result in a maximum recovery of 70% of timber volume (Piponiot et al., 2019). Harvest intensity appears to play an important role in forest recovery rates, thus heavily logged forests recover biomass at a slower rate than moderately logged forests (0.25 and $0.32\,m^3/year/ha$, respectively; Bonnell et al., 2011). Furthermore, full structural recovery takes longer in heavily logged forests than those that have been moderately logged (112 years and 95 years, respectively; Bonnell et al., 2011), and recovery is also delayed on abandoned skid trails and tracks (Pinard et al., 2000). What is clear is that full recovery of timber stocks requires 40 years or more (Roopsind et al., 2017), while carbon stocks can take more than 60 years to recover to pre-logging levels (Roopsind et al., 2017; Ruslandi, et al., 2017a).

As for other ecosystem services (evapotranspiration, watershed services etc.), our knowledge is limited. Such long recovery periods are economically unattractive and therefore rarely implemented. Restoration efforts and post-logging interventions therefore represent a method to improve post-logging recovery rates, though it is worth noting that these should be used as a supplement rather than as a replacement for improved pre-logging practises.

8.2 Enhanced recovery via post-logging interventions

The current political climate has ushered in enormous incentives for large-scale restoration of degraded tropical forest to improve ecosystem services and biodiversity in human-impacted landscapes (Brancalion et al., 2017). Initiatives such as the Bonn Challenge and New York Declaration on Forests have been widely embraced, with over 40 countries and institutions so far committing to the recovery of 350 million hectares of degraded land by 2030 (Bonn Challenge, 2018). Such is the enthusiasm for these restoration programmes that 10% of countries appear to have committed more area for restoration under the Bonn Challenge than they have available (Bastin et al., 2019). Despite this, global forest restoration potential is estimated at 0.9 billion hectares, creating ample opportunity to restore the biodiversity of degraded landscapes while storing an extra 205 gT of carbon (Bastin et al., 2019). The restoration of logged forests through post-logging interventions thus represents a method to combine political goals of landscape restoration with sustainable long-term timber production.

Post-logging interventions, such as the cutting or poisoning of vines and competing non-commercial tree species, can more than double the natural regrowth of commercial tree species (Peña-Claros et al., 2008), while enrichment planting of commercial species can allow full recovery of timber stocks within 40 years (Ruslandi, et al., 2017b). Such interventions are also capable of improving the recovery of carbon stocks. Lianas restrict tree growth and can reduce large-tree carbon stocks by up to 50%, since large trees store a disproportionately large amount of carbon, removal of lianas after logging allows improved growth and carbon sequestration (Villegas et al., 2009). Such interventions are capable of improved recovery of timber and carbon stocks, as well as the provisioning of local communities with non-timber forest products and stable employment (Cerullo and Edwards, 2019).

Post-logging interventions can, however, cause adverse biodiversity impacts when timber recovery or carbon sequestration are the key aims (Putz and Redford, 2009). The planting of a limited number of fast-growing tree species can reduce both species and functional diversity, and the removal of vines and understory plants can reduce biodiversity due to the loss of food, foraging substrate, nesting sites and protection (Edwards et al., 2009). The evidence base of the impacts of post-logging interventions on biodiversity is, however, more limited than carbon or timber stock recovery (Cerullo and Edwards, 2019). Removal of lianas in Panama increased community-level canopy tree reproduction (García León et al., 2018) and logging gap tree species diversity by 65% (Schnitzer and Carson, 2010), whereas non-timber tree thinning in the Congo had little effect on tree diversity (Gourlet-Fleury et al., 2013).

Impacts of post-logging interventions on animal diversity appear equally unclear. In Borneo, enrichment planting increases vegetation cover (see chapter "Monitoring tropical forest degradation and restoration with satellite remote sensing: A test using Sabah Biodiversity Experiment" by Wu et al., this issue). However, enrichment planting and liberation cutting has little effect on invertebrate (Edwards et al., 2010c) or IUCN red-listed bird abundance (Ansell et al., 2011). Liberation cutting also results in negative impacts on fruit-eating birds likely due to the loss of fruit sources, but positive impacts on insectivorous birds presumably due to a more open understory (Edwards et al., 2009). Dung beetle communities in restored forests are similar to those in unlogged old-growth forest, and the community composition of treated forest more closely resembles old-growth forest than does naturally regenerating forest in terms of effective diversity (Cerullo et al., 2019). By contrast, bird communities in restored forest had fewer specialists with a more truncated evolutionary history in comparison to bird communities within naturally regenerating forest (Cosset and Edwards, 2017).

Timber enhancement could provide two additional beneficial biodiversity outcomes. First, by maintaining timber harvest potential and the associated needs to protect these resources, enhancement may reduce the potential for logged forests to be reclassified for agricultural clearance or be illegally encroached upon, with enormous environmental benefits. Second, in either land-sparing or land-sharing logging scenarios, enhancement may restrict the temptation for landowners to increase exploitation levels (Cerullo and Edwards, 2019). Under a land-sparing scenario, post-logging interventions in areas under high-intensity timber extraction could improve timber restocking rates to restrict the incursion of harvest activities into 'spared' areas of primary forest. Similarly, under land-sharing scenarios, timber-restocking efforts could prevent adoption of higher-intensity timber extraction throughout concessions.

The range of both negative and beneficial impacts of post-logging interventions on different aspects of biodiversity suggests a complex interaction between interventions and the species-specific traits that make species either vulnerable to or benefit from them. How forests intensively treated with timber enhancing activities like liberation cutting will function under climate change is also an underexplored area. Evidence suggests that droughts during El Niño years can undermine the positive impacts of liana cutting on forest regeneration (O'Brien et al., 2019). Due to this, more extensive research is required to identify any general long-term consequences of post-logging interventions and to improve our understanding of the finer points of biodiversity responses to these interventions. In the meantime, post-logging interventions that centre on timber or carbon recovery must make attempts to limit negative consequences for local biodiversity by targeting activities towards easily accessible, small forested areas.

9. Logged forests under climate change

Climate change in the 21st century represents a huge threat for biodiversity globally, and particularly so in the tropics. Amazonia is currently warming at a rate of 0.25 °C per decade, with the potential for a rise of 3.3 °C by the end of the century (Malhi et al., 2008). Accompanying the predicted warming is significant drying across large regions, with enhanced duration and intensity of dry seasons, and permanent disruption of El Niño events (Cochrane and Laurance, 2008), which are capable of increasing pantropical fires by 133% (Chen et al., 2017).

Additionally, selective logging of primary forests increases fire vulnerability through increased desiccation, fuel provisioning, opening of the canopy and drying of understory (Cochrane and Laurance, 2008; Malhi et al., 2008). It follows that in a warming climate where drought frequency and intensity is expected to increase, selectively logged forests and the biodiversity they contain are under serious threat of increasing fire vulnerability and the destruction that accompanies forest fires (see above).

Synergies between climate change and rapid land-use change thus put tropical forests under particular threat (Brodie et al., 2012). However, despite large increases in local temperatures after conversion of primary forest to agricultural land, selectively logged forests exhibit minimal local temperature increases, and could therefore provide important refugia for thermally restricted species (Senior et al., 2017). In intensively selectively logged forests in Borneo where large structural differences remain after 9–12 years, logging activity has very little impact on both macroclimate and microclimate thermal buffering (Jucker et al., 2018; Senior et al., 2018; Fig. 2B), with rapid recovery of the primary forest thermal environment in the Amazon within 5 years post-selective harvest (Mollinari et al., 2019). Only when focusing on pre-clearance salvage logged forest, in which regulations on minimum felling diameter, repeat logging incursions, etc., have been removed making structural damage extreme (Blonder et al., 2018; Hardwick et al., 2015; Jucker et al., 2018), and when using a single 1-ha plot of tall primary forest as a control (Blonder et al., 2018), are microclimate temperatures warmer than primary forest after ~10 years of forest recovery.

Compared to primary forests, logged forests in Borneo maintained a similar density of understorey microhabitat features, including tree holes, leaf litter and deadwood, which offer favourable microclimates (Senior et al., 2018). The maintenance of thermally buffered microhabitats in logged forests may prove crucial, given their importance in providing species with effective refuge from extreme temperature events (Scheffers et al., 2014). However, whether selectively logged forests really can equal primary forest in their thermal buffering potential, and by extension their ability to retain temperature-sensitive species under climate change scenarios (Senior et al., 2018) remains contentious. In particular, attempts to simultaneously capture longer-term temporal variation in microhabitat temperatures alongside information on the distribution and availability of microhabitats in multidimensional forest space, from the undergrowth through to the canopy, are now needed.

Drought exposure poses an additional largely unquantified risk to logged forests. Natural drought and rainfall exposure experiments have demonstrated decreased tree growth and increased mortality, with large trees affected most severely (Corlett, 2016). This could pose particular problems for logged forests, since large trees are favoured by commercial timber extraction. Investigation into 14 commercial tree species in the Guiana Shield has found variability in their ability to cope with drought-induced water stress, with some species proving significantly more vulnerable than others (Fargeon et al., 2016). In light of this, expanding these predictions to important and valuable commercial species, such as Mahogany and Ipe, is a research priority. Such information would allow for conservation plans specific for valuable species vulnerable to water stress to maintain economic productivity of logged forests under a changing climate. An element of conservation plans for commercial species could involve active gene transfer from model plants to be grown in timber plantations, with increasing commercial tree drought tolerance using such methods (Zhang et al., 2013).

Solid understanding of how selectively logged forests will respond to the threats posed by climate change through increased temperatures and drought frequency is still lacking. Selectively logged forests may be capable of rapidly recovering the thermal climate of primary forests (Mollinari et al., 2019; Senior et al., 2017, 2018), but how this recovery will be affected by the extreme temperatures expected under various climate scenarios remains to be seen. Similarly, the increasing frequency and severity of droughts will exacerbate water-stress in large merchantable trees as well as fire risk, and we are still some way from knowing how selectively logged forests will cope under such conditions. The scant attention paid to the potential effects of climate change on logged forests and rates of timber extraction poses a real concern for our understanding of how logged forests and associated industry will cope under climate change, with immediate attention required to better inform future management practise.

10. How to drive improved logging management—Legal levers and monitoring, certification, REDD+, and FLR

10.1 Legal levers and monitoring

A solid starting point for improving the management of logging is through legal means that restrict illegal logging, reduce corruption and enforce the law. Illegal logging is pervasive across the tropics, representing between 50–90% of tropical timber products (Nellemann, 2012). The large uncertainty is due to the secretive nature of illegal logging and that the impacts are often difficult to distinguish from legal logging (Tacconi, 2012). Such impacts can be direct via reductions in carbon stocks, species richness, and NFTP for local communities, or indirect through influencing political systems, timber markets, and economic returns (Pacheco et al., 2016). For instance, according to a spatially explicit simulation of the Amazonian timber industry, illegal logging drastically reduces the profitability of legal logging (Santos de Lima et al., 2018). Under scenarios where illegal logging was operating outside of concessions, Santos de Lima et al., observed a 59% drop in the concession area under legal reduced-impact logging across a thirty-year cycle (Fig. 2C), with an incurred estimated loss of $1.3 billion from timber concessions in National Forests. Alongside its other manifest social and ecological harms, illegal logging ultimately undercuts the viability of a sustainable forest sector.

Steps to reduce corruption and illegal logging in the tropics have been taken by large consumers of tropical timber (e.g. The EU's FLEGT [Forest Law Enforcement, Governance and Trade] Action Plan), which aim to support timber producing nations, promote legal markets and tackle systemic corruption (http://www.euflegt.efi.int/flegt-action-plan). However, effective policies must also be implemented throughout the timber producing regions themselves, and be backed by a step-change in the timber sourcing practises of emerging markets in China, Vietnam, and India, whose less-discerning imports are impeding the possibility for consumer-oriented measures to curtail illegal logging.

Greater monitoring of logging could serve as a critical component for detecting illegal harvesting as well as enforcing legally prescribed logging limits. Currently, most monitoring of logging focuses on the use of Landsat imagery, but this is often only capable of detecting areas logged at intensities $>20\,m^3/ha$ with large canopy gaps, thus complicating the detection of the widespread lower-intensity logging deployed throughout the Amazon and Congo. However, improvements in the detectability of low-intensity logging are being made through innovative applications of new technology. For example, a machine-learning algorithm that uses Landsat imagery has been applied to selectively logged forests in the Brazilian Amazon and was capable of detecting 40% of low-intensity logging activity (Fig. 2D).

Since many of the impacts of logging are not visible through changes in canopy structure, satellite-derived monitoring should, where possible, be complimented by below-canopy survey techniques. Bioacoustic monitoring is emerging as a scalable way of coupling observed changes in forest management and structure with on-the-ground information on biodiversity impacts and human disturbance. Bioacoustic devices are already being deployed as a means of detecting the noises of chainsaws and gunshots in real-time

(Fig. 2A ii). As accompanying softwares advance, acoustic data could be used as a tool for testing the effectiveness of forest sustainability schemes and as a method of corporate accountability (Burivalova et al., 2019b).

To further promote transitions towards timber management that prioritises the maintenance of sustainable yields in a socially and environmentally acceptable manner, strong institutions are needed to generate funding, enable policies and influence business practises (Cerullo and Edwards, 2019). The prevailing 'cut-and-run' outlook of large logging corporations chasing immediate profits requires a significant overhaul. Current political volatility in a number of tropical governments and the awarding of short-term timber licences reduces commercial incentives to plan for the long term and invest in sustainable harvesting practises and post-logging interventions. Extending the length of logging contracts and giving preference to those with a track record of good environmental practise could rectify the short-term aims that permeate the industry. Furthermore, certification schemes can play a role by including long-term sustainable practises within their requirement framework, and initiatives such as the Bonn Challenge could direct large sources of funding towards climate-smart logging activities (Cerullo and Edwards, 2019).

10.2 Certification

Certification schemes are a market-based mechanism that can help to fund sustainable logging, where certified status is granted for adopting environmentally conscious practise. The principle example is the Forest Stewardship Council (FSC); initiated in 1994, it is the largest scheme of its kind, yet it has had limited reach in tropical forest (<2%). FSC criteria are based on environmentally sustainable logging while providing wider social benefits (Bennett, 2000). Despite the higher costs involved in FSC-certified logging, the financial benefits arising from certification has created a market that financially rewards more sustainable practise. High-quality certified wood can fetch price premiums of 27–57% compared to uncertified products (Kollert and Lagan, 2007), and FSC-certified organisations can expect additional revenue as large as $4.98/m^3 of certified production (Breukink et al., 2015).

FSC-certified concessions can provide societal (Miteva et al., 2015) and environmental (Sheil and Meijaard, 2010) benefits beyond their financial incentives, but this is not always the case and certification does not guarantee more positive environmental outcomes. In the Peruvian

Amazon, for example, there was similar avian diversity and composition in FSC-certified and uncertified concessions (Campos-Cerqueira et al., 2019). In the Congo, deforestation rates within FSC-certified and uncertified concessions did not differ (Panlasigui et al., 2018), and FSC-certified concessions can exhibit equal or greater loss of intact forest landscapes compared to non-certified concessions (Kleinschroth et al., 2017, 2019b).

In both Indonesia (Griscom et al., 2014) and the Congo (Umunay et al., 2019), certified concessions emitted similar levels of carbon compared to non-certified concessions (Griscom et al., 2014). However, incorporating specific emission reductions techniques into FSC requirements (such as RIL-C) could significantly reduce carbon emissions from certified concessions. Broader-scale evidence on the effectiveness of certification schemes in delivering environmental, social, and economic sustainability is meagre despite some being in operation for over two decades (Burivalova et al., 2019a). Identifying management features of certified concessions that effectively retain biodiversity and reduce emissions will allow wider application of these successful aspects of concession management and inclusion within certification criteria frameworks.

10.3 REDD+

Payments for ecosystem services (PES) have become an important part of the conservation toolkit, the largest scheme of which is 'reducing emissions from deforestation and forest degradation' (REDD+). The scheme financially compensates landowners for maintaining their forests and avoiding carbon emissions via deforestation and degradation (Angelsen, 2009). REDD+ was initially envisaged to create a dynamic private carbon market, but this has since failed to materialise, with bilateral agreements providing the majority of current REDD+ funding (Fletcher et al., 2016). However, REDD+ still has potential to provide favourable conservation outcomes, by acquiring large areas of land as set-asides for conservation and funding sustainable management of privately owned tropical forest.

The suitability of carbon payments is highly dependent on the condition of land considered for protection. Attempting to protect unlogged, primary forest from logging would require high carbon prices due to the large opportunity costs of maintaining unharvested forest. In Southeast Asia, for instance, high densities of medium-value tree species results in high carbon prices (~\$28/tCO_2) to offset these opportunity costs (Fisher et al., 2011a). In the Amazon, where logging targets rare, high-value species, the opportunity cost of protecting other tree species drops steeply (Fisher et al., 2014). Because logged forests retain the majority of primary forest biodiversity and have significantly reduced opportunity costs of protection, they can represent a good investment for conservation. In Borneo, for example, logged forests have lost 60–80% of the wood value of unlogged forests, but retain over 75% of bird and dung beetle biodiversity (Fisher et al., 2011b).

REDD+ could also fund activities to reduce the carbon emissions associated with logging. Widespread implementation of RIL across Indonesia has the potential to reduce carbon emissions by 638 million tonnes at a cost of \$23.77/tCO_2, while simultaneously providing the biodiversity and environmental benefits associated with RIL (Graham et al., 2017). Similarly, the adoption of RIL-C techniques in the Congo Basin can reduce logging-related emissions by 50% without jeopardising yield (Umunay et al., 2019). Maintaining yields while reducing carbon emissions is particularly favourable under payment schemes, such as REDD+, as it guarantees no leakage to other areas, since there is no displacement of demand and thus logging activity. Furthermore, concessions adhering to RIL-C guidelines could secure funding from REDD+ as well as certification under FSC, receiving economic benefits in the form of both carbon payments and market premiums for certified products (Kollert and Lagan, 2007). However, FSC certification does not result in any reduction in carbon emissions compared to conventional practise (Umunay et al., 2019). FSC requirements need adjusting accordingly to ensure reduced carbon emissions in certified concessions; the easiest pathway to achieve this is through the inclusion of RIL-C practises into certification criteria.

Criticisms of REDD+—The value of market-based schemes, including REDD+, has been questioned, notably being described as the latest 'fad' in conservation thinking by Redford et al. (2013). Much criticism of REDD+ is based on the failed development of the voluntary market (private investors as opposed to governments and NGOs). In the years 2006–2014, \$9.8 billion was pledged to REDD+, but only 10% of funding came through the voluntary market (Fletcher et al., 2016). The majority of the money was provided through bilateral agreements worth over \$7 billion (Boucher, 2015). Future pledges to REDD+ come from similar sources: ~\$10 billion of cumulative international pledges have been made by the Norwegian and German governments. It therefore appears that REDD+ is moving away from a market-based mechanism and becoming increasingly reliant on pledges (Angelsen, 2017). This movement is largely the result of a failure to place binding national caps on emissions by the 2015 Paris

Agreement, but this could change with continued multilateral climate discussions. The introduction of binding emission caps in the near future could revive the voluntary carbon market.

Being a market-driven force, a common criticism of REDD+ is that projects will prioritise cost-effective emissions reductions, neglecting biodiversity co-benefits in favour of cheap carbon emission offsets (Venter and Koh, 2012). However, tangible evidence of this criticism is scant, and research suggests the opposite is occurring. Murray et al. (2015) show that current REDD+ projects in Indonesia target high biodiversity areas with lower carbon densities, and similar evidence exists for Tanzania (Lin et al., 2014) and Brazil (De Barros et al., 2014). The focus on protection of biodiverse areas is probably a result of the heavy influence of NGOs in current REDD+ projects (Murray et al., 2015). On the other hand, only 25% of projects fell within areas of high deforestation risk. These areas have high opportunity costs associated with them and are therefore unattractive for carbon payments, leaving them in danger of being neglected by REDD+ in favour of cheaper land. It is thus also likely that REDD+ will fail to prevent the logging of primary forest.

REDD+ may also displace logging to areas unprotected by carbon payments (Lim et al., 2017). Such leakage can occur at both local and international scales (Kuik, 2014) and undermines the effectiveness of REDD+, but knowledge on the factors that drive leakage and its extent are lacking (Atmadja and Verchot, 2012). At the wider scale, tracking international trade flows of timber may help to identify transnational leakage (Lim et al., 2017), with closer collaboration between REDD+ nations required to restrict large-scale leakage.

REDD+ may also marginalise (or continue the marginalisation of) vulnerable people (Thompson et al., 2011). Despite some successful social impacts, such as promoting the rights of indigenous communities (Jagger and Rana, 2017) and positive impacts on livelihood assets (Atela et al., 2015), REDD+ projects have not been found to consistently benefit local people. Evidence from Nigeria and Vietnam shows restriction of forest access compromises livelihoods (Asiyanbi, 2016; McElwee et al., 2016), and in Indonesia and Tanzania the land-use practises promoted by projects were not financially sustainable for local communities (Lounela, 2015; Svarstad and Benjaminsen, 2017). Furthermore, REDD+ projects can negatively impact those that are in greater need of project benefits, such as in Nepal, where tightened forest control restricted access to alternative livelihood resources for the poorest households despite benefit-sharing mechanisms (Poudel et al., 2015). To ensure the marginalisation of local people by

REDD+ within logged forest is avoided, greater care should be made to create equity frameworks that avoid generalising across regions, involve locals in decision making and consider local context.

10.4 Forest and landscape restoration—Promoting active management of selectively logged forests

The emerging Forest and Landscape Restoration (FLR) agenda, with its high-level political buy-in and dual goals of enhancing human well-being and ecological integrity could direct substantial funding and resources towards improving management of logged forests. Within production forests, a spectrum of potential interventions could qualify for FLR-based support, with the types of intervention prioritised contingent on restoration goals and the management regime forests fall under (Cerullo and Edwards, 2019). For instance, within community-run production forests, FLR payments could support employment-generating silvicultural interventions for sustaining or increasing yields across logging rotations (e.g. post-logging activities such as liberation cutting). Within ex-production forests, particularly on public lands straddling relatively forest poor communities, FLR activities might involve planting interventions aimed at counteracting the impacts of previously damaging logging operations (e.g. planting on slopes or riversides to reduce erosion), or recovering non-timber forest products of livelihood importance (Cerullo and Edwards, 2019).

Two things are critical for expanding the potential for FLR to deliver positive ecological and livelihood outcomes within selectively logged forests. First is the creation of fair multi-stakeholder processes for determining restoration goals, and which interventions to target where. This is particularly important given that the relative costs and benefits associated with improved management of relatively intact degraded forests will have to be balanced against restoration alternatives, including of establishing agro-forestry systems, woodlot plantations and silvo-pastoral systems on deforested lands (Brancalion et al., 2017). Further quantifying the socio-economic values of different restoration options (Budiharta et al., 2014) and developing transparent, scientifically sound decision-making frameworks for optimising sectoral planning and targeting of restoration interventions will form a central component.

Second, to extend the socio-economic role of logged forest interventions within wider landscapes, planned restorative activities must be buttressed by robust tenure (McLain et al., 2018) and community involvement (Holl, 2017). Besides typically limited silvicultural capacity, a major constraint often preventing communities from investing in optimal pre- and post-logging management is either tenure insecurity or the high cost of forest management licences. Clarifying and securing tenure, either by granting ownership of forest parcels to local communities directly or by developing hybrid public-private initiatives and company-community models, such as timber outgrower schemes, offers a way forward. Breaking down licence structures to award smaller licences that enable community management of heterogeneous landscapes is critical if communities are to invest and derive benefit from restorative activities in logged forest under FLR (Cerullo and Edwards, 2019).

Concessions of mostly timber-depleted forest, granted for the purpose of restoration, should be more widely explored as a potential livelihood enhancing strategy under FLR, particularly since such forests are often left unmanaged and thus vulnerable to clearance (Murdiyarso et al., 2011). In part, this is what Indonesia's Ecosystem Restoration Concession model hopes to achieve. Sixty-year forest restoration licences are granted to Indonesian companies seeking to generate sustainable businesses from payments for recovering ecosystem services, ecotourism and the marketisation of non-timber forest products (Walsh et al., 2011), in theory providing an impetus for halting deforestation. Thus far, the ERC model has yet to emerge as a fully-fledged financially viable instrument for managing degraded forest. However, they are attracting corporate interest, with 573,455 ha out of 2.7 million ha allocated for ERCs currently under management.

The potential for other restoration-concession licencing frameworks to support logged forest conservation and livelihood-enhancing strategies should be piloted elsewhere under the emerging FLR agenda. Priority might be given to existing community-run or smallholder concessions with a successful history of supporting forest conservation. For instance, there are community-managed forests in Zambia where a single chiefdom can now muster more forest guards than the entire national forest service (FAO, 2018).

11. Conclusions and conservation recommendations

As we enter the Anthropocene, a large scientific consensus underscores the critical importance that well-managed selectively logged forests can play in tackling the challenges of biodiversity loss and climate change. Despite prevailing poor logging practises and slow adoption of well-validated strategies for limiting the ecological and social impacts of logging, production forests continue to maintain an array of vital ecosystem services and functions, and are second only to primary forests in their capacity to support tropical biodiversity. Fully unlocking the high-conservation value of logged forests will require tropical governments, industries and emerging climate and forest conservation and restoration agendas to support and enforce the policies, norms and behaviours that ensure that production forests widely employ best-practises and avoid clearance and burning.

Priorities for improving the management of logged forests will vary locally according to management targets as well as broadly across tropical regions. For example, particularly in Southeast Asia, historically intensive timber extraction means large portions of tropical forest are fast approaching timber depletion, with some areas entering their third or even fourth logging rotations. In such regions, a key priority is to explore and implement ways of recovering resilience, and actively protecting economically imperilled forests that have undergone multiple logging rotations (Cerullo and Edwards, 2019; Fisher et al., 2011b). Elsewhere, especially in Amazonia and emerging logging frontiers in Papua New Guinea, precedent should instead be given to ensuring that legal logging follows prescribed reduced-impact techniques and that significant tracts of old-growth forest are left undisturbed. Low-hanging fruits for improving logging outcomes pan-tropically lie in better spatial planning and post-logging management of logging roads, in cost-effective efforts to sustain timber yields, and in harnessing legal levers and certification standards to invoke transitions in the tropical timber sector.

The role that science must play in nurturing such transitions has changed. After decades of quantifying in increasingly finer detail the impacts of logging on biodiversity and forest functions and services, now tropical forest scientists must prioritise solutions-focused and applied research. Critical frontiers include how to best arrange logging operations to maximise landscape-level biodiversity and ecosystem services, the relative costs, benefits and scalability of post-logging silvicultural interventions, and the tractability of approaches for detecting and tackling illegal logging. There is also a dire need to examine the conditions under which differing logging concession licence agreements (e.g. size, length and ownership structure of concessions) help or hinder forest conservation goals and desirable management practises.

In summary, there is enormous potential for logged tropical forests to deliver numerous vital services to humanity and to play a central role in diverting catastrophic environmental outcomes in the Anthropocene. Ultimately, their capacity to do so rests on consumers, policies, and payments fully recognising that a forest is worth far more than just its timber.

References

Alamgir, M., Campbell, M.J., Sloan, S., Goosem, M., Clements, G.R., Mahmoud, M.I., Laurance, W.F., 2017. Economic, socio-political and environmental risks of road development in the tropics. Curr. Biol. 27, R1130–R1140.

Angelsen, A. (Ed.), 2009. Realising REDD+: National Strategy and Policy Options. CIFOR, https://doi.org/10.17528/cifor/002871.

Angelsen, A., 2017. REDD+ as result-based aid: general lessons and bilateral agreements of Norway. Rev. Dev. Econ. 21 (2), 237–264.

Anitha, K., Joseph, S., Chandran, R.J., Ramasamy, E.V., Prasad, S.N., 2010. Tree species diversity and community composition in a human-dominated tropical forest of Western Ghats biodiversity hotspot, India. Ecol. Complex. 7, 217–224.

Ansell, F.A., Edwards, D.P., Hamer, K.C., 2011. Rehabilitation of logged rain forests: avifaunal composition, habitat structure, and implications for biodiversity-friendly REDD+. Biotropica 43, 504–511.

Asiyanbi, A.P., 2016. A political ecology of REDD+: property rights, militarised protectionism, and carbonised exclusion in Cross River. Geoforum 77, 146–156.

Asner, G.P., Broadbent, E.N., Oliveira, P.J.C., Keller, M., Knapp, D.E., Silva, J.N.M., 2006. Condition and fate of logged forests in the Brazilian Amazon. Proc. Natl. Acad. Sci. U. S. A. 103, 1294712950.

Atela, J.O., Minang, P.A., Quinn, C.H., Duguma, L.A., 2015. Implementing REDD+ at the local level: assessing the key enablers for credible mitigation and sustainable livelihood outcomes. J. Environ. Manag. 157, 238–249.

Atmadja, S., Verchot, L., 2012. A review of the state of research, policies and strategies in addressing leakage from reducing emissions from deforestation and forest degradation (REDD+). Mitig. Adapt. Strateg. Glob. Chang. 17, 311–336.

Bacha, C.J.C., Rodriguez, L.C.E., 2007. Profitability and social impacts of reduced impact logging in the Tapajós National Forest, Brazil—a case study. Ecol. Econ. 63, 70–77.

Barber, C.P., Cochrane, M.A., Souza Jr., C.M., Laurance, W.F., 2014. Roads, deforestation, and the mitigating effect of protected areas in the Amazon. Biol. Conserv. 177 (2014), 203–209.

Barlow, J., Peres, C.A., 2004. Ecological responses to El Niño–induced surface fires in central Brazilian Amazonia: management implications for flammable tropical forests. Philos. Trans. R. Soc. Lond. Ser. B Biol. Sci. 359, 367–380.

Barlow, J., Peres, C.A., 2008. Fire-mediated dieback and compositional cascade in an Amazonian forest. Philos. Trans. R. Soc., B 363, 1787–1794.

Barlow, J., Haugaasen, T., Peres, C.A., 2002. Effects of ground fires on understorey bird assemblages in Amazonian forests. Biol. Conserv. 105, 157–169.

Bastin, J.F., Finegold, Y., Garcia, C., Mollicone, D., Rezende, M., Routh, D., Zohner, C.M., Crowther, T.W., 2019. The global tree restoration potential. Science 365, 76–79.

Bennett, E.L., 2000. Timber certification: where is the voice of the biologist? Conserv. Biol. 14 (4), 921–923.

Berry, N.J., Phillips, O.L., Lewis, S.L., Hill, J.K., Edwards, D.P., Tawatao, N.B., Ahmad, N., Magintan, D., Khen, C.V., Maryati, M., Ong, R.C., Hamer, K.C., 2010. The high value of logged tropical forests: lessons from northern Borneo. Biodivers. Conserv. 19, 985–997.

Bertault, J.G., Sist, P., 1997. An experimental comparison of different harvesting intensities with reduced-impact and conventional logging in East Kalimantan, Indonesia. For. Ecol. Manag. 94, 209–218.

Bicknell, J.E., Struebig, M.J., Edwards, D.P., Davies, Z.G., 2014. Improved timber harvest techniques maintain biodiversity in tropical forests. Curr. Biol. 24, R1119–R1120.

Bicknell, J.E., Gaveau, D.L., Davies, Z.G., Struebig, M.J., 2015. Saving logged tropical forests: closing roads will bring immediate benefits: peer-reviewed letter. Front. Ecol. Environ. 13, 73–74.

Blaser, J., Sarre, A., Poore, D., Johnson, S., 2011. Status of Tropical Forest Management 2011. 38, ITTO, pp. 376–384, Technical Series.

Blonder, B., Both, S., Coomes, D., Elias, D., Jucker, T., Kvasnica, J., Majalap, N., Malhi, Y.S., Milodowski, D., Riutta, T., Svátek, M., 2018. Extreme and highly heterogeneous microclimates in selectively logged tropical forests. Front. For. Glob. Change 1, 5.

Bonn Challenge, 2018. http://www.bonnchallenge.org/.

Bonnell, T.R., Reyna-Hurtado, R., Chapman, C.A., 2011. Post-logging recovery time is longer than expected in an East African tropical forest. For. Ecol. Manag. 261, 855–864.

Boucher, D.H., 2015. The REDD/carbon market offsets debate: big argument, small potatoes. J. Sustain. For. 34, 547–558.

Brais, S., 2001. Persistence of soil compaction and effects on seedling growth in northwestern Quebec. Soil Sci. Soc. Am. J. 65, 1263–1271.

Brancalion, P.H., Lamb, D., Ceccon, E., Boucher, D., Herbohn, J., Strassburg, B., Edwards, D.P., 2017. Using markets to leverage investment in forest and landscape restoration in the tropics. Forest Policy Econ. 85, 103–113.

Brandão Jr., A.O., Souza Jr., C.M., 2006. Mapping unofficial roads with Landsat images: a new tool to improve the monitoring of the Brazilian Amazon rainforest. Int. J. Remote Sens. 27, 177–189.

Brando, P.M., Balch, J.K., Nepstad, D.C., Morton, D.C., Putz, F.E., Coe, M.T., Silvério, D., Macedo, M.N., Davidson, E.A., Nóbrega, C.C., Alencar, A., 2014. Abrupt increases in Amazonian tree mortality due to drought–fire interactions. Proc. Natl. Acad. Sci. 111, 6347–6352.

Breukink, G., Levin, J., Mo, K., 2015. Profitability and sustainability in responsible forestry. Economic impacts of FSC certification on forest operators, World Wide Fund for Nature, Gland, Switzerland, (also available at http://wwf.panda.org/wwf_news, 250330.).

Broadbent, E.N., Asner, G.P., Keller, M., Knapp, D.E., Oliveira, P.J., Silva, J.N., 2008. Forest fragmentation and edge effects from deforestation and selective logging in the Brazilian Amazon. Biol. Conserv. 141, 1745–1757.

Brodie, J., Post, E., Laurance, W.F., 2012. Climate change and tropical biodiversity: a new focus. Trends Ecol. Evol. 27, 145–150.

Brown, K.A., Gurevitch, J., 2004. Long-term impacts of logging on forest diversity in Madagascar. Proc. Natl. Acad. Sci. 101 (16), 6045–6049.

Bruijnzeel, L.A., 2004. Hydrological functions of tropical forests: not seeing the soil for the trees? Agric. Ecosyst. Environ. 104, 185–228.

Buckley, D.S., Crow, T.R., Nauertz, E.A., Schulz, K.E., 2003. Influence of skid trails and haul roads on understory plant richness and composition in managed forest landscapes in Upper Michigan, USA. For. Ecol. Manag. 175, 509–520.

Budiharta, S., Meijaard, E., Erskine, P.D., Rondinini, C., Pacifici, M., Wilson, K.A., 2014. Restoring degraded tropical forests for carbon and biodiversity. Environ. Res. Lett. 9 (11), 114020.

Burivalova, Z., Şekercioğlu, Ç.H., Koh, L.P., 2014. Thresholds of logging intensity to maintain tropical forest biodiversity. Curr. Biol. 24, 1893–1898.

Burivalova, Z., Hua, F., Koh, L.P., Garcia, C., Putz, F., 2017. A critical comparison of conventional, certified, and community management of tropical forests for timber in terms of environmental, economic, and social variables. Conserv. Lett. 10, 4–14.

Burivalova, Z., Allnutt, T.F., Rademacher, D., Schlemm, A., Wilcove, D.S., Butler, R.A., 2019a. What works in tropical forest conservation, and what does not: effectiveness of four strategies in terms of environmental, social, and economic outcomes. Conserv. Sci. Pract. 1, e28.

Burivalova, Z., Game, E.T., Butler, R.A., 2019b. The sound of a tropical forest. Science 363, 28–29.

Butsic, V., Kuemmerle, T., 2015. Using optimization methods to align food production and biodiversity conservation beyond land sharing and land sparing. Ecol. Appl. 25, 589–595.

Campos-Cerqueira, M., Mena, J.L., Tejeda-Gómez, V., Aguilar-Amuchastegui, N., Gutierrez, N., Aide, T.M., 2019. How does FSC forest certification affect the acoustically active fauna in Madre de Dios, Peru? Remote Sens. Ecol. Conserv.

Cardillo, M., Mace, G.M., Jones, K.E., Bielby, J., Bininda-Emonds, O.R., Sechrest, W., Orme, C.D.L., Purvis, A., 2005. Multiple causes of high extinction risk in large mammal species. Science 309, 1239–1241.

Cazzolla Gatti, R., Castaldi, S., Lindsell, J.A., Coomes, D.A., Marchetti, M., Maesano, M., Di Paola, A., Paparella, F., Valentini, R., 2015. The impact of selective logging and clearcutting on forest structure, tree diversity and above-ground biomass of African tropical forests. Ecol. Res. 30, 119–132.

Cerullo, G.R., Edwards, D.P., 2019. Actively restoring resilience in selectively logged tropical forests. J. Appl. Ecol. 56, 107–118.

Cerullo, G.R., Edwards, F.A., Mills, S.C., Edwards, D.P., 2019. Tropical forest subjected to intensive post-logging silviculture maintains functionally diverse dung beetle communities. For. Ecol. Manag. 444, 318–326.

Chan, N.W., Parker, D.J., 1996. Response to dynamic flood hazard factors in peninsular Malaysia. Geogr. J. 162, 313–325.

Chaudhary, A., Burivalova, Z., Koh, L.P., Hellweg, S., 2016. Impact of forest management on species richness: global meta-analysis and economic trade-offs. Sci. Rep. 6, 23954.

Chazdon, R.L., Peres, C.A., Dent, D., Sheil, D., Lugo, A.E., Lamb, D., Stork, N.E., Miller, S.E., 2009. The potential for species conservation in tropical secondary forests. Conserv. Biol. 23, 1406–1417.

Chen, Y., Morton, D.C., Andela, N., Van Der Werf, G.R., Giglio, L., Randerson, J.T., 2017. A pan-tropical cascade of fire driven by El Niño/Southern Oscillation. Nat. Clim. Chang. 7, 906–911.

Cleary, D.F., 2017. Impact of logging on tree, liana and herb assemblages in a Bornean forest. J. Sustain. For. 36, 806–817.

Cochrane, M.A., 2003. Fire science for rainforests. Nature 421, 913–919.

Cochrane, M.A., Laurance, W.F., 2008. Synergisms among fire, land use, and climate change in the Amazon. Ambio 522–527.

Cochrane, M.A., Alencar, A., Schulze, M.D., Souza, C.M., Nepstad, D.C., Lefebvre, P., Davidson, E.A., 1999. Positive feedbacks in the fire dynamic of closed canopy tropical forests. Science 284, 1832–1835.

Cole, L.E., Bhagwat, S.A., Willis, K.J., 2014. Recovery and resilience of tropical forests after disturbance. Nat. Commun. 5, 3906.

Coleman, E.A., Fleischman, F.D., 2012. Comparing forest decentralization and local institutional change in Bolivia, Kenya, Mexico, and Uganda. World Dev. 40, 836–849.

Corlett, R.T., 2016. The impacts of droughts in tropical forests. Trends Plant Sci. 21, 584–593.

Cosset, C.C.P., Edwards, D.P., 2017. The effects of restoring logged tropical forests on avian phylogenetic and functional diversity. Ecol. Appl. 27, 1932–1945.

Costa, F., Magnusson, W., 2002. Selective logging effects on abundance, diversity, and composition of tropical understory herbs. Ecol. Appl. 12 (3), 807–819.

Costantini, D., Edwards, D.P., Simons, M.J., 2016. Life after logging in tropical forests of Borneo: a meta-analysis. Biol. Conserv. 196, 182–188.

Dantzer, B., Fletcher, Q.E., Boonstra, R., Sheriff, M.J., 2014. Measures of physiological stress: a transparent or opaque window into the status, management and conservation of species? Conserv. Physiol. 2, 023.

De Barros, A.E., Macdonald, E.A., Matsumoto, M.H., Paula, R.C., Nijhawan, S., Malhi, Y., Macdonald, D.W., 2014. Identification of areas in Brazil that optimize conservation of forest carbon, jaguars, and biodiversity. Conserv. Biol. 28, 580–593.

Santos de Lima, L., Merry, F., Soares-Filho, B., Rodrigues, H.O., dos Santos Damaceno, C., Bauch, M.A., 2018. Illegal logging as a disincentive to the establishment of a sustainable forest sector in the Amazon. PloS One 13, e0207855.

DeFries, R.S., Rudel, T., Uriarte, M., Hansen, M., 2010. Deforestation driven by urban population growth and agricultural trade in the twenty-first century. Nat. Geosci. 3, 178–181.

Dessie, A., Bredemeier, M., 2013. The effect of deforestation on water quality: a case study in Cienda Micro Watershed, Leyte, Philippines. Resour. Environ. 3, 1–9.

Dobson, A., Lodge, D., Alder, J., Cumming, G.S., Keymer, J., McGlade, J., Mooney, H., Rusak, J.A., Sala, O., Wolters, V., Wall, D., Winfree, R., Xenopoulos, M.a., 2006. Habitat loss, trophic collapse, and the decline of ecosystem services. Ecology 87, 1915–1924.

Douglas, I., 1999. Hydrological investigations of forest disturbance and land cover impacts in South-East Asia: a review. Philos. Trans. R. Soc. Lond. B Biol. Sci. 354, 1725–1738.

Edwards, D.P., Ansell, F.A., Ahmad, A.H., Nilus, R., Hamer, K.C., 2009. The value of rehabilitating logged rainforest for birds. Conserv. Biol. 23, 1628–1633.

Edwards, D.P., Larsen, T.H., Docherty, T.D.S., Ansell, F.A., Hsu, W.W., Derhé, M.A., Hamer, K.C., Wilcove, D.S., 2010a. Degraded lands worth protecting: the biological importance of Southeast Asia's repeatedly logged forests. Proc. R. Soc. B Biol. Sci. *278*, 82–90.

Edwards, D.P., Hodgson, J.A., Hamer, K.C., Mitchell, S.L., Ahmad, A.H., Cornell, S.J., Wilcove, D.S., 2010c. Wildlife-friendly oil palm plantations fail to protect biodiversity effectively. Conserv. Lett. 3 (4), 236–242.

Edwards, F.A., Edwards, D.P., Hamer, K.C., Davies, R.G., 2013a. Impacts of logging and conversion of rainforest to oil palm on the functional diversity of birds in Sundaland. Ibis 155, 313–326.

Edwards, D.P., Woodcock, P., Newton, R.J., Edwards, F.A., Andrews, D.J., Docherty, T.D., Mitchell, S.L., Ota, T., Benedick, S., Bottrell, S.H., Hamer, K.C., 2013b. Trophic flexibility and the persistence of understory birds in intensively logged rainforest. Conserv. Biol. 27, 1079–1086.

Edwards, D.P., Tobias, J.A., Sheil, D., Meijaard, E., Laurance, W.F., 2014a. Maintaining ecosystem function and services in logged tropical forests. Trends Ecol. Evol. 29, 511–520.

Edwards, D.P., Magrach, A., Woodcock, P., Ji, Y., Lim, N.T.L., Edwards, F.A., Larsen, T.H., Hsu, W.W., Benedick, S., Khen, C.V., Chung, A.Y., 2014b. Selective-logging and oil palm: multitaxon impacts, biodiversity indicators, and trade-offs for conservation planning. Ecol. Appl. 24, 2029–2049.

Edwards, F.A., Edwards, D.P., Larsen, T.H., Hsu, W.W., Benedick, S., Chung, A., Vun Khen, C., Wilcove, D.S., Hamer, K.C., 2014c. Does logging and forest conversion to oil palm agriculture alter functional diversity in a biodiversity hotspot? Anim. Conserv. 17, 163–173.

Edwards, D.P., Gilroy, J.J., Woodcock, P., Edwards, F.A., Larsen, T.H., Andrews, D.J., Derhé, M.A., Docherty, T.D., Hsu, W.W., Mitchell, S.L., Ota, T., 2014d. Land-sharing versus land-sparing logging: reconciling timber extraction with biodiversity conservation. Glob. Chang. Biol. 20, 183–191.

Edwards, F.A., Finan, J., Graham, L.K., Larsen, T.H., Wilcove, D.S., Hsu, W.W., Chey, V.K., Hamer, K.C., 2017. The impact of logging roads on dung beetle assemblages in a tropical rainforest reserve. Biol. Conserv. 205, 85–92.

Eisenberg, J.N., Cevallos, W., Ponce, K., Levy, K., Bates, S.J., Scott, J.C., Hubbard, A., Vieira, N., Endara, P., Espinel, M., Trueba, G., 2006. Environmental change and infectious disease: how new roads affect the transmission of diarrheal pathogens in rural Ecuador. Proc. Natl. Acad. Sci. 103, 19460–19465.

Ekroos, J., Ödman, A.M., Andersson, G.K., Birkhofer, K., Herbertsson, L., Klatt, B.K., Olsson, O., Olsson, P.A., Persson, A.S., Prentice, H.C., Rundlöf, M., 2016. Sparing land for biodiversity at multiple spatial scales. Front. Ecol. Evol. 3, 145.

Elkin, C., Possingham, H., 2008. The role of landscape-dependent disturbance and dispersal in metapopulation persistence. Am. Nat. 172, 563–575.

Ellis, P.W., Gopalakrishna, T., Goodman, R.C., Putz, F.E., Roopsind, A., Umunay, P.M., Zalman, J., Ellis, E.A., Mo, K., Gregoire, T.G., Griscom, B.W., 2019. Reduced-impact logging for climate change mitigation (RIL-C) can halve selective logging emissions from tropical forests. For. Ecol. Manag. 438, 255–266.

Ewers, R.M., Boyle, M.J.W., Gleave, R.A., Plowman, N.S., Benedick, S., Bernard, H., Bishop, T.R., Bakhtiar, E.Y., Chey, V.K., Chung, A.Y.C., Davies, R.G., Edwards, D.P., Eggleton, P., Fayle, T.M., Hardwick, S.R., Homathevi, R., Kitching, R.L., Khoo, M., Luke, S.H., March, J.J., Nilus, R., Pfeifer, M., Rao, S.V., Sharp, A.C., Snaddon, J.L., Stork, N.E., Struebig, M.J., Wearn, O.R., Yusah, K.M., Turner, E.C., 2015. Logging cuts the functional importance of invertebrates in tropical rainforest. Nat. Commun. 6, 6836.

Fargeon, H., Aubry-Kientz, M., Brunaux, O., Descroix, L., Gaspard, R., Guitet, S., Rossi, V., Hérault, B., 2016. Vulnerability of commercial tree species to water stress in logged forests of the Guiana shield. Forests 7, 105.

Fayle, T.M., Edwards, D.P., Foster, W.A., Yusah, K.M., Turner, E.C., 2015. An ant–plant by-product mutualism is robust to selective logging of rain forest and conversion to oil palm plantation. Oecologia 178, 441–450.

Fimbel, R.A., Grajal, A., Robinson, J.G. (Eds.), 2001. The Cutting Edge: Conserving Wildlife in Logged Tropical Forest. Columbia University Press, New York.

Fisher, B., Edwards, D.P., Giam, X., Wilcove, D.S., 2011a. The high costs of conserving Southeast Asia's lowland rainforests. Front. Ecol. Environ. 9, 329–334.

Fisher, B., Edwards, D.P., Larsen, T.H., Ansell, F.A., Hsu, W.W., Roberts, C.S., Wilcove, D.S., 2011b. Cost-effective conservation: calculating biodiversity and logging trade-offs in Southeast Asia. Conserv. Lett. 4, 443–450.

Fisher, B., Edwards, D.P., Wilcove, D.S., 2014. Logging and conservation: Economic impacts of the stocking rates and prices of commercial timber species. Forest Policy Econ. 38, 65–71.

Fletcher, R., Dressler, W., Büscher, B., Anderson, Z.R., 2016. Questioning REDD+ and the future of market-based conservation. Conserv. Biol. 30, 673–675.

Food and Agriculture Organization of the United Nations, 2012. State of the World's Forests. Rome.

Food and Agriculture Organization of the United Nations, 2014. State of the World's forests. Rome.

Food and Agriculture Organization of the United Nations, 2015. Global Forest Resources Assessment 2015. Main report.

Food and Agriculture Organization of the United Nations, 2018. Rethinking Forest Concessions.

Foster, W.A., Snaddon, J.L., Turner, E.C., Fayle, T.M., Cockerill, T.D., Ellwood, M.F., Broad, G.R., Chung, A.Y., Eggleton, P., Khen, C.V., Yusah, K.M., 2011. Establishing the evidence base for maintaining biodiversity and ecosystem function in the oil palm landscapes of South East Asia. Philos. Trans. R. Soc., B 366, 3277–3291.

França, F., Barlow, J., Araújo, B., Louzada, J., 2016. Does selective logging stress tropical forest invertebrates? Using fat stores to examine sublethal responses in dung beetles. Ecol. Evol. 6, 8526–8533.

García León, M.M., Martínez Izquierdo, L., Mello, F.N.A., Powers, J.S., Schnitzer, S.A., 2018. Lianas reduce community-level canopy tree reproduction in a Panamanian forest. J. Ecol. 106 (2), 737–745.
Gardner, T., 2010. Monitoring Forest Biodiversity: improving Conservation Through Ecologically Responsible Management. Earthscan, London.
Ghazoul, J., Sheil, D., 2010. Tropical Rain Forest Ecology, Diversity, and Conservation. Oxford University Press.
Gibson, L., Lee, T.M., Koh, L.P., Brook, B.W., Gardner, T.A., Barlow, J., Peres, C.A., Bradshaw, C.J., Laurance, W.F., Lovejoy, T.E., Sodhi, N.S., 2011. Primary forests are irreplaceable for sustaining tropical biodiversity. Nature 478 (7369), 378.
Gourlet-Fleury, S., Mortier, F., Fayolle, A., Baya, F., Ouédraogo, D., Bénédet, F., Picard, N., 2013. Tropical forest recovery from logging: a 24 year silvicultural experiment from Central Africa. Philos. Trans. R. Soc., B 368 (1625), 20120302.
Graham, V., Laurance, S.G., Grech, A., Venter, O., 2017. Spatially explicit estimates of forest carbon emissions, mitigation costs and REDD+ opportunities in Indonesia. Environ. Res. Lett. 12, 044017.
Griscom, B.W., Goodman, R.C., 2015. Reframing the sharing vs sparing debate for tropical forestry landscapes. J. Trop. For. Sci. 27, 145–147.
Griscom, B., Ellis, P., Putz, F.E., 2014. Carbon emissions performance of commercial logging in East Kalimantan, Indonesia. Glob. Chang. Biol. 20, 923–937.
Griscom, B.W., Goodman, R.C., Burivalova, Z., Putz, F.E., 2017. Carbon and biodiversity impacts of intensive versus extensive tropical forestry. Conserv. Lett. 11, e12362.
Guariguata, M.R., García-Fernández, C., Sheil, D., Nasi, R., Herrero-Jauregui, C., Cronkleton, P., Ingram, V., 2010. Compatibility of timber and non-timber forest product management in natural tropical forests: perspectives, challenges, and opportunities. For. Ecol. Manag. 259, 237–245.
Hamer, K.C., Hill, J.K., Benedick, S., Mustaffa, N., Sherratt, T.N., Maryati, M., Chey, V.K., 2003. Ecology of butterflies in natural and selectively logged forests of northern Borneo: the importance of habitat heterogeneity. J. Appl. Ecol. 40, 150–162.
Hamer, K.C., Newton, R.J., Edwards, F.A., Benedick, S., Bottrell, S.H., Edwards, D.P., 2015. Impacts of selective logging on insectivorous birds in Borneo: the importance of trophic position, body size and foraging height. Biol. Conserv. 188, 82–88.
Hamilton, A.J., Basset, Y., Benke, K.K., Grimbacher, P.S., Miller, S.E., Novotný, V., Samuelson, G.A., Stork, N.E., Weiblen, G.D., Yen, J.D., 2010. Quantifying uncertainty in estimation of tropical arthropod species richness. Am. Nat. 176, 90–95.
Hardus, M.E., Lameira, A.R., Menken, S.B., Wich, S.A., 2012. Effects of logging on orangutan behavior. Biol. Conserv. 146, 177–187.
Hardwick, S.R., Toumi, R., Pfeifer, M., Turner, E.C., Nilus, R., Ewers, R.M., 2015. The relationship between leaf area index and microclimate in tropical forest and oil palm plantation: forest disturbance drives changes in microclimate. Agric. For. Meteorol. 201, 187–195.
Hartanto, H., Prabhu, R., Widayat, A.S.E., Asdak, C., 2003. Factors affecting runoff and soil erosion: plot-level soil loss monitoring for assessing sustainability of forest management. For. Ecol. Manag. 180, 361–374.
Hattori, D., Kenzo, T., Irino, K.O., Kendawang, J.J., Ninomiya, I., Sakurai, K., 2013. Effects of soil compaction on the growth and mortality of planted dipterocarp seedlings in a logged-over tropical rainforest in Sarawak, Malaysia. For. Ecol. Manag. 310, 770–776.
Haugaasen, T., Barlow, J., Peres, C.A., 2003. Effects of surface fires on understorey insectivorous birds and terrestrial arthropods in central Brazilian Amazonia. Anim. Conserv. 6 (4), 299–306. https://doi.org/10.1017/S1367943003003366.
Hawthorne, W.D., Sheil, D., Agyeman, V.K., Juam, M.A., Marshall, C.A.M., 2012. Logging scars in Ghanaian high forest: towards improved models for sustainable production. For. Ecol. Manag. 271, 27–36.

Hendrison, J., 1990. Damage-Controlled Logging in Managed Tropical Rain Forest in Suriname. University of Wageningen, Thesis.
Hill, J.K., Hamer, K.C., 2004. Determining impacts of habitat modification on diversity of tropical forest fauna: the importance of spatial scale. J. Appl. Ecol. 41 (4), 744–754.
Holl, K.D., 2017. Restoring tropical forests from the bottom up. Science 355, 455–456.
Holmes, T.P., Blate, G.M., Zweede, J.C., Pereira Jr., R., Barreto, P., Boltz, F., Bauch, R., 2002. Financial and ecological indicators of reduced impact logging performance in the eastern Amazon. For. Ecol. Manag. 163, 93–110.
Hosonuma, N., Herold, M., De Sy, V., De Fries, R.S., Brockhaus, M., Verchot, L., Angelsen, A., Romijn, E., 2012. An assessment of deforestation and forest degradation drivers in developing countries. Environ. Res. Lett. 7, 044009.
Houghton, R.A., 2012. Carbon emissions and the drivers of deforestation and forest degradation in the tropics. Curr. Opin. Environ. Sustain. 4, 597–603.
Iponga, D.M., Mikolo-Yobo, C., Lescuyer, G., Assoumou, F.M., Levang, P., Tieguhong, J.C., Ngoye, A., 2018. The contribution of NTFP-gathering to rural people's livelihoods around two timber concessions in Gabon. Agrofor. Syst. 92, 157–168.
Jagger, P., Rana, P., 2017. Using publicly available social and spatial data to evaluate progress on REDD+ social safeguards in Indonesia. Environ. Sci. Pol. 76, 59–69.
Ji, Y., Ashton, L., Pedley, S.M., Edwards, D.P., Tang, Y., Nakamura, A., Kitching, R., Dolman, P.M., Woodcock, P., Edwards, F.A., Larsen, T.H., 2013. Reliable, verifiable and efficient monitoring of biodiversity via metabarcoding. Ecol. Lett. 16, 1245–1257.
Jucker, T., Hardwick, S.R., Both, S., Elias, D.M.O., Ewers, R.M., Milodowski, D.T., Swinfield, T., Coomes, D.A., 2018. Canopy Structure and Topography Jointly Constrain the Microclimate of Human-Modified Tropical Landscapes.
Jumbe, C.B.L., Angelsen, A., 2006. Do the poor benefit from devolution policies? Evidence from Malawi's forest co-management program. Land Econ. 82, 562–581.
Katovai, E., Sirikolo, M., Srinivasan, U., Edwards, W., Laurance, W.F., 2016. Factors influencing tree diversity and compositional change across logged forests in the Solomon Islands. For. Ecol. Manag. 372, 53–63.
Keenan, R.J., Reams, G.A., Achard, F., Freitas, J.V.D., Grainger, A., Lindquist, E., 2015. Forest ecology and management dynamics of global forest area: results from the FAO Global Forest Resources Assessment 2015. For. Ecol. Manag. 352, 9–20.
Kleinschroth, F., Healey, J.R., 2017. Impacts of logging roads on tropical forests. Biotropica 49, 620–635.
Kleinschroth, F., Healey, J.R., Sist, P., Mortier, F., Gourlet-Fleury, S., 2016a. How persistent are the impacts of logging roads on Central African forest vegetation? J. Appl. Ecol. 53, 1127–1137.
Kleinschroth, F., Healey, J.R., Gourlet-Fleury, S., 2016b. Sparing forests in Central Africa: re-use old logging roads to avoid creating new ones. Front. Ecol. Environ. 14, 9–10.
Kleinschroth, F., Healey, J.R., Gourlet-Fleury, S., Mortier, F., Stoica, R.S., 2017. Effects of logging on roadless space in intact forest landscapes of the Congo Basin. Conserv. Biol. 31, 469–480.
Kleinschroth, F., Laporte, N., Laurance, W.F., Goetz, S.J., Ghazoul, J., 2019a. Road expansion and persistence in forests of the Congo Basin. Nat. Sustain. 2, 628–634.
Kleinschroth, F., Garcia, C., Ghazoul, J., 2019b. Reconciling certification and intact forest landscape conservation. Ambio 48, 153–159.
Kollert, W., Lagan, P., 2007. Do certified tropical logs fetch a market premium?: a comparative price analysis from Sabah, Malaysia. Forest Policy Econ. 9, 862–868.
Kremen, C., 2015. Reframing the land-sparing/land-sharing debate for biodiversity conservation. Ann. N. Y. Acad. Sci. 1355, 52–76.
Kuik, O., 2014. REDD+ and international leakage via food and timber markets: a CGE analysis. In: Mitigation and Adaptation Strategies for Global Change. Vol. 19. pp. 641–655.

Laurance, W.F., Croes, B.M., Tchignoumba, L., Lahm, S.A., Alonso, A., Lee, M.E., Campbell, P., Ondzeano, C., 2006a. Impacts of roads and hunting on central African rainforest mammals. Conserv. Biol. 20, 1251–1261.

Laurance, W.F., Nascimento, H.E., Laurance, S.G., Andrade, A., Ribeiro, J.E., Giraldo, J.P., Lovejoy, T.E., Condit, R., Chave, J., Harms, K.E., D'Angelo, S., 2006b. Rapid decay of tree-community composition in Amazonian forest fragments. Proc. Natl. Acad. Sci. 103, 19010–19014.

Laurance, W.F., Sayer, J., Cassman, K.G., 2014. Agricultural expansion and its impacts on tropical nature. Trends Ecol. Evol. 29, 107–116.

Leck, M.A., Parker, V.T., Simpson, R.L., 2008. Seedling Ecology and Evolution. Cambridge University Press.

Legras, S., Martin, E., Piguet, V., 2018. Conjunctive implementation of land sparing and land sharing for environmental preservation. Ecol. Econ. 143, 170–187.

Lele, S., 2009. Watershed services of tropical forests: from hydrology to economic valuation to integrated analysis. Curr. Opin. Environ. Sustain. 1, 148–155.

Lewis, S.L., Edwards, D.P., Galbraith, D., 2015. Increasing human dominance of tropical forests. Science 349, 827–832.

Lim, F.K., Carrasco, L.R., McHardy, J., Edwards, D.P., 2017. Perverse market outcomes from biodiversity conservation interventions. Conserv. Lett. 10, 506–516.

Lin, L., Sills, E., Cheshire, H., 2014. Targeting areas for reducing emissions from deforestation and forest degradation (REDD+) projects in Tanzania. Glob. Environ. Chang. 24, 277–286.

Lounela, A., 2015. Climate change disputes and justice in Central Kalimantan, Indonesia. Asia Pac. Viewp. 56, 62–78.

Lowood, H.E., 1990. The calculating forester: quantification, cameral science, and the emergence of scientific forestry management in Germany. In: The Quantifying Spirit in the 18th Century. , pp. 315–342.

MacDicken, K.G., Sola, P., Hall, J.E., Sabogal, C., Tadoum, M., de Wasseige, C., 2015. Global progress toward sustainable forest management. For. Ecol. Manag. 352, 47–56.

Magrach, A., Senior, R., Rogers, A., Nurdin, D., Benedick, S., Laurance, W.F., Santamaria, L., Edwards, D.P., 2016. Selective logging in tropical forests decreases the robustness of liana-tree interaction networks to the loss of host tree species. Proc. R. Soc. B 283, 20153008.

Mahmood, R., Pielke Sr., R.A., Hubbard, K.G., Niyogi, D., Dirmeyer, P.A., McAlpine, C., Carleton, A.M., Hale, R., Gameda, S., Beltrán-Przekurat, A., Baker, B., 2014. Land cover changes and their biogeophysical effects on climate. Int. J. Climatol. 34, 929–953.

Makarieva, A.M., Gorshkov, V.G., 2007. Biotic pump of atmospheric moisture as driver of the hydrological cycle on land. Hydrol. Earth Syst. Sci. Discuss. 11, 1013–1033.

Makarieva, A.M., Gorshkov, V.G., Sheil, D., Nobre, A.D., Bunyard, P., Li, B.L., 2014. Why does air passage over forest yield more rain? Examining the coupling between rainfall, pressure, and atmospheric moisture content. J. Hydrometeorol. 15, 411–426.

Malhi, Y., Roberts, J.T., Betts, R.A., Killeen, T.J., Li, W., Nobre, C.A., 2008. Climate change, deforestation, and the fate of the Amazon. Science 319, 169–172.

Matricardi, E.A., Skole, D.L., Pedlowski, M.A., Chomentowski, W., Fernandes, L.C., 2010. Assessment of tropical forest degradation by selective logging and fire using Landsat imagery. Remote Sens. Environ. 114, 1117–1129.

Matricardi, E.A.T., Skole, D.L., Pedlowski, M.A., Chomentowski, W., 2013. Assessment of forest disturbances by selective logging and forest fires in the Brazilian Amazon using Landsat data. Int. J. Remote Sens. 34, 1057–1086.

McElwee, P., Van Nguyen, T., Nguyen, D., Tran, N., Le, H., Nghiem, T., Thi Vu, H., 2016. Using REDD+ policy to facilitate climate adaptation at the local level: synergies and challenges in Vietnam. Forests 8, 11.

McLain, R., Lawry, S., Guariguata, M.R., Reed, J., 2018. Toward a tenure-responsive approach to forest landscape restoration: a proposed tenure diagnostic for assessing restoration opportunities. Land Use Policy. https://doi.org/10.1016/j.landusepol.2018.11.053.

Meijaard, E., Abram, N.K., Wells, J.A., Pellier, A.S., Ancrenaz, M., Gaveau, D.L., Runting, R.K., Mengersen, K., 2013. People's perceptions about the importance of forests on Borneo. PloS One 8 (9), e73008.

Menton, M., Merry, F., Lawrence, A., Brown, N., 2009. Company–community logging contracts in Amazonian settlements: impacts on livelihoods and NTFP harvests. Ecol. Soc. 14, 39.

Messina, S., Edwards, D.P., Eens, M., Costantini, D., 2018. Physiological and immunological responses of birds and mammals to forest degradation: a meta-analysis. Biol. Conserv. 224, 223–229.

Miller, S.D., Goulden, M.L., Hutyra, L.R., Keller, M., Saleska, S.R., Wofsy, S.C., Figueira, A.M.S., Da Rocha, H.R., De Camargo, P.B., 2011. Reduced impact logging minimally alters tropical rainforest carbon and energy exchange. Proc. Natl. Acad. Sci. 108, 19431–19435.

Miteva, D.A., Loucks, C.J., Pattanayak, S.K., 2015. Social and environmental impacts of forest management certification in Indonesia. PLoS One 10 (7).

Mollinari, M.M., Peres, C.A., Edwards, D.P., 2019. Rapid recovery of thermal environment after selective logging in the Amazon. Agric. For. Meteorol. 278, 107637.

Montejo-Kovacevich, G., Hethcoat, M.G., Lim, F.K., Marsh, C.J., Bonfantti, D., Peres, C.A., Edwards, D.P., 2018. Impacts of selective logging management on butterflies in the Amazon. Biol. Conserv. 225, 1–9.

Murdiyarso, D., Dewi, S., Lawrence, D., Seymour, F., 2011. Indonesia's forest moratorium: a stepping stone to better forest governance? Working Paper 76. CIFOR, Bogor, Indonesia.

Murray, J.P., Grenyer, R., Wunder, S., Raes, N., Jones, J.P., 2015. Spatial patterns of carbon, biodiversity, deforestation threat, and REDD+ projects in Indonesia. Conserv. Biol. 29, 1434–1445.

Nabe-Nielsen, J., Severiche, W., Fredericksen, T., Nabe-Nielsen, L.I., 2007. Timber tree regeneration along abandoned logging roads in a tropical Bolivian forest. New For. 34, 31–40.

Ndoye, O., Tieguhong, J.C., 2004. Forest resources and rural livelihoods: the conflict between timber and non-timber forest products in the Congo Basin. Scand. J. For. Res. 19, 36–44.

Nellemann, C., 2012. Green carbon, black trade: illegal logging, tax fraud and laundering in the world's tropical forests. In: A Rapid Response Assessment. United Nations Environment Programme. GRID-Arendal.

Nelson, P.N., Gabriel, J., Filer, C., Banabas, M., Sayer, J.A., Curry, G.N., Koczberski, G., Venter, O., 2014. Oil palm and deforestation in Papua New Guinea. Conserv. Lett. 7, 188–195.

Noutcheu, R., Snook, L.K., Tchatat, M., Taedoumg, H., Tchingsabe, O., Tieguhong, J.C., 2016. Do logging concessions decrease the availability to villagers of foods from timber trees? A quantitative analysis for Moabi (*Baillonella toxisperma*), Sapelli (*Entandrophragma cylindricum*) and Tali (*Erythrophleum suaveolens*) in Cameroon. For. Ecol. Manag. 381, 279–288.

O'Brien, M.J., Philipson, C.D., Reynolds, G., Dzulkifli, D., Snaddon, J.L., Ong, R., Hector, A., 2019. Positive effects of liana cutting on seedlings are reduced during El Niño-induced drought. J. Appl. Ecol. 56 (4), 891–901.

Pacheco, P., Cerutti, P.O., Edwards, D.P., Lescuyer, G., Mejia, E., Navarro, G., Obidzinski, K., Pokorny, B., Sist, P., 2016. Multiple and Intertwined Impacts of Illegal Forest Activities. IUFRO.

Pan, Y., Birdsey, R.A., Fang, J., Houghton, R., Kauppi, P.E., Kurz, W.A., Phillips, O.L., Shvidenko, A., Lewis, S.L., Canadell, J.G., Ciais, P., 2011. A large and persistent carbon sink in the world's forests. Science 333, 988–993.

Panlasigui, S., Rico-Straffon, J., Pfaff, A., Swenson, J., Loucks, C., 2018. Impacts of certification, uncertified concessions, and protected areas on forest loss in Cameroon, 2000 to 2013. Biol. Conserv. 227, 160–166.

Pearson, T.R., Brown, S., Murray, L., Sidman, G., 2017. Greenhouse gas emissions from tropical forest degradation: an underestimated source. Carbon Bal. Manag. 12 (1), 3.

Peña-Claros, M., Fredericksen, T.S., Alarcón, A., Blate, G.M., Choque, U., Leaño, C., Licona, J.C., Mostacedo, B., Pariona, W., Villegas, Z., Putz, F.E., 2008. Beyond reduced-impact logging: silvicultural treatments to increase growth rates of tropical trees. For. Ecol. Manag. 256, 1458–1467.

Pereira Jr., R., Zweede, J., Asner, G.P., Keller, M., 2002. Forest canopy damage and recovery in reduced-impact and conventional selective logging in eastern Para, Brazil. For. Ecol. Manag. 168, 77–89.

Peres, C.A., Barlow, J., Haugaasen, T., 2003. Vertebrate responses to surface wildfires in a central Amazonian forest. Oryx 37, 97–109.

Peres, C.A., Gardner, T.A., Barlow, J., Zuanon, J., Michalski, F., Lees, A.C., Vieira, I.C., Moreira, F.M., Feeley, K.J., 2010. Biodiversity conservation in human-modified Amazonian forest landscapes. Biol. Conserv. 143, 2314–2327.

Pillay, R., Fletcher, R.J., Sieving, K.E., Udell, B.J., Bernard, H., 2016. Bioacoustic monitoring reveals shifts in breeding songbird populations and singing behaviour with selective logging in tropical forests. J. Appl. Ecol. 56, 2482–2492.

Pinard, M.A., Putz, F.E., 1996. Retaining forest biomass by reducing logging damage. Biotropica 28, 278–295.

Pinard, M., Howlett, B., Davidson, D., 1996. Site conditions limit pioneer tree recruitment after logging of dipterocarp forests in Sabah, Malaysia. Biotropica 28, 2–12.

Pinard, M.A., Barker, M.G., Tay, J., 2000. Soil disturbance and post-logging forest recovery on bulldozer paths in Sabah, Malaysia. For. Ecol. Manag. 130, 213–225.

Piponiot, C., Rödig, E., Putz, F.E., Rutishauser, E., Sist, P., Ascarrunz, N., Blanc, L., Derroire, G., Descroix, L., Guedes, M.C., Coronado, E.H., 2019. Can timber provision from Amazonian production forests be sustainable? Environ. Res. Lett. 14, 064014.

Plumptre, A.J., 1995. The importance of" seed trees" for the natural regeneration of selectively logged tropical forest. Commonw. For. Rev. 253–258.

Pokorny, B., Sabogal, C., Silva, J.N.M., Bernardo, P., Souza, J., Zweede, J., 2005. Compliance with reduced-impact harvesting guidelines by timber enterprises in terra firme forests of the Brazilian Amazon. Int. For. Rev. 7, 9–20.

Poudel, M., Thwaites, R., Race, D., Dahal, G.R., 2015. Social equity and livelihood implications of REDD+ in rural communities—a case study from Nepal. Int. J. Commons 9, 177–208.

Poulsen, J.R., Clark, C.J., Mavah, G., Elkan, P.W., 2009. Bushmeat supply and consumption in a tropical logging concession in northern Congo. Conserv. Biol. 23, 1597–1608.

Poulsen, J.R., Clark, C.J., Bolker, B.M., 2011. Decoupling the effects of logging and hunting on an Afrotropical animal community. Ecol. Appl. 21, 1819–1836.

Putz, F.E., Pinard, M.A., 1993. Reduced-impact logging as a carbon-offset method. Conserv. Biol. 7 (4), 755–757.

Putz, F.E., Redford, K.H., 2009. Dangers of carbon-based conservation. Glob. Environ. Chang. 4, 400–401.

Putz, F.E., Dykstra, D.P., Heinrich, R., 2000. Why poor logging practices persist in the tropics. Conserv. Biol. 14, 951–956.

Putz, F.E., Blate, G.M., Redford, K.H., Fimbel, R., Robinson, J., 2001. Tropical forest management and conservation of biodiversity: an overview. Conserv. Biol. 15, 7–20.

Putz, F.E., Sist, P., Fredericksen, T., Dykstra, D., 2008. Reduced-impact logging: challenges and opportunities. For. Ecol. Manag. 256, 1427–1433.
Putz, F.E., Zuidema, P.A., Synnott, T., Peña-Claros, M., Pinard, M.A., Sheil, D., Vanclay, J.K., Sist, P., Gourlet-Fleury, S., Griscom, B., Palmer, J., 2012. Sustaining conservation values in selectively logged tropical forests: the attained and the attainable. Conserv. Lett. 5, 296–303.
Putz, F.E., Baker, T., Griscom, B.W., Gopalakrishna, T., Roopsind, A., Umunay, P.M., Zalman, J., Ellis, E.A., Ellis, P.W., 2019. Intact forest in selective logging landscapes in the tropics. Front. For. Glob. Chang. 2, 30.
Ramage, B.S., Marshalek, E.C., Kitzes, J., Potts, M.D., 2013. Conserving tropical biodiversity via strategic spatiotemporal harvest planning. J. Appl. Ecol. 50, 1301–1310.
Ramdani, F., Moffiet, T., Hino, M., 2014. Local surface temperature change due to expansion of oil palm plantation in Indonesia. Clim. Chang. 123, 189–200.
Redford, K.H., Padoch, C., Sunderland, T., 2013. Fads, funding, and forgetting in three decades of conservation. Conserv. Biol. 27, 437–438.
Richter, F., 2001. Financial and Economic Assessment of Timber Harvesting Operations in Sarawak, Malaysia. Forest Harvesting Case-Study (FAO).
Rist, L., Shanley, P., Sunderland, T., Sheil, D., Ndoye, O., Liswanti, N., Tieguhong, J., 2012. The impacts of selective logging on non-timber forest products of livelihood importance. For. Ecol. Manag. 268, 57–69.
Romero, C., 1999. Reduced-impact logging effects on commercial non-vascular pendant epiphyte biomass in a tropical montane forest in Costa Rica. For. Ecol. Manag. 118, 117–125.
Roopsind, A., Wortel, V., Hanoeman, W., Putz, F.E., 2017. Quantifying uncertainty about forest recovery 32-years after selective logging in Suriname. For. Ecol. Manag. 391, 246–255.
Runting, R.K., Griscom, B.W., Struebig, M.J., Satar, M., Meijaard, E., Burivalova, Z., Cheyne, S.M., Deere, N.J., Game, E.T., Putz, F.E., Wells, J.A., 2019. Larger gains from improved management over sparing–sharing for tropical forests. Nat. Sustain. 2 (1), 53–61.
Ruslandi, Romero, C., Putz, F.E., 2017b. Financial viability and carbon payment potential of large-scale silvicultural intensification in logged dipterocarp forests in Indonesia. Forest Policy Econ. 85, 95–102.
Ruslandi, Cropper Jr., W.P., Putz, F.E., 2017a. Tree diameter increments following silvicultural treatments in a dipterocarp forest in Kalimantan, Indonesia: a mixed-effects modelling approach. For. Ecol. Manag. 396, 195–206.
Salati, E., Vose, P.B., 1984. Amazon basin: a system in equilibrium. Science 225 (4658), 129–138.
Saner, P., Loh, Y.Y., Ong, R.C., Hector, A., 2012. Carbon stocks and fluxes in tropical lowland dipterocarp rainforests in Sabah, Malaysian Borneo. PloS One 7 (1), e29642.
Santika, T., Wilson, K.A., Budiharta, S., Law, E.A., Poh, T.M., Ancrenaz, M., Struebig, M.J., Meijaard, E., 2019. Does oil palm agriculture help alleviate poverty? A multidimensional counterfactual assessment of oil palm development in Indonesia. World Dev. 120, 105–117.
Scheffers, B.R., Edwards, D.P., Diesmos, A., Williams, S.E., Evans, T.A., 2014. Microhabitats reduce animal's exposure to climate extremes. Glob. Chang. Biol. 20, 495–503.
Schneider, J.E., 2014. Guest editor's introduction: energy homeostasis in context. Horm. Behav. **66**, 1–6.
Schnitzer, S.A., Carson, W.P., 2010. Lianas suppress tree regeneration and diversity in treefall gaps. Ecol. Lett. 13, 849–857.
Senior, R.A., Hill, J.K., González del Pliego, P., Goode, L.K., Edwards, D.P., 2017. A pantropical analysis of the impacts of forest degradation and conversion on local temperature. Ecol. Evol. 7, 7897–7908.

Senior, R.A., Hill, J.K., Benedick, S., Edwards, D.P., 2018. Tropical forests are thermally buffered despite intensive selective logging. Glob. Chang. Biol. 24, 1267–1278.
Senior, R.A., Hill, J.K., Edwards, D.P., 2019. Global loss of climate connectivity in tropical forests. Nat. Clim. Chang. 9 (8), 623–626.
Sheil, D., Meijaard, E., 2010. Purity and prejudice: deluding ourselves about biodiversity conservation. Biotropica 42, 566–568.
Siegert, F., Ruecker, G., Hinrichs, A., Hoffmann, A.A., 2001. Increased damage from fires in logged forests during droughts caused by El Nino. Nature 414, 437–440.
Sloan, S., Campbell, M.J., Alamgir, M., Lechner, A.M., Engert, J., Laurance, W.F., 2019. Trans-national conservation and infrastructure development in the Heart of Borneo. PLoS One 14 (9).
Spracklen, D.V., Arnold, S.R., Taylor, C.M., 2012. Observations of increased tropical rainfall preceded by air passage over forests. Nature 489, 282.
Struhsaker, T.T., et al., 1997. Ecology of an African Rain Forest: Logging in Kibale and the Conflict Between Conservation And Exploitation. University Press of Florida.
Sudarshana, P., Nageswara-rao, M., Soneji, J.R., 2012. Tropical Forests. InTech, Rijeka.
Svarstad, H., Benjaminsen, T.A., 2017. Nothing succeeds like success narratives: a case of conservation and development in the time of REDD. J. East. Afr. Stud. 11, 482–505.
Swaine, M.D., Agyeman, V.K., 2008. Enhanced tree recruitment following logging in two forest reserves in Ghana. Biotropica 40, 370–374.
Tacconi, L. (Ed.), 2012. Illegal Logging: "Law Enforcement, Livelihoods and the Timber Trade". Routledge.
Tan-Soo, J.S., Pattanayak, S.K., 2019. Seeking natural capital projects: forest fires, haze, and early-life exposure in Indonesia. Proc. Natl. Acad. Sci. 116, 5239–5245.
Terborgh, J., Nuñez-Iturri, G., Pitman, N.C., Valverde, F.H.C., Alvarez, P., Swamy, V., Pringle, E.G., Paine, C.E.T., 2008. Tree recruitment in an empty forest. Ecology 89, 1757–1768.
Thompson, M.C., Baruah, M., Carr, E.R., 2011. Seeing REDD+ as a project of environmental governance. Environ. Sci. Pol. 14, 100–110.
Tittler, R., Messier, C., Goodman, R.C., 2016. Triad forest management: local fix or global solution. In: Laroque, G.R. (Ed.), Ecological Forest Management Handbook. CRC Press, pp. 33–45. No. 45.
Umunay, P.M., Gregoire, T.G., Gopalakrishna, T., Ellis, P.W., Putz, F.E., 2019. Selective logging emissions and potential emission reductions from reduced-impact logging in the Congo Basin. For. Ecol. Manag. 437, 360–371.
UNEP-WCMC, 2009. UNEP World Conservation Monitoring Centre: Annual Report 2009. Technical report, UNEP-WCMC.
UNFF, 2007. Non-legally Binding Instrument on All Types of Forests. Technical report, United Nations Forum on Forests.
United Nations, 2015. Adoption of the Paris Agreement (United Nations, New York, 2015).
United Nations, 2018. The Sustainable Development Goals Report 2018 (United Nations, New York, 2018).
Van Gardingen, P.R., McLeish, M.J., Phillips, P.D., Fadilah, D., Tyrie, G., Yasman, I., 2003. Financial and ecological analysis of management options for logged-over Dipterocarp forests in Indonesian Borneo. For. Ecol. Manag. 183, 1–29.
Venter, O., Koh, L.P., 2012. Reducing emissions from deforestation and forest degradation (REDD+): game changer or just another quick fix? Ann. N. Y. Acad. Sci. 1249, 137–150.
Villegas, Z., Peña-Claros, M., Mostacedo, B., Alarcón, A., Licona, J.C., Leaño, C., Pariona, W., Choque, U., 2009. Silvicultural treatments enhance growth rates of future crop trees in a tropical dry forest. For. Ecol. Manag. 258, 971–977.

Walsh, R.P.D., Bidin, K., Blake, W.H., Chappell, N.A., Clarke, M.A., Douglas, I., Ghazali, R., Sayer, A.M., Suhaimi, J., Tych, W., Annammala, K.V., 2011. Long-term responses of rainforest erosional systems at different spatial scales to selective logging and climatic change. Philos. Trans. R. Soc., B 366, 3340–3353.
Wang, K., Dickinson, R.E., 2012. A review of global terrestrial evapotranspiration: observation, modeling, climatology, and climatic variability. Rev. Geophys. 50, RG2005.
Webb, T.J., Woodward, F.I., Hannah, L., Gaston, K.J., 2005. Forest cover rainfall relationships in a biodiversity hotspot: the Atlantic forest of Brazil. Ecol. Appl. 15, 1968–1983.
Wei, L., Chevalier, R., Archaux, F., Gosselin, F., 2015. Influence of stand attributes and skid trail area on stand-scale ground flora diversity. Can. J. For. Res. 45, 1816–1826.
West, T.A., Vidal, E., Putz, F.E., 2014. Forest biomass recovery after conventional and reduced-impact logging in Amazonian Brazil. For. Ecol. Manag. 314, 59–63.
Wilkie, D., Shaw, E., Rotberg, F., Morelli, G., Auzel, P., 2000. Roads, development, and conservation in the Congo Basin. Conserv. Biol. 14, 1614–1622.
Wilson, K.A., Meijaard, E., Drummond, S., Grantham, H.S., Boitani, L., Catullo, G., Dennis, R., Dutton, I., Falcucci, A., Maiorano, L., Possingham, H.P., Rondinini, C., Venter, O., Watts, M., 2010. Conserving biodiversity in production landscapes. Ecol. Appl. 20, 1721–1732.
Xu, H., Li, Y., Liu, S., Zang, R., He, F., Spence, J.R., 2015. Partial recovery of a tropical rain forest a half-century after clear-cut and selective logging. J. Appl. Ecol. 52, 1044–1052.
Zhang, L., Liu, M., Qiao, G., Jiang, J., Jiang, Y., Zhuo, R., 2013. Transgenic poplar "NL895" expressing CpFATB gene shows enhanced tolerance to drought stress. Acta Physiol. Plant. 35, 603–613.

Further reading

Abdul Rahim, N., Harding, D., 1992. Effects of selective logging methods on water yield and streamflow parameters in Peninsular Malaysia. J. Trop. For. Sci. 5, 130–154.
Asner, G.P., Rudel, T.K., Aide, T.M., DeFries, R., Emerson, R., 2009. A contemporary assessment of change in humid tropical forests. Conserv. Biol. 23, 1386–1395.
Duchelle, A.E., Simonet, G., Sunderlin, W.D., Wunder, S., 2018. What is REDD+ achieving on the ground? Curr. Opin. Environ. Sustain. 32, 134–140.
Edwards, D.P., Fisher, B., Boyd, E., 2010b. Protecting degraded rainforests: enhancement of forest carbon stocks under REDD+. Conserv. Lett. 3, 313–316.
Gaveau, D.L., Sheil, D., Salim, M.A., Arjasakusuma, S., Ancrenaz, M., Pacheco, P., Meijaard, E., 2016. Rapid conversions and avoided deforestation: examining four decades of industrial plantation expansion in Borneo. Sci. Rep. 6, 32017.
Gilroy, J.J., Woodcock, P., Edwards, F.A., Wheeler, C., Baptiste, B.L., Uribe, C.A.M., Haugaasen, T., Edwards, D.P., 2014. Cheap carbon and biodiversity co-benefits from forest regeneration in a hotspot of endemism. Nat. Clim. Chang. 4, 503–507.
Graham, V., Laurance, S.G., Grech, A., McGregor, A., Venter, O., 2016. A comparative assessment of the financial costs and carbon benefits of REDD+ strategies in Southeast Asia. Environ. Res. Lett. 11, 114022.
Green, R.E., Cornell, S.J., Scharlemann, J.P., Balmford, A., 2005. Farming and the fate of wild nature. Science 307, 550–555.
Huang, M., Asner, G.P., 2010. Long-term carbon loss and recovery following selective logging in Amazon forests. Glob. Biogeochem. Cycles 24, 1–15.
Kauffman, J.B., Uhl, C., 1990. Interactions of anthropogenic activities, fire, and rain forests in the Amazon Basin. In: Goldammer, J.G. (Ed.), Fire in the Tropical Biota. Ecological Studies, Vol. 84. Springer, Berlin, pp. 117–134.

Phalan, B., Onial, M., Balmford, A., Green, R.E., 2011. Reconciling food production and biodiversity conservation: land sharing and land sparing compared. Science 333, 1289–1291.
Rajamani, L., Brunnée, J., 2017. The legality of downgrading nationally determined contributions under the Paris Agreement: lessons from the US disengagement. J. Environ. Law 29, 537–551.
Rautner, M., Leggett, M., Davis, F., 2013. The Little Book of Big Deforestation Drivers. Global Canopy Programme, Oxford.
Ruslandi, Venter, O., Putz, F.E., 2011. Overestimating conservation costs in Southeast Asia. Front. Ecol. Environ. 9, 542–544.
Sitch, S., Hayes, D., 2011. A large and persistent carbon sink in the world's forests. Science 333, 988–993.

CHAPTER TWO

Novelty in the tropical forests of the 21st century

Ariel E. Lugo[a,*], Oscar J. Abelleira Martínez[b], Ernesto Medina[a,c], Gerardo Aymard[d,†], Tamara Heartsill Scalley[a]

[a]International Institute of Tropical Forestry, USDA Forest Service, Río Piedras, Puerto Rico
[b]Department of Agroenvironmental Sciences, University of Puerto Rico, Mayagüez, Puerto Rico
[c]Centro de Ecología, Instituto Venezolano de Investigaciones Científicas, Caracas, Venezuela
[d]UNELLEZ-Guanare, Programa de Ciencias del Agro del Mar, Herbario Universitario (PORT), Mesa de Cavacas, estado Portuguesa, Venezuela
[*]Corresponding author: e-mail address: aelugo1234@gmail.com

Contents

† Current address: Compensation International Progress S.A., Ciprogress Greenlife, Bogotá, DC Colombia.

Advances in Ecological Research, Volume 62
ISSN 0065-2504
https://doi.org/10.1016/bs.aecr.2020.01.008

Abstract

We review literature relevant to assessing the future of tropical forests and supplement the review with new data from the lowlands of Venezuela. Compared to today, future tropical forests will have a higher level of novelty, defined as the degree of dissimilarity of a system relative to a historical baseline. Processes of succession and evolution generate novelty in forests and have done so for millennia. Under increasing human activity and climate change, the rate of generation of novelty has increased and the resulting forests are termed novel forests to distinguish them from historical forests. Historical forests are less exposed to anthropogenic disturbances and operate at slower levels of novelty generation. Acclimation, adaptation, changes in species composition and dominance, and changes in the proportions of species in communities are the responses of the biota to climate change and anthropogenic disturbances. Therefore, novelty contributes to the persistence of tropical forests in spite of increasing levels of human activity. Novel forests are similar to historical forests in terms of structure but they are younger, they have a faster turnover of mass and chemical elements, and different species composition. Historical species assemblages cannot cope with the altered environments that result from chronic anthropogenic disturbances. The dominant species in novel forests tend to be, and function as, pioneer species. High levels of species dominance in novel forests influence the proportions of chemical elements, which when coupled to species traits and attributes, help explain how novel forests cope with the conditions that result from anthropogenic activities. Novelty is more common in the tropics than in other latitudinal regions, and within the tropics, it is more common in islands where human activity is more intense than in continents. Novel tropical forests in islands have greater representation and dominance of introduced species than novel forests in continents, where native species with wide geographic distributions dominate. Regardless of geography, novel tropical forests share similar attributes and functioning. The adaptability of novel forests to extreme conditions created by human activity signals a future for tropical forests that is different from predictions of constant degradation, homogenization, and loss of biodiversity. Instead, a process of recombination of species (all taxa) into new species assemblages maintains structure, function, physiognomy, species richness, and ecological services. This remixing initially involves loss of large organisms, certain groups of species, and loss of old-growth attributes of forests. Some of these losses can be reversed through succession, assuming there is sufficient time to restore depleted stores such as soil organic matter. Continued environmental change will stimulate continued remixing of species, loss of vulnerable species, gains of less vulnerable ones, and more dissimilarity with historical forests. Novel forests are an answer to the changes induced by climate change and other anthropogenic disturbances, and as such require conservation measures, because as they mature, novel forests usually diversify and help restore lost biodiversity. We also review strategies to conserve biodiversity and optimize ecological services using novel forest succession.

It is impossible to have any idea as to which will be the future dominants, but it is clear that plenty of combinations are possible and that the advance towards a climax (if such a thing can be visualized for equatorial formations) is a very indirect one.

(Aubréville, 1938, p. 531).

1. Introduction

When André Aubréville, a French botanist, faced the complexity of the equatorial closed forest of the Ivory Coast (Aubréville, 1938), he saw and described the paradoxical order and stochasticity displayed by the many tree species that formed the magnificent forest. He also pondered the future of the tropical forest and expressed his point of view summarized in the quote above. At the time, he could not anticipate all the changes that would ensue in the region as a result of human influence and climate change. He did comment about the presence of species associated with humans, noting that anyone experiencing the novelty of running into an oil palm in the middle of the forest was witness to the scarce but irrefutable evidence of past human influence on a forest stand in early 20th century Africa. These days, in highly modified tropical islands such as Puerto Rico, botanists might instead find novelty in encountering a once common, but now rare, old-growth native tree species in a secondary forest site.

As did Aubréville almost a century ago, we take a critical look at the degree of transformation of tropical forests, but this time emphasizing the effects of climate change and other anthropogenic disturbances on these ecosystems. Our aim is to project current trends into the future when we expect that novel tropical forests, which we define later, will be more common than they are today. To establish the structural and compositional characteristics of novel forests in the continental tropics, we present new data from the llands of Venezuela and compare those data with similar information from the montane tropics and Caribbean islands. The Appendix contains the methods and justifications for these analyses. We also explore the ecophysiological characteristics of the successional tree species that predominate under disturbed conditions, the functioning of tropical novel forests, review criteria on how to intervene in novel forests for optimizing their ecological services, and illustrate the implications of varying levels of intervention for the future successional trajectories of novel forests. Finally, we address novel forest conservation as a strategy for assuring the persistence of threatened tropical biodiversity.

2. Novelty in ecosystems

Today, the world is characterized by constant change in almost all of its components, including the location of land masses, the climate, and the biogeography of the biota, to name a few. The magnitude, variety, and longevity of human-induced changes has led to the suggestion that we should refer to the present, not as the Holocene Epoch, but as the Anthropocene Epoch (Lewis and Maslin, 2015; Waters et al., 2016). These changes influence the global environment and promote novelty in ecological systems.

Jackson (2013) reflecting on these changes concluded that ecological novelty is not new. Humans have interacted with tropical forests for at least 100 millennia (Roberts et al., 2017), influenced the global environment, including land surface transformation, and changed the composition of the atmosphere as well (Lewis and Maslin, 2015). While the type and scale of these activities has varied across the world, sometimes leading to significant landscape effects (van der Kaars et al., 2017), the oldest evidence of human occupation in the tropical Americas dates back to 13,000 years BP (Roosevelt et al., 2002). Montoya et al. (2020) hypothesized that early scattered human populations, even in low densities, may have been responsible for some of the current patterns of forest biodiversity and structure in Amazonia. What has changed is the scale of human activity, which increasingly exerts greater influence over ecosystems and the environmental conditions of the world. This explains why when Hobbs et al. (2009) defined novel ecosystems, they included 'human agency' as one of the necessary components of the definition. There are three foundational principles that support the establishment of novelty in ecosystems (Mascaro et al., 2013). These are: (1) the individualistic concept of H. Gleason, which states that the flora and fauna move independently of one another in response to the environment. (2) The notion that the biotic and abiotic characteristics of an ecosystem are tethered; they interact and influence each other. (3) The fact that humans cause changes in ecosystems that are directional and permanent.

Mascaro et al. (2013) defined a novel ecosystem (p. 55) as: ... *one of abiotic, biotic, and social components (and their interactions) that, by virtue of human influence, differ from those that prevailed historically, having a tendency to self-organize and manifest novel qualities without intensive human management.*

This definition restricts novelty to that generated through human influence. As we discuss below, novelty can also be generated by ecological processes in the absence of human intervention, but at this historic moment, human influence is accelerating the generation of novelty in ecosystems. This definition also requires a base of comparison, which Radeloff et al. (2015) addressed and we discuss below. Finally, this definition does not make a distinction on the geographic origin of the species that become part of novel ecosystems and is consonant with the Anthropocene Epoch that we live under today. We anticipate that novel ecosystems will continue to increase in abundance into the future as human control over the world reaches a peak, and it behoves scientists to study and understand their structuring and functioning to assure proper conservation policies and activities. Radeloff et al. (2015) developed a formal and quantifiable definition of abiotic and biotic novelty in ecosystems and mapped global novelty, finding that the tropics emerged as the region with the greatest level of novelty in the world.

The definition of novelty proposed by Radeloff et al. (2015) was (p. 2052): *the degree of dissimilarity of a system measured in one or more dimensions relative to a reference baseline, usually defined either as the present or a time window to the past.* The definition requires a baseline of comparison, results in a continuum of novelty, both in time and space, and allows for the comparison of multiple dimensions, both abiotic and biotic. Novelty can develop in almost all abiotic and biotic parameters or ecosystem attributes. Those used by Radeloff et al. (2015) included climate change, atmospheric nitrogen deposition, and changes associated with human population density. Equally important to the biota and the environment is the carbon dioxide concentration in the atmosphere. The concentration of this gas has oscillated over millennia but has recently increased at a fast rate powered by human activity. Because there is a limitation of tropical data availability to conduct the kinds of analyses in Radeloff et al. (2015), we will use a biotic criterion to assess novelty in tropical forests. Specifically, we focus on the species composition, which changes in response to climatic and other environmental disturbances. The tropics have been sufficiently explored taxonomically to allow the exploration of biotic novelty under the influence of human agency, even if we lack sufficient details on the abiotic changes associated with biotic novelty. Our review of tropical forests emphasizes biotic novelty with inferences about abiotic causes, when those causes are not quantified.

Novelty in the form of new species, new species assemblages within communities, rapid turnover of species, or new abiotic conditions is a natural product of species evolution and forest succession. Novelty has existed and will continue to exist as long as evolution and succession occur on the planet. Novelty is a measure of the degree of adaptation to changing conditions under which species evolve and species assemblages change over time. However, the pace of novelty generation has increased in the Anthropocene due to increased human activity and the heterogeneity in environmental conditions that it creates (Lugo et al., 2018). Novelty is revealed in the changes of the relative importance of species within species assemblages, the gains and losses of species from assemblages, and by the remixing of species into different rank order of species irrespective of geographic origin.

3. Change in tropical climates

No one questions the importance of climate to vegetation functioning and physical appearance, and one can safely assume that if the climate changes, the vegetation will respond. The unanswered question at this historic moment is: What is the projected magnitude of climate change into the near future, and what level of persistence can we expect of historical ecological systems given the projected level of climate change? We also do not know the relative roles in the generation of novelty in ecosystems played by climate change and other human disturbances.

A review by Bailey (2009) suggested that tropical rain and dry climates would increase in geographic area while boreal and snow climates would decrease (see also Belda et al., 2014). These predictions are based on global climate models that have a poor performance when dealing with tropical latitudes. For example, the range in the level of the 2000–06 trend in net ecosystem productivity predicted by seven climate models of the Coupled Model Intercomparison Project Phase 5 was about $80\,gC\,m^{-2}\,year^{-1}$ for most latitudes except the tropics. For the tropics, the range of the output among the models was greater than $220\,gC\,m^{-2}\,year^{-1}$ or almost three times more uncertain than for temperate and boreal latitudes (Cavaleri et al., 2015). Worse yet, in terms of the prediction for tropical latitudes, some models describe the region as a net source of carbon, while others do so as a net sink. Global climate models do not recognize the diversity of climatic conditions in the tropics as revealed by the Holdridge Life Zone System (Holdridge, 1947). The Holdridge system depicts climate in general, and that of the tropics and subtropics in particular, with greater geographic precision and repeatability than the Köppen-Trewartha system (Lugo et al., 1999a). The application of the system is limited by the availability of empirical climate data, which the Köppen-Trewartha system was designed to overcome (Bailey, 2009).

We define the tropics and subtropics as those geographic locations without frost in the lowlands (0–500 m elevation; Holdridge, 1967). If climate change reduces the frost occurrence in the lowlands of continents and islands, it is possible that the geographic extent of areas defined as tropical and subtropical will expand. A faster atmospheric warming rate at the higher northern latitudes of the tropics and subtropics, i.e., the Tropic of Cancer at 23.5° north latitude, can have an effect on the northern distributions of tropical and subtropical species. Low temperature sensitive species in tropical, subtropical, and warm temperate life zones can expand their northern range if the frequency of frost declines and if they can overcome other ecological barriers associated with latitudinal changes and migration. The same is true, but in the opposite direction, at the southern limits of the Tropic of Capricorn.

The depiction of the geographic distribution of tropical and subtropical forests by life zone has been used to illustrate the effects on those forests of the accelerating climate change (e.g. Cramer and Leemans, 1993; Emanuel et al., 1985; Smith et al., 1992). One would expect that climate change shifts the geographic influence of life zone conditions. Some areas will shift to warmer and drier life zones while others will shift to wetter and cooler ones, and some might not change. A shift in life zone distribution, coupled with the effects of human activity and extreme events, are likely to induce changes in the composition and the proportions of species in species assemblages. Moreover, the relative importance of a life zone shift and level of anthropogenic disturbances may not be homogeneous geographically. One could expect that some geographic regions may be more vulnerable to a given force of change, or there might be locations where both forces of change (climatic and anthropogenic) converge. An example is the Amazon forests where anthropogenic changes such as deforestation are not distributed homogeneously within climatic life zones (Laurance et al., 2006; Phillips et al., 2008).

A problem when anticipating climate change over specific locations is the resolution of global climate change models. For example, when addressing climate change over montane tropical areas, Helmer et al. (2019) observed that the grid size of global climate change models was too coarse (0.5°) to assume that any of those projections would reflect the conditions that local vegetation would face. Moreover, they indicated that within one grid cell of a global climate model, *climate changes with elevation over distances shorter than each cell* (p. 4/34). Thus, the predicted ecophysiological effects of climate change based on global climate models must be interpreted with caution and with consideration to the difficulties in downscaling global models to local scales, and the ecological and environmental variation within local geographic and ecological scales.

In a study that overcomes the resolution limitations of global climate models, Helmer et al. (2019) predicted that the climatic conditions that lead to the familiar tropical montane cloud forests and *páramo* vegetation in the Caribbean and throughout the Neotropics, will mostly disappear in 25–45 years under a worst-case scenario that projects the Earth-atmosphere energy balance change as a result of continuing increases in greenhouse gas emissions and land use changes. As discussed by Helmer et al. (2019), the implications to the persistence of the unique biodiversity of this rich and extensive region of the Neotropics are dire.

Henareh et al. (2016) used a different approach than used by Helmer et al. (2019) to project into the future climate change conditions in Puerto Rico. They downscaled the output of 12 general circulation climate models to assess the distribution of the Holdridge life zones of Puerto Rico from 1960–90 to 2071–99. Life zones shifted from wetter to drier life zones and they projected the loss of the wettest life zone in the island, the subtropical rain forest life zone. In their worse case scenarios, the island loses its subtropical wet, and rain forest life zones, and gains five life zones, including four tropical ones (tropical wet forest, tropical moist forest, tropical dry forest, and tropical very dry forest), and a subtropical thorn woodland. The predicted trend is for warmer and drier life zone conditions. In all scenarios, the model predicts a shift in the most geographically extensive life zone of Puerto Rico from subtropical moist forest to tropical dry forest within 100 years. The application of similar methods yielded similar results for continental Central America; a decrease in the extent of cloud forests at high elevations and an extension of drier climates into higher elevations (Kamalkar et al., 2008). The heterogeneous topography of insular montane landscapes, combined with anthropogenic barriers to dispersal and migration, including abandoned fields and urban development, are likely to affect the expected movement of organisms in response to climate change, more so than in continental areas where dispersal and migration of vegetation may be more feasible.

Cloud forests and other life zones at high elevations may be more likely to be affected by climate change than lowland life zones given the differences in areal extent of the life zones. A lowland life zone is more likely to cover a larger area than a montane one given the steepness of the climatic gradients in mountains relative to lowlands. Therefore, climate change is more likely to affect montane forests much quicker than lowland forests (but see Esquivel-Muelbert et al. (2018) and Edwards et al. (2019), for examples, where the synergy between climate change and other anthropogenic disturbances create greater effects on lowland Amazonian forests). As an insular example, Campos Cerqueira et al. (2017) found bird populations expanding their home ranges in the Luquillo Mountains of Puerto Rico. They found that of 29 bird species in these mountains, eight species, or 28%, showed significant shifts in their range limits, mostly towards higher elevation, between 1998 and 2015. A continental example is Morueta Holme et al. (2015) study of 210 years of vegetation change on the Chimborazo volcano in Ecuador. They compared today's vegetation with the description of von Humboldt and Bonpland (1807). They found an upward shift in the distribution of vegetation zones as well as increases of >500 m in the maximum elevation limits of individual plant taxa. These range shifts are consistent with increased temperatures and glacier retreat on Chimborazo since Humboldt's visit.

The expected climatic changes can fundamentally shift the moisture and temperature conditions under which most organisms live today. This can lead to shifts in the dominance of species within species assemblages to those with ecological response strategies associated to faster rates of processes and smaller size organisms. Corresponding spatial-temporal lags in dispersal and migration-mediated responses may also become more evident with time. A single or a small subset of species from the total species pool already dominates primary and secondary tropical forests in some regions of the tropics (Hubbell, 2001; Kazmierczak et al., 2016; ter Steege et al., 2013). We anticipate that species dominance in novel forests will shift to a subset of those species with favoured traits such as faster growth, larger leaves, fistulose branches, and significant amount of seeds (Gómez-Pompa, 1971; Ramos Prado et al., 1982; Vázquez Yánez, 1980). The shift involves responses to both climatic and edaphic (soil condition) changes. A shift to higher species abundance is implicit in the notion of hyperdominance observed in the trees of the Amazon by ter Steege et al. (2019). They assessed 10,071 tree species and found that over half the stems belong to just 227 species.

A similar imbalance was also observed in species to genus ratios. Over half of all Amazonian tree species belong to genera with 100 or more species, while 52% of the genera have 10 or fewer species (Dexter and Chave, 2016). Some of these hyper dominants are tree species with some evidence of domestication that dominate large forest areas, especially in those locations near archaeological sites (Levis et al., 2017).

4. Three case studies regarding novelty in tropical forests

Novelty in the tropics is revealed in changes in the composition of floras. We reviewed tropical floras from 28 continental and 33 insular locations (Appendix) and found that the percentage of naturalized species in the published floras averaged 9.9 (SE 0.1) and 17.5 (SE 0.1), respectively. The level of species naturalization is more obvious in islands where species from other geographic locations are easier to record. Moreover, synanthropic taxa, i.e., taxa associated with humans, were introduced since the colonial and exploration times, due the Europeanization of their colonies (Correoso Rodríguez, 2005). In Puerto Rico, the species richness of most taxonomic groups has increased as a result of species introductions and naturalization being greater than species extinctions (Lugo et al., 2012a,b; Torres and Snelling, 1997). All these increases in faunal and floral composition strongly suggest that the species assemblages in these tropical locations contain novel combinations of species different from historical combinations of species.

Novel tropical forests appear to be more abundant in the insular tropics than the continental tropics (Ewel et al., 2013) because human population and activity are concentrated in islands compared to continental regions. As a result, a greater fraction of insular territory is affected by human activity when compared to continental territory. This means that the insular biota faces more environmental challenges and anthropogenic disturbances than continental biota, thus generating more novelty. Thus, while islands may appear different when compared to continents because of the intensity of change in their species assemblages, they are reflecting particular circumstances while following the same ecological principles that govern continental systems. Our review of the tropical literature shows that novel forests occur on the montane and lowland continental tropics and are beginning to be recognized as such (Aymard, 2017; Aymard and González, 2014). Nevertheless, we need to examine the outcomes of novelty generation in islands vs. continents to firmly establish what they have in common and where they diverge.

In this section we compare novel forests in Caribbean islands with those in the South American continent. We use Puerto Rico as representative of Caribbean islands, and the Venezuelan Andes (from now on *los Andes*) and Venezuelan Llanos (from now on, *los Llanos*) as representatives of montane and lowland continental areas, respectively. Like in the forests of Caribbean islands, the floras of *los Andes* and *los Llanos* have experienced intense human activity, which has led to changes in the species composition and structural complexity of its vegetation (Aymard, 2017; Aymard and González, 2014; Sarmiento and Llambí, 2011). By comparing continental vegetation with Puerto Rican novel forests, we can advance the understanding of the emergence of novelty in tropical forests and gain insights about the potential future species composition, structuring, and functioning of tropical forests. We start by summarizing the current understanding of novel forests in Puerto Rico, followed by information about vegetation in *los Andes* and *los Llanos*.

4.1 Island novel forests: A case study from the Caribbean

The vegetation survey of Puerto Rico and the US Virgin Islands of Gleason and Cook (1926) documented the importance of native species in the locations that they visited. Detailed succession studies, such the one by Smith (1970) following ionizing radiation and experimental cutting of forests in the Luquillo Mountains, also established a succession of primary and secondary native species with no obvious deviation in terms of species composition from what was expected in the absence of human intervention, even if the experiment themselves (ionizing radiation and clearcutting) were human-induced. The expression of novelty in this succession involved the substitution of the floristic composition of the primary forest with secondary forest ones, and changes in the proportions of species in seedling populations. Four individuals of *Phytolacca icosandra* colonized and expanded on the radiation and the cut sites in locations subjected to heavy trampling. Over the next 23 years the radiation site generated high levels of novelty through a high turnover of herbs, vines, ferns, saplings, seedlings, and graminoids. This succession led to a closed canopy forest with the same species described by Smith (Taylor et al., 1995) but assembled with a different proportion of importance measured by the species Importance Values, i.e., the sum of a species relative density and relative basal area, expressed in percent (Appendix). Notably, all taxa involved in this study were native species. The generation of novelty in this succession was low from the point of view of floristic composition, but the rate of biomass accumulation in this radiation-stressed system was different from similar forest stands responding to hurricane effects (fig. 19.5 in Lugo and Heartsill Scalley, 2014). In this review we use the term 'historical succession' and 'historical secondary forests' to designate successions and their outcomes (respectively) where the native species composition involved deviates very little from what is normally expected in the absence of anthropogenic effects.

Studies of succession on abandoned pastures in northeastern Puerto Rico revealed a higher level of novelty generation than observed in the experimental succession studied by Smith (1970) and Taylor et al. (1995). The findings of these studies (Aide et al., 1995, 1996; Lugo, 1992; Zimmerman et al., 1995, 2000) included: the initial failure of native pioneer tree species to colonize the pasture; over one decade of high levels of dominance by a group of shrubs and treelets, including high elevation sites with a single species (*Miconia prasina*) with 100% Importance Value; the eventual establishment of native tree species; and the emergence of introduced species in the succession (10 of 112 woody species). Aide et al. (1996) noted that the original species composition had not developed after 40 years of succession and that these secondary forests could not be distinguished from undisturbed sites in terms of tree density, basal area, number of species, species diversity, or physiognomy, i.e., the shape, height, and physical appearance of vegetation. A statistical analysis of the succession trajectories resulted in the expected importance of age, elevation, and time as correlates with stand characteristics and species composition. However, those analyses and canonical ordinations could not explain the observed high degree of floristic variation. We argue that this floristic variation is a measure of the high degree of novelty generated by the successions under study. We term this type of succession novel succession and its outcome novel secondary forests (or just novel forests), because both the path and the outcome of succession deviate significantly from historical successions. In Puerto Rico, the deviation includes the presence of introduced species, but as we will see in the example from Venezuela, the deviation from historical patterns may or not involve introduced species.

Aide et al. (1995, 1996) and Zimmerman et al. (2000), still working in northeastern Puerto Rico, explored the causes that prevented native pioneer species from colonizing abandoned pasture sites. They identified biotic and abiotic barriers that impeded forest regeneration and seedling establishment, including depleted seed banks, lack of advanced regeneration, the composition and abundance of seed in seed rain, distance to seed sources, loss of seed dispersers, and species traits. Soil compaction, loss of soil structure, and erosion were identified as abiotic factors that gave advantages to introduced tree species over native ones. Silander (1979) found that the pioneer *Cecropia schreberiana*, a Caribbean endemic element, could not colonize pastures because insects reduced seed viability to 1 or 2 months, and high and fluctuating soil temperatures and/or low and fluctuating soil moisture reduced germination rates. Clearly there is a strong selection at the point of tree establishment (ecesis) that can regulate the species composition of developing forests.

Aide et al. (2000) studied the succession of 71 abandoned pastures in four geographic locations within Puerto Rico, and confirmed the results discussed above while adding an additional level of novelty to the ecological outcome of their observations. They found more introduced tree and shrub species present in the successional trajectories of abandoned pastures compared to sites previously subject to other forms of agriculture, such as shade coffee and subsistence farming. The presence of introduced species, coupled to a predominance of native tree species regenerating in the understory, resulted in forests with species composition that contrasted with the species composition of old forest sites in the vicinity of the emerging forests on abandoned pastures. The differences were sufficient to lead the authors to conclude (p. 333) that: *...it will take many years for the species composition to converge with that of the old forests*. Alternatively, what if the species composition does not converge? Or better yet, why should it be presumed that it has to converge? Aubréville (1938) made this point clear in his seminal paper: there can be as many successional trajectories as there are biotic, abiotic, and historical (i.e. anthropogenic) contexts.

The repetitive inventory of Puerto Rico's forests, which began in 1982 (Birdsey and Weaver, 1982), and continues today (Marcano Vega, 2019), provided sufficient information at the insular scale to show that 75% of the forest area of Puerto Rico were novel forests (Lugo and Helmer, 2004; Martinuzzi et al., 2013). These novel forests are shaped by the climate (Brandeis et al., 2009) as is the other 25% of the forest area with historical forest cover. An extensive literature documents the structural and species composition of these novel forests (e.g. Abelleira Martínez and Lugo, 2008; Abelleira Martínez et al., 2010; Lugo et al., 2011a, 2012a), the novel animal communities associated with them (e.g. Abelleira Martínez, 2008; Lugo et al., 2012b) and other associated plants (e.g. Pérez et al., 2012), and temporal and spatial structural variation in dry climates (e.g. Molina Colón and Lugo, 2006; Molina Colón et al., 2011; Ramjohn et al., 2012).

These novel forests do not show evidence of homogenization of the biota because the very high dominance of canopy species early in the succession decreases over time and also decreases vertically within stands with increasing species richness in the lower forest strata (Abelleira Martínez, 2010). In fact, fig. 20.3 in Lugo and Brandeis (2005) shows that a hectare of novel forest has a greater number of tree species than a hectare of mature historical forests dominated by native species.

4.2 Montane continental novel communities: A case study from the Andes

The mountaintops in *los Andes* comprise ecosystems known as the *páramos* (Vareschi, 1970). Their vegetation includes a considerable number of endemic species with a particular physiognomy. Standing out is the Espeletiinae group, with a remarkable degree of morphological diversity, including trees with profusely branched, dichotomous or unbranched stems, shrubs, and rosettes that can be sessile, short-branched or, notably, giant caulescent rosettes (Cuatrecasas, 2013). The *páramo* vegetation also comprises numerous representatives of Poaceae, Asteraceae (*Espeletia* sensu *lato*), and Ericaceae (Jiménez Rivillas et al., 2018).

For about 400 years, the *páramos* of the state of Mérida have been dedicated to wheat farming with no apparent efforts to conserve soil. When fields are abandoned, the historical vegetation is slow to return. Sarmiento and Smith (2011) studied the vegetation of this region in relation to past and present land use and found four community variants with different combinations of species: *páramo* (assumed to be undisturbed, but less developed than *páramos* under more favourable conditions), degraded shrublands, transitional vegetation, and degraded communities dominated by a matrix of Poaceae elements called *pajonales* in Peru. They found that after 40 years of abandonment, there is no evidence to show that the combination of plant species that constituted the *páramo* was returning to the abandoned lands. The results of a linearized correspondence analysis (their figure 7), showed that the vegetation aligned along a soil degradation axis with the *páramo* on one extreme and the degraded shrublands in the other. Transitional vegetation overlapped the *páramo* and degraded shrublands but was closer to the *páramo* than to the degraded shrublands.

The *pajonales* separated from the other three vegetation types and fell outside the main disturbance gradient. *Pajonales* exhibited a dramatic change in species composition including the dominance of a taxon that ranges from Argentina to the state of Utah in North America; the grass *Jarava ichu*.

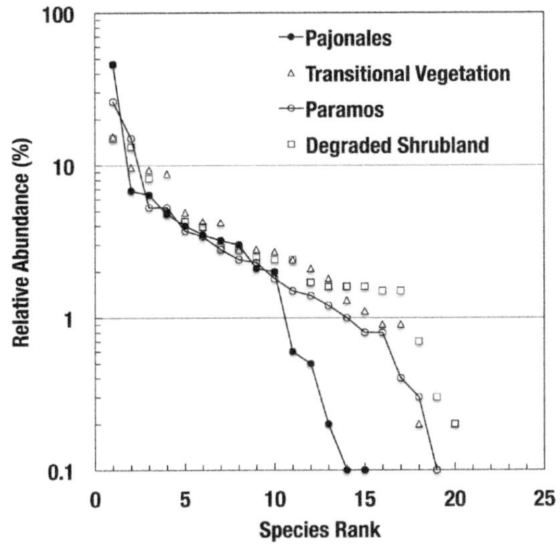

Fig. 1 Ranked-abundance curves for four vegetation types in the wheat-growing region of the Venezuelan Andes. *Data are from Sarmiento, L., and J. K. Smith. 2011. Degradación de laderas durante el ciclo triguero en los Andes venezolanos y factores que limitan su restauración. 17-34 in F. Herrera and I. Herrera, Eds. La restauración ecológica en Venezuela: fundamentos y experiencias. Ediciones IVIC, Instituto Venezolano de Investigaciones Científicas, Caracas, Venezuela.*

When all the species in all four communities were ranked by abundance (Fig. 1), the graph shows lower number of species and higher abundance for the top ranked species of the *pajonales* compared to *páramo*. The community attributes of the *pajonal* compared with those of the historical *páramo* (increased abundance, lower number of species, different common species, and different mix of species) deviate considerably from the historical vegetation, reflecting a high degree of novelty.

As discussed above for Puerto Rico, the shift in community structure and composition is associated with deteriorating soil conditions due to anthropogenic disturbance. Sarmiento and Smith (2011) found that *pajonales* were associated with significant change in soil texture (more sand and less clay), over 60% reduction in soil organic matter, and lower soil nitrogen relative to the *páramos*. As in Puerto Rico, soil erosion changed the path of succession. The biotic response involved changes in species composition and dominance that led to a biological capability that could deal with the novel conditions of anthropogenic disturbance. The main difference was that in Puerto Rico, introduced species prevailed through all the succession but in *los Andes*, native species prevailed and a grass with a wider geographic distribution emerged as the dominant species. Moreover, the remixing of species through succession in Puerto Rico included both native and non-native species, but in *los Andes*, the emergence of introduced species was only observed for the first 2 years, and from there on, succession was dominated by native species (Sarmiento and Llambí, 2011).

4.3 Lowland Continental Novel Forests: A case study from the Venezuelan Llanos

The 90 0.1 ha-plots (100 by 10 m) from *los Llanos* that we analysed (Fig. 2) had a total of 872 taxonomically resolved morphospecies from at least 96 plant families. However, only 52 plant families (226 species) were contributing more than 50% dominance as measured by the species Importance Value. Further, only 169 species were represented in the grouping of the top three species of each plot arranged by Importance Value. When we focused on the 15 plots with the highest values of Importance Value observed among their dominant species, we found 44 species, and from those then selected the 15 species with the highest Importance Value among them (Table 1). These species are represented by only ten families and are all native species, including trees, palms, and bamboo. Only two of the tree species

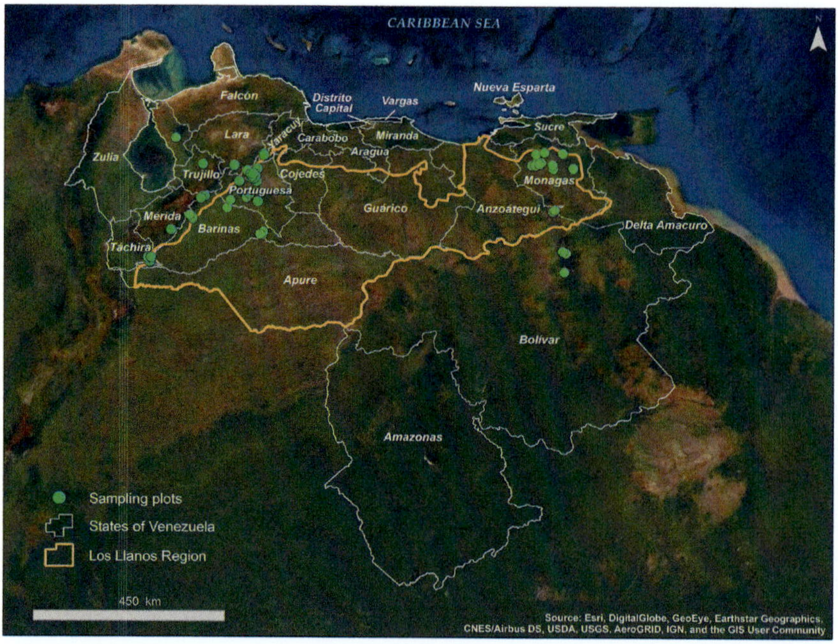

Fig. 2 Map of Venezuela showing the location of 90 vegetation plots in the Llanos. *Data are from G. Aymard and the cartography is by Olga Ramos González.*

Table 1 Species with the highest importance values (based on stem density and basal area) in novel forests of *los Llanos* of Venezuela.

Species	Family	Importance value (%)
Inga oerstediana	Fabaceae	64.29
Guazuma ulmifolia	Malvaceae	62.57
Cochlospermum vitifolium	Bixaceae	56.51
Virola elongata	Myristicaceae	51.04
Tapirira guianensis	Anacardiaceae	46.39
Guarea guidonia	Meliaceae	45.88
*Attalea maripa**	Arecaceae	44.92
Protium tenuifolium	Burseraceae	43.46
*Guadua angustifolia**	Poaceae	39.91
Anacardium excelsum	Anacardiaceae	39.59
Inga nobilis	Fabaceae	39.51
Senegalia polyphylla	Fabaceae	37.39
Rudgea crassiloba	Rubiaceae	37.12
Ceiba pentandra	Malvaceae	35.69
Protium stevensonii	Burseraceae	34.75

The list includes woody (tree, shrub, liana), and non-woody (*) standing (bamboo and palm).

(*Ceiba pentandra* and *Guazuma ulmifolia*) appear in the list of 29 pioneer species of the Neotropics published by Chazdon (2014), which is a list influenced by forests with greater rainfall than those in *los Llanos*.

High dominance in forests, measured by the sum of the Importance Value of the top three species in a plot, was associated with lower species per plot and lower complexity index (Fig. 3). The complexity index of the community tended to increase with a reduction in dominance, but this relationship was not statistically significant at $P < 0.05$. It appears that the structural attributes of communities (height, basal area, stem density) upon which the complexity index is based are not dependent on the floristic composition of the community, as was described by Aide et al. (1995, 1996, 2000) in Puerto Rico. In other words, communities are seldom novel from the point of view of stand structural parameters. The source of novelty emerges from floristic composition and the effects of species on ecosystem functioning that we discuss below.

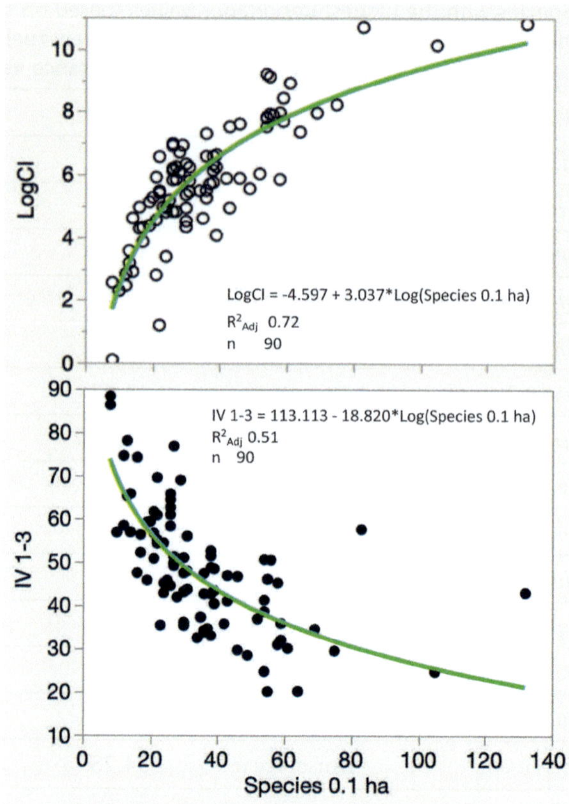

Fig. 3 Non-linear relationship between the number of species per 0.1 ha with the Complexity Index of vegetation (open circles) and sum of the Importance Value of the three-most dominant species in 90 plots in the Venezuelan Llanos (filled black circles). *Data are from G. Aymard.*

The dominant species in the 90 plots have in common a wide geographic distribution throughout the Neotropics, including from Argentina to Mexico, or from Brazil through Colombia to Panama, and also from the Caribbean islands. They all function as pioneer species but have different traits. For example, some are shade intolerant with rapid growth (*Senegalia, Handroanthus, Tabebuia, Hura*), some have seeds dispersed by animals (*Inga, Guazuma*), or by wind (*Ceiba, Heliocarpus, Cochlospermum*). The genera *Inga, Spondias*, and *Erythrina* are examples of taxa that can grow in mature forests favoured by human activity because of their dispersal by animals, and human use for their products or for shade. Species of the genera *Virola, Tapirira, Guarea, Anacadium, Protium, Annona*, and *Ficus* have fleshy seeds dispersed by animals and generally occur in later stages of succession where shade tolerance favours them. *Heliocarpus americanus* grows in pre-montane forests above 400 m elevation. Palms form mature forests on sites with soils with poor drainage as is also the case with *Anacardium, Guarea,* and *Zygia.* Leguminosae (sensu *lato*) and species of *Inga* and *Pterocarpus* have associations with nitrogen-fixing bacteria.

The analysis of secondary forests in *los Llanos* established that these forests had high levels of species dominance, with a few species accounting for over 50% of the Importance Value of the communities. Moreover, these dominant species were native species with wide geographic distributions throughout the Neotropics. We now ask: How different are these communities from those described by Veillon in the 1950s to the 1970s?

Jean Pierre Veillon made an extraordinary effort to describe what he designated the natural forests of Venezuela (Veillon, 1995, 1997), which we identify as historical forests. Between 1953 and 1978 he set up 62 large plots (>1 ha) in undisturbed forested areas or in areas in the process of being altered by logging activities. He focused on trees larger than 10 cm diameter at breast height and described in detail the structure and floristic composition of forests throughout nine states in Venezuela, including those of *los Llanos*. Because he was interested in the primary vegetation under various climatic, edaphic, and geomorphological conditions, he also collected information on the level of human activity and intervention within the forests that he sampled. Veillon studied six locations (Table 2) in *los Llanos* where, in his view, the primary and secondary vegetation could be compared because of their similar climate and close proximity. His main findings were that human intervention changed the floristical composition and structure of the stands. Stands with notable human influence were considered secondary vegetation. He also found that these secondary forest stands had not recovered their original species composition when compared with stands that he defined as primary forest stands. We compared species dominance in those same six stands using indices of species relative Importance Value measured by species Importance Values, and also compared them with those studied by Aymard as discussed above (Table 1).

As suggested by Veillon, there was little overlap in the top three dominant species of primary and secondary forests in *los Llanos* (Table 2). Moreover, the Importance Value of individual species and the sum of the Importance Value indices of the top three species did not differ between the two groups. The Importance Value in Veillon's plots were generally low, with the exception

Table 2 Species with the highest importance values in mature, climax, and secondary forests of *los Llanos* of Venezuela sampled by Veillon (1997).

Site and location	Dominant species (IV)	Importance value
Primary forests		
Alto Tucupido (24.7)	***Bombacopsis quinata*** (***Pochota fendleri***)	10.6
230 m elevation. Guanare, Portuguesa	*Unknown*	7.6
	Chrysophyllum caracasanum (*Pradosia caracasana*)	6.8
Ticoporo Aereopuerto (44.2)	***Symmeria paniculata***	16.1
120 m elevation. Pedraza, Barinas	***Bombacopsis quinata (Pochota fendleri)***	15.4
	Pouteria anibaefolia (Pouteria reticulata)	12.7
Mijagual (47.7)	*Anacardium excelsum*	26.5
150 m elevation. Pedraza, Barinas	***Pouteria anibaefolia (Pouteria reticulata)***	16.1
	Sorocea sprucei	5.1
Secondary forests		
Caimital (33.5)	***Symmeria paniculata***	13.7
150 m elevation. Obispos, Barinas	***Maquira coriacea***	10.5
	Bombacopsis quinata (Pochota fendleri)	9.2
Hato Sansun (28.21)	*Bravaisia integerrima*	13.0
100 m elevation. Guanare, Portuguesa	*Spondias mobin*	9.4
	Maquira coriacea	5.8
La Ricaldera (41.9)	*Piptadenia peregrina* (*Anadenanthera peregrina*)	18.8
200 m elevation, Obispos, Barinas	*Cochlospermum vitifolium*	12.6
	Protium heptaphyllum	10.5

All sites are in the tropical dry forest life zone sensu Veillon (1974). Importance Value (IV) is in percent, based on tree density and basal area. The IV for three dominant species is in parenthesis next to the site location. Species in bold appear in more than one site. Scientific names are updated in parenthesis.

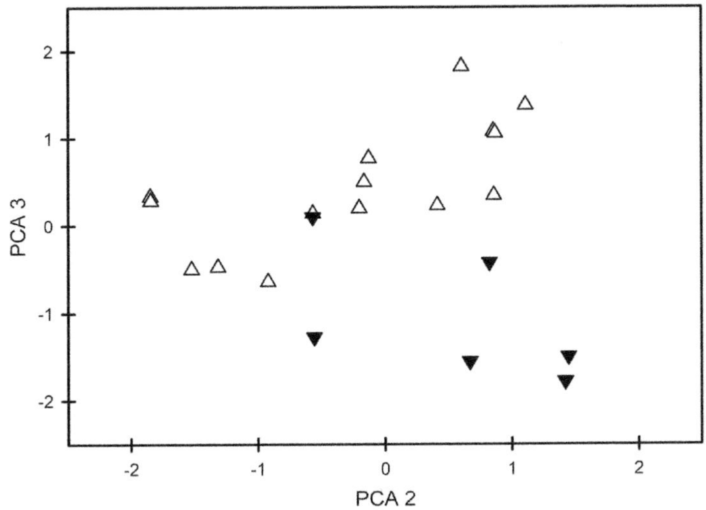

Fig. 4 Relationship between the number of species per 0.1 ha (X axis) and the annual basal area growth in m^2 ha^{-1} in 90 vegetation plots in the Venezuelan Llanos. *Data for plots from G. Aymard are symbolized with open triangles and from Veillon, J. P. 1997. Los bosques naturales de Venezuela. Parte III. Los bosques tropóficos o veraneros de la zona de vida de bosque seco tropical. Universidad de Los Andes- Consejo de Publicaciones, Instituto Forestal Latinoamericano, Facultad de Ciencias Forestales y Ambientales, Mérida, Venezuela, p. 127 with filled inverted triangles.*

of *Anacardium excelsum* with an Importance Value of 26.5. This was due to the large size of these trees (high basal area), which are among the largest trees in *los Llanos*. Although the stands were different floristically, they were not different in terms of species dominance; the level of novelty in the successional forests was low compared to the historical primary forest stands. In contrast, species dominance in the subsequent Aymard inventory in Table 1 was higher than all six Veillon stands by a factor of 2. This is illustrated in the distribution of plots in ordination space as defined by the Importance Value contribution of individual species (Fig. 4), which has shifted from the proportions observed in the Veillon plots to those in the Aymard plot inventory. The horizontal axis separates plots by their second highest Importance Value species ($r^2 = 0.705$), while the vertical axis separates plots based on their highest Importance Value ($r^2 = 0.987$).

The secondary species in Veillon plots, like those in Aymard plots, had wide geographic distribution. The range of *Symmeria paniculata*, *Maquira coriacea*, and *Protium heptaphyllum* extended from Bolivia to Venezuela. *Spondias mombin* extended from Bolivia to Mexico and the Caribbean, while the distribution of *Anadenanthera peregrina* is accelerated by humans because its seeds are the source of a powerful hallucinogenic snuff called *yopo*. Currently, the distribution of this species ranges from Argentina to Venezuela, the Guianas, Brazil and the Caribbean. *Cochlospermum vitifolium* extends from South America, Central America, Mexico, and the Caribbean, and *Bravaisia integerrima* from Colombia, Venezuela, the Caribbean, Central America, and Mexico.

The suite of species dominating the historical forests of *los Llanos* had little overlap with those dominating Aymard plots, i.e., 4 species in Veillon plots appeared among the list of the top 15 dominant species in Aymard plots. Notably, one species that dominated secondary forests in Veillon plots was among the top dominants in Aymard's plots (e.g. *Cochlospermum vitifolium*), which raises questions about the level of disturbance in Veillon's primary forest plots. These comparisons reflect a high level of novelty in the forests that now prevail in *los Llanos* compared to historical forests studied by Veillon.

5. The functioning of novel tropical forests

In the preceding text we have argued that some anthropogenic disturbances, such as soil erosion, can affect the conditions of plant establishment (ecesis) for a long time, essentially acting as chronic events. The novel conditions for ecesis that result from these anthropogenic disturbances can lead to novel successional trajectories and novel forests that are different from those associated with the historical forests that they replace. In this section we focus attention on functional comparisons between novel and historical forests. Such comparisons are affected by the age of stands, particularly when dealing with successional trajectories. Lugo et al. (2018) took age into consideration when comparing the functioning and related ecological attributes of novel and historical forests (their table 1). They compared young novel with young historical, and mature novel with mature historical forests, and found that mature novel forests were more similar to young novel forests than they were to mature historical forests. The age range of available stands limited comparisons because the age difference between mature historical and mature novel was high. The oldest novel forests available for comparison were less than 100 years old, while mature historical stands can be twice that age.

Age also has an influence on other comparisons that involve species and system attributes. At the species level, Grime and Pierce (2012) analysed the role of species within ecosystems by what they called the CSR strategies. These involved the maximization of resource acquisition (C), maintenance of metabolic performance in variable and unpredictable conditions (S), and regeneration or the rapid completion of the life cycle (R). Lugo et al. (2018) compared novel with historical forests in the subtropical moist forest life zone of Puerto Rico using the CSR strategy as the measure of comparison. They found that historical forests were consistently S-selected, while novel forests had both C-selected and R-selected traits. In no comparison did novel forest species fall on the S-selected strategies.

At the ecosystem-level novel and historical forests differentiate by what Walker and Salt (2006) defined as slow variables. In any system, a slow variable is a storage or compartment, such as soil or wood in forests, with a slow turnover rate. The turnover rate is calculated by dividing a flux (mass per area and time) into the compartment by the storage (mass per area), resulting in X turnovers per time. In a forest, for example, state variables can be arranged by decreasing turnover rate: leaves < roots < aboveground wood < soil organic matter. In this example, soil organic matter, with the slowest turnover rate is a slow variable. Slow variables are important for ecosystem function because their turnover is associated with resilience. Slow variables take more time to develop, but they provide buffering to disturbances. Conversely, fast variables recover quickly but are less resistant to change. When comparing novel with historical forests, Lugo et al. (2018) found that slow variables had larger storages in historical forests than in novel forests, where fast variables predominate over slow ones.

The three case studies that we presented above have shown that novel forests, in contrast to historical forests, have higher species dominance and novel species assemblages with introduced and/or native species with wide geographic distribution. The question that we address next is: What are the implications of high species dominance and novel species composition to ecosystem-level functional attributes of forests?

We do not expect to find novel ecophysiological characteristics at the species level when comparing novel and historical secondary forests. The reason is that the types of species in secondary novel forests are mostly pioneer species with similar ecophysiological traits as those in historical secondary forests. The common ecophysiological traits of pioneer species occur regardless of whether they are part of a novel or a historical forest. However, the ecophysiological traits of pioneer species contrast with those of species classified as non-pioneer or late-succession species (Table 3). Exceptional cases emerge from remnant agricultural or ornamental trees that have

Table 3 Properties of secondary forests and the dynamics of ecosystem reconstruction (Bazzaz, 1979).

Attribute	Early successional plants	Late-successional plants
Seeds		
Dispersal in time	Long	Short
Wind dispersed or animal dispersed?	Wind	Animal
Seed germination enhanced by:		
Light	Yes	No
Fluctuating temperatures	Yes	No
High NO_3^- concentrations	Yes	No?
Inhibited by far-red light	Yes	No
High CO_2 concentrations	Yes	No?
Photosynthesis		
Light saturation intensity	High	Low
Light compensation point	High	Low
Efficiency at low light	Low	High
Photosynthetic rates	High	Low
Respiration rates	High	Low
Transpiration rates	High	Low
Stomatal and mesophyll resistances	Low	High
Resistance to water transport	Low	High
Acclimation potential	High	Low
Recovery from resource limitation	Fast	Slow
Physiological response breadth	Broad	Narrow
Resource acquisition rates	Fast	Slow?
Material allocation flexibility	Fast	Slow?

naturalized and do not necessarily conform entirely to an early-successional set of plant traits as are the cases of *Delonix regia* and *Mangifera indica*. Nevertheless, these differences are expected to hold in most examples of species belonging to these two groupings, regardless of whether they are growing in novel or historical forests. Similarly, the ecophysiological response of tropical tree species and communities to the environmental challenges associated with climate change and deforestation are beginning to be understood (Table 4), and it is unlikely that the ecophysiological responses to those challenges will be different in novel vs. historical forests. The same would apply to systems level comparisons of mass and nutrient fluxes between those two types secondary forests (but see table 1 in Lugo et al., 2018).

We do expect that individual species traits could influence the comparison of novel vs. historical secondary forests, if the species in question through its dominance in the community could influence the behaviour of the whole forest. Given the differences in the degree of species dominance

Table 4 Likely ecophysiological challenges to tropical plants and other predicted changes that can affect tropical forests as a result of environmental changes associated with the Anthropocene Epoch.

Environmental challenge	Effects on plants or forests	Plant or forest responses	References
At the plant population level			
Increased in atmospheric carbon dioxide concentration	Response is species-specific; transient increase in photosynthesis rates of C-3 plants. Decrease of stomatal conductance	Increased nitrogen use efficiency, non-structural carbohydrates, and water use efficiency. Frequency of lianas related to CO_2 and wind dispersed seeds	Körner (2003), Long et al. (2004), Cernusak et al. (2013), Phillips et al. (2002), Wright et al. (2004), and Schnitzer and Bongers (2011)
Increase in air temperature	Respiration rates increase faster than photosynthesis rates	Acclimation, hot nights increase carbon loss. Woody lianas can acclimate to nighttime warming. Leaves can acclimate to gradual warming	Lloyd and Farquhar (2008), Clark (2004), Slot and Kitajima (2015), and Slot and Winter (2017)

Continued

Table 4 Likely ecophysiological challenges to tropical plants and other predicted changes that can affect tropical forests as a result of environmental changes associated with the Anthropocene Epoch.—cont'd

Environmental challenge	Effects on plants or forests	Plant or forest responses	References
Changes in rainfall	Recurrent severe drought, uncertainty on local amounts and distribution. Water stress and loss of local species	Migration to less water stressed locations. Expansion of water stress resistant species into drier locations. Selection of wind dispersed over animal dispersed seeds	Huang et al. (2013) and Zhang et al. (2018)
At the community and forest level			
Loss of forest area	Potential loss of species, loss of biotic connectivity, productivity, and level of biogeochemical cycles	Fragmentation of forests and their functions	Hansen et al. (2013) and Austin et al. (2017)
Increased in atmospheric carbon dioxide concentration	Some forests increase growth rates, others do not as they are not limited by carbon	Increase in growth might be temporary, average minimum temperature might drive growth	Clark et al. (2003) and Körner (2003)
Increase in air temperature	Increase plant and soil respiration Increased decomposition. Increased demand for water	Faster mineralization of elements, unclear effect on net productivity, possibility of an unknown threshold from carbon sink to source	Doughty and Goulden (2008) and Wood et al. (2012)
Changes in rainfall	Increased drought frequency Stronger effects of seasonal fires Reduction in carbon sequestration	Changes in species composition Increase in fast growing, short leaved tree species	Lewis (2006) and Huang et al. (2013)

between novel and historical secondary forests, it may be reasonable to expect differences when comparisons involve high levels of dominance of species whose attributes or traits are sufficiently unique to influence the comparison. For example, due to relatively small leaf size and short leaf longevity (deciduousness) of leguminous trees, novel stands dominated by these species allow for higher inputs of solar radiation, throughfall, and litterfall to the forest floor compared with stands dominated by large-leaved evergreen species with a larger Leaf Area Index (Abelleira Martínez, 2011). A difficulty with this expectation is the absence of sufficient ecophysiological and functional data to allow for valid statistical differences to surface. Also, we know that forest structure and physiognomy are similar in novel and historical forests (Aide et al., 2000), and it is also possible that systems dynamics such as mass and nutrient turnovers might also be similar. Therefore, the following review on functioning is exploratory and preliminary rather than conclusive.

5.1 Organic matter dynamics

Novel forests in the subtropical moist forest life zone of Puerto Rico have a high turnover of wood mass and volume as expected of young secondary forests (Table 5). The low standing biomass and volume in these forests' turnover twice and five times a year, respectively. Tree mortality is low (1.6% year^{-1}). Most of the volume turnover (3.7% year^{-1}) is due to tree growth.

Novel forests in *los Llanos* also exhibited rapid growth rates. The rate of basal area accumulation (growth) in the *los Llanos* averaged 1.6 m^2 ha^{-1} year^{-1} with a wide range of variation (0.01–8.5). On average, the annual turnover of basal area in *los Llanos* was 3.5% (range of 2.2–6.7), a value similar to the turnover of wood volume in Puerto Rican forests (Table 5). This average rate of growth is high and was not related to the age of stands (range of 15–45 years). Values are higher than those in Aide et al. (2000) for novel forests on abandoned pastures in Puerto Rico. Species density explained 30% of the variation in basal area growth at *los Llanos* (Fig. 5). The variation in growth can be attributed to contrasting environmental conditions and levels of disturbance as was reported in Puerto Rico by Aide et al. (1996) when they also encountered differences in the rates of biomass accumulation among different successional sequences.

Erickson et al. (2014) found a low leaf litter mass on the forest floor of secondary forests (both novel and historical) of the subtropical moist forest life zone in Puerto Rico. This low accumulation of leaf litter, coupled to high leaf litterfall rates, results in high leaf mass turnover on the forest floor (Table 5). The turnover of leaf litter mass was 100 times faster than the turnover of wood mass.

Table 5 Standing volume, total biomass (above and below ground), volume growth, volume mortality, volume removals, mass accumulation, and mass and volume turnover of novel forests of the subtropical forest life zone of Puerto Rico.

Parameter and units	Mass or volume	Annual rates	Annual turnover (%)
Standing volume ($m^3\ ha^{-1}$) and rate of volume growth ($m^3\ ha^{-1}\ year^{-1}$)	78.8	2.95 (growth)	3.7
		1.25 (mortality)	1.6
		0.25 (removals)	0.3
		4.45 (total)	5.6
Total mass ($Mg\,ha^{-1}$) and rate of mass production ($Mg\,ha^{-1}\ year^{-1}$)	94.6	2.0	2.1
Leaf litter mass ($Mg\,ha^{-1}$) and leaf fall rate ($Mg\,ha^{-1}\ year^{-1}$)	2.3	6.8	296

Data covers 230,985 ha of forests or 49% of insular forests. Data for stand volume and mass are from Marcano Vega (2019) and apply to trees with a diameter at breast height ≥ 12.5 cm. Data for leaf litter mass are from Erickson et al. (2014) and for leaf fall rate we used the mean of six annual measurements in Lugo et al. (2011b), Abelleira Martínez (2011), and Fonseca da Silva (2011).

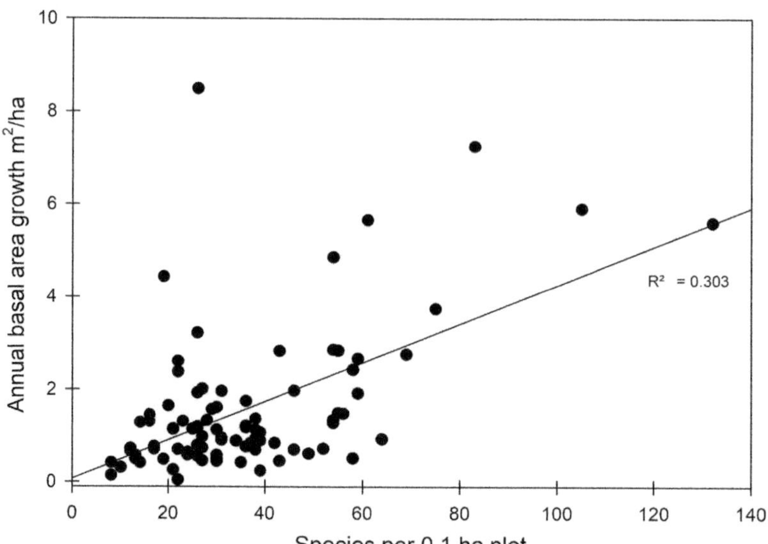

Fig. 5 Principal Components Analyses ordination based on Importance Values of species per plot. *Data are from G. Aymard.*

Novel forests, like most secondary forests, are carbon sinks given their high accumulation of mass and basal area over periods that exceed 40 years. At the insular scale, for which we have robust data, forests are accumulating 1 $Mg\,C\,ha^{-1}\,year^{-1}$ (Marcano Vega, 2019). The carbon sink function of these novel forests is confirmed at the stand level, when detailed ecophysiological measurements are conducted in tandem with stand structural ones (Lugo et al., 2008), or when isotopes are used to trace the accumulation of soil organic matter through succession, coupled with long-term tree growth data (Silver et al., 2004).

5.2 Cycling of elements

The high turnover of wood mass and wood volume in novel forests is accompanied by an even faster turnover of chemical elements in the leaf litter (Fig. 6). Element return via leaf litterfall was larger than element accumulation in leaf litter (Table 6). Novel forest stands had faster turnover than historical secondary forests with the exception of iron (Fig. 6). The flux and turnover of elements through leaf litter is not uniform but varies with the element, with potassium and manganese being the most mobile followed by nitrogen and magnesium. Aluminium is particularly immobile in the

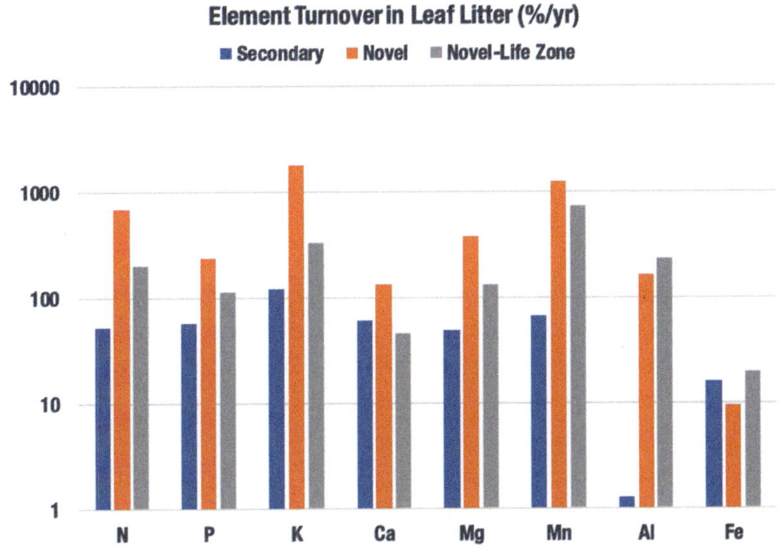

Fig. 6 Annual element turnover in leaf litter of a historical secondary forest and a novel secondary forest in the subtropical moist forest life zone of Puerto Rico. Estimates of turnover rate are based on data in Table 6. For the estimate of novel element flux-life zone element storage, we used the chemical storage data for the life zone in Erickson et al. (2014).

Table 6 Elemental mass accumulated in leaf litter and element return in leaf fall of novel and historical secondary forests in the subtropical moist forest life zones.

Forest type	N	P	K	Ca	Mg	Mn	Al	Fe	S
Storage (kg ha^{-1})									
Moist Forest Life Zone	71	3	9	104	12	1	8	8	14
Novel on Alluvial	17	2	2	44	5	1	15	22	2
Novel on Alluvial	8	1	1	13	2	0	4	6	1
Novel on Alluvial	37	2	3	48	6	1	14	20	5
Secondary on Volcanic 1	145	5	25	121	31	2	63	48	
Secondary on Volcanic 2	168	6	22	125	29	2	35	35	
Flux (kg ha^{-1} year^{-1})									
Novel on Volcanic 98–99	181	4	33	55	19	6	24	2	27
Novel on Volcanic 99–00	105	3	29	38	13	4	11	1	15
Secondary on Volcanic 1	93	3	28	77	17	2	1	4	
Secondary on Volcanic 1	77	3	27	80	12	1	1	11	
Secondary on Volcanic 2	87	3	29	71	15	2	1	4	
Secondary on Volcanic 2	72	3	28	74	14	1	1	9	

Data for the life zone as a whole are from Erickson et al. (2014). Data for the historical secondary forest on volcanic substrate are from Lugo et al. (1999b) and data for novel forests on alluvial and volcanic substrate are, respectively, from Abelleira Martínez and Lugo et al. (2008) and Lugo et al. (2011b). In this comparison, the difference between novel and historical is the presence of introduced species in novel and not in historical forests. Values are rounded to the nearest kg, and empty cells mean there are no data.

historical secondary forest. For discussion of the role of litterfall in nutrient cycling in historical forests, see chapter "Revisiting nutrient cycling by litterfall—Insights from 15 years of litter manipulation in old-growth lowland tropical forest" by Sayer et al., this issue.

The analysis of Erickson et al. (2014) of 11 elements in fallen leaf litter and 143 historical and novel secondary forest stands systematically located across Puerto Rico, revealed that the stoichiometry of leaf litter was influenced by climate (life zone), geologic substrate (karst vs. non-karst), stand age (up to 60 years), and species composition. They found that leaf litter chemistry on the island was spatially heterogeneous and that elements behaved independently, i.e., responding differentially to various control factors or to tree species. For example, the basal area of leguminous trees correlated with leaf litter nitrogen concentration. Erickson et al. (2014) suggested that the availability of nitrogen and phosphorus on insular forest floors was greater than before deforestation, because nutrient availability increased with decreasing forest stand age and novel species composition. Litter with higher nitrogen and phosphorus concentration, i.e., more chemically labile, would favour the rapid chemical turnover and support faster uptake and productivity.

The contribution of the introduced *Spathodea campanulata* to rapid element turnover in leaf litter was confirmed in a decomposition experiment in three geologic formations in the subtropical moist forest life zone (Lugo and Abelleira Martínez, 2018). They found the fastest leaf decomposition rates in the Neotropics, and faster decomposition rates for leaves from novel forests compared to leaves from historical forests. These differences may be reflecting species differences in terms of the species-specific chemical quality of leaf litter, rather than the novelty or lack of it of forests.

Lugo and Erickson (2017) related nutrient concentration anomalies in fallen leaf litter in the subtropical dry forest life zone to species dominance, both native and non-native (introduced). Species groups, such as nitrogen-fixing species and other naturalized species that dominate novel dry forests, have a disproportional influence on forest element stoichiometry. For example, historical stands dominated by *Citharexylum fruticosum* had significantly lower carbon concentration anomalies than other species assemblages. Novel dry forests, compared to the mean of all forest species assemblages island-wide, tend to have fallen leaf litter with lower than average manganese and sodium concentrations and lower than average C/N and C/P ratios. After accounting for significant differences in stand age, geology, and or precipitation, novel dry forests compared to historical dry forests have higher carbon anomalies, lower calcium and sodium anomalies, and lower C/N ratio anomalies. Nitrogen, sulphur, phosphorus, magnesium, potassium, manganese, aluminium, and iron concentration anomalies did not differ among the forest assemblages or between novel dry and historical dry forests. The overarching pattern in the comparisons of element concentration anomalies between novel and historical dry forests and with island-wide forests is the absence of a consistent pattern among elements (Lugo and Erickson, 2017).

The stoichiometric characteristics of species assemblages can influence forest functioning through the speed of recycling and accumulation of nutrients, not only in leaf litter dynamics as discussed above, but also at

Table 7 Traits of introduced tree species that dominate novel forests in the moist life zones of north and western Puerto Rico (Little et al., 1974; Little and Wadsworth, 1964).

	Leaf		Fruit		Nitrogen
Species	Phenology	Size	Flesh	Size	Fixer
Albizia procera[a]	D	S	No	L	Yes
Castilla ellastica	D	L	Yes	S	No
Delonix regia[a]	D	S	No	L	No
Mangifera indica	E	L	Yes	L	No
Peltophorum pterocarpum[a]	SD[b]	S	No	S	Yes
Senna siamea[a]	SD[b]	S	No	L	No
Spathodea campanulata	D	L	No	L	No
Terminalia catappa	SD	L	Yes	S	No

[a]Leguminous tree species (i.e. Fabaceae).
[b]Species classified as evergreen by Little and Wadsworth (1964) but that exhibit deciduous behaviour in some areas of moist Puerto Rico (Abelleira Martínez, personal observation).
For phenology: D is deciduous, SD is semi-deciduous, and E is evergreen. For size, small is S, large is L, and medium is M. For leaves, size corresponds to leaflet size.

the level of whole stands (Lugo et al., 2011a, 2012a). In addition, different leaf chemistry traits can affect species composition, diversity, and food web dynamics of organisms in litter and soil as was shown by Barberena Arias (2008). She related biotic responses at the animal species level to the leaf litter stoichiometry of individual tree species.

The traits of species exhibiting high dominance (Tables 7 and 8) make a difference not only on biotic interactions among organisms as shown in those tables, but also in the flux and turnover of nutrients through the forest compartments that they dominate, be they novel or historical. Dominant species not only capture the resources of sites, but they have the capacity to modify site conditions through their cover, differential use of resources, and regulating nutrient storages and fluxes. Moreover, each dominant species has a particular stoichiometry requirement and capability that allows it to function and influence site conditions. The best examples are those associated with nitrogen fixation, which clearly influence the nitrogen cycle of stands. But their varied ecological traits allow each species to optimize functioning under the conditions that match their traits (Vitousek, 1990; Vitousek and Walker, 1989). Such functioning has implications to management in terms of matching species to site conditions and using species to restore degraded sites.

Table 8 Structural and functional attributes of large (≥10 cm diameter at breast height [DBH]) trees in 250 m^2 plots and small (≥2.5 < 10 cm DBH) trees in concentric 100 m^2 plots established on novel forest sites in the subtropical moist forest life zone of Puerto Rico.

	Large trees						Small trees	
Dominant introduced tree species	**Sites (N)**	**Tree density (trees/ha)**	**Basal area ($m^2\ ha^{-1}$)**	**Dominant species importance value (%)**	**Leaf Area Index**	**Introduced species importance value (%)**	**Bat-dispersed species importance value (%)**	**Community-weighed mean of fruit size (cm)**
Albizia procera	6[a]	670	21.6	64.2	2.6	50.0	0.9	9.6
Castilla elastica	3	613	43.6	68.3	4.8	38.1	26.7	3.4
Delonix regia	2	520	31.3	59.6	N/A	67.3	0	30.8
Mangifera indica	2	520	42.2	43.9	4.2	1.7	0	2.3
Peltophorum pterocarpum	2	540	26.8	75.7	4.0	48.0	3.4	6.1
Senna siamea	4[a]	873	27.7	51.6	4.4	42.4	5.6	9.8
Spathodea campanulata	5	888	38.0	67.2	5.2	30.4	0.8	7.3
Terminalia catappa	3[a]	329	46.2	65.4	3.8	13.1	51.6	6.7

[a]Large and small trees were sampled in 250 and 100 m^2 plots except in sites denoted with an asterisk where 5–9 quarter points per size class were used at 1–2 sites. Relative frequency was also considered when estimating species Importance Values on sites sampled with quarter points (Cottam and Curtis, 1956).

Tree density, basal area, and dominant species Importance Values correspond to the mean estimated across the sites sampled. Species Importance Values were based on relative tree density and basal area. Leaf Area Index corresponds to the mean of a subset of sites (1–4) sampled during March–May 2017. Importance Values of introduced species and bat-dispersed tree species for small trees are based on pooled values for introduced species (Little and Wadsworth, 1964; Little et al., 1974) and for six species typically dispersed by bats in Puerto Rico: *Andira inermis*, *Calophyllum calaba*, *Melicoccus bijugatus*, *Terminalia catappa*, *Thespesia grandiflora*, and *Syzigium jambos* (Rodríguez Durán, 2005). *Andira inermis* and *C. calaba* are native, *T. grandiflora* is endemic, and *M. bijugatus*, *T. catappa* and *S. jambos* are introduced. The community weighted mean of fruit size was estimated according to maximum fruit length.

6. Tropical forests of the future

Viewed from the point of view of biodiversity and conservation biology, the future of tropical forests depends on what conservationists can accomplish before current forests collapse or are converted to other cover (Sodhi et al., 2007). Sodhi et al. (2007) stated (p. ix): *Habitat loss and fragmentation, overexploitation of species and environmental degradation are operating on a massive scale worldwide, and these are the processes fueling the current extinction crisis. These human impacts on the biosphere predict a grim future for global biodiversity, especially considering that in many regions the rates of destruction and modification are accelerating....the present-day and future biodiversity crisis loom largest in the tropics.* The future of tropical forests, although dire, is not as grim as expressed from the perspective of a particular academic point of view. For example, Wright and Muller Landau (2006) examined future scenarios of population growth and deforestation in the tropics and found that as those processes slow down, the area of remaining tropical forests is much larger than most studies suggest. Any credible discussion of the future of tropical forests has to consider scale and temporal effects on the predictions; small scale, short-term phenomena are difficult if not impossible to extrapolate to global and long-term scales. The global generalization in Sodhi et al. (2007) may be true at smaller geographic scales but not necessarily at the global scale. As an example, the future condition of forests is associated with their area. Forest area is related to the number of species and the exchange and accumulation of matter, which influence biogeochemical cycles. All modelling, regardless of scale, requires knowledge of the area involved to estimate extinctions rates, potential carbon loss or accumulation, or importance of fragmentation. Small changes in area greatly affect model outputs.

For the tropics, reliable knowledge about the area of tropical forests at global and even national scales is illusive. Since the first global inventory by Zon and Sparhawk (1923), the perception of tropical forest area increased until 1990, when the Food and Agriculture Organization (FAO, 1993) began standardized estimates of global forest areas. Since 1990 the perception of the area of tropical forests has decreased until 2015, but at rates that have been adjusted downward from earlier estimates by the FAO (see fig. 1 in Keenan et al., 2015). Satellite technology also results in variable estimates, depending on the definition of 'forest' and the resolution of images. Due to high uncertainty in the basic information available, the larger the temporal or spatial scale used for predictions or estimates, the greater the uncertainty of the outcome.

The problem is that anticipating the future of tropical forests is a lot more complex than generally realized, particularly if the reality of the

Anthropocene Epoch and the creativity of humans and of evolutionary and successional processes are ignored. As an example, Keenan et al. (2015) identify 19 countries (mostly tropical) that have transitioned from forest loss to forest gain between 1990 and 2015. Forest and landscape fragmentation are also reversible processes as is species endangerment. Only extinction is irreversible, but the current and future rates of the process are subject of debate (Briggs, 2017), whose outcome does not affect the problem of predicting the future. Therefore, the future of tropical forests can at best be imagined or inferred given the temporal and spatial complexity of landscapes and the uncertainty of the Anthropocene environment. For Chazdon (2014), the secondary forests of today will be tropical forests of the future. We agree, but ask: Will they conform to historical species combinations or will the mixing of species lead to novel secondary forests? Will they function at similar rates as historical secondary forests do? We believe that the answers to these questions require the contributions of many disciplines because of the complexity introduced by human activities and the ever-changing environmental conditions.

The documented Holocene history of human activity on the Neotropical lowlands was the subject of analysis by scientists representing many disciplines who came together to address the historical ecology of the region (Balée and Erickson, 2006). They concluded after comprehensive analysis of historical trajectories that human nature per se is not the culprit of the calamities described by conservation biologists. Moreover, they describe cases where human activity results in situations favourable to biodiversity and abiotic conditions. In the view of Balée and Erickson (2006, p. 9) *...causality can be addressed to historically defined configurations of interrelationships over time between specific societies and their economies, on the one hand, and given environments, on the other.*

Our perspective is cemented on Anthropocene conditions, which are more variable and less predictable than those of the Holocene. These conditions select for ecological systems and organisms with fast metabolic rates, low biomass, and shorter-lived and small-sized organisms (Brown, 1995; Enquist et al., 1998; West et al., 1999), a process that is already in motion (Galetti and Dirzo, 2013; Wilkie et al., 2011; Wright et al., 2007). Thus, the novel forests of the future may lack large, long-lived organisms unless humans intervene, through conservation-oriented management, to prevent so. Some taxa will be missing or be more affected than others, particularly those with large sized organisms. These reductions in the numbers of large organisms reflect intense human use and require and equally intense or greater effort to overcome it.

If we limit the prediction to the ecological point of view and to current trends of environmental change, we anticipate that tropical forest ecosystems will continue to change towards high turnover of nutrients, mass, and species. Greater levels of disturbances (anthropogenic and non-anthropogenic) and constantly changing environmental conditions power the increases in turnover. Those conditions also favour a change in forest physiognomy from tree to shrub or shrublands. Shrublands are best suited for coping with extreme environmental conditions given their physiognomy, plant traits, and ecophysiological characteristics (Lugo et al., 2019). The structure of these future forests and shrublands and their plant species richness (not their species composition) will be similar to the forests and shrublands of today as will be the range of speeds of their processes. We base this prediction on the case studies presented above, which showed that after 30–100 years of succession, those parameters did not change even though the species composition did. Species composition and the ecological services associated to particular species are obvious differences between novel and historical forests. Also, the secondary forests of *los Llanos* studied so far do not harbour naturalized species like the Puerto Rican ones. This difference reinforces the notion that as time proceeds, species will mix regardless of geographic origin. It is possible that introduced species have a greater potential to dominate when the local flora is less diverse as it happens in Puerto Rico relative to *los Llanos*. Forest fragmentation should continue to vary temporally in response to human activity with particular effects to particular taxa that are favoured or not by fragmentation. Moreover, the expansion of agricultural, mining, and urban areas will likely continue to reduce the area under forest cover.

6.1 Conservation of novel tropical forests and interventions for ecological services

The conservation of novel forests is important to the conservation of biodiversity as they represent natural responses to prevailing environmental conditions (Ellis, 2015; Radeloff et al., 2015). We reject the notion that only the primary forests, as important as they are, are irreplaceable for sustaining tropical biodiversity (Gibson et al., 2011). Novel forests can provide sites where active conservation interventions are implemented for the sake of native, primary, threatened, or endangered species (see chapter "Protecting environmental and socio-economic values of selectively logged tropical forests in the Anthropocene" by Bousfield et al., this issue). Novel forests are further advanced in successional time compared to non-forest land cover under environmental conditions that can support forests, which means they can be more effectively managed for a variety of purposes that require time to develop. These purposes include increased levels of biodiversity, rehabilitating soil, replenishment of slow variables that provide resilience, for food and wood production, for restoring lost species, and for watershed protection (Table 9; Chapin et al., 2009; Light et al., 2013).

In this section, we review the literature and work in progress in Puerto Rico on passive and active interventions in novel tropical forests. These terms (passive and active) reflect the range in the intensity level of human intervention required to conserve forests of all types. Passive interventions are at the lower range of human intervention, while active interventions are at the upper level of human intervention. The actions that we discuss, even the active ones, strive to minimize human interventions while

Table 9 Potential tradeoffs in supporting, regulating, provision, and cultural ecosystem services (*sensu* Chapin et al., 2009) in novel forests undergoing hands-off conservation (i.e. status quo), and agroforestry and silviculture conservation interventions.

Ecosystem service	Status quo	Agroforestry	Silviculture
Supporting			
Carbon storage	0/+	+	++
Biodiversity	−/0	+	++
Regulating			
Microclimate	0/+	0/+	+
Soil rehabilitation	0/+	0/+	+
Provision			
Food	+	++	0
Wood	0/+	+	++
Cultural			
Identity	−/0	+	+
Recreation	0/+	+	+

Positive (+), negative (−) and neutral (0) effects indicate potential changes in ecosystem services capacity relative to historical forests in the case of novel forests under status quo, and relative to novel forests under status quo for agroforestry and silviculture interventions. Double signs (++) indicate superior effects by intervention. For some ecosystem service and conservation strategy combinations, the potential range in relative changes is separated by a backslash.

maximizing the ecological work of forest organisms. We address the successional trajectories and faunal composition of novel forests passively managed for biodiversity conservation in comparison with those under active conservation interventions. In doing so, we report on on-going experiments aimed at designing and testing active interventions that couple food and wood production with the conservation of biodiversity and the restoration of native species into novel forests through agroforestry and silvicultural practises (Table 9 and Fig. 7).

6.2 Passive interventions

Insular novel forests are depleted in late-successional species, possibly due to their restricted distribution, lack of dispersers, or failure to regenerate and grow with competing secondary vegetation (Abelleira Martínez, 2010; Ewel et al., 2013; Mascaro et al., 2008; Zimmerman et al., 2000). Humans can unknowingly alter the ecological situation and favour the conservation of late-successional species. For example, the tree species *Guaiacum officinale* is native to the dry forests of the Caribbean and Mesoamerica, and its prized wood has resulted in it being currently listed as threatened by IUCN. Nevertheless, this species is planted in urban areas in Puerto Rico for its ornamental value. Within 30 years of its urban planting and without human assistance, seeds of *Guaiacum* had dispersed, germinated, and became

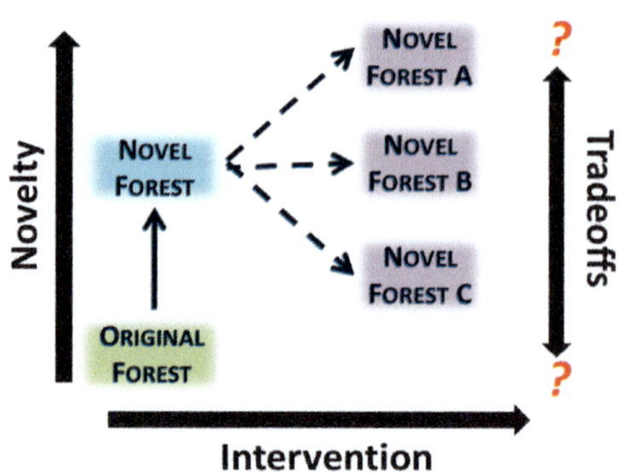

Fig. 7 Deforestation, agricultural land use, land abandonment, and species introductions have converted most original native forest cover into novel forests in Puerto Rico. Interventions can result in novel forests that (A) increase, (B) remain the same, or (C) decrease in novelty due to positive, neutral, or negative feedbacks between traits of introduced species (i.e. flora and fauna) and disturbance levels associated to agroforestry or silviculture. Identifying interventions that help reduce ecosystem service tradeoffs in novel forests can contribute towards a resilient social-ecological-technological system.

established on a novel forest at a site with greater humidity than found in its native environment (Abelleira Martínez et al., 2015). Although this is a highly contextual local-scale example, it nevertheless illustrates that when given the opportunity, species forming novel forests contribute to the ex situ conservation of late-successional species without direct human assistance.

Humans do set the initial conditions under which novelty develops and sustains diverse species assemblages. This is done by influencing the functional trait composition of novel forests by virtue of the traits of introduced or domesticated species that are favoured by anthropogenic activities such as agriculture, landscaping, and trade (Tables 7 and 8). Regardless of the factors influencing the functional traits in novel forests at the time of their establishment (ecesis), it is the successional processes operating outside of human influence what shapes the structure and functioning of communities over subsequent temporal scales.

Social factors that mediate succession of traits include the use of introduced species in agroforestry combined with the myriad of environmental conditions that can be the outcome of site-level management practises. Although species that colonize abandoned agricultural fields are typically wind dispersed, small seeded species; animals may also be the agents of colonization. An example is that of *Terminalia catappa*, a species introduced for ornamental purposes throughout the tropics and that is typically dispersed by bats in Puerto Rico (Rodríguez Durán, 2005). These animals may be attracted to coconut palms and dispersing *Terminalia* in the process, shaping succession and adding novelty. On the other hand, the dominant species may also be remnant fruit trees planted by humans, such as *Mangifera indica*, akin to remnant late-successional trees present in continental secondary forests following deforestation and fragmentation (Guevara et al., 2005). Given the variety of traits of dominant tree species in insular novel forests, the ecological services that are potentially provided by these forests, such as carbon storage, biodiversity conservation, microclimatic regulation and soil rehabilitation, are likely to vary as well (Table 9).

Changes in the functional trait composition of forest cover can modify ecosystem function and the capacity to provide ecological services (Finegan et al., 2015; Lavorel et al., 2011). For example, the difference in the community-weighed mean traits (Grime, 1998) of insular novel forests relative to historical forests may be larger than in continents due to higher monodominance in insular forests (i.e. Importance Value of dominant species in continental forests of 35–64% vs. 44–76% in insular forests based on Tables 1 and 8).

Given the above, one can expect that the traits of dominant introduced species can influence the species and trait composition of juvenile trees that get established and grow in the understory of novel forests. They do so by attracting particular seed dispersal agents or by creating microclimatic conditions that favour the growth and establishment of species with particular traits or trait values. Therefore, passive interventions benefit from an understanding of species traits and their influence on community development. Application of this understanding allows better alignment between conservation objectives and favoured forest species composition.

6.3 Active interventions

Some sectors of society place higher value on native and endemic species, and historical species assemblages than they do for introduced species or taxa with high abundance. This normative gives social licence for active interventions with novel forests (Davis et al., 2011; Hobbs et al., 2014; Hulvey et al., 2013). These interventions can be designed and implemented in tandem with activities that generate socio-economic benefits, such as the establishment of agroforestry systems in existing novel forests. Interventions can also be aimed towards the restoration and assisted migration of rare native species that were previously common or present in historical forests, but that were decimated due to deforestation, wood exploitation, and agricultural expansion (Jackson and Hobbs, 2009; Wadsworth, 1950; Wright and Muller Landau, 2006).

The relatively greater extent of novel forests in islands, coupled with scarce land area, places greater demand for innovative conservation strategies that favour ecological services (Brauman et al., 2007; Ewel et al., 2013; Gould et al., 2017).

Gould et al. (2017) suggest devising and incentivizing interventions that use existing forest cover for agroforestry production of shade crops and for silvicultural stand enrichment with late-successional hardwoods in a way that existing forest cover can be conserved in tandem with increased food and, perhaps, wood production. Ecosystem service tradeoffs (*sensu* Foley et al., 2005) need to be assessed in the design and implementation of interventions such that tradeoffs are minimized, and interventions act to synergistically increase societal benefits and value of novel forests across multiple management objectives (Table 9). For example, agroforestry interventions aimed at the enrichment of novel forest stands with fruit trees that are also late-successional or native, such as *Genipa americana* and *Pouteria multiflora*, can couple food production with increased capacity for carbon storage (e.g. dense wood and recalcitrant leaves), native species conservation, cultural identity, and recreation. The same strategy applies to stand enrichment with late-successional native or endemic tree species.

Agroforestry systems are as diverse as environmental conditions and the evolution of cultures and peoples and are considered a feasible alternative to agricultural intensification for supplying food while conserving forests (Cernansky, 2018; Jose, 2009; Weaver, 1979). Agroforestry practises are ancient and have moulded the composition of forests that were once considered primary and pristine, such as the Mayan forests of the Yucatán peninsula, *terra preta* forests in the Amazon, the *taungya* systems of tropical Asia, and the ancient terrace systems of Sri Lanka and Indonesia (Diemont and Martin, 2009; Weaver, 1979; Willis et al., 2004). The social-ecological-technological context and cultural value of agroforestry also plays a role in species selection and local practises in agroforestry systems. For example, breadfruit (*Arthocarpus atillis*) is planted in subsistence farms and commercial mono-cultures in Puerto Rico and is rarely used as a shade tree, whereas in Pacific islands, from where breadfruit originates, it is used for shade of introduced cacao, coffee, bananas, and other shade crops (Elevicht and Ragone, 2018). The species assemblages that result from agroforestry practises are novel within the timeframe in which they are implemented, but as intervention ceases and time elapses, these systems change through self-organization and become so wild and complex as to elicit pristine attributes by those that describe them (Jackson and Hobbs, 2009; Willis et al., 2004).

Some elements of novelty generated by past agricultural practises in Puerto Rico's forests, such as dominance and presence of *Mangifera indica*, could be reduced with time and ecological succession, especially if these species lack adequate seed dispersal agents. The novelty that has been added by the addition of introduced species with large fruits and no analogue native seed dispersal agents will eventually be reduced unless introduced fauna begin to disperse these fruits, which may be already happening (below). Novelty can also be reduced by active conservation interventions (e.g. Gbètoho et al., 2017). However, we focus on experimental conservation interventions designed to increase the capacity of novel forests to provide ecosystem services while reducing tradeoffs (Table 9). Although interventions for ecosystem services has certainly occurred in the past and contemporary times in other areas, we focus on experimental interventions that explicitly consider novelty as an aspect of the system to be conserved in Puerto Rico.

6.4 Experiments

Experiments in progress in Puerto Rico are based on the results of studies of novel forests and they aim to match the structural and functional characteristics of novel forests with the requirements of forest understory crops or trees deemed important to conservation objectives such as maintaining forest cover and native canopy species. The high nutrient concentration and decomposition rates of leaf litter in novel forests can act to increase nutrient turnover rates and productivity of crops planted in these sites (Erickson et al., 2014; Lugo and Abelleira Martínez, 2018). Plant traits such as leaf longevity and recalcitrance, canopy leaf area, and the capacity to fix nitrogen can also influence microclimate (i.e. incoming light levels, relative humidity, and temperature; Abelleira Martínez, 2011), soil conditions (i.e. organic matter and nutrient availability; Lugo and Abelleira Martínez, 2018), and regulate understory tree growth. Forest stands dominated by different species, functional groups, or trait values, may favour growth and productivity of one or another crop tree planted in their understory (Abelleira Martínez, 2019). In *Spathodea campanulata* forests, for example, full deciduousness and high leaf nutrient content coupled with fast decomposition rates act to make stoichiometric relationships (i.e. C/N and C/P) in standing litter and soil more favourable for plant growth (Abelleira Martínez, 2011; Lugo and Abelleira Martínez, 2018; Lugo et al., 2011a, b). Experiments can also test for the effect of dominance of any one of the most common introduced tree species in Puerto Rico's novel forests, particularly the effect of their species traits, on the growth, survival and, eventually, fruit production of crop trees planted in the understory (Tables 7 and 8).

One experiment aimed at incorporating agroforestry production of shade crops, commercial fruit trees, and native yet relatively rare edible fruit trees, into existing novel forests in the subtropical moist forest life zone of Puerto Rico (Abelleira Martínez, 2019; Túa Ayala and Abelleira Martínez, 2019). The planted crop trees consisted of deep shade tolerant cacao (*Theobroma cacao*), mid-shade tolerant coffee (*Coffea arabica* var. bourbon), the native mid-shade tolerant forest species jagua (*Genipa americana*) and jácana (*Pouteria multiflora*), and light-demanding commercial fruit species avocado (*Persea americana*) and breadfruit (*Artocarpus altilis*).

The unexpected passage of hurricane María over the 31 research sites, allowed for testing survivability of crops after such an event (see also chapter "Soil biogeochemical responses of a tropical forest to warming and hurricane disturbance" by Reed et al. (this issue) for discussion of the effects of hurricane María on an experimentally warmed tropical forest). The event caused an unplanned thinning and pruning of trees and branches across the vertical strata of the canopy except for the low understory that was protected from winds and falling limbs due to low stature and high flexibility

Fig. 8 Juvenile tree of *Pouteria multiflora* planted 2 weeks before hurricane María on a volcanic novel forest site dominated by *Albizia procera* in western Puerto Rico (A), and juvenile trees regenerating naturally in the understory of a novel forest karst site dominated by *Spathodea campanulata* in northern Puerto Rico (B). On both sites, canopy trees of dominant introduced species were heavily affected, resulting in high amounts of coarse woody debris and litter on the forest floor among juvenile trees that survived and that were and fully foliated in some cases. Both photos were taken a month or so after the hurricane's passage.

of juvenile trees (Fig. 8). Virtually all fruit trees planted before the hurricane survived the disturbance (Túa Ayala and Abelleira Martínez, 2019). A year and a half after planting at 10 of the sites established before the hurricane, high (>70%) survival was observed for all planted species except for avocado, breadfruit, and coffee on sites dominated by *Albizia procera* and

Delonix regia. On the other hand, cacao and coffee were setting flower in some sites, and some breadfruit, jacana, and jagua trees reached more than twice the initial (1 m) height; in the case of breadfruit reaching up to 3 m. The success of these plantings in withstanding a category 4 hurricane and reaching reproductive maturity in times equalling those of traditional agroforestry shade crop systems (i.e. cacao and coffee; Haggar et al., 2011; Somarriba and Beer, 2011) illustrate the potential of novel forests to harbour productive agroforestry systems in Puerto Rico and the potential of agroforestry systems to be resilient to future stressors (Fig. 7).

A second experiment had the objective of restoring a late-succession wetland tree species to the understory of novel forests on riparian floodplain sites. The experiment was established in 2010 on novel forests dominated by *Spathodea campanulata* on abandoned sugar cane fields in riparian wetlands (Abelleira Martínez and Lugo, 2008). Initial results show higher (>50%) survival of *Pterocarpus officinalis* trees planted in floodplain areas away from levees, where they are more susceptible to flood waters and debris (Cruz and Abelleira Martínez, personal communication, 2019). The plantings were protected by the forest canopy during hurricane María, and canopy opening may eventually accelerate species turnover.

In a third experiment, the rare late-succession tree species *Dacryodes excelsa* and *Manilkara bidentata*, and the endangered *Coccoloba pubescens* and *Stahlia monosperma* were planted in a line arrangement in novel forests dominated by leguminous and non-leguminous species across the subtropical moist forest life zone. The traits of leguminous vs. non-leguminous species often contrast, with the former group typically capable of fixing nitrogen and possessing deciduous leaves of low laminar leaf size (Tables 7 and 8). These traits, when combined, may favour the growth of species with early-successional traits (i.e. light and resource demanders) in novel forests dominated by legumes, and make non-legume dominated novel forests better sites for late-successional species establishment. Some of these rare native tree species are typical of historical forests in wetter (i.e. *Dacryodes excelsa* and *Coccoloba pubescens*) and drier (i.e. *Stahlia monosperma*) life zones, and their successful establishment can help populations expand and migrate. Due to anthropogenic barriers, dispersal of some species that lack effective natural seed dispersal mechanisms in contemporary times benefit from human-mediated dispersal to keep up with land conversion and climate change. The establishment of these species in novel forests can increase their value for biodiversity conservation, cultural identity and aesthetics, and improve the capacity of novel forests to support socio-economic activities, such as ecotourism, outdoor education, and recreation, that benefit neighbouring communities (Tavárez and Elbakidze, 2019). Coupled with agroforestry production and silviculture (Table 9; Lugo, 2012), these activities can empower communities in tandem with increasing their perspective and appreciation towards novel and historical ecosystems.

The three examples of experimental interventions that are being implemented in novel forests of Puerto Rico can help other island nations cope with expanding novel forest cover. In Fiji, for example, novel forests dominated by *Spathodea campanulata* seem to result in negative effects to socio-economic activities such as agricultural production and tourism, affecting the capacity of these systems to provide ecological services (Brown and Daigneault, 2014). Alternatively, Fijian institutions and stakeholders may be compelled to accept that these systems are a natural, inevitable response of forests to human intervention, and turn their attention towards the design and implementation of interventions to conserve these forests in the way they deem more valuable; be that accelerating species turnover towards more historical states or incorporating agroforestry crops. After all, Polynesian islands conserve agroforestry systems dominated by breadfruit, which was originally introduced from one island to others for a variety of purposes, making these agroforestry systems, although culturally engrained, also highly novel. The same reasoning in ecosystem service tradeoffs may be applied to the conservation of novel forests dominated by *Swietenia* spp. in islands as far apart as Dominica in the Caribbean and the Philliphine Archipelago (Cernansky, 2018; Norghauer et al., 2011).

The realization that novel forests are not 'invasive', 'thrash', or 'weed-dominated' homogenized forests but that they are heterogeneous and diverse forests intricately related to human activities is important for the understanding of tropical landscapes. Such realization can help island communities adopt a perspective that can facilitate the inclusion of novel forests into conservation strategies aimed at better adapting to increased number of disturbances and lowering vulnerability to extreme events (Hobbs et al., 2014). Although the examples from Puerto Rico can be illustrative, interventions should be context-specific and designed for taking into consideration the needs of peoples, communities, societies, and cultures.

6.5 Historical and novel fauna

While certain environmental conditions favour small organisms over large ones, humans create conditions, either purposefully or by neglect, which favour large sized organisms and even vulnerable ones that without human intervention would not survive. For example, in Puerto Rico, *Rhesus* monkeys were introduced for medical trials at the medical research facilities of the US Navy base in Sabana Seca, and for experiments in the Magueyes island Experiment Station of the University of Puerto Rico at Lajas. In the dawn of the 21st century, these monkeys extended in range and population size throughout Puerto Rico's coastal lowlands, consuming crops and scaring people. Monkeys have begun to alter the composition of forests by dispersing tree species with large sized fruits and seeds, such as the Caribbean native mammee apple (*Mammea americana),* and endemic and native cacti (Augusto Carvajal, personal communication; Omar Monsegur, personal communication; S.A. Sloan, personal communication; Watlington, 2009). The persistence of these monkeys in Puerto Rico may eventually modify the functional (i.e. reproductive) trait composition of novel and historical forests (Table 8).

The success and outcomes of assisted migration of endangered native flora and fauna in Puerto Rico, although equally intriguing, is less controversial and has received positive reception and support from the public, government, and private entities. For example, the endangered Puertorrican Parrot (*Amazona vitatta*) and Sapo Concho frog (*Bufo lemur*) have been re-introduced with modest success in areas outside of the 20th century ranges of populations, which was restricted to protected areas in the Luquillo Mountains for the parrot and the Guánica dry forest for the frog. As their populations spread, these species will benefit from resources, such as food and habitat, found in novel forests. For example, the Dominican Parakeet and the Scarlet Macaw are two parrot species (i.e. Psittaciformes) that actively forage pods and consume seeds of *Spathodea campanulata* trees in urban novel forests (Abelleira Martínez, 2008). Moreover, the protected areas where many native and endangered fauna are now found, are delimited and restricted in size and surrounded by a matrix of novel forests. Humans can intervene to make these novel forests better resources for this fauna.

The examples above are just a few among many that illustrate the natural mixing of species and the ensuing assembly of novel communities of plants and animals functioning in response to their traits and life history requirements irrespective of their geographic origin. As discussed in detail in Lugo et al. (2012b), these assemblages self-organize, experience succession, and reach maturity and sustainability in tropical landscapes. They are the harbingers of the future tropical forests.

7. Conclusion

This literature review on the role of climate change in shaping the tropical forests of the future focused its attention on changes already in progress that anticipate the future state of tropical forests. We used three case studies supplemented with new data to enrich information in the literature. Tropical forests will be shaped not only by climate change but also by other anthropogenic disturbances that interact in synergy with a changing climate. We anticipate that tropical forests will continue to change as they have done in the past. They will change and adapt to an increasing level of disturbances through the generation of novelty. Novelty is measured by the degree of dissimilarity of a system measured in one or more dimensions relative to a reference baseline, usually a historical one. Novelty is the product of evolutionary and successional processes; whose rate of emergence increases in proportion to anthropogenic disturbances. In the current Anthropocene Epoch, novelty in ecosystems will increase dramatically because the environment itself is increasingly changing and uncertain. We used a species composition criterion to assess the current level of novelty in tropical forests and to make judgement about the implications to the future of these forests, known as novel forests. In contrast to novel forests, historical forests are exposed to lower anthropogenic stressors, and their generation of novelty through succession and evolution is much lower.

Climate models predict a dire future for tropical forests, particularly those in montane regions. Land use changes also affect tropical forests and in combination with climate change cause greater stress to forests. We found that changes in species composition was a major response of vegetation to disturbances associated with anthropogenic activities. Moreover, that response is usually accompanied by an increased level of dominance by a few species. In Caribbean islands the change in species composition and high dominance favours introduced species but in the continental lowland and montane regions, native species with wide geographic distributions predominate. In both locations (tropical islands and continents) the floras are enriched by the introduction and naturalization of species.

Novel forests behaved as historical forests in terms of many functional attributes such as having high rates of mass, volume, and nutrient turnover. We could not establish differences in ecophysiological and functional traits with historical secondary forests. The importance of dominant species in influencing the behaviour of forest stands was well established, whether the species was introduced or native. The level of dominance is important for establishing differences among stands as was well demonstrated for the nitrogen dynamics in relation to basal area of nitrogen-fixing species. Highly dominant species like *Spathodea campanulata* (an introduced species) or *Citharexylum fruticosum* (a native species) in Puerto Rico were influential in affecting nutrient and carbon dynamics, respectively, in their species assemblages. While much research remains to be done in the functional characteristics of novel forests, the results available suggest their functional behaviour is similar to the behaviour of historical secondary forests, and thus there is no apparent reason to expect novel forests to be somehow abnormal or detrimental to the biogeochemical fluxes of landscapes or the delivery of ecological services.

Given the complexity of the tropical biome and our poor understanding of its functioning, we caution against the extrapolation of local events to global scales. Thus, our prediction of the future of tropical forests focus on the current trend towards smaller organisms, faster rates of turnover of organic mass and chemical elements, greater species dominance, and different combinations of species. Because these emerging novel forests are doing so under the forces of natural selection and self-organization that have always regulated ecosystem functioning and development, they represent adaptive solutions to emerging environmental challenges. These forests support biodiversity and provide services that need conservation, as is normally done with historical forests.

We reviewed different examples of passive and active interventions on novel forests with the objective of providing social and ecological services to people. In so doing the importance of species traits, particularly the traits of highly dominant species, emerged as important for conservation. Knowing species traits and incorporating those traits into conservation schemes reduce the need for active interventions and promote passive interventions, because the ecological work is done by the organisms rather than people. Similarly, tropical forests will have a positive outcome into the future if humans give their organisms the space and time to respond with novelty to the challenges of the Anthropocene.

Acknowledgements

This study was conducted in collaboration with the University of Puerto Rico. Olga Ramos prepared the location map for the vegetation plots in the Venezuelan Llanos. This study was supported by USDA-NIFA Grants 1009339, 1012459 and 21878230 to the University of Puerto Rico Agricultural Experiment Station and by the International Institute of Tropical Forestry, USDA Forest Service. The following students of the Department of Agroenvironmental Sciences, University of Puerto Rico at Mayagüez, contributed to this manuscript by collecting data in support of Table 8: Génesis Túa Ayala, Jennifer Rivera San Antonio, Natalia Quiñones, Gabriel Báez Rivera and Anthony Pérez Méndez. Rey E. Cruz Aguilar and Alejandro J. Marengo Casul conducted the analysis of tropical floras.

Appendix

This appendix contains material developed in support of this literature review. The material is presented in two sections. First, a description of the information collected in *los Llanos* of Venezuela by Gerardo Aymard. This section contains background information about *los Llanos*, its flora, and the methods used to analyse the data collected in the field. Second, a table with the list of tropical locations were we found published flora that itemized introduced and naturalized species. The table includes the data extracted from the published flora.

A.1 Los Llanos of Venezuela

Description of the los Llanos Region. The broad lowland zone that extends from northeastern Colombia to eastern Venezuela, occupies an area of ca. 532,000 km^2 (Huber et al., 2006). *Los Llanos* comprise 26.6% (about 240,000 km^2) of the continental area of Venezuela (Fig. 2). It is the largest savannah area in north-western South America—with rich sedimentary deposits of Quaternary origin (Schargel, 2015). The tropical dry forest life zone covers most of the *los Llanos* geography, but within this climate, vegetation varies depending on edaphic conditions, local climate, periodic flooding, and anthropogenic activity. The region also has smaller areas with other life zones and transitions between life zones such as the transition from tropical dry to very dry thorn forest, and less frequent, from dry semi-deciduous to evergreen forests (Ewel and Madriz, 1968).

The flora of *los Llanos* has a measurable affinity with the Caribbean flora (Takhtajan, 1986, DRYFLOR et al., 2016). About 60% of the *los Llanos* species belong to the Caribbean bioregion (Aymard, 2017). Currently, the vegetation of *los Llanos* is composed of a mosaic of savannahs mixed with pastures, shrubby vegetation, gallery forests; an array of communities composed by thorn dry forest, and a transition of dry semi-deciduous to evergreen forests highly transformed by human activity. The region contains extensive tropical savannahs whose origins are controversial because of the relative influences of anthropogenic activity, fire, seasonal flooding, edaphic factors, and climate. We do not address the extensive literature on the anthropogenic vs. non- anthropogenic origin of savannah vegetation. Instead, we focus on existing forests in *los Llanos*.

Background and Methods. Veillon (1986, 1995, 1997) described the structure, species composition, and dominance of *los Llanos* forests. We used those descriptions as the base of comparison to assess the degree of novelty in the current forest cover of the region. To do so, we also analysed unpublished raw data from one of the authors (G. Aymard). He had systematically sampled vegetation that had been affected by anthropogenic activity with the eventual objective of restoring historical forests. However, the floristic composition of stands, much like the work in the Andes discussed above, did not reflect evidence that the combinations of species of the historical vegetation was present in the developing mature forests in *los Llanos*.

Aymard's data comprises 90 0.1 ha-plots (100 m by 10 m) established in *los Llanos* and the seasonally dry forests of Bolívar, Trujillo and Zulia states, the latter *fide* (Figueroa, 2011; Fig. 2). In these plots, all free-standing stems of woody (trees, shrubs, lianas) vegetation, and non-woody (tree ferns, bamboo, palms) vegetation with a diameter at breast height >2.5 cm were identified to species and their height and diameter measured. From those data, we calculated species Importance Values based on the relative density and relative basal area of each species, and the complexity index for stands sensu Holdridge (1967). The complexity index is calculated by multiplying the height of the three tallest trees (m), number of tree species per 0.1 ha, stand basal area (m^2 per 0.1 ha), stem density per 0.1 ha, and 10^{-3}. We calculated the basal area growth rate ($m^2\ ha^{-1}\ year^{-1}$) of forest stands by dividing their basal area by their age. The basal area growth rate was then expressed as a percent of the standing basal area (% $year^{-1}$). We used stand species composition and structural information to identify the most dominant species by plot, the relationship between stand dominance and complexity index, and the relationship between stand dominance and number of species. We further explored Importance Values with multivariate ordination analyses (principal component analyses, PCA) to illustrate the differences in distribution and proportion of Importance Values by species per plot. The PCA was done on a correlation matrix (PC ORD v 7). Using a subset of the 90 plots inventoried by G. Aymard, we compared the 15 plots with the highest levels of dominance to the six historical forest plots from Veillon to ascertain difference in parameters that could be attributed to the generation of novelty. Geographically, the Veillon plots were between 100 and 650 km from those selected for comparison.

For descriptive statistics and regression analysis we used JMP version 14.2 (SAS Institute Inc., Cary NC, USA).

A.2 Analysis of published floras

This work involved compiling bibliographical references that document absolute numbers of native and naturalized species of vascular plants in tropical floras. Texts of the early 20th century floras seldom recognize introduced (or naturalized) species; those that do often group native and naturalized species together and do not provide separate enumerations. However, in the late 20th century we observed an increasing trend of reporting and quantifying introduced species and naturalizations. This compendium consists mainly of books but includes some online databases and scientific articles of floristic inventories. An annotated bibliography of the flora's covered by this analysis is available from the senior author. The following table contains the countries or regions covered by the survey. Multiple reports for the same country reflect different floras or area coverage.

Country	Native	Naturalized	Sum	Percent naturalized
Angola	304	20	324	6.2
Antioquia, Colombia	8302	612	8914	6.9
Bermuda	165	303	468	64.7
Bolivia	2709	24	2733	0.9
Brazil	580	44	624	7.1
British Guiana	536	65	601	10.8
Cayman Islands	24,030	250	24,280	1.0
Colombia	25,585	168	25,753	0.7
Colombia	33,849	85	33,934	0.3
Cuba	308	117	425	27.5
Cuba	63	1	64	1.6
Cuba	6504	516	7020	7.4
Cuba	75	5	80	6.3
Cuba	67	10	77	13.0
Cuba	12	2	14	14.3
Cuba	94	1	95	1.1
Cuba	15	9	24	37.5

Continued

—cont'd

Country	Native	Naturalized	Sum	Percent naturalized
Cuba	47	2	49	4.1
El Salvador	486	97	583	16.6
El Salvador	205	86	291	29.6
El Salvador	257	75	332	22.6
El Salvador	1757	313	2070	15.1
Ethiopia	309	19	328	5.8
French Guiana	5900	165	6065	2.7
Galapagos Islands	480	264	744	35.5
Galapagos Islands		185	185	100.0
Galapagos Islands	85	1	86	1.2
Guam	402	64	466	13.7
Guatemala	8194	55	8249	0.7
Guiana Shield	13,074	293	13,367	2.2
Hawaii	860	139	999	13.9
Hawaii	60	92	152	60.5
Jawaii	1358	1366	2724	50.1
Hawaii (montane)		349	349	100.0
India	17,212	730	17,942	4.1
India	18,115	471	18,586	2.5
Indonesia		651	651	100.0
Kenya	779	29	808	3.6
Kruger NP, South Africa		159	159	100.0
Fiji Islands	288	128	416	30.8
Margarita Island, Venezuela	642	2	644	0.3
Martinique	1668	1111	2779	40.0
Netherlands Antilles	418	49	467	10.5
Nicaragua	1548	408	1956	20.9

—cont'd

Country	Native	Naturalized	Sum	Percent naturalized
Panama	9140	380	9520	4.0
Philippines	9465	582	10,047	5.8
Puerto Rico	2337	575	2912	19.7
Puerto Rico	2900	300	3200	9.4
Puerto Rico	3126	583	3709	15.7
Puerto Rico and the US Virgin Islands	176	74	250	29.6
Rwanda	486	240	726	33.1
Rwanda	157	25	182	13.7
Somalia	34	7	41	17.1
South Florida		187	187	100.0
South Florida and Virgin Islands	41	28	69	40.6
South Sudan	17	5	22	22.7
St. Bartholomew	367	125	492	25.4
St. Croix	920	26	946	2.7
St. John	642	105	747	14.1
Taiwan	3910	340	4250	8.0
Tanzania	869	41	910	4.5
Trinidad & Tobago	267	56	323	17.3
Uganda	226	24	250	9.6
Virgin Gorda, British Virgin Islands	367	36	403	8.9
Zambia	415	25	440	5.7
Zimbabwe	5533	391	5924	6.6

References

Abelleira Martínez, O.J., 2008. Observations on the fauna that visit *Spathodea campanulata* forests in Puerto Rico. Acta Cient. 22, 37–42.

Abelleira Martínez, O.J., 2010. Invasion by native tree species prevents biotic homogenization in novel forests of Puerto Rico. Plant Ecol. 211, 49–64.

Abelleira Martínez, O.J., 2011. Flooding and profuse flowering result in high litterfall in novel *Spathodea campanulata* forests in northern Puerto Rico. Ecosphere 2, 105. https://doi.org/10.1890/ES1811-00165.00161.

Abelleira Martínez, O.J., 2019. Geographic distribution and spatial attributes of African tulip tree forests in north-central Puerto Rico: implications for social-ecological resilience. J. Agric. Univ. P. R. 103.

Abelleira Martínez, O., Lugo, A.E., 2008. Post sugar cane succession in moist alluvial sites in Puerto Rico. In: Myster, R.W. (Ed.), Post-Agricultural Succession in the Neotropics. Springer, New York, pp. 73–92.

Abelleira Martínez, O.J., Rodríguez, M.A., Rosario, I., Soto, N., López, A., Lugo, A.E., 2010. Structure and species composition of novel forests dominated by an introduced species in northcentral Puerto Rico. New For. 39, 1–18.

Abelleira Martínez, O.J., Meléndez Ackerman, E.J., García Montiel, D., Parrotta, J.A., 2015. Seed dispersal turns an experimental plantation on degraded land into a novel forest in urban northern Puerto Rico. For. Ecol. Manage. 357, 68–75.

Aide, T.M., Zimmerman, J.K., Herrera, L., Rosario, M., Serrano, M., 1995. Forest recovery in abandoned tropical pastures in Puerto Rico. For. Ecol. Manage. 77, 77–86.

Aide, T.M., Zimmerman, J.K., Rosario, M., Marcano, H., 1996. Forest recovery in abandoned cattle pastures along an elevational gradient in northeastern Puerto Rico. Biotropica 28, 537–548.

Aide, T.M., Zimmerman, J.K., Pascarella, J.B., Rivera, L., Vega, H.M., 2000. Forest regeneration in a chronosequence of tropical abandoned pastures: implications for restoration. Restor. Ecol. 8, 328–338.

Aubréville, A., 1938. Regeneration patterns in the closed forest of Ivory Coast. In: Chazdon, R.L., Whitmore, T.C. (Eds.), Foundations of Tropical Forest Biology: Classic Papers With Commentaries. University of Chicago, pp. 523–537. 2001. English translation.

Austin, K.G., González Roglich, M., Schaffer Smith, D., Schwantes, A.M., Swenson, J.J., 2017. Trends in size of tropical deforestation events signal increasing dominance of industrial-scale drivers. Environ. Res. Lett. 12, 054009.

Aymard, G., 2017. Adiciones a la flora vascular de los Llanos de Venezuela: nuevos registros y estados taxonómicos. Biollania (Edición Especial) 15, 1–296.

Aymard, G., González, V., 2014. Los bosques de los llanos de Venezuela: aspectos de su estructura, composición florística y estado actual de conservación. In: Rangel, J.O. (Ed.), Colombia diversidad biótica XIV. La región de la Orinoquia de Colombia. Universidad Nacional de Colombia, Instituto de Ciencias Naturales, Bogotá, Colombia, pp. 483–532.

Bailey, R.G., 2009. Ecosystem Geography. From Ecoregions to Sites, Second ed. Springer, New York, NY.

Balée, W., Erickson, C.L., 2006. Time, complexity, and historical ecology. In: Balée, W., Erickson, C.L. (Eds.), Time and Complexity in Historical Ecology: Studies in the Neotropical Lowlands. Columbia University Press, New York, NY, pp. 1–17.

Barberena Arias, M.F., 2008. Single Tree Species Effects on Temperature, Nutrients and Arthropod Diversity in Litter and Humus in the Guánica Dry Forest. Dissertation, University of Puerto Rico, Río Piedras.

Bazzaz, F.A., 1979. The physiological ecology of plant succession. Annu. Rev. Ecol. Syst. 10, 351–371.

Belda, M., Holtanová, E., Halenka, T., Kalvová, J., 2014. Climate classification revisited: from Köppen to Trewartha. Climate Res. 59, 1–13.

Birdsey, R.A., Weaver, P.L., 1982. The forest resources of Puerto Rico. In: Resource Bulletin SO-85. USDA Forest Service Southern Forest Experiment Station, New Orleans, Louisiana.

Brandeis, T.J., Helmer, E., Marcano Vega, H., Lugo, A.E., 2009. Climate shapes the novel plant communities that form after deforestation in Puerto Rico and the U.S. Virgin Islands. For. Ecol. Manage. 258, 1704–1718.
Brauman, K.A., Daily, G.C., Ka'eo Duarte, T., Mooney, H.A., 2007. The nature and value of ecosystem services: an overview highlighting hydrologic services. Annu. Rev. Env. Resour. 32, 67–98.
Briggs, J.C., 2017. Emergence of a sixth mass extinction? Biol. J. Linn. Soc. 20, 1–6.
Brown, J.H., 1995. Macroecology. University of Chicago Press.
Brown, P., Daigneault, A., 2014. Cost–benefit analysis of managing the invasive African tulip tree (*Spathodea campanulata*) in the Pacific. Environ. Sci. Policy 39, 65–76.
Campos Cerqueira, M., Arendt, W.J., Wunderle Jr., J.M., 2017. Have bird distributions shifted along an elevational gradient on a tropical mountain? Ecol. Evol. 2017, 1–11.
Cavaleri, M.A., Reed, S.C., Smith, W.K., Wood, T.E., 2015. Urgent need for warming experiments in tropical forests. Glob. Chang. Biol. 21, 1–11. https://doi.org/10.1111/gcb.12860.
Cernansky, R., 2018. How to build a forest. Nature 560, 542–544.
Cernusak, L.A., Winter, K., Dalling, J.W., Holtum, J.A.M., Jaramillo, C., Körner, C., Leakey, A.D.B., Norby, R.J., Poulter, B., Turner, B.L., Wright, S.J., 2013. Tropical forest responses to increasing atmospheric CO_2: current knowledge and opportunities for future research. Funct. Plant Biol. 40, 531–551.
Chapin III, F.S., Kofinas, G.P., Folke, C., 2009. Principles of Ecosystem Stewardship: Resilience-Based Natural Resource Management in a Changing World. Springer, New York, NY.
Chazdon, R.L., 2014. Second Growth: The Promise of Tropical Forest Regeneration in an Age of Deforestation. The University of Chicago Press, Chicago, IL, p. 445.
Clark, D.A., 2004. Sources or sinks? The responses of tropical forests to current and future climate and atmospheric composition. Phil. Trans. R. Soc. London 359, 477–491.
Clark, D.A., Piper, S.C., Keeling, C.D., Clark, D.B., 2003. Tropical rain forest tree growth and atmospheric carbon dynamics linked to interannual temperature variation during 1984-2000. Proc. Natl. Acad. Sci. 100, 5852–5857.
Correoso Rodríguez, M., 2005. La introducción de especies. Lineamientos generales. ¿Enriquecimiento de la biodiversidad o problemas ecológicos? Geospacial 2, 64–75.
Cottam, G., Curtis, J.T., 1956. The use of distance measures in phytosociological sampling. Ecology 37, 451–460.
Cramer, W.P., Leemans, R., 1993. Assessing impacts of climate change on vegetation using climate classification systems. In: Solomon, A.M., Shugart, H.H. (Eds.), Vegetation Dynamics & Global Change. Springer, Boston, MA, pp. 190–217. https://doi.org/10.1007/978-1-4615-2816-6_101993.
Cuatrecasas, J., 2013. A systematic study of the subtribe Espeletiinae (Heliantheae, Asteraceae). Mem. N. Y. Bot. Gard. 107, 14–689.
Davis, M.A., Chew, M.K., Hobbs, R.J., Lugo, A.E., Ewel, J.J., Vermeij, G.J., Brown, J.H., Rosenzweig, M.L., Gardner, M.R., Carroll, S.P., Thompson, K., Pickett, S.T.A., Stromberg, J.C., Del Tredici, P., Suding, K.N., Ehrenfeld, J.G., Grime, J.P., Mascaro, J., Briggs, J.C., 2011. Don't judge species on their origins. Nature 474, 153–154.
Dexter, K., Chave, J., 2016. Evolutionary patterns of range size, abundance and species richness in Amazonian angiosperm trees. Peer J. 4, e2402.
Diemont, S.A.W., Martin, J.F., 2009. Lacandon Maya ecosystem management: sustainable design for subsistence and environmental restoration. Ecol. Appl. 19, 254–266.
Doughty, C.E., Goulden, M.L., 2008. Are tropical forests near a high temperature threshold? J. Geophys. Res. 113, G00B07. https://doi.org/10.1029/2007JG000632.

DRYFLOR, Banda-R, K., Delgado-Salinas, A., Dexter, K.G., Linares-Palomino, R., Oliveira-Filho, A., Prado, D., Pullan, M., Quintana, C., Riina, R., Rodríguez M., G.M., Weintritt, J., Acevedo-Rodríguez, P., Adarve, J., Álvarez, E., Aranguren B., A., Arteaga, J.C., Aymard, G., Castaño, A., Ceballos-Mago, N., Cogollo, Á., Cuadros, H., Delgado, F., Devia, W., Dueñas, H., Fajardo, L., Fernández, Á., Franklin, J., Freid, E.H., Galetti, L.A., Gonto, R., González-M., R., Graveson, R., Helmer, E.H., Idárraga, Á., López, R., Marcano-Vega, H., Martínez, O.G., Maturo, H.M., McDonald, M., McLaren, K., Melo, O., Mijares, F., Mogni, V., Molina, D., del Pilar Moreno, N., Nassar, J.M., Neves, D.M., Oakley, L.J., Oatham, M., Olvera-Luna, A.R., Pezzini, F.F., Dominguez, O.J.R., Ríos, M.E., Rivera, O., Rodríguez, N., Rojas, A., Särkinen, T., Sánchez, R., Smith, M., Vargas, C., Villanueva, B., Pennington, R.T., 2016. Plant diversity patterns and their conservation implications in neotropical dry forests. Science 353, 1383–1387.

Edwards, D.P., Socolar, J.B., Mills, S.C., Burivalova, Z., Koh, L.P., Wilcove, D.S., 2019. Conservation of tropical forests in the Anthropocene. Curr. Biol. 29, R1008–R1020. https://doi.org/10.1016/j.cub.2019.08.026.

Elevicht, C.R., Ragone, D., 2018. Breadfruit agroforestry guide: Planning and implementation of regenerative organic methods. Breadfruit Institute of the National Tropical Botanical Garden and Permanent Agriculture Resources, Kalaheo, Hawaii, Holualoa, Hawaii. www.ntbg.org/breadfruit and www.agroforestry.net.

Ellis, E.C., 2015. Ecology in an anthropogenic biosphere. Ecol. Monogr. 85, 287–331.

Emanuel, W.R., Shugart, H.H., Stevenson, M.P., 1985. Climatic change and the broad-scale distribution of terrestrial ecosystem complexes. Clim. Change 7, 29–43.

Enquist, B.J., Brown, J.H., West, G.B., 1998. Allometric scaling of plant energetics and population density. Nature 395, 163–165.

Erickson, H.E., Helmer, E.H., Brandeis, T.J., Lugo, A.E., 2014. Controls on fallen leaf chemistry and forest floor element masses in native and novel forests across a tropical island. Ecosphere 5, 48, Article.

Esquivel-Muelbert, A., Baker, T.R., Dexter, K.G., Lewis, S.L., Brienen, R.J.W., Feldpausch, T.R., Lloyd, J., Monteagudo-M., A., Arroyo, L., Álvarez-D., E., Higuchi, N., Marimon, B.S., Marimon-Junior, B.H., Silveira, M., Vilanova, E., Gloor, E., Malhi, Y., Chave, J., Barlow, J., Bonal, D., Davila, N., Erwin, T., Fauset, S., Hérault, B., Laurance, S., Poorter, L., Qie, L., Stahl, C., Sullivan, M.J.P., ter Steege, H., Vos, V.A., Zuidema, P.A., Almeida, E., de Oliveira, E.A., Andrade, A., Vieira, S.A., Aragão, L., Araujo-Murakami, A., Arets, E., Aymard C., G.A., Barbosa Camargo, P., Barroso, J.G., Bongers, F., Boot, R., Camargo, J.L., Castro, W., Chama-Moscoso, V., Comiskey, J., Cornejo-Valverde, F., Lola da Costa, A.C., del Aguila Pasquel, J., Di Fiore, T., Duque, L.F., Elias, F., Engel, J., Flores- Llampazo, G., Galbraith, D., Herrera-F., R., Honorio-Coronado, E., Hubau, W., Jimenez-R., E., Nogueira L., A.J., Umetsu, R.K., Laurance, W., Lopez-G., G., Lovejoy, T., Melo-Cruz, O.A., Morandi, P.S., Neill, D., Núñez-V., P., Pallqui-Camacho, N.C., Parada G., A., Pardo, G., Peacock, J., Peña-Claros, M., Peñuela-M., M.C., Petronelli, P., Pickavance, G.C., Pitman, N., Prieto, A., Quesada, C., Ramírez-A., H., Réjou-M., M., Restrepo-C., Z., Roopsind, A., Rudas, A., Salomão, R., Silva, N., Silva-E., J., Singh, J., Stropp, J., Terborgh, J., Thomas, R., Toledo, M., Torres-L., A., Valenzuela-G., L., van de Meer, P.J., van der Heijden, G., van der Hout, P., Vasquez-M., R., Vela, C., Guimarães-V., I.C., Phillips, O.L., 2018. Compositional response of Amazon forests to climate change. Glob. Chang. Biol. 24, 1–18. https://doi.org/10.1111/gcb.14413.

Ewel, J.J., Madriz, A., 1968. Zonas de vida de Venezuela. In: Memoria explicativa sobre el mapa ecológico. Ministerio de Agricultura y Cria, Dirección de Investigación, Republica de Venezuela, Caracas, Venezuela, p. 265.

Ewel, J.J., Mascaro, J., Kueffer, C., Lugo, A.E., Lach, L., Gardener, M.R., 2013. Islands: where novelty is the norm. In: Hobbs, R.J., Higgs, E.S., Hall, C.M. (Eds.), Novel Ecosystems: Intervening in the New Ecological World Order. Wiley-Blackwell, New Jersey, pp. 29–44.

FAO, 1993. Forest resources assessment 1990: tropical countries. In: Food and Agriculture Organization Forestry Paper 112, Rome, Italy.

Figueroa, V., 2011. Los bosques secos de la costa oriental del lago de Maracaibo, estado Zulia. Venezuela. Biollania (Edicion Especial) 10 (Edición Especial), 189–196.

Finegan, B., Peña Claros, M., de Oliveira, A., Ascarrunz, N., Bret Harte, M.S., Carreño Rocabado, G., Casanoves, F., Díaz, S., Velepucha, P.E., Fernández, F., Licona, J.C., Lorenzo, L., Salgado Negret, B., Vaz, M., Poorter, L., 2015. Does functional trait diversity predict above-ground biomass and productivity of tropical forests? Testing three alternative hypotheses. J. Ecol. 103, 191–201.

Foley, J.A., DeFries, R., Asner, G.P., Barford, C., Bonan, G., Carpenter, S.R., Chapin, F.S., Coe, M.T., Daily, G.C., Gibbs, H.K., Helkowski, J.H., Holloway, T., Howard, E.A., Kucharik, C.J., Monfreda, C., Patz, J.A., Prentice, C., Ramankutty, N., Snyder, P.K., 2005. Global consequences of land use. Science 309, 570–574.

Fonseca da Silva, J., 2011. Ecophysiology and Productivity of *Castilla elastica*, an Introduced Tropical Tree Species. University of Puerto Rico, Rio Piedras, Puerto Rico.

Galetti, M., Dirzo, R., 2013. Ecological and evolutionary consequences of living in a defaunated world. Biol. Conserv. 163, 1–6.

Gbètoho, A.J., Aoudji, A., Roxburgh, L., Ganglo, J., 2017. Assessing the suitability of pioneer species for secondary forest restoration in Benin in the context of global climate change. Bois For. Tro. 332, 43–55.

Gibson, L., Lee, T.M., Koh, L.P., Brook, B.W., Gardner, T.A., Barlow, J., Peres, C.A., Bradshaw, C.J.A., Laurance, W.F., Lovejoy, T.E., Sodhi, N.S., 2011. Primary forests are irreplaceable for sustaining tropical biodiversity. Nature 478, 378–381.

Gleason, H.A., Cook, M.T., 1926. Plant ecology of Porto Rico. In: Scientific Survey of Porto Rico and the Virgin Islands. Academy of Sciences, New York, NY, New York, pp. 1–173.

Gómez-Pompa, A., 1971. Posible papel de la vegetación secundaria en la evolución de la flora tropical. Biotropica 3, 125–135.

Gould, W.A., Wadsworth, F.H., Quiñones, M., Fain, S.J., Álvarez Berríos, N., 2017. Land use, conservation, forestry, and agriculture in Puerto Rico. Forests 8, 242.

Grime, J.P., 1998. Benefits of plant diversity to ecosystems: immediate, filter and founder effects. J. Ecol. 86, 902–910.

Grime, J.P., Pierce, S., 2012. The Evolutionary Strategies That Shape Ecosystems. Wiley-Blackwell, Chichester, West Sussex, UK.

Guevara, S., Laborde, J., Sánchez-Ríos, G., 2005. Los árboles que la selva dejo atrás. Interciencia 30, 595–601.

Haggar, J., Barrios, M., Bolaños, M., Merlo, M., Moraga, P., Munguia, R., Ponce, A., Romero, S., Soto, G., Staver, C., Virginio, E.d.M.F., 2011. Coffee agroecosystem performance under full sun, shade, conventional and organic management regimes in Central America. Agrofor. Syst. 82, 285–301.

Hansen, M.C., Potapov, P.V., Moore, R., Hancher, M., Turubanova, S.A., Tyukavina, A., Thau, D., Stehman, S.V., Goetz, S.J., Loveland, T.R., Kommareddy, A., Egorov, A., Chini, L., Justice, C.O., Townshend, J.R.G., 2013. High-resolution global maps of 21st-century forest cover change. Science 342, 850–853.

Helmer, E.H., Gerson, E.A., Baggett, L.S., Bird, B.J., Ruzycki, T.S., Voggesser, S.M., 2019. Neotropical cloud forests and páramo to contract and dry from declines in cloud immersion and frost. PLoS One 14, e0213155.

Henareh, A.K., Gould, W.A., Harmsen, E., Terando, A., Quiñones, M., Collazo, J.A., 2016. Climate change implications for tropical islands: interpolating and interpreting statistically downscaled GCM projections for management and planning. J. Appl. Meteorol. Climatol. 55, 265–282.

Hobbs, R.J., Higgs, E., Harris, J.A., 2009. Novel ecosystems: implications for conservation and restoration. Trends Ecol. Evol. 24, 599–605.

Hobbs, R.J., Higgs, E., Hall, C.M., Bridgewater, P., Chapin III, F.S., Ellis, E.C., Ewel, J.J., Hallett, L.M., Harris, J., Hulvey, K.B., Jackson, S.T., Kennedy, P.L., Kueffer, C., Lach, L., Lantz, T.C., Lugo, A.E., Mascaro, J., Murphy, S.D., Nelson, C.R., Perring, M.P., Richardson, D.M., Seastedt, T.R., Standish, R.J., Starzomski, B.M., Suding, K.N., Tognetti, P.M., Yakob, L., Yung, L., 2014. Managing the whole landscape: historical, hybrid, and novel ecosystems. Front. Ecol. Environ. 12, 557–564.

Holdridge, L.R., 1947. Determination of the world plant formations from simple climatic data. Science 105, 367–368.

Holdridge, L.R., 1967. Life Zone Ecology. Tropical Science Center, San José, Costa Rica.

Huang, P., Xie, S.P., Hu, K., Huang, G., Huang, R., 2013. Patterns of the seasonal response of tropical rainfall to global warming. Nat. Geosci. 6, 357–361. https://doi.org/10.1038/NGEO1792, AOP April 14.

Hubbell, S.P., 2001. The Unified Neutral Theory of Biodiversity and Biogeography. Princeton University, New Jersey.

Huber, O., Duno de Stefano, R., Aymard, G., Riina, R., 2006. Flora and vegetation of the Venezuelan Llanos: a review. In: Pennington, T.R., Lewis, G.P., Ratter, J.A. (Eds.), Neotropical Savannas and Seasonal Dry Forests: Plant Diversity, Biogeography, and Conservation. Taylor & Francis, Boca Raton, FL, pp. 95–120.

Hulvey, K.B., Standish, R.J., Hallett, L.M., Starzomsky, B.M., Murphy, S.D., Nelson, C.R., Gardener, M.R., Kennedy, P.L., Seastedt, T.R., Suding, K.N., 2013. Incorporating novel ecosystems into management frameworks. In: Hobbs, R.J., Higgs, E.S., Hall, C.M. (Eds.), Novel Ecosystems: Intervening in the New Ecological World Order. Wiley-Blackwell, New Jersey, pp. 157–171.

Jackson, S.T., 2013. Perspective: ecological novelty is not new. In: Hobbs, R.J., Higgs, E.S., Hall, C.M. (Eds.), Novel Ecosystems: Intervening in the New Ecological World Order. John Wiley & Sons, Chichester, Wet Sussex, UK, pp. 63–65.

Jackson, S.T., Hobbs, R., 2009. Ecological restoration in the light of ecological history. Science 325, 567–569.

Jiménez Rivillas, C., García, J.J., Quijano Abril, M.A., Daza, J.M., Morrone, J.J., 2018. A new biogeographical regionalisation of the Páramo biogeographic province. Aust. Syst. Bot. 31, 296–310.

Jose, S., 2009. Agroforestry for ecosystem services and environmental benefits: an overview. Agrofor. Syst. 76, 1–10.

Kamalkar, A.V., Bradley, R.S., Diaz, H.F., 2008. Climate change scenario for Costa Rican montane forests. Geophys. Res. Lett. 35, L11702. https://doi.org/10.1029/2008GL033940.

Kazmierczak, M., Backmann, P., Fedriani, J.M., Fischer, R., Hartmann, A.K., Huth, A., May, F., Müller, M.S., Taubert, F., Grimm, V., Groeneveld, J., 2016. Monodominance in tropical forests: modelling reveals emerging clusters and phase transitions. J. R. Soc. Interface 13, 201601223.

Keenan, R.J., Reams, G.A., Achard, F., de-Freitas, J.V., Grainger, A., Lindquist, E., 2015. Dynamics of global forest area: results from the FAO Global Forest Resources Assessment 2015. For. Ecol. Manage. 352, 9–20.

Körner, C., 2003. Carbon limitation in trees. J. Ecol. 91, 4–17.

Laurance, W.F., Nascimento, H.E.M., Laurance, S.G., Andrade, A.C., Fearnside, P.M., Ribeiro, J.E.L., Capretz, R.L., 2006. Rain forest fragmentation and the proliferation of successional trees. Ecology 87, 469–482.

Lavorel, S., Grigulis, K., Lamarque, P., Colace, M.P., Garden, D., Girel, J., Pellet, G., Douzet, R., 2011. Using plant functional traits to understand the landscape distribution of multiple ecosystem services. J. Ecol. 99, 135–147.

Levis, C., Costa, F.R.C., Bongers, F., Peña-Claros, M., Clement, C.R., Junqueira, A.B., Neves, E.G., Tamanaha, E.K., Figueiredo, F.O.G., Salomão, R.P., Castilho, C.V., Magnusson, W.E., Phillips, O.L., Guevara, J.E., Sabatier, D., Molino, J.-F., Cárdenas López, D., Mendoza, A.M., Pitman, N.C.A., Duque, A., Núñez Vargas, P., Zartman, C.E., Vasquez, R., Andrade, A., Camargo, J.L., Feldpausch, T.R., Laurance, S.G.W., Laurance, W.F., Killeen, T.J., Mendonça Nascimento, H.E., Montero, J.C., Mostacedo, B., Amaral, I.L., Guimarães Vieira, I.C., Brienen, R., Castellanos, H., Terborgh, J., de Jesus Veiga Carim, M., da Silva Guimarães, J.R., de Souza Coelho, L., de Almeida Matos, F.D., Wittmann, F., Mogollón, H.F., Damasco, G., Dávila, N., García-Villacorta, R., Coronado, E.N.H., Emilio, T., de Andrade Lima Filho, D., Schietti, J., Souza, P., Targhetta, N., Comiskey, J.A., Marimon, B.S., Marimon Jr., B.-H., Neill, D., Alonso, A., Arroyo, L., Carvalho, F.A., de Souza, F.C., Dallmeier, F., Pansonato, M.P., Duivenvoorden, J.F., Fine, P.V.A., Stevenson, P.R., Araujo-Murakami, A., Aymard C., G.A., Baraloto, C., do Amaral, D.D., Engel, J., Henkel, T.W., Maas, P., Petronelli, P., Revilla, J.D.C., Stropp, J., Daly, D., Gribel, R., Paredes, M.R., Silveira, M., Thomas-Caesar, R., Baker, T.R., da Silva, N.F., Ferreira, L.V., Peres, C.A., Silman, M.R., Cerón, C., Valverde, F.C., Di Fiore, A., Jimenez, E.M., Mora, M.C.P., Toledo, M., Barbosa, E.M., de Matos Bonates, L.C., Arboleda, N.C., de Sousa Farias, E., Fuentes, A., Guillaumet, J.-L., Jørgensen, P.M., Malhi, Y., de Andrade Miranda, I.P., Phillips, J.F., Prieto, A., Rudas, A., Ruschel, A.R., Silva, N., von Hildebrand, P., Vos, V.A., Zent, E.L., Zent, S., Cintra, B.B.L., Nascimento, M.T., Oliveira, A.A., Ramírez-Angulo, H., Ramos, J.F., Rivas, G., Schöngart, J., Sierra, R., Tirado, M., van der Heijden, G., Torre, E.V., Wang, O., Young, K.R., Baider, C., Cano, A., Farfan-Rios, W., Ferreira, C., Hoffman, B., Mendoza, C., Mesones, I., Torres-Lezama, A., Medina, M.N.U., van Andel, T.R., Villarroel, D., Zagt, R., Alexiades, M.N., Balslev, H., García-Cabrera, K., Gonzales, T., Hernandez, L., Huamantupa-Chuquimaco, I., Manzatto, A.G., Milliken, W., Cuenca, W.P., Pansini, S., Pauletto, D., Arevalo, F.R., Reis, N.F.C., Sampaio, A.F., Giraldo, L.E.U., Sandoval, E.H.V., Gamarra, L.V., Vela, C.I.A., ter Steege, H., 2017. Persistent effects of pre-Columbian plant domestication on Amazonian forest composition. Science 355, 925–931.

Lewis, S.L., 2006. Tropical forests and the changing earth system. Philos. Trans. R. Soc. Lond. B Biol. Sci. 361, 195–210.

Lewis, S.L., Maslin, M.A., 2015. Defining the Anthropocene. Nature 519, 171–180. https://doi.org/10.1038/nature14258.

Light, A., Thompson, A., Higgs, E.S., 2013. Valuing novel ecosystems. In: Hobbs, R.J., Higgs, E.S., Hall, C.M. (Eds.), Novel Ecosystems: Intervening in the New Ecological World Order. Wiley-Blackwell, New Jersey, pp. 257–268.

Little, E.L., Wadsworth, F.H., 1964. Common trees of Puerto Rico and the Virgin Islands. Agriculture Handbook. USDA Forest Service, Washington DC, U.S.A, p. 249.

Little, E.L., Woodbury, R.O., Wadsworth, F.H., 1974. Trees of Puerto Rico and the Virgin Islands. In: USDA Agriculture Handbook 449. USDA Forest Service, Washington, DC.

Lloyd, J., Farquhar, G.D., 2008. Effect of rising temperatures and [CO_2] on the physiology of tropical forest trees. Philos. Trans. R. Soc. B 363, 1811–1817.

Long, S.P., Ainsworth, E.A., Rogers, A., Ort, D.R., 2004. Rising atmospheric carbon dioxide: plants FACE the future. Annu. Rev. Plant Biol. 55, 591–628.

Lugo, A.E., 1992. Comparison of tropical tree plantations with secondary forests of similar age. Ecol. Monogr. 62, 1–41.

Lugo, A.E., 2012. Social silviculture: a new paradigm in the search for sustainable land conservation in the tropics? Bois For. Trop. 314, 3–5.

Lugo, A.E., Abelleira Martínez, O.L., 2018. Stoichiometry of decomposing *Spathodea campanulata* leaves in novel puertorrican forests. For. Ecol. Manage. 430, 176–187. https://doi.org/10.1016/j.foreco.2018.07.059.

Lugo, A.E., Brandeis, T.A., 2005. New mix of alien and native species coexist in Puerto Rico's landscapes. In: Burslem, D., Pinard, M., Hartley, S. (Eds.), Biotic Interactions in the Tropics. Their Role in the Maintenance of Species Diversity. Cambridge University Press, Cambridge, pp. 484–509.

Lugo, A.E., Erickson, H.E., 2017. Novelty and its ecological implications to dry forest functioning and conservation. Forests 8, 161. https://doi.org/10.3390/f8050161.

Lugo, A.E., Heartsill Scalley, T., 2014. Research in the Luquillo Experimental Forest has advanced understanding of tropical forests and resolved management issues. In: Hayes, D.C., Stout, S.L., Crawford, R.H., Hoover, A.P. (Eds.), USDA Forest Service Experimental Forests and Ranges: Research for the Long-Term. Springer, New York, NY, pp. 435–461. 672 p.

Lugo, A.E., Helmer, E., 2004. Emerging forests on abandoned land: Puerto Rico's new forests. For. Ecol. Manage. 190, 145–161.

Lugo, A.E., Brown, S.L., Dodson, R., Smith, T.S., Shugart, H.H., 1999a. The Holdridge life zones of the conterminous United States in relation to ecosystem mapping. J. Biogeogr. 26, 1025–1038.

Lugo, A.E., Domínguez Cristóbal, C., Santos, A., Torres Morales, E., 1999b. Nutrient return and accumulation in litter of a secondary forest in the coffee region of Puerto Rico. Acta Cient. 13, 43–74.

Lugo, A.E., Fonseca da Silva, J., Sáez Uribe, A., 2008. Balance de carbono en un bosque de *Castilla elastica:* resultados preliminares. Acta Cient. 22, 13–28.

Lugo, A.E., Abelleira, O.J., Collado, A., Viera, C.A., Santiago, C., Vélez, D.O., Soto, E., Amaro, G., Charón, G., Colón Jr., H., Santana, J., Morales, J.I., Rivera, K., Ortíz, L., Rivera, L., Maldonado, M., Rivera, N., Vázquez, N.J., 2011a. Allometry, biomass, and chemical content of novel African tulip tree (*Spathodea campanulata)* forests in Puerto Rico. New For. 42, 267–283. https://doi.org/10.1007/s11056-011-9258-8.

Lugo, A.E., Domínguez Cristóbal, C., Méndez, N., 2011b. Hurricane Georges accelerated litterfall fluxes of a 26-yr-old novel secondary forest in Puerto Rico. In: Lupo, A.R. (Ed.), Recent Hurricane Research: Climate, Dynamics, and Societal Impacts. InTech, Rijeka, Croatia, pp. 535–554.

Lugo, A.E., Abelleira Martínez, O., Fonseca da Silva, J., 2012a. Aboveground biomass, wood volume, nutrient stocks and leaf litter in novel forests compared to native forests and tree plantations in Puerto Rico. Bois For. Trop. 314, 7–16.

Lugo, A.E., Carlo, T.A., Wunderle Jr., J.M., 2012b. Natural mixing of species: novel plant-animal communities on Caribbean Islands. Anim. Conserv. 15, 233–241.

Lugo, A.E., Winchell, K.M., Carlo, T.A., 2018. Novelty in ecosystems. In: DellaSala, D.A., Goldstein, M.I. (Eds.), The Encyclopedia of the Anthropocene. Elsevier, Oxford, pp. 259–271.

Lugo, A.E., Medina, E., Cuevas, E., Ramos González, O., 2019. Ecological and physiological aspects of Caribbean shrublands. Caribbean Nat. 58, 1–35.

Marcano Vega, H., 2019. Los bosques de Puerto Rico, 2014. USDA Forest Service, Southern Research Station, Resource Bulletin SRS-224, Asheville, NC.

Martinuzzi, S., Lugo, A.E., Brandeis, T.J., Helmer, E.H., 2013. Geographic distribution and level of novelty of Puerto Rican forests. In: Hobbs, R.J., Higgs, E.S., Hall, C. (Eds.), Novel Ecosystems: Intervening in the New Ecological World Order. Wiley, Oxford, pp. 81–87.

Mascaro, J., Becklund, K.K., Hughes, R.F., Shnitzer, S.A., 2008. Limited native plant regeneration in novel, exotic-dominated forests on Hawaii. For. Ecol. Manage. 256, 593–606.

Mascaro, J., Harris, J.A., Lach, L., Thompson, A., Perring, M.P., Richardson, D.M., Ellis, E.C., 2013. Origins of the novel ecosystems concept. In: Hobbs, R.J., Higgs, E.S., Hall, C.M. (Eds.), Novel Ecosystems: Intervening in the New Ecological World Order. John Wiley & Sons, Chichester, West Sussex, UK, pp. 45–57.

Molina Colón, S., Lugo, A.E., 2006. Recovery of a subtropical dry forest after abandonment of different land uses. Biotropica 38, 354–364.

Molina Colón, S., Lugo, A.E., Ramos, O., 2011. Novel dry forests in southwestern Puerto Rico. For. Ecol. Manage. 262, 170–177.

Montoya, E., Lombardo, U., Levis, C., Aymard, G.A., Mayle, F.E., 2020. Human contribution to amazonian diversity: pre-Columbian legacy to current plant communities. In: Rull, V., Carnaval, A. (Eds.), Neotropical Diversification. Springer, Berlin. https://doi.org/10.1007/978-3-030-31167-4.

Morueta Holme, N., Engemanna, K., Sandoval Acuña, P., Jonas, J.D., Segnitz, R.M., Svenninga, J.C., 2015. Strong upslope shifts in Chimborazo's vegetation over two centuries since Humboldt. Proc. Natl. Acad. Sci. 112, 12741–12745.

Norghauer, J.M., Martin, A.R., Mycroft, E.E., James, A., Thomas, S.C., 2011. Island Invasion by a threatened tree species: evidence for natural enemy release of mahogany (*Swietenia macrophylla*) on Dominica, Lesser Antilles. PLoS One 6, e18790. https://doi.org/10.1371/journal.pone.0018790.

Pérez, M., Sastre De Jesús, I., Lugo, A.E., Abelleira Martínez, O.J., 2012. Bryophyte species diversity in secondary forests dominated the introduced species *Spathodea campanulata* Beauv. in Puerto Rico. Biotropica 44, 763–770.

Phillips, O.L., Martínez, R.V., Arroyo, L., Baker, T.R., Killeen, T., Lewis, S.L., Malhi, Y., Mendoza, A.M., Neill, D., Núãez Vargas, P., Alexiades, M., Cerón, C., Di Fiore, A., Erwin, T., Jardim, A., Palacios, W., Saldias, M., Vinceti, B., 2002. Increasing dominance of large lianas in Amazonian forests. Nature 418, 770–773.

Phillips, O.L., Lewis, S.L., Baker, T.R., Chao, K.J., Higuchi, N., 2008. The changing Amazon forest. Philos. Trans. R. Soc. B 363, 1819–1827.

Radeloff, V.C., Williams, J.W., Bateman, B.L., Burke, K.D., Carter, S.K., Childress, E.S., Cromwell, K.J., Gratton, C., Hasley, A.O., Kraemer, B.M., Latzka, A.W., Marin-Spiotta, E., Meine, C.D., Munoz, S.E., Neeson, T.M., Pidgeon, A.M., Rissman, A.R., Rivera, R.J., Szymanski, L.M., Usinowicz, J., 2015. The rise of novelty in ecosystems. Ecol. Appl. 25, 2051–2068.

Ramjohn, I.A., Murphy, P.G., Burton, T.M., Lugo, A.E., 2012. Survival and rebound of Antillean dry forests: role of forest fragments. For. Ecol. Manage. 284, 124–132.

Ramos Prado, J.M., Delgado-rueda, M., Del Amo, S., Fernández, E., 1982. Análisis estructural de un área de vegetación secundaria en Uxpanapa, Veracruz. Biotica 7, 7–29.

Roberts, P., Hunt, C., Arroyo-Kalin, M., Boivin, N., 2017. The deep human prehistory of global tropical forests and its relevance for modern conservation. Nat. Plants 3, 1–9. https://doi.org/10.1038/nplants.2017.93.

Rodríguez Durán, A., 2005. Murciélagos. In: Joglar, R. (Ed.), Biodiversidad de Puerto Rico: vertebrados y ecosistemas. Editorial del Instituto de Cultura de Puerto Rico, San Juan, PR, pp. 241–274.

Roosevelt, A.C., Douglas, J., Brown, L., 2002. Migrations and adaptations of the first Americans: Clovis and pre-Clovis viewed from South America. Mem. Calif. Acad. Sci. 27, 159–236.

Sarmiento, L., Llambí, L.D., 2011. Regeneración del páramo después de un disturbio agrícola: síntesis de dos décadas de investigación en sistemas con descansos largos de la cordillera de Mérida. In: Herrera, F., Herrera, I. (Eds.), La restauración ecológica en Venezuela: fundamentos y experiencias. Ediciones IVIC, Instituto Venezolano de Investigaciones Científicas, Caracas, Venezuela, pp. 123–145.

Sarmiento, L., Smith, J.K., 2011. Degradación de laderas durante el ciclo triguero en los Andes venezolanos y factores que limitan su restauración. In: Herrera, F., Herrera, I. (Eds.), La restauración ecológica en Venezuela: fundamentos y experiencias. Ediciones IVIC, Instituto Venezolano de Investigaciones Científicas, Caracas, Venezuela, pp. 17–34.

Schargel, R., 2015. Suelos llaneros: geomorfología, distribución geográfica y caracterización. In: López-F, R., Hétier, J.-M., López-H, D., Schargel, R., Zinck, A. (Eds.), Tierras llaneras de Venezuela. IRD-CIDIAT, Mérida, Venezuela, pp. 63–153.

Schnitzer, S.A., Bongers, F., 2011. Increasing liana abundance and biomass in tropical forests: emerging patterns and putative mechanisms. Ecol. Lett. 14, 397–406.

Silander, S.R., 1979. A study of the ecological life history of *Cecropia peltata* L., an early secondary successional species in the rain forest of Puerto Rico. Masters. University of Tennessee, Knoxville, TN.

Silver, W.L., Kueppers, L.M., Lugo, A.E., Ostertag, R., Matzek, V., 2004. Carbon sequestration and plant community dynamics following reforestation of tropical pasture. Ecol. Appl. 14, 1115–1127.

Slot, M., Kitajima, K., 2015. General patterns of acclimation of leaf respiration to elevated temperatures across biomes and plant types. Oecologia 177, 885–900.

Slot, M., Winter, K., 2017. In situ temperature response of photosynthesis of 42 tree and liana species in the canopy of two Panamanian lowland tropical forests with contrasting rainfall regimes. New Phytol. 214, 1103–1117.

Smith, R.F., 1970. The vegetation structure of a Puerto Rican rain forest before and after short-term gamma irradiation. In: Odum, H.T., Pigeon, R.F. (Eds.), A Tropical Rain Forest. National Technical Information Service, Springfield, VA, pp. D103–D140.

Smith, T.M., Shugart, H.H., Bonan, G.B., Smith, J.B., 1992. Modeling the potential response of vegetation to global climate change. Adv. Ecol. Res. **22**, 93–98. 98a, 98b, 99–116. https://doi.org/10.1016/S0065-2504(08)60134-8.

Sodhi, N.S., Brook, B.W., Bradshaw, C.J.A., 2007. Tropical Conservation Biology. Blackwell Publishing, Malden, MA.

Somarriba, E., Beer, J., 2011. Productivity of *Theobroma cacao* agroforestry systems with timber or legume service shade trees. Agrofor. Syst. 81, 109–121.

Takhtajan, A., 1986. Floristic Regions of the World. University of California Press, Berkeley, p. 522.

Tavárez, H.S., Elbakidze, L., 2019. Valuing recreational enhancements in the San Patricio urban forest of Puerto Rico: a choice experiment approach. Forest Policy Econ. 109, 102004.

Taylor, C.M., Silander, S., Waide, R.B., Pfeiffer, W.J., 1995. Recovery of a tropical forest after gamma irradiation: a 23-year chronicle. In: Lugo, A.E., Lowe, C. (Eds.), Tropical Forests: Management and Ecology. Springer-Verlag, New York, NY, pp. 258–285.

ter Steege, H., Pitman, N.C.A., Sabatier, D., Baraloto, C., Salomão, R.P., Guevara, J.E., Phillips, O.L., Castilho, C.V., Magnusson, W.E., Molino, J.-F., Monteagudo, A., Núñez Vargas, P., Montero, J.C., Feldpausch, T.R., Honorio C., E.N., Killeen, T.J., Mostacedo, B., Vasquez, R., Assis, R.L., Terborgh, J., Wittmann, F., Andrade, A., Laurance, W.F., Laurance, S.G.W., Marimon, B.S., Marimon Jr., B.-H., Guimarães Vieira, I.C., Amaral, I.L., Brienen, R., Castellanos, H., Cárdenas López, D., Duivenvoorden, J.F., Mogollón, H.F., de Almeida Matos, F.D., Dávila, N., García-Villacorta, R., Stevenson D., P.R., Costa, F., Emilio, T., Levis, C., Schietti, J., Souza, P., Alonso, A., Dallmeier, F., Duque, A., Fernández Piedade, M.T., Araujo-Murakami, A., Arroyo, L., Gribel, R., Fine, P.V.A., Peres, C.A., Toledo, M., Aymard C., G.A., Baker, T., Cerón, C., Engel, J., Henkel, T.W., Maas, P., Petronelli, P., Stropp, J., Zartman, C.E., Daly, D., Neill, D., Silveira, M., Paredes, M.R., Chave, J., de Andrade Lima F., D., Jørgensen, P.M., Fuentes, A., Schöngart, J., Cornejo, F., Di Fiore, A., Jimenez, E.M., Peñuela M., M.C., Phillips, J.F., Rivas, G., van Andel, T.R., von Hildebrand, P., Hoffman, B., Zent, E.L., Malhi, Y., Prieto, A., Rudas, A.,

Ruschell, A.R., Silva, N., Vos, V., Zent, S., Oliveira, A.A., Cano S., A., Gonzales, T., de Nascimento, M.T., Ramirez-Angulo, H., Sierra, R., Tirado, M., Umaña M., M.N., van der Heijden, G., Vela, C.I.A., Vilanova, E., Vriesendorp, C., Wang, O., Young, K.R., Baider, C., Balslev, H., Ferreira, C., Mesones, I., Torres-Lezama, A., Urrego, L.E., Zagt, R., Alexiades, M.N., Hernández, L., Huamantupa-C., I., Milliken, W., Palacios, W., Pauletto, D., Valderrama S., E., Valenzuela G., L., Dexter, K.G., Feeley, K., Lopez-González, G., Silman, M.I.R., 2013. Hyperdominance in the Amazonian tree flora. Science 342, 1243092. https://doi.org/10.1126/science.1243092.

ter Steege, H., Mota de Oliveira, S., Pitman, N.C.A., Sabatier, D., Antonelli, A., Guevara Andino, J.E., Aymard, G.A., Salomão, R.P., 2019. Towards a dynamic list of Amazonian tree species. In: Scientific Reports. February, http://www.nature.com/srep.

Torres, J.A., Snelling, R.R., 1997. Biogeography of Puerto Rican ants: a non-equilibrium case? Biodivers. Conserv. 6, 1103–1121.

Túa Ayala, G.Z., Abelleira Martínez, O.J., 2019. Interventions for agroforestry and species restoration in novel forests of Puerto Rico: enrichment planting success before and after Hurricane María. Braz. J. For. Res. 39, e201902043.

van der Kaars, S., Miller, G.M., Turney, C.S.M., Cook, E.J., Nürnberg, D., Schönfeld, J., Kershaw, A.P., Lehman, S.J., 2017. Humans rather than climate the primary cause of Pleistocene megafaunal extinction in Australia. Nat. Commun. 8, 141–142.

Vareschi, V., 1970. Flora de los páramos de Venezuela. Ediciones del Rectorado, Universidad de Los Andes, Mérida, Venezuela.

Vázquez Yánez, C., 1980. Notas sobre la autoecología de los árboles pioneros de rápido crecimiento de la selva tropical lluviosa. Trop. Ecol. 21, 103–112.

Veillon, J.P., 1974. Relaciones entre los bosques naturales y el clima en diferentes zonas ecológicas del noroccidente de Venezuela. Instituto de Silvicultura. Facultad de Ciencias Forestales, Universidad de los Andes, Mérida. Venezuela, p. 184.

Veillon, J.P., 1986. Especies forestales autoctonas de los boques naturales de Venezuela. Instituto Forestal Latinoamericano, Instituto de Silvicultura, Universidad de los Andes, Mérida, Venezuela, p. 199.

Veillon, J.P., 1995. Los bosques naturales de Venezuela. Parte II. Los bosques xerófilos zonas de vida: bosque espinoso tropical y bosque muy seco tropical. Instituto de Silvicultura, Universidad de los Andes-MARNR-SEFORVEN, Mérida, Venezuela, p. 127.

Veillon, J.P., 1997. Los bosques naturales de Venezuela. Parte III. Los bosques tropóficos o veraneros de la zona de vida de bosque seco tropical. Universidad de Los Andes- Consejo de Publicaciones, Instituto Forestal Latinoamericano, Facultad de Ciencias Forestales y Ambientales, Mérida, Venezuela, p. 127.

Vitousek, P.M., 1990. Biological invasions and ecosystem processes: towards an integration of population biology and ecosystem studies. Oikos 57, 7–13.

Vitousek, P.M., Walker, L.R., 1989. Biological invasion by *Myrica faya* in Hawaii: plant demography, nitrogen fixation, ecosystem effects. Ecol. Monogr. 59, 247–265.

von Humboldt, A., Bonpland, A., 1807. Essai sur la géographie des plantes; Accompagné d'un tableau physique des régions equinoxiales. Levrault, Schoell et Compagnie, Paris, France.

Wadsworth, F.H., 1950. Notes on the climax forests of Puerto Rico and their destruction and conservation prior to 1900. Caribbean Forest. 11, 38–47.

Walker, B., Salt, D., 2006. Resilience Thinking: Sustaining Ecosystems and People in a Changing World. Island Press, Washington, DC.

Waters, C.N., Zalasiewicz, J., Summerhayes, C., Barnosky, A.D., Poirier, C., Gałuszka, A., Cearreta, A., Edgeworth, M., Ellis, E.C., Ellis, M., Jeandel, C., Leinfelder, R., McNeill, J.R., Richter, D.d., Steffen, W., Syvitski, J., Vidas, D., Wagreich, M., Williams, M., Zhisheng, A., Grinevald, J., Odada, E., Oreskes, N., Wolfe, A.P., 2016. The Anthropocene is functionally and stratigraphically distinct from the Holocene. Science 351, aad2622. https://doi.org/10.1126/science.aad2622.

Watlington, F., 2009. Guaraguaos, ardillas, cotorras y cuervos en el "Central Park" de Puerto Rico. Acta Cient. 23, 69–72.
Weaver, P.L., 1979. La agri-silvicultura en la América tropical. Unasylva 31, 2–12.
West, G.B., Brown, J.H., Enquist, B.J., 1999. A general model for the structure and allometry of plant vascular systems. Nature 400, 664–667.
Wilkie, D.S., Bennett, E.L., Peres, C.A., Cunningham, A.A., 2011. The empty forest revisited. Ann. N. Y. Acad. Sci. 1223, 120–128.
Willis, K.J., Gillson, L., Brncic, T.M., 2004. How "virgin" is virgin rainforest? Science 304, 402.
Wood, T.E., Cavaleri, M.A., Reed, S.C., 2012. Tropical forest carbon balance in a warmer world: a critical review spanning microbial- to ecosystem-scale processes. Biol. Rev. 87, 912–927. https://doi.org/10.1111/j.1469-185X.2012.00232.x.
Wright, S.J., Muller Landau, H.C., 2006. The future of tropical forest species. Biotropica 38, 287–301.
Wright, S.J., Calderon, O., Hernandez, A., Paton, S., 2004. Are lianas increasing in importance in tropical forests? A 17-year record from Panama. Ecology 85, 484–489.
Wright, S.J., Stoner, K.E., Beckman, N., Corlett, R.T., Dirzo, R., Muller Landau, H.C., Nuñez-lturri, G., Peres, C.A., Wang, B.C., 2007. The plight of large animals in tropical forests and the consequences for plant regeneration. Biotropica 38, 289–291.
Zhang, T., Niinemets, Ü., Justin Sheffield, J., Lichstein, J.W., 2018. Shifts in tree functional composition amplify the response of forest biomass to climate. Nature 556, 99–102.
Zimmerman, J.K., Aide, T.M., Rosario, M., Herrera, L., 1995. Effects of land management and a recent hurricane on forest structure and composition in the Luquillo Experimental Forest, Puerto Rico. For. Ecol. Manage. 77, 65–76.
Zimmerman, J.K., Pascarella, J.B., Aide, T.M., 2000. Barriers to forest regeneration in an abandoned pasture in Puerto Rico. Restor. Ecol. 8, 350–360.
Zon, R., Sparhawk, W.N., 1923. Forest Resources of the World. McGraw-Hill Book Co., New York, NY, USA.

Further reading

Barger, et al., 2002. Nutrient Limitation to primary productivity in a secondary savanna in Venezuela. Biotropica 34, 493–501.
Burgos Rodríguez, J.A., Avilés Rodríguez, K.A., Kolbe, J.J., 2016. Effects of invasive Green Iguanas (*Iguana iguana*) on seed germination and seed dispersal potential in southeastern Puerto Rico. Biol. Invasions 18, 2775–2782.
Lugo, A.E., Abelleira, O.J., Borges, S., Colón, L.L., Meléndez, S., Rodríguez, M.A., 2006. Preliminary estimate of earthworm abundance and species richness in *Spathodea campanulata* Beauv. forests in northern Puerto Rico. Caribb. J. Sci. 42, 325–330.

CHAPTER THREE

Monitoring tropical forest degradation and restoration with satellite remote sensing: A test using Sabah Biodiversity Experiment

Jinhui Wu[a,b,]*, Bin Chen[d], Glen Reynolds[e], Jun Xie[f], Shunlin Liang[c], Michael J. O'Brien[e,g], Andrew Hector[b,]*

[a]State Key Laboratory of Remote Sensing Science, Faculty of Geographical Science, Beijing Normal University, Beijing, China
[b]Department of Plant Sciences, University of Oxford, Oxford, United Kingdom
[c]Department of Geographical Sciences, University of Maryland College Park, College Park, MD, United States
[d]Department of Land, Air and Water Resources, University of California, Davis, CA, United States
[e]Southeast Asia Rainforest Research Partnership (SEARRP), Kota Kinabalu, Malaysia
[f]School of Environmental Sciences, University of Hull, Hull, United Kingdom
[g]Área de Biodiversidad y Conservación, Universidad Rey Juan Carlos, Móstoles, Spain
*Corresponding authors: e-mail address: jwu@mail.bnu.edu.cn; andrew.hector@plants.ox.ac.uk

Contents

Advances in Ecological Research, Volume 62
ISSN 0065-2504
https://doi.org/10.1016/bs.aecr.2020.01.005

Abstract

Selective logging has been so extensive that harvested forest now exceeds unlogged areas in most tropical forest regions outside of the Amazon. In response, in Southeast Asia, enrichment planting with dipterocarp tree species is carried out in an attempt to accelerate restoration of forest structure and functioning. However, assessing the impacts of degradation (from selective logging and other causes) and subsequent restoration with field measurements is expensive and time-consuming. There is therefore a need to develop methods for the assessment of forest quality using remote sensing. Here, we use high spatial resolution satellite imagery and advanced remote sensing products to monitor the pattern and dynamics of estimated vegetation cover, Leaf Area Index (LAI), and the biomass of plots within a field-scale (500 ha) replicated and randomized manipulation that compares different forest restoration treatments with naturally regenerating controls within the Sabah Biodiversity Experiment (SBE). We also compare the biodiversity experiment plots with the surrounding area of the Malua Forest Reserve that was selectively logged for the second time in 2007. In general, satellite remote sensing detected differences in degradation between the once- and twice-logged areas as well as between the different experimental restoration treatments. We found that approximately 70% of the Malua Forest Reserve experienced a decrease of vegetation cover after the selective relogging in 2007, while the Sabah Biodiversity Experiment area that was not relogged showed increasing vegetation cover. Within the experiment, we found that plots restored using Enrichment line planting, had higher remote sensed vegetation cover (Mean $\pm$ SE: 66.90 $\pm$ 0.06 vs. 61.96 $\pm$ 0. 16) and LAI (Mean $\pm$ SE: 5.09 $\pm$ 0.03 vs. 4.61 $\pm$ 0.11) than that of unenriched plots. Among the enrichment planted plots, those planted with mixtures of (4 or 16) species exhibited higher vegetation cover (Mean $\pm$ SE: 67.72 $\pm$ 0.06 vs. 65.35 $\pm$ 0. 09) and LAI (Mean $\pm$ SE: 5.29 $\pm$ 0.04 vs. 4.82 $\pm$ 0.06) than that of monoculture plots. Overall, when our test case of the Sabah Biodiversity Experiment was viewed through the lens of remote sensing indicators, satellite imagery was able to detect changes in forest quality due to selective logging and restoration enrichment planting. Furthermore, our results suggest that diverse mixtures of planted tree seedlings enhance restoration of forest canopies compared to planting with single species. Confirmation with ground data will be needed to validate these results and to better understand the biological processes determining tropical forest degradation and restoration.

1. Introduction

In most zones of tropical forest outside of the Amazon, the area that has been selectively logged now exceeds the area of old growth affecting their diversity, functioning and stability (Alvarado and Sandberg, 2001; Asner et al., 2004b; Laporte et al., 2007; Nepstad et al., 1999; Tuck et al., 2016) and sometimes predisposing them to conversion to other land uses such as oil palm plantation. In southeast Asia, enrichment planting with dipterocarp tree species is carried out to accelerate restoration of forest structure and functioning (Millet et al., 2013; Tigabu et al., 2010). Enrichment planting is practiced in a number of forms but the basic idea is to increase establishment and recovery of forests by planting seedlings of the timber tree species that have been selectively logged out (Hector, 2011). However, assessing the impacts of degradation on forest quality (from selective logging and other causes) and on the efficacy of restoration is expensive and time-consuming using field measurements. There is therefore a need to develop complementary methods for the assessment of forest quality using remote sensing data that can be applied at larger spatial scales relevant to management and conservation.

Remote sensing has proven useful in large-scale vegetation monitoring because it enables consistent observations of vegetation across time and space (Chen et al., 2019; Liang, 2004; Liang et al., 2018; Turner, 2014; Zhu et al., 2016). Many studies have used remote sensing data to assess the impacts of logging on forest dynamics (Foody and Cutler, 2003; Matricardi et al., 2010; Perez et al., 2016; Rocchini et al., 2007; Turner, 2014; Turner et al., 2003). These previous studies have contributed to substantial improvement in remote sensing-based techniques to assess various aspects of forest diversity and structure and the ecosystem functions and services that they support. However, significant gaps exist in attempts to better link field ecology and remote sensing. Firstly, coupling remote sensing data and in-situ ecological field data is challenging, since the scale of the data from remote sensing and field ecology can differ substantially (Chambers et al., 2007; Kerr and Ostrovsky, 2003; Muraoka and Koizumi, 2009; Turner, 2014) with the spatial resolution of remote sensing data typically much coarser than that of data from measurements taken in the field. Secondly, in tropical regions, the obscuring effects of cloud cover are serious which leads to a large uncertainty of remote sensing data (de Souza et al., 2010; Gao and Li, 2000; Ju and Roy, 2008; Liou, 1992; Simpson et al., 2000; Xu et al., 2016). In recent years, with the progress that has been made on satellite technology and remote sensing inversion models, a number of high-quality and long-term quantitative remote sensing products (e.g. Leaf Area Index (LAI), Fraction of Absorbed Photosynthetically Active Radiation (FAPAR), Gross Primary Productivity (GPP) and Land Cover) have been developed to monitor the distribution and variation of terrestrial vegetation

(Claverie et al., 2016; Friedl et al., 2002; Liang, 2004; Liang et al., 2012; Myneni et al., 2002; Running et al., 2000; Xiao et al., 2016; Yuan et al., 2010; Zhu et al., 2013). In addition, some of these quantitative remote sensing products exhibit very high spatial resolution (Chen et al., 2015; Feng et al., 2016; Hansen et al., 2013; Liang et al., 2013; Sexton et al., 2013b; Song et al., 2016). This progress provides a great opportunity for comprehensively assessing the impacts of both degradation and restoration on forest quality which we do here using the Sabah Biodiversity Experiment as a test case.

The Sabah Biodiversity Experiment (SBE) covers 500 ha of Malua Forest Reserve, an area of selectively logged forest bordering primary forest located in the Malaysian state of Sabah, Borneo (Hector et al., 2011). Malua Forest Reserve was selectively logged in the early 1980s and again in the mid-2000s, although the 500 ha areas of the Sabah Biodiversity Experiment site was excluded from the second round of selective logging (in 2007) (Saner et al., 2012). We can therefore examine differences between the treatment plots within the experiment and compare the Sabah Biodiversity Experiment as a whole with the surrounding relogged forest in the rest of Malua Forest Reserve. Sabah Biodiversity Experiment, manipulates the identity, composition and diversity of enrichment planted dipterocarps to assess their impacts on the functioning and stability of selectively logged rainforests during restoration and compare them with those of unrestored plots within the same design (Tuck et al., 2016). The experiment contains 124 four-hectare (200 × 200 m) plots which include five different treatment groups that we compare here, namely unplanted plots, monoculture plots, 4-species mixture plots, 16-species mixture plots, and 16-species mixtures plots with enhanced climber cutting. Sabah Biodiversity Experiment is also part of a global network of tree diversity experiments and is one of only two representatives in the palaeo-tropics.

In this study, we use high spatial resolution satellite imagery and advanced remote sensing products to assess degradation of the once-logged area of the biodiversity experiment with the twice-logged area of the surrounding Malua Forest Reserve and to compare different forest restoration treatments within the Sabah Biodiversity Experiment (SBE). Specifically, we use Sabah Biodiversity Experiment as a test case to assess what differences in vegetation canopy attributes satellite remote sensing can detect between:

(1) The areas of Malua forest reserve that was relogged in 2007 vs. the Sabah Biodiversity Experiment site that was not relogged.

(2) The unplanted vs. enrichment planted plots within Sabah Biodiversity Experiment 15 years after the start of planting.

(3) Sabah Biodiversity Experiment plots that have been enrichment planted with seedling mixtures of different diversity.

(4) Sabah Biodiversity Experiment plots that have had lianas removed (enhanced climber cutting) vs. matched plots subject to standard Enrichment line planting with the same mixtures of 16 dipterocarp seedlings.

2. Methodology

2.1 Location

The Sabah Biodiversity Experiment is located in the southern part of the Malua Forest Reserve (Fig. 1). The Sabah Biodiversity Experiment covers 500 ha of the 35,000 ha Malua Forest Reserve. The climate is typical wet tropical with an average temperature of 27 °C and an annual rainfall of >3000 mm, distributed over two distinct wet seasons (Lussetti et al., 2016; Marsh and Greer, 1992; Saner, 2009). Malua Forest Reserve was logged in the 1980s and relogged in 2007, while the Sabah Biodiversity Experiment was excluded from the second round of logging since this area had been protected since the establishment of the experiment in 2002.

The experiment contains 124 four-hectare (200 × 200 m) plots, split between two blocks that are north and south of an old logging road. There are 64 plots in the south block and 60 plots in the north block. The experiment manipulates the identity, composition and diversity of the enrichment planted dipterocarp seedlings and forest management technique of liana removal (enhanced climber cutting). Ninety-six of these plots form a gradient in the diversity of enrichment planted seedlings with four levels: 1, 4, 16 enrichment planted species and 16 enrichment planted species with liana removal. The remaining plots are comprised of 12 unenriched naturally regenerating controls, and an additional 16 plots enrichment planted with the 16-species mixtures and subject to 2 rounds of liana removal (O'Brien et al., 2019). In this enhanced climber cutting treatment, climbers are cut throughout the whole plot, which is expected to improve recovery time during restoration. The design also manipulates the composition of the enrichment planted seedlings with, at the 2 extremes, 16 different single-species treatments vs. a single 'full mixture' of all 16 species combined (Table S1 in the online version at https://doi.org/10.1016/bs.aecr.2020.01.005). At the intermediate diversity level, species compositions within the 4-species mixtures provide a gradient of generic richness and are designed to produce a range of canopy structures once the planted seedlings

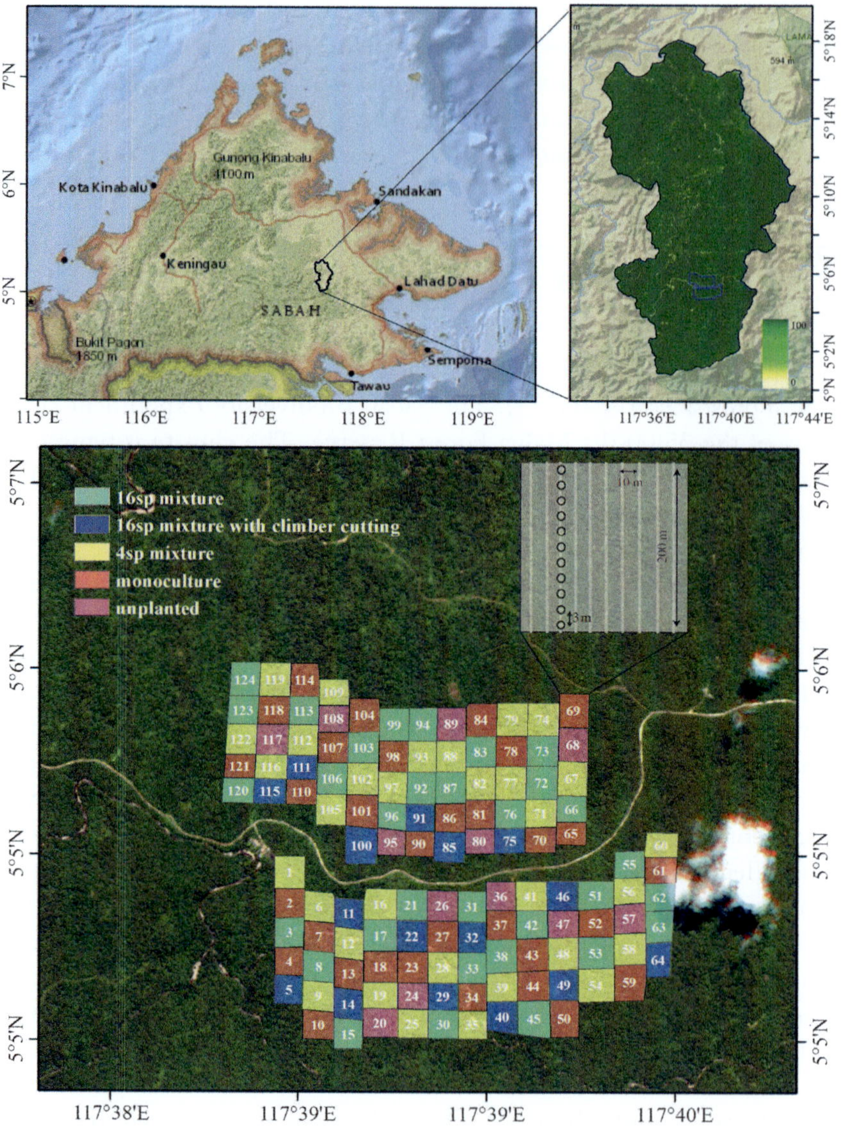

Fig. 1 Location of the Malua Forest Reserve (top) and Sabah Biodiversity Experiment with details of the Enrichment line planting.

mature but these treatments are not addressed further here. Each planted species richness level has 32 plots: the 1- and 4-species richness levels contain 16 different compositions, each replicated twice, plus 32 replicates of the 16-species mixture, all balanced equally across the 2 blocks.

Following standard enrichment planting practice, the initial cohort of seedlings (cohort 1 planted January 2002–September 2003) were supplemented with a second cohort (cohort 2 planted September 2008–August 2011) that replaced initial mortality. Since establishment, the plots (sometimes all of them, sometimes only a subset) have been periodically surveyed. Further details of Sabah Biodiversity Experiment and Malua Forest Reserve can be found in previous publications (Hector et al., 2011; Saner et al., 2012; Tuck et al., 2016).

2.2 Remote sensing data

Landsat Vegetation Continuous Fields tree cover, MODIS MCD15A3H Leaf Area Index (LAI), Estimated Deforested Area Biomass, and RapidEye image were selected to monitor the vegetation canopy structure in this study based on considerations of data accuracy, the area of study zone and the time period of the experiment (Table 1).

2.2.1 Landsat vegetation continuous fields tree cover

The Landsat Vegetation Continuous Fields tree cover (Landsat tree cover) layers contain estimates of the percentage of horizontal ground in each 30 m pixel covered by vegetation greater than 5 m in height globally (Sexton et al., 2013a). In this study, we rename the Landsat tree cover as Landsat vegetation cover, since in Sabah, tropical rainforest, almost all plants Landsat can detect are higher than 5 m. The product is derived from all seven bands of Landsat-5 Thematic Mapper (TM) and/or Landsat Enhanced Thematic Mapper Plus (ETM+). The spatial resolution of Landsat tree cover dataset is 30 m, which is suitable for comparing the vegetation canopy structure among different kinds of the 200-by-200 m plots within Sabah

Table 1 Remote sensing datasets.

Data	Spatial resolution	Temporal resolution	Time cover
Landsat VCF vegetation cover	30 m	5-year	2000, 2005, 2010
MODIS MCD15A3H LAI	500 m	4-day	2001–2017
Estimated deforested area aboveground biomass	30 m	12-year	2000–2012
RapidEye-based vegetation cover	5 m	/	Aug. 2012
RapidEye-based leaf area index (LAI)	5 m	/	Aug. 2012
RapidEye-based aboveground biomass	5 m	/	Aug. 2012

Biodiversity Experiment. To date, this dataset has four epochs—2000, 2005, 2010, and 2015—each of which is a composite of several years. For example, vegetation cover 2005 shows the average condition of tree cover from 2003 to 2008, because vegetation cover 2005 was calculated from satellite images which are collected from 2003 to 2008 (Gutman et al., 2008, 2013), while vegetation cover 2000 shows the average vegetation cover from 1999 to 2002 and vegetation cover 2010 shows the average vegetation cover from 2008 to 2012. To avoid confusion it is important to point out that in this study we use vegetation cover 2005 to monitor the vegetation cover of study zone after the 2007 logging, since the input satellite images of vegetation cover for the 2005 epoch in Sabah were mostly collected from 2008. Vegetation cover for the 2000, 2005 and 2010 epochs were therefore selected for this study.

2.2.2 MODIS MCD15A3H LAI

Leaf Area Index (LAI: one-sided green leaf area per unit ground area) is an important vegetation biophysical variable (Bréda, 2003; Watson, 1947; Xiao et al., 2014, 2016). MODIS MCD15A3H LAI has been widely used in forest monitoring, which exhibits very high accuracy (Brede et al., 2018; Li et al., 2018; Miller et al., 2019; Ranga Myneni and Yuri, 2015). The temporal range of this dataset is from 2003 to 2017 and the temporal resolution of MCD15A3H LAI is 4-day. The spatial resolution of MCD15A3H LAI is 500 m, which although larger than the 200 m plots could be used to compare the vegetation canopy structure between the selectively relogged area (Malua Forest Reserve) and the area protected from the 2007 relogging (the 500 ha area of Sabah Biodiversity Experiment as a whole). In the process of LAI calculation, the cloud contamination of albedo, one of the main input data for LAI calculation has been effectively removed. In addition, this dataset provides a layer of quality control (QC), which contains the cloud state pixel-by-pixel. That is very important for removing cloud effect, one of the serious effects in tropical area (Myneni et al., 2015). In this study, the maximum value composite (MVC) procedure was took to generate monthly LAI from 4-day LAI. By doing so, cloud contamination, shadow effects, aerosol and water vapour effects can be minimized (Knyazikhin, 1999). The yearly LAI was generated by averaging the monthly LAI.

2.2.3 Estimated deforested area biomass

This dataset provides estimates of aboveground biomass (AGB) loss from 2000 to 2012 at 30 m resolution (Baccini et al., 2016). The biomass estimates are only for areas where deforestation occurred during the period 2000–2012. Estimates are derived using in-situ measurements and a RandomForest statistical model which correlated Geoscience Laser Altimeter System (GLAS) LiDAR data, Landsat 7 Enhanced Thematic Plus (ETM+) satellite imagery, reflectance, elevation, and biophysical variables. A separate RandomForest model is trained for each continent using a stratified sampling design of roughly 13,000 GLAS-based biomass estimates for each model. This dataset was selected to measure the biomass loss in Sabah Biodiversity Experiment and Malua Forest Reserve.

2.2.4 *RapidEye image*

In this study, the RapidEye satellite image of the Sabah Biodiversity Experiment site was acquired in August 2012. RapidEye satellite images exhibit a temporal resolution of 5.5 days and a high-resolution of 5 m (Jung-Rothenhäusler et al., 2007; Tyc et al., 2005). RapidEye's high spatial resolution provides a chance to monitor the forest canopy structure among different plots in a detailed way. The multi-spectral scanner on the satellites acquires data in five spectral bands. The blue (0.44–0.51 μm), green (0.52–0.59 μm), red (0.63–0.68 μm) and near-infrared (0.76–0.85 μm) bands are very similar to the corresponding Landsat spectral bands, while the sensor has an additional red-edge band (0.69–0.73 μm), giving additional sensitivity to changes in the reflectance of vegetation (Chander et al., 2013; Pfeifer et al., 2016).

2.3 Methods

A group of satellite imagery and remote sensing products were pre-processed (calibration, clip, cloud and cloud shadowing-masking) using ArcGIS and ENVI.

2.3.1 *Vegetation metrics inversion from RapidEye image*

We applied atmospheric correction to the RapidEye image using the FLAASH atmospheric correction model. The vegetation cover, Leaf Area Index (LAI), and aboveground biomass (AGB) were calculated from the empirical formulas which were developed in Pfeifer et al. (2016). In the development of these empirical formulas, the in-situ data used for model fitting were collected from tropical rainforest in Sabah, Borneo near to our site, and the remote sensing data they used were RapidEye images. Thus, we replicated the inversion method to calculate the vegetation cover, LAI, and biomass. These inversed metrics exhibited high accuracy, since the accuracy had been validated. In addition, vegetation cover, Leaf Area Index, and biomass data exhibit very high spatial resolution since the spatial resolution of RapidEye images is 5 m. Further details of the inversion method and accuracy assessment can be found in the previous publication (Pfeifer et al., 2016).

$$\begin{aligned}\text{Vegetation cover} = 2.66 - 0.66{*}\text{Red} + 0.3{*}\text{RedEdge} - 0.08{*}\text{NearIR} \\ + 1.48{*}\text{DissB4} - 0.42{*}\text{DissB5} - 0.17{*}\text{DissB3}\end{aligned} \tag{1}$$

$$\begin{aligned}\text{LAI} = 0.9 - 0.59{*}\text{Red} + 0.41{*}\text{RedEdge} - 0.11{*}\text{NearIR} \\ -0.53{*}\text{DissB3} + 1.08{*}\text{DissB4} - 0.36{*}\text{DissB5}\end{aligned} \tag{2}$$

$$\begin{aligned}\text{Biomass} = 19.45 - \exp(\text{MSAVI2}) - 2.39{*}\text{Green} + 1.08{*}\text{RedEdge} \\ + 2.65{*}\text{DissB2} - 0.28{*}\text{DissB3} - 0.13{*}\text{DissB5} + 0.09{*}\text{DissB4}\end{aligned} \tag{3}$$

where Vegetation cover is the RapidEye-based vegetation cover, LAI is the RapidEye-based Leaf Area Index, Biomass is the RapidEye-based biomass, MSAVI2 is the Modified Soil-Adjusted Vegetation Index 2 (Qi et al., 1994), Green is the green band of RapidEye image, Red is the red band of RapidEye image, NearIR is the near-infrared band of RapidEye image, RedEdge is the red-edge band of RapidEye image, DissB2, DissB3, DissB4, and DissB5 are grey-level dissimilarities of green band, red band, near-infrared band, and red-edge band, respectively (Gallardo-Cruz et al., 2012).

2.3.2 Statistical analysis

Our focus for the analysis of the satellite indicators of vegetation cover, Leaf Area Index (LAI), and biomass was on the estimation of means and variabilities, so we derived pixel-level estimates with standard errors (mean $\pm$ SE), for each of the key questions of interest listed at the end of the introduction: unenriched plots vs. enriched plots; monoculture plots vs. mixture plots; 4-species mixture plots vs. 16-species mixture; and, 16-species mixture with climber cutting vs. 16-species mixture without climber cutting. We focus on graphical presentation of means and confidence intervals where degree of (non-)overlap can be used to judge statistical significance (e.g. means with 95% CIs that are at the point of overlap are significantly different at $P < 0.01$

approximately). Assessing the effects of the 2007 relogging by comparing the Sabah Biodiversity Experiment with the surrounding Malua Forest Reserve is complicated by the difference in size of the two areas. We addressed this in two ways. First, we randomly selected 100 samples in Malua Forest Reserve with the same 500-ha area as Sabah Biodiversity Experiment and calculated the mean and standard error of vegetation cover and LAI of these as a representative sample of the whole area. However, the intensity of the 2007 relogging varied across the Malua Forest Reserve so we also selected 10 samples (of the same area as the biodiversity experiment) in the most heavily logged area of Malua Forest Reserve as a representative sample of the most heavily relogged areas (worst case scenario).

3. Results

In our study, we first divided the Malua Forest Reserve into two parts to compare the once-logged (1984–86) 500-ha area of Sabah Biodiversity Experiment, with the remaining surrounding area of the Malua Forest Reserve that was relogged in 2007 before comparing different treatments within the biodiversity experiment design.

3.1 Effects of relogging on vegetation canopy attributes

After the 2007 relogging the vegetation cover of the Malua Forest Reserve declined by 3% on average and by 7% in its most severely logged parts, while the vegetation cover of Sabah Biodiversity Experiment that was protected from the relogging exhibited a 5% increase (Fig. 2).

Although the change of mean vegetation cover was not large in Malua Forest Reserve, the proportion of the area affected was high. To examine the spatial variability in relogging intensity we divided Malua Forest Reserve into six parts: 56.96% (19,936 ha) of the area of Malua Forest Reserve showed 0–9% decrease of vegetation cover, 3.99% (1396.5 ha) of the area of the reserve showed 9–18% decrease of vegetation cover, and 1.54% (539 ha) of the area of the reserve showed over 18% decrease of vegetation cover. 36.18% (12,633 ha), 0.94% (329 ha), and 0.39% (136.5 ha) of the reserve showed 0–9%, 9–18%, and over 18% increase of vegetation cover (Fig. 3). After the 2007 relogging, there was a considerable decrease of vegetation cover in the centre of Malua Forest Reserve, while the 500 ha of the Sabah Biodiversity Experiment protected from the

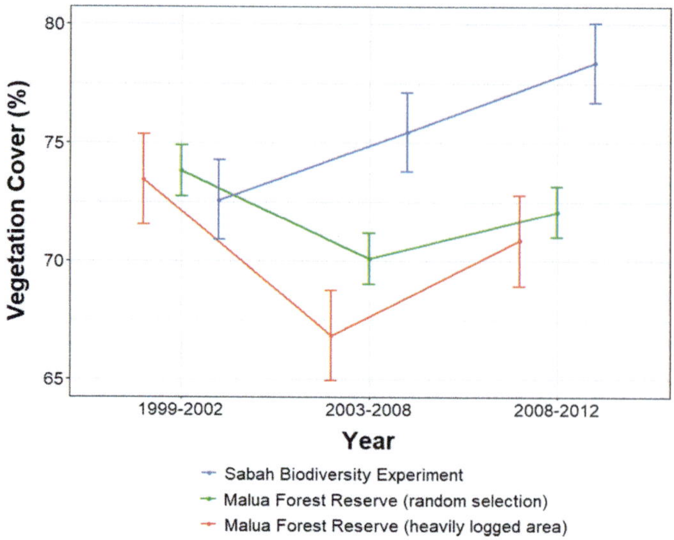

Fig. 2 Landsat vegetation cover with 95% confidence intervals of Sabah Biodiversity Experiment and Malua Forest Reserve before, during and after the selective relogging in 2007 (values jittered to avoid overlap).

relogging mostly exhibited an increasing trend of vegetation cover. Among the areas of Malua Forest Reserve where vegetation cover increased, we found that some areas were caused by rivers drying up followed by vegetation regrowth which locally offset the decline in cover due to relogging. We cannot tell whether the drying up of small water courses was associated with the relogging or not.

The Leaf Area Index (LAI) of Sabah Biodiversity Experiment was higher than that of the surrounding Malua Forest Reserve, particularly its most heavily relogged areas (Fig. 4). The LAI of Sabah Biodiversity Experiment and Malua Forest Reserve remained relatively stable from 2004 to 2007. After the 2007 relogging, the LAI of Malua Forest Reserve exhibited a significant decrease (Fig. 4). In heavily logged areas, the LAI began to decrease earlier than whole Malua Forest Reserve, suggesting the most heavily relogged areas were those where the relogging started earliest and went on longest. The average LAI of Malua Forest Reserve reached its lowest levels in 2009 shortly after the relogging and then recovered. In comparison, the LAI of Sabah Biodiversity Experiment did not show significant fluctuation from 2007 to 2013 consistent with its protection from relogging. The LAI of both the Sabah Biodiversity Experiment

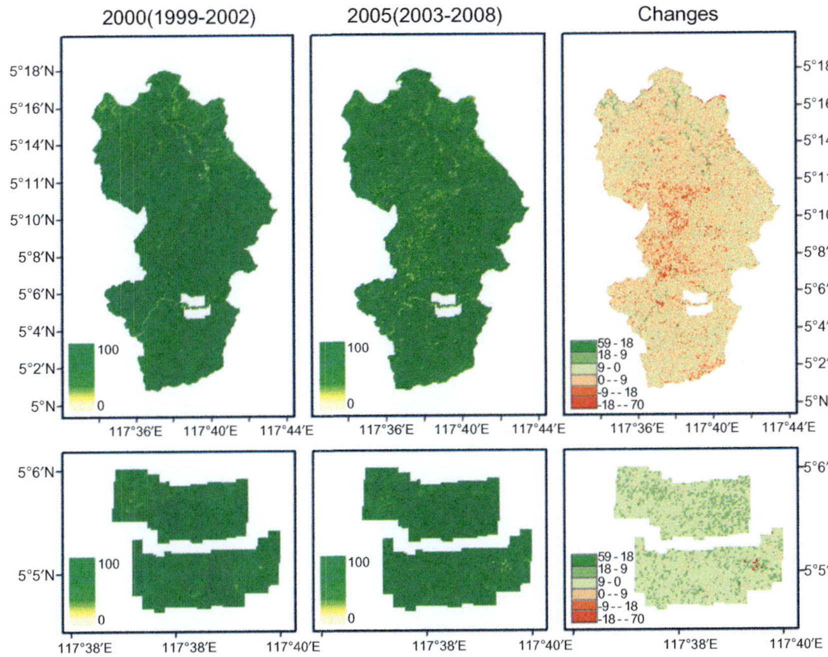

Fig. 3 Landsat vegetation cover of Sabah Biodiversity Experiment and Malua Forest Reserve with the change in cover after the selective relogging in 2007 (right).

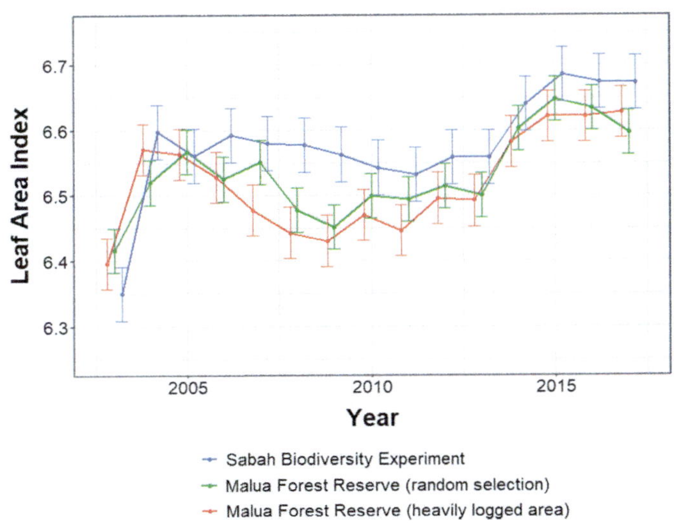

Fig. 4 Variation of mean annual MODIS Leaf Area Index (LAI) with 95% Confidence Intervals (values jittered for clarity).

site and the rest of the surrounding Malua Forest Reserve showed significant increases from 2013 to 2015. The selective relogging in 2007 led to differences between Malua Forest Reserve and Sabah Biodiversity Experiment in both LAI (Fig. 4) and vegetation cover (Fig. 2), so that the two satellite remote sensing indicators agreed and provided a consistent picture of changes in forest quality.

The area with biomass loss was mainly in the middle part of Malua Forest Reserve. In total, we estimate that the Malua Forest Reserve lost approximate 2.7 million Mg of aboveground biomass from the 11,433 ha of deforested area during the period 2000 through 2012.

3.2 Effects of enrichment planting on vegetation canopy attributes

The Landsat vegetation cover showed that the unenriched plots exhibited lower vegetation cover and less change from 1999 to 2012 (Fig. 5). There were no consistent differences in the vegetation cover of planted and unplanted plots before the enrichment planting. After the enrichment planting in 2002/03, the unenriched plots exhibited ~4% higher vegetation cover than the unplanted plots, and this difference in vegetation cover then

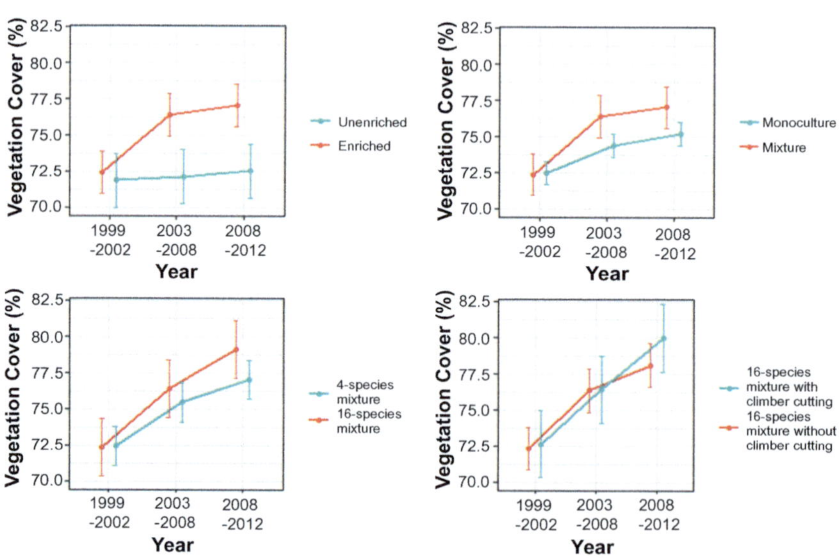

Fig. 5 Landsat vegetation cover of different enrichment planting groups with 95% confidence intervals. Note that the first round of enrichment planting was only completed in 2003 and of climber cutting in 2011.

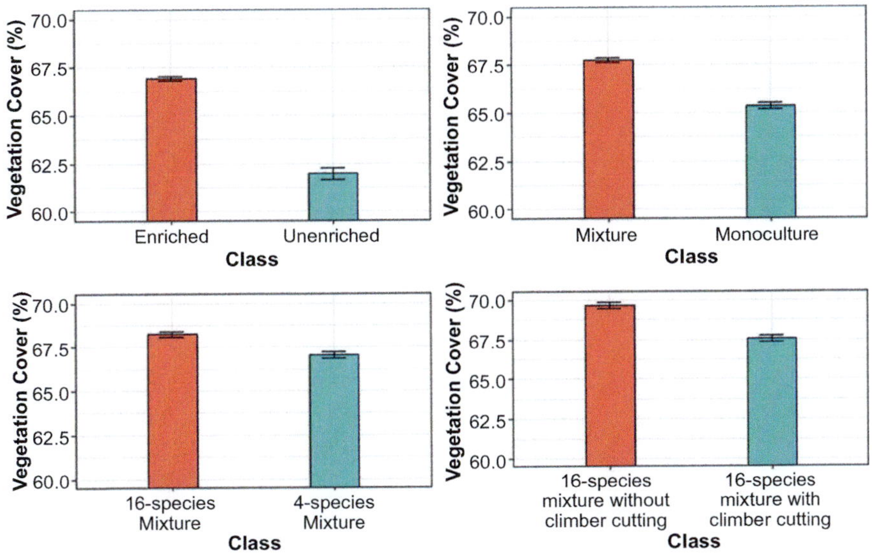

Fig. 6 RapidEye-based vegetation cover of different enrichment planting groups with 95% confidence intervals in August 2012.

increased over time. The value of RapidEye-based vegetation cover was lower than that of Landsat vegetation cover, and the RapidEye-based vegetation cover showed that the gap between enriched plots and unenriched plots was slightly larger than that in Landsat vegetation cover (Fig. 6). In both of these two datasets, the standard errors of vegetation cover in planted plots was relatively lower than that in unplanted plots (consistent with standard deviations). Moreover, the standard errors of vegetation cover in enrichment planted plots and unenriched plots did not change substantially during the period of 1999–2012.

For the RapidEye-based LAI, we found the enriched plots ($N=110$, two plots were excluded since the cloud cover) exhibited LAI values 0.48 units higher on average than that of unenriched plots ($N=12$) (Fig. 7). For RapidEye-based biomass, we found the same pattern as found with the LAI dataset (Fig. 8).

3.3 Effects of planted tree diversity on vegetation canopy attributes

After the enrichment planting was carried out, the vegetation cover of mixture plots became 2% higher than that of monoculture plots, and this difference in vegetation cover has persisted to the most recent

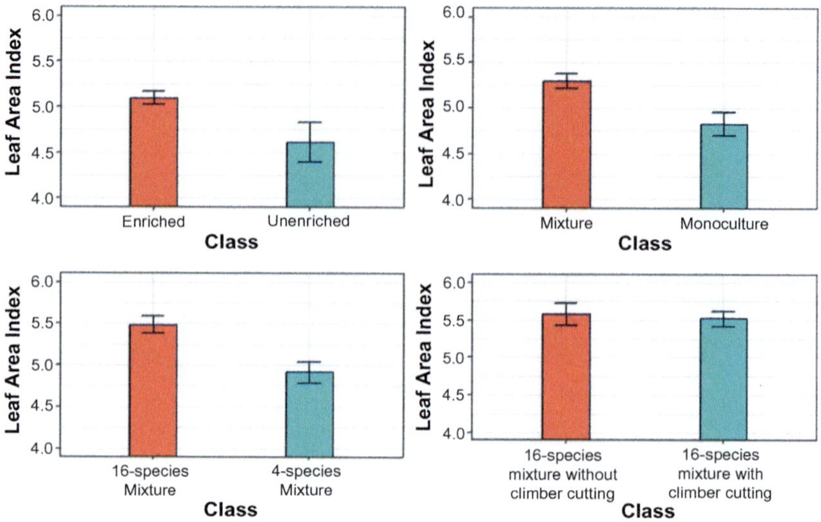

Fig. 7 RapidEye-based leaf area index (LAI) of different enrichment planting groups with 95% confidence intervals in August 2012.

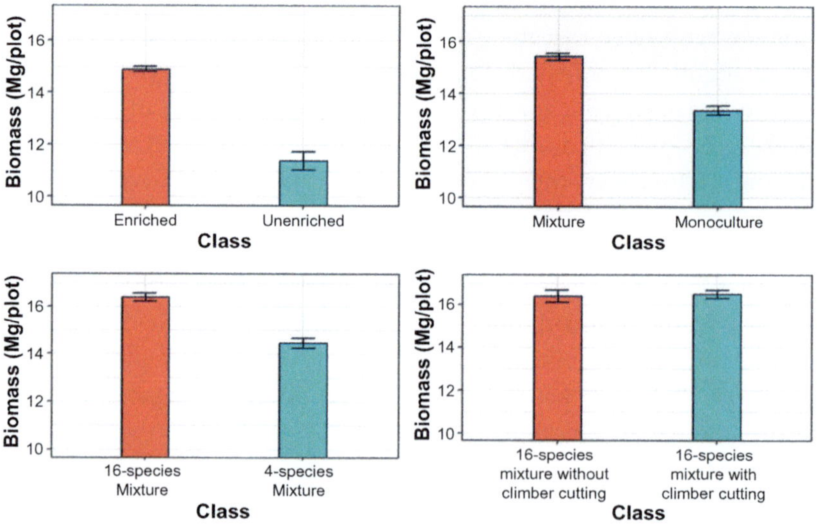

Fig. 8 RapidEye-based biomass of different enrichment planting groups with 95% confidence intervals in August 2012.

measurements (Fig. 5). The RapidEye-based vegetation cover also showed that the vegetation cover of mixture plots was higher than that of monoculture plots (Fig. 6). Moreover, we found mixture plots exhibited differences of 0.47 LAI units and 2 Mg/plot biomass higher than monoculture plots (Figs 7 and 8).

In the Landsat data, the vegetation cover of 16-species mixture plots became higher than that of 4-species mixture plots after the enrichment planting, a difference that became wider over time (Fig. 5). For the RapidEye-based vegetation cover, LAI, and biomass, the 16-species mixture plots also generally exhibit higher cover estimates (Figs 6–8).

3.4 Effects of enhanced climber cutting on vegetation canopy attributes

For the Landsat vegetation cover data, with the different enrichment planting strategies, the plots of 16-species mixtures with enhanced climber cutting exhibited similar vegetation cover to that of 16-species mixture plots without climber cutting (Fig. 5). The vegetation cover of the 16-species mixtures both with and without enhanced climber cutting exhibited a 6% increase from 1999 to 2012. For the RapidEye-based vegetation cover data, the plots of 16-species mixture with climber cutting exhibited lower vegetation cover than 16-species mixture plots without climber cutting, which was not consistent to the result we get from Landsat vegetation cover (Fig. 6). However, there was no clear effect of climber cutting on the RapidEye estimates of biomass and LAI (Figs 7 and 8). Overall, the effects of enhanced climber cutting were more mixed that those for enrichment planting - probably due to the inability to distinguish between tree cover and liana cover.

4. Discussion

In this study, we used Sabah Biodiversity Experiment as a test case to assess to what degree easily-accessible satellite remote sensing imagery could assess the effects of degradation and restoration on indicators of forest canopy quality. In general, our results confirm the ability of the satellite remote sensing to detect both the effects of forest degradation and of different restoration practices (although the ability of different satellite remote sensing products to distinguish the climber cutting treatments was inconsistent). In particular, our results suggest that enrichment planting enhances restoration and that diverse mixtures of tree seedlings are more effective than using single-species treatments.

4.1 Impacts of selective logging

Previous studies have assessed the process of forest degradation using a variety of remote sensing approaches to estimate the extent and impacts of selective logging on forest quality (Asner et al., 2002, 2004a, 2005; de Wasseige and Defourny, 2004; Matricardi et al., 2010; Souza Jr and Barreto, 2000). In this study, we used Sabah Biodiversity Experiment as a test case to assess to what degree easily available satellite remote sensing imagery could assess the effects of degradation and restoration on indicators of forest quality. We found the vegetation cover of the Malua Forest Reserve and its most heavily logged area exhibited 3% and 7% decreases, respectively, in the years after the 2007 relogging, while the vegetation cover of Sabah Biodiversity Experiment exhibited a 5% increase during the same period confirming that satellite remote sensing was able to distinguish between areas of selectively logged forest that had been subjected to a second round of selective logging or not. Although these numerical differences between Malua Forest Reserve and Sabah Biodiversity Experiment are consistent with what we would expect following relogging (or not) we raise a note of caution—because visible light does not penetrate leaves (Rees, 2013) optical remote sensing data may underestimate effects of selective logging on the multilayered structure of tropical forests. Furthermore, with selective logging, it is typically only the larger, highly valuable trees that are removed from the forest (Asner et al., 2004a, 2005; Edwards et al., 2014; Lambert, 1992; Laporte et al., 2007; Verissimo et al., 1995) and the indicators of canopy cover and related variables used in our study may be reduced less than other properties that the satellite data available to us could not address as well—stocks of carbon for example. It is known that in tropical rainforests, selective logging can result in smaller changes in cover than clear-cut logging because the logged areas remain covered by leaves of surrounding trees when the harvested trees are removed (Gatti et al., 2015; Souza Jr et al., 2005; Thiollay, 1992). Thus, in this study, the effects of selective logging detected using the 30-by-30 m and 500-by-500 m resolution remote sensing data reflects only the places where no trees are found around or under the selectively logged trees. While the vegetation cover and LAI data in this study generally give consistent results it is important to note that these measures are only expected to agree under limited circumstances. For example, if an area of forest were (hypothetically) covered by a single layer of leaves it would have a vegetation cover of 100% and an LAI of one. However, tropical forests are often characterized by tall, multilayered canopies meaning that LAI can range up to several LAI units while vegetation cover reaches its maximum at 100%.

We estimate that Malua Forest Reserve lost approximately 2.7 million Mg of biomass from the 11,433 ha deforested area from 2000 to 2012 due to the relogging in 2007, while the Sabah Biodiversity Experiment lost none during the same period (as expected since it was protected from the relogging). This loss of biomass through selective logging is qualitatively consistent with the result from field estimates (although at the higher end of previous estimates), which found that the initial selective logging in the 1980s decreased the dipterocarp stock of the Sabah Biodiversity Experiment by 55–66% in Malua Forest Reserve compared to the old growth forest in nearby Danum Valley conservation area even 22 years after logging (Saner et al., 2012).

The differences in tree cover and LAI after the relogging become smaller within a few years, presumably due to growth of unlogged vegetation and new recruits (dipterocarps do not re-sprout from the stump when cut). This is consistent with previous studies that have found rapid recovery of canopy cover estimates in some tropical rainforests, where, for example, half of the logging trails could no longer be seen from satellite remote sensing data about 5 years after the logging activities because of a decrease in the spectral difference between the forest and the trails (de Wasseige and Defourny, 2004; Gatti et al., 2015). Overall, while our remote sensing analysis was able to distinguish canopy characteristics of the area of Malua Forest Reserve that had been relogged from the areas that had not, the approach may underestimate effects of selective logging on other aspects of forest quality (e.g. carbon stocks).

4.2 The effects of enrichment planting

Enrichment planting is intended to increase seedling establishment and recovery of forests by planting seedlings of the timber tree species that have been selectively logged. Enrichment planting is generally done in poorly-stocked areas of logged-over forests and associated degraded area (abandoned logging roads and log-yards, abandoned shifting cultivation area, etc.) (Chan et al., 2008). A total of 25,857 ha of forest plantation had been created using enrichment planting techniques in peninsular Malaysia (Safa et al., 2004). Previous field studies have found that enrichment planting improved the soil biological status where biomass C and N content were significantly higher than in unenriched areas (Karam et al., 2012). However, few studies have used remote sensing data to assess the effects of enrichment planting. In this study, we found enrichment planting improved vegetation canopy attributes where the vegetation cover, LAI, and biomass were higher than unenriched forest, and the difference in vegetation cover increased over time. This is probably the first regional-scale study utilizing satellite observations to quantify the effects of enrichment planting in Sabah, Malaysia.

4.3 The potential value of diversification of enrichment planting schemes

Most enrichment planting restores areas with single-species planting or low-diversity mixtures. However, the demonstration of a general positive relationship between biodiversity and ecosystem functioning raises the potential value of diversifying enrichment planting (Cardinale et al., 2006, 2007, 2011; César et al., 2016; Duffy, 2009; Forshed et al., 2008; Gravel et al., 2011; Hector, 2009; Hector et al., 2010; Hector and Bagchi, 2007; Hector and Hooper, 2002; Lussetti et al., 2016; Schnitzer et al., 2004; Yachi and Loreau, 1999). In this study, we found that multi-species mixtures exhibited higher vegetation cover, LAI, and biomass than that of monoculture plots. In addition, the LAI of multi-species mixture plots increased more rapidly over time than that of monocultures. That could be because an ecosystem with a high species richness functions more effectively in terms of resource capture and cycling, and higher levels of productivity (Huang et al., 2018; Liang et al., 2016). Indeed, our results provide some support for higher vegetation cover, LAI and biomass values of 16-species mixture plots, compared with 4-species mixture plots (although the degree of differences varied with response variable and source of remote sensing data—see figures).

The assessment of changes in vegetation canopy attributes also depends on the resolution of the remote sensing data (Benediktsson et al., 2012). We found 5 m resolution remote sensing data provided more three-dimensional structure information of LAI, compared with 30 m or 500 m resolution data. For example, the 5 m resolution RapidEye-based metrics, produced a larger difference in LAI between mixture and monoculture plots than the smaller difference in vegetation cover from the coarser resolution data.

4.4 The consequences of liana removal

Some studies have found that climber cutting can speed-up restoration of selectively logged forests (César et al., 2016; Forshed et al., 2008; Lussetti et al., 2016; Schnitzer et al., 2004). In this study, there was some inconsistency between the different remote sensing data: the Landsat vegetation cover and RapidEye-based vegetation cover showed inconsistent and sometimes opposing results about the effects of climber cutting. This is probably because Landsat vegetation cover 2010 shows the vegetation cover average condition from 2008 to 2012, while the vegetation cover RapidEye image is the condition of that day in August 2012. The image quality for a single day is uncertain, since it may be affected by the weather, tree condition, sensor viewing and sun illumination angles of that day, while these effects can be minimized by integrating long-term images (e.g. Landsat vegetation cover 2010). In addition, under normal climate conditions (excluding ENSO events) climber cutting can decrease the mortality of dipterocarp species and increase the compensatory regrowth of their canopies (Lussetti et al., 2016).

4.5 Limitations and future research

Although the high accuracy and high-resolution quantitative remote sensing datasets used here lead to better performance on terrestrial vegetation monitoring compared with previous moderate-resolution or qualitative remote sensing datasets (Hansen et al., 2010, 2013; Hansen and DeFries, 2004; Wilson and Jetz, 2016; Xie et al., 2008), they still face two known limitations. Firstly, compared with the in-situ ecology data, the spatial resolution of current remote sensing datasets still requires improvements, especially for specific tree species monitoring. Secondly, the performance of current quantitative remote sensing datasets is relatively limited when being used to detect three-dimension tree structure, such as tree height and diameter at breast height which are important factors for monitoring vegetation structure (Carlson and Ripley, 1997; Lefsky et al., 2002; Turner et al., 1999; Waring et al., 1995). For example, the monoculture plots and the 4-species mixture plots both contain different compositions (Table S1 in the online version at https://doi.org/10.1016/bs.aecr.2020.01.005), but each is replicated only twice and we therefore did not compare them using the satellite data available for this study. To detect the performance of each composition, it is necessary to add airborne LiDAR data for tree structure monitoring. However, there are few airborne LiDAR data continuously available for vegetation monitoring on a large scale. During recent years, there are some studies that integrate airborne LiDAR data and optical remote sensing data to develop high-quality products (Lefsky, 2010; Lim et al., 2003; Simard et al., 2011). Progress has been made by ecologists and geographers in Malaysian Borneo (Asner et al., 2018). The advantage of these methods is that they can get more structural information from earth's surface, though the spatial-temporal coverage of these products may be limited. Future research will consider the development of long-term high-resolution remote sensing datasets when coupled with the LiDAR data to improve the ability of comprehensive vegetation monitoring. This remote sensing data will need appropriate validation by in-situ field measurements, something that is absent in this study. The replication and randomization of treatments within the blocked design of Sabah Biodiversity Experiment should ensure that the ability of the satellite remote sensing to discriminate among the different plots is reliable but validation with field measurements will provide additional assurance as well as being essential in order to understand the biological processes underlying the different treatment effects.

5. Conclusion

In this study, we used a large scale, long-term field manipulation (Sabah Biodiversity Experiment) as a test of the ability of high spatial resolution satellite imagery and advanced remote sensing products to assess the effects of selective logging and of different restoration treatments on forest canopy quality. In general, our study was able to detect differences in estimated vegetation cover, Leaf Area Index (LAI), and biomass due to selective logging as well as differences among the different forest restoration treatments, namely enrichment planting, climber cutting and natural (unassisted) recovery. In particular, our results suggest that enrichment planting enhances restoration and that diverse mixtures of tree species are more effective than single-species plantings. However, the monitoring ability of optical remote sensing data to detect these impacts is still limited. In the longer-term, LiDAR will allow more comprehensive assessments of forest restoration. Remote sensing measurements of forest quality will also need to be accompanied by field measurements to validate the treatment effects and to better understand the biological processes affecting forest degradation and restoration.

Glossary

Landsat Landsat refers a series of artificial satellites that monitor the earth's resources by photographing the surface at different wavelengths and resolutions. Since 1972, eight Landsat satellites have been launched. The Landsat program is the longest-running enterprise for acquisition of satellite imagery of Earth.

MODIS (Moderate Resolution Imaging Spectroradiometer) MODIS (or Moderate Resolution Imaging Spectroradiometer) is a key instrument aboard the Terra and Aqua satellites. The instrument capture data in 36 spectral bands ranging in wavelength from 0.4 to 14.4 μm and at varying spatial resolutions (2 bands at 250 m, 5 bands at 500 m and 29 bands at 1 km).

RapidEye RapidEye refers to the constellation of five earth observation satellites owned and operated by Planet Labs. Each sensor capture data in five distinct bands of the electromagnetic spectrum: Blue (440–510 nm), Green (520–590 nm), Red (630–690 nm), Red-Edge (690–730 nm) and Near-Infrared (760–880 nm), and at varying spatial resolution of 5 m.

Leaf Area Index (LAI) The total one-sided green leaf area per unit ground surface area. Leaves are the major eco-physiological parts of a plant that interact with the atmosphere, including absorbing and assimilating carbon dioxide, intercepting light necessary for photosynthesis, releasing oxygen that is formed as a byproduct of photosynthesis and so on. LAI is a reliable parameter for plant growth.

Vegetation cover The fraction or percentage of the unit ground surface covered by green leaves.

Gross Primary Productivity (GPP) GPP refers the rate of how an ecosystem's producers or autotrophs collect and save a certain amount of chemical energy referred to as biomass at a specific time. In other words, it's the rate at which energy is stored as biomass by plants or other primary producers and made available to the consumers in the ecosystem.

Spatial resolution Spatial resolution measures the smallest angular separation between two objects. For satellite images, this is represented in pixels and the spatial resolution for a given image is noted as how many metres that pixel represents.

Temporal resolution Temporal resolution refers to the frequency at which imagery is recorded for a particular area.

Atmospheric correction Atmospheric correction refers the process of removing the effects of the atmosphere on the reflectance values of images taken by satellite or airborne sensors.

ArcGIS ArcGIS is a geographic information system (GIS) for working with maps and geographic information maintained by Esri.

ESRI Esri is an international supplier of geographic information system (GIS) software, web GIS and geodatabase management applications.

ENVI ENVI is an image analysis software for working with remote sensing and geographic information maintained by L3Harris Geospatial.

Acknowledgements

Thanks to all the many people who have supported and worked on Sabah Biodiversity Experiment. Our thanks also to Tommaso Jucker and two anonymous reviewers for their constructive criticism of earlier versions of this chapter. J.W. was supported by the

National Key Research and Development Program of China (2016YFA0600101) and the China-UK Joint Research and Innovation Partnership Fund (China Scholarship Council). M.J.O'B. was supported by the Atracción de Talento Investigador Modalidad I Fellowship from the Comunidad de Madrid (2018-T1/AMB-11095) and the University of Zurich URPP GCB. A.H. was supported by the NERC Human Modified Tropical Forests thematic programme (Biodiversity and Land-use Impacts on tropical ecosystem function consortium). This chapter is a product of the South East Asia Rainforest Research Partnership programme and Sabah Biodiversity Experiment publication 21.

References

Alvarado, E., Sandberg, D., 2001. Logging in Tropical Forest: Literature Review on Ecological Impacts. University of Washington, USDA Forest Service.

Asner, G.P., Keller, M., Pereira Jr., R., Zweede, J.C., 2002. Remote sensing of selective logging in Amazonia: assessing limitations based on detailed field observations, Landsat ETM+, and textural analysis. Remote Sens. Environ. 80, 483–496.

Asner, G.P., Keller, M., Pereira Jr., R., Zweede, J.C., Silva, J.N., 2004a. Canopy damage and recovery after selective logging in Amazonia: field and satellite studies. Ecol. Appl. 14, 280–298.

Asner, G.P., Keller, M., Silva, J.N., 2004b. Spatial and temporal dynamics of forest canopy gaps following selective logging in the eastern Amazon. Glob. Chang. Biol. 10, 765–783.

Asner, G.P., Knapp, D.E., Broadbent, E.N., Oliveira, P.J., Keller, M., Silva, J.N., 2005. Selective logging in the Brazilian Amazon. Science 310, 480–482.

Asner, G.P., Brodrick, P.G., Philipson, C., Vaughn, N.R., Martin, R.E., Knapp, D.E., Heckler, J., Evans, L.J., Jucker, T., Goossens, B., 2018. Mapped aboveground carbon stocks to advance forest conservation and recovery in Malaysian Borneo. Biol. Conserv. 217, 289–310.

Baccini, A., Walker, W., Farina, M., Houghton, R.A., 2016. CMS: Estimated Deforested Area Biomass, Tropical America, Africa, and Asia, 2000. ORNL Distributed Active Archive Center.

Benediktsson, J.A., Chanussot, J., Moon, W.M., 2012. Very high-resolution remote sensing: challenges and opportunities [point of view]. Proc. IEEE 100, 1907–1910.

Bréda, N.J., 2003. Ground-based measurements of leaf area index: a review of methods, instruments and current controversies. J. Exp. Bot. 54, 2403–2417.

Brede, B., Gastellu-Etchegorry, J.-P., Lauret, N., Baret, F., Clevers, J., Verbesselt, J., Herold, M., 2018. Monitoring forest phenology and leaf area index with the autonomous, low-cost transmittance sensor PASTiS-57. Remote Sens. (Basel) 10, 1032.

Cardinale, B.J., Srivastava, D.S., Duffy, J.E., Wright, J.P., Downing, A.L., Sankaran, M., Jouseau, C., 2006. Effects of biodiversity on the functioning of trophic groups and ecosystems. Nature 443, 989.

Cardinale, B.J., Wright, J.P., Cadotte, M.W., Carroll, I.T., Hector, A., Srivastava, D.S., Loreau, M., Weis, J.J., 2007. Impacts of plant diversity on biomass production increase through time because of species complementarity. Proc. Natl. Acad. Sci. 104, 18123–18128.

Cardinale, B.J., Matulich, K.L., Hooper, D.U., Byrnes, J.E., Duffy, E., Gamfeldt, L., Balvanera, P., O'connor, M.I., Gonzalez, A., 2011. The functional role of producer diversity in ecosystems. Am. J. Bot. 98, 572–592.

Carlson, T.N., Ripley, D.A., 1997. On the relation between NDVI, fractional vegetation cover, and leaf area index. Remote Sens. Environ. 62, 241–252.

César, R.G., Holl, K.D., Girão, V.J., Mello, F.N., Vidal, E., Alves, M.C., Brancalion, P.H., 2016. Evaluating climber cutting as a strategy to restore degraded tropical forests. Biol. Conserv. 201, 309–313.

Chambers, J.Q., Asner, G.P., Morton, D.C., Anderson, L.O., Saatchi, S.S., Espírito-Santo, F.D., Palace, M., Souza Jr., C., 2007. Regional ecosystem structure and function: ecological insights from remote sensing of tropical forests. Trends Ecol. Evol. 22, 414–423.

Chan, H.T., Shamsudin, I., Ismail, P., 2008. An In-Depth Look at Enrichment Planting (Malaysian Forest Records No. 47). Forest Research Institute Malaysia, Kuala Lumpur, Malaysia.

Chander, G., Haque, M.O., Sampath, A., Brunn, A., Trosset, G., Hoffmann, D., Roloff, S., Thiele, M., Anderson, C., 2013. Radiometric and geometric assessment of data from the RapidEye constellation of satellites. Int. J. Remote Sens. 34, 5905–5925.

Chen, J., Chen, J., Liao, A., Cao, X., Chen, L., Chen, X., He, C., Han, G., Peng, S., Lu, M., 2015. Global land cover mapping at 30 m resolution: a POK-based operational approach. ISPRS J. Photogramm. Remote Sens. 103, 7–27.

Chen, C., Park, T., Wang, X., Piao, S., Xu, B., Chaturvedi, R.K., Fuchs, R., Brovkin, V., Ciais, P., Fensholt, R., Tømmervik, H., Bala, G., Zhu, Z., Nemani, R.R., Myneni, R.B., 2019. China and India lead in greening of the world through land-use management. Nat. Sustain. 2, 122–129.

Claverie, M., Matthews, J., Vermote, E., Justice, C., 2016. A 30+ year AVHRR LAI and FAPAR climate data record: algorithm description and validation. Remote Sens. (Basel) 8, 263.

de Souza, E.G., Scharf, P.C., Sudduth, K.A., 2010. Sun position and cloud effects on reflectance and vegetation indices of corn. Agron. J. 102, 734–744.

de Wasseige, C., Defourny, P., 2004. Remote sensing of selective logging impact for tropical forest management. For. Ecol. Manage. 188, 161–173.

Duffy, J.E., 2009. Why biodiversity is important to the functioning of real-world ecosystems. Front. Ecol. Environ. 7, 437–444.

Edwards, D.P., Tobias, J.A., Sheil, D., Meijaard, E., Laurance, W.F., 2014. Maintaining ecosystem function and services in logged tropical forests. Trends Ecol. Evol. 29, 511–520.

Feng, M., Sexton, J.O., Channan, S., Townshend, J.R., 2016. A global, high-resolution (30-m) inland water body dataset for 2000: first results of a topographic–spectral classification algorithm. Int. J. Digital Earth 9, 113–133.

Foody, G.M., Cutler, M.E., 2003. Tree biodiversity in protected and logged Bornean tropical rain forests and its measurement by satellite remote sensing. J. Biogeogr. 30, 1053–1066.

Forshed, O., Karlsson, A., Falck, J., Cedergren, J., 2008. Stand development after two modes of selective logging and pre-felling climber cutting in a dipterocarp rainforest in Sabah, Malaysia. For. Ecol. Manage. 255, 993–1001.

Friedl, M.A., McIver, D.K., Hodges, J.C., Zhang, X.Y., Muchoney, D., Strahler, A.H., Woodcock, C.E., Gopal, S., Schneider, A., Cooper, A., 2002. Global land cover mapping from MODIS: algorithms and early results. Remote Sens. Environ. 83, 287–302.

Gallardo-Cruz, J.A., Meave, J.A., González, E.J., Lebrija-Trejos, E.E., Romero-Romero, M.A., Pérez-GarcíA, E.A., Gallardo-Cruz, R., Hernández-Stefanoni, J.L., Martorell, C., 2012. Predicting tropical dry forest successional attributes from space: is the key hidden in image texture? PLoS One 7, e30506.

Gao, B.-C., Li, R.-R., 2000. Quantitative improvement in the estimates of NDVI values from remotely sensed data by correcting thin cirrus scattering effects. Remote Sens. Environ. 74, 494–502.

Gatti, R.C., Castaldi, S., Lindsell, J.A., Coomes, D.A., Marchetti, M., Maesano, M., DI Paola, A., Paparella, F., Valentini, R., 2015. The impact of selective logging and clearcutting on forest structure, tree diversity and above-ground biomass of African tropical forests. Ecol. Res. 30, 119–132.

Gravel, D., Bell, T., Barbera, C., Bouvier, T., Pommier, T., Venail, P., Mouquet, N., 2011. Experimental niche evolution alters the strength of the diversity–productivity relationship. Nature 469, 89.

Gutman, G., Byrnes, R.A., Masek, J., Covington, S., Justice, C., Franks, S., Headley, R., 2008. Towards monitoring land-cover and land-use changes at a global scale: the global land survey 2005. Photogramm. Eng. Remote. Sens. 74, 6–10.

Gutman, G., Huang, C., Chander, G., Noojipady, P., Masek, J.G., 2013. Assessment of the NASA–USGS global land survey (GLS) datasets. Remote Sens. Environ. 134, 249–265.

Hansen, M.C., DeFries, R.S., 2004. Detecting long-term global forest change using continuous fields of tree-cover maps from 8-km advanced very high resolution radiometer (AVHRR) data for the years 1982–99. Ecosystems 7, 695–716.

Hansen, M.C., Stehman, S.V., Potapov, P.V., 2010. Quantification of global gross forest cover loss. Proc. Natl. Acad. Sci. 107, 8650–8655.

Hansen, M.C., Potapov, P.V., Moore, R., Hancher, M., Turubanova, S., Tyukavina, A., Thau, D., Stehman, S., Goetz, S., Loveland, T., 2013. High-resolution global maps of 21st-century forest cover change. Science 342, 850–853.

Hector, A., 2009. Darwin's 'principle of divergence' and the link between biodiversity and ecosystem functioning. In: Stocklin, J., Hoxterman, E. (Eds.), Darwin und die botanik. Vierteljahrsschrift der Naturforschenden Gesellschaft in Zürich, pp. 182–191.

Hector, A., 2011. Ecology: diversity favours productivity. Nature 472, 45.

Hector, A., Bagchi, R., 2007. Biodiversity and ecosystem multifunctionality. Nature 448, 188.

Hector, A., Hooper, R., 2002. Darwin and the first ecological experiment. Science 295, 639–640.

Hector, A., Hautier, Y., Saner, P., Wacker, L., Bagchi, R., Joshi, J., Scherer-Lorenzen, M., Spehn, E.M., Bazeley-White, E., Weilenmann, M., 2010. General stabilizing effects of plant diversity on grassland productivity through population asynchrony and overyielding. Ecology 91, 2213–2220.

Hector, A., Philipson, C., Saner, P., Chamagne, J., Dzulkifli, D., O'Brien, M., Snaddon, J.L., Ulok, P., Weilenmann, M., Reynolds, G., 2011. The Sabah biodiversity experiment: a long-term test of the role of tree diversity in restoring tropical forest structure and functioning. Philos. Trans. R. Soc. Lond. B Biol. Sci. 366, 3303–3315.

Huang, Y., Chen, Y., Castro-Izaguirre, N., Baruffol, M., Brezzi, M., Lang, A., Li, Y., Härdtle, W., Von Oheimb, G., Yang, X., 2018. Impacts of species richness on productivity in a large-scale subtropical forest experiment. Science 362, 80–83.

Ju, J., Roy, D.P., 2008. The availability of cloud-free Landsat ETM+ data over the conterminous United States and globally. Remote Sens. Environ. 112, 1196–1211.

Jung-Rothenhäusler, F., Weichelt, H., Pach, M., 2007. RapidEye—a novel approach to space borne geo-information solutions. In: Proceedings of ISPRS Hannover Workshop, pp. 1–4.

Karam, D., Arifin, A., Radziah, O., Shamshuddin, J., Majid, N., Hazandy, A., Zahari, I., Nor Halizah, A., Rui, T., 2012. Impact of long-term forest enrichment planting on the biological status of soil in a deforested dipterocarp forest in Perak, Malaysia. Scientific World Journal 2012, 641346.

Kerr, J.T., Ostrovsky, M., 2003. From space to species: ecological applications for remote sensing. Trends Ecol. Evol. 18, 299–305.

Knyazikhin, Y., 1999. MODIS Leaf Area Index (LAI) and Fraction of Photosynthetically Active Radiation Absorbed by Vegetation (FPAR) Product (MOD 15) Algorithm Theoretical Basis Document. http://eospso.gsfc.nasa.gov/atbd/modistabls.heml.

Lambert, F., 1992. The consequences of selective logging for Bornean lowland forest birds. Philos. Trans. R. Soc. Lond. B Biol. Sci. 335, 443–457.

Laporte, N.T., Stabach, J.A., Grosch, R., Lin, T.S., Goetz, S.J., 2007. Expansion of industrial logging in Central Africa. Science 316, 1451.

Lefsky, M.A., 2010. A global forest canopy height map from the moderate resolution imaging spectroradiometer and the geoscience laser altimeter system. Geophys. Res. Lett. 37, 15.

Lefsky, M.A., Cohen, W.B., Parker, G.G., Harding, D.J., 2002. Lidar remote sensing for ecosystem studies: lidar, an emerging remote sensing technology that directly measures the three-dimensional distribution of plant canopies, can accurately estimate vegetation structural attributes and should be of particular interest to forest, landscape, and global ecologists. AIBS Bull. 52, 19–30.

Li, C., Wulf, H., Schmid, B., He, J.-S., Schaepman, M.E., 2018. Estimating plant traits of alpine grasslands on the Qinghai-Tibetan plateau using remote sensing. IEEE J. Sel. Top. Appl. Earth Obs. Remote Sens. 11, 2263–2275.

Liang, S., 2004. Quantitative Remote Sensing of Land Surfaces. John Wiley & Sons.

Liang, S., Li, X., Wang, J., 2012. Advanced Remote Sensing: Terrestrial Information Extraction and Applications. Academic Press.

Liang, S., Zhao, X., Liu, S., Yuan, W., Cheng, X., Xiao, Z., Zhang, X., Liu, Q., Cheng, J., Tang, H., 2013. A long-term Global LAnd Surface Satellite (GLASS) data-set for environmental studies. Int. J. Digital Earth 6, 5–33.

Liang, J., Crowther, T.W., Picard, N., Wiser, S., Zhou, M., Alberti, G., Schulze, E.-D., McGuire, A.D., Bozzato, F., Pretzsch, H., 2016. Positive biodiversity-productivity relationship predominant in global forests. Science 354, aaf8957.

Liang, S., Shi, J., Yan, G., 2018. Recent Progress in Quantitative Land Remote Sensing in China. Multidisciplinary Digital Publishing Institute.

Lim, K., Treitz, P., Wulder, M., St-Onge, B., Flood, M., 2003. LiDAR remote sensing of forest structure. Prog. Phys. Geogr. 27, 88–106.

Liou, K.-N., 1992. Radiation and Cloud Processes in the Atmosphere. Theory, Observation, and Modeling. Oxford University Press.

Lussetti, D., Axelsson, E.P., Ilstedt, U., Falck, J., Karlsson, A., 2016. Supervised logging and climber cutting improves stand development: 18 years of post-logging data in a tropical rain forest in Borneo. For. Ecol. Manage. 381, 335–346.

Marsh, C.W., Greer, A.G., 1992. Forest land-use in Sabah, Malaysia: an introduction to Danum Valley. Philos. Trans. R. Soc. Lond. B Biol. Sci. 335, 331–339.

Matricardi, E.A., Skole, D.L., Pedlowski, M.A., Chomentowski, W., Fernandes, L.C., 2010. Assessment of tropical forest degradation by selective logging and fire using Landsat imagery. Remote Sens. Environ. 114, 1117–1129.

Miller, P., Kumar, A., Mote, T., Moraes, F., Mishra, D., 2019. Persistent hydrological consequences of hurricane Maria in Puerto Rico. Geophys. Res. Lett. 46 (3), 1413–1422.

Millet, J., Tran, N., Ngoc, N.V., Thi, T.T., Prat, D., 2013. Enrichment planting of native species for biodiversity conservation in a logged tree plantation in Vietnam. New For. 44, 369–383.

Muraoka, H., Koizumi, H., 2009. Satellite ecology (SATECO)—linking ecology, remote sensing and micrometeorology, from plot to regional scale, for the study of ecosystem structure and function. J. Plant Res. 122, 3–20.

Myneni, R., Knyazikhin, Y., Park, T., 2015. MCD15A3H MODIS/Terra+Aqua Leaf Area Index/FPAR 4-day L4 Global 500m SIN Grid V006. [Dataset], NASA EOSDIS Land Processes DAAC. https://doi.org/10.5067/MODIS/MCD15A3H.006. (Accessed 9 February 2020).

Myneni, R.B., Hoffman, S., Knyazikhin, Y., Privette, J., Glassy, J., Tian, Y., Wang, Y., Song, X., Zhang, Y., Smith, G., 2002. Global products of vegetation leaf area and fraction absorbed PAR from year one of MODIS data. Remote Sens. Environ. 83, 214–231.

Nepstad, D.C., Verssimo, A., Alencar, A., Nobre, C., Lima, E., Lefebvre, P., Schlesinger, P., Potter, C., Moutinho, P., Mendoza, E., 1999. Large-scale impoverishment of Amazonian forests by logging and fire. Nature 398, 505.

O'Brien, M.J., Philipson, C.D., Reynolds, G., Dzulkifli, D., Snaddon, J.L., Ong, R., Hector, A., 2019. Positive effects of liana cutting on seedlings are reduced during El Niño-induced drought. J. Appl. Ecol. 56, 891–901.

Perez, L., Nelson, T.A., Coops, N.C., Fontana, F., Drever, C.R., 2016. Characterization of spatial relationships between three remotely sensed indirect indicators of biodiversity and climate: a 21years' data series review across the Canadian boreal forest. Int. J. Digital Earth 9 (7), 676–696.

Pfeifer, M., Kor, L., Nilus, R., Turner, E., Cusack, J., Lysenko, I., Khoo, M., Chey, V., Chung, A., Ewers, R., 2016. Mapping the structure of Borneo's tropical forests across a degradation gradient. Remote Sens. Environ. 176, 84–97.

Qi, J., Chehbouni, A., Huete, A., Kerr, Y., Sorooshian, S., 1994. A modified soil adjusted vegetation index. Remote Sens. Environ. 48, 119–126.

Ranga Myneni, K., Yuri, T.P., 2015. MOD15A3H MODIS/Combined Terra + Aqua Leaf Area Index/FPAR Daily L4 Global 500m SIN Grid.

Rees, W.G., 2013. Physical Principles of Remote Sensing. Cambridge University Press.

Rocchini, D., Ricotta, C., Chiarucci, A., 2007. Using satellite imagery to assess plant species richness: the role of multispectral systems. Appl. Veg. Sci. 10, 325–331.

Running, S.W., Thornton, P.E., Nemani, R., Glassy, J.M., 2000. Global terrestrial gross and net primary productivity from the earth observing system. In: Methods in Ecosystem Science. Springer.

Safa, M.S., Ibrahim, Z., Abdu, A., 2004. Potentialities of New Line Planting Technique of Enrichment Planting in Peninsular Malaysia: a Review of Resource Sustainability and Economic Feasibility. In: KUSTEM 3rd Annual Seminal, Kuala Terenganu, Terenganu Darul Iman, 4–5 May, Malaysia.

Saner, P., 2009. Ecosystem Carbon Dynamics in Logged Forest of Malaysian Borneo. PhD thesis, University of Zurich, Zurich, Switzerland.

Saner, P., Loh, Y.Y., Ong, R.C., Hector, A., 2012. Carbon stocks and fluxes in tropical lowland dipterocarp rainforests in Sabah, Malaysian Borneo. PLoS One 7, e29642.

Schnitzer, S.A., Parren, M.P., Bongers, F., 2004. Recruitment of lianas into logging gaps and the effects of pre-harvest climber cutting in a lowland forest in Cameroon. For. Ecol. Manage. 190, 87–98.

Sexton, J.O., Song, X.-P., Feng, M., Noojipady, P., Anand, A., Huang, C., Kim, D.-H., Collins, K.M., Channan, S., Dimiceli, C., 2013a. Global, 30-m resolution continuous fields of tree cover: landsat-based rescaling of MODIS vegetation continuous fields with lidar-based estimates of error. Int. J. Digital Earth 6, 427–448.

Sexton, J.O., Song, X.-P., Huang, C., Channan, S., Baker, M.E., Townshend, J.R., 2013b. Urban growth of the Washington, DC–Baltimore, MD metropolitan region from 1984 to 2010 by annual, Landsat-based estimates of impervious cover. Remote Sens. Environ. 129, 42–53.

Simard, M., Pinto, N., Fisher, J.B., Baccini, A., 2011. Mapping forest canopy height globally with spaceborne lidar. J. Geophys. Res. Biogeo. 2005–2012, 116.

Simpson, J.J., Jin, Z., Stitt, J.R., 2000. Cloud shadow detection under arbitrary viewing and illumination conditions. IEEE Trans. Geosci. Remote Sens. 38, 972–976.

Song, X.-P., Sexton, J.O., Huang, C., Channan, S., Townshend, J.R., 2016. Characterizing the magnitude, timing and duration of urban growth from time series of Landsat-based estimates of impervious cover. Remote Sens. Environ. 175, 1–13.

Souza Jr., C., Barreto, P., 2000. An alternative approach for detecting and monitoring selectively logged forests in the Amazon. Int. J. Remote Sens. 21, 173–179.

Souza Jr., C.M., Roberts, D.A., Cochrane, M.A., 2005. Combining spectral and spatial information to map canopy damage from selective logging and forest fires. Remote Sens. Environ. 98, 329–343.

Thiollay, J.-M., 1992. Influence of selective logging on bird species diversity in a Guianan rain forest. Conserv. Biol. 6, 47–63.

Tigabu, M., Savadogo, P., Odén, P.C., Xayvongsa, L., 2010. Enrichment planting in a logged-over tropical mixed deciduous forest of Laos. J. For. Res. 21, 273–280.

Tuck, S.L., O'Brien, M.J., Philipson, C.D., Saner, P., Tanadini, M., Dzulkifli, D., Godfray, H.C.J., Godoong, E., Nilus, R., Ong, R.C., 2016. The value of biodiversity for the functioning of tropical forests: insurance effects during the first decade of the Sabah biodiversity experiment. Proc. R. Soc. B Biol. Sci. 283, 20161451.

Turner, W., 2014. Sensing biodiversity. Science 346, 301–302.

Turner, D.P., Cohen, W.B., Kennedy, R.E., Fassnacht, K.S., Briggs, J.M., 1999. Relationships between leaf area index and Landsat TM spectral vegetation indices across three temperate zone sites. Remote Sens. Environ. 70, 52–68.

Turner, W., Spector, S., Gardiner, N., Fladeland, M., Sterling, E., Steininger, M., 2003. Remote sensing for biodiversity science and conservation. Trends Ecol. Evol. 18, 306–314.

Tyc, G., Tulip, J., Schulten, D., Krischke, M., Oxfort, M., 2005. The RapidEye mission design. Acta Astronaut. 56, 213–219.

Verissimo, A., Barreto, P., Tarifa, R., Uhl, C., 1995. Extraction of a high-value natural resource in Amazonia: the case of mahogany. For. Ecol. Manage. 72, 39–60.

Waring, R.H., Way, J., Hunt, E.R., Morrissey, L., Ranson, K.J., Weishampel, J.F., Oren, R., Franklin, S.E., 1995. Imaging radar for ecosystem studies. Bioscience 45, 715–723.

Watson, D., 1947. Variation in net assimilation rate and leaf area between species and varieties and within and between years. Ann. Bot. N. S. XI (41), 41–76.

Wilson, A.M., Jetz, W., 2016. Remotely sensed high-resolution global cloud dynamics for predicting ecosystem and biodiversity distributions. PLoS Biol. 14, e1002415.

Xiao, Z., Liang, S., Wang, J., Chen, P., Yin, X., Zhang, L., Song, J., 2014. Use of general regression neural networks for generating the GLASS leaf area index product from time-series MODIS surface reflectance. IEEE Trans. Geosci. Remote Sens. 52, 209–223.

Xiao, Z., Liang, S., Wang, J., Xiang, Y., Zhao, X., Song, J., 2016. Long-time-series global land surface satellite leaf area index product derived from MODIS and AVHRR surface reflectance. IEEE Trans. Geosci. Remote Sens. 54, 5301–5318.

Xie, Y., Sha, Z., Yu, M., 2008. Remote sensing imagery in vegetation mapping: a review. J. Plant Ecol. 1, 9–23.

Xu, M., Jia, X., Pickering, M., Plaza, A.J., 2016. Cloud removal based on sparse representation via multitemporal dictionary learning. IEEE Trans. Geosci. Remote Sens. 54, 2998–3006.

Yachi, S., Loreau, M., 1999. Biodiversity and ecosystem productivity in a fluctuating environment: the insurance hypothesis. Proc. Natl. Acad. Sci. 96, 1463–1468.

Yuan, W., Liu, S., Yu, G., Bonnefond, J., Chen, J., Davis, K.J., Desai, A.R., Goldstein, A.H., Gianelle, D., Rossi, F., 2010. Global estimates of evapotranspiration and gross primary production based on MODIS and global meteorology data. Remote Sens. Environ. 114, 1416–1431.

Zhu, Z., Bi, J., Pan, Y., Ganguly, S., Anav, A., Xu, L., Samanta, A., Piao, S., Nemani, R.R., Myneni, R.B., 2013. Global data sets of vegetation leaf area index (LAI) 3g and fraction of photosynthetically active radiation (FPAR) 3g derived from global inventory modeling and mapping studies (GIMMS) normalized difference vegetation index (NDVI3g) for the period 1981 to 2011. Remote Sens. (Basel) 5, 927–948.
Zhu, Z., Piao, S., Myneni, R.B., Huang, M., Zeng, Z., Canadell, J.G., Ciais, P., Sitch, S., Friedlingstein, P., Arneth, A., Cao, C., Cheng, L., Kato, E., Koven, C., Li, Y., Lian, X., Liu, Y., Liu, R., Mao, J., Pan, Y., Peng, S., Peñuelas, J., Poulter, B., Pugh, T.A.M., Stocker, B.D., Viovy, N., Wang, X., Wang, Y., Xiao, Z., Yang, H., Zaehle, S., Zeng, N., 2016. Greening of the Earth and its drivers. Nat. Clim. Chang. 6, 791.

CHAPTER FOUR

The long and short of it: A review of the timescales of how fire affects soils using the pulse-press framework

Adam F.A. Pellegrini[a,b,*], **Robert B. Jackson**[a,c,d]

[a]Department of Earth System Science, Stanford University, Stanford, CA, United States
[b]Department of Plant Sciences, University of Cambridge, Cambridge, United Kingdom
[c]Woods Institute for the Environment, Stanford University, Stanford, CA, United States
[d]Precourt Institute for Energy, Stanford University, Stanford, CA, United States
[*]Corresponding author: e-mail address: afapellegrini@gmail.com

Contents

Abstract

Understanding factors that regulate ecosystem responses to disturbances is a long-standing ecological goal and has been useful in predicting the effects of global change. The response of an ecosystem to a single perturbation does not necessarily predict its response to repeated perturbations—highlighting the 'pulse' vs. 'press' framework to predict ecosystem responses to single or recurring perturbations, respectively. This framework can help guide hypotheses around how ecosystems may, or may not, respond to changes in disturbance regimes. Here we apply the pulse-press framework to understand fire effects on soils and plant communities. Specifically, we combine a quantitative meta-analysis on the response of soil carbon and nutrients with a literature review of the response of microbial and plant communities. We argue that fire is a

Advances in Ecological Research, Volume 62
ISSN 0065-2504
https://doi.org/10.1016/bs.aecr.2020.01.010

cumulative disturbance driving a 'press' response and can be both coupled and decoupled from the 'pulse' response to a single burn depending on whether the response variable is regulated by the direct effects of heating or the indirect effects of changing plant and microbial communities. Our meta-analysis demonstrates that carbon and nutrient content in mineral soils responded differently because carbon and nitrogen, which are volatilized to the atmosphere in a fire and turn over relatively slowly in soil, are regulated by the long-term losses of plant biomass inputs more so than heating-induced volatilization, and thus change relatively little in a single fire. Contrastingly, the elements calcium, potassium, and phosphorus, which are enriched in ash because of their resistance to volatilization, are regulated by the intermittent pulses into the soil following a fire. A literature review suggests microbial biomass and decomposition activity tended to decline after a single fire because of heat-induced mortality, which was consistent with the effects of repeated burning, likely due to combined effects of repeated heat-induced mortality and lower soil organic matter. Another important 'press' response was how nutrient losses shape the nutrient use strategies of plants (tissue stoichiometry and nutrient acquisition strategy) and how changes in plant strategies influence soil responses. Past studies have generally found that species with low nutrient concentrations in biomass, high nutrient recycling before tissue senescence, and a greater reliance on mycorrhizal symbioses tend to increase in abundance with frequent burning. Many of these traits likely lead to slower decomposition and nutrient availability, but few studies have rigorously linked fire-driven shifts in traits with carbon and nitrogen cycling. The processes we have identified are likely to be especially important in tropical ecosystems, where disturbance by fire is shifting rapidly, attributable to changes in land use and climate. Tropical savannas are some of the most responsive landscapes to altered fire regimes, with large changes in soil carbon and nutrients, suggesting that shifts in fire regimes may change the carbon balance of these ecosystems substantially. We propose that more fire-manipulation experiments are needed in tropical forests, especially in moist forests where plants are not as well adapted to fire.

1. Introduction

The response of ecosystems to altered fire regimes will be a key factor determining the trajectory of the global carbon (C) sink (Andela et al., 2017; van der Werf et al., 2017) and the preservation of biodiversity (Bond and Parr, 2010). Some of the largest changes in fire regimes are occurring in tropical ecosystems, such as savannas and forests (Andela et al., 2017; Trumbore et al., 2015), which also play critical roles in the global carbon cycle (Dixon et al., 1994). Shifts in fire activity caused by changes in land use and climate (Abatzoglou and Williams, 2016; Andela et al., 2017) are hypothesized to have long-term effects on ecosystem functioning because of the potential for past fires to influence processes like nutrient availability for decades

(Johnstone et al., 2016; Zarin et al., 2005). Such long-lasting effects of fire are hypothesized to emerge through a variety of mechanisms, such as changes in soil carbon and nutrients, microbial communities and plant physiological traits.

The response of soils to fire is complex because the immediate effects of a single fire are not always consistent with the long-term effects of repeated burning. For example, inorganic nitrogen (N) in the mineral soil may increase immediately following a single fire (Wan et al., 2001), but can be significantly depleted with repeated burning as the availability of mineralizable organic matter declines (Guinto et al., 2001; Norris et al., 2013; Pellegrini et al., 2014; Reich et al., 2001). Similarly, plant species colonizing a landscape immediately after a fire can have nutrient-rich tissues, high leaf area, and rapid biomass turnover, while those persisting in frequently burned landscapes often have relatively nutrient poor tissues and long-lived biomass (Cavender-Bares and Reich, 2012; Hoffmann et al., 2012; Pellegrini, 2016). Consequently, the effect of fire on ecosystems that emerges after decadal alterations in fire frequencies may remain undetected when only single burns are examined.

Here, we evaluate the factors that determine the response of soils from short- to long-term changes in fire regimes, through combining pulse-press and hierarchical response frameworks. To accomplish this, we review the literature on how fire impacts soils (both carbon and nutrient content and microbial communities) by contrasting the data from the effects of a single fire with those of repeated burning. To test the hierarchical scales at which fire effects emerge, we explore how plant and microbial communities respond to fire and how their compensatory responses lead to potential changes in soil nutrient cycling. Our aim is to provide a better understanding of how fire affects soils and why the effects of single fires may sometimes differ from those of repeated burning. To accomplish this aim, we combine a quantitative meta-analysis on the response of soil carbon and nutrients with a literature review of the response of microbial and plant communities. We first take a global approach to explore the applicability of these frameworks and then explicitly discuss the relevance to tropical ecosystems.

2. The pulse-press framework

The effect of burning on ecosystems can potentially be conceptualized using the pulse-press framework, which has been developed to reconcile the different effects of a discrete 'pulse' change in resource supply with the continued 'press' of chronic changes in resource supply (Ives, 1995; Ives and Carpenter, 2007). This framework has been applied to many topics such as flooding in streams (Lake, 2000), precipitation and drought in terrestrial ecosystems (Sala et al., 2012; Schwinning et al., 2004), and eutrophication in aquatic systems (Ives and Carpenter, 2007). With drought, for example, the 'pulse' effects are determined by intraseasonal dynamics of how individual rainfall events affect ecosystems; the 'press' effects focus on interseasonal dynamics of how longer-term changes in rainfall across multiple growing seasons affect ecosystems (Schwinning and Sala, 2004). The response to 'pulse' declines in precipitation is not always predictive of the 'press' declines due to prolonged changes in precipitation (Schwinning and Sala, 2004). For example, different timescales of physiological responses (rapid enzyme production vs. slow biomass production) can result in a short change in rainfall eliciting a change in these rapid response variables, while a prolonged change induces shifts in slower response variables, such as plant community biomass and composition (Harris et al., 2018).

Similar analogies exist for fire. For example, in tropical savannas, plant communities can rapidly regrow and resprout after a single fire (Hoffmann et al., 2009), but repeated burning changes community composition by limiting the ability of juvenile plants to become adults, suppressing woody plant abundance and biomass over decades of repeated burning (Higgins et al., 2007; Moreira, 2000). The effects of single vs. repeated fires on belowground properties may also differ because a single fire may result in pulsed increases due to leaching of ash into soils (which would occur on daily to annual timescales) whereas repeated burning may result in losses due to lower aboveground plant biomass inputs into soils due to repeated combustion (which would occur on decadal timescales) (Fig. 1). In ecosystems with intense canopy wildfires that can directly combust soil organic matter, such as in boreal forests, the pulse and press effects may converge because organic matter losses from soils occur in both scenarios of single and repeated burns (Johnson and Curtis, 2001; Nave et al., 2011; Walker et al., 2019). Yet in many ecosystems that experience surface fires, such as tropical savannas and forests, soil responses can be determined by changes in plant communities on decadal timescales, as either succession when fire is excluded or slow losses of biomass and changing plant composition when fires recur (Higgins et al., 2007; Moreira, 2000). As a result of declining plant biomass inputs, soil organic matter can decline (e.g. in tropical savannas (Coetsee et al., 2010, Pellegrini et al., 2015)).

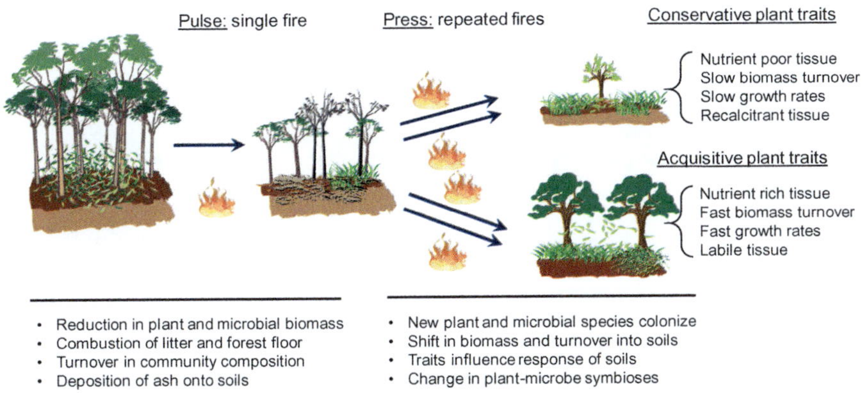

Fig. 1 Hypothesized conceptual schematic illustrating the potential responses of the ecosystem to alterations in burning. A single surface fire tends to reduce plant and microbial biomass as well as combust biomass in live plants (in wood and leaves) and the forest floor and litter layers (Kauffman et al., 1994). Elements that are not volatilized can be redeposited as ash that can be leached into the soils. Species with traits allowing them to resist mortality or re-grow rapidly after fire increase in abundance. Repeated burning restricts the turnover of aboveground biomass into the soils and reduces microbial activity (Metcalfe et al., 2018; Pellegrini et al., 2014), which can reduce soil nutrients (Guinto et al., 2001). Changes in the plant community can shift plant functional traits (Higgins et al., 2012), which generally group plant species into two broad 'strategies' of species that are conservative because they grow slowly and have nutrient poor and recalcitrant tissue and low biomass turnover (top scenario) (Wigley et al., 2016) or acquisitive because they grow quickly, have nutrient rich and labile tissue, and rapid biomass turnover (bottom scenario) (Pellegrini, 2016).

3. The hierarchical response framework

Theory suggests that pulse effects can scale to press effects through a hierarchical response framework via shifts in individual traits and species composition (Smith et al., 2009). Here we apply such a framework through the lens of plant-soil interactions (Bardgett et al., 2013); as repeated burning changes nutrient availability, plants and microbes may respond by shifting how they use and acquire nutrients (e.g., tissue stoichiometry and symbioses) (Fig. 2). In turn, changes in the physiological traits and composition of plants and microbes may further influence the rate at which nutrients become available through decomposition (Hobbie, 1992, 2015) (Fig. 2). Empirical studies have demonstrated that fire can change plant traits relevant to carbon and nutrient turnover in soils (Cavender-Bares and Reich, 2012;

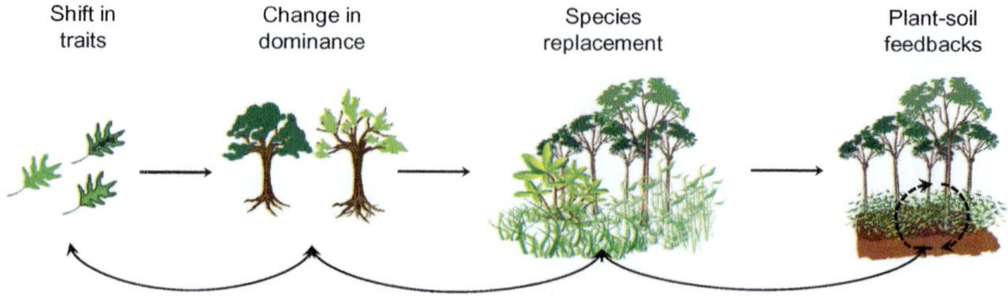

Fig. 2 Conceptual schematic of response hierarchies to changes in fire. The responses scale from shifts in traits, to turnover in the dominant species, to immigration of new species and loss of existing species in the ecosystem, to the restructuring of plant soil feedbacks. The arrows indicate that fire-driven changes emerge over the course of years to decades, but the key theme is that fire can alter the physiology of the plant community at these different scales, and that changes in certain physiological traits can influence soil biogeochemistry. Consequently, the observed changes in soil biogeochemistry may partly be due to the response of plant traits.

Dantas et al., 2013a; Hoffmann et al., 2005). For example, species favoured by fire can invest in resource acquisition such as N fixation (Newland and DeLuca, 2000) or have high nutrient resorption, thick and long-lived leaves, low tissue nutrient content, and high C:N ratios (Maracahipes et al., 2018; Pellegrini, 2016; Wigley et al., 2016), which can influence decomposition (Norris et al., 2013).

The response of plant communities may help explain the pulse vs. press effects on soils because the turnover of the plant community tends to be larger under prolonged changes in fire regimes. Plant communities can recover to pre-fire states after a single fire, even if it takes several decades (Clement and Touffet, 1990; Hanes, 1971); however, changes in fire frequency or intensity tends to alter plant species composition over the course of several years to decades (Glitzenstein et al., 2012; Lewis and Debuse, 2012; Peterson and Reich, 2001). Given that the press response of the plant community to fire is different from the pulse response, repeated burning may have distinct effects on the physiological traits of the plant community that could influence soil C and nutrient storage and turnover.

4. The pulse response: single fire events

4.1 Changes in elemental concentrations in soils

Elemental losses during fire are regulated by the heat of the fire combined with the physical properties of the elements and their compounds.

Combustion usually occurs around 650 °C, which is above the temperature that volatilization begins to take place for carbon (C) and nitrogen (N) (200 °C) and sulfur (S) (375 °C). Potassium (K) and phosphorus (P) have slightly higher thresholds of 774 °C, whereas magnesium (Mg) and calcium (Ca) have even higher thresholds of 1107 °C and 1484 °C, respectively (Hosking, 1938; Raison et al., 1985; Tiedemann, 1987). These temperature thresholds relative to the intensity of the fire determine whether elements in combusted biomass are volatilized to the atmosphere vs. pyromineralized into simpler compounds and retained in the remaining ash and char after the fire (Raison et al., 1985; Seiler and Crutzen, 1980). Over longer timescales, the 'fate' of the elements remaining in the system can be highly variable as they can be lost via erosion and runoff or taken up by regrowing plants and microbes and/or sorbed to minerals (DeBano and Conrad, 1978; Raison et al., 1985; Seiler and Crutzen, 1980). One hypothesis is that fire leads to belowground enrichment of elements with higher volatilization thresholds than combustion (K, P, Mg, and Ca) because they are redistributed from being stored in aboveground plant biomass into ash that is transported into the mineral soil (Butler et al., 2017). Carbon and N, on the other hand, are hypothesized to be lost from soils either through direct combustion of organic matter or a reduction in plant biomass inputs (Pellegrini et al., 2015).

Synthesis studies have found that concentrations of C and N can be enriched multiple years after a fire (Johnson and Curtis, 2001), reduced in particular ecosystems and after intense wildfires (Nave et al., 2011; Wang et al., 2012), or change relatively little beneath the top few centimeters of soil (Wan et al., 2001). For example, Wan et al. (2001) synthesized data from 87 studies across many ecosystems and concluded that total C and N concentrations in mineral soils are relatively insensitive to fire, except in the top few centimetres where direct combustion may be occurring. In contrast, Nave et al. (2011) synthesized data from 57 studies in temperate forests and found fire depleted mineral soil C and N concentrations but not stocks because of a concurrent increase in bulk density. In both studies, wildfires (which they interpreted to have higher intensities than prescribed fires because wildfires often occur in drier and/or windier conditions) tended to have larger negative effects on mineral soil concentrations. Consequently, losses are sensitive to the intensity of burning because fire intensity correlates with combustion of organic matter from the mineral horizon (Baird et al., 1999). Nonetheless, for many surface fires, which are usually low in intensity, mineral soil C and N do not change much in a single fire.

The pulse response of other nutrients, such as P and cations, differs from C and N. Several studies have shown that single fires tend to enrich soil P concentrations (Butler et al., 2018). Changes in soil cations are also apparent, but to our knowledge there has yet to be a meta-analysis of how single fires can affect cations (although see Pellegrini et al., 2018 for an analysis of long-term effects of repeated burning). Consequently, to evaluate the effect of fire on Ca concentrations (given that it has a higher temperature threshold), we compiled data from several sites ($n=21$) included in the other meta-analyses for comparability (Wan et al., 2001) and tested for effects using mixed-effects models weighting the contribution of studies by their variance and replication as is standard in other meta-analysis approaches (Hedges et al., 1999; Pellegrini et al., 2018) (see Supplementary Table S1 in the online version at https://doi.org/10.1016/bs.aecr.2020.01.010). We found that soil Ca tended to be enriched after a single fire (+59%, $P<0.0001$), and that the enrichment increased with the time since fire (time since fire effect, $Q_M=11.08$, $P<0.001$). The temporal trends suggest that the redistribution of Ca into soil pools continues to occur over multiple years.

Comparing across elements, the volatilization thresholds of elements can be important in determining the pulse response of a single fire. Elements with higher thresholds are not lost to the atmosphere and thus are redistributed from ash into soils relatively rapidly; in contrast, elements with low thresholds tend to change relatively little in the soils because despite large aboveground losses mineral soils tend to not reach sufficiently high temperatures for volatilization to occur.

Unfortunately, most studies are in temperate ecosystems, especially temperate forests. It is likely that the response of the different elements to single fires are similar in tropical ecosystems, given that physical properties of the elements determine their potential volatility. For example, the slash and burn fires in tropical forests tend to support conclusions of C and N in aboveground biomass being volatilized to the atmosphere in a fire (Kauffman et al., 1993).

4.2 Changes in microbial communities

Understanding the pulse response of microbial communities holds promise for improving our ability to predict how fire influences soil C and nutrient turnover and availability. Multiple syntheses have evaluated the effect of single fires on soil microbial communities, which suggest fire can (i) suppress fungi, especially ectomycorrhizal types, more than bacteria (Dove and Hart,

2017), (ii) reduce microbial biomass (Dooley and Treseder, 2012; Pressler et al., 2018), and (iii) change decomposition activity (Wan et al., 2001; Wang et al., 2012), particularly after high intensity wildfires (Dooley and Treseder, 2012; Wang et al., 2012).

The implications of changes in microbial community composition and abundance can be assessed through measurements of heterotrophic respiration and N mineralization rates. For example, respiration can decline immediately following a fire, often attributed to reduced microbial abundance rather than substrate availability, but respiration can also increase, potentially due to an increase in soluble sugars (DeLuca and Zouhar, 2000, Pietikäinen and Fritze, 1995). Mineralization of organic N behaves in similarly complex ways, as there tends to be an increase in nitrification rates from a few days to months following fire. The increase in nitrification is hypothesized to be due to increases in exchangeable cations through ash deposition promoting nitrification (DeLuca and Sala, 2006). The effects of mineralization rates on inorganic N concentrations are clearer: a meta-analysis demonstrated that ammonium concentrations increase immediately (days to months) after a fire, while nitrate concentrations peak 6–12 months after a fire (Wan et al., 2001). But over long timescales of repeated burning, inorganic N declines (Pellegrini et al., 2018, 2020).

Extracellular enzymes involved in decomposition produced by microbes offer another avenue to evaluate fire effects on soil carbon and nutrient cycling but have received little attention in past syntheses. After a single fire, activity of decomposing enzymes can increase (Boerner et al., 2006; Rietl and Jackson, 2012). This result suggests fire may elevate nutrient turnover, which could then help explain the increase in inorganic nutrients observed following prescribed burning (Wan et al., 2001). Contrastingly, fire can reduce the activity of enzymes in decomposing litter layers (Köster et al., 2016), as well as soil pools (Pourreza et al., 2014). Shifts in enzyme activity may result from lower substrate availability due to reduced inputs, lower microbial biomass resulting from heat-induced mortality, turnover in species composition, or other unexplored mechanisms.

5. The press response: repeated burning

5.1 Elemental concentrations in soils: Divergence of pulse-press

The pulse effect of fire on soil C and N concentrations can be minimal and vary across ecosystems, climates, and fire intensities. In contrast, repeated burning over decades can have more consistent effects on soil C and N, suggesting that the pulse effect of a single fire does not correspond with the press effect of repeated burning (i.e., the two are decoupled). For example, we updated our previous global meta-analysis (Pellegrini et al., 2018) of 48 sites (citations included below and in the references in Pellegrini et al., 2018) with data from four additional sites (Burns, 1952; Godwin et al., 2017; Guénon et al., 2013). The sites included in the analysis have (i) experienced altered fire frequencies (either due to natural variability or intentional manipulation) for 9 or more years, and (ii) the burned plots in all the sites have experienced more than one fire event. Re-analysis including the new sites confirmed the previous findings of a strong depletion effect of repeated burning on total soil C and N in savanna-grasslands and broadleaf forests (on average, frequently burned plots had −22% ($P<0.001$) and −23% ($P<0.001$) less C and N than fire-protected plots, respectively), but enrichment in needleleaf forests (on average, frequently burned plots had +24% ($P=0.001$) and + 15% ($P=0.03$) more C and N than fire protected plots, respectively).

Moreover, the effect of frequent burning increased through time in savanna-grasslands and broadleaf forests, with the largest losses of soil C and N occurring in sites that experienced the longest alteration of fire frequencies. Consequently, the response of mineral soil C and N to a single fire event can be decoupled from the response to decades of repeated burning (Fig. 3). The different responses according to ecosystem type illustrate that the direction of the decoupling can depend on plant community composition, which may be due to a variety of processes that we do not explore here.

Further analyses demonstrated that tropical and temperate ecosystems responded in similar ways. In analyses of soil C and N, there were no significant differences in the response ratios between tropical (we aggregated sub-tropical and tropical regions) and temperate regions that were savanna-grassland and broadleaf forest ecosystems (region effect for C: $Q_M=0.3$, $P>0.50$, for N: $Q_M=1.1$, $P=0.30$). In the cases of P and Ca, all ecosystems converged on similar responses (region effect for P: $Q_M=0.03$, $P>0.50$, for Ca: $Q_M=0.37$, $P>0.50$). Most tropical sites are tropical savannas. However, many of the tropical savannas have enough rainfall and extensive wet season to allow closed-canopy forests to form in the absence of fire (Pellegrini et al., 2014). Consequently, the fire experiments in these wet savannas provide a useful baseline for how drier tropical forests may respond to fire.

The potential divergence between pulse and press responses of soil C and N are hypothesized to occur because changes in soil C and N are regulated

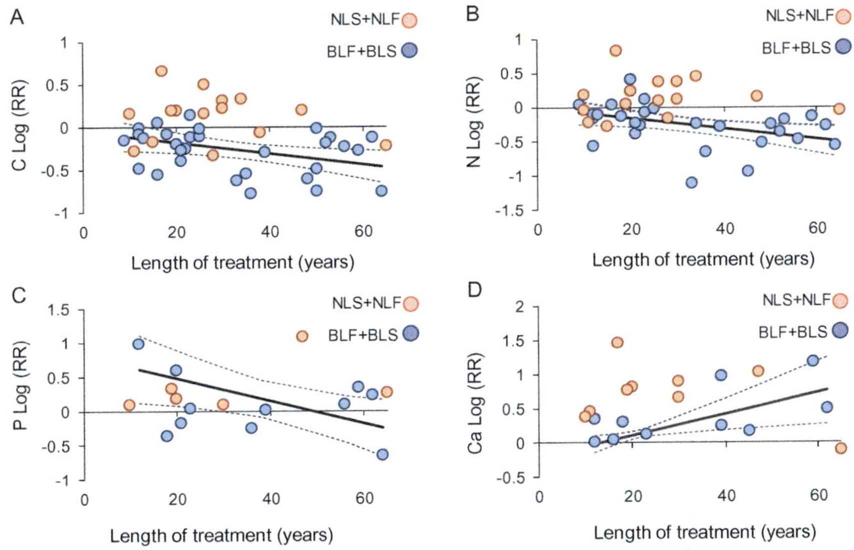

Fig. 3 Meta-regression between the length of fire treatment for sites that experienced at least 10 years of altered fire frequencies and the log response ratio of soil carbon (C, panel A), nitrogen (N, panel B), phosphorus (P, panel C), and calcium (Ca, panel D). The x-axis 'length of fire treatment' represents the number of years that fire frequencies differed between the high vs. low fire frequency treatments (in some cases due to natural variability in fire regime but in others due to intentional manipulation). Values shown are split into sites classified as either broadleaf forests and broadleaf savannas (BLF + BLS) or needleleaf forests and needleleaf savannas (NLF + NLS). Lines indicate significant meta-regressions, which were present only for BLF + BLS, C: $P = 0.019$; N: $P = 0.015$; P: $P = 0.021$; Ca: $P = 0.010$ for significant slopes. The responses of soil carbon and nutrients were not correlated with length of fire manipulation treatment in the NLF + NLS sites ($P > 0.05$).

by the long turnover time of soil organic matter and the rate at which plant biomass inputs (e.g., litterfall or root production and turnover) into soil changes with fire. Through successive burns, plant biomass inputs into soils declines due to repeated combustion of leaf litter, tree mortality, and losses of root biomass (Pellegrini et al., 2015, 2020; Raison et al., 1993; Reich et al., 2001). Consequently, repeated burning is predicted to change soils over decades as lower biomass inputs have an increasingly noticeable effect, which is consistent with the observations that repeated burning reduces soil C and N over decadal timescales (observed in Pellegrini et al. (2018) and updated with additional data in Fig. 3). Consequently, fire-driven losses of plant biomass inputs can change mineral soil C and N storage over decadal timescales as repeated burning depletes plant biomass storage and inputs into soils.

In contrast to C and N, the responses of Ca and P to single and repeated fires are similar (Fig. 3). For P, our previous meta-analysis found that across all sites the initial enrichment of P in soils persisted even after 30 years of repeated burning, but then became insignificant as frequent burning continues (Pellegrini et al., 2018). Our updated analysis illustrates that the enrichment and then decline of soil P emerges only in savanna-grasslands and broadleaf forests (study length, $P=0.02$, Fig. 3). Consequently, the initial redistribution of P into soils is sustained over multiple decades, but over time the enrichment decreases, which is likely due to various loss processes, such as erosion and leaching (Bêche et al., 2005; DeBano and Conrad, 1978). Soil P in needleleaf forests responded differently because it remains unchanged through time regardless of the number of fires. However, more data are needed to verify the lack of temporal change in needleleaf ecosystems because one of the longest studies used a relatively weak P extraction method (Mehlich 3) that likely does not capture changes in the more recalcitrant soil P fractions (Cross and Schlesinger, 1995); if that study is removed from the meta-analysis, soil P in needleleaf ecosystems declined.

Calcium enrichment with more frequent burning is more stable through time (Fig. 3) but with two nuances: (i) the enrichment is significantly greater in needleleaf ecosystems than in savanna-grasslands and broadleaved forests, and (ii) the enrichment in savanna-grasslands and broadleaf forests increases with the length of time plots experienced higher fire frequencies. The more consistent enrichment of Ca than P with fire is intuitive given its higher volatilization temperature.

We propose that the relative coupling or decoupling of various elements can be linked to the timescale of the mechanism regulating their input into soils. In the cases of Ca and P, the primary factor driving changes is the redistribution of ash, which occurs over days to months to a few years. However, the enrichment can change after multiple decades of increased burning potentially because of sustained leaching and erosion, suggesting that competing pathways reduce the enrichment but only on long timescales (e.g., >30 years). For C and N, the primary mechanism determining the effect of fire on soil pools is the reduction of biomass inputs into soils. Lower inputs reduce soil C and N slowly because the turnover of organic matter in soils occurs on decadal to centennial timescales in many systems (Jackson et al., 2017); consequently the lack of an observable pulse effect is likely because of insufficient time and losses to detect significant effects.

5.2 Microbial community enzymes: convergence of pulse-press

Only a few studies document the effects of changes in fire frequency on microbial community composition and activity (Bastias et al., 2006; Buscardo et al., 2015; Campbell et al., 2008; Oliver et al., 2015). These studies suggest that frequent burning generally shifts both microbial composition and activity (Ajwa et al., 1999; Buscardo et al., 2015). For example, repeatedly burned plots can have lower beta glucosidase activity relative to unburned controls (Ajwa et al., 1999; Boerner et al., 2005), suggesting slower organic matter turnover. Similarly, in an oak-hickory forest, repeated annual and periodic burning reduced the activity of acid phosphatase, alpha and beta glucosidase, sulfatase, and urease, relative to unburned plots, reductions that were all correlated with lower microbial biomass (Eivazi and Bayan, 1996). Consequently, preliminary evidence points to enzymes responding to persistent changes in fire frequency, but the causes of these changes are unclear. Mechanisms leading to shifts in enzymes could include losses of microbial biomass, shifts in C and nutrient availability, and/or changes in microbial composition. From the studies that do exist, the pulse and press response of microbes appear to be similar, where biomass and activity changes after single and multiple fires. Perhaps one reason for the convergence is the high sensitivity of microbes to mortality from heat during fire and the potential for relatively rapid compositional turnover.

6. Hierarchical responses of plants to fire-driven changes in nutrients

Fire-driven changes in soil nutrients have the potential to interact with changes in plant composition and their physiological traits. For example, fire-driven N losses may stimulate the colonization of fast growing N-fixing plants that produce a lot of nutrient rich litter, which can decompose into soils and replenish the N lost via fire; resultantly, the response of plants to low N conditions 'compensates' for the N losses at the level of the ecosystem. On the other hand, if the colonizing plants have traits such as high N resorption and lignin content and long-lived leaves, they can reduce the inputs of N from plant biomass into soils; resultantly, the of the plant 'escalates' N limitation at the landscape scale (Fig. 4).

Multiple studies have demonstrated that fire history can determine the composition of species with different, typically more conservative, traits (e.g., low specific leaf area, low concentrations of N, and high N resorption

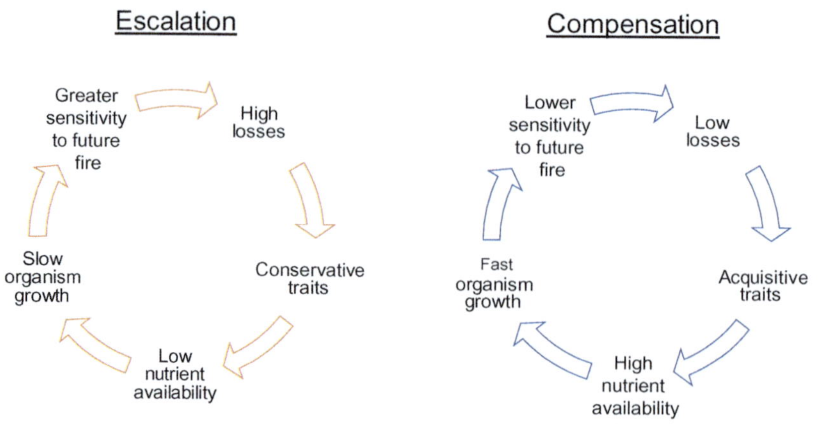

- High nutrient resorption and lower quality litter
- Species with lower leaf area
- Reduction in enzyme activity
- Reduction in mycorrhizal symbioses

- Colonization by nitrogen-fixing plants
- Greater investment in nutrient acquiring enzymes
- Increased production of herbaceous species

Fig. 4 Conceptual schematic of possible feedbacks between changes in soil nutrients, plant and microbial responses, and the impact of the responses on turnover in nutrients. Evidence for the examples are cited in the text.

efficiency) (Cavender-Bares and Reich, 2012; Dantas et al., 2013b; Pellegrini, 2016; Wigley et al., 2016). In Brazilian tropical savannas, for example, tree species specializing in savannas that burn frequently and have low soil N tend to have low foliar N and leaf area, and thicker, longer-lived leaves than species in forests that tend to not burn and have higher N (Dantas et al., 2013a; Hoffmann et al., 2005). Similarly, in temperate oak savannas, where frequent burning reduces soil N stocks and mineralization (Reich et al., 2001), trees from frequently burned plots withdrew a greater proportion of nutrients from leaves before senescence, producing lower quality litter that decomposed more slowly relative to individuals in fire protected plots (Hernández and Hobbie, 2008; Norris and Reich, 2009). In subtropical ecosystems such as eucalypt forests, frequent burning can reduce the relative concentrations of nutrients in foliar biomass (Butler et al., 2017). Consequently, evidence suggests that species with traits that further escalate nutrient limitation tend to increase in abundance with more frequent fire.

In addition to changes in the tissue traits of plants, fire-driven shifts in the abundance of plant species with different symbioses are thought to be important in regulating long-term effects of fire on soils (Hart et al., 2005a,b). For example, colonization by N-fixing plants may recover N lost in fire (Newland and DeLuca, 2000), and some studies have proposed that the persistence of N fixers in the landscape is a critical factor buffering against N losses from soils (Johnson and Curtis, 2001). Subsequently, if the abundance of plants capable of N fixation increases with frequent burning, fixers may provide a critical pathway for replacing N losses (Yelenik et al., 2012). In temperate ecosystems that contain both herbaceous legumes and woody actinorhizal plants, species capable of fixing N tend to rapidly colonize burned landscapes (Newland and DeLuca, 2000; Peterson et al., 2007; Yelenik et al., 2012). Recent work has questioned the degree to which N fixers can balance N losses even when they increase in abundance following fire (Tierney et al., 2019). Moreover, the colonization of N-fixing plants in frequently burned areas is not necessarily generalizable across biomes: the relative abundance of woody plant N fixers does not appear to be influenced by fire in tropical savannas and forests but rather depends on water availability (Pellegrini et al., 2015, 2016).

Changes in the abundance of plants that form mycorrhizal symbioses may also impact the formation and turnover of soil organic matter (Taylor et al., 2016) because mycorrhizal symbionts can be strongly sensitive to fires (Cairney and Bastias, 2007). Single fires tend to have strong negative effects on fungal activity on root tips (Dove and Hart, 2017), potentially reducing the nutrient uptake capacity of plants. Studies from areas burned repeatedly over the course of multiple decades find similar negative effects of fire on fungi (Bastias et al., 2006). In ectomycorrhizal communities, such as those dominated by pines, oaks and eucalypts, frequent burning can reduce fungal species richness on root tips (Buscardo et al., 2010) and lower the abundance of colonized root tips (Tuininga and Dighton, 2004). However, there does not seem to be as clear of an effect of fire on arbuscular mycorrhizal (AM) communities, potentially because of the resilience of grasses to fire in many frequently burned ecosystems. For example, in a North American temperate grassland, 10 years of repeated burning did not reduce AM fungal colonization (Eom et al., 1999). In contrast, in African tropical savanna, $\sim$40 years of repeated burning reduced AM colonization (Hartnett et al., 2004), suggesting a lower capacity for plants to acquire inorganic nutrients. The broad patterns of how fire changes the composition of plant species with different symbiosis strategies remains elusive, limiting inference into the role of changes in the composition of species with different symbiosis strategies in determining belowground responses to fire. However, preliminary data suggest fire can regulate the potential formation of mycorrhizal symbioses, either by directly killing the symbiont or changing the composition of plants with different symbiotic strategies.

7. Relevance to tropical ecosystems

Connecting short- and long-term effects of disturbances on tropical ecosystems has been a focus of research for several decades (Brando et al., 2014; Moreira, 2000; Ryan and Williams, 2011; Uhl and Jordan, 1984). Extensive research has evaluated the effect of fire on tropical rainforests given their large extent and exposure to increased fire frequencies (Andela et al., 2017; Morton et al., 2006; Van der Werf et al., 2009). However, much research on the effects of repeated burning on tropical rainforests focuses on the response of vegetation, primarily trees, which tend to decline rapidly with more frequent and intense fires (Balch et al., 2013; Brando et al., 2014; Uhl and Jordan, 1984). In contrast to vegetation responses, very few studies have evaluated the effects of repeated burning on tropical rainforest soils and nutrient cycling. Notable exceptions have found that historical fire frequency can be important in regulating ecosystem recovery because of fire-driven N losses (Zarin et al., 2005) and potentially also through the suppression of decomposition (Metcalfe et al., 2018). In contrast to the limited work on repeated burning effects on tropical rainforest soils, dozens of studies have investigated the effect of single fires on soil C and nutrients (Døckersmith et al., 1999; Kauffman et al., 2009; Ryan et al., 2011; Uhl and Jordan, 1984), which have found that slash-and-burn fires can result in a transient enrichment of soil nutrients, but intense fires can deplete soil nutrients in the top few centimetres. Our pulse-press framework leads to the hypothesis that the long-term effects of repeated burning on soils likely results in losses of C and N if the fire-return interval is short enough to prohibit the balancing of nutrient losses by atmospheric deposition and/or N fixation inputs, which can provide the necessary N for regrowth of secondary tropical rainforests (Batterman et al., 2013; Brookshire et al., 2019). Consequently, a large research gap exists between the multitude of studies on single fire effects with the reality that rainforests are being increasingly exposed to repeated burning.

In contrast to the limited work on the effect of repeated burning on tropical rainforest soils and nutrient cycling, there has been extensive work in tropical savannas and seasonal forests (Guinto et al., 2001; Holdo et al., 2012; Nardoto and Bustamante, 2003; Pellegrini et al., 2014; Turner et al., 2008; Verma and Jayakumar, 2018). Generally, fire exclusion increases the storage and turnover of soil C and nutrient stocks, whereas increased fire frequencies decreases C and nitrogen storage and turnover (Pellegrini et al.,

2018). Changes in soils arise due to combustion losses of C and nutrients from grass and tree biomass, despite savannas being a highly fire adapted system. Analyses of plant functional traits (e.g., leaf physiology) find that species adapted to repeatedly burned environments tend to have conservative nutrient use traits that may encourage lower soil nutrient availability (Dantas et al., 2013a; Hoffmann et al., 2005; Pellegrini, 2016). Aside from studies on soils and plants, little is known about how microbial communities respond to fire in tropical savanna ecosystems, limiting the understanding of mechanisms contributing to changes in biogeochemical cycling.

Even less is known about the responses of different forest and savanna types (such as seasonally flooded and peatland ecosystems). Fire effects in these ecosystem types are largely driven by weather variability resulting in dry enough conditions for fire to ignite and spread (van der Werf et al., 2010). For example, studies of peat combustion in boreal forests found moisture to be the critical factor regulating combustion (Rein et al., 2008). Seasonally inundated ecosystems also may respond differently than well-drained tropical ecosystems because of their thick organic horizons, which can be combustible as evidenced from tropical and boreal peatland ecosystems (Turetsky et al., 2015; Usup et al., 2004). Consequently, a single fire can hypothetically have large immediate effects on soils in seasonally flooded and peatland ecosystems, potentially causing the press and pulse effects to be similar.

8. Conclusions

Fire effects on ecosystems can be conceptualized through a pulse vs. press framework by considering the timescales of the processes changed by fire. Carbon and nitrogen, which are lost primarily aboveground to the atmosphere in a fire, turn over slowly in soil and change because of the long-term losses of plant biomass inputs with frequent burning. In contrast, elements enriched in ash because of their resistance to volatilization, such as calcium and phosphorus, are regulated by the intermittent pulses into the soil following a fire, resulting in a convergence of pulse and press responses. Soil microbes are sensitive to heating effects in a single fire, which tend to become magnified with repeated burning (i.e., generally larger losses of biomass and decomposition and changes in species composition), connecting pulse and press responses. Plant-soil interactions are likely critical for regulating the long-term effects of repeated burning given that species with nutrient-poor tissues with low decomposability and a greater reliance on mycorrhizal symbioses tend to increase in abundance with frequent burning. These traits can decrease the turnover of carbon and nitrogen, creating a fire-nutrient feedback that may be a key characteristic of ecosystems that burn. However, repeated burning experiments are largely lacking in some of the ecosystems most rapidly experiencing changes in fire regimes such as tropical rainforests. In tropical rainforests, where plants are not as well adapted to fire and thus are more prone to mortality in a fire, repeated burning may have even more dramatic effects on soils and plant-soil interactions because of the rapid declines in biomass inputs and change in species composition.

References

Abatzoglou, J.T., Williams, A.P., 2016. Impact of anthropogenic climate change on wildfire across western US forests. Proc. Natl. Acad. Sci. U. S. A. 113, 11770–11775.

Ajwa, H.A., Dell, C.J., Rice, C.W., 1999. Changes in enzyme activities and microbial biomass of tallgrass prairie soil as related to burning and nitrogen fertilization. Soil Biol. Biochem. 31, 769–777.

Andela, N., Morton, D.C., Giglio, L., Chen, Y., van der Werf, G.R., Kasibhatla, P.S., DeFries, R.S., Collatz, G.J., Hantson, S., Kloster, S., Bachelet, D., Forrest, M., Lasslop, G., Li, F., Mangeon, S., Melton, J.R., Yue, C., Randerson, J.T., 2017. A human-driven decline in global burned area. Science 356, 1356–1362.

Baird, M., Zabowski, D., Everett, R.L., 1999. Wildfire effects on carbon and nitrogen in inland coniferous forests. Plant Soil 209, 233–243.

Balch, J.K., Massad, T.J., Brando, P.M., Nepstad, D.C., Curran, L.M., 2013. Effects of high-frequency understorey fires on woody plant regeneration in southeastern Amazonian forests. Philos. Trans. R. Soc. Lond. B Biol. Sci. 368, 20120157.

Bardgett, R.D., Manning, P., Morriën, E., De Vries, F.T., 2013. Hierarchical responses of plant-soil interactions to climate change: consequences for the global carbon cycle. J. Ecol. 101, 334–343.

Bastias, B.A., Huang, Z.Q., Blumfield, T., Xu, Z., Cairney, J.W.G., 2006. Influence of repeated prescribed burning on the soil fungal community in an eastern Australian wet sclerophyll forest. Soil Biol. Biochem. 38, 3492–3501.

Batterman, S.A., Hedin, L.O., van Breugel, M., Ransijn, J., Craven, D.J., Hall, J.S., 2013. Key role of symbiotic dinitrogen fixation in tropical forest secondary succession. Nature 502, 224–227.

Bêche, L.A., Stephens, S.L., Resh, V.H., 2005. Effects of prescribed fire on a Sierra Nevada (California, USA) stream and its riparian zone. For. Ecol. Manage. 218, 37–59.

Boerner, R.E.J., Brinkman, J.A., Smith, A., 2005. Seasonal variations in enzyme activity and organic carbon in soil of a burned and unburned hardwood forest. Soil Biol. Biochem. 37, 1419–1426.

Boerner, R.E., Waldrop, T.A., Shelburne, V.B., 2006. Wildfire mitigation strategies affect soil enzyme activity and soil organic carbon in loblolly pine (Pinus taeda) forests. Can. J. For. Res. 36, 3148–3154.

Boerner, R.E.J., Huang, J., Hart, S.C., 2009. Impacts of fire and fire surrogate treatments on forest soil properties: a meta-analytical approach. Ecol. Appl. 19, 338–358.

Bond, W.J., Parr, C.L., 2010. Beyond the forest edge: ecology, diversity and conservation of the grassy biomes. Biol. Conserv. 143, 2395–2404.

Brando, P.M., Balch, J.K., Nepstad, D.C., Morton, D.C., Putz, F.E., Coe, M.T., Silvério, D., Macedo, M.N., Davidson, E.A., Nóbrega, C.C., Alencar, A., Soares-Filho, B.S., 2014. Abrupt increases in Amazonian tree mortality due to drought–fire interactions. Proc. Natl. Acad. Sci. U. S. A. 111, 6347–6352.

Brookshire, E.N.J., Wurzburger, N., Currey, B., Menge, D.N.L., Oatham, M.P., Roberts, C., 2019. Symbiotic N fixation is sufficient to support net aboveground biomass accumulation in a humid tropical forest. Sci. Rep. 9, 7571.

Burns, P.Y., 1952. Effect of Fire on Forest Soils in the Pine Barren Region of New Jersey. Yale University.

Buscardo, E., Rodríguez-Echeverría, S., Martín, M.P., De Angelis, P., Pereira, J.S., Freitas, H., 2010. Impact of wildfire return interval on the ectomycorrhizal resistant propagules communities of a Mediterranean open forest. Fungal Biol. 114, 628–636.

Buscardo, E., Rodríguez-Echeverría, S., Freitas, H., De Angelis, P., Pereira, J.S., Muller, L.A.H., 2015. Contrasting soil fungal communities in Mediterranean pine forests subjected to different wildfire frequencies. Fungal Divers. 70, 85–99.

Butler, O.M., Lewis, T., Chen, C., 2017. Prescribed fire alters foliar stoichiometry and nutrient resorption in the understorey of a subtropical eucalypt forest. Plant Soil 410, 181–191.

Butler, O.M., Elser, J.J., Lewis, T., Mackey, B., Chen, C., 2018. The phosphorus-rich signature of fire in the soil-plant system: a global meta-analysis. Ecol. Lett. 21, 335–344.

Cairney, J.W.G., Bastias, B.A., 2007. Influences of fire on forest soil fungal communities. Can. J. For. Res. 37, 207–215.

Campbell, C.D., Cameron, C.M., Bastias, B.A., Chen, C., Cairney, J.W.G., 2008. Long term repeated burning in a wet sclerophyll forest reduces fungal and bacterial biomass and responses to carbon substrates. Soil Biol. Biochem. 40, 2246–2252.

Cavender-Bares, J., Reich, P.B., 2012. Shocks to the system: community assembly of the oak savanna in a 40-year fire frequency experiment. Ecology 93, S52–S69.

Choromanska, U., DeLuca, T.H., 2001. Prescribed fire alters the impact of wildfire on soil biochemical properties in a ponderosa pine forest. Soil Sci. Soc. Am. J. 65, 232.

Choromanska, U., DeLuca, T.H., 2002. Microbial activity and nitrogen mineralization in forest mineral soils following heating: evaluation of post-fire effects. Soil Biol. Biochem. 34, 263–271.

Clement, B., Touffet, J., 1990. Plant strategies and secondary succession on Brittany heathlands after severe fire. J. Veg. Sci. 1, 195–202.

Coetsee, C., Bond, W.J., February, E.C., 2010. Frequent fire affects soil nitrogen and carbon in an African savanna by changing woody cover. Oecologia 162, 1027–1034.

Cross, A.F., Schlesinger, W.H., 1995. A literature review and evaluation of the Hedley fractionation: applications to the biogeochemical cycle of soil phosphorus in natural ecosystems. Geoderma 64, 197–214.

Dantas, V.L., Batalha, M.A., Pausas, J.G., 2013a. Fire drives functional thresholds on the savanna-forest transition. Ecology 94, 2454–2463.

Dantas, V., Pausas, J., Batalha, M., Paula Loiola, P., Cianciaruso, M., 2013b. The role of fire in structuring trait variability in Neotropical savannas. Oecologia 171, 487–494.

DeBano, L.F., Conrad, C.E., 1978. The effect of fire on nutrients in a chaparral ecosystem. Ecology 59, 489–497.

DeLuca, T., Zouhar, K., 2000. Effects of selection harvest and prescribed fire on the soil nitrogen status of ponderosa pine forests. For. Ecol. Manage. 138, 263–271.

DeLuca, T.H., Sala, A., 2006. Frequent fire alters nitrogen transformations in ponderosa pine stands of the inland northwest. Ecology 87, 2511–2522.

Dixon, R.K., Solomon, A.M., Brown, S., Houghton, R.A., Trexier, M.C., Wisniewski, J., 1994. Carbon pools and flux of global forest ecosystems. Science 263, 185–190.

Døckersmith, I., Giardina, C., Sanford Jr., R., 1999. Persistence of tree related patterns in soil nutrients following slash-and-burn disturbance in the tropics. Plant Soil 209, 137–157.

Dooley, S.R., Treseder, K.K., 2012. The effect of fire on microbial biomass: a meta-analysis of field studies. Biogeochemistry 109, 49–61.

Dove, N.C., Hart, S.C., 2017. Fire reduces fungal species richness and in situ mycorrhizal colonization: a meta-analysis. Fire Ecol. 13, 37–65.

Eivazi, F., Bayan, M.R., 1996. Effects of long-term prescribed burning on the activity of select soil enzymes in an oak-hickory forest. Can. J. For. Res. 26, 1799–1804.

Eom, A.-H., Hartnett, D.C., Wilson, G.W.T., Figge, D.A.H., 1999. The effect of fire, mowing and fertilizer amendment on arbuscular mycorrhizas in tallgrass prairie. Am. Midl. Nat. 142, 55–70.

Glitzenstein, J.S., Streng, D.R., Masters, R.E., Robertson, K.M., Hermann, S.M., 2012. Fire-frequency effects on vegetation in North Florida pinelands: another look at the long-term Stoddard fire research plots at tall timbers research station. For. Ecol. Manage. 264, 197–209.

Godwin, D., Kobziar, L., Robertson, K., 2017. Effects of fire frequency and soil temperature on soil CO_2 efflux rates in old-field pine-grassland forests. Forests 8, 274.

Grady, K.C., Hart, S.C., 2006. Influences of thinning, prescribed burning, and wildfire on soil processes and properties in southwestern ponderosa pine forests: a retrospective study. For. Ecol. Manage. 234, 123–135.

Groeschl, D., Johnson, J., Smith, D., 1993. Wildfire effects on forest floor and surface soil in a table mountain pine-pitch pine forest. Int. J. Wildland Fire 3, 149–154.

Grove, T.S., O'Connell, A.M., Dimmock, G.M., 1986. Nutrient changes in surface soils after an intense fire in jarrah (Eucalyptus marginata Donn ex Sm.) forest. Austral Ecol. 11, 303–317.

Guénon, R., Vennetier, M., Dupuy, N., Roussos, S., Pailler, A., Gros, R., 2013. Trends in recovery of Mediterranean soil chemical properties and microbial activities after infrequent and frequent wildfires. Land Degrad. Dev. 24, 115–128.

Guinto, D.F., Xu, Z.H., Saffigna, P.G., House, A.P.N., Saffigna, P.G., House, A.P.N., 2001. Soil chemical properties and forest floor nutrients under repeated prescribed-burning in eucalypt forests of South-East Queensland, Australia. N. Z. J. For. Sci. 31, 170–187.

Hanes, T.L., 1971. Succession after fire in the chaparral of southern California. Ecol. Monogr. 41, 27–52.

Harris, R.M.B., Beaumont, L.J., Vance, T.R., Tozer, C.R., Remenyi, T.A., Perkins-Kirkpatrick, S.E., Mitchell, P.J., Nicotra, A.B., McGregor, S., Andrew, N.R., Letnic, M., Kearney, M.R., Wernberg, T., Hutley, L.B., Chambers, L.E., Fletcher, M.S., Keatley, M.R., Woodward, C.A., Williamson, G., Duke, N.C., Bowman, D.M.J.S., 2018. Biological Responses to the Press and Pulse of Climate Trends and Extreme Events. Nature Publishing Group.

Hart, S.C., Classen, A.T., Wright, R.J., 2005a. Long-term interval burning alters fine root and mycorrhizal dynamics in a ponderosa pine forest. J. Appl. Ecol. 42, 752–761.

Hart, S., DeLuca, T., Newman, G., 2005b. Post-fire vegetative dynamics as drivers of microbial community structure and function in forest soils. For. Ecol. Manage. 220, 166–184.

Hartnett, D.C., Potgieter, A.F., Wilson, G.W.T., 2004. Fire effects on mycorrhizal symbiosis and root system architecture in southern African savanna grasses. Afr. J. Ecol. 42, 328–337.

Hedges, L.V., Gurevitch, J., Curtis, P.S., 1999. The meta-analysis of response ratios in experimental ecology. Ecology 80, 1150–1156.

Hernández, D.L., Hobbie, S.E., 2008. Effects of fire frequency on oak litter decomposition and nitrogen dynamics. Oecologia 158, 535–543.

Higgins, S.I., Bond, W.J., February, E.C., Bronn, A., Euston-Brown, D.I.W., Enslin, B., Govender, N., Rademan, L., O'Regan, S., Potgieter, A.L.F., 2007. Effects of four decades of fire manipulation on woody vegetation structure in savanna. Ecology 88, 1119–1125.

Higgins, S.I., Bond, W.J., Combrink, H., Craine, J.M., February, E.C., Govender, N., Lannas, K., Moncreiff, G., Trollope, W.S.W., 2012. Which traits determine shifts in the abundance of tree species in a fire-prone savanna? J. Ecol. 100, 1400–1410.

Hobbie, S.E., 1992. Effects of plant species on nutrient cycling. Trends Ecol. Evol. 7, 336–339.

Hobbie, S.E., 2015. Plant species effects on nutrient cycling: revisiting litter feedbacks. Trends Ecol. Evol. 30, 357–363.

Hoffmann, W.A., Franco, A.C., Moreira, M.Z., Haridasan, M., 2005. Specific leaf area explains differences in leaf traits between congeneric savanna and forest trees. Funct. Ecol. 19, 932–940.

Hoffmann, W.A., Adasme, R., Haridasan, M., de Carvalho, M.T., Geiger, E.L., Pereira, M.A.B., Gotsch, S.G., Franco, A.C., 2009. Tree topkill, not mortality, governs the dynamics of savanna-forest boundaries under frequent fire in Central Brazil. Ecology 90, 1326–1337.

Hoffmann, W.A., Geiger, E.L., Gotsch, S.G., Rossatto, D.R., Silva, L.C.R., Lau, O.L., Haridasan, M., Franco, A.C., 2012. Ecological thresholds at the savanna-forest boundary: how plant traits, resources and fire govern the distribution of tropical biomes. Ecol. Lett. 15, 759–768.

Holdo, R.M., Mack, M.C., Arnold, S.G., 2012. Tree canopies explain fire effects on soil nitrogen, phosphorus and carbon in a savanna ecosystem. J. Veg. Sci. 23, 352–360.

Hosking, J.S., 1938. The ignition at low temperatures of the organic matter in soils. J. Agric. Sci. 28, 393–400.

Ives, A.R., 1995. Predicting the response of populations to environmental change. Ecology 76, 926–941.

Ives, A.R., Carpenter, S.R., 2007. Stability and diversity of ecosystems. Science 317, 58–62.

Jackson, R.B., Lajtha, K., Crow, S.E., Hugelius, G., Kramer, M.G., Piñeiro, G., 2017. The ecology of soil carbon: pools, vulnerabilities, and biotic and abiotic controls. Annu. Rev. Ecol. Evol. Syst. 48, 419–445.

Jiménez Esquilín, A.E., Stromberger, M.E., Shepperd, W.D., 2008. Soil scarification and wildfire interactions and effects on microbial communities and carbon. Soil Sci. Soc. Am. J. 72, 111.

Johnson, D.W., Curtis, P.S., 2001. Effects of forest management on soil C and N storage: meta analysis. For. Ecol. Manage. 140, 227–238.

Johnstone, J.F., Allen, C.D., Franklin, J.F., Frelich, L.E., Harvey, B.J., Higuera, P.E., Mack, M.C., Meentemeyer, R.K., Metz, M.R., Perry, G.L., Schoennagel, T., Turner, M.G., 2016. Changing disturbance regimes, ecological memory, and forest resilience. Front. Ecol. Environ. 14, 369–378.

Kauffman, J.B., Sanford Jr., R.L., Cummings, D.L., Salcedo, I.H., Sampaio, E.V.S.B., 1993. Biomass and nutrient dynamics associated with slash fires in neotropical dry forests. Ecology 74, 140–151.

Kauffman, J.B., Cummings, D.L., Ward, D., 1994. Relationships of fire, biomass and nutrient dynamics along a vegetation gradient in the Brazilian Cerrado. J. Ecol. 82, 519–531.

Kauffman, J.B., Hughes, R.F., Heider, C., 2009. Carbon pool and biomass dynamics associated with deforestation, land use, and agricultural abandonment in the neotropics. Ecol. Appl. 19, 1211–1222.

Kitzberger, T., Raffaele, E., Heinemann, K., Mazzarino, M.J., 2005. Effects of fire severity in a north Patagonian subalpine forest. J. Veg. Sci. 16, 5–12.

Köster, K., Berninger, F., Heinonsalo, J., Lindén, A., Köster, E., Ilvesniemi, H., Pumpanen, J., 2016. The long-term impact of low-intensity surface fires on litter decomposition and enzyme activities in boreal coniferous forests. Int. J. Wildland Fire 25, 213–223.

Lake, P.S., 2000. Disturbance, patchiness, and diversity in streams. J. N. Am. Benthol. Soc. 19, 573–592.

LeDuc, S.D., Rothstein, D.E., 2007. Initial recovery of soil carbon and nitrogen pools and dynamics following disturbance in jack pine forests: a comparison of wildfire and clearcut harvesting. Soil Biol. Biochem. 39, 2865–2876.

Lewis, T., Debuse, V.J., 2012. Resilience of a eucalypt forest woody understorey to long-term (34-55 years) repeated burning in subtropical Australia. Int. J. Wildland Fire 21, 980–991.

Maracahipes, L., Carlucci, M.B., Lenza, E., Marimon, B.S., Marimon, B.H., Guimarães, F.A.G., Cianciaruso, M.V., 2018. How to live in contrasting habitats? Acquisitive and conservative strategies emerge at inter- and intraspecific levels in savanna and forest woody plants. Perspect. Plant Ecol. Evol. Syst. 34, 17–25.

Martí-Roura, M., Casals, P., Romanyà, J., 2013. Long-term retention of post-fire soil mineral nitrogen pools in Mediterranean shrubland and grassland. Plant Soil 371, 521–531.

Metcalfe, D.B., Rocha, W., Balch, J.K., Brando, P.M., Doughty, C.E., Malhi, Y., 2018. Impacts of fire on sources of soil CO_2 efflux in a dry Amazon rain forest. Glob. Chang. Biol. 24, 3629–3641.

Monleon, V.J., Cromack Kermit, J., Landsberg, J.D., 1997. Short-and long-term effects of prescribed underburning on nitrogen availability in ponderosa pine stands in Central Oregon. Can. J. For. Res. 27, 369–378.

Moreira, A.G., 2000. Effects of fire protection on savanna structure in Central Brazil. J. Biogeogr. 27, 1021–1029.

Morton, D.C., DeFries, R.S., Shimabukuro, Y.E., Anderson, L.O., Arai, E., del Bon Espirito-Santo, F., Freitas, R., Morisette, J., 2006. Cropland expansion changes deforestation dynamics in the southern Brazilian Amazon. Proc. Natl. Acad. Sci. U.S. A. 103, 14637–14641.

Nardoto, G.B., Bustamante, M.M.D.C., 2003. Effects of fire on soil nitrogen dynamics and microbial biomass in savannas of Central Brazil. Pesq. Agrop. Brasileira 38, 955–962.

Nave, L.E., Vance, E.D., Swanston, C.W., Curtis, P.S., 2011. Fire effects on temperate forest soil C and N storage. Ecol. Appl. 21, 1189–1201.

Newland, J.A., DeLuca, T.H., 2000. Influence of fire on native nitrogen-fixing plants and soil nitrogen status in ponderosa pine—Douglas-fir forests in western Montana. Can. J. For. Res. 30, 274–282.

Norris, M.D., Reich, P.B., 2009. Modest enhancement of nitrogen conservation via retranslocation in response to gradients in N supply and leaf N status. Plant Soil 316, 193–204.

Norris, M.D., Avis, P.G., Reich, P.B., Hobbie, S.E., 2013. Positive feedbacks between decomposition and soil nitrogen availability along fertility gradients. Plant Soil 367, 347–361.

Oliver, A.K., Callaham, M.A., Jumpponen, A., 2015. Soil fungal communities respond compositionally to recurring frequent prescribed burning in a managed southeastern US forest ecosystem. For. Ecol. Manage. 345, 1–9.

Pellegrini, A.F.A.F.A., 2016. Nutrient limitation in tropical savannas across multiple scales and mechanisms. Ecology 97, 313–324.

Pellegrini, A.F.A., Hoffmann, W.A., Franco, A.C., 2014. Carbon accumulation and nitrogen pool recovery during transitions from savanna to forest in Central Brazil. Ecology 95, 342–352.

Pellegrini, A.F.A., Hedin, L.O., Staver, A.C., Govender, N., Henry, H.A.L., 2015. Fire alters ecosystem carbon and nutrients but not plant nutrient stoichiometry or composition in tropical savanna. Ecology 96, 1275–1285.

Pellegrini, A.F.A., Staver, A.C., Hedin, L.O., Charles-Dominique, T., Tourgee, A., 2016. Aridity, not fire, favors nitrogen-fixing plants across tropical savanna and forest biomes. Ecology 97, 2177–2183.

Pellegrini, A.F.A., Ahlström, A., Hobbie, S.E., Reich, P.B., Nieradzik, L.P., Staver, A.C., Scharenbroch, B.C., Jumpponen, A., Anderegg, W.R.L., Randerson, J.T., Jackson, R.B., 2018. Fire frequency drives decadal changes in soil carbon and nitrogen and ecosystem productivity. Nature 553, 194–198.

Pellegrini, A.F., McLauchlan, K.K., Hobbie, S.E., Mack, M.C., Marcotte, A.L., Nelson, D.M., Perakis, S.S., Reich, P.B., Whittinghill, K., 2020. Frequent burning causes large losses of carbon from deep soil layers in a temperate savanna. J. Ecol. https://doi.org/10.1111/1365-2745.13351.

Peterson, D.W., Reich, P.B., 2001. Prescribed fire in oak savanna: fire frequency effects on stand structure and dynamics. Ecol. Appl. 11, 914–927.

Peterson, D.W., Reich, P.B., Wrage, K.J., Franklin, J., 2007. Plant functional group responses to fire frequency and tree canopy cover gradients in oak savannas and woodlands. J. Veg. Sci. 18, 3–12.

Pietikäinen, J., Fritze, H., 1995. Clear-cutting and prescribed burning in coniferous forest: comparison of effects on soil fungal and total microbial biomass, respiration activity and nitrification. Soil Biol. Biochem. 27, 101–109.

Ponder, F., Tadros, M., Loewenstein, E.F., 2009. Microbial properties and litter and soil nutrients after two prescribed fires in developing savannas in an upland Missouri Ozark Forest. For. Ecol. Manage. 257, 755–763.

Pourreza, M., Hosseini, S.M., Safari Sinegani, A.A., Matinizadeh, M., Dick, W.A., 2014. Soil microbial activity in response to fire severity in Zagros oak (Quercus brantii Lindl.) forests, Iran, after one year. Geoderma 213, 95–102.

Pressler, Y., Moore, J.C., Cotrufo, M.F., 2018. Belowground community responses to fire: meta-analysis reveals contrasting responses of soil microorganisms and mesofauna. Oikos 128, 309–327.

Prieto-Fernández, A., Acea, M.J., Carballas, T., 1998. Soil microbial and extractable C and N after wildfire. Biol. Fertil. Soils 27, 132–142.

Raison, R.J., Khanna, P.K., Woods, P.V., 1985. Transfer of elements to the atmosphere during low-intensity prescribed fires in three Australian subalpine eucalypt forests. Can. J. For. Res. 15, 657–664.

Raison, R.J., O'connell, A.M., Khanna, P.K., Keith, H., 1993. Effects of repeated fires on nitrogen and phosphorus budgets and cycling processes in forest ecosystems. In: Fire in Mediterranean ecosystems. Commission of the European countries, Ecosystem Research Report. vol. 5. Commission of the European Communities, Brussels, pp. 347–363.

Reich, P.B., Peterson, D.W., Wedin, D.A., Wrage, K., 2001. Fire and vegetation effects on productivity and nitrogen cycling across a forest-grassland continuum. Ecology 82, 1703–1719.

Rein, G., Cleaver, N., Ashton, C., Pironi, P., Torero, J.L., 2008. The severity of smouldering peat fires and damage to the forest soil. Catena 74, 304–309.

Rietl, A.J., Jackson, C.R., 2012. Effects of the ecological restoration practices of prescribed burning and mechanical thinning on soil microbial enzyme activities and leaf litter decomposition. Soil Biol. Biochem. 50, 47–57.

Ryan, C.M., Williams, M., 2011. How does fire intensity and frequency affect miombo woodland tree populations and biomass? Ecol. Appl. 21, 48–60.

Ryan, C.M., Williams, M., Grace, J., 2011. Above-and belowground carbon stocks in a miombo woodland landscape of Mozambique. Biotropica 43, 423–432.

Sala, O.E., Gherardi, L.A., Reichmann, L., Jobbágy, E., Peters, D., 2012. Legacies of precipitation fluctuations on primary production: theory and data synthesis. Philos. Trans. R. Soc. Lond. Ser. B Biol. Sci. 367, 3135–3144.

Schwinning, S., Sala, O.E., 2004. Hierarchy of responses to resource pulses in arid and semi-arid ecosystems. Oecologia 141, 211–220.

Schwinning, S., Sala, O.E., Loik, M.E., Ehleringer, J.R., 2004. Thresholds, memory, and seasonality: understanding pulse dynamics in arid/semi-arid ecosystems. Oecologia 141, 191–193.

Seiler, W., Crutzen, P.J., 1980. Estimates of gross and net fluxes of carbon between the biosphere and the atmosphere from biomass burning. Clim. Change 2, 207–247.

Smith, M.D., Knapp, A.K., Collins, S.L., 2009. A framework for assessing ecosystem dynamics in response to chronic resource alterations induced by global change. Ecology 90, 3279–3289.

Taylor, M.K., Lankau, R.A., Wurzburger, N., 2016. Mycorrhizal associations of trees have different indirect effects on organic matter decomposition. J. Ecol. 104, 1576–1584.

Tiedemann, A.R., 1987. Combustion losses of sulfur from forest foliage and litter. For. Sci. 33, 216–223.

Tierney, J.A., Hedin, L.O., Wurzburger, N., 2019. Nitrogen fixation does not balance fire-induced nitrogen losses in longleaf pine savannas. Ecology 100, e02735.

Trumbore, S., Brando, P., Hartmann, H., 2015. Forest health and global change. Science 349, 814–818.

Tuininga, A.R., Dighton, J., 2004. Changes in ectomycorrhizal communities and nutrient availability following prescribed burns in two upland pine–oak forests in the New Jersey pine barrens. Can. J. For. Res. 34, 1755–1765.

Turetsky, M.R., Benscoter, B., Page, S., Rein, G., van der Werf, G.R., Watts, A., 2015. Global vulnerability of peatlands to fire and carbon loss. Nat. Geosci. 8, 11–14.

Turner, J., Lambert, M., Jurskis, V., Bi, H., 2008. Long term accumulation of nitrogen in soils of dry mixed eucalypt forest in the absence of fire. For. Ecol. Manage. 256, 1133–1142.

Uhl, C., Jordan, C.F., 1984. Succession and nutrient dynamics following forest cutting and burning in Amazonia. Ecology 65, 1476–1490.

Usup, A., Hashimoto, Y., Takahashi, H., Hayasaka, H., 2004. Combustion and thermal characteristics of peat fire in tropical peatland in Central Kalimantan, Indonesia. Tropics 14, 1–19.

Van de Vijver, C.A.D.M., Poot, P., Prins, H.H.T., 1999. Causes of increased nutrient concentrations in post-fire regrowth in an East African savanna. Plant Soil 214, 173–185.

Van der Werf, G.R., Morton, D.C., DeFries, R.S., Olivier, J.G.J., Kasibhatla, P.S., Jackson, R.B., Collatz, G.J., Randerson, J.T., 2009. CO_2 emissions from forest loss. Nat. Geosci. 2, 737–738.

van der Werf, G.R., Randerson, J.T., Giglio, L., Collatz, G.J., Mu, M., Kasibhatla, P.S., Morton, D.C., DeFries, R.S., van Jin, Y., van Leeuwen, T.T., 2010. Global fire emissions and the contribution of deforestation, savanna, forest, agricultural, and peat fires (1997–2009). Atmos. Chem. Phys. 10, 11707–11735.

van der Werf, G.R., Randerson, J.T., Giglio, L., van Leeuwen, T.T., Chen, Y., Rogers, B.M., Mu, M., van Marle, M.J.E., Morton, D.C., Collatz, G.J., Yokelson, R.J., Kasibhatla, P.S., 2017. Global fire emissions estimates during 1997–2016. Earth Syst. Sci. Data 9, 697–720.

Verma, S., Jayakumar, S., 2018. Effect of recurrent fires on soil nutrient dynamics in a tropical dry deciduous forest of Western Ghats, India. J. Sustain. For. 37, 678–690.

Walker, X.J., Baltzer, J.L., Cumming, S.G., Day, N.J., Ebert, C., Goetz, S., Johnstone, J.F., Potter, S., Rogers, B.M., Schuur, E.A.G., Turetsky, M.R., Mack, M.C., 2019. Increasing wildfires threaten historic carbon sink of boreal forest soils. Nature 572, 520–523.

Wan, S., Hui, D., Luo, Y., 2001. Fire effects on nitrogen pools and dynamics in terrestrial ecosystems: a meta-analysis. Ecol. Appl. 11, 1349–1365.

Wang, Q., Zhong, M., Wang, S., 2012. A meta-analysis on the response of microbial biomass, dissolved organic matter, respiration, and N mineralization in mineral soil to fire in forest ecosystems. For. Ecol. Manage. 271, 91–97.

Wigley, B.J., Slingsby, J.A., Díaz, S., Bond, W.J., Fritz, H., Coetsee, C., 2016. Leaf traits of African woody savanna species across climate and soil fertility gradients: evidence for conservative versus acquisitive resource-use strategies. J. Ecol. 104, 1357–1369.

Yelenik, S.G., Perakis, S.S., Hibbs, D.E., 2012. Regional constraints to biological nitrogen fixation in post-fire forest communities. Ecology 94, 739–750.

Zarin, D.J., Davidson, E.A., Brondizio, E., Vieira, I.C.G., Sá, T., Feldpausch, T., Schuur, E.A.G., Mesquita, R., Moran, E., Delamonica, P., 2005. Legacy of fire slows carbon accumulation in Amazonian forest regrowth. Front. Ecol. Environ. 3, 365–369.

Data for the single fire studies are in the supplemental tables

Baird et al., 1999; Boerner et al., 2009; Choromanska and DeLuca, 2001, 2002; Grady and Hart, 2006; Groeschl et al., 1993; Grove et al., 1986; Jiménez Esquilín et al., 2008; Kauffman et al., 1993, 1994; Kitzberger et al., 2005; LeDuc and Rothstein, 2007; Martí-Roura et al., 2013; Monleon et al., 1997; Ponder et al., 2009; Prieto-Fernández et al., 1998; Van de Vijver et al., 1999.

CHAPTER FIVE

Revisiting nutrient cycling by litterfall—Insights from 15 years of litter manipulation in old-growth lowland tropical forest

Emma J. Sayer[a,b,*], Chadtip Rodtassana[c], Merlin Sheldrake[b,d], Laëtitia M. Bréchet[e,f], Oliver S. Ashford[g], Luis Lopez-Sangil[h], Deirdre Kerdraon-Byrne[a], Biancolini Castro[b], Benjamin L. Turner[b], S. Joseph Wright[b], Edmund V.J. Tanner[d]

[a]Lancaster Environment Centre, Lancaster University, Lancaster, United Kingdom
[b]Smithsonian Tropical Research Institute, Panama, Republic of Panama
[c]Department of Botany, Faculty of Science, Chulalongkorn University, Bangkok, Thailand
[d]Department of Plant Sciences, University of Cambridge, Cambridge, United Kingdom
[e]Department of Biology, Centre of Excellence PLECO (Plant and Ecosystems), University of Antwerp, Wilrijk, Belgium
[f]INRA, UMR Ecology of Guiana Forests, AgroParisTech, Cirad, CNRS, Université des Antilles, Université de Guyane, Kourou, French Guiana
[g]Integrative Oceanography Division, Scripps Institution of Oceanography, University of California, San Diego, CA, United States
[h]Teagasc, Johnstown Castle, Wexford, Ireland
*Corresponding author: e-mail address: e.sayer@lancaster.ac.uk

Contents

Advances in Ecological Research, Volume 62
ISSN 0065-2504
https://doi.org/10.1016/bs.aecr.2020.01.002

Abstract

The crucial role of tropical forests in the global carbon balance is underpinned by their extraordinarily high biomass and productivity, even though the majority of tropical forests grow on nutrient-poor soils. Nutrient cycling by litterfall has long been considered essential for maintaining high primary productivity in lowland tropical forests but few studies have tested this assumption experimentally. We review and synthesise findings from the Gigante Litter Manipulation Project (GLiMP), a long-term experiment in lowland tropical forest in Panama, Central America, in which litter has been removed from or added to large-scale plots for 15 years. We assessed changes in soil and litter nutrient concentrations in response to the experimental treatments and estimated nutrient return and nutrient use efficiency to indicate changes in nutrient cycling. The soil concentrations of most nutrients increased with litter addition and declined with litter removal. Litter removal altered nitrogen, potassium, manganese and zinc cycling, demonstrating the importance of litter inputs for maintaining the availability of these elements to plants. By contrast, litter addition only altered nitrogen cycling and, despite low concentrations of available soil phosphorus, the effects of litter manipulation on phosphorus cycling were inconsistent. We discuss potential mechanisms underlying the observed changes, and we emphasise the importance of decomposition processes in the forest floor for retaining nutrient elements, which partially decouples nutrient cycling from the mineral soil. Finally, by synthesising GLiMP studies conducted during 15 years of litter manipulation, we highlight key knowledge gaps and avenues for future research into tropical forest nutrient cycling.

1. Introduction

1.1 The dual role of litterfall in tropical forest nutrient cycling

The importance of tropical forests in global biogeochemical cycles is undisputed: tropical forests represent the largest store of terrestrial biomass carbon (Brown and Lugo, 1982), regulate continental scale hydrological cycles (Gloor et al., 2013) and contribute to global atmospheric circulation (Malhi et al., 2014). Unfortunately, our understanding of carbon (C) and nutrient dynamics in tropical forest ecosystems is deficient, which hampers our ability to predict the effects of global changes (Bonan, 2008; Wieder et al., 2015). Long-term shifts in temperature (e.g. Corlett, 2011), precipitation (Meir and Woodward, 2010), and atmospheric deposition (Hietz et al., 2011) are likely to have major impacts on tropical forest ecosystem functioning by influencing tree growth and species composition (Wright, 2005) as well as elemental cycling. In this context, tropical forest nutrient cycling is of considerable interest because nutrient cycles regulate forest C storage (Finzi et al., 2011) and many tropical forests maintain extraordinarily high biomass and productivity, despite growing on nutrient-poor soils (Vitousek and Sanford, 1986).

Litterfall and litter decomposition play a central role in the cycling of C and nutrients in tropical forests (Sayer and Tanner, 2010). Although leaf litter only contains a small proportion of the nutrient content in above-ground forest biomass (typically $<10\%$; Proctor, 1987), it dominates the intra-annual cycle of nutrients through rapid turnover (Schlesinger, 1991); litterfall therefore represents a major pathway for the transfer of mineral nutrients and organic matter from vegetation to soil (Vitousek and Sanford, 1986). Many lowland tropical forests are characterised by conservative or "tight" nutrient cycling with low losses of nutrients (Vitousek, 1984), which is facilitated by substantial direct uptake of macronutrients from decomposing leaf litter by plant roots (Medina and Cuevas, 1989; Tobón et al., 2004). The processes of microbial immobilisation and mineralisation of nutrients from organic material provide a "slow-release" mechanism of nutrients to plants, which limits nutrient losses in gaseous forms or via leaching (Sayer et al., 2012), and fine roots concentrated in organic horizons and the upper layers of the mineral soil intercept nutrients as they are released from decomposing litter (Attiwill and Adams, 1993; Yavitt et al., 2011). Indeed, due to the large amounts of nutrients contained in litter and other decomposing organic material on the soil surface ("forest floor"; Box 1), nutrient cycling between plants and soil can become partially decoupled from the mineral soil (Leuschner et al., 2001; Sayer et al., 2006a) and cycling of some nutrients in organic matter can meet much of the annual demand by plants (Attiwill and Adams, 1993). In addition to acting as a nutrient reservoir, the forest floor also intercepts nutrients in throughfall (Tobón et al., 2004) and can retain nutrients that would otherwise be lost through leaching or, in tropical soils with low base saturation and high aluminium mobility, through sorption to metal oxides (Medina and Cuevas, 1989; Tobón et al., 2004). Hence, litter inputs play a dual role in tropical forest nutrient cycling: as a primary source of nutrients and as an important mechanism for nutrient retention within the ecosystem.

BOX 1 Glossary—Guiding definitions of key terms used in this work.

Term	Definition and context
Direct cycling/uptake (of nutrients)	The capture and uptake of nutrients by roots and mycorrhizal fungi upon mineralisation and release from decomposing organic matter in the forest floor; regarded as an important nutrient conservation mechanism in highly weathered, infertile soils
Forest floor	The organic horizon on the surface of the mineral soil, comprising litter, roots, and organic material in various stages of decay (Sayer, 2006); often referred to as the "O Horizon"
Immobilisation (of nutrients)	Microbial conversion of nutrients from inorganic to organic forms (i.e. incorporation into cells), making them unavailable to plants; immobilisation is the reverse of mineralisation
Leaky nutrient cycle	Inefficient nutrient cycle characterised by high losses of a given element through leaching or in gaseous forms; nutrient cycling within an ecosystem can be characterised as leaky for one nutrient but as "tight" or efficient for another
Litter layer	The surface layer of the forest floor, which is not in an advanced stage of decomposition and comprises fallen leaves, needles, fruits, flowers, and twigs (Sayer, 2006); often referred to as the L or O_L horizon
Mineralisation (of nutrients)	Microbial conversion of nutrients in organic forms into soluble inorganic forms that may be available to plants; usually occurs during the decomposition of organic matter
Nutrient limitation	Nutrient limitation is inferred when additions of an essential element in biologically available forms cause an increase in the rate of an ecosystem process (sensu Tanner et al., 1998)
Nutrient return	The amount of a given nutrient cycled annually in litterfall, calculated as the product of litterfall mass and litter nutrient concentrations; also referred to as nutrient accession

BOX 1 Glossary—Guiding definitions of key terms used in this work.—cont'd

Nutrient use efficiency	A measure of how well plants use available mineral nutrients; defined for forests as the amount of organic matter lost from plants or permanently stored in plants per unit of nutrients lost or stored; calculated here as the inverse of nutrient concentrations in aboveground litterfall (sensu Vitousek, 1982)
Tight (conservative) nutrient cycle	Nutrient cycle characterised by high nutrient use efficiency and minimal losses of mineral nutrients, e.g. by rapid uptake by plants, mycorrhizas or decomposers (sensu Vitousek, 1984); also referred to as "efficient" or "conservative"

We know relatively little about nutrient and C cycling in tropical forests compared to temperate wooded systems (Townsend et al., 2011), in particular about the specific role of litterfall (Sayer, 2006). Fertilisation experiments in tropical forests have demonstrated that various ecosystem processes are limited by the availability of different nutrients (Cleveland and Townsend, 2006; Kaspari et al., 2008; Ostertag, 2010; Vitousek et al., 1993; Wright et al., 2011; Wright et al., 2018), but explicit consideration of nutrient return by litterfall is still crucial for understanding elemental cycling. Leaf litter can essentially be viewed as a natural complete fertiliser (Sayer et al., 2012) not only because it contains all the elements required for plant growth, but also because those nutrients are cycled with substantial amounts of C (*c.* 50% leaf dry mass). The stoichiometric balance of C and nutrients is critical to a large number of processes underpinning ecosystem C and nutrient cycling (Elser et al., 2010; Sardans et al., 2012; Sinsabaugh et al., 2008). Importantly, elemental stoichiometry drives the decomposition of organic material (Manzoni et al., 2010), because microbial decomposers rely on organic C as a source of energy, and usually process C and nutrients together (Finzi et al., 2011; Sinsabaugh et al., 2008). Carbon inputs also modify the rate at which nutrients are processed and released, as well as the forms in which nutrients are present in the soil (Palm et al., 1997), and there is evidence that nutrient availability can influence the storage or release of soil C in tropical forests (Cleveland et al. 2006; Nottingham et al., 2015; Wieder et al., 2015). Despite the clear importance of coupled

C and nutrient cycling by litterfall, we still lack an integrated understanding of nutrient pathways from leaf litter to decomposer organisms and soil through to re-uptake by roots. The gaps in our knowledge stem partly from the difficulties of accounting for the heterogeneity of tropical forest ecosystems (Townsend et al., 2011) and the extraordinarily high diversity of plants (Sayer and Banin, 2016), but also because experimental assessments of the role of leaf litter in tropical forest elemental dynamics are scarce.

1.2 Assessing the role of litterfall using manipulative experiments

Experimental manipulation of litter dates back to the 1800s and has provided key insights into the role of litterfall in forests (Sayer, 2006). Historically, studies have focussed on litter removal experiments to elucidate the importance of nutrients cycled in litterfall for maintaining forest productivity (Sayer, 2006). Recognition of important knowledge gaps in soil organic C dynamics has also given rise to numerous litter manipulation experiments, most notably the Detritus Input and Removal Treatments (DIRT) network (Nadelhoffer et al., 2004; Lajtha et al., 2018), which include litter removal and litter addition treatments, but also consider belowground carbon inputs using "no roots" and "no inputs" treatments. The DIRT network has focussed principally on C dynamics, demonstrating that aboveground litter inputs may be just as important for soil C sequestration as root inputs, and that site fertility plays a key role in the contribution of aboveground litter to soil C stocks (Lajtha et al., 2018). By contrast, there are few experimental studies of nutrient cycling by litterfall and most of our knowledge about the long-term effects of enhanced or reduced litter inputs on forest nutrient status is derived from temperate systems (Sayer, 2006). Although it seems reasonable to assume that leaf litter will play a similar role in temperate and tropical forest ecosystem C and nutrient dynamics, there are three crucial distinctions: (1) rapid decomposition in a warm humid tropical climate results in fast turnover of organic matter and high rates of element cycling; (2) in many tropical forests, growth occurs all year round and litterfall seasonality is much less pronounced than in temperate deciduous forests; (3) the vast majority of tropical forests are highly diverse, and the diversity of species contributing litter also influences decomposition processes by shaping soil fauna communities (Laird-Hopkins et al., 2017).

We aimed to provide a comprehensive overview of the role of litterfall in tropical forest nutrient cycling by establishing a large-scale, long-term litter manipulation experiment. The Gigante Litter Manipulation Project

(GLiMP) has been running continuously since 2003 and has afforded numerous insights into the role of leaf litter in tropical forest ecosystem processes. Here, we synthesise the results of 15 years of research within the GLiMP experimental plots and compare key findings with other studies in both temperate and tropical forests. We complement the results of published studies with new datasets to investigate the trajectory of change in nutrient cycles in response to long-term litter addition and removal treatments and assess the consequences for broader ecosystem functioning. Finally, we aim to motivate future research by identifying persisting and emerging knowledge gaps about the role of litterfall in lowland tropical forests (Box 2).

2. The Gigante Litter Manipulation Project (GLiMP)

2.1 Study site and experimental design

The long-term litter manipulation experiment is located on Gigante Peninsula (9° 06′ N, 79° 54′ W), part of the Barro Colorado Nature Monument in Panama, Central America. Mean annual rainfall is *c.* 2600 mm with a strong dry season from January to April and the mean annual temperature is 26 °C (Leigh, 1999). The soil is a moderately acidic Oxisol with pH *c.* 5.5, a total C content of *c.* 4%, and low concentrations of extractable phosphorus (*c.* 0.5 $mg\,kg^{-1}$) and potassium (*c.* 40 $mg\,kg^{-1}$) but high concentrations of calcium (*c.* 1500 $mg\,kg^{-1}$) and magnesium (*c.* 400 $mg\,kg^{-1}$) at 0–10 cm depth (Ashford et al., 2013; Sayer and Tanner, 2010; Sheldrake et al., 2017; Tanner et al., 2016; Yavitt et al., 2009).

Between 2000 and 2002, 15 plots were established within a 40-ha area of old-growth lowland tropical forest. Each plot measures 45-m × 45-m and is bordered by 0.5-m deep trenches to limit nutrient and water transfer; the trenches were double-lined with construction plastic and back-filled with soil (Sayer and Tanner, 2010). All measurements and samples were taken in the inner 30-m × 30-m of each plot to limit potential effects of trenching. Starting in January 2003, the litter has been removed every 1–2 months from five plots and added to five plots, where it is spread out as evenly as possible, leaving five plots as undisturbed controls (Fig. 1). The treatments were initially assigned to the plots in a stratified random design based on the annual litterfall in 2002 (Sayer and Tanner, 2010), but as the litter treatments were paired geographically for logistical reasons, each replicate block comprises one pair of litter removal (L−) and litter addition (L+) plots and the closest control plot.

BOX 2 Major knowledge gaps and future research directions in tropical forest nutrient cycling.

Collectively, the GLiMP studies demonstrate that ecosystem responses to declining nutrient availability involve multiple mechanisms, including plant physiological adaptations to reduce nutrient demand, shifts in the species composition of microbial communities, mining of organic compounds, and direct uptake of nutrients from the forest floor. Hence, future research to address the following major knowledge gaps will greatly advance our understanding of tropical forest nutrient and carbon dynamics.

Knowledge Gap 1: Plant adaptations to nutrient availability

Expanding current theory of nutrient limitation to account for plant adaptation to nutrient availability could greatly improve our understanding of tropical forest nutrient cycling and ecosystem responses to global change. To achieve this, we need to assess changes in plant productivity and foliar nutrient concentrations against a variety of other mechanisms that control nutrient demand. Studies of nutrient allocation to different structures and physiological or metabolic processes could be particularly informative for understanding plant responses to altered nutrient supply. A combination of gradient studies and controlled experiments would then link differences in plant productivity and nutrient allocation to adaptation mechanisms such as resorption, prolonged leaf retention, and shifts in biomass allocation.

Knowledge Gap 2: Micronutrient cycles

Micronutrients are key to understanding decomposition food-webs but we still know very little about how most trace elements are cycled through the ecosystem, or how their availability influences the dynamics of other elements. A more complete picture of trace elements in tropical forest functioning could be gained from broad studies and networks assessing the effects of micronutrient additions along a gradient of macronutrient availability, followed by targeted experiments to investigate specific processes or groups of organisms (Kaspari and Powers, 2016). Experiments assessing the impact of both macro- and micronutrient availability on microbial processes and plant-microbe interactions could reveal particularly important insights into ecosystem functioning.

Knowledge Gap 3: The relative contribution of "direct nutrient cycling" and nutrients stored in soil organic matter

We provide multiple lines of evidence for nutrient uptake from the forest floor and the potential for soil organic matter to act as a major alternative source of scarce nutrients. We still lack detailed studies tracking the flow of nutrients and carbon in litterfall through various stages of decomposition to re-uptake by plants, and a greater focus on the formation and turnover of soil organic matter in tropical forests could contribute to quantifying the relative importance of organic compounds as potential sources of nutrients. Molecular and isotopic techniques developed since the initial characterisation of direct nutrient uptake could facilitate a new wave of research into microbial processing and transfer of nutrients in organic materials at different stages of decomposition and by distinct functional groups.

BOX 2 Major knowledge gaps and future research directions in tropical forest nutrient cycling.—cont'd

Knowledge Gap 4: Links between plant and microbial nutrient requirements

Microbial mineralisation of nutrients in organic matter provides substantial amounts of nutrients for plant growth but microbial resource requirements vary widely among different functional groups (Camenzind et al., 2018). Our experiment provides an excellent platform for studies of microbial communities and functions linked to elemental cycling via litterfall, but very little work in this area has been achieved to date. The GLiMP studies demonstrate direct effects of litter quantity on heterotrophic respiration via the addition of fresh organic carbon (Sayer et al., 2007, 2011), and indirect effects on mycorrhizal associations via altered plant nutrient demand (Sheldrake et al., 2017, 2018). Shifts in microbial communities may also underpin many of the observed patterns of nutrient accumulation and release from litter, and it is conceivable that other crucial microbial processes, such as N-fixation, are also heavily influenced by changes in litterfall. New molecular techniques facilitate numerous avenues for research into microbial communities and processes (Camenzind et al., 2018), which could contribute to addressing the current knowledge gaps on micronutrient and forest floor processes.

Fig. 1 The Gigante Litter Manipulation Project (GLiMP), showing adjacent litter removal and control plots separated by plastic-lined trenches (A) during the first year of treatments in 2003; (B) the soil surface after the first application of treatments in February 2003; and (C) the soil surface after 5 years of treatments in 2008.

2.2 Field and laboratory measurements

Soil samples for analysis of nutrient concentrations and pH were taken annually from 2004 to 2007 and in 2009, 2010, 2012 and 2017 (see Table 1 for full details of analysis in original data sources). Each year, four to nine soil cores were collected from 0 to 10 cm depth in each plot; samples in 2004 were extracted individually but in all other years, the cores were mixed to give one composite sample per plot. Briefly, inorganic nitrogen

Table 1 Datasets used to analyse patterns of change in nutrient cycling in a lowland tropical forest in Panama, Central America, in response to 15 years of litter addition and litter removal treatments, giving the citation for previously published data, where N is nitrogen, N_{TOT} is total soil N, NO_3^--N is soil nitrate-N, P is phosphorus, P_{EXTR} is extractable soil P, K is potassium, Ca is calcium, Mg is magnesium, B is boron, Mn is manganese, Na is sodium, Zn is zinc, pH is soil pH, SOC is soil organic carbon, C_{MIC} is microbial biomass carbon and N_{MIC} is microbial biomass N.

Source	Year(s)	Soil properties (0–10 cm depth)	Litter nutrients
Sayer et al. (2007)	2004, 2006	N_{TOT}[a], SOC, C_{MIC}, N_{MIC},	
Sayer and Tanner (2010)	2004–2007	NO_3^--N, P_{EXTR}, K, Ca, Mg[b], B, Mn, Na[b], Zn, pH	N, P, K, Ca, Mg, B, Mn, Na, Zn
Sayer et al. (2011)	2007	SOC	
Sayer et al. (2012)	2007	C_{MIC}, N_{MIC}	
Ashford et al. (2013)	2010	NO_3^--N, P_{EXTR}, K, Ca, Mg, B, Mn, Na, pH	
Tanner et al. (2016)	2009	N_{TOT}, SOC, NO_3^--N, P_{EXTR}, K, Ca, Mg, pH	
Sheldrake et al. (2017)	2012	N_{TOT}, SOC, NO_3^--N, P_{EXTR}, K, Ca, Mg, Mn, Na, pH	
Previously unpublished data	2013		N, P, K, Ca, Mg, B, Mn, Na, Zn
	2016	N_{TOT}, SOC, C_{MIC}, N_{MIC}, NO_3^--N, pH	
	2017		N, P, K, Ca, Mg, B, Mn, Na, Zn
	2018	P_{EXTR}, K, Ca, Mg, B, Mn, Zn	

[a]2004 only.
[b]No data for 2006.

(N; ammonium-N and nitrate-N) was determined by colourimetry after extraction with potassium chloride. Soils were stored for 24 h before extraction from 2004 to 2007 and extracted immediately in 2010, 2012 and 2017. This difference in storage times can affect the relative concentrations of ammonium-N and nitrate-N (Turner and Romero, 2009). Issues with contamination affected the results for ammonium-N in several years, which precluded analysis of long-term trends and we therefore only present the results for nitrate-N. Phosphorus, potassium, magnesium, calcium and micronutrients were determined by inductively-coupled plasma spectrometry (ICP-MS) after Mehlich III extraction, except in 2010 and 2012, when extractable soil P (P_{EXTR}) was determined by resin extraction (Turner and Romero, 2009). Where relevant, values for soil boron were corrected for iron interference according to Turner et al. (2016). Soil pH was measured on fresh soils in deionised water. Total N was determined by Kjeldahl extraction in 2004, 2009, 2012 and 2016, total soil organic C was determined by combustion oxidation in 2004, 2007, 2009, 2012, 2015 and 2016, and microbial biomass C and N were assessed by chloroform fumigation in 2004, 2006–2008, and 2016.

Litter nutrients and nutrient return in litterfall were assessed using dried litter samples collected from 10 traps per plot in September or October annually from 2005 to 2007 (Table 1) and from 5 traps per plot in December 2013 and October 2017. The samples (excluding the woody fraction) were pooled by plot and year, shredded, mixed, and a subsample was finely ground for nutrient analysis. Nitrogen concentrations in the litter were determined by complete combustion gas chromatography and all other nutrient concentrations were determined by ICP-MS after nitric acid digest.

After 15 years of litter manipulation, we used ion exchange resins to assess short-term changes in nutrient exchange rates in the topsoil. Nutrient exchange rates during the rainy season were determined using plant root simulator probes (PRS probes™, WesternAg, Canada), which consist of ion exchange resins held in plastic support frames. We installed four pairs of anion and cation exchange probes vertically at 0–10 cm in each plot in June 2017 (early rainy season) and November 2017 (late rainy season). The probes remained in place for 42 days on each occasion. After retrieval, the probes were cleaned with deionised water and returned to the manufacturer for analysis. Nitrate-N was determined by colourimetry and all remaining nutrient ions were measured by ICP-MS after acid digest. Nutrient exchange rates are based on the surface area of the resin and the exposure time and thus units are μg nutrient $10\,cm^{-2}\,42\,d^{-1}$; all data analyses were based on mean values per plot ($n = 5$ per treatment).

2.3 Data analysis and synthesis

We performed all analyses in R v.3.5.1 (R Development Core Team, 2018) using the lme4 package for linear mixed effects models (Bates et al., 2015) and the lmerTest package (Kuznetsova et al., 2017) to derive significance values for individual treatment levels, using Satterthwaite's approximation to estimate degrees of freedom. We estimated the annual nutrient return by litter for 2005–2007, 2013 and 2017 by multiplying annual total small litter mass per plot ($g\,m^{-2}\,year^{-1}$) by the nutrient concentrations of the litter ($mg\,kg^{-1}$) collected in September and October (2005–2007 and 2017) or December (2013) each year. We then calculated nutrient use efficiency sensu Vitousek (1982) as the ratio of annual litterfall dry mass to the nutrient return by litter. We also estimated the mean and total amount of nutrients transferred between L− and L+ plots based on mean annual litterfall and nutrient concentrations per treatment, using plot-level data for 2005–2007, 2013 and 2017 (Table 1), and nutrient concentrations from one composite sample per treatment for 2003 and 2009–2012 (Sayer and Tanner, 2010; Rodtassana, 2016; Table 2).

To assess patterns of change in nutrient cycling in the GLiMP plots, we collated data on nutrient concentrations in topsoil and litter from 2004 to

Table 2 Estimated annual nutrient return in litterfall in control (CT) litter addition (L+) and litter removal (L−) treatments in lowland tropical forest in Panama, Central America, showing means and standard errors for $n = 15$ years, and the total amount of each nutrient transferred from L− to L+ plots over the duration of the experiment.

	Nutrient return in litterfall ($kg\,ha^{-1}\,year^{-1}$)			
	CT	**L+**	**L−**	**Transferred ($kg\,ha^{-1}$)**
Nitrogen (N)	191 ± 24	234 ± 28	149 ± 15	2246
Phosphorus (P)	6.30 ± 0.6	7.28 ± 0.7	5.14 ± 0.4	77
Potassium (K)	59.5 ± 8.3	73.0 ± 12	47.3 ± 7.2	710
Calcium (Ca)	207 ± 34	218 ± 33	169 ± 23	2532
Magnesium (Mg)	46.5 ± 6.9	50.7 ± 7.5	36.8 ± 4.7	553
Boron (B)	0.57 ± 0.1	0.65 ± 0.1	0.48 ± 0.07	7.2
Manganese (Mn)	4.05 ± 0.6	4.01 ± 0.4	2.45 ± 0.3	37
Sodium (Na)	6.40 ± 1.1	7.04 ± 1.2	5.3 ± 0.9	80
Zinc (Zn)	0.42 ± 0.06	0.31 ± 0.03	0.31 ± 0.03	4.6

2018. Published studies describe the datasets from 2004 through 2012 (Table 1) and are supplemented here with new data collected between 2013 and 2018 to evaluate longer-term trends and predictions made when the experiment began. There were small methodological differences among years (e.g. sampling month, sample processing and laboratory analyses) that could potentially confound trends over time. These methodological differences will have similar effects on all three treatments (L+, L− and control) and we therefore analysed log response ratios, which minimise variation associated with methodological differences among years by standardising values from the L+ and L− treatments by values from the control treatment. We calculated log response ratios (RR) for each variable, block and year as follows:

$$\mathrm{RR} = \ln(\mathrm{R_X}/\mathrm{R_{CT}}) \tag{1}$$

where R_X is the measured value in treatment X and R_{CT} is the measured value in the control.

We first assessed the influence of litter manipulation on soil nutrients, litter nutrients, nutrient return in litterfall and nutrient use efficiency using linear mixed effects models (*lmer* function) with treatment as a fixed effect, and block and year as random effects. We then tested for trajectories in treatment responses (RR) over time using separate linear models for each treatment. We quantified nutrient exchange rates once in 2017, present their absolute values for each treatment and assess the effect of litter manipulation using linear mixed effects models, with treatment as a fixed effect, and block and sampling time as random effects.

For all linear mixed effects models, we dropped terms sequentially to determine their significance, using AIC and *P*-values to check for model improvement (Pinheiro and Bates, 2000). We tested final models against corresponding null models using likelihood ratio tests and assessed the fit of each model with diagnostic plots. For all significant terms, we report the effect size *d* and significance at $P < 0.05$ in the text. We also report marginally significant trends for $P < 0.1$ and $|d| > 0.6$. Statistics for the best fit model for each response variable are given in Supplementary Table S1 in the online version at https://doi.org/10.1016/bs.aecr.2020.01.002.

We calculated Cohen's *d* effect size (Cohen, 1988) for each year with blocks pooled as follows:

$$d = (\mathrm{M_X} - \mathrm{M_{CT}})/\mathrm{SD_{pooled}} \tag{2}$$

where M_X is the mean of a given treatment, M_{CT} is the mean of the control and SD_{pooled} is the pooled standard deviation:

$$SD_{pooled} = \sqrt{\left(\left(SD_X{}^2 + SD_{CT}{}^2\right)/2\right)} \quad (3)$$

where SD_X and SD_{CT} are the standard deviations of treatment and control, respectively. We present mean Cohen's d (± S.E.) across years for each response variable. Cohen's d represents standard deviations of difference, where $d \geq 0.8$ is a large effect (Cohen, 1988) and generally corresponded to significance at $P < 0.05$ in our analyses.

3. Macronutrient cycling in litter

Biogeochemical theory hypothesises that concentrations of rock-derived nutrients such as phosphorus (P) decline over geological time as a result of erosion and leaching, whereas nitrogen (N), which is derived mainly from biotic processes or atmospheric deposition, accumulates over time (Walker and Syers, 1976). Consequently, P is more likely to be limiting than N in old, highly weathered soils in the tropics, whereas N is more likely to be limiting in young soils in the temperate zone, but there are numerous lines of evidence for N and P co-limitation in many lowland tropical forests (see, e.g. Sayer and Banin, 2016; Townsend et al., 2011; Wright et al., 2018). Litterfall is the dominant pathway for N and P cycling (Schlesinger, 1991), and leaf litter also contains substantial amounts of potassium (K), calcium (Ca) and magnesium (Mg). Indeed, foliar concentrations of these nutrients often reflect soil fertility in tropical forests (Vitousek and Sanford, 1986) and hence 15 years of litter removal and addition treatments are likely to have a substantial impact on macronutrient cycling. We present and discuss the patterns for each nutrient in turn.

3.1 Nitrogen

Nitrogen (N) is the most extensively studied nutrient with respect to forest productivity and nutrient cycling. Foliar N is largely present in chlorophylls, proteins, and enzymes (Loomis, 1997), and is therefore the main nutrient element cycled in litterfall (Medina and Cuevas, 1989). Decades of research suggest that N is relatively more available to plants in the lowland tropics compared to temperate forests but there is also ample evidence for N-limitation of multiple ecosystem processes in tropical forests (e.g. Kaspari et al., 2008; Wright et al., 2018).

The forest soils in the Panama Canal area are relatively rich in N (Hietz et al., 2011; Santiago et al., 2005). High rates of N mineralisation at our study site (Yavitt, 2000) indicate rapid N cycling (Adams, 1986) and a "leaky" N-cycle characterised by substantial N losses in gaseous forms or via leaching (Corre et al., 2010; Erickson et al., 2001; Koehler et al., 2009). As the declines in soil N with litter removal in temperate forests were greater in soils with high initial N content (Sayer, 2006), we also expected substantial losses of N in our L− plots. Accordingly, during the first 5 years of treatments (2004–2007) soil inorganic N concentrations declined rapidly with litter removal, and there was a sizeable increase in nitrate-N with litter addition (Sayer and Tanner, 2010). A decade later (2016–2018), litter manipulation has clearly affected total N concentrations in the topsoil, as well as nitrate-N and litter N concentrations (Fig. 2A and C).

The largest and most rapid changes in response to litter manipulation were observed for nitrate-N concentrations in the soil (Fig. 2A). Within 2 years of the start of treatments, soil nitrate-N concentrations had declined with litter removal and increased with litter addition (Sayer and Tanner, 2010). After 7 years, nitrate-N had declined by 57% in L− plots, increased threefold in L+ plots (Sayer et al., 2012) and the higher nitrate-N concentrations with litter addition were apparent to 15 cm depth (Tanner et al., 2016). The differences in soil nitrate-N concentrations between L− and L+ plots have persisted throughout the experiment, with a greater overall effect of litter removal ($d = -2.38 \pm 0.34$, $P < 0.001$) compared to litter addition ($d = 1.83 \pm 0.31$, $P < 0.001$) and soil nitrate-N increased in the L+ plots over time ($F_{1,38} = 8.1$; $P = 0.007$). After 15 years of treatments, the exchange rate of nitrate-N in L+ plots was at least two-fold higher than in control plots in both the early and late rainy season ($P < 0.001$; Fig. 3A). We hypothesise that these changes in nitrate-N in response to litter manipulation occur because heterotrophic nitrification is stimulated by organic matter (Adams, 1986; Verstraete and Focht, 1977) and uses organic N compounds instead of ammonium as a substrate (Pedersen et al., 1999; Schimel et al., 1984).

Total soil N did not respond to litter manipulation during the first year of the experiment (Sayer et al., 2012) but concentrations of total N in the L− plots declined after 2009 ($F_{1,21} = 4.8$; $P = 0.04$; Fig. 2B), with a large effect size ($d = -2.21 \pm 0.31$, $P < 0.001$). We expected the total soil N pool to respond relatively slowly to litter manipulation because a large fraction of total soil N is bound in soil organic matter (Knicker et al., 1993). Long-term reductions in N inputs from leaf litter and decreased availability of inorganic

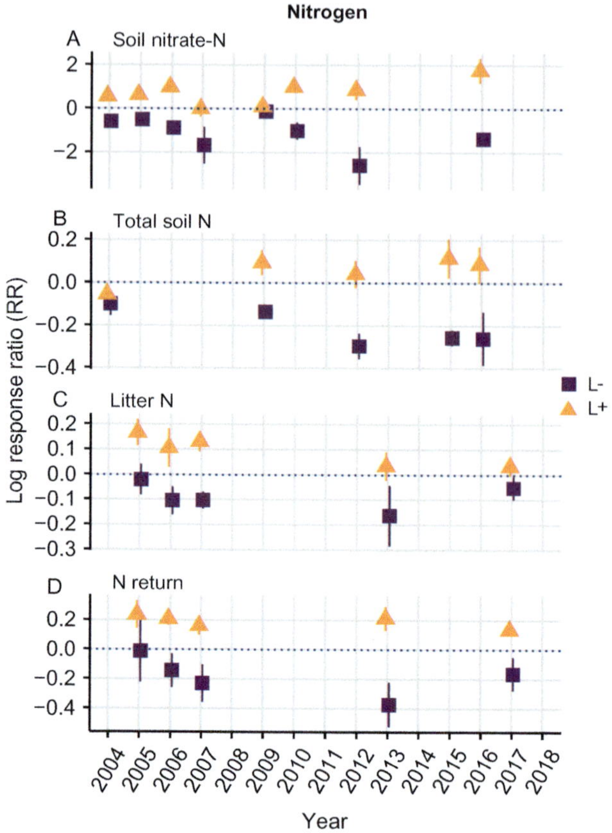

Fig. 2 Changes in nitrogen (N) cycling in response to 15 years of litter addition (orange triangles) and litter removal (purple squares) treatments in a lowland tropical forest in Panama, Central America. Differences in (A) nitrate-N and (B) total N concentrations in the soil at 0–10cm depth, (C) N concentrations in litter (litter N) and (D) the estimated annual N return by litterfall (N return) are given as log response ratios, where the dashed blue line (y=0) indicates no difference relative to controls; dots and whiskers are means and standard errors for $n=5$.

N in soils are likely to have stimulated microbial degradation of soil organic matter as a source of N (Nottingham et al., 2012, 2015). Consistent with this hypothesis, the L− treatment reduced soil organic C concentrations ($d=-2.29\pm0.34$, $P<0.001$; Fig. 4A), microbial biomass C ($d=-1.27\pm 0.24$; $P=0.025$; Fig. 4A) and microbial biomass N ($d=-1.28\pm0.24$; $P=0.026$; Fig. 4C). The negligible effect of litter addition on total soil N concentrations and microbial biomass corresponds to the smaller increase in soil organic C in the L+ plots ($d=0.75\pm0.20$, $P=0.005$; Figs 3B and 4A–C).

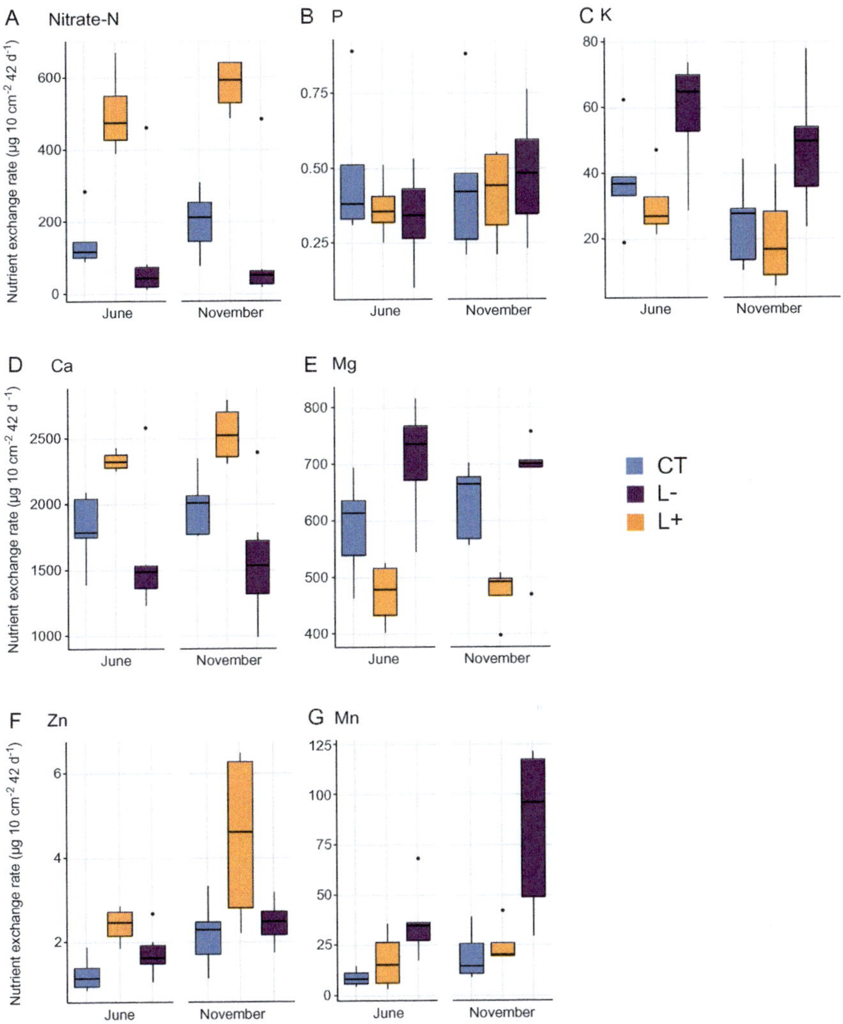

Fig. 3 Exchange rates of (A) soil nitrate-N, (B) phosphorus (P), (C) potassium (K), (D) calcium (Ca), (E) magnesium (Mg), (F) zinc (Zn), and (G) manganese (Mn) after 15 years of litter addition (L+, orange) and litter removal (L−, purple) treatments compared to controls (CT, blue) during the early rainy season (June) and towards the end of the rainy season (November) in a lowland tropical forest in Panama, Central America; nutrient exchange rates were measured by ion exchange resin strips at 0–10 cm depth during 42 days; boxes denote the 25th and 75th percentiles and median lines are given for $n = 5$, whiskers indicate values up to 1.5 × the interquartile range, and dots indicate outliers.

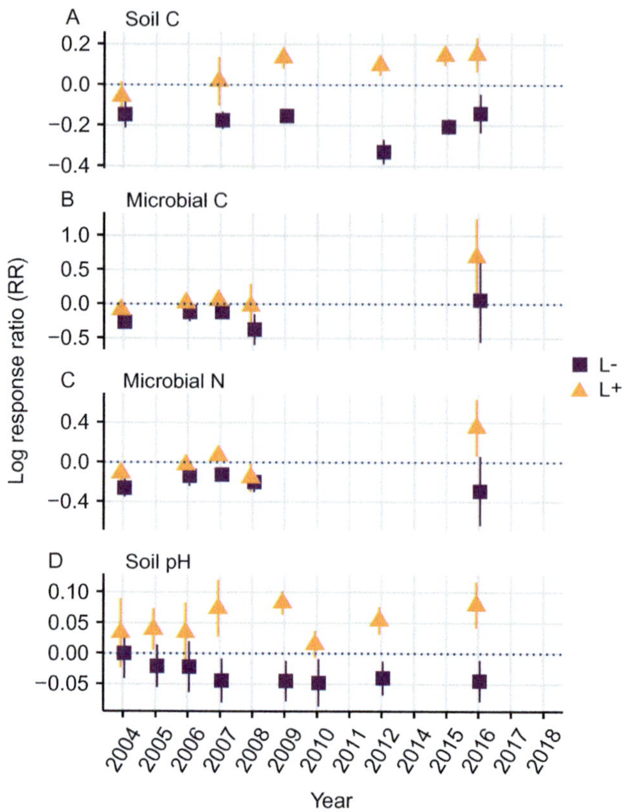

Fig. 4 Changes in (A) soil carbon concentrations (soil C), (B) microbial biomass carbon (microbial C), (C) microbial biomass nitrogen (microbial N) and (D) soil pH at 0–10 cm depth in response to 15 years of litter addition (orange triangles) and litter removal (purple squares) treatments in a lowland tropical forest in Panama, Central America. Differences are given as log response ratios, where the dashed blue line ($y = 0$) indicates no difference relative to controls; dots and whiskers are means and standard errors for $n = 5$.

The N concentrations in leaves (Sayer and Tanner, 2010), roots (Rodtassana and Tanner, 2018) and litter (Sayer et al., 2012) declined in the L− plots, with corresponding increases in the L+ plots and particularly rapid increases with litter addition in the first years of the experiment (Sayer and Tanner, 2010; Fig. 2C). However, the effect of litter addition declined over time ($F_{1,23} = 5.0$; $P = 0.034$) and by 2017, the mean effects on litter N concentrations were similar for litter addition ($d = 1.13 \pm 0.21$, $P < 0.001$) and litter removal ($d = -0.90 \pm 0.11$, $P < 0.001$). The changes in litter N concentrations likely reflect the differences in nitrate-N availability among treatments because microbial retention of nitrate-N is low, plants are better competitors for nitrate-N than microorganisms, and rates of N mineralisation are usually closely related to plant uptake (Attiwill and Adams, 1993). Lower availability of N in the L− plots could have contributed to lower abundance of arbuscular mycorrhizal (AM) fungi and the significant changes in AM fungal communities observed after 9 years of litter removal (Sheldrake et al., 2017). AM fungal hyphae have much higher N concentrations than plant tissues (Hodge et al., 2010) and hence the disruption of the N cycle by the removal of litter could have a negative impact on the abundance of some AM fungal taxa (Sheldrake et al., 2017).

Nitrogen return in litterfall also declined substantially with litter removal ($d = -1.72 \pm 0.44$, $P = 0.025$) and increased with litter addition ($d = 1.40 \pm 0.02$, $P < 0.001$). We transferred *c.* 149 kg N ha^{-1} year^{-1} from L− to L+ plots (Table 2) and we initially predicted that higher N concentrations in plant material would create a positive feedback to litter addition by boosting photosynthetic capacity (Sayer and Tanner, 2010) and contributing to increased litter production in the L+ plots (Sayer et al., 2012; Rodtassana, 2016; Rodtassana et al. unpublished data). However, there was no trajectory of change in N return by litterfall over time (Fig. 2D) and hence, the increasing levels of nitrate-N in the soil suggest that nitrate-N supply has exceeded plant demand in the L+ plots. Indeed, high concentrations of nitrate-N (Fig. 2A) and increased nitrous oxide emissions from the soil in the L+ plots (Welch et al., 2019) indicate an increasingly leaky N-cycle and the potential for substantial N losses from the system after 15 years of litter addition.

There is a strong relationship between N use efficiency and N availability across temperate and tropical forests (Vitousek, 1982). Accordingly, after 15 years of litter manipulation, N use efficiency had increased in the L− plots ($d = 0.90 \pm 0.16$; $P < 0.001$) and declined in the L+ plots ($d = -1.16 \pm 0.27$; $P = 0.002$) relative to the controls (Fig. 5A). The N use efficiency we estimated for L− plots (79), L+ plots (64), and controls (70) places our study site among N-rich tropical forests (NUE < 80; Vitousek, 1982), but the enhanced N use efficiency in the L− plots is perhaps the first sign of N-limitation after 15 years of litter removal. Hence, despite the initial moderate availability of soil N at our study site and local atmospheric N deposition of *c.* 9 kg ha^{-1} year^{-1} (Hietz et al., 2011), we observed substantial changes in N cycling in response to 15 years of litter manipulation, which illustrates the importance of N cycling by litterfall and N retention in organic matter.

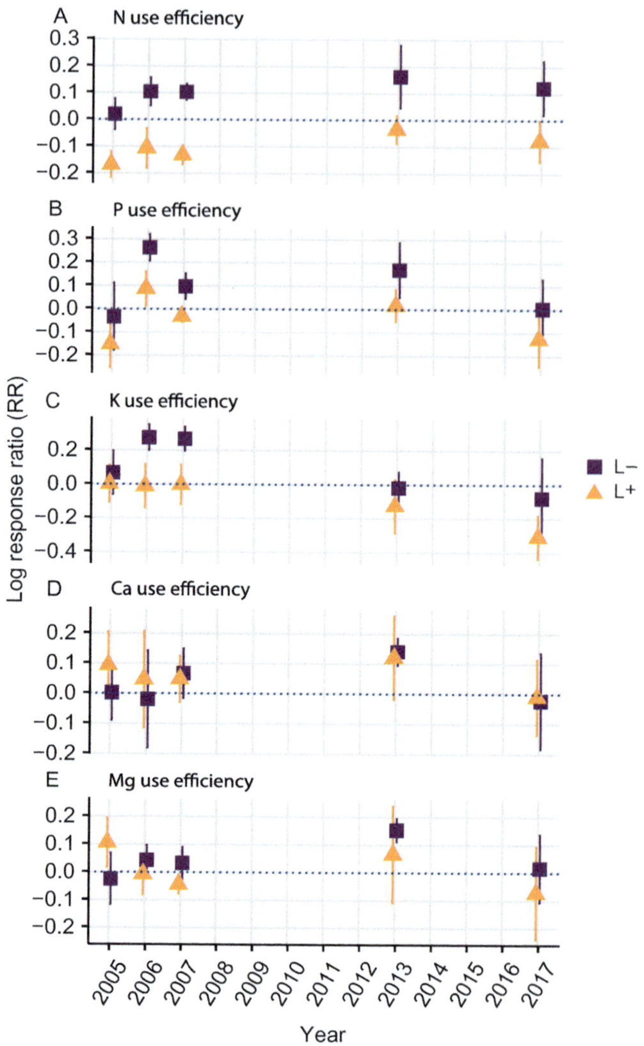

Fig. 5 Changes in nutrient use efficiency for (A) nitrogen (N), (B) phosphorus (P), (C) potassium (K), (D) calcium (Ca) and (E) magnesium (Mg) in response to 15 years of litter addition (orange triangles) and litter removal (purple squares) treatments in a lowland tropical forest in Panama, Central America. Differences are given as log response ratios, where the dashed blue line ($y=0$) indicates no difference relative to controls; dots and whiskers are means and standard errors for $n=5$.

The manifest changes in soil and litter N concentrations and N use efficiency during 15 years of litter manipulation not only highlight the importance of litterfall for N cycling but are also characteristic of an ecosystem adapted to high N availability. The rapid declines in soil and litter N concentrations in the L− plots indicate that conservation mechanisms for N are lacking, and the changes in N use efficiency support Wright's (2019) hypothesis that tropical trees adapted to high levels of N cycling also have relatively high demands for N.

3.2 Phosphorus

Phosphorus (P) is a key component of numerous compounds in leaves, including cell walls, nucleic acids, ATP and co-enzymes. Unlike N, a large proportion of foliar P is present in inorganic forms (Sinclair and Vadez, 2002). Much of the demand for P in natural forests can be met by the cycling of P in organic matter (Attiwill and Adams, 1993), especially in old, highly weathered tropical soils with physical and chemical properties that promote strong P sorption (Townsend et al., 2011; Vitousek et al., 2010). Hence, many lowland tropical forests have low soil concentrations of available P (John et al., 2007; Vitousek, 1984; Vitousek et al., 2010) and a large proportion of P cycling can be restricted to organic horizons in soils with high content of iron and aluminium oxides (Attiwill and Adams, 1993). Lowland tropical tree species have likely evolved strategies to adapt to low soil P availability (Mo et al., 2019; Zalamea et al., 2016), resulting in a "tight" or highly efficient P cycle (Vitousek, 1984). Numerous lines of evidence indicate that P cycling at our study site is conservative and that various ecosystem processes are limited by P availability, including decomposition (Kaspari et al., 2008), seedling growth and photosynthesis (Pasquini and Santiago, 2012; Santiago et al., 2012), and tree growth (Wright et al., 2011).

Litter inputs play a dual role in maintaining P cycling in soils with high sorption capacity. First, the P contained in litter is much more mobile than that held in the soil, as the latter is rapidly sorbed to iron, aluminium and calcium in soil minerals (Yavitt, 2000) and hence, litterfall provides much of the P required for plant growth (Attiwill and Adams, 1993; Vitousek, 1982). Second, compounds released during litter decomposition can increase the overall availability of inorganic P in soils by forming complexes with aluminium and iron (Ehrenfeld et al., 2005) and reducing P sorption (Palm et al., 1997). Our litter manipulation treatments clearly have the capacity to influence both of these aspects of P cycling. Using a radioactive phosphate tracer (^{32}P), Schreeg et al. (2013) demonstrated lower P sorption in L+ soils and higher P sorption in L− soils compared to controls after 5 years of litter manipulation treatments. Hence, given the low relative availability of inorganic P in the soil, we expected that litter addition would enhance P availability, whereas P concentrations in soil and litter would decline rapidly with litter removal (Sayer and Tanner, 2010), resulting in greater P use efficiency or reduced forest productivity.

The patterns of change we observed for P concentrations in soil and litter are at odds with our initial expectations. During the first 5 years of treatments, we measured lower soil extractable P (P_{EXTR}) concentrations in the L− plots compared to controls at 0–2 cm depth (Sayer and Tanner, 2010; Vincent et al., 2010), lower P_{EXTR} in the L− plots compared to L+ plots to 20-cm depth (Tanner et al., 2016), and lower litter P concentrations in the L− plots (Sayer and Tanner, 2010). We had expected to observe a gradual decline in soil and litter P concentrations in response to chronic removal of P with litter and it is surprising that there was no clear trajectory of change in soil or litter P concentrations in the L− plots over the 15 years of treatments (Fig. 6A and B). Despite a large overall effect of litter removal on P_{EXTR} ($d=-1.13\pm0.28$, $P=0.008$), the effect varied widely from year to year (Fig. 6A). The difference between treatments can be explained by a small but more consistent trend towards increasing soil P_{EXTR} concentrations with litter addition ($P=0.084$), which increased over time ($F_{1,38}=5.0$, $P=0.03$; $d=0.68\pm0.29$; Fig. 6A). However, even after almost 15 years of treatments, phosphate exchange rates in the topsoil were unaffected by litter manipulation (Fig. 3B). The decrease in litter P concentrations was also smaller than expected ($d=-0.70\pm0.29$; $P=0.06$; Fig. 6B), and although there was a significant treatment effect on P return by litterfall, neither treatment differed significantly from the controls.

Various studies in the GLiMP plots provide two plausible explanations for the apparent lack of substantial changes in P cycling with litter removal: (1) declining plant productivity has reduced the overall requirement for P uptake by plants and (2) plants have access to other sources of P. Although the growth of mature trees appears to be unaffected by litter manipulation to date (Sayer and Tanner, 2010; Edmund Tanner et al., unpublished data), lower annual litterfall in the L− plots (Rodtassana, 2016; C. Rodtassana et al., unpublished data) and lower P return in leaf litter (Fig. 6C) indicates that decreased leaf production has contributed to reduced plant P demand. Nonetheless, the slight decline in P return in litterfall only accounts for *c.* 13% of the P transferred annually from the L+ to L− plots (*c.* 5 kg P ha^{-1}; Table 2) and it is likely that alternative sources of P play a key role in maintaining plant P supply at our study site.

Organic P (P_{ORG}) is a likely alternative source of P for plants because it is abundant in tropical mineral soils (Harrison, 1987; Turner and Engelbrecht, 2011) and turnover of P_{ORG} could supply substantial amounts

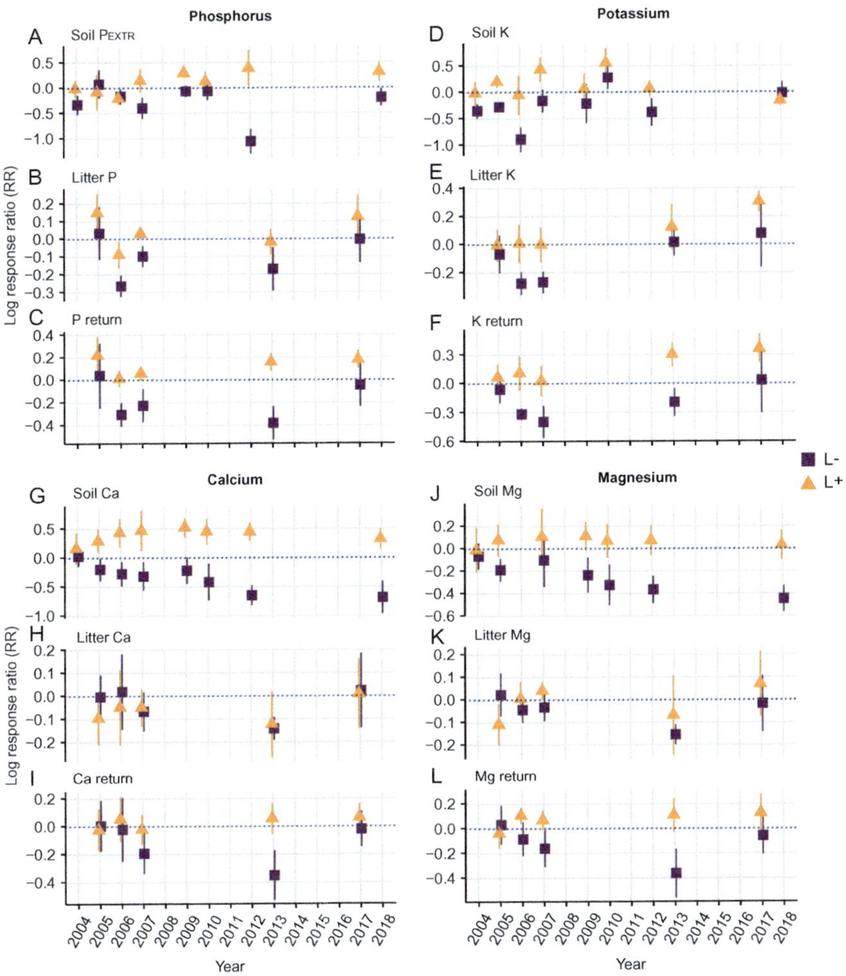

Fig. 6 Changes in (A–C) phosphorus (P), (D–F) potassium (K), (G–I) calcium (Ca) and (J–L) magnesium (Mg) cycling in response to 15 years of litter addition (orange triangles) and litter removal (purple squares) treatments in a lowland tropical forest in Panama, Central America. Differences in soil extractable nutrient concentrations at 0–10 cm depth (A, D, G, J), litter nutrient concentrations (B, E, H, K) and the estimated annual nutrient return by litterfall (C, E, I, L) are given as log response ratios, where the dashed blue line (y = 0) indicates no difference relative to controls; dots and whiskers are means and standard errors for $n = 5$.

of bioavailable P (Tiessen et al., 1992). In 2006, after 3 years of treatments, P_{ORG} in the surface soils of the L− plots had already declined by 23% relative to the controls and increased turnover of P_{ORG} provided an estimated *c.* 1.4 kg P ha^{-1} $year^{-1}$ in the L− treatment (Vincent et al., 2010), representing almost a third of the P removed with litter. After 9 years of litter removal, P_{ORG} was *c.* 18% lower in the L− plots than the controls and microbial P had declined by two thirds (Sheldrake et al., 2017). Hence, increased turnover of P_{ORG} has likely compensated for the lower availability of inorganic P to some extent (Vincent et al., 2010) but it is uncertain how long this alternative supply can meet demands. It will be a matter of great interest to monitor changes in distinct P pools to understand the potential for P_{ORG} to sustain forest P-status in future.

Phosphorus availability is also influenced by AM fungi, as they represent a crucial mechanism for the uptake of P and its transport to plants (Smith and Read, 2010). Sheldrake et al. (2017, 2018) observed shifts in AM fungal communities in the L− plots that potentially signal a change in P acquisition strategies, and which could help plants to maintain their P status. Although both litter manipulation treatments altered AM fungal communities in the soil, only litter removal altered fungal communities in roots, indicating a potential change in plant-mycorrhizal relationships (Sheldrake et al., 2018). In addition, AM fungal community composition in the L− plots was similar to adjacent N-fertilised plots (Sheldrake et al., 2018), which is noteworthy because greater availability of N can create or exacerbate P-limitation (Vitousek et al., 2010). Hence, the similarity of AM fungal communities in the L- and N-fertilised plots in our study forest indicates a potentially important role for AM fungi in plant P-acquisition when the demand for readily available P outstrips the supply (Sheldrake et al., 2018).

It is conceivable that, despite the removal of large amounts of P, declining litter production reduced the demand for P in the L− plots, while increased turnover of P_{ORG} replenished soil supplies, and greater P-acquisition by altered mycorrhizal associations helped meet plant P requirements (Sheldrake et al., 2018). Although we cannot rule out other mechanisms of P supply, such as the transport of P reserves from subsoil horizons to the topsoil by deep roots (Grubb, 1989), it is nonetheless surprising that the annual removal of P ($77\,kg\,P\,ha^{-1}$ in total over 15 years; Table 2) has not resulted in more pronounced changes in P cycling after 15 years. Given the substantial decline in organic P reserves after 9 years of litter removal (Sheldrake et al., 2017), it is possible that the P status of trees in the L− plots will eventually decline to levels that substantially impede productivity.

The effects of litter addition on P cycling were also minor and inconsistent. Although there was no discernible change in soil P concentrations with litter addition for several years, this coincided with increasing litter production during the rainy season (Sayer et al., 2012; Sayer and Tanner, 2010), which indicates that the extra nutrients added with the litter were used to boost productivity. If this was the case, soil P_{EXTR} only started to increase once litter production had reached a new equilibrium and the nutrient demands for greater litter production were met. However, we also observed no consistent increase in litter P concentrations and the trend towards greater P return by litterfall in the L+ plots ($d=0.88\pm0.16$, $P=0.05$; Fig. 6C) can be attributed to greater litter production (Sayer et al., 2012; Rodtassana et al. unpublished data).

Some of the temporal variation in P cycling with litter addition might be due to the first application of the treatment in January 2003, which comprised the entire litter standing crop from the L− plots. Based on nutrient concentrations in the litter standing crop of the control plots in 2005 (Sayer et al., 2010), the first application of litter to the L+ plots would have doubled the amount of P in the litter standing crop (*c.* 11 kg P ha^{-1}; Sayer et al., 2010), which is twice the mean amount of P returned annually with litterfall (Table 2). The large increase in P return in litterfall in the L+ plots in 2005 could therefore signal enhanced litter production and higher foliar P concentrations in response to the large pulse of nutrients added with litter at the start of the experiment. A similar substantial and rapid change to litter P inputs, with no corresponding change in soil P concentrations, was observed after a one-off fourfold increase in litter inputs in a wet tropical forest in Costa Rica (Wood et al., 2009). Hence, much of the P cycled in litterfall is likely to be taken up by plants (Sayer et al., 2012). Furthermore, higher soil P_{ORG} concentrations at the surface after 3 years of litter addition (Vincent et al., 2010) and a *c.* 11% increase in P_{ORG} in topsoil after 9 years of litter addition (Sheldrake et al., 2017) indicate that aboveground litter also plays a role in maintaining P_{ORG} stocks. This seems likely because litter is a source of organic compounds such as DNA (Turner and Engelbrecht, 2011) and because more readily accessible P inputs from litter would reduce the turnover rates for P_{ORG} in the soil (Vincent et al., 2010).

The P use efficiency at our study site does not indicate severe P-limitation (Vitousek, 1982), but we nonetheless observed slightly higher P use efficiency in L− plots compared to the controls ($d=0.79\pm0.45$; $P=0.036$), although there was no consistent trajectory of change in P use efficiency over time (Fig. 5B). Taken together, GLiMP studies support the concept of tight P cycling in lowland tropical forest, but we found little evidence for P-limitation with litter removal. Although it is possible that alternative sources of P are able to compensate for the substantial removal of nutrients with litter, it is also conceivable that co-limitation of different biological processes by multiple nutrients (Kaspari et al., 2008; Wright et al., 2011) has masked changes in P cycling. The balance of nutrient elements in plant tissues is largely determined by the requirements of essential biological processes, as well as adaptation to local conditions, and is thus less flexible than previously thought (Elser et al., 2010). Based on the premise that the N-cycle in our forest is leaky, but P-cycling is conservative, rapid and substantial losses of N with litter removal could have reduced P requirements, which would explain the contrasting results for shifts in N and P cycling in response to litter removal. Relatively constant litter N:P ratios in both treatments provide some support for stoichiometric constraints of plant responses to the litter manipulation treatments. Indeed, the simultaneous removal of multiple nutrients with the litter could constrain the cycles of individual elements and their potential responses to the L− treatment.

All things considered, the surprisingly minor response of P cycling to long-term litter removal can be attributed to a combination of P conservation mechanisms, as well as exploitation of alternative P sources and reduced P demand with the removal of multiple other elements, in particular N. New research to determine the relative importance of different processes in maintaining tropical forest P status could provide valuable information for tropical forest restoration on P-poor soils.

3.3 Potassium

Potassium regulates stomatal activity, enzyme activation, and the transport of water in plants (Prajapati and Modi, 2012) but as K is rapidly leached from the canopy (Likens et al., 1994), throughfall is the major pathway for K cycling in tropical forests (Vitousek and Sanford, 1986). Nonetheless, we transferred *c.* 47 kg K ha^{-1} $year^{-1}$ between L− and L+ plots (Table 2), which is ¾ of the K inputs from throughfall at the study site (Vitousek and Sanford, 1986). In addition, there are strong biotic controls on K cycling (Tripler et al., 2006) and the forest floor contributes substantially to the retention of K (Tobón et al., 2004).

The concentrations of K in the soils at our study site (84–131 mg kg^{-1}) are similar to soils across Mesoamerica (Barthold et al., 2008; Wright et al., 2011) but concentrations of exchangeable K are lower than expected (Yavitt et al., 2009) and previous work in the study forest has demonstrated K-limitation of plant growth (Santiago et al., 2012; Wright et al., 2011;

Yavitt et al., 2011) and decomposition processes (Kaspari et al., 2008). We expected that availability of K to plants in the L+ plots would be enhanced by greater K retention in the forest floor, indicated by substantially reduced K release from litter in the L+ plots during the early stages of decomposition (Sayer et al., 2006b). We further expected that the impact of litter removal on K availability would be amplified by higher leaching losses (Sayer et al., 2012). It is thus surprising that the initial strong responses of soil K concentrations to both litter manipulation treatments appear to have attenuated over time (Fig. 6D). We initially measured a rapid and sizeable decline in soil K concentrations in the L− plots between 2004 and 2006, but although soil K concentrations were lower in the L− plots compared to the controls across all years, the effects were moderate ($d = -0.76 \pm 0.15$, $P = 0.041$) and inconsistent (Fig. 6D). In the L+ plots, soil K concentrations increased markedly in 2005 and 2007, and although soil K generally increased in response to litter addition ($P = 0.039$) the pattern over time was so inconsistent that the mean effect across all years was negligible ($d = 0.38 \pm 0.15$; Fig. 6D).

High temporal variability in leaching losses of K provides one explanation for the lack of a clear trajectory of change in soil K concentrations over time. Potassium is highly soluble and readily leached from litter (Schreeg et al., 2013) and soil (Likens et al., 1994); hence, even small shifts in sampling times combined with variable inter- and intra-annual precipitation patterns are likely to influence our results. We sampled soils at the end of the rainy season each year from 2004 to 2007 (Sayer et al., 2012; Sayer and Tanner, 2010) but during or at the end of the dry season in 2009, 2010, 2012 and 2018 (Ashford et al., 2013; Sheldrake et al., 2017; Tanner et al., 2016), which complicates the interpretation of long-term patterns. Previous work at our study site, and nearby, demonstrated accumulation of K in soils during the dry season (Yavitt and Wright, 1996) and rapid release of K from decomposing litter at the start of the rainy season (Cornejo et al., 1994; Sayer et al., 2006b). Accordingly, our measurements of nutrient exchange rates using resin probes showed a clear decline in K supply over the course of the rainy season. Greater exchange rates of K in the L− plots ($P = 0.037$; Fig. 3C) could suggest that more K is reaching the mineral soil via throughfall and greater movement of K through the soil in the absence of a litter layer (Sayer et al., 2012).

Treatment differences for litter K concentrations are less variable than those for soil because we determined nutrients in litter samples collected at the same time each year. Although K leaches rapidly from leaves after abscission (Schreeg et al., 2013), and substantial losses of K can occur from litter traps when samples are collected monthly (Cuevas and Medina, 1986), it is reasonable to assume that relative losses of K from litter-traps were the same across all three treatments.

The initial decline in litter K concentrations in the L− plots largely mirrors the patterns for soil K, with substantial declines in 2006 and 2007 but only a minor and inconsistent effect across all years (litter: $P=0.036$, $d=-0.54\pm0.26$; Fig. 6E). The overall decline in K return by litterfall ($d=-1.11\pm0.38$, $P=0.005$) suggests that the declining K supply in the L− plots is unable to meet plant demands (Fig. 6F). However, interannual variability was high and in 2017, after 15 years of litter removal, we measured similar litter K concentrations and K return by litterfall to the controls (Fig. 6E and F). By contrast, although litter K concentrations and K return by litter were unaffected by litter addition in the first 5 years of the experiment, litter K concentrations increased over time in response to litter addition ($F_{1,23}=5.3$, $P=0.03$) and K return was higher in later years ($F_{1,23}=4.4$, $P=0.048$; Fig. 6F). Direct uptake of K by roots from decomposing litter (Burghouts et al., 1992) in the L+ plots could explain the higher litter K concentrations and increased K return by litterfall in 2013 and 2017. We measured greater K stocks in the forest floor (Sayer et al., 2010) as well as higher fine root biomass in the forest floor within 2 years of litter addition (Sayer et al., 2006a), and the latter persisted even after 10 years of treatments (Rodtassana and Tanner, 2018). However, given the high interannual variability in soil and litter K concentrations, it remains to be seen if this trend persists in future.

Surprisingly, there was some indication that litter manipulation altered K use efficiency (treatment effect: $P=0.008$, Fig. 5C), although neither treatment differed significantly from the controls. Nonetheless, differences in K use efficiency between litter removal and addition treatments support various lines of evidence that processes at our study site are limited by low K availability (Kaspari et al., 2008; Wright et al., 2011; Wright et al., 2018). Hence, despite the importance of K inputs by throughfall and the high variability in soil and litter K concentrations among years, the GLiMP studies nonetheless demonstrate that litterfall contributes substantially to K cycling in our study forest. Importantly, litter inputs also appear to play a role in reducing leaching losses of K via retention in the forest floor and possible direct uptake by roots from decomposing organic matter.

3.4 Calcium and magnesium

Calcium (Ca) and magnesium (Mg) are two of the most abundant elements in litter (Tobón et al., 2004) and we transferred *c.* 169 kg Ca ha^{-1} $year^{-1}$ and *c.* 37 kg Mg ha^{-1} $year^{-1}$ between L− and L+ plots (Table 2). Calcium is one of the main constituents of cell walls in leaves (Kirkby and Pilbeam, 1984), whereas Mg plays a key role in photosynthesis (Cakmak and Kirkby, 2008). Litterfall can contribute 80–90% of the Ca and 65–85% of the Mg being cycled in tropical forests (Campo et al., 2000; Parker, 1983) and although Ca or Mg can be limiting to productivity in some tropical forests (Cuevas and Medina, 1988), the soils at our study site are relatively rich in exchangeable Mg (553 mg kg^{-1}) and particularly rich in Ca (1690 mg kg^{-1}) compared to other highly weathered tropical soils (Yavitt et al., 2009).

Despite high Ca concentrations in the topsoil, we observed sizeable and rapid changes in soil Ca (Ca^{2+}) to litter manipulation (Sayer and Tanner, 2010). As substantial amounts of Ca cycle through litterfall (Table 2), soil Ca concentrations increased with litter addition ($d = 1.28 \pm 0.12$, $P < 0.001$), decreased with litter removal ($d = -1.21 \pm 0.31$, $P < 0.001$), and the effect of litter removal became more pronounced over time ($F_{1,37} = 7.6$, $P = 0.009$; Fig. 6G). Our study of nutrient exchange rates also demonstrates the importance of litterfall and decomposition for Ca cycling. As Ca is a structural component of leaves and is not soluble (Schreeg et al., 2013), its release during decomposition strongly mirrors mass loss (Gosz et al., 1976) and thus Ca exchange rates were higher in the L+ plots ($P = 0.034$), especially towards the end of the rainy season, but remained unchanged in the L− plots (Fig. 3D). Litter Ca concentrations did not change significantly with either treatment, but this is not surprising given the high soil Ca concentrations at our study site (Yavitt et al., 2009), which likely represent the main source of Ca for plants. Thus, although the concentrations of exchangeable Ca in the soil were altered by litter manipulation, litter Ca concentrations and Ca return by litter were unaffected by litter manipulation (Fig. 6H and I).

The patterns of change for Mg in soil and litter are broadly similar to those for Ca (Fig. 6J–L), and differences in the cycling of the two elements can largely be explained by the greater mobility of Mg and its crucial role in various processes involved in photosynthesis (Cakmak and Kirkby, 2008). Both high and low concentrations of Mg in plant cells can have adverse effects on photosynthesis (Shaul, 2002) and the maintenance of Mg homeostasis in plant tissues could explain why litter Mg concentrations remained relatively constant, even though soil Mg concentrations were significantly lower in L− plots compared to controls ($d = -1.09 \pm 0.16$, $P < 0.001$) and declined over time ($F_{1,32} = 5.7$, $P = 0.02$; Fig. 6J). Although litter addition had no discernible effect on soil Mg concentrations, Mg exchange rates in the topsoil were lower in L+ plots compared to the controls ($P < 0.001$), especially in the late growing season, and there was a non-significant trend towards higher Mg exchange rates in the L− plots ($P = 0.12$; Fig. 3E). However, although Mg is relatively mobile and prone to leaching, it is also retained in organic matter in the forest floor (Gosz et al., 1973; Gransee and Führs, 2013). We hypothesise that the slight increase in Mg exchange rates in the L− plots reflects greater leaching losses with litter removal and the lower exchange rates of Mg with litter addition reflect accumulation of Mg in the thick forest floor of the L+ plots (Sayer et al., 2010) and slower release of Mg during decomposition (Sayer et al., 2006b).

Calcium and Mg are important for exchange reactions that buffer acidity in many soils, so the observed changes in soil Ca and Mg concentrations probably resulted in the sizeable increase in soil pH with litter addition ($d = 1.12 \pm 0.22$, $P < 0.001$) and the parallel decline with litter removal ($d = -0.83 \pm 0.15$, $P = 0.004$; Fig. 4D). Although soil acidification with increased nitrification is a common issue in fertilisation experiments (Wright, 2019), it is unlikely to explain the changes in pH in response to litter manipulation because we would expect lower rates of nitrification in the L− plots. High spatial variation in forest soil pH is associated with the concentrations of exchangeable Ca and Mg (Finzi et al., 1998) and a litter manipulation study in temperate forest observed a strong relationship between changes in Ca and Mg concentrations and soil pH (Tóth et al., 2011). At our study site, the response of soil pH to litter addition and removal treatments was strongly related to the magnitude of change in the combined concentrations of Ca and Mg ($R^2 = 0.73$, $P < 0.001$; Fig. 7). The initial rapid changes in soil pH reflect the low buffering capacity of the soil at our study site (Yavitt et al., 2009) and the considerable influence of organic matter inputs on soil pH (McFee et al., 1977). Hence, nutrient shortages with litter removal could be exacerbated by the combination of lower soil pH and declining soil organic C concentrations.

Collectively, the GLiMP studies demonstrate that litterfall plays an important role in maintaining soil Ca and Mg concentrations and regulating soil pH, despite high availability of Ca and Mg in the soils at our study site. The marked changes in soil Ca, Mg and pH we observed with litter removal in particular suggest that litterfall may be crucial for buffering pH changes and maintaining the availability Ca and Mg in the majority of tropical forests.

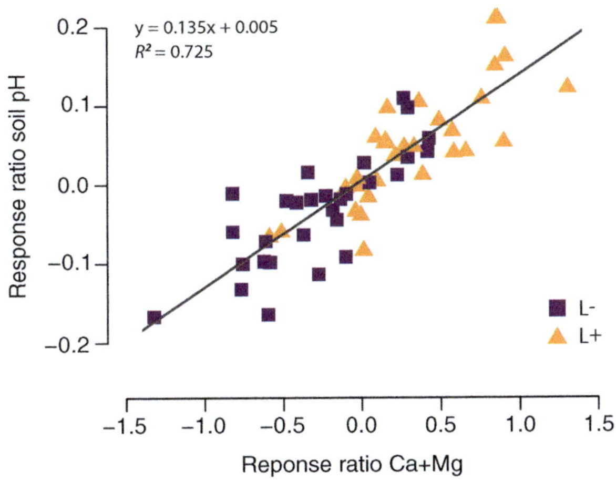

Fig. 7 The relationship between changes in soil calcium and magnesium concentrations (Ca + Mg) and changes in soil pH at 0–10 cm depth during 15 years of litter addition (yellow triangles) and litter removal (purple squares) treatments in a lowland tropical forest in Panama, Central America. Changes relative to the controls are given as log response ratios calculated for each of $n = 5$ replicate blocks and $n = 6$ years (2004, 2005, 2007, 2009, 2010, 2012).

4. Micronutrients and decomposition processes

Our knowledge of micronutrient cycling in natural ecosystems is limited and the roles of biological processes in the availability of trace elements in soils has yet to be fully established (Li et al., 2008). There is a particular dearth of information about trace element cycling in tropical forest ecosystems, even though they are essential for biological processes from the cellular level to the ecosystem scale (Kaspari and Powers, 2016). The role of trace elements as enzyme co-factors makes them particularly crucial for nutrient cycling, as well as decomposition (Kaspari et al., 2008, 2014; Powers and Salute, 2011) and soil food webs (Kaspari et al., 2017) in tropical forests. Here, we present the evidence for biotic cycling of four micronutrients and discuss the specific links to decomposition processes and the nutrient requirements of decomposer organisms.

4.1 Boron, sodium, zinc and manganese

We measured concentrations of five micronutrients (boron, iron, manganese, sodium and zinc) in multiple years. The concentrations of boron

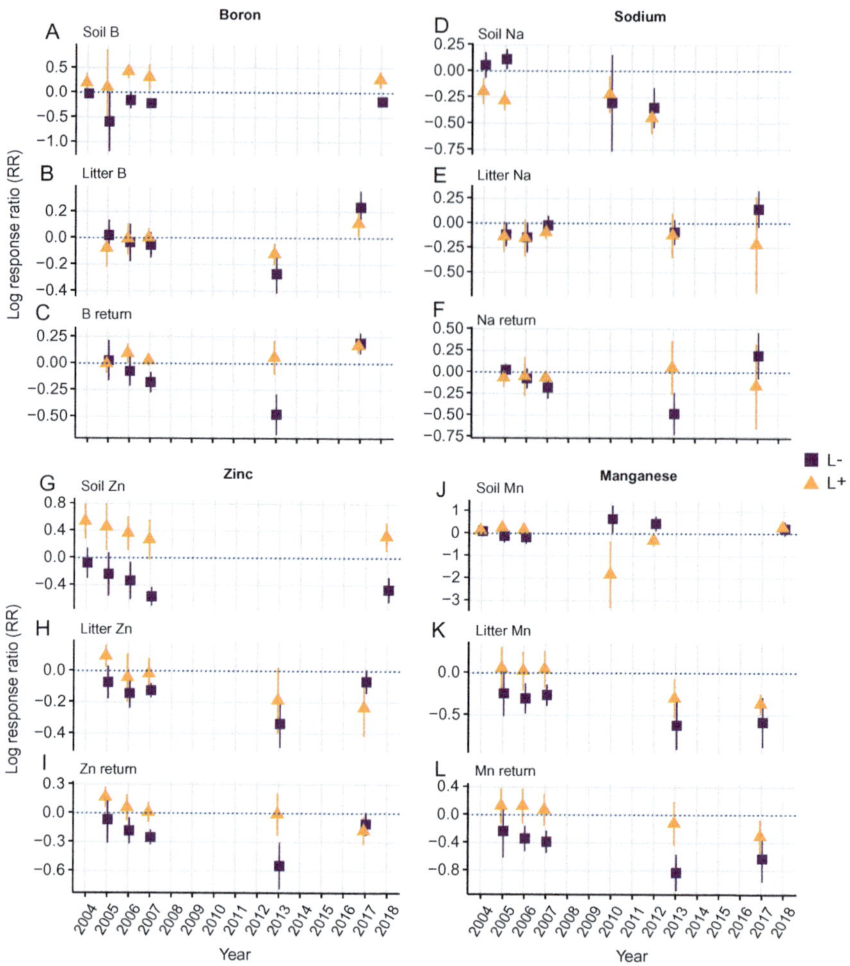

Fig. 8 Changes in (A–C) boron (B), (D–F) sodium (Na), (G–I) zinc (Zn) and (J–L) manganese (Mn) cycling in response to 15 years of litter addition (orange triangles) and litter removal (purple squares) treatments in a lowland tropical forest in Panama, Central America. Differences in soil extractable nutrient concentrations at 0–10 cm depth (A, D, G, J), litter nutrient concentrations (B, E, H, K) and the estimated annual nutrient return by litterfall (C, E, I, L) are given as log response ratios, where the dashed blue line (y = 0) indicates no difference relative to controls; dots and whiskers are means and standard errors for $n = 5$.

(B) sodium (Na), and zinc (Zn) in the soil, and the concentration of manganese (Mn) in litter responded to at least one litter treatment (Fig. 8), suggesting that the cycles of these elements are influenced by litterfall—most likely via decomposition processes.

Boron (B) is critical for biological N cycling because it is involved in various processes that mediate interactions between legumes and rhizobia (Bolaños et al., 2004). Boron is only present in trace amounts in the soil and litterfall, and hence we only transferred *c.* 0.5 kg B ha^{-1} $year^{-1}$ between L− and L+ plots (Table 2). Nonetheless, the increase in soil B concentrations with litter addition indicate that litterfall could be important for maintaining B availability at our study site ($d = 1.0 \pm 0.14$, $P < 0.001$; Fig. 8A). However, the differences in soil B among treatments did not increase over time and were not reflected in litter B concentrations, B return or B use efficiency (Fig. 8B and C and 9A).

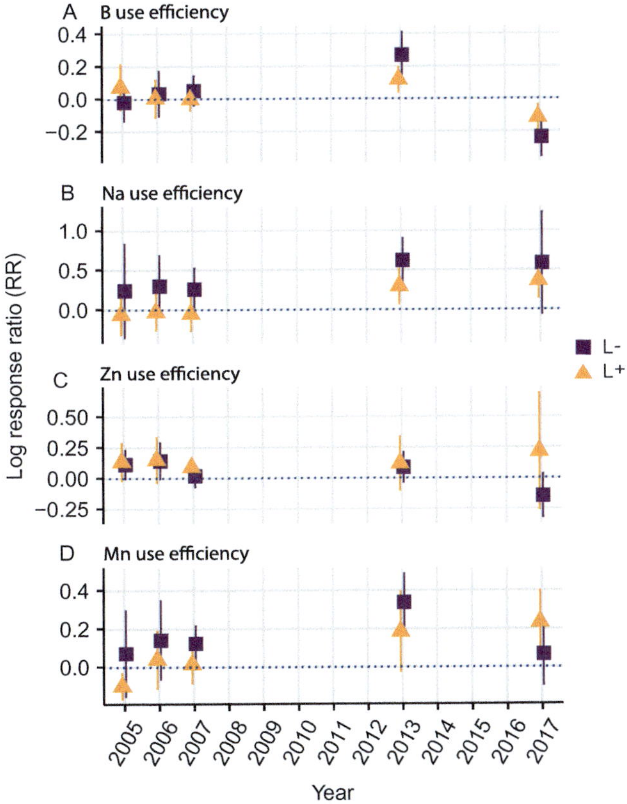

Fig. 9 Changes in nutrient use efficiency for (A) boron (B), (B) sodium (Na), (C) zinc (Zn), and (D) manganese (Mn) in response to 15 years of litter addition (orange triangles) and litter removal (purple squares) treatments in a lowland tropical forest in Panama, Central America. Differences are given as log response ratios, where the dashed blue line ($y = 0$) indicates no difference relative to controls; dots and whiskers are means and standard errors for $n = 5$.

Low variation in foliar B concentrations is expected because the range of B concentrations that is beneficial to plants is extremely narrow (Goldberg, 1997).

Sodium plays an important role in decomposition food-webs because it is essential to numerous arthropod taxa (Kaspari et al., 2014) and has been linked to arthropod abundance in the forest floor at our study site (Sayer et al., 2010). We transferred *c.* 5 kg Na ha^{-1} $year^{-1}$ between L− and L+ plots (Table 2) but we did not expect soil or litter Na concentrations to be affected by litter manipulation because our site is located within 30 km of the coast and likely receives sufficient Na inputs from marine deposition (Kaspari et al., 2014). Accordingly, litter removal did not affect soil Na concentrations (Fig. 8D) and neither treatment influenced litter Na concentrations, Na return or Na use efficiency (Figs 8E and F and 9B) but surprisingly, soil Na concentrations were lower in the L+ plots ($d = -1.41 \pm 0.30$, $P < 0.001$, Fig. 8D). It is likely that lower Na concentrations in the mineral soil with litter addition are the result of effective retention of mobile nutrients by the thick forest floor because Na concentrations increased with forest floor dry mass in the L+ plots (Sayer et al., 2010).

Zinc is an essential component of thousands of plant proteins but around 90% of the Zn in soils is insoluble and unavailable for plant uptake (Broadley et al., 2007). We only transferred *c.* 0.3 kg Zn ha^{-1} $year^{-1}$ between L− and L+ plots (Table 2), but there was a consistent pattern of higher Zn concentrations in the soil with litter addition ($d = 0.85 \pm 0.06$, $P < 0.001$), and lower concentrations with litter removal ($d = 0.76 \pm 0.16$, $P = 0.03$, Fig. 8G). We also measured greater Zn exchange rates in the L+ plots ($P = 0.005$), particularly at the end of the rainy season (Fig. 3F). Lower litter Zn concentrations in the L− plots ($d = 0.70 \pm 0.11$, $P = 0.004$; Fig. 8H) also resulted in lower Zn return by litter ($d = -0.81 \pm 0.24$, $P = 0.004$; Fig. 8I) and greater Zn use efficiency in response to litter removal ($d = 0.63 \pm 0.23$, $P = 0.012$; Fig. 9C).

Although chemical and physical processes are thought to be more important for Na and Zn cycling than biological processes (Laskowski et al., 1995), the changes in soil Na and Zn with litter manipulation are most likely linked to nutrient retention during decomposition processes in the forest floor. Litter decomposition can be enhanced by addition of Na (Kaspari et al., 2014) and Zn (Powers and Salute, 2011), which is likely related to high decomposer requirements for these elements (Gosz et al., 1976). The concentration of Na in arthropod biomass can be much higher than in the surrounding forest floor (Cromack et al., 1977), and tropical forest fungi accumulate both nutrients (Stark, 1972). Interestingly, Na concentrations in decomposing boles were also higher in the L+ plots compared to the controls (Gora et al., 2018) and hence, the lower Na concentrations in the mineral soil could indicate active transport of Na by fungi into decomposing woody litter. The higher Zn supply in the L+ plots at the end of the wet season indicates substantial release of Zn during the later stages of decomposition (Fig. 3F). Hence, accumulation of Na and Zn in the forest floor is likely to accelerate litter turnover, whereas lower Zn availability in the L− plots could contribute to slower rates of decomposition (Sayer et al., 2006b).

Manganese (Mn) cycling is of particular interest because of the strong relationship between litter Mn concentrations and decomposition rates in multiple ecosystems (Keiluweit et al., 2015) and the effect of Mn in reducing tree growth on nearby Barro Colorado Island (Zemunik et al., 2018). Leaf litter is a major source of Mn (Marschner, 1986) and we transferred *c.* 2.5 $kg\ Mn\ ha^{-1}\ year^{-1}$ between L− and L+ plots (Table 2), so it is somewhat surprising that the concentrations of Mn in the soil showed no clear response to litter manipulation (Fig. 8J). However, $Mn^{3+/4+}$ oxides in the mineral soil are insoluble and unavailable for plant uptake (Keiluweit et al., 2015), so their concentrations are less likely to be affected by litter manipulation. By contrast, the higher exchange rates of Mn with litter removal, especially during the rainy season, ($P = 0.002$; Fig. 3G) indicate greater amounts of water-soluble Mn in the soil of the L− plots, possibly as a result of lower pH (Laskowski et al., 1995; Page, 1962). Increased leaching of water-soluble Mn could explain the lower concentrations of Mn in litter in the L− plots after the first 4 years of treatments ($d = -0.94 \pm 0.06$; $P = 0.003$; Fig. 8K), and the decline in Mn return in litter ($d = -1.02 \pm 0.17$; $P = 0.002$; Fig. 8L). Greater Mn use efficiency with litter removal ($d = 0.97 \pm 0.10$, $P < 0.001$; Fig. 9D) suggests tight cycling of Mn in leaf litter, and the decline in litter but not soil Mn concentrations in the L− plots could indicate the importance of Mn cycling with organic matter and its direct uptake by roots from the forest floor. Decomposer fungi actively cycle Mn from plant material, maintaining it in forms available for plant uptake (Keiluweit et al., 2015). Hence, fungal accumulation of Mn in the forest floor would reduce leaching losses of water-soluble Mn and might also explain why we observed no increase in soil Mn with litter addition. Indeed, given the central role of Mn in fungal decomposition of organic material, greater concentrations of Mn in the forest floor could explain higher decay rates of fine woody litter (Sayer et al., 2006b) and leaf litter (Sayer and Tanner, 2010) in the L+ plots.

Overall, our findings reveal potentially important links between decomposer nutrient requirements and the observed changes in Mn, Na, and Zn with litter manipulation, and hence that litterfall and decomposition processes are crucial for the retention and cycling of multiple micronutrients in tropical forests. Given the combined influence of site fertility and litter inputs on soil C dynamics in forest ecosystems (Lajtha et al., 2018), further research into the links between micronutrient availability and decomposition processes could potentially improve our understanding of soil C storage in tropical forests.

4.2 Links between nutrients, decomposition, and organisms in the soil and forest floor

The GLiMP treatments demonstrate a key role of litterfall in the cycling or retention of most of the nutrients we investigated (Fig. 10). However, just as litterfall is one of the main pathways for the transfer of C and nutrients between plants and soil, decomposer organisms make those mineral elements available to plants. The nutrient requirements of decomposers contribute substantially to nutrient cycling but are often overlooked. Hence, the response of decomposition processes to altered litter inputs is likely to be influenced by the abundance and community composition of arthropods in litter and soil, as well as the availability of nutrients. As such, the effects of litter manipulation on organisms in the soil and forest floor deserve a separate mention.

Decaying litter on the forest floor not only represents a source of nutrients and substrate to decomposers, but also buffers variation in humidity and temperature (Sayer, 2006) and provides habitat space for a wide range of organisms (Gill, 1969; Poser, 1990; Sayer et al., 2010). It is therefore not surprising that changes in the abundance and diversity of decomposer organisms and arthropods are generally greater in response to litter removal than litter addition (Sayer, 2006). At our study site, arthropod abundance in both the litter and the mineral soil was best explained by the depth and mass of the forest floor (Ashford et al., 2013; Sayer et al., 2010) and differences in litter decomposition rates were at least partially explained by mesoarthropod abundance (Sayer et al., 2006b). Arthropod community composition was strongly related to P, Ca and Na concentrations in the forest floor (Sayer et al., 2010) and to P availability in the mineral soil (Ashford et al., 2013). Decreased arthropod abundance and lower soil nutrients likely contributed to lower decay rates of leaf litter in the L− plots (Rodtassana, 2016; Sayer et al., 2006b). By contrast, greater retention of nutrients (particularly K,

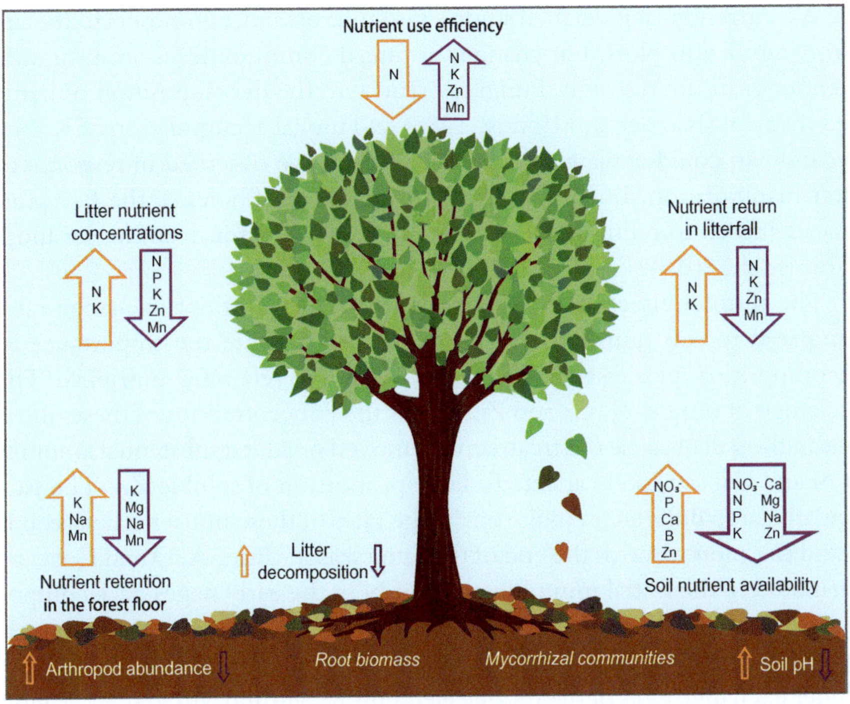

Fig. 10 Conceptual diagram illustrating the effects of litter addition (orange arrows) and litter removal (purple arrows) on different components of the forest nutrient cycle, where upward and downward arrows denote an increase or decrease, respectively, in nutrient concentrations in soil and litter, nutrient retention by the forest floor, stand-level nutrient use efficiency, soil pH, arthropod abundance or litter decomposition rates. Abbreviations inside arrows indicate nutrient elements significantly affected by litter manipulation treatments, following standard chemical nomenclature; labels in italics indicate other variables influenced by litter manipulation without a clear directional change.

Mn, Na and Zn) by decomposer organisms in the forest floor and rapid turnover of the litter standing crop (Rodtassana, 2016) likely accelerated the cycling of nutrients from litter while minimising losses from the system in the L+ plots (Gosz et al., 1976; Tobón et al., 2004). Termites contribute to the break-down of woody debris and their abundance is strongly linked to the availability of Na (Kaspari et al., 2014), but their distribution is patchy (Salick et al., 1983) and we observed no consistent differences among treatments in the abundance of termites in the forest floor or soil (Ashford et al., 2013; Sayer et al., 2010).

We currently lack detailed studies into microbial decomposers in the litter manipulation plots, but changes in fungal communities is an intriguing avenue for future research. Fungi are crucial to the decomposition of lignified material (Rayner and Boddy, 1988) and fungal accumulation of K, Na, Zn and Mn could explain some of the patterns we observed in response to litter manipulation. In addition, the slower decay of boles in the L− plots was attributed to reduced availability of N, P and K for decomposer fungi (Gora et al., 2018).

The distinct effects of the litter treatments on nutrient exchange rates compared to soil nutrient concentrations demonstrate the importance of decomposition processes in the forest floor for retaining nutrients. The exchange of nitrate-N, Ca and Zn mirror the concentrations of these nutrients in the soil because the treatments removed or added substantial amounts of N and Ca, and likely affected a large proportion of soluble Zn. The particularly large differences in the exchange rates of these nutrients between L+ and L− plots towards the end of the rainy season (Fig. 3A, D and F) can be attributed to the initial immobilisation of N in the early stages of decomposition (Sayer et al., 2006b) and the release of Ca and Zn during the late stages of decomposition (Laskowski et al., 1995) in the L+ plots. By contrast, the higher exchange rates of the mobile elements K, Mg and Mn in the L− plots (Fig. 3C, E and G), despite lower or unchanged concentrations in the soil, indicate increased leaching of mobile elements in the absence of an intact litter layer. Hence, more explicit consideration of the links between nutrient availability, decomposer organisms and the turnover of organic matter in the forest floor could shed light on many open questions about tropical forest nutrient cycling.

5. Synthesis and conclusions

The Gigante Litter Manipulation Project has contributed greatly to our understanding of nutrient cycling by litterfall. By disrupting or augmenting nutrient inputs in litter, the experimental treatments allow us to tease apart the relative importance of nutrient supply in the soil or litter, and test some of the basic principles of tropical forest nutrition and nutrient use efficiency (Fig. 10). Fifteen years of continuous litter removal and addition treatments in large experimental plots have expanded the mechanistic basis for the efficient use of several nutrients and highlighted numerous avenues for future research (Box 2).

5.1 Nutrient availability and nutrient use efficiency

We assessed overall changes in forest nutrient cycling in response to long-term litter manipulation treatments by calculating nutrient use efficiency, which provides an ecosystem-level measure of nutrient availability to plants (Vitousek, 1982). Nutrient use efficiency can be regulated by the allocation of a given element to different tissues or functions, or via resorption of foliar nutrients prior to leaf abscission (Vitousek, 1982). Hence, changes in nutrient use efficiency in response to litter manipulation underscore the importance of nutrient cycling by litterfall. Although we observed no clear trajectory of change in nutrient use efficiency over time in our experimental plots, the continuous addition or removal of nutrients with litter still had an overall effect on N, P, K, Mn and Zn cycling, as well as Ca, Mg and B concentrations in the soil (Fig. 10).

The extent of changes in N and P status of litter and soils at our study site conforms to the widely held view of a leaky N but tight P cycle. Nonetheless, we found that cycling of N in litter is critical for sustaining N availability to plants, and that N return in litterfall helps maintain the N-status of the forest, even on soils with moderate or high N-availability. By contrast, the maintenance of plant P-status despite 15 years of litter removal suggests that plants are less dependent on P cycling by litterfall than previously presumed. Many tree species at our study site occur on low-P soils across central Panama (Condit et al., 2013; Wright et al., 2018) and might maintain their P status by accessing alternative sources of P (Liu et al., 2018; Steidinger et al., 2015), via increased resorption (Chapin, 1980) or other physiological adaptations to reduce their P-demand (Mo et al., 2019). Multiple conservation strategies for P, but not N, would explain why the relationship between soil nutrient availability and litter nutrient concentrations is much stronger for N than for P (Vitousek, 1982; Wright, 2019). It is possible that plants adapted to high levels of N cycling also have relatively high demands for N (Wright, 2019), which would explain why we observed rapid changes in N but not P use efficiency in response to litter removal (Fig. 5B). Increased Mn and Zn use efficiency in response to litter removal (Fig. 9C and D), as well as the potential shift in K use efficiency indicate strong biotic controls on these elemental cycles via decomposition processes in the forest floor. Indeed, major declines in nutrient availability in the mineral soil in response to litter removal were largely restricted to the most abundant elements (N, Ca, Mg), which provides strong support for the biotic retention and cycling of scarce and potentially limiting nutrients in lowland tropical forests (Box 2; Fig. 10).

Clearly, extractable nutrients in soil samples only provide a snapshot of potential nutrient availability to plants, and many of the discrepancies we observed in the responses of soil and litter nutrient concentrations to the litter manipulation treatments can be explained by the retention of nutrients in the forest floor. Our study of nutrient exchange rates gives a good indication of how decomposing plant matter in the forest floor influences measurements of nutrient availability from the mineral soil. Higher exchange rates of nitrate-N, Ca, and Zn in the mineral soil of the L+ plots compared to the L− plots (Fig. 3A, D and F) suggest that large amounts of these nutrients are being released from the forest floor to the mineral soil before they can be taken up by plants. By contrast, higher exchange rates of mobile nutrients such as K, Mg and Mn with litter removal compared to litter addition indicates increased leaching of those elements in the absence of the forest floor (Fig. 3C, E and G).

Of particular note are the changes we observed in trace element cycling. The building blocks of life involve at least 25 different elements, but we know relatively little about the role of micronutrients in ecosystem function (Kaspari and Powers, 2016). Our data only included measurements of five trace elements, but we present strong evidence for changes in four micronutrients (B, Na, Mn, Zn) in response to litter manipulation and an increase in nutrient use efficiency for Mn and Zn. There is clearly much more work to be done to characterise micronutrient cycling and the relative importance of different trace elements in tropical ecosystem function (Box 2). The GLiMP studies indicate that a more comprehensive view of tropical forest nutrient cycling would need to consider specific plant adaptations to shortages of different elements, alternative sources of scarce nutrients, and detailed patterns of nutrient retention and release from decomposing organic matter.

5.2 The importance of forest floor processes

As decomposition is rapid in moist lowland tropical forests, the depth of the superficial organic horizons can vary seasonally at our study site, from a thick layer of fresh litter and decomposing organic material at the end of the dry season to a thin and patchy covering of partially decomposed leaves towards the end of the rainy season (Wieder and Wright, 1995). The dynamic nature of the tropical forest floor requires high spatial and temporal resolution for process studies. Many key steps of multiple elemental cycles appear to bypass the mineral soil, and although direct nutrient uptake from the forest floor was demonstrated over 40 years ago (Herrera et al., 1978; Stark and Jordan, 1978), the relevant processes are still relatively poorly characterised (Box 2).

Collectively, our results give a strong indication that a substantial proportion of multiple nutrient cycles occur within the forest floor (Fig. 10). First, altered fine root distribution and greater proliferation of fine roots (Rodtassana and Tanner, 2018; Sayer et al., 2006a) and fungal hyphae (Sheldrake et al., 2017) into the thick forest floor in the L+ plots demonstrate that plants perceive the added litter as a "nutrient hotspot", where nutrients can be taken up as they are mineralised and released from the litter. Second, slower release of K and Mg during decomposition with litter addition (Sayer et al., 2006b) and contrasting mobility of these elements in the mineral soil in the L+ and L− plots (Fig. 3C and E) point towards substantial retention of K and Mg in the forest floor. Third, the shifts in AM fungal communities and depletion of organic P in the L− plots signal substantial changes in P-acquisition and cycling with litter removal, demonstrating the importance of organic matter for maintaining P-status in the forest. Finally, the increase in microbial biomass in the mineral soil of L+ plots was surprisingly small, despite substantial inputs of C and nutrients with the added litter (Fig. 4B and C). Given the cumulative evidence for nutrient retention and uptake in the forest floor, we propose that measurements of processes in the mineral soil only represent a small part of the picture. It follows that explicit consideration of processes in the forest floor could greatly improve our understanding of nutrient and C dynamics in tropical forests (Box 2).

5.3 Microbial pathways of nutrient cycling

We could begin to address many of the open questions about long-term effects of altered litter inputs with in-depth assessments of microbial communities (Box 2). As our results demonstrate that some processes in the litter layer and mineral soil are at least partly decoupled, current discrepancies may be resolved by linking changes in microbial functional groups to specific substrates and processes. Substantially higher rates of soil respiration in the L+ plots are a strong indication that microbial activity in the mineral soil is influenced by aboveground litter inputs (Sayer et al., 2007, 2011). However, the effects of litter addition on soil microbial biomass and soil C were much smaller than expected (Fig. 4B and C) and a recent assessment of broad microbial groups represented by phospho-lipid fatty acid biomarkers revealed only minor effects of both litter treatments on microbial community composition and the abundance of bacterial functional groups in the mineral soil (E. Sayer et al., unpublished data).

Although others have investigated the responses of soil microbial communities to litter manipulation, their findings vary widely. Several studies noted only limited responses in temperate (Leitner et al., 2016; Wang et al., 2017) and tropical forests (Cantrell et al., 2014; Chen et al., 2014), whereas others noted increased abundance of Acidobacteria in response to litter removal, and greater abundance of Alphaproteobacteria with litter addition in a tropical forest, which were related to changes in soil C and N pools (Nemergut et al., 2010). Analysis of phospho-lipid fatty acid biomarkers has also revealed shifts in the relative abundances of broad microbial functional groups, with greater abundance of stress-tolerant Gram-positive bacteria in response to litter removal (Pisani et al., 2016; Wang et al., 2013) and increased abundance of actinomycetes, Gram-positive, and Gram-negative bacteria in response to litter addition (Pisani et al., 2016). However, changes in microbial function (Leff et al., 2012) and extracellular activity (Weintraub et al., 2013; Zhao et al., 2017) with litter manipulation are not always consistent, even within individual studies. Hence, there is still a dearth of information on microbial pathways of nutrient cycling in tropical forests, especially over the long-term and at the ecosystem scale.

The results from studies of AM fungi in the GLiMP plots demonstrate that aboveground litter inputs can shape belowground microbial communities via changes in nutrient availability (Sheldrake et al., 2017, 2018). Although AM fungi are not directly involved in decomposition processes, substantially greater AM fungal colonisation of roots in the forest floor in L+ and control plots indicate that AM fungi nonetheless play an important role in plant uptake of nutrients from organic matter (Sheldrake et al., 2017). Consequently, research into interactions between roots, AM fungi and saprobes in the forest floor could shed new light on tropical nutrient cycling and the concept of direct nutrient uptake. Indeed, fungal decomposition of litter might be particularly crucial to the cycling and retention of several elements in the forest floor. Fungi actively transport and accumulate N, K, Mn and Zn during decomposition (Cromack et al., 1975; Stark, 1972), and these processes are severely disrupted by the litter removal treatments. It is therefore noteworthy that the availability of these nutrients to plants in the L− plots declined to such an extent that nutrient use efficiency increased (Figs. 5 and 9). Shifts in the composition and activity of other microbial groups are likely to elucidate some of the mechanisms underpinning the observed changes in nutrient cycling with litter manipulation, and we call for future work using high-resolution techniques to assess changes in microbial community composition and function in response to altered litter inputs (Box 2).

Our review demonstrates the need for a more holistic view of elemental cycling and nutrient co-limitation. Our insights into tropical forest nutrient cycling arise from studies spanning the molecular to the ecosystem scale, and incorporate aspects of plant, soil, and microbial ecology with biogeochemistry (Fig. 10). Hence, the GLiMP studies not only provide novel insights into the importance of litterfall in tropical forest nutrient cycling, they also collectively demonstrate the importance of "integrative biogeochemical ecology" (Kaspari and Powers, 2016) via the wide range of different processes involved in elemental cycling and the timescales within which they operate.

Acknowledgements

We thank J. Bee, L. Hayes, S. Queenborough, R. Upson, M. Vorontsova, S. Valdez and S. Valdez who established the experimental plots, J.A. Valdez and G. Perez for starting the treatments, and the many Cambridge students who have helped with sampling and measurements. We are grateful to the people who have contributed to research within the Gigante Litter Manipulation Project over the years: A. Cheesman, E. Gora, A. Lacey, S.F. Leitman, J.S. Powers, R. Ross, L. Schreeg, L. Sutcliffe, D. Ureña, A. Vincent, R. Wiederholt, and F. Zeugin. We further thank W. Foster, M. Garcia, M. Kaspari, J. Yavitt for their expert advice, as well as O. Acevedo, B. Jimenez and M. Cano for logistical help and D. Agudo and T. Romero for laboratory analyses. Finally, we are particularly indebted to our "raker", Francisco Valdez, who has moved litter around the forest for over a decade. The project was established with a Mellon Fund Grant to E.V.J.T. and has been funded by a Gates Cambridge Scholarship to E.J.S., a European Union Marie-Curie Fellowship (MOIF-CT-2005-21728) to E.J.S., a Smithsonian Institution Postdoctoral Fellowship to E.J.S., a European Research Council Starting Grant (FP/2007-2013/ERC Grant Agreement No. 307888) to E.J.S., and the Smithsonian Tropical Research Institute.

References

Adams, J.A., 1986. Identification of heterotrophic nitrification in strongly acid larch humus. Soil Biol. Biochem. 18, 324–339.

Ashford, O.S., Foster, W.A., Turner, B.L., Sayer, E.J., Sutcliffe, L., Tanner, E.V.J., 2013. Litter manipulation and the soil arthropod community in a lowland tropical rainforest. Soil Biol. Biochem. 62, 5–12.

Attiwill, P.M., Adams, M.A., 1993. Nutrient cycling in forests. New Phytol. 124, 561–582.

Barthold, F.K., Stallard, R.F., Elsenbeer, H., 2008. Soil nutrient-landscape relationship in a lowland tropical rainforest in Panama. For. Ecol. Manag. 225, 1135–1148.

Bates, D., Mächler, M., Bolker, B., Walker, S., 2015. Fitting linear mixed-effects models using lme4. J. Stat. Softw. 67, 1–48.

Bolaños, L., Lukaszewski, K., Bonilla, I., Blevins, D., 2004. Why boron? Plant Physiol. Biochem. 42, 907–912.

Bonan, G.B., 2008. Forests and climate change: forcings, feedbacks, and the climate benefits of forests. Science 320, 1444–1449.
Broadley, M.R., White, P.J., Hammond, J.P., Zelko, I., Lux, A., 2007. Zinc in plants. New Phytol. 173, 677–702.
Brown, S., Lugo, A.E., 1982. The storage and production of organic matter in tropical forests and their role in the global carbon cycle. Biotropica 14, 161–187.
Burghouts, T., Ernsting, G., Korthals, G., De Vries, T., 1992. Litterfall, leaf litter decomposition and litter invertebrates in primary and selectively logged dipterocarp forest in Sabah, Malaysia. Philos. Trans. R. Soc., B 335, 407–416.
Cakmak, I., Kirkby, E.A., 2008. Role of magnesium in carbon partitioning and alleviating photooxidative damage. Physiol. Plant. 133, 692–704.
Camenzind, T., Hättenschwiler, S., Treseder, K.K., Lehmann, A., Rillig, M.C., 2018. Nutrient limitation of soil microbial processes in tropical forests. Ecol. Monogr. 88, 4–21.
Campo, J., Maass, J.M., Jaramillo, V.J., Yrízar, A.M., 2000. Calcium, potassium, and magnesium cycling in a Mexican tropical dry forest ecosystem. Biogeochemistry 49, 21–36.
Cantrell, S.A., Molina, M., Lodge, D.J., Rivera-Figueroa, F.J., Ortiz-Hernández, M.L., Marchetti, A.A., Cyterski, M.J., Pérez-Jiménez, J.R., 2014. Effects of a simulated hurricane disturbance on forest floor microbial communities. For. Ecol. Manag. 332, 22–31.
Chapin III, F.S., 1980. The mineral nutrition of wild plants. Annu. Rev. Ecol. Syst. 11, 233–260.
Chen, H., Gurmesa, G.A., Liu, L., Zhang, T., Fu, S., Liu, Z., Dong, S., Ma, C., Mo, J., 2014. Effects of litter manipulation on litter decomposition in a successional gradient of tropical forests in southern China. PLoS One 9, e99018.
Cleveland, C.C., Townsend, A.R., 2006. Nutrient additions to a tropical rain forest drive substantial soil carbon dioxide losses to the atmosphere. Proc. Natl. Acad. Sci. 103, 10316–10321.
Cohen, J., 1988. Statistical Power Analysis for the Behavioral Sciences. Lawrence Earlbaum Associates, Hillsdale, New Jersey, USA.
Condit, R., Engelbrecht, B.M.J., Pino, D., Pérez, R., Turner, B.L., 2013. Species distributions in response to individual soil nutrients and seasonal drought across a community of tropical trees. Proc. Natl. Acad. Sci. U. S. A. 110, 5064–5068.
Corlett, R.T., 2011. Impacts of warming on tropical lowland rainforests. Trends Ecol. Evol. 26, 606–613.
Cornejo, F.H., Varela, A., Wright, S.J., 1994. Tropical forest litter decomposition under seasonal drought: nutrient release, fungi and bacteria. Oikos 70, 183–190.
Corre, M.D., Veldkamp, E., Arnold, J., Wright, S.J., 2010. Impact of elevated N input on soil N cycling and losses in old-growth lowland and montane forests in Panama. Ecology 91, 1715–1729.
Cromack, K., Todd, R.L., Monk, C.D., 1975. Patterns of basidiomycete nutrient accumulation in conifer and deciduous forest litter. Soil Biol. Biochem. 7, 265–268.
Cromack, K., Sollins, P., Todd, R.L., Crossley, D.A., Fender, W.M., Fogel, R., Todd, A.W., 1977. Soil microorganism—arthropod interactions: fungi as major calcium and sodium sources. In: Mattson, W.J. (Ed.), The Role of Arthropods in Forest Ecosystems. Springer, Berlin, Heidelberg, pp. 78–84.
Cuevas, E., Medina, E., 1986. Nutrient dynamics within Amazonian forests. I. Nutrient flux in fine litter fall and efficiency of nutrient utilization. Oecologia 68, 466–472.
Cuevas, E., Medina, E., 1988. Nutrient dynamics within Amazonian forests. Oecologia 76, 222–235.
Ehrenfeld, J.G., Ravit, B., Elgersma, K., 2005. Feedback in the plant-soil system. Annu. Rev. Env. Resour. 30, 75–115.

Elser, J.J., Fagan, W.F., Kerkhoff, A.J., Swenson, N.G., Enquist, B.J., 2010. Biological stoichiometry of plant production: metabolism, scaling and ecological response to global change. New Phytol. 186, 593–608.

Erickson, H., Keller, M., Davidson, E.A., 2001. Nitrogen oxide fluxes and nitrogen cycling during postagricultural succession and forest fertilization in the humid tropics. Ecosystems 4, 67–84.

Finzi, A.C., Canham, C.D., Van Breemen, N., 1998. Canopy tree–soil interactions within temperate forests: species effects on pH and cations. Ecol. Appl. 8, 447–454.

Finzi, A.C., Austin, A.T., Cleland, E.E., Frey, S.D., Houlton, B.Z., Wallenstein, M.D., 2011. Responses and feedbacks of coupled biogeochemical cycles to climate change: examples from terrestrial ecosystems. Front. Ecol. Environ. 9, 61–67.

Gill, R.W., 1969. Soil microarthropod abundance following old-field litter manipulation. Ecology 50, 805–816.

Gloor, M.R.J.W., Brienen, R.J., Galbraith, D., Feldpausch, T.R., Schöngart, J., Guyot, J.L., Espinoza, J.C., Lloyd, J., Phillips, O.L., 2013. Intensification of the Amazon hydrological cycle over the last two decades. Geophys. Res. Lett. 40, 1729–1733.

Goldberg, S., 1997. Reactions of boron with soils. Plant and Soil 193, 35–48.

Gora, E.M., Sayer, E.J., Turner, B.L., Tanner, E.V.J., 2018. Decomposition of coarse woody debris in a long-term litter manipulation experiment: a focus on nutrient availability. Funct. Ecol. 32, 1128–1138.

Gosz, J.R., Likens, G.E., Bormann, F.H., 1973. Nutrient release from decomposing leaf and branch litter in the Hubbard Brook Forest, New Hampshire. Ecol. Monogr. 43, 173–191.

Gosz, J.R., Likens, G.E., Bormann, F.H., 1976. Organic matter and nutrient dynamics of the forest and forest floor in the Hubbard Brook Forest. Oecologia 22, 305–320.

Gransee, A., Führs, H., 2013. Magnesium mobility in soils as a challenge for soil and plant analysis, magnesium fertilization and root uptake under adverse growth conditions. Plant and Soil 368, 5–21.

Grubb, P.J., 1989. The role of mineral nutrients in the tropics: a plant ecologist's view. In: Proctor, J. (Ed.), Mineral Nutrients in Tropical Forest and Savanna Ecosystems. Blackwell, Oxford, pp. 417–439.

Harrison, A.F., 1987. Soil Organic Phosphorus. A Review of World Literature. CAB International, Oxford.

Herrera, R., Merida, T., Stark, N., Jordan, C.F., 1978. Direct phosphorus transfer from leaf litter to roots. Naturwissenschaften 65, 208–209.

Hietz, P., Turner, B.L., Wanek, W., Richter, A., Nock, C.A., Wright, S.J., 2011. Long-term change in the nitrogen cycle of tropical forests. Science 334, 664–666.

Hodge, A., Helgason, T., Fitter, A.H., 2010. Nutritional ecology of arbuscular mycorrhizal fungi. Fungal Ecol. 3, 267–273.

John, R., Dalling, J.W., Harms, K.E., Yavitt, J.B., Stallard, R.F., Mirabello, M., Hubbell, S.P., Valencia, R., Navarrete, H., Vallejo, M., Foster, R.B., 2007. Soil nutrients influence spatial distributions of tropical tree species. Proc. Natl. Acad. Sci. U. S. A. 104, 864–869.

Kaspari, M., Powers, J.S., 2016. Biogeochemistry and geographical ecology: embracing all twenty-five elements required to build organisms. Am. Nat. 188, S62–S73.

Kaspari, M., Garcia, M.N., Harms, K.E., Santana, M., Wright, S.J., Yavitt, J.B., 2008. Multiple nutrients limit litterfall and decomposition in a tropical forest. Ecol. Lett. 11, 35–43.

Kaspari, M., Clay, N.A., Donoso, D.A., Yanoviak, S.P., 2014. Sodium fertilization increases termites and enhances decomposition in an Amazonian forest. Ecology 95, 795–800.

Kaspari, M., Bujan, J., Weiser, M.D., Ning, D., Michaletz, S.T., Zhili, H., Enquist, B.J., Waide, R.B., Zhou, J., Turner, B.L., Wright, S.J., 2017. Biogeochemistry drives

diversity in the prokaryotes, fungi, and invertebrates of a Panama forest. Ecology 98, 2019–2028.
Keiluweit, M., Nico, P., Harmon, M.E., Mao, J., Pett-Ridge, J., Kleber, M., 2015. Long-term litter decomposition controlled by manganese redox cycling. Proc. Natl. Acad. Sci. U. S. A. 112, 5253–5260.
Kirkby, E.A., Pilbeam, D.J., 1984. Calcium as a plant nutrient. Plant Cell Environ. 7, 397–405.
Knicker, H., Fründ, R., Lüdemann, H.D., 1993. The chemical nature of nitrogen in native soil organic matter. Naturwissenschaften 80, 219–221.
Koehler, B., Corre, M.D., Veldkamp, E., Wullaert, H., Wright, S.J., 2009. Immediate and long-term nitrogen oxide emissions from tropical forest soils exposed to elevated nitrogen input. Glob. Chang. Biol. 15, 2049–2066.
Kuznetsova, A., Brockhoff, P.B., Christensen, R.H.B., 2017. lmerTest package: tests in linear mixed effects models. J. Stat. Softw. 82 (13), article 13.
Laird-Hopkins, B.C., Bréchet, L.M., Trujillo, B.C., Sayer, E.J., 2017. Tree functional diversity affects litter decomposition and arthropod community composition in a tropical forest. Biotropica 49, 903–911.
Lajtha, K., Bowden, R.D., Crow, S., Fekete, I., Kotroczó, Z., Plante, A., Simpson, M.J., Nadelhoffer, K.J., 2018. The detrital input and removal treatment (DIRT) network: insights into soil carbon stabilization. Sci. Total Environ. 640, 1112–1120.
Laskowski, R., Niklinska, M., Maryanski, M., 1995. The dynamics of chemical elements in forest litter. Ecology 76, 1393–1406.
Leff, J.W., Wieder, W.R., Taylor, P.G., Townsend, A.R., Nemergut, D.R., Grandy, A.S., Cleveland, C.C., 2012. Experimental litterfall manipulation drives large and rapid changes in soil carbon cycling in a wet tropical forest. Glob. Chang. Biol. 18, 2969–2979.
Leigh, E.G., 1999. Tropical Forest Ecology. Oxford University Press, Oxford, UK.
Leitner, S., Sae-Tun, O., Kranzinger, L., Zechmeister-Boltenstern, S., Zimmermann, M., 2016. Contribution of litter layer to soil greenhouse gas emissions in a temperate beech forest. Plant and Soil 403, 455–469.
Leuschner, C., Hertel, D., Coners, H., Büttner, V., 2001. Root competition between beech and oak: a hypothesis. Oecologia 126, 276–284.
Li, J., Richter, D.D., Mendoza, A., Heine, P., 2008. Four-decade responses of soil trace elements to an aggrading old-field forest: B, Mn, Zn, Cu, and Fe. Ecology 89, 2911–2923.
Likens, G.E., Driscoll, C.T., Buso, D.C., Siccama, T.G., Johnson, C.E., Lovett, G.M., Ryan, D.F., Fahey, T., Reiners, W.A., 1994. The biogeochemistry of potassium at Hubbard brook. Biogeochemistry 25, 61–125.
Liu, X., Burslem, D.F., Taylor, J.D., Taylor, A.F., Khoo, E., Majalap-Lee, N., Helgason, T., Johnson, D., 2018. Partitioning of soil phosphorus among arbuscular and ectomycorrhizal trees in tropical and subtropical forests. Ecol. Lett. 21, 713–723.
Loomis, R.S., 1997. On the utility of nitrogen in leaves. Proc. Natl. Acad. Sci. U. S. A. 94, 13378–13379.
Malhi, Y., Gardner, T.A., Goldsmith, G.R., Silman, M.R., Zelazowski, P., 2014. Tropical forests in the Anthropocene. Annu. Rev. Env. Resour. 39, 125–159.
Manzoni, S., Trofymow, J.A., Jackson, R.B., Porporato, A., 2010. Stoichiometric controls dynamics of carbon, nitrogen, and phosphorus in decomposing litter. Ecol. Monogr. 80, 89–106.
Marschner, H., 1986. Mineral Nutrition of Higher Plants. Academic, Amsterdam.
McFee, W.W., Kelly, J.M., Beck, R.H., 1977. Acid precipitation effects on soil pH and base saturation of exchange sites. Water Air Soil Pollut. 7, 401–408.
Medina, E., Cuevas, E., 1989. Patterns of nutrient accumulation and release in Amazonian forests of the upper Rio Negro basin. In: Proctor, J. (Ed.), Mineral Nutrition in Tropical Forest and Savanna Ecosystems. Blackwell Scientific, Oxford, UK, pp. 217–240.

Meir, P., Woodward, F.I., 2010. Amazonian rain forests and drought: response and vulnerability. New Phytol. 187, 553–557.
Mo, Q., Li, Z., Sayer, E.J., Yu, S., Zou, B., Li, Y., Tang, J., Heskel, M., Zheng, X., Ding, Y., Mao, R., Wang, F., 2019. Foliar phosphorus fractions reveal how tropical plants maintain photosynthetic rates despite low soil phosphorus availability. Funct. Ecol. 33, 503–513.
Nadelhoffer, K.J., Boone, R.D., Bowden, R.D., Canary, J., Kaye, J., Micks, P., Ricca, A., McDowell, W., Aitkenhead, J., 2004. The DIRT experiment. Litter and root influences on forest soil organic matter stocks and function. In: Foster, D.R., Aber, J.D. (Eds.), Forests in Time. Yale University Press, New Haven, pp. 300–315.
Nemergut, D.R., Cleveland, C.C., Wieder, W.R., Washenberger, C.L., Townsend, A.R., 2010. Plot-scale manipulations of organic matter inputs to soils correlate with shifts in microbial community composition in a lowland tropical rain forest. Soil Biol. Biochem. 42, 2153–2160.
Nottingham, A.T., Turner, B.L., Chamberlain, P.M., Stott, A.W., Tanner, E.V.J., 2012. Priming and microbial nutrient limitation in lowland tropical forest soils of contrasting fertility. Biogeochemistry 111, 219–237.
Nottingham, A.T., Turner, B.L., Stott, A.W., Tanner, E.V.J., 2015. Nitrogen and phosphorus constrain labile and stable carbon turnover in lowland tropical forest soils. Soil Biol. Biochem. 80, 26–33.
Ostertag, R., 2010. Foliar nitrogen and phosphorus accumulation responses after fertilization: an example from nutrient-limited Hawaiian forests. Plant and Soil 334, 1–14.
Page, E.R., 1962. Studies in soil and plant manganese. Plant and Soil 16, 247–257.
Palm, C.A., Myers, R.J., Nandwa, S.M., 1997. Combined use of organic and inorganic nutrient sources for soil fertility maintenance and replenishment. In: Buresh, R.J., Sanchez, P.A., Calhoun, F. (Eds.), Replenishing Soil Fertility in Africa. Soil Science Society of America, pp. 193–217.
Parker, G.G., 1983. Throughfall and stemflow in the forest nutrient cycle. Adv. Ecol. Res. 13, 57–133.
Pasquini, S.C., Santiago, L.S., 2012. Nutrients limit photosynthesis in seedlings of a lowland tropical forest tree species. Oecologia 168, 311–319.
Pedersen, H., Dunkin, K.A., Firestone, M.K., 1999. The relative importance of autotrophic and heterotrophic nitrification in a conifer forest soil as measured by ^{15}N tracer and pool dilution techniques. Biogeochemistry 44, 125–150.
Pinheiro, J.C., Bates, D.M., 2000. Mixed-Effects Models in S and S-PLUS. Springer, New York.
Pisani, O., Lin, L.H., Lun, O.O., Lajtha, K., Nadelhoffer, K.J., Simpson, A.J., Simpson, M.J., 2016. Long-term doubling of litter inputs accelerates soil organic matter degradation and reduces soil carbon stocks. Biogeochemistry 127, 1–14.
Poser, T., 1990. The influence of litter manipulation on the centipedes of a beech wood. In: Minelli, A. (Ed.), 7th International Congress on Myriapods. Brill, Leiden, pp. 235–245.
Powers, J.S., Salute, S., 2011. Macro-and micronutrient effects on decomposition of leaf litter from two tropical tree species: inferences from a short-term laboratory incubation. Plant and Soil 346, 245–257.
Prajapati, K., Modi, H.A., 2012. The importance of potassium in plant growth–a review. Indian J. Plant Sci. 1, 177–186.
Proctor, J., 1987. Nutrient cycling in primary and old secondary rainforests. Appl. Geogr. 7, 135–152.
R Development Core Team, 2018. R: A Language and Environment for Statistical Computing. R Foundation for Statistical Computing. Vienna, Austria.
Rayner, A.D., Boddy, L., 1988. Fungal Decomposition of Wood. Its Biology and Ecology. John Wiley & Sons Ltd., N.J., Sussex

Rodtassana, C., 2016. Litter Manipulation Effects on Fine Root and Litterfall Dynamics in a Tropical Forest. Ph.D. Thesis, University of Cambridge, UK.

Rodtassana, C., Tanner, E.V.J., 2018. Litter removal in a tropical rain forest reduces fine root biomass and production but litter addition has few effects. Ecology 99, 735–742.

Salick, J., Herrera, R., Jordan, C.F., 1983. Termitaria: nutrient patchiness in nutrient-deficient rain forests. Biotropica 15, 1–7.

Santiago, L.S., Schuur, E.A., Silvera, K., 2005. Nutrient cycling and plant–soil feedbacks along a precipitation gradient in lowland Panama. J. Trop. Ecol. 21, 461–470.

Santiago, L.S., Wright, S.J., Harms, K.E., Yavitt, J.B., Korine, C., Garcia, M.N., Turner, B.L., 2012. Tropical tree seedling growth responses to nitrogen, phosphorus and potassium addition. J. Ecol. 100, 309–316.

Sardans, J., Rivas-Ubach, A., Peñuelas, J., 2012. The elemental stoichiometry of aquatic and terrestrial ecosystems and its relationships with organismic lifestyle and ecosystem structure and function: a review and perspectives. Biogeochemistry 111, 1–39.

Sayer, E.J., 2006. Using experimental litter manipulation to assess the roles of leaf litter in the functioning of forest ecosystems. Biol. Rev. 81, 1–31.

Sayer, E.J., Banin, L.F., 2016. Tree nutrient status and nutrient cycling in tropical forest—lessons from fertilization experiments. In: Goldstein, G., Santiago, L.S. (Eds.), Tropical Tree Physiology. Springer, Switzerland, pp. 275–297.

Sayer, E.J., Sutcliffe, L.M., Ross, R.I., Tanner, E.V.J., 2010. Arthropod abundance and diversity in a lowland tropical forest floor in Panama: the role of habitat space vs. nutrient concentrations. Biotropica 42, 194–200.

Sayer, E.J., Tanner, E.V.J., 2010. Experimental investigation of the importance of litterfall in lowland semi-evergreen tropical forest nutrient cycling. J. Ecol. 98, 1052–1062.

Sayer, E.J., Tanner, E.V.J., Cheesman, A.W., 2006a. Increased litterfall changes fine root distribution in a moist tropical forest. Plant and Soil 281, 5–13.

Sayer, E.J., Tanner, E.V.J., Lacey, A.L., 2006b. Effects of litter manipulation on early-stage decomposition and meso-arthropod abundance in a moist tropical forest. For. Ecol. Manag. 229, 285–293.

Sayer, E.J., Powers, J.S., Tanner, E.V.J., 2007. Increased litterfall in tropical forests boosts the transfer of soil CO_2 to the atmosphere. PLoS One 2, e1299.

Sayer, E.J., Heard, M.S., Grant, H.K., Marthews, T.R., Tanner, E.V.J., 2011. Soil carbon release enhanced by increased tropical forest litterfall. Nat. Clim. Chang. 1, 304–307.

Sayer, E.J., Tanner, E.V.J., Wright, S.J., Yavitt, J.B., Harms, K.E., Powers, J.S., Kaspari, M., Garcia, M.N., Turner, B.L., 2012. Comparative assessment of lowland tropical forest nutrient status in response to fertilization and litter manipulation. Ecosystems 15, 387–400.

Schimel, J.P., Firestone, M.K., Killham, K.S., 1984. Identification of heterotrophic nitrification in a Sierran forest soil. Appl. Environ. Microbiol. 48, 802–806.

Schlesinger, W.H., 1991. Biogeochemistry: An Analysis of Global Change. Academic press, Oxford.

Schreeg, L.A., Mack, M.C., Turner, B.L., 2013. Leaf litter inputs decrease phosphate sorption in a strongly weathered tropical soil over two time scales. Biogeochemistry 113, 507–524.

Shaul, O., 2002. Magnesium transport and function in plants: the tip of the iceberg. Biometals 15, 307–321.

Sheldrake, M., Rosenstock, N., Revillini, D., Olsson, P.A., Mangan, S., Sayer, E.J., Turner, B.L., Tanner, E.V.J., 2017. Arbuscular mycorrhizal fungal community composition is altered by long-term litter removal but not litter addition in a lowland tropical forest. New Phytol. 214, 455–467.

Sheldrake, M., Rosenstock, N., Mangan, S., Revillini, D., Sayer, E.J., Olsson, P.A., Wallander, H., Tanner, E.V.J., Wright, S.J., Turner, B.L., 2018. Responses of arbuscular

mycorrhizal fungi tolong-term inorganic and organic nutrient addition in a lowland tropical forest. ISME J. 12, 2433–2445.
Sinclair, T.R., Vadez, V., 2002. Physiological traits for crop yield improvement in low N and P environments. Plant and Soil 245, 1–15.
Sinsabaugh, R.L., Lauber, C.L., Weintraub, M.N., Ahmed, B., Allison, S.D., Crenshaw, C., Contosta, A.R., Cusack, D., Frey, S., Gallo, M.E., Gartner, T.B., Hobbie, S.E., Holland, K., Keeler, B.L., Powers, J.S., Stursova, M., Takacs-Vesbach, C., Waldrop, M.P., Wallenstein, M.D., Zak, D.R., Zeglin, L.H., 2008. Stoichiometry of soil enzyme activity at global scale. Ecol. Lett. 11, 1252–1264.
Smith, S.E., Read, D.J., 2010. Mycorrhizal Symbiosis. Academic press.
Stark, N.M., 1972. Nutrient cycling pathways and litter fungi. Bioscience 22, 355–360.
Stark, N.M., Jordan, C.F., 1978. Nutrient retention by the root mat of an Amazonian rain forest. Ecology 59, 434–437.
Steidinger, B.S., Turner, B.L., Corrales, A., Dalling, J.W., 2015. Variability in potential to exploit different soil organic phosphorus compounds among tropical montane tree species. Funct. Ecol. 29, 121–130.
Tanner, E.V.J., Vitousek, P.M., Cuevas, E., 1998. Experimental investigation of nutrient limitation of forest growth on wet tropical mountains. Ecology 79, 10–22.
Tanner, E.V.J., Sheldrake, M.W., Turner, B.L., 2016. Changes in soil carbon and nutrients following 6 years of litter removal and addition in a tropical semi-evergreen rain forest. Biogeosciences 13, 6183–6190.
Tiessen, H., Salcedo, I.H., Sampaio, E.V.S.B., 1992. Nutrient and soil organic matter dynamics under shifting cultivation in semiarid northeastern Brazil. Agric. Ecosyst. Environ. 38, 139–151.
Tobón, C., Sevink, J., Verstraten, J.M., 2004. Litterflow chemistry and nutrient uptake from the forest floor in northwest Amazonian forest ecosystems. Biogeochemistry 69, 315–339.
Tóth, J.A., Nagy, P.T., Krakomperger, Z., Veres, Z., Kotroczó, Z., Kincses, S., Fekete, I., Papp, M., 2011. Effect of litter fall on soil nutrient content and pH, and its consequences in view of climate change (Síkfőkút DIRT project). Acta Silv. Lign. Hung. 7, 75–86.
Townsend, A.R., Cleveland, C.C., Houlton, B.Z., Alden, C.B., White, J.W.C., 2011. Multi-element regulation of the tropical forest carbon cycle. Front. Ecol. Environ. 9, 9–17.
Tripler, C.E., Causal, S.S., Likens, G.E., Walter, M.T., 2006. Patterns in potassium dynamics in forest ecosystems. Ecol. Lett. 9, 451–466.
Turner, B.L., Engelbrecht, B.M., 2011. Soil organic phosphorus in lowland tropical rain forests. Biogeochemistry 103, 297–315.
Turner, B.L., Romero, T.E., 2009. Short-term changes in extractable inorganic nutrients during storage of tropical rain forest soils. Soil Sci. Soc. Am. J. 73, 1972–1979.
Turner, B.L., Bielnicka, A.W., Dalling, J.W., Wolf, J.A., 2016. Interference by iron in the determination of boron by ICP-OES in Mehlich-III extracts and total element digests of tropical forest soils. Commun. Soil Sci. Plant Anal. 47, 2378–2386.
Verstraete, W., Focht, D.D., 1977. Biochemical ecology of nitrification and denitrification. In: Alexander, M. (Ed.), Advances in Microbial Ecology. Springer, Boston, pp. 135–214.
Vincent, A.G., Turner, B.L., Tanner, E.V.J., 2010. Soil organic phosphorus dynamics following perturbation of litter cycling in a tropical moist forest. Eur. J. Soil Sci. 61, 48–57.
Vitousek, P.M., 1982. Nutrient cycling and nutrient use efficiency. Am. Nat. 119, 553–572.
Vitousek, P.M., 1984. Litterfall, nutrient cycling and nutrient limitation in tropical forests. Ecology 65, 285–298.
Vitousek, P.M., Sanford Jr., R.L., 1986. Nutrient cycling in moist tropical forest. Annu. Rev. Ecol. Syst. 17, 137–167.

Vitousek, P.M., Walker, L.R., Whiteaker, L.D., Matson, P.A., 1993. Nutrient limitations to plant growth during primary succession in Hawaii volcanoes National Park. Biogeochemistry 23, 197–215.

Vitousek, P.M., Porder, S., Houlton, B.Z., Chadwick, O.A., 2010. Terrestrial phosphorus limitation: mechanisms, implications, and nitrogen–phosphorus interactions. Ecol. Appl. 20, 5–15.

Walker, T.W., Syers, J.K., 1976. The fate of phosphorus during pedogenesis. Geoderma 15, 1–19.

Wang, Q., He, T., Wang, S., Liu, L., 2013. Carbon input manipulation affects soil respiration and microbial community composition in a subtropical coniferous forest. Agric. For. Meteorol. 178, 152–160.

Wang, J.J., Pisani, O., Lin, L.H., Lun, O.O., Bowden, R.D., Lajtha, K., Simpson, A.J., Simpson, M.J., 2017. Long-term litter manipulation alters soil organic matter turnover in a temperate deciduous forest. Sci. Total Environ. 607, 865–875.

Weintraub, S.R., Wieder, W.R., Cleveland, C.C., Townsend, A.R., 2013. Organic matter inputs shift soil enzyme activity and allocation patterns in a wet tropical forest. Biogeochemistry 114, 313–326.

Welch, B., Gauci, V., Sayer, E.J., 2019. Tree stem bases are sources of CH_4 and N_2O in a tropical forest on upland soil during the dry to wet season transition. Glob. Chang. Biol. 25, 361–372.

Wieder, R.K., Wright, S.J., 1995. Tropical forest litter dynamics and dry season irrigation on Barro Colorado Island, Panama. Ecology 76, 1971–1979.

Wieder, W.R., Cleveland, C.C., Smith, W.K., Todd-Brown, K., 2015. Future productivity and carbon storage limited by terrestrial nutrient availability. Nat. Geosci. 8, 441.

Wood, T.E., Lawrence, D., Clark, D.A., Chazdon, R.L., 2009. Rain forest nutrient cycling and productivity in response to large-scale litter manipulation. Ecology 90, 109–121.

Wright, S.J., 2005. Tropical forests in a changing environment. Trends Ecol. Evol. 20, 553–560.

Wright, S.J., 2019. Plant responses to nutrient addition experiments conducted in tropical forests. Ecol. Monogr. 89, e01382.

Wright, S.J., Yavitt, J.B., Wurzburger, N., Turner, B.L., Tanner, E.V., Sayer, E.J., Santiago, L.S., Kaspari, M., Hedin, L.O., Harms, K.E., Garcia, M.N., 2011. Potassium, phosphorus, or nitrogen limit root allocation, tree growth, or litter production in a lowland tropical forest. Ecology 92, 1616–1625.

Wright, S.J., Turner, B.L., Yavitt, J.B., Harms, K.E., Kaspari, M., Tanner, E.V., Bujan, J., Griffin, E.A., Mayor, J.R., Pasquini, S.C., Sheldrake, M., 2018. Plant responses to fertilization experiments in lowland, species-rich, tropical forests. Ecology 99, 1129–1138.

Yavitt, J.B., 2000. Nutrient dynamics of soil derived from different parent material on Barro Colorado Island, Panama. Biotropica 32, 198–207.

Yavitt, J.B., Wright, S.J., 1996. Temporal patterns of soil nutrients in a Panamanian moist forest revealed by ion-exchange resin and experimental irrigation. Plant and Soil 183, 117–129.

Yavitt, J.B., Harms, K.E., Garcia, M.N., Wright, S.J., He, F., Mirabello, M.J., 2009. Spatial heterogeneity of soil chemical properties in a lowland tropical moist forest, Panama. Aust. J. Soil. Res. 47, 674–687.

Yavitt, J.B., Harms, K.E., Garcia, M.N., Mirabello, M.J., Wright, S.J., 2011. Soil fertility and fine root dynamics in response to 4 years of nutrient (N, P, K) fertilization in a lowland tropical forest, Panama. Austral Ecol. 36, 433–445.

Zalamea, P.C., Turner, B.L., Winter, K., Jones, F.A., Sarmiento, C., Dalling, J.W., 2016. Seedling growth responses to phosphorus reflect adult distribution patterns of tropical trees. New Phytol. 212, 400–408.

Zemunik, G., Davies, S.J., Turner, B.L., 2018. Soil drivers of local-scale tree growth in a lowland tropical forest. Ecology 99, 2844–2852.

Zhao, Q., Classen, A.T., Wang, W.W., Zhao, X.R., Mao, B., Zeng, D.H., 2017. Asymmetric effects of litter removal and litter addition on the structure and function of soil microbial communities in a managed pine forest. Plant and Soil 414, 81–93.

Further reading

Jordan, C.F., Herrera, R., 1981. Tropical rain forests: are nutrients really critical? Am. Nat. 117, 167–180.

CHAPTER SIX

Soil biogeochemical responses of a tropical forest to warming and hurricane disturbance

Sasha C. Reed[a,*], Robin Reibold[a], Molly A. Cavaleri[b], Aura M. Alonso-Rodríguez[c], Megan E. Berberich[c], Tana E. Wood[c]
[a]U.S. Geological Survey, Southwest Biological Science Center, Moab, UT, United States
[b]College of Forest Resources and Environmental Science, Michigan Technological University, Houghton, MI, United States
[c]U.S.D.A. Forest Service, International Institute of Tropical Forestry, Río Piedras, PR, United States
*Corresponding author: e-mail address: screed@usgs.gov

Contents

Abstract

Tropical forests represent <15% of Earth's terrestrial surface yet support >50% of the planet's species and play a disproportionately large role in determining climate due to the vast amounts of carbon they store and exchange with the atmosphere. Currently, disturbance patterns in tropical ecosystems are changing due to factors such as increased land use pressure and altered patterns in hurricanes. At the same time, these regions are expected to experience unprecedented warming before 2100. Despite the importance of these ecosystems for forecasting the global consequences of multiple stressors, our understanding of how projected changes in climate and

Advances in Ecological Research, Volume 62
ISSN 0065-2504
https://doi.org/10.1016/bs.aecr.2020.01.007

disturbance will affect the biogeochemical cycling of tropical forests remains in its infancy. Until now, no studies to our knowledge have evaluated forest recovery following hurricane disturbance within the context of concurrent climatic change. Here, we present soil biogeochemical results from a tropical forest field warming experiment in Puerto Rico where, a year after experimental warming began, Hurricanes Irma and María greatly altered the forest, allowing a unique opportunity to explore the interacting effects of hurricane disturbance and warming. We tracked post-hurricane forest recovery for a year without warming to assess legacy effects of prior warming on the disturbance response, and then reinitiated warming treatments to further evaluate interactions between forest recovery and warmer temperatures. The data showed that warming affected multiple aspects of soil biogeochemical cycling even in the first year of treatment, with particularly large positive effects on soil microbial biomass pools (e.g., increases of 54%, 33%, and 38% relative to the control plots were observed for microbial biomass carbon, nitrogen, and phosphorus, respectively after 6 months of warming). We also observed significant effects of the hurricanes on soil biogeochemical cycling, as well as interactive controls of warming and disturbance. Taken together, our results showed dynamic soil responses that suggest the future of biogeochemical cycling in this tropical wet forest will be strongly shaped by the directional effects of warming and the episodic effects of hurricanes.

1. Introduction

Disturbances, such as hurricanes, play a key role in shaping forest landscape patterns and processes over years, decades, and centuries (Johnstone et al., 2016; Lugo, 2008; Lugo and Scatena, 1995). Most ecosystems are commonly recovering from past events, and thus interpretation of current forest dynamics should ideally be considered within the context of the historical disturbance regime (Bormann and Likens, 1979; Cooper, 1926; Turner, 2010; White, 1979). However, we are now entering a time when directional changes in climate, such as increasing temperatures, are likely to disrupt key ecosystem processes, which can increase vulnerability to other disturbances and endanger the persistence of the affected ecosystems (Folke et al., 2004). A strong interaction between disturbance and climate change severely limits our ability to predict how ecosystems respond to perturbation, as the outcomes of these interactions may not be apparent until a disturbance event has occurred (Johnstone et al., 2016). Despite the large potential for interactive effects between climate change and disturbance in tropical ecosystems, our understanding of the consequences of these combined effects for ecosystem function remains poor, in part because disturbance events, such as strong hurricanes, are relatively infrequent. Here, we took advantage of an ongoing field warming experiment in

Puerto Rico that was struck by two major hurricanes to ask the questions: (1) Does prior warming affect soil biogeochemical responses to hurricane disturbance? and (2) Does increased temperature affect the speed and magnitude of recovery of biogeochemical processes following disturbance?

Tropical forests are ideal ecosystems for addressing these questions, given the high rate of biogeochemical cycling and the speed of forest recovery. For example, within 15 years of forest regrowth following hurricanes, tropical forest canopy structure can be nearly indistinguishable from old growth forests (Heartsill-Scalley et al., 2010). Further, tropical forests play a disproportionately large role in Earth's biogeochemical cycling relative to the spatial extent of their land cover (Townsend et al., 2011). Lowland tropical forests make up only 15% of the planet's terrestrial surface, yet these productive ecosystems account for over 2/3 of our live terrestrial plant biomass (Pan et al., 2013), nearly one-third of all soil carbon (C) (Jobbagy and Jackson, 2000; Tarnocai et al., 2009), and exchange more carbon dioxide with the atmosphere than any other biome (Beer et al., 2010; Foley et al., 2003). Tropical forests also represent large uncertainties in our capacity to forecast future C cycling and thus climate (Cavaleri et al., 2015). For example, although data suggest tropical forest ecosystems determine the size of the land C sink (Ahlström et al., 2015), our understanding of how tropical forests will respond to global changes, such as increased temperature and altered precipitation, is lacking (Wood et al., 2012). Nevertheless, the implications for such uncertainty related to the biogeochemical consequences of tropical forest responses are particularly important. If the massive amount of C stored in tropical forest soils increases or decreases in response to change, even seemingly subtle alterations could have large implications for the amount of CO_2 in the atmosphere (see also "Protecting environmental and socio-economic values of selectively logged tropical forests in the Anthropocene" by Bousfield et al., this issue).

Most ecosystems are adapted to some form of disturbance regime. Hurricanes, for example, affect nearly every continent in the world and are among the most intense weather disturbances in forested ecosystems (McDowell, 2011). Strong hurricanes are known to cause significant disruption to forest structure and reductions in canopy height (Heartsill-Scalley et al., 2010; Fig. 1), which has cascading consequences for biogeochemical and hydrologic cycling in forests (Comita et al., 2009; Heartsill-Scalley et al., 2007; Herbert et al., 1999; Lodge et al., 1991; Lugo et al., 1983; McDowell, 2011; McDowell and Liptzin, 2014; Sánchez-Sánchez and Islebe, 1999; Silver and Vogt, 1993). For example, green leaves stripped from vegetation and deposited on the forest floor can deliver a substantial

Fig.1 (A) Photographs showing canopy cover before and after the 2017 hurricanes at the warming experiment site in the Luquillo Experimental Forest, Puerto Rico. (B) Percent canopy openness before the hurricanes and following Hurricane Irma and Hurricane María. Dashed lines indicate dates the 2017 hurricanes in Puerto Rico. These data underscore both the large effects of hurricanes on the structure of tropical forests, as well as the rapid pace of recovery.

pulse of C and nutrients that are equivalent to more than a year's worth of litter inputs, resulting in 1.3 and 2.4 years' worth of phosphorus (P) and nitrogen (N), respectively deposited on the forest floor in less than 1 week (Lodge et al., 1991). The opening of the forest canopy also affects a range of processes via large changes in forests' microclimate. As much as 40% of rainfall can be intercepted by an intact forest canopy, and rainfall that does pass through the canopy (i.e., throughfall) is chemically altered due to leaf nutrient leaching and the washing off of dust and debris (Heartsill-Scalley et al.,

2007). Immediately following hurricanes, canopy rainfall interception is negligible, and the chemistry of throughfall most closely reflects that of rainfall (Heartsill-Scalley et al., 2007). Canopy disturbance also significantly alters understory microclimate and light availability (Comita et al., 2009; Scatena et al., 1996), driving a much more variable microclimate without the buffering effect of the forest canopy, which is also likely to alter patterns of soil biogeochemical cycling (González et al., 2014; Lodge et al., 2014; Silver and Vogt, 1993). However, in places where such events are a long-term component of the disturbance regime, plant communities that experience frequent hurricanes have often developed adaptation strategies that provide both resistance and resilience to these forms of disturbance (Lugo et al., 1983; Sánchez-Sánchez and Islebe, 1999).

Research also suggests that tropical forests are facing a significant shift toward a novel temperature regime in the coming decades, with temperatures in the tropics and subtropics increasing by 3–5 °C within the next 20 years (Anderson, 2011; Diffenbaugh and Scherer, 2011). Given the narrow temperature range under which tropical forests typically operate, combined with evidence that these forests may be near high temperature thresholds, there is potential for tropical forest C cycling to respond dramatically to rising temperatures (Clark et al., 2003; Doughty and Goulden, 2008; Wood et al., 2012, 2019). Responses from warming experiments in other types of ecosystems show the likelihood of soil biogeochemical responses to increased temperatures (e.g., Bai et al., 2013; Rustad et al., 2001), but a lack of such experiments in tropical forests has hindered our capacity to explore similarities or differences in tropical environments (Cavaleri et al., 2015). Natural temperature gradients (e.g., elevational gradients) in the tropics offer important opportunities for observational studies of temperature effects on plant and ecosystem functioning (Malhi et al., 2010; Nottingham et al., 2019; Selmants et al., 2014; Townsend et al., 1995), yet the warmer temperature regimes predicted for the already-warm lowland tropics do not currently exist in tropical forests today (Wright, 2010). Thus, warming experiments are needed to achieve the projected temperatures and to assess how tropical forests change as temperatures rise (Cavaleri et al., 2015).

Given this potential for increased temperatures to "stress" tropical forested ecosystems, the combined effects of hurricane disturbance and warmer temperatures could fundamentally alter the functioning and the resilience of these ecosystems, resulting in direct and legacy effects of prior warming on soil biogeochemical cycling in the months following disturbance.

Furthermore, as the forest progresses through the expected stages of post-hurricane recovery, increased temperature could alter the trajectory and duration of forest recovery patterns, resulting in shifted ecosystem states (Johnstone et al., 2016) and changes to coupled biogeochemical cycles. An improved understanding of how hurricane disturbance and temperature interact to regulate tropical forest biogeochemical cycling is essential not only for considering the overall fate of these forests, but for improving predictions of future global C cycling and climate (Ahlström et al., 2015; Cavaleri et al., 2015; Reed et al., 2012; Van Gestel et al., 2018). Using the Tropical Responses to Altered Climate Experiment (TRACE) in situ manipulation and an opportunity to assess interactions of this experimental warming with forest responses to 2017's Hurricanes Irma and María, we hypothesized that prior warming would significantly alter soil C and nutrient cycling in the months following hurricane disturbance. We further predicted that warming in a post-hurricane environment would drive trajectories of recovery divergent from those in the unwarmed plots, resulting in fundamentally altered coupled C and nutrient cycles and a system with altered resilience to future disturbance (Johnstone et al., 2016).

2. Methods

2.1 Site description

The study site is a lowland subtropical wet forest (sensu Holdridge et al., 1971) located near the U.S. Forest Service Sabana Field Research Station, which is within the Luquillo Experimental Forest in northeastern Puerto Rico (18° 19′ 28.56"N, −65° 43' 50.16"W). The study area is ~100 m.a.s.l. and is a secondary forest that has regenerated from pasture since the early 1950s (Kimball et al., 2018). The site has a long history of hurricane disturbance, including Hurricanes Hugo (1989) and Georges (1998), which both occurred within the timeframe of this forest's regeneration. The mean annual temperature is 24 °C, and mean annual precipitation is ~3500 mm. Rainfall occurs throughout the year and, on average, no month receives less than 200 mm of precipitation (Heartsill-Scalley et al., 2007). Temperature exhibits relatively low seasonal variation, with mean monthly temperatures varying just 4 °C throughout the year (Kimball et al., 2018). Soils are classified as Ultisols and are deep and rich in clay (Scatena, 1989). The site has relatively steep slopes that range from 15 to 26 degrees with an average slope of 21 degrees (Kimball et al., 2018).

2.2 Experimental design

The Tropical Responses to Altered Climate Experiment (TRACE) is made up of six 4-m diameter hexagonal plots: three control and three heated. All plots were at least 10 m apart. For the heated plots, six infrared (IR) heaters (Model Raymax 1010, Watlow Electric Manufacturing Co., St. Louis, MO) were installed on crossbars at ~2.6 m height from the ground. Heating from this height warms the understory canopy and the soils. The same infrastructure (crossbars, posts, and concrete footings) used for the warming plots was also constructed around the unwarmed control plots to control for any effects of treatment installation and/or shading. Instead of IR heaters for these plots, aluminium panels were installed on crossbars as false heaters. Multiple soil microclimate sensors were installed (Campbell Scientific CS655; Logan, UT): Each plot has one soil temperature and moisture sensor at 40–50 cm depth, one sensor at 20–30 cm depth, and three surface sensors at 0–10 cm depth. The surface sensors are at the edge of the plot, midway to the center of the plot, and at the center of the plot. The deeper soil sensors are in the center of the plot. Warming treatments began on September 28, 2016. The IR heaters were designed and controlled so heated plots maintained temperatures 4.0 °C higher than the mean of control plot temperatures ±0.1 °C, as sensed by IR thermometers (Kimball et al., 2018). During the first year of heating prior to the hurricanes, heating was homogeneous, and there was little observable difference among temperatures of the 0–10 cm depth soil at the plots' centers, edges, and midway between. During this first year with the warming treatments, soil temperatures at 40–50 cm depth increased ~3 °C compared to the control plots. Details about the warming treatment's design, installation, and performance can be found in Kimball et al. (2018).

In 2017, the coastal U.S. and Puerto Rico were struck by a series of major hurricanes that will have long-lasting implications for these tropical forests. On September 5, 2017, after 11.5 months of experimental warming, the IR heaters were turned off in preparation for the hurricanes. On September 7, 2017 the eye of category 5 Hurricane Irma passed just 15 miles north of Puerto Rico. Less than 2 weeks later, Hurricane María made landfall on the island as a Category 4 storm with sustained winds up to 250 km/h, and 500 mm of precipitation fell in 24 h. Hurricanes Irma and María strongly affected forest structure across the island. This disturbance completely defoliated all vegetation, severed large branches, and drove a large number of treefalls, all of which substantially increased canopy openness (Uriarte et al., 2019; Fig. 1), with significant consequences for forest structure and function (e.g., Heartsill-Scalley et al., 2007; Lodge et al., 1991; Silver and Vogt, 1993). Although the majority of branches and leaves were lost from the canopy, most trees survived, and forest regeneration began quickly (Fig. 1).

While TRACE infrastructure was designed to withstand hurricane-force winds and precipitation, falling trees and branches during Hurricanes Irma and María damaged portions of the heating and plot infrastructure. Tree trunks and large branches were removed from the plots following the hurricanes, but leaf litter and small woody debris (≤3 cm diameter) were left in place. Warming treatments were fully re-initiated approximately 1 year later, on September 28, 2018. Following the reinstatement of warming, changes to forest structure significantly altered the way warming occurred in the soils (Fig. 2). Hurricane-induced increases in understory growth

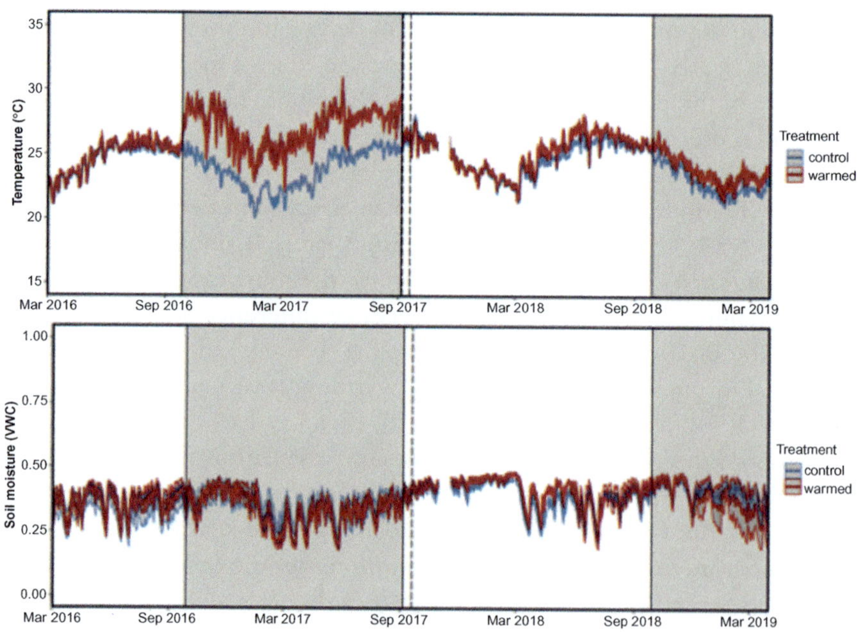

Fig. 2 (A) Soil temperature (°C) and (B) soil moisture (volumetric water content, unitless) values for the top 10 cm of soil in the control (blue) and warmed (red) plots ($n = 3$ for each treatment). The data are daily means with standard error for among plot variation shown with coloured shading. There is a gap in the soil microclimate data in November, 2017 due to problems with the data logger battery system. The gray shading blocks show the times when the warming treatments were active and the vertical dashed lines show when Hurricanes Irma and María struck the site. Results show that the warming increased soil temperatures at all measurement times, but the warming effects were larger prior to the hurricanes. The warming also caused significant but small reductions in volumetric water content.

intercepted more energy from the IR lamps, and as a result the soils warmed on average + 1 °C at 0–10 cm depth as opposed to the +3.2 °C observed at the same soil depth prior to the hurricanes.

The hurricanes that struck Puerto Rico were, of course, not a planned treatment and because the entire forest was affected, the disturbance left an experimental design without a disturbance control (i.e., all plots, warmed and unwarmed, were affected by the hurricanes). A lack of power throughout Puerto Rico following Hurricane María and damage to warming infrastructure also meant that warming treatments were not possible for several months following the hurricanes. With that in mind, the warming treatments were stopped for a full year so that we could fully assess the legacy effects of warming on a recovering tropical forest without the confounding effects of active warming. While the need to stop the warming did inhibit assessment of the consequences of active warming during the first year of regrowth, the in situ warming experiment combined with the potentially once-in-a-century Category 4 hurricane offered a unique opportunity to assess the effects of prior warming on forest responses to hurricane disturbance and to explore the interacting effects of warming and disturbance on the trajectory of forest recovery. Taken together, the data presented here show pre-treatment data, nearly a year of warming, a year of forest regrowth without warming following large hurricane disturbance, and then recovery from hurricane disturbance with warming treatments reinitiated.

2.3 Soil sampling and analysis

At each sampling timepoint, soil samples were taken from 0 to 10 cm depth and were analysed for total C and total N, KCl-extractable inorganic N (NH_4^+ + NO_3^-), Bray-extractable PO_4^{3-}, K_2SO_4-extractable organic C and N, and microbial biomass C, N, and P concentrations. Unfortunately, unanticipated problems with the chloroform methods resulted in unusable soil microbial biomass data for the pre-warming timepoint of March 10, 2016. Chloroform methods were refined and were successful at all other timepoints. Soils were harvested from each plot and were returned to the on-site laboratory where roots and visible organic matter were manually removed from each sample. All soil processing and extractions were conducted on the same day soils were collected. For all extractions, extracts were filtered through Whatman #1 filter paper (GE Healthcare, Chicago, IL), frozen immediately, and stored frozen until analysis.

To assess total C and N concentrations, soils were dried at 60 °C, ground, and were analysed for total C and N using an elemental analyser (Elementar Vario Micro Cube, Elementar Inc., Langenelsbold, Germany). For extractable inorganic N concentrations, fresh soil was shaken with 2 N KCl for 1 h, allowed to sit overnight, and filtered (as described in Reed et al., 2008). Ammonium (NH_4^+) and nitrate (NO_3^-) concentrations were analysed using colorimetry with the indophenol blue method and a cadmium reduction-Greis Ilosvay method on an autoanalyser (Unity Scientific Smartchem 200 Discrete autoanalyser, Milford, MA; Reed et al., 2008). To assess Bray-extractable PO_4^{3-}, fresh soil was extracted with Bray solution (dilute HCl + NH_4F) and shaken for 1 min (Bray and Kurtz, 1945). Samples were filtered immediately and PO_4^{3-} concentrations were measured colorimetrically using an ascorbic acid molybdate analysis on the SmartChem autoanalyser (Sparks et al., 1996). For TOC and TDN analyses, fresh soil was extracted with 0.5 M K_2SO_4. Samples were shaken for 1 h, filtered, and analysed on a Shimadzu TOC/TDN (TOC-V_{CPN} with TNM-1 attachment, Shimadzu Corporation, Kyoto, Japan).

A chloroform addition method was used to assess microbial biomass C, N, and P. Briefly, the method is a modification of the chloroform fumigation method (Brookes et al., 1985; Vance, 1987), and ethanol-free chloroform is directly added to soil instead of fumigating with chloroform in a sealed chamber headspace. For microbial C, N, and P, fresh soil was placed into an Erlenmeyer flask, and 1 mL of chloroform was added before sealing the flask with a neoprene stopper. The flasks were left in the dark for 14–16 h before being unstoppered and aired for 1 h. Samples for microbial biomass C and N concentrations were extracted with K_2SO_4 and were analysed for TOC and TDN concentrations as described above. Samples for microbial biomass P were extracted with Bray and were analysed for PO_4^{3-} concentrations as described above. Initial K_2SO_4 extract values (for C and N) or initial Bray extract values (for P) were subtracted from the K_2SO_4 or Bray values from the samples with chloroform addition to estimate microbial biomass C, N, and P concentrations. The data provided here have not been converted using the proportionality constants sometimes used for C, N, and P (Brookes et al., 1985, Vance et al., 1987; e.g., $Kc = 0.45$ for C). Canopy openness was measured from the center of each of the six experimental plots on the dates shown in Fig. 1 using a spherical densiometer (Forestry Suppliers, Inc.). Measurements were taken following the manufacturer's instructions.

All data were tested for normality and homoscedasticity before analysis. Variation in all soil biogeochemical pools were assessed over time, by treatment, and with associated interactions using a repeated measures general linear model (GLM). Plots were treated as a random variable to account for native differences among them. For single time points, significant differences between treatments were explored with multi-variate GLM. Significance was determined at $\alpha < 0.05$ and all analyses were performed using SPSS statistical software (IBM Inc., Armonk, NY).

3. Results

3.1 Warming treatment effects on soil moisture

The volumetric water content was an average of 1.45% lower in the warmed plots relative to the control plot for the first round of warming (10/4/2016–9/5/2017), and an average of 3.79% lower in the warmed plots relative to the control plots for the second round of warming (9/29/2018–8/27/2019; Fig. 2).

3.2 Total carbon and nitrogen

Total soil C and N concentrations in the top 10 cm of soil varied significantly with time ($P < 0.001$), showing significant increases in C and N concentrations in the year following the hurricanes and subsequent declines after that time (Fig. 3). Although there was some suggestion of warming-induced increases in total soil C and N during the first year of warming before the hurricanes struck, these trends were not significant and there were no interactive effects of warming on the patterns of soil C and N over the course of post-hurricane forest recovery. The C:N ratios of the soil did not change through time or with the warming treatment.

3.3 Extractable carbon, nitrogen, and phosphorus

Similarly to total soil C and N, there were not significant warming treatment effects for extractable organic C or total dissolved N concentrations at any timepoint. Extractable organic C and N concentrations did increase through time. In contrast, the concentrations of inorganic forms of extractable soil N and P did respond significantly to warming (for NH_4^+, NO_3^-, and PO_4^{3-} there was a significant effect of warming ($P < 0.05$) for at least one time point of every extractable pool; Figs. 4 and 5). Concentrations of these highly available extractants were also affected by the hurricane and showed interactive effects of hurricane disturbance and warming. For example, patterns in inorganic N showed strong responses to these global change

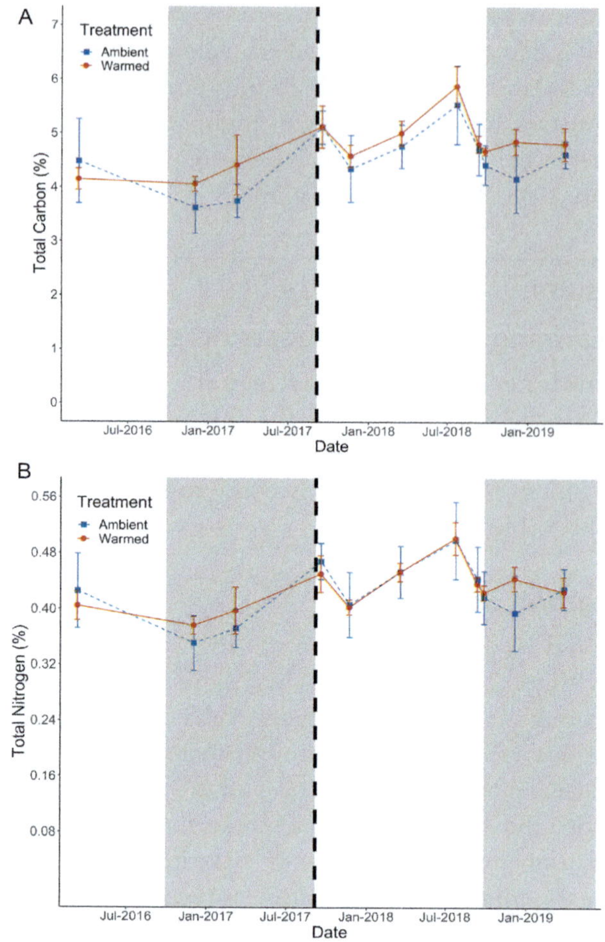

Fig. 3 Total soil (A) carbon and (B) nitrogen concentrations from 0–10 cm depth in the ambient (blue squares) and warmed (red circles) plots. Measurements were made through time and values are means ± 1 standard error. Gray shading shows the times when the warming treatments were active, and the dashed vertical line represents when Hurricane Irma occurred. Total carbon and nitrogen concentrations varied through time, most notably changing after the hurricanes, and there were no significant warming treatment effects.

pressures and different forms of N responded in different ways. Extractable ammonium concentrations increased significantly following the hurricane, with values that remained elevated above all pre-hurricane values for all post-hurricane sampling points (Fig. 4A; $P < 0.01$). In addition to hurricane effects, warming significantly decreased ammonium concentrations prior to the hurricanes, but although concentrations of ammonium were consistently lower in the warmed plots relative to control plots after the

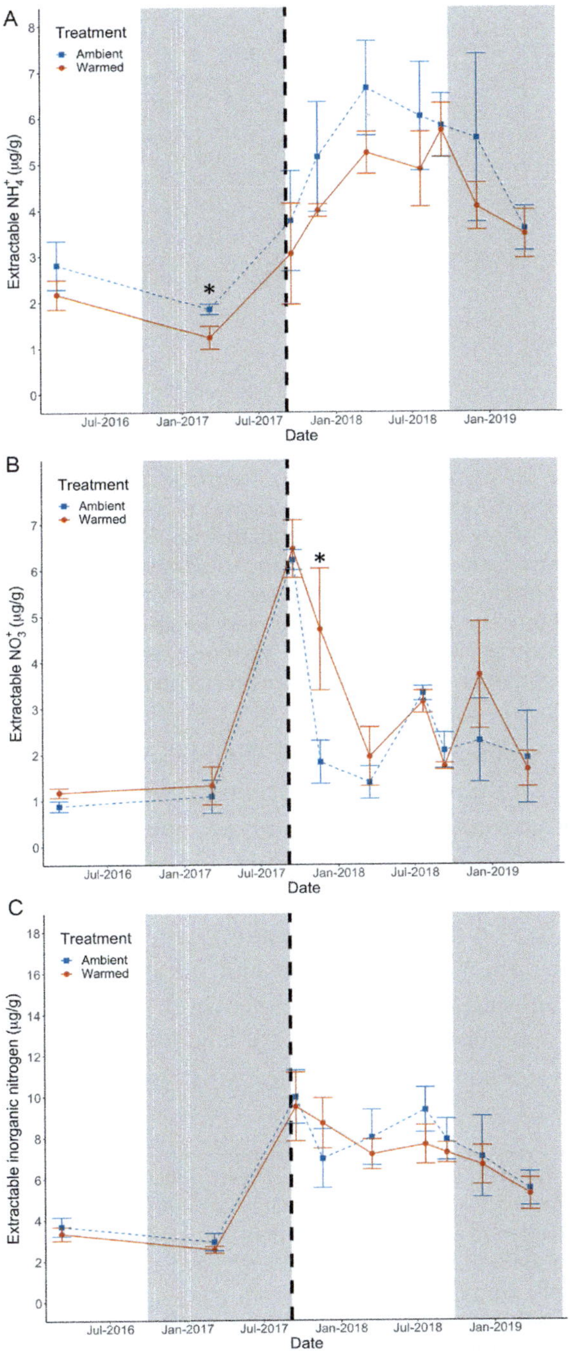

Fig. 4 Extractable (A) soil ammonium (NH_4^+), (B) nitrate (NO_3^-), and (C) inorganic nitrogen concentrations from 0–10 cm depths in the ambient (blue squares) and *(Continued)*

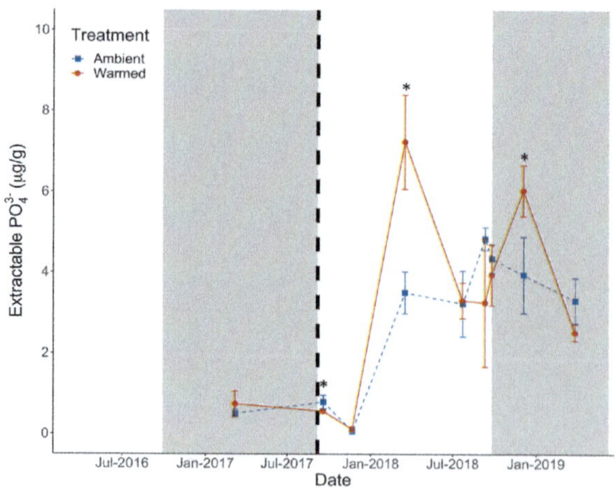

Fig. 5 Extractable soil phosphorus (PO_4^{3-}) concentrations from 0–10 cm depths in the ambient (blue squares) and warmed (red circles) plots. Measurements were made through time and values are means ± 1 standard error. Gray shading shows the times when the warming treatments were active, and the dashed vertical line represents the occurrence of Hurricane Irma. Data showed concentrations varied through time, especially following hurricane activity, with warming treatment effects at specific times (depicted with asterisks). Results show that both hurricane disturbance and warming significantly affected dynamic soil phosphorus concentrations.

hurricane, these differences were not significant. For nitrate, warming alone did not significantly affect concentrations prior to the hurricanes, but following hurricane disturbance, nitrate concentrations were significantly higher in the warmed plots during the year of recovery before warming was reinitiated ($P=0.012$). The hurricane did not have as sustained an effect on nitrate concentrations as it did ammonium concentrations. When summing ammonium and nitrate to explore changes in inorganic N concentrations, we found that the hurricanes significantly increased the extractable inorganic N concentrations and that these concentrations remained elevated nearly 2 years after the hurricanes

Fig. 4—Cont'd warmed (red circles) plots. Measurements were made through time and values are means ± 1 standard error. Gray shading shows the times when the warming treatments were active, and the dashed vertical line represents the occurrence of Hurricane Irma. Data show that extractable nitrogen concentrations varied through time, especially following hurricane activity, with warming treatment effects at specific times (significant differences among treatments depicted with asterisks). Results show that both hurricane disturbance and warming significantly affected dynamic soil nitrogen concentrations of this tropical wet forest.

(e.g., inorganic N concentrations were >50% higher at all timepoints post hurricane relative to all pre-hurricane measures).

Extractable soil P concentrations varied greatly with time ($P < 0.001$), with large increases in P observed following the hurricanes (extractable PO_4^{3-} concentrations were over four times higher post-hurricane relative to values observed prior to the hurricanes). Overall, the amount of extractable P remained elevated relative to any of the pre-hurricane measurements almost 2 years after the hurricanes occurred (Fig. 5). In contrast, warming treatments significantly reduced extractable P prior to Hurricane Maria ($P < 0.05$). Warming also interacted with the effects of hurricane disturbance such that extractable P was more than doubled in the plots that had been previously warmed 6 months after the hurricanes occurred than in the controls which had received no warming (Fig. 5). The extractable P concentrations were also significantly higher in the warmed plots during the second round of warming in December 2018 for one measurement timepoint.

3.4 Microbial biomass carbon, nitrogen, and phosphorus

Surprisingly, the hurricanes did not significantly affect microbial biomass C or N in the year following the events (Fig. 6). In contrast to microbial biomass C and N, microbial biomass P did significantly increase during the year following the hurricanes and remained higher than pre-treatment concentrations for all post-hurricane timepoints ($P < 0.001$). Warming affected microbial biomass C, N, and P concentrations only during times when the warming treatments were active (Fig. 6). Although we did not have pre-treatment data for microbial biomass due to methodological issues, none of the other pre-treatment values, including those for the highly dynamic extractable N pools, showed significant differences prior to the warming treatment, supporting the idea that microbial biomass concentrations likely did not have spurious differences between treatment plots prior to the initiation of warming. Further, for all microbial biomass element pools, the differences between the warming and the control plots disappeared when the warming was not active (i.e. the timepoints during the year post-hurricane when warming was ceased; Fig. 6), lending further support for the suggestion that the warming treatment, when active, caused strong differences in microbial biomass C, N, and P. In all cases, when warming had a significant effect on microbial biomass concentrations of any element, the effect was a warming-induced increase in concentrations (Fig. 6).

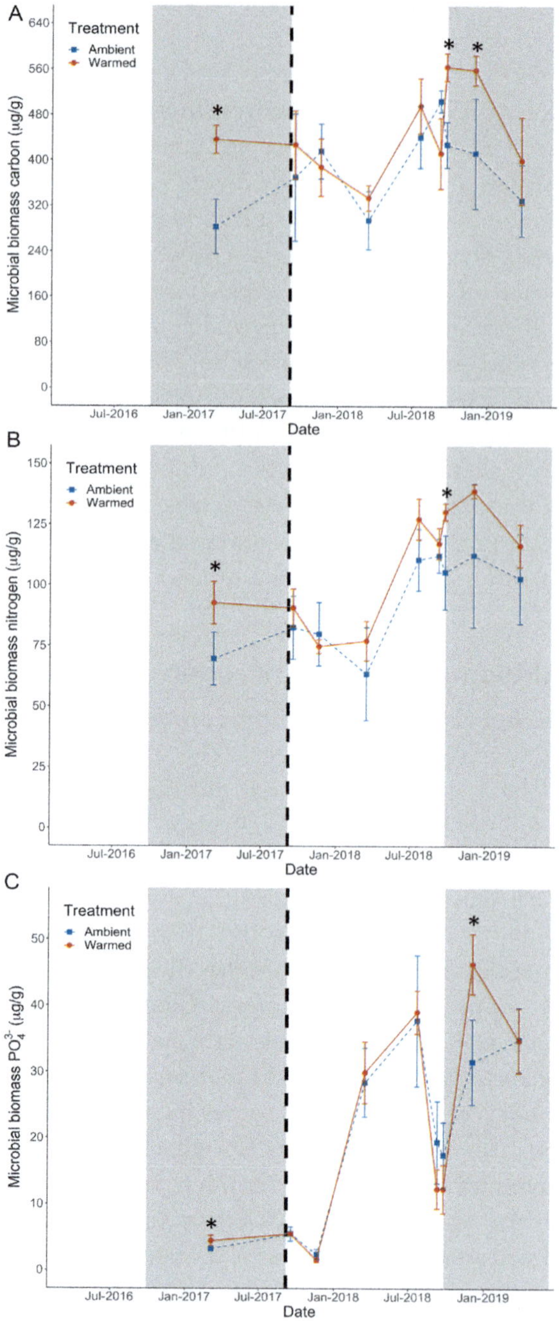

Fig. 6 Soil microbial (A) carbon, (B) nitrogen, and (C) phosphorus concentrations from 0–10 cm depths in the ambient (blue squares) and warmed (red circles) plots. Measurements were made through time and values are means ± 1 standard error. Gray shading shows the times when the warming treatments were active, and the dashed vertical line represents the occurrence of Hurricane Irma. Data showed that concentrations varied through time, especially with warming, and significant warming treatment effects are depicted with asterisks.

4. Discussion

This study assessed the soil biogeochemical responses of a tropical wet forest to an in situ warming treatment and subsequent interactions with disturbance following a once-in-a-century hurricane (i.e., Hurricane María). Like the majority of tropical forest sites (Chazdon, 2003; Houghton and Nassikas, 2018), the study area is a secondary forest (regenerating from pasture for ∼70 years). While no tropical forest can be a proxy to predict the responses of all tropical forests, the data described here are relevant to the large number of tropical forests that have legacies of land use history and disturbance (see chapter "Novelty in the tropical forests of the 21st century" by Lugo et al., this issue), which can help us begin to constrain tropical responses to concurrent global changes. We observed a high responsiveness of this forest's soil biogeochemical cycles to both warming and hurricane disturbance, as well as their interactive effects. The data from the warming experiment demonstrated a rapid response of tropical soil's faster-cycling C and nutrient pools to increased temperature. Multi-timepoint biogeochemical data following Hurricanes Irma and María showed the dynamism with which tropical forest soils responded to and the new trajectories of change following large-scale hurricane disturbance. The interactions between the effects of warming and hurricanes lend insight into the consequences of hurricanes in a warmer world and provide a novel look into a potential future for some tropical forest soils.

4.1 Warming effects

To our knowledge, the soil biogeochemical data from the year of in situ experimental warming prior to the hurricanes represent the first such published values for any tropical wet forest and lend unique insight into tropical forest belowground responses to increasing temperatures. While no warming experiment can perfectly mimic the manner, extent, or variation that will occur with actual warming (Amthor et al., 2010; Aronson and McNulty, 2009), in situ warming experiments can provide a mechanistic understanding of how systems interact with, are driven by, and respond to temperature change (Aronson and McNulty, 2009; Cavaleri et al., 2015; Hanson et al., 2017; Kimball et al., 2008).

In this tropical wet forest, warming treatments did not affect total soil C and N concentrations in a single year (Fig. 3). The hurricanes restructured the forest before we could assess longer-term total C and N pool responses to warming in the absence of recent disturbance, nevertheless, we did observe significant effects of warming on soil extractable and microbial C and nutrient pools in the first year of warming (Figs. 4–6), with different responses from different pools and elements. In contrast to our predictions, extractable organic C and total dissolved N did not vary significantly with warming, however, warming decreased extractable P by 30% (Fig. 5). Effects on extractable P are of particular interest as P is the element often believed most limiting to the activity of lowland tropical forest plants and microbes (Cleveland and Townsend, 2006, Elser et al., 2007, Cleveland et al., 2011, Camenzind et al., 2018; but see Wright, 2019). Determining the underlying mechanisms controlling a dynamic response of P to warming would be of great value, and many components of P cycling could be at play. For example, clay-rich Ultisols, which are common in lowland tropical forests (Reed and Wood, 2016), have a high affinity for P (Olander and Vitousek, 2004) and these interactions can be strongly mediated by soil moisture and soil redox conditions (Chacon et al., 2006; Lin et al., 2018; Wood et al., 2016). Thus, in addition to competition between roots and microbes, there is also competition with the soil mineral surface for available P pools. In addition to warming the soils, our treatments also resulted in a small reduction of soil moisture in the surface soils (Fig. 2), which could affect P adsorption to mineral soils. A wealth of data suggest that understanding how P availability responds to altered climate and disturbance is critical for our understanding and prediction of how tropical forests store and cycle C (Camenzind et al., 2018; Cleveland and Townsend, 2006; Fleischer et al., 2019; Reed et al., 2015; Turner et al., 2018; Wright, 2019; Yang et al., 2016), and the data presented here suggest that warming has strong potential to affect the P pools believed most biologically accessible (Fig. 5).

Microbial biomass C, N, and P all significantly increased with the warming during both sets of warming events (i.e., pre- and post-hurricane; Fig. 6). As in most ecosystems, tropical forest microbes are biogeochemical engines that drive myriad ecosystem processes and that can determine globally-relevant functions, such as soil C storage and CO_2 exchange with the atmosphere (Cleveland et al., 2003; Cleveland and Townsend, 2006; Sayer et al., 2007; Wood et al., 2012). An increase in the size of the microbial biomass pool has numerous implications for the cycling of nutrients and for the fate of the large amounts of C that enter tropical forest soils (e.g., Cleveland et al., 2010). For example, a larger microbial biomass pool could immobilize limiting nutrients, alter competition between plants and heterotrophs, and shift the proportion of C that is released to the atmosphere versus stored in soil. It is also noteworthy that microbial biomass C, N, and

P concentrations were affected by the warming treatments only during times when the warming was active, as there were no legacy effects of the warming treatment in the year following Hurricanes Irma and María when there was no active warming. Further, for microbial biomass C and N, it was warming alone that showed significant effects on soil concentrations in the year following the hurricanes (i.e., there was no effect of hurricanes on microbial biomass C or N in the unwarmed plots the year after hurricane disturbance). These results suggest that microbial communities are responding quickly to the temperature treatments: when temperatures rose with the treatments, so did microbial biomass, and when the treatments were stopped, microbial biomass pools quickly declined and the differences among treatments disappeared. As would be expected with real-world warming, the warming treatments resulted in altered soil moisture and, although these changes were small (Fig. 2), further research at the site is underway to disentangle the direct effects of temperature from any indirect effects of warming-induced changes to soil hydrology.

4.2 Hurricane effects

Most soil biogeochemical measurements showed significant change with hurricane disturbance and, for all measured values except microbial biomass C and N concentrations, the effects of the hurricanes were larger than the effects of warming (although we cannot yet know how the relative effects of warming and disturbance will compare with the effects of longer-term warming). While assessments of soil biogeochemistry through time in tropical forests recovering from hurricanes are rare, our results are in line with the observational data, experimental results, and model predictions that exist for lowland tropical forests (González et al., 2014; Lodge et al., 1991; Lugo, 2008; McDowell and Liptzin, 2014; Sanford et al., 1991; Silver et al., 2014). Specifically, we observed an increase in C and N concentrations in the year following the hurricanes, followed by a decline back to original values (Fig. 3). These changes may seem minor (e.g. soil total C concentrations ranged between ~,4% to ~6%), but they are highly relevant differences from a C cycle and climate feedback perspective. This responsiveness of total soil C and N concentrations has implications for understanding the potential for rapid terrestrial biosphere effects on the global C cycle and for model representation of these productive ecosystems. An improved understanding of the mechanisms behind these changes remains important. For example, the build-up of soil total C and N in the year following the hurricanes likely comes from the large inputs of leaves, particularly green leaves, that accompany defoliation and represent important inputs into tropical forest soils (Lodge et al., 1991). Partitioning these inputs into C's fate as stabilized organic matter versus leached losses into waterways versus gaseous CO_2 and CH_4 emissions is critical for modelling potential feedbacks to climate change (Cavaleri et al., 2015; McDowell, 2011; Sanford et al., 1991).

The concentrations of extractable soil N pools were significantly affected by hurricane disturbance and changed as the forest recovered (Fig. 4). Extractable ammonium (NH_4^+) concentrations rose consistently following the hurricane. Increased concentrations of this biologically available pool of N could in part be an effect of the large reductions in root biomass often observed following hurricanes (Parrotta and Lodge, 1991; Silver and Vogt, 1993). Fewer roots to immobilize NH_4^+ could result in higher concentrations in the soil, as could increased mineralization of the elevated litter inputs observed with hurricane disturbance (Lodge et al., 1991). See also chapter "Revisiting nutrient cycling by litterfall—Insights from 15 years of litter manipulation in old-growth lowland tropical forest" Sayer et al. (this issue), for further discussion of the impacts of increased litterfall on nutrients. However, elevated soil NH_4^+ is not a response observed consistently with hurricane disturbance, as NH_4^+ concentrations have also been shown to decline (Erickson and Ayala, 2004). These discrepancies among results could be related to forest land use history, soil type, sampling frequency, and/or differences in the nature of the hurricane disturbance. More data are needed to improve our understanding of the patterns and mechanisms behind soil biogeochemical responses to hurricanes. In contrast to NH_4^+, extractable nitrate (NO_3^-) concentrations rose immediately after the hurricanes, but then declined. The relatively rapid decline in NO_3^- could be due in part to increased denitrification and/or leaching losses, which are loss pathways typically expected for NO_3^-. Hurricanes have been shown to stimulate N_2O losses (Erickson and Ayala, 2004), which could come from increased nitrification and/or denitrification rates. Elevated soil solution NO_3^- has been observed with other hurricanes and with simulated hurricane disturbance (Shiels et al., 2015), however, these patterns are not universal across hurricane events and sites (Erickson and Ayala, 2004). Nevertheless, long-term assessment of stream water chemistry following hurricanes supports the idea of sustained inputs in NO_3^- moving from soils into waterways for years following large hurricane events (McDowell, 2011; Schaefer et al., 2000).

Inorganic N concentrations ($NH_4^+ + NO_3^-$) are commonly considered a highly available N pool for plants and microbes, and the patterns in the concentrations of extractable inorganic N observed here suggest that the availability of these pools varies through time following a hurricane event (Fig. 4C). There are relatively few data of which we are aware that assess these dynamic pools through time with hurricane disturbance, in part because the events themselves are infrequent, and those presented here show not only the magnitude of hurricane effects on soil N cycling, but also how the patterns of NH_4^+ and NO_3^- can vary. This is particularly interesting because results from other ecosystem types suggest that the form of N (i.e., NH_4^+ vs. NO_3^-) can be as or more important in determining the ecological consequences of N cycling changes as the total amount of N (Austin et al., 2006).

The hurricane also significantly affected soil P cycling: extractable P declined by over 50% immediately following the hurricanes, but then increased dramatically—increasing 4 times in the control plots and over 10 times in the warmed plots—and remained significantly higher than all pre-hurricane values (Fig. 5). Increases in these available forms of P have important implications for forest recovery. For example, research using experimental hurricane effects found that adding debris to forests with intact canopies could increase forest basal area by 10% for at least 7 years after applying the treatments (Zimmerman et al., 2014). In the context of forest canopies that close after hurricanes on timelines of less than 2 years (e.g., Fig. 1), this suggests that hurricane-induced increases in litter resources are likely to fuel forest recovery. Movement of P from the large amounts of debris that is deposited from the canopy to the forest floor during hurricanes has strong potential to increase the availability of limiting nutrients (Lodge et al., 1991, 2014; Wright, 2019). For example, in a litter manipulation experiment in Panama, results showed the litter layer was important in supplying P to biota: the addition of litter not only quickly stimulated nutrient cycling, but also had longer-term effects on soil microbes and C dynamics (Sayer et al., 2012; Sayer and Tanner, 2010). Lodge et al. (2014) showed that P from hurricane-induced litterfall was both immobilized in the forest floor and moved to the soil below, likely through litter leaching. Their calculations suggested that most of the P lost from the green leaves deposited after a hurricane likely enters the soil or is taken up directly by plant roots (Lodge et al., 2014). The soil extractable and microbial P data shown here support the idea that this P is moving into and is maintained in surface soils in multiple P forms. In light of the many studies suggesting P plays a critical role in regulating C and N cycling in terrestrial ecosystems, particularly in highly weathered tropical forest soils such as those described here (Camenzind et al., 2018; McGlynn et al., 2007; Reed et al., 2007; Reed and Wood, 2016), it will be important to determine how these observed effects change in concert with changes to plant and microbial communities through time. For example, the increases in extractable P occurred concurrently with large increases in microbial biomass P (Fig. 6C), which is in contrast to the more tightly-cycled P suggested by the data for the pre-hurricane timepoints (Figs. 5 and 6). Such increases could be due not only to changes in microbial stoichiometric ratios (Cleveland and Liptzin, 2007), but also to luxury consumption or microbial community composition shifts. Overall, hurricane effects on the aspects of the P cycle measured here were of a higher magnitude and were more sustained than for patterns observed in C or N.

4.3 Interactive effects of warming and hurricanes

We saw significant interactions between warming and hurricane disturbance, including legacy effects of warming even during times when the warming treatment was not active. For example, extractable NO_3^- and PO_4^{3-} concentrations were significantly higher in the warmed plots months after the warming treatments had ceased following Hurricane María (Figs. 4B and 5). Such changes to the N and P economies of the soils suggest the potential for longer-term effects for a range of forest organisms (Cleveland and Townsend, 2006; Elser et al., 2007; McGlynn et al., 2007). Of note, microbial biomass C, N, and P concentrations, which showed such strong warming effects during times when warming treatments were active, did not show significant legacy effects of warming (i.e., biomass concentrations were not different between control and warmed soils post-hurricane during the year when warming was not occurring). However, when we look at times when warming and recovery from hurricane co-occurred (i.e. the second warming event), we see significant effects of warming on the trajectory of recovery and interactions between warming and disturbance. For example, microbial biomass C, N, and P concentrations, as well as those of soil NO_3^- and PO_4^{3-}, are higher in the warmed plots relative to the control plots and higher than the concentrations observed before the hurricanes. These data suggest some potential increases in soil fertility for a warmer recovering forest. Over the longer-term, coupling the soil changes with plant growth and community composition will help determine if these changes to soil biogeochemistry help support ecosystem resilience or, instead, shift the system into a new state (Johnstone et al., 2016). It is important to mention that the concurrent effects of warming and hurricane disturbance in many cases would not have been predictable from summing the individual effects of a single change.

For example, the effects of warming on extractable soil PO_4^{3-} in a recovering forest were many times larger and opposite in direction than the effects of warming on that P pool prior to the Hurricane María. As in other systems, disturbance and climate change interactions may alter not only the magnitude of the effect of a single driver, but also the sign.

Taken together, the results from this study point to a highly dynamic soil biogeochemical response to warming and hurricane disturbance, with significant interactions between the two change factors. In the context of understanding how multiple global change stressors will affect terrestrial ecosystems (Johnstone et al., 2016; Turner, 2010), the data from this study point to the likelihood that disturbance and climate change will have profound effects on tropical rain forest soils and their function. In particular, these results suggest tropical secondary forests recovering from hurricane disturbance in a warmer environment will have a larger soil microbial biomass pool and greater access to available nutrients such as the N in NO_3^- and the P in PO_4^{3-} (Figs. 4–6), at least in the short-term. Opposite to our predictions, increases in these pools could result in increased productivity of tropical forests, with more fertility fueling faster response. However, increases in the size of these pools, and their large variation among sampling times, also suggest faster cycling and larger nutrient losses from the system (e.g., via leaching and denitrification). In this way, the faster turnover and decomposition that warming may stimulate when large inputs of litter fall from the canopy could result in an overall reduction in ecosystem fertility and thus lowered resilience over the long-term. Future studies that further elucidate the mechanisms and consequences of interactive changes for tropical wet forests in a warmer world will be of the utmost importance for forecasting the fate of these diverse, productive, and globally important ecosystems.

Acknowledgements

This research was supported by the U.S. Department of Energy Office of Science Terrestrial Ecosystem Sciences Program (Awards Number 89243018S-SC-000014 and DE-SC-0011806), the National Science Foundation (DEB-1754713), the USDA Forest Service International Institute of Tropical Forestry (IITF), the University of Puerto Rico-Río Piedras, and the USGS Ecosystems Mission Area. All research at IITF is done in collaboration with the University of Puerto Rico. We are grateful to Armin Howell for help with soil collection and preparation, to Nick Melone for soil processing, and to Michala Phillips for help with figure creation. We are also indebted to the TRACE staff and to the numerous interns and volunteers who made this work possible. Any use of trade, firm, or product names is for descriptive purposes only and does not imply endorsement by the U.S. Government.

References

Ahlström, A., Raupach, M.R., Schurgers, G., Smith, B., Arneth, A., Jung, M., Reichstein, M., Canadell, J.G., Friedlingstein, P., Jain, A.K., 2015. The dominant role of semi-arid ecosystems in the trend and variability of the land CO_2 sink. Science 348, 895–899.

Amthor, J.S., Hanson, P.J., Norby, R.J., Wullschleger, S.D., 2010. A comment on "appropriate experimental ecosystem warming methods by ecosystem, objective, and practicality" by Aronson and McNulty. Agric. For. Meteorol. 150, 497–498.

Anderson, B.T., 2011. Near-term increase in frequency of seasonal temperature extremes prior to the 2 C global warming target. Clim. Change 108, 581.

Aronson, E.L., McNulty, S.G., 2009. Appropriate experimental ecosystem warming methods by ecosystem, objective, and practicality. Agric. For. Meteorol. 149, 1791–1799.

Austin, A., Sala, O., Jackson, R., 2006. Inhibition of nitrification alters carbon turnover in the Patagonian steppe. Ecosystems 9, 1257–1265.

Bai, E., Li, S., Xu, W., Li, W., Dai, W., Jiang, P., 2013. A meta-analysis of experimental warming effects on terrestrial nitrogen pools and dynamics. New Phytol. 199, 441–451.

Beer, C., Reichstein, M., Tomelleri, E., Ciais, P., Jung, M., Carvalhais, N., Rödenbeck, C., Arain, M.A., Baldocchi, D., Bonan, G.B., 2010. Terrestrial gross carbon dioxide uptake: global distribution and covariation with climate. Science 329, 834–838.

Bormann, F.H., Likens, G.E., 1979. Catastrophic disturbance and the steady state in northern hardwood forests: a new look at the role of disturbance in the development of forest ecosystems suggests important implications for land-use policies. Am. Sci. 67, 660–669.

Bray, R.H., Kurtz, L.T., 1945. Determination of total, organic and available forms of phosphorus in soils. Soil Sci. 59, 39–45.

Brookes, P.C., Kragt, J.F., Powlson, D.S., Jenkinson, D.S., 1985. Chloroform fumigation and the release of soil nitrogen: the effects of fumigation time and temperature. Soil Biol. Biochem. 17, 831–835.

Camenzind, T., Hättenschwiler, S., Treseder, K.K., Lehmann, A., Rillig, M.C., 2018. Nutrient limitation of soil microbial processes in tropical forests. Ecol. Monogr. 88, 4–21.

Cavaleri, M.A., Reed, S.C., Smith, W.K., Wood, T.E., 2015. Urgent need for warming experiments in tropical forests. Glob. Chang. Biol. 21, 2111–2121.

Chacon, N., Silver, W.L., Dubinsky, E.A., Cusack, D.F., 2006. Iron reduction and soil phosphorus solubilization in humid tropical forests soils: the roles of labile carbon pools and an electron shuttle compound. Biogeochemistry 78, 67–84.

Chazdon, R.L., 2003. Tropical forest recovery: legacies of human impact and natural disturbances. Perspect. Plant Ecol. Evol. Syst. 6, 51–71.

Clark, D.A., Piper, S., Keeling, C., Clark, D.B., 2003. Tropical rain forest tree growth and atmospheric carbon dynamics linked to interannual temperature variation during 1984–2000. Proc. Natl. Acad. Sci. 100, 5852–5857.

Cleveland, C.C., Liptzin, D., 2007. C: N: P stoichiometry in soil: is there a "Redfield ratio" for the microbial biomass? Biogeochemistry 85, 235–252.

Cleveland, C.C., Townsend, A.R., 2006. Nutrient additions to a tropical rain forest drive substantial soil carbon dioxide losses to the atmosphere. Proc. Natl. Acad. Sci. U. S. A. 103, 10316–10321.

Cleveland, C.C., Townsend, A.R., Schmidt, S.K., Constance, B.C., 2003. Soil microbial dynamics and biogeochemistry in tropical forests and pastures, southwestern Costa Rica. Ecol. Appl. 13, 314–326.

Cleveland, C.C., Wieder, W.R., Reed, S.C., Townsend, A.R., 2010. Experimental drought in a tropical rain forest increases soil carbon dioxide losses to the atmosphere. Ecology 91, 2313–2323.

Cleveland, C.C., Townsend, A.R., Taylor, P., Alvarez-Clare, S., Bustamante, M.M., Chuyong, G., Dobrowski, S.Z., Grierson, P., Harms, K.E., Houlton, B.Z., 2011. Relationships among net primary productivity, nutrients and climate in tropical rain forest: a pan-tropical analysis. Ecol. Lett. 14, 939–947.

Comita, L.S., Uriarte, M., Thompson, J., Jonckheere, I., Canham, C.D., Zimmerman, J.K., 2009. Abiotic and biotic drivers of seedling survival in a hurricane-impacted tropical forest. J. Ecol. 97, 1346–1359.

Cooper, W.S., 1926. The fundamentals of vegetational change. Ecology 7, 391–413.

Diffenbaugh, N.S., Scherer, M., 2011. Observational and model evidence of global emergence of permanent, unprecedented heat in the 20th and 21st centuries. Clim. Change 107, 615–624.

Doughty, C.E., Goulden, M.L., 2008. Are tropical forests near a high temperature threshold? J. Geophys. Res. Biogeo. 113, G00B007.

Elser, J., Bracken, M.E.S., Cleland, E.E., Gruner, D.S., Harpole, W.S., Hillebrand, H., Ngai, J.T., Seabloom, E.W., Shurin, J.B., Smith, J.E., 2007. Global analysis of nitrogen and phosphorus limitation of primary producers in freshwater, marine and terrestrial ecosystems. Ecol. Lett. 10, 1135–1142.

Erickson, H.E., Ayala, G., 2004. Hurricane-induced nitrous oxide fluxes from a wet tropical forest. Glob. Chang. Biol. 10, 1155–1162.

Fleischer, K., Rammig, A., De Kauwe, M.G., Walker, A.P., Domingues, T.F., Fuchslueger, L., Garcia, S., Goll, D.S., Grandis, A., Jiang, M., 2019. Amazon forest response to CO_2 fertilization dependent on plant phosphorus acquisition. Nat. Geosci. 12, 736–741.

Foley, J.A., Costa, M.H., Delire, C., Ramankutty, N., Snyder, P., 2003. Green surprise: how terrestrial ecosystems could affect earth's climate. Front. Ecol. Environ. 1, 38–44.

Folke, C., Carpenter, S., Walker, B., Scheffer, M., Elmqvist, T., Gunderson, L., Holling, C.S., 2004. Regime shifts, resilience, and biodiversity in ecosystem management. Annu. Rev. Ecol. Evol. Syst. 35, 557–581.

González, G., Lodge, D.J., Richardson, B.A., Richardson, M.J., 2014. A canopy trimming experiment in Puerto Rico: the response of litter decomposition and nutrient release to canopy opening and debris deposition in a subtropical wet forest. For. Ecol. Manage. 332, 32–46.

Hanson, P.J., Riggs, J.S., Nettles IV, W.R., Phillips, J.R., Krassovski, M.B., Hook, L.A., Gu, L., Richardson, A.D., Aubrecht, D.M., Ricciuto, D.M., 2017. Attaining whole-ecosystem warming using air and deep-soil heating methods with an elevated CO_2 atmosphere. Biogeosciences 14, 861–883.

Heartsill-Scalley, T., Scatena, F.N., Estrada, C., McDowell, W., Lugo, A.E., 2007. Disturbance and long-term patterns of rainfall and throughfall nutrient fluxes in a subtropical wet forest in Puerto Rico. J. Hydrol. 333, 472–485.

Heartsill-Scalley, T., Scatena, F.N., Lugo, A.E., Moya, S., Estrada Ruiz, C.R., 2010. Changes in structure, composition, and nutrients during 15 yr of hurricane-induced succession in a subtropical wet forest in Puerto Rico. Biotropica 42, 455–463.

Herbert, D.A., Fownes, J.H., Vitousek, P.M., 1999. Hurricane damage to a Hawaiian forest: nutrient supply rate affects resistance and resillience. Ecology 80, 908–920.

Holdridge, L.R., Grenke, W.C., Hatheway, W.H., Liang, T., Tosi, J.A., 1971. Forest Environments in Tropical Life Zones: A Pilot Study. Pergamon Press, Oxford.

Houghton, R.A., Nassikas, A.A., 2018. Negative emissions from stopping deforestation and forest degradation, globally. Glob. Chang. Biol. 24, 350–359.

Jobbagy, E.G., Jackson, R.B., 2000. The vertical distribution of soil organic carbon and its relation to climate and vegetation. Ecol. Appl. 10, 423–436.

Johnstone, J.F., Allen, C.D., Franklin, J.F., Frelich, L.E., Harvey, B.J., Higuera, P.E., Mack, M.C., Meentemeyer, R.K., Metz, M.R., Perry, G.L., 2016. Changing disturbance regimes, ecological memory, and forest resilience. Front. Ecol. Environ. 14, 369–378.

Kimball, B.A., Conley, M.M., Wang, S., Lin, X., Luo, C., Morgan, J., Smith, D., 2008. Infrared heater arrays for warming ecosystem field plots. Glob. Chang. Biol. 14, 309–320.

Kimball, B.A., Alonso-Rodríguez, A.M., Cavaleri, M.A., Reed, S.C., González, G., Wood, T.E., 2018. Infrared heater system for warming tropical forest understory plants and soils. Ecol. Evol. 8, 1932–1944.

Lin, Y., Bhattacharyya, A., Campbell, A.N., Nico, P.S., Pett-Ridge, J., Silver, W.L., 2018. Phosphorus fractionation responds to dynamic redox conditions in a humid tropical forest soil. J. Geophys. Res. Biogeo. 123, 3016–3027.

Lodge, D.J., Scatena, F., Asbury, C., Sanchez, M., 1991. Fine litterfall and related nutrient inputs resulting from Hurricane Hugo in subtropical wet and lower montane rain forests of Puerto Rico. Biotropica 23, 336–342.

Lodge, D.J., Cantrell, S.A., González, G., 2014. Effects of canopy opening and debris deposition on fungal connectivity, phosphorus movement between litter cohorts and mass loss. For. Ecol. Manage. 332, 11–21.

Lugo, A.E., 2008. Visible and invisible effects of hurricanes on forest ecosystems: an international review. Austral Ecol. 33, 368–398.

Lugo, A.E., Scatena, F., 1995. Ecosystem-level properties of the Luquillo Experimental Forest with emphasis on the tabonuco forest. In: Tropical Forests: Management and Ecology. Springer, pp. 59–108.

Lugo, A.E., Applefield, M., Pool, D.J., McDonald, R.B., 1983. The impact of Hurricane David on the forests of Dominica. Can. J. For. Res. 13, 201–211.

Malhi, Y., Silman, M., Salinas, N., Bush, M., Meir, P., Saatchi, S., 2010. Introduction: elevation gradients in the tropics: laboratories for ecosystem ecology and global change research. Glob. Chang. Biol. 16, 3171–3175.

McDowell, W.H., 2011. Impacts of hurricanes on forest hydrology and biogeochemistry. In: Forest Hydrology and Biogeochemistry. Springer, pp. 643–657.

McDowell, W.H., Liptzin, D., 2014. Linking soils and streams: response of soil solution chemistry to simulated hurricane disturbance mirrors stream chemistry following a severe hurricane. For. Ecol. Manage. 332, 56–63.

McGlynn, T.P., Salinas, D.J., Dunn, R.R., Wood, T.E., Lawrence, D., Clark, D.A., 2007. Phosphorus limits tropical rain Forest litter Fauna. Biotropica 39, 50–53.

Nottingham, A.T., Bååth, E., Reischke, S., Salinas, N., Meir, P., 2019. Adaptation of soil microbial growth to temperature: using a tropical elevation gradient to predict future changes. Glob. Chang. Biol. 25, 827–838.

Olander, L.P., Vitousek, P.M., 2004. Biological and geochemical sinks for phosphorus in soil from a wet tropical forest. Ecosystems 7, 404–419.

Pan, Y., Birdsey, R.A., Phillips, O.L., Jackson, R.B., 2013. The structure, distribution, and biomass of the world's forests. Annu. Rev. Ecol. Evol. Syst. 44, 593–622.

Parrotta, J.A., Lodge, D.J., 1991. Fine root dynamics in a subtropical wet forest following hurricane disturbance in Puerto Rico. Biotropica 23, 343–347.

Reed, S.C., Wood, T.E., 2016. Soil phosphorus cycling in tropical soils: an ultisol and oxisol perspective. In: Lal, Stewart, (Eds.), Soil Phosphorus. CRC Press.

Reed, S.C., Cleveland, C.C., Townsend, A.R., 2007. Controls over free-living nitrogen fixation in a lowland tropical rain forest. Biotropica 39, 585–592.

Reed, S.C., Cleveland, C.C., Townsend, A.R., 2008. Tree species control rates of free-living nitrogen fixation in a tropical rain forest. Ecology 89, 2924–2934.

Reed, S.C., Wood, T., Cavaleri, M.A., 2012. Tropical forests in a warming world. New Phytol. 193, 27–29.

Reed, S.C., Yang, X., Thornton, P.E., 2015. Incorporating phosphorus cycling into global modeling efforts: a worthwhile, tractable endeavor. New Phytol. 208, 324–329.

Rustad, L., Campbell, J., Marion, G., Norby, R., Mitchell, M., Hartley, A., Cornelissen, J., Gurevitch, J., 2001. A meta-analysis of the response of soil respiration, net nitrogen mineralization, and aboveground plant growth to experimental ecosystem warming. Oecologia 126, 543–562.

Sánchez-Sánchez, O., Islebe, G., 1999. Hurricane Gilbert and structural changes in a tropical forest in South-Eastern Mexico. Glob. Ecol. Biogeogr. 8, 29–38.

Sanford Jr., R.L., Parton, W.J., Ojima, D.S., Lodge, D.J., 1991. Hurricane effects on soil organic matter dynamics and forest production in the Luquillo experimental Forest, Puerto Rico: results of simulation modeling. Biotropica 23, 364–372.

Sayer, E.J., Tanner, E.V., 2010. Experimental investigation of the importance of litterfall in lowland semi-evergreen tropical forest nutrient cycling. J. Ecol. 98, 1052–1062.

Sayer, E.J., Powers, J.S., Tanner, E.V., 2007. Increased litterfall in tropical forests boosts the transfer of soil CO2 to the atmosphere. Plos One 2, e1299.

Sayer, E.J., Wright, S.J., Tanner, E.V., Yavitt, J.B., Harms, K.E., Powers, J.S., Kaspari, M., Garcia, M.N., Turner, B.L., 2012. Variable responses of lowland tropical forest nutrient status to fertilization and litter manipulation. Ecosystems 15, 387–400.

Scatena, F.N., 1989. An introduction to the Physiography and History of the Bisley Experimental Watersheds in the Luquillo Mountains of Puerto Rico. General Technical Report SO-72. vol. 72 US Dept of Agriculture, Forest Service, Southern Forest Experiment Station, New Orleans, LA, p. 22.

Scatena, F., Moya, S., Estrada, C., Chinea, J., 1996. The first five years in the reorganization of aboveground biomass and nutrient use following Hurricane Hugo in the Bisley Experimental Watersheds, Luquillo Experimental Forest, Puerto Rico. Biotropica 28, 424–440.

Schaefer, D.A., McDowell, W.H., Scatena, F.N., Asbury, C.E., 2000. Effects of hurricane disturbance on stream water concentrations and fluxes in eight tropical forest watersheds of the Luquillo Experimental Forest, Puerto Rico. J. Trop. Ecol. 16, 189–207.

Selmants, P.C., Litton, C.M., Giardina, C.P., Asner, G.P., 2014. Ecosystem carbon storage does not vary with mean annual temperature in Hawaiian tropical montane wet forests. Glob. Chang. Biol. 20, 2927–2937.

Shiels, A.B., Gonzalez, G., Lodge, D.J., Willig, M.R., Zimmerman, J.K., 2015. Cascading effects of canopy opening and debris deposition from a large-scale hurricane experiment in a tropical rain forest. Bioscience 65, 871–881.

Silver, W.L., Vogt, K.A., 1993. Fine root dynamics following single and multiple disturbances in a subtropical wet forest ecosystem. J. Ecol. 81, 729–738.

Silver, W.L., Hall, S.J., González, G., 2014. Differential effects of canopy trimming and litter deposition on litterfall and nutrient dynamics in a wet subtropical forest. For. Ecol. Manage. 332, 47–55.

Sparks, D.L., Helmke, P., Page, A., 1996. Methods of Soil Analysis: Chemical Methods. SSSA.

Tarnocai, C., Canadell, J., Schuur, E.A., Kuhry, P., Mazhitova, G., Zimov, S., 2009. Soil organic carbon pools in the northern circumpolar permafrost region. Global Biogeochem. Cycles 23, GB2023.

Townsend, A.R., Vitousek, P.M., Trumbore, S.E., 1995. Soil organic matter dynamics along gradients in temperature and land use on the island of Hawaii. Ecology 76, 721–733.

Townsend, A.R., Cleveland, C.C., Houlton, B.Z., Alden, C.B., White, J.W., 2011. Multi-element regulation of the tropical forest carbon cycle. Front. Ecol. Environ. 9, 9–17.

Turner, M.G., 2010. Disturbance and landscape dynamics in a changing world. Ecology 91, 2833–2849.

Turner, B.L., Brenes-Arguedas, T., Condit, R., 2018. Pervasive phosphorus limitation of tree species but not communities in tropical forests. Nature 555, 367.

Uriarte, M., Thompson, J., Zimmerman, J.K., 2019. Hurricane María tripled stem breaks and doubled tree mortality relative to other major storms. Nat. Commun. 10, 1362.

Van Gestel, N., Shi, Z., Van Groenigen, K.J., Osenberg, C.W., Andresen, L.C., Dukes, J.S., Hovenden, M.J., Luo, Y., Michelsen, A., Pendall, E., 2018. Predicting soil carbon loss with warming. Nature 554, E4.

Vance, E.D., 1987. Microbial biomass measurements in forest soils: the use of the chloroform fumigation-incubation method in strongly acid soils. Soil Biol. Biochem. 19, 697–702.

Vance, E.D., Brookes, P.C., Jenkinson, D.S., 1987. An extraction method for measuring soil microbial biomass C. Soil Biol. Biochem. 19, 703–707.

White, P.S., 1979. Pattern, process, and natural disturbance in vegetation. Bot. Rev. 45, 229–299.

Wood, T.E., Cavaleri, M.A., Reed, S.C., 2012. Tropical forest carbon balance in a warmer world: a critical review spanning microbial-to ecosystem-scale processes. Biol. Rev. 87, 912–927.

Wood, T.E., Matthews, D., Vandecar, K., Lawrence, D., 2016. Short-term variability in labile soil phosphorus is positively related to soil moisture in a humid tropical forest in Puerto Rico. Biogeochemistry 127, 35–43.

Wood, T.E., Cavaleri, M.A., Giardina, C.P., Khan, S., Mohan, J.E., Nottingham, A.T., Reed, S.C., Slot, M., 2019. Soil warming effects on tropical forests with highly weathered soils. In: Ecosystem Consequences of Soil Warming. Elsevier, pp. 385–439.

Wright, S.J., 2010. The future of tropical forests. Ann. N. Y. Acad. Sci. 1195, 1–27.

Wright, S.J., 2019. Plant responses to nutrient addition experiments conducted in tropical forests. Ecol. Monogr. 89, e01382.

Yang, X., Thornton, P.E., Ricciuto, D.M., Hoffman, F.M., 2016. Phosphorus feedbacks constraining tropical ecosystem responses to changes in atmospheric CO2 and climate. Geophys. Res. Lett. 43, 7205–7214.

Zimmerman, J.K., Hogan, J.A., Shiels, A.B., Bithorn, J.E., Carmona, S.M., Brokaw, N., 2014. Seven-year responses of trees to experimental hurricane effects in a tropical rainforest, Puerto Rico. For. Ecol. Manage. 332, 64–74.

CHAPTER SEVEN

Critical role and collapse of tropical mega-trees: A key global resource

Bruno X. Pinho[a,*], Carlos A. Peres[b,c], Inara R. Leal[a], Marcelo Tabarelli[a]

[a]Departamento de Botânica, Universidade Federal de Pernambuco, Recife, Pernambuco, Brazil
[b]School of Environmental Sciences, University of East Anglia, Norwich Research Park, Norwich, United Kingdom
[c]Departamento de Sistemática e Ecologia, Universidade Federal da Paraíba, João Pessoa, Paraíba, Brazil
*Corresponding author: e-mail address: bxpinho@hotmail.com

Contents

Abstract

Very large tropical forest trees ('mega-trees') represent an irreplaceable habitat associated with large benefits in terms of biodiversity and ecosystem services. Here we provide a comprehensive overview of the importance of tropical mega-trees relative to biodiversity persistence, ecosystem services, and sociocultural value. We describe all contemporary threats to mega-trees as a result of the conversion of old-growth forests into human-degraded landscapes exposed to climate change. We reveal a myriad of taxa, functional groups and ecosystem services that are directly or indirectly supported by tropical mega-trees, such as highly illuminated crowns for epiphytes and large boles and branches sheltering many vertebrate and invertebrate taxa. Mega-trees also amplify the resource spectra associated with non-redundant niche space, including thermal

Advances in Ecological Research, Volume 62
ISSN 0065-2504
https://doi.org/10.1016/bs.aecr.2020.01.009

buffering and extended vertical microclimates above the forest canopy. Mega-trees also make the largest contribution to forest productivity, aboveground biomass, and timber. However, protecting this resource is a major challenge due to the natural rarity of mega-trees, illegal logging, old-growth forest conversion and climate change. We argue for 'productive forest landscapes' as a strategy to protect mega-trees and highlight their connections with sustainable development goals and other global targets. This strategy poses economic, political, technological and social challenges, as sparing tropical forests still incurs high opportunity costs that few stakeholders can realistically afford. Despite these challenges, the future of tropical forests mega-trees and local livelihoods are inextricably linked, and productive forest landscapes would bring clear unanticipated benefits for future generations and tropical biodiversity.

1. Introduction

Tropical forests play a prominent role at local, regional and global scales in relation to biodiversity persistence, provision of livelihood goods (e.g. timber, fibre, fuelwood, medicinal plants) and ecosystem services (e.g. climate regulation, carbon sequestration and storage) essential for human well-being (Beer et al., 2010; Bonan, 2008; Houghton et al., 2015; Malhi, 2012). Tropical forests thus represent a key component of global sustainability (Malhi et al., 2014), but old-growth tropical forests worldwide continue to be relentlessly converted into degraded human-modified landscapes (Laurance et al., 2014; Lewis et al., 2015).

In this context, it is speculated that a large portion of the direct or indirect benefits provided by tropical forests are disproportionately contributed by very large trees or 'mega-trees' (see Fig. 1 for definition of a 'mega-tree' and key features). This is particularly the case of lowland tropical forests, in which mega-tree species attain the highest levels of species richness and biomass contribution (Balzotti et al., 2017; Corlett and Primack, 2011; Gonmadje et al., 2017). To provide a single example, Southeast Asian forests and their flora contain over 500 mega-tree species (Ghazoul, 2016). Although mega-trees usually represent a small fraction of the overall tree abundance, they account for most of the forest biomass, thereby contributing the lion's share of effects on biogeochemical cycles and ecosystem processes such as carbon storage, water cycling and nutrient dynamics (Bastin et al., 2018; Lutz et al., 2018; Nepstad et al., 1994; Sist et al., 2014; Slik et al., 2013). Moreover, by over-towering the prevailing forest canopy and interacting with the hotter, drier and more open conditions of the emergent stratum, mega-trees add non-redundant complexity to the vertical vegetation structure and control forest microclimate, providing

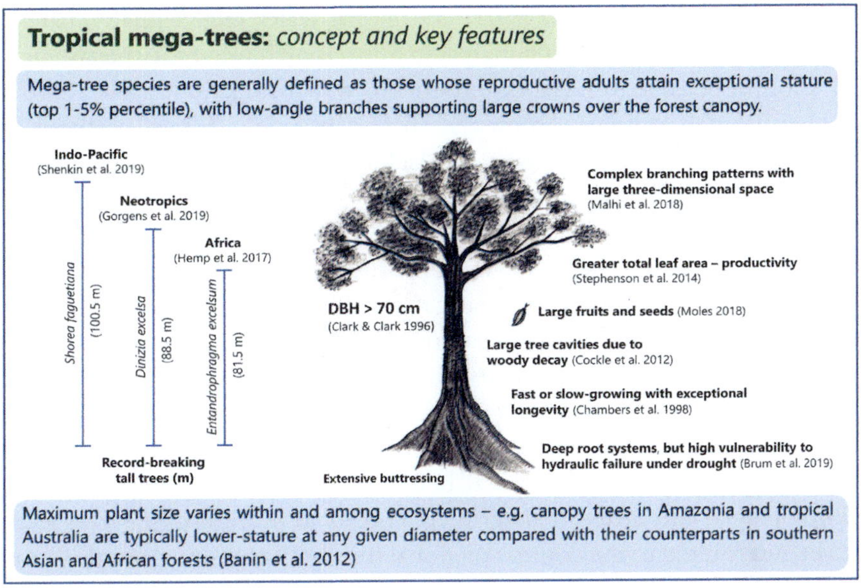

Fig. 1 Definition and key attributes of tropical mega-tree species. *Hand-drawn illustration by Jannah Oliveira.*

unique microhabitats for myriad of organisms, many of which are obligate mega-tree attendants and/or commensals (Cockle et al., 2011a,b; Woods et al., 2015). Mega-tree species can also be exceptionally long-lived (i.e. over hundreds of years), enhancing their long-term value and the ecological benefits they provide (Chambers et al., 1998; Laurance et al., 2004). This brief description supports the notion that mega-tree species represent a unique physical, functional and ecological life-history strategy that cannot be replaced by other plant functional groups.

However, exceptional stature also imposes costs (see Falster and Westoby, 2003), such as higher investment in supporting stem and root structures in relation to photosynthetic foliage, thereby reducing reproductive output (Kawecki, 1993); increasing risk of hydraulic failure (i.e. lower resistance to cavitation) under extreme drought conditions (Brum et al., 2019); and higher breakage rates due to strong winds (Laurance et al., 2000). Many of these features partly explain why mega-trees are highly sensitive to natural and human disturbances. Currently, mega-trees are exposed to an increasing myriad of threats operating from local (e.g. selective logging; Sist et al., 2014) to regional (e.g. forest fragmentation, Laurance et al., 2000; Santos et al., 2008; severe droughts and wildfires, Barlow et al., 2003; Nepstad et al., 2007) and global scales (e.g. climate change; Bennett et al., 2015), with unanticipated impacts on biodiversity persistence, provision of ecosystem services and human well-being.

Despite such a disproportionately important role, a robust synthesis of the importance of mega-trees is still lacking. Although some reviews on the ecology and conservation of large ancient trees are available (Lindenmayer and Laurance, 2016, 2017), these provide little emphasis on the ecological role and conservation status of mega-trees in tropical forest environments, thus leaving several gaps to be properly described and/or filled. A more comprehensive examination on the extent to which mega-trees support tropical biodiversity through countless direct and indirect interactions with a myriad of taxa and ecological groups is still required. Moreover, tropical mega-trees support several ecosystem functions and services, forest products and cultural benefits, most of which are yet to be acknowledged and properly described. Threats to tropical mega-trees must also be interpreted in the context of global trends/changes, to provide sound guidelines for mega-tree conservation action embedded in global agendas. By underappreciating the multiple contribution of tropical mega-trees and their requirements, we risk not only the persistence of this global heritage but also the opportunity to develop tropical forest regions sustainably.

Here we provide a comprehensive overview of the importance of tropical mega-trees in relation to biodiversity persistence, ecosystem services, and economic and cultural values. We particularly focus on their biodiversity benefits across eight categories of supporting mechanisms. We also describe and integrate all contemporary threats to mega-trees as an expected consequence of the conversion of old-growth forests into degraded human-modified landscapes that are increasingly exposed to climate change. To tackle this alarming trajectory, we argue in favour of 'productive forest landscapes' as a strategy to protect mega-tree species and highlight their positive connections with the sustainable development goals and other global targets. To achieve this ambitious goal, we highlight (1) key gaps to be filled, such as the relative contribution of mega-tree species to the diversity of tropical floras, NTFP opportunities and an acknowledgement of mega-trees as priceless 'spiritual entities' in contemporary and ancient human cultures, and (2) public forestry policies as a key strategy of socioeconomic development. Our narrative is based on a review of over a thousand papers systematically compiled through the Web of Science and Google Scholar covering many aspects of the ecology, evolution, distribution and conservation of mega-trees (for Methods, see Box 1). We acknowledge this contribution is far from

BOX 1 Details on the search methods employed in this study.
In this chapter, we compiled, organized and synthesized relevant literature to provide a comprehensive overview of the importance of tropical mega-trees in supporting biodiversity and ecosystem functioning, their economic and cultural value, and contemporary threats to their persistence. We did not conduct a formal systematic review, as this chapter is essentially qualitative and aims to review all relevant information, identify key knowledge gaps and propose conservation strategies. Many of the reported topics have very limited empirical evidence, and thus need further detailed investigation. Firstly, we searched for articles in Web of Knowledge and Google Scholar, constrained to studies conducted in the tropics but with no restriction on year of publication. Our search was based on commonly used terms to define mega-tree species (i.e. 'big trees', 'large trees', 'large old trees', 'giant trees') in association with 'tropical rainforest', OR 'tropical forest' OR only 'tropics', AND terms related to specific functions, ecological groups or threats, as described in each section below, such as 'productivity', 'biomass', 'carbon', 'nutrients', 'microclimate', 'vertical stratification', 'biodiversity', 'epiphytes', 'animals', 'birds', 'cavity-nesting birds', 'bats', 'primates', 'reptiles', 'amphibians', 'insects', 'ants', 'beetles', 'economic value', 'NTFP', 'timber extraction', 'carbon market', 'climate change', 'fire', 'forest fragmentation', using different combinations of search terms. We also considered additional references cited in the articles compiled, resulting in the examination of over a thousand titles and abstracts, from which we selected and carefully examined ca. 300 articles containing direct or indirect evidence for the specific role of tropical mega-trees and contemporary threats to this group. We then summarized information on the following key topics on tropical mega-trees: 1. Ecological roles (a—Supporting biodiversity, b—Supporting ecosystem functions and services), 2. Economic value (a—Timber, b—NTFPs, c—Carbon market), 3. Contemporary threats (a—Landscape modification, b—Timber extraction, c—Climate change), and 4. Synthesis and way forward.

exhaustive, but we hope to stimulate debate on sustainable use of tropical forests and the required research and policy agendas that consider mega-trees as highly sensitive keystone species.

2. Key ecological roles of tropical mega-trees

2.1 Supporting biodiversity

Mega-trees per se contribute greatly to tropical biodiversity, but the extent to which this operates and the biotic diversity involved remains a knowledge

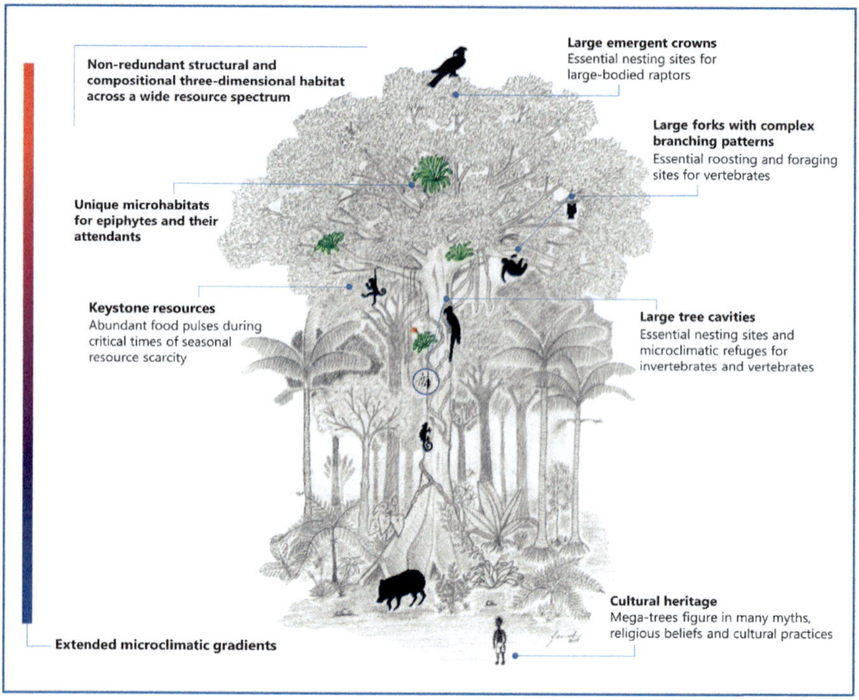

Fig. 2 Schematic illustration of key structures and support to biodiversity provided by tropical mega-tree species. *Hand-drawn illustration by Jannah Oliveira.*

gap. We refer to hundreds to thousands of tree species with direct and indirect effects on tropical forest biodiversity. Biodiversity-supporting services provided by mega-tree species are intrinsically associated with their complex and colossal three-dimensional structure (Fig. 2). This includes large crowns, low-angled bifurcations and natural cavities that directly support particular substrates/habitats or food resources for a myriad of taxa. Immense crowns filtering direct sunlight above the forest canopy also amplify the resource spectra associated with non-redundant niche space, including thermal buffering and extended vertical microclimatic gradients (a form of indirect support). Many of the directly supported taxa interact with others through trophic cascades and plant-animal interactions, potentially extending the role exerted by mega-trees. Although most of these services are yet to be examined (at least in terms of beneficiary taxa), it is now possible to provide a preliminary overview by integrating the available evidence into a framework considering both direct and indirect support, driving processes and beneficiary taxa or functional groups (Table 1).

Table 1 Summary of key direct and indirect support for biodiversity provided by mega-tree species in tropical forests, including driving processes and beneficiary taxa or ecological groups, and associated ecosystem services.

	Structures/resources	Support type	Taxa/ecological groups that benefit	Ecosystem functions and services
Direct support	Large emergent crowns and forks	Unique microhabitat conditions and resources (e.g. canopy humus)	Epiphytes	Carbon sequestration, biomass retention
		Nesting sites	Large-bodied raptors (e.g. Falconiformes)	Top-down control
		Nesting sites	Birds, stingless bees	Pollination, seed dispersal
		Roosting sites	Bats, birds, lizards	Pollination, seed dispersal, top-down control, pest control
		Sleeping sites	Primates	Seed dispersal
	Tree cavities	Microrefuges to extreme microclimatic conditions	Arthropods, frogs, lizards	Pathogen population control
		Breeding sites, peccary wallows	Peccaries, amphibians, reptiles	Pathogen population control
	Uprooting—temporary or permanent upland pools	Abundant food resources during critical times of low resource availability ('keystone resources')	Entire communities of frugivores and pollinators	Pollination, seed dispersal
	Masting flowering/ fruits	Microclimatic buffering	Shade-tolerant flora and their attendants	Carbon sequestration and storage
Indirect support	Large trunks with emergent crowns	Extended vertical physical support and microclimatic gradients	Microclimate-specialist species, entire communities	Maintaining trophic structure, preventing cascading extinctions and community collapse

2.1.1 Unique microhabitats for epiphytes

Epiphytes represent more than one-third of all vascular plant species in some tropical forests (Benzing, 2008; Nieder et al., 2001). Although heliophilic epiphytes exhibit low host-species specificity (Wagner et al., 2015), most species are strongly associated with very large trees, particularly large and highly illuminated crowns; e.g. several bark ferns, orchids and atmospheric bromeliads (Gonzalez et al., 2017; Woods et al., 2015). For instance, almost half of all bromeliad species occurring in the Atlantic Forest of northeast Brazil are found in the illuminated and desiccated crows supported by emergent tree species (Siqueira-Filho and Tabarelli, 2006).

The largest tree individuals within a population also tend to be the oldest, and epiphyte abundance and diversity are expected to increase with tree size as a result of increasing habitat area and colonization time as host trees become larger and older. Indeed, the positive correlations between tree size and epiphyte abundance, biomass and diversity are widely reported for different tropical forest realms (Ding et al., 2016; Gonzalez et al., 2017; Woods et al., 2015; Zotz and Vollrath, 2003). Using rigorous null models, Woods et al. (2015) tested the role of tree size on epiphyte assemblages and concluded that microhabitat heterogeneity also increases with tree size (e.g. wider variation of vapour pressure deficit from the inner to the outer crown), and that most epiphyte species are specialized to specific conditions and resources (e.g. canopy humus), provided exclusively by trees >70 cm DBH. For instance, only 6% of epiphyte species (mostly bark ferns) were significantly associated with small trees, whereas 57% were associated with mega-trees (Woods et al., 2015). Beyond mere physical support, mega-trees thus amplify microhabitat heterogeneity, thereby promoting more diverse and abundant epiphyte communities.

2.1.2 Nesting sites for large raptors

Colossal crowns and their huge bifurcations are essential if not irreplaceable platforms for nesting sites for several large raptor species (González and Vargas, 2011; Jenny, 2012; Lewis, 1986), most of which are at risk of extinction (Bildstein et al., 1998). For instance, the largest and rarest tropical eagle species on Earth, the Philippine Eagle (*Pithecophaga jefferyi*), rely on large emergent trees for nesting, as they provide natural platforms such as epiphytic ferns on their large high crowns (Lewis, 1986) and a panoramic view for the eagles to scan the landscape. Due to increasing human pressure, including forest loss and selective logging, this species is on the brink of extinction (Lewis, 1986).

Another emblematic extant mega-raptor is the Harpy Eagle (*Harpia harpyja*), the largest Neotropical forest raptor (Jenny, 2012). Harpies nest on several species of emergent trees, virtually all of which are targeted by the logging industry (Alvarez-Cordero, 1996). It is therefore unsurprising that the current distribution of Harpies and very large trees is roughly congruent across the Neotropics (Miranda et al., 2019). Raptors are important apex predators, playing a key role in top-down control of arboreal herbivore populations, such as sloths (Sergio et al., 2008). However, it remains unclear to what extent the collapse of mega-trees could lead to a trophic cascade via arboreal folivore release in the absence of large canopy raptors induced by nesting habitat limitation.

2.1.3 Tree cavities as nesting sites for invertebrates and vertebrates

Mega-trees also provide cavities that are essential for nesting and shelter of many vertebrate and invertebrate taxa (Cockle et al., 2011a; Gibbons and Lindenmayer, 2002; Scheffers et al., 2014), as many animal populations are limited by natural tree cavities (Cockle et al., 2010; Newton, 1994). Nests of stingless bees (Apidae, Meliponini) in tropical rainforests are almost invariably (80–100%) found in branches, trunks and roots of mega-trees, and nest density is typically correlated with the density of mega-trees (Eltz et al., 2002, 2003; Samejima et al., 2004). Studies also suggest that cavity-nesting stingless bees are limited by nest-site availability. For instance, the addition of artificial nest boxes increases bee nest density (Inoue et al., 1993) and many colonies may be found within the same tree, particularly the largest specimens (Eltz et al., 2003; Samejima et al., 2004). Stingless bees are one of the most important groups of pollinating insects in Neotropical forests, using a wide variety of floral resources (Momose et al., 1998; Nagamitsu and Inoue, 2005). There is evidence that height partitioning may occur even within the same tree, with larger bees selecting the upper crowns for nesting (Kajobe, 2015), likely indicating competition for higher cavities. Nest tree selection by stingless bees may also depend on the availability of resin for nest construction (Leonhardt and Blüthgen, 2009), which is largely found in large, late-successional trees (Samejima et al., 2004). In some temperate forests more than 40% of all bird and mammal species depend on cavity availability for nesting or roosting (Gibbons and Lindenmayer, 2002; Newton, 1994). Tropical forests, however, host the greatest diversity of cavity-nesting birds, which account for a substantial component of the overall avian diversity, 678 species in the Neotropical realm alone (van der Hoek et al., 2017). For instance, cavity-nesting bird species account for 22% of the Atlantic

Forest avifauna (Cockle et al., 2011a). Yet the ecology of tree cavity supply, distribution and selection by birds and other vertebrates is, at best, embryonic in tropical forests compared to a vast analogous literature from temperate forests (but see Cockle et al., 2011a).

In addition to larger, deeper cavities, mega-trees are also likely to be a preferred resource for cavity-nesting birds, because they can avoid both terrestrial and scansorial predators by selecting cavities in isolated crowns (Brightsmith, 2005; Cockle et al., 2011a). Suitable avian nest cavities are more likely to occur in very large emergent trees (>100 cm DBH), and nest density increases where artificial nest boxes are supplemented in tropical forests, indicating that cavity supply for cavity-nesting birds is limiting (Cockle et al., 2010; Newton, 1994). This is particularly true for secondary (non-excavating) cavity-nesters, which rely on existing tree cavities and represent the most threatened and species-rich group of cavity-nesting birds (van der Hoek et al., 2017). This group preferentially selects large live trees with high cavities produced by wood decay, while excavators (i.e. primary cavity-nesters) tend to select smaller-girthed dead trees ('snags') to construct their cavities (Cockle et al., 2011a, 2012). Among secondary cavity-nesters, some have more specific requirements and are particularly suppressed by tree cavity scarcity. For instance, tropical falcons are frequently found defending their nests against other cavity-nesting bird species such as parrots (Cornelius et al., 2008), as they depend on larger cavities with large openings, which represent a small fraction of the available tree cavities in tropical forests (Boyle et al., 2008). In a study in Mesoamerica, cavities used by Barred Forest Falcon (*Micrastur ruficollis*), Collared Falcon (*Micrastur semitorquatus*) and Laughing Falcon (*Herpetotheres cachinnans*) were all in very large (mean DBH >95 cm) emergent trees, which represented only 5% of all surveyed trees (Gerhardt, 2004).

2.1.4 Promoting habitat heterogeneity for amphibians and reptiles

The abundance and richness of herpetofauna in tropical forests are typically higher in areas with elevated densities of mega-trees, particularly those with extensive buttressing and large cavities that promote habitat heterogeneity (Voris, 1977; Whitfield and Pierce, 2005). For instance, amphibians and reptiles often benefit from temporary pools in cavities in mega-trees, which are used as breeding sites (Savage, 2002; Wells, 2010). Uprooting of large trees also generates large amounts of spatial heterogeneity at the ground level, including semi-permanent upland water bodies such as sizeable peccary wallows, which are critical sites for pond-breeding frogs in Neotropical forests

(Beck et al., 2010; Gascon, 1991). Additionally, the huge and complex structure of large tree forks provide essential shelter and foraging sites for other herpetofauna, particularly lizards (Ellinger et al., 2001; Vitt, 1991).

2.1.5 Roosting sites for mammals

Mega-tree cavities also form critical roosting habitat for bats, which play diverse ecological roles in tropical forests, including seed dispersal, pollination and predation (Kunz et al., 2011). Bat foraging requirements are often very general, but they have specific roosting requirements, which are essential for their survival and reproduction (Kunz and Lumsden, 2003). Tree cavities are one of the main roosting habitats used by tropical forest bats, especially those in the families Phyllostomidae and Emballonuridae in the Neotropics (Kunz and Lumsden, 2003). Furthermore, most species select large cavities in large-diameter emergent trees (Evelyn and Stiles, 2003; Kunz and Lumsden, 2003). Bat roosting sites for a common frugivorous bat species (*Sturnira lilium*, Phyllostomidae) located using radiotelemetry in Mexico were always in trees larger than 50 cm DBH. Gleaning faunivore bats, which account for ~20% of all bat species in Neotropical forests, are also highly restricted to very large trees in mature forests (Simmons and Voss, 1998). Large-girthed trees are selected by bats because they are more likely to host cavities with a large internal volume. Emergents rising above surrounding canopy trees also provide reduced predation risk and the appropriate microclimate; for instance, higher solar radiation provides energetic benefits for bats (Kunz and Lumsden, 2003). The vertical forest canopy heterogeneity also determines the composition and diversity of arboreal vertebrate assemblages, so that higher-stature old-growth forests are associated with greater taxonomic and phylogenetic diversity (e.g. Martins et al., 2017).

Several tropical forest primate species also select large emergent trees as safe sleeping sites, which is critical for their fitness, as they may spend over half of their lives within these natural shelters (Cheyne et al., 2012; Lutermann et al., 2010; Phoonjampa et al., 2010). In particular, several gibbon species (Hylobatidae) commonly overnight in the thickest, tallest large-crowned emergent trees (Cheyne et al., 2012; Phoonjampa et al., 2010). As gibbons do not build nests, they are very vulnerable to predator attacks, including by raptors, snakes and felids. By selecting large emergent trees as sleeping sites, primates not only avoid competition with other frugivores, but also minimize predation risk (Cheyne et al., 2012). Moreover, mega-trees provide essential and often super-abundant resources, including flowers, seeds and mature fruit pulp for many arboreal and terrestrial consumers (Dunn et al., 2009), and their vast crowns can accommodate large feeding aggregations and group sizes.

2.1.6 Keystone plant resources for frugivores

Mega-trees provide an important resource contribution to entire assemblages of forest frugivores (Diaz-Martin et al., 2014; Peres, 2000; Shanahan and Compton, 2001; Terborgh, 1986a,b). Some large-crowned species in tropical forests (e.g. *Ficus* spp.) provide 'keystone plant resources' (sensu Terborgh, 1986a), such as fruits, seeds and flowers that become available in large amounts during critical times of low resource availability over the annual cycle, thereby having a disproportionately large effect on frugivores. For instance, only 12 plant species were found sustain up to 80% of the animal biomass during a period of ripe fruit scarcity at Cocha Cashu, Peru, thereby determining the carrying capacity of the forest for the frugivore community (Terborgh, 1986b). Several mega-tree species, particularly in the Moraceae and Sapotaceae, provide keystone resources. Loss of such keystone resources—for example, via selective timber extraction—can lead to the functional collapse of whole forest ecosystems due to impacts on a range of trophic levels.

2.1.7 Extended structural and microclimatic gradients

Although mega-trees are often scattered at low densities in an otherwise more homogeneous forest canopy, they account for a disproportionately large amount of aboveground forest biomass and carbon storage (Ali et al., 2019; Bastin et al., 2018; Lutz et al., 2018; Slik et al., 2013) and break up the forest canopy by projecting their crowns well above the canopy (Clark and Clark, 1996). This creates singularity in upper canopy conditions by enhancing structural complexity, while also affecting the forest understorey, promoting cooler and wetter microclimates, and increasing water and nutrient resource availability (Hardwick et al., 2015; Jucker et al., 2018; Manning et al., 2006). Mega-trees therefore account for much additional microclimatic buffering and extended vertical microclimatic gradients (Hardwick et al., 2015; Jucker et al., 2018), which benefit specialists such as shade-tolerant flora and their attendants. This is particularly important in tropical forests, where approximately one-third of all woody plant species (from shrubs to subcanopy trees) are shade-tolerant, inhabit the forest understorey (Gentry et al., 1982), and where many species operate near their thermal tolerance limits (Doughty and Goulden, 2008). Moreover, the diversity of many animal taxa is vertically stratified in tropical rainforests, including arthropods (Charles and Basset, 2005), birds (Chmel et al., 2016), primates (Peres, 1993) and other mammals (Vieira and Monteiro-Filho, 2003). Therefore, the collapse of the emergent layer could potentially lead to the complete reorganization of animal communities, as species are compressed downwards and interact competitively with others.

2.1.8 Supporting extensive foodwebs

By directly supporting several groups of plants and animals, mega-trees are indirectly connected to extensive foodwebs and plant-animal interactions. For example, epiphytes such as bird's nest ferns and tank bromeliads sustained by the emergent forest layer (Benzing, 2008; Ellwood and Foster, 2004) host a diverse fauna, with different epiphyte species typically associated with different animal assemblages (Stuntz et al., 2002a). Ellwood and Foster (2004) showed that a single large epiphyte can support an invertebrate biomass comparable to that of the entire tree crown in which they occur, which would double estimates of invertebrate biomass in tropical forest canopies. Like large tree cavities, epiphytes can buffer extreme weather conditions, thereby providing 'climatic microrefugia' for sensitive animal taxa (Scheffers et al., 2014; Stuntz et al., 2002b). Although the importance of these *microrefuges* is more evident in temperate ecosystems with pronounced climatic variation, they are also critical in tropical forests, where elevated temperatures can be detrimental to ectothermic fauna with low thermal tolerance (e.g. arthropods, frogs and lizards), particularly under future climate change scenarios (Scheffers et al., 2014). Specifically, microhabitat temperature buffering for several species represents the only way to escape lethal conditions resulting from extreme climatic events (Scheffers et al., 2014).

Overall, mega-trees can be described as 'keystone structures' (sensu Tews et al., 2004) or 'ecosystem engineers' (sensu Jones et al., 1994) and as facilitators for multiple species and ecological groups, thereby imparting disproportionate influence on biodiversity persistence, by their effect in augmenting structural heterogeneity and widening the resource spectrum of an ecosystem. However, investigations into the direct and indirect roles of mega-trees for their dependent biodiversity remain embryonic. The upper canopy of tropical forests remains one of the least studied environments across all terrestrial biomes, and is often referred to as 'the last biotic frontier' (Erwin, 1983) or a 'black-box' (Nakamura et al., 2017).

2.2 Supporting ecosystem functions and services

In addition to supporting biodiversity, mega-tree species are intrinsically linked to key ecosystem services that tropical forests provide from local to global scales (Fig. 3).

2.2.1 Forest productivity, aboveground biomass and carbon stocks

Tropical forests account for one-third of the productivity and one-quarter of the carbon stock of the Earth's terrestrial biosphere (Beer et al., 2010; Bonan, 2008). Beyond its clear importance in regulating climate change (Houghton et al., 2015), aboveground forest biomass (AGB) is perhaps the main vegetation driver of several ecosystem processes, such as litter production and decomposition (see Lohbeck et al., 2015 for an example from tropical secondary forests). Understanding the drivers of variation in AGB is thus key to predicting ecosystem functioning and adequately managing ecosystems.

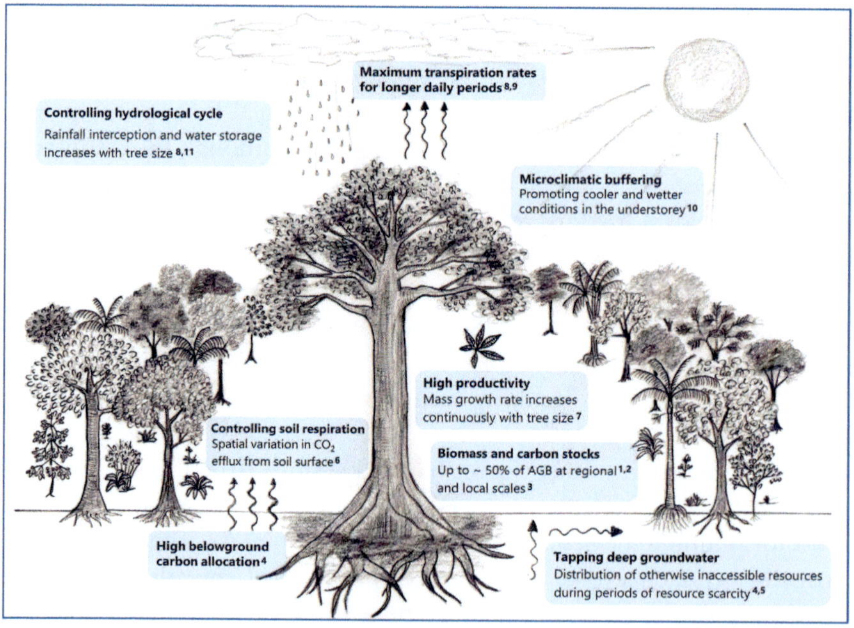

Fig. 3 Schematic illustration of key structures and support to ecosystem services by tropical mega-tree species. Superscript numbers indicate associated references: 1—Slik et al. (2013); 2—Lutz et al. (2018); 3—Brown et al. (1995), 4—Nepstad et al. (1994); 5—Brum et al. (2019); 6—Katayama et al. (2009); 7—Stephenson et al. (2014); 8—Goldstein et al. (1998); 9—Jordan and Kline (1977); 10—Jucker et al. (2018); 11—Herwitz (1985). *Hand-drawn illustration by Jannah Oliveira.*

Forest productivity and AGB are driven both by direct effects of abiotic conditions on tree growth/survival and by vegetation properties, including taxonomic (e.g. species richness), structural (e.g. basal area) and functional (e.g. wood density) attributes (Poorter et al., 2015, 2017; van der Sande et al., 2017). Tree species diversity may increase tropical forest AGB, probably through niche complementarity effects on resource use efficiency, which are apparently limited to small spatial scales (~0.1 ha; Poorter et al., 2015). On the other hand, forest structure attributes markedly determine variation in AGB at local (Poorter et al., 2015), regional (Malhi et al., 2006) and continental scales (Slik et al., 2013). Specifically, tropical forest AGB tends to increase markedly with increasing mean stem diameter, stand basal area and density of mega-trees (Lutz et al., 2018; Poorter et al., 2015; Slik et al., 2013). For instance, in a pan-tropical analysis on the drivers of AGB variation across 120 lowland tropical forests, Slik et al. (2013) demonstrated that density of mega-trees (DBH >70 cm) explains ca. 70% of the site-level variation in aboveground biomass. Furthermore, higher density of mega-trees is the sole cause of the higher forest AGB in Palaeotropical compared to Neotropical forests (Slik et al., 2013). The positive effects of mega-tree attributes on tropical forest AGB override the effects of species richness and other tree assemblage attributes (Ali et al., 2019).

Mega-trees have extensive well-lit crowns, and thus continuously sequester carbon at high rates, making the largest contribution to forest productivity (Stephenson et al., 2014). Formerly, it was thought that individual mass growth rate should decline with tree size, but on the basis of ~700,000 trees across 403 tropical and temperate species, Stephenson et al. (2014) showed that, despite the decline in leaf-level productivity, mass growth rate increases continuously with tree size, probably as a result of increasing total leaf area. To give an idea of the disproportionate productivity of a single mega-tree, trees with 100 cm DBH will each year add the mean mass equivalent of an entire conspecific tree of 10–20 cm DBH (Stephenson et al., 2014).

Although tropical mega-trees account for a small fraction of the overall tree abundance, they store a substantial fraction of forest biomass, which explains their marked effects on AGB at different spatial scales (Bastin et al., 2018; Lutz et al., 2018; Slik et al., 2013). At continental scales, on average they account for only 1.5%, 2.4% and 3.8% of stems larger than 10 cm DBH across forest sites in South America, Southeast Asia and Africa, respectively, but store approximately 25%, 40% and 45% of AGB in these three regions (Slik et al., 2013). Regional gradients in AGB can also be explained by variation in the density of mega-trees. For instance, ~60% of the variation in AGB across Amazonia is explained by forest basal area alone (Baker et al., 2004) and the higher AGB of Eastern Amazonian forests compared to Western Amazonia probably results from the higher abundance of very large trees >100 cm DBH. In this region, trees larger than 60 cm DBH account for less than 10% of the total tree density, but store half of the forest biomass (Sist et al., 2014). At local scales, Brown et al. (1995) found that only 3% of trees (emergent and large canopy species) accounted for 50% of AGB in the southwestern Amazon. Individual biomass increases exponentially with stem diameter, which explains the disproportionate contribution of large trees to forest biomass in comparison with small trees (Poorter et al., 2015). Finally, the relative contribution of large trees to carbon stocks is likely to still be underestimated, as their very large crowns with complex branching patterns are not taken into account by conventional estimates of forest biomass based on general allometric equations (Malhi et al., 2018).

2.2.2 Controlling resource dynamics and climatic conditions

The pool and magnitude of resource fluxes (i.e. water, carbon and nutrients) are strongly influenced by variation in organismal size, as suggested by the Metabolic Scaling Theory (Enquist et al., 2015). The high vegetative and reproductive biomass (i.e. wood, foliage, flowers, fruits) associated with mega-trees not only provides support for biodiversity, but also determines patterns of ecosystem functioning from local to regional scales. For instance, species with larger woody surface area may store greater amounts of water under heavy rainfall (Herwitz, 1985), and deep rooting systems allow mega-trees to tap otherwise inaccessible groundwater (Brum et al., 2019), thus potentially influencing hydrological regimes (Nepstad et al., 1994). Such features allow mega-trees to maintain maximum transpiration rates for longer daily periods (Goldstein et al., 1998). For instance, every 0.1 m^2 increase in basal sapwood area allows an increment of 10 kg in diurnal water storage capacity in forest trees in Panama (Goldstein et al., 1998). This is also reflected in transpiration rates, with rates varying from 0.7 L in small understorey trees to 1180 L in mega-trees in Amazonia (Jordan and Kline, 1977).

The large root systems and extensive foliage of tropical mega-trees also support continuous photosynthetic activity and distribution of deep water (among other resources) during periods of extreme resource scarcity (Goldstein et al., 1998; Nepstad et al., 1994). This potentially sustains a number of understorey tree species with shallow roots, thus avoiding ecosystem collapse. Moreover, the larger belowground carbon associated with mega-trees largely controls spatial variation in tropical forest soil respiration (i.e. efflux of CO_2 from the soil surface; Katayama et al., 2009). Tropical mega-trees may therefore largely control ecosystem functioning, by altering spatial-temporal patterns of resource availability, microclimatic and hydro-ecological conditions, and regional rainfall regimes. Yet, our current understanding of the role of tropical mega-trees in controlling resource dynamics and climatic conditions is still very limited, despite their global conservation significance.

2.2.3 Mega-trees as spiritual/religious entities or deities

Sacred trees frequently figure in indigenous cosmologies, myths, religious beliefs, cultural practices and historical events for a myriad of both ancient and contemporary cultures or ethnic groups worldwide (Cusack, 2011). This immemorial phenomenon is not restricted to places of worship centred on large trees, but also includes sites at which sacred trees become central locations for rituals, including weddings, rain-making ceremonies, funerals (e.g. burial sites), parades, sacrifices, judgements or social gatherings (Dafni, 2007). Frequently, sacred trees represent a crucial component of sacred forest groves (Gadgil, 2018). In other traditions, parts of a sacred tree are considered a form of talisman (Dafni, 2007). Such a diverse relationship with sacred trees highlights the spiritual depths that noteworthy trees have evoked in human societies.

Many sacred trees include tropical mega-tree species such as the sacred Maya Ceiba or 'Yaaxche' (*Ceiba pentandra*), a crucial element of the Mayan cosmovision which connects the upper, middle, and underworlds (Zidar and Elisens, 2009). This mega-tree species is still a dominant element of the emergent layer across forests from Amazonia to southern Mexico. Baobab trees (*Adansonia* spp.) are sacred entities and a cosmovision element for several African cultures (Vieilledent et al., 2013), as well as magnificent sightseeing targets for ecotourism. In India, strangler figs (*Ficus religiosa, F. benghalensis*), which begin life as epiphytes but achieve the stature of mega-trees (Kandari et al., 2014) are also of cultural significance. Buddha achieved enlightenment beneath the crown of a fig tree (*F. religiosa*), which is now named the Bodhi tree (Niroula and Singh, 2015). Other mega-tree species figure heavily in Hinduism (Niroula and Singh, 2015) and a comprehensive search across the ethnographic literature is likely to reveal the existence of several tropical mega-tree species considered sacred, as is the case in temperate regions (Cusack, 2011). In addition to their significance to local communities, several sacred trees, groves and forests have been spared from agricultural conversion, thereby becoming important conservation sites (Bhagwat and Rutte, 2006). Sacred mega-trees also increasingly attract tourism and pilgrimages, rendering this a key topic in both conservation and sustainable development science (Haberman, 2013; Vieilledent et al., 2013).

3. Economic value of mega-trees

Tropical forests provide many commercially valuable products, including both timber and non-timber forest products. They also provide globally important ecosystem services, the market value of which is still poorly quantified, such as carbon sequestration/storage and water cycling. Timber and non-timber forest products have been exploited and traded by humans for centuries, while the carbon market is a recent and promising economic avenue for remaining forest stands (Box 2). Although these assets are provided by trees in general, mega-trees often contribute the most in terms of forest biomass, as they exhibit specific features, such as large size and high wood density (Lutz et al., 2018; Poorter et al., 2015), that make them most attractive to the logging industry (Cannon et al., 1998; Gérard et al., 2017). Therefore, large canopy trees make a disproportionately large contribution to human well-being, especially in the timber and carbon markets.

3.1 Timber

Since colonial times, timber and wood products have remained the main source of revenue from tropical forests. Indeed, many tropical countries began their economic development in the aftermath of colonial independence, by exploiting their natural timber stocks (Ghazoul, 2016). Annual exports of primary and secondary wood products from tropical forests exceeded US$20 billion in the first decade since 2000, with further predicted increases as more countries focus their exports on high-value secondary wood products (FAO, 2018). This tropical timber trade is dominated by Southeast Asia, with emergent dipterocarp tree species—that often reach heights >80 m—the mainstay. During the 1980s, tropical hardwoods from Southeast Asia accounted for 84% of the export market from developing countries, while only 14% came from Africa and 1% from Latin America (Laird, 1995). This leading role has not changed much over the years; for instance, more roundwood is harvested from Borneo alone than from the whole of Africa and the Amazon combined (Curran et al., 2004). Dipterocarps contribute over 85% of Indonesia's timber exports and

BOX 2 Most important economic uses of tropical mega-trees.

Timber

For centuries, timber from mega-trees supported the naval enterprise and urban infrastructure worldwide. Many colonial capitals such as Havana, Salvador and Goa exhibited active shipyards sustained by mega-trees such as teak (*Tectona grandis*), rosewood (*Dalbergia nigra*) and sissoo (*Dalbergia sissoo*) (Tripati et al., 2005). Operational caravels and galleons are long gone, but hardwoods from almost extinct mega-trees can still be admired in colossal baroque churches and administrative buildings of many colonial capitals, and in staircases, massive wardrobes and other furniture made from rosewood in many churches and monasteries across the Brazilian coastal regions (Bernal et al., 2011). Timber exploitation reduced the number and size of populations of the most valuable mega-trees, such as the big-leaf mahogany (*Swietenia macrophylla*), and facilitated access for conversion of tropical forest into agriculture and pastureland (Nepstad et al., 2004).

Non-timber forest products

Non-timber forest products (NTFP) including medicinal plants, fibres, resins, latex, fruits, seeds, construction materials, and bushmeat have been also exploited by forest-dwellers for centuries, for both subsistence and trade. For example, the Borneo camphor (*Dryobalanopsis aromatica*) mega-tree has been traded by Arab merchants since the 6th century (Shiva and Jantan, 1998). Recent studies provide evidence that some Amazonian mega-tree species, such as the Brazil nut (*Bertholletia excelsa*), were domesticated by palaeo-Indians and subsequently increased in density over the last 5000 years (Levis et al., 2017). Although the exploitation of NTFP can be sustainable, chronic use of some resources as seeds can lead to the senescence of mega-tree populations, due to systematic failure in seedling recruitment (Peres et al., 1997; Peres et al., 2003).

Carbon market

To reduce tropical forest carbon emissions into the atmosphere as carbon dioxide, the United Nations Framework Convention on Climate Change created economic incentives to assist developing countries in voluntarily reducing emissions from deforestation and forest degradation, or 'REDD'. Countries that demonstrate emission reductions may be able to trade those carbon credits on the international carbon market or elsewhere. Most of the key REDD guidelines were completed by 2013, with the final book finished in 2015 (with sustainable management of forests, conservation of forest carbon stocks, and enhancement of forest carbon stocks constituting the '+' in 'REDD+'). As mega-trees account for up to half of all aboveground forest biomass (Slik et al., 2013), they have a disproportionate contribution to the carbon market.

generate as much as US$6 billion in annual export revenues, and the Kalimantan region alone provides 60% of Indonesia's exports (Ghazoul, 2016).

The sheer abundance of large timber trees in Asia's tropical forests continues to make timber harvesting very lucrative. Asian dipterocarp rainforests have the highest stature of any tropical forest, showing the large contribution of mega-trees to the timber trade. However, the leading role of mega-trees in tropical timber markets is a global pattern, rather than a prerogative of Asia. From the 300 commercially valuable tree species assembled in the Tropical Timber Atlas (Gérard et al., 2017), 278 are tropical angiosperms, 256 (92%) of which are mega-tree species of >70 cm DBH (Table 2). Among these species, 155 (60%) can also be defined as hardwoods (i.e. wood density >0.7 $g\,cm^{-3}$), again demonstrating the timber quality of mega-trees. For example, the five most commercially valuable timber species in Brazil (*Tabebuia* sp., *Hymenaea courbaril*, *Astronium lecointei*, *Manilkara* sp. and *Dinizia excelsa*) are very large, hardwood trees, all of which are worth over US$300 per m^3 of sawnwood (ITTO, 2018). Despite the economic and environmental impact posed by timber exploitation, transforming timber markets to a more sustainable trajectory still requires basic science, such as better taxonomic nomenclature of 'commercial species'. For example, most of the information on wood proprieties and grades is still organized by each species' commercial name, with several taxonomic species or even genera grouped into a single name (Table 2).

3.2 Non-timber forest products

Non-timber forest products (NTFPs) generate significant income across local to regional tropical markets, not to mention their domestic use for subsistence (Laird, 1995). For instance, NTFP markets in Indonesia were worth US$125 million annually in the early 1980s (de Jong and Mendelsohn, 1992). Mega-tree NTFPs, such as rosewood oil (*Dalbergia nigra*), chicle (*Manilkara zapota*), and Brazil nuts (*Bertholletia excelsa*), are traded widely in international markets (Laird, 1995). Like timber, NTFP exploited from Asian mega-trees are concentrated among dipterocarps, including resins and oleoresins such as *Dipterocarpus* spp., camphor waxy (*Cinnamomum camphora*, but also *Dryobalanops aromatica*), and Illipe nut and butter fat (a group of 15 species of *Shorea*). Of these, dammar and Illipe nut and butter fat are widely traded and continue to support local economies in Sumatra and Kalimantan (see review on the use and threats to dipterocarp NTFPs in Ghazoul, 2016).

Table 2 Important families, genera and species of tropical mega-trees used in the timber industry per continent.

	Africa	Asia and Oceania	Latin America
Number of families	32	25	26
Most important families[a]	Fabaceae (37) Malvaceae (10) Sapotaceae (8)	Dipterocarpaceae (9) Fabaceae (8) Burseraceae (5)	Fabaceae (26) Lecythidaceae (5) Moraceae (5)
Number of genera	98	58	77
Most important genera[a]	*Dacryodes* (12) *Beilschmiedia* (11) *Bombax* (8)	*Shorea* (84)[b] *Hopea* (15) *Dipterocarpus* (12)	*Peltogyne* (8) *Ocotea* (7) *Andira* (6), *Copaifera* (6) and *Swietenia* (6)
Number of species	117	59	80
Most important species[c]	*Letestua durissima* *Klainedoxia gabonensis* *Lophira alata* *Desbordesia insignis* *Bobgunnia fistuloides* *Coula edulis* *Manilkara mabokeensis* *Cynometra ananta* *Pentaclethra macrophylla* *Autranella congolensis*	*Diospyros ebenum* *Dialium platysepalum* *Dalbergia oliveri* *Chloroxylon swietenia* *Eusideroxylon zwageri* *Neobalanocarpus heimii* *Shorea laevis* *Eucalyptus diversicolor* *Dalbergia latifólia* *Hopea forbesii*	*Bocoa prouacensis* *Brosimum rubescens* *Manilkara bidentata* *Dipteryx odorata* *Dinizia excelsa* *Handroanthus impetiginosa* *Mora excelsa* *Chlorocardium rodiei* *Sacoglottis guianensis* *Lecythis idatimon*

[a]Most important families and genera were based on the number of genera and species, respectively.
[b]The Tropical Timber Atlas is organized by the commercial name and for some speciose genera (e.g. *Shorea*) several species are known by the same name.
[c]Most important species were based on their wood density.
Source: Tropical Timber Atlas (Gérard et al., 2017).

Afrotropical mega-trees also play an important role as NTFPs. For instance, from 38 species used by communities in Cameroon, 12 are mega-trees. These include African walnut (*Coula edulis*), erimado (*Ricinodendron heudelotii*) and moabi (*Baillonella toxisperma*) used as spices and condiments, resins collected from okoumé (*Aucoumea klaineana*), and medicines from alstonia (*Alstonia boonei*) and bubinga (*Guibourtia tessmannii*) (Dijk, 1999). Some species produce both timber and non-timber products, such as the okoumé, which yields a resin tapped prior to tree felling (Laird, 1995). In southern Cameroon, 19 of 31 timber species (representing 86% of total volume) exploited by a Dutch logging company (GWZ) are also used by local communities as NTFPs (Dijk, 1999; Laird, 1995). Due to unsustainable exploitation, some NTFPs extracted from mega-trees reach high prices in international markets, such as the bark of the African cherry (*Prunus africana*), which provides a compound exported to Europe and the United States, used to treat prostate disorders (Waterman, 1994). As supplies have begun to diminish elsewhere, large quantities of *Prunus* bark have begun to be exploited in Equatorial Guinea for a lucrative export market of US$220 million/year (Sunderland and Obama, 1999).

In Latin America many of the major NTFPs are also extracted from mega-tree species, such as rosewood oil and Brazil nut, as well as andiroba (*Carapa guianensis*) and copaiba (*Copaifera langsdorffii*). These species support a fast-growing market of medicinal and cosmetic products, opening a new opportunity for the sustainable use of Amazonian forests (Klimas et al., 2012; Newton et al., 2012). For instance, 580 tons of copaiba oleoresin was produced in the Brazilian Amazon alone in 2010, generating US$2,416,186 (IBGE, 2010). In Amazonian countries, including Peru, Brazil and Ecuador, NTFPs often yield greater returns per hectare than timber extraction, cattle ranching, and plantation forestry (Balick and Mendelsohn, 1992; Grimes et al., 1994; Peters et al., 1989). Among 335 timber species used in the Brazilian Amazon, one-third have also been valued for their fruits, medicinal properties, and/or gums and resins. In some cases, such as rosewood, non-timber values exceed timber values (Clay and Clement, 1993). For example, Brazil traded over 30,000 tons of Brazil nuts in 2007, generating over US$23 million (IBGE, 2007).

The extent of the potential contribution of non-timber products sourced from mega-trees to the economy of tropical countries is therefore significant. However, there is little information on species used as NTFPs and their commercial value, as most studies consider local markets and fail to consider quantitative measures of NTFP natural production rates (de Jong and

Mendelsohn, 1992). Therefore, there is a need to compile information on production and harvesting regimes, as well as their ecological impacts, in order to sustainably manage wild NTFPs. As demonstrated by the emerging cosmetic market based on fruits and seeds from Amazonian forests (Funasaki et al., 2016), NTFPs represent an immense opportunity for sustainable development. However, this is still neglected globally, particularly if we consider that tropical floras contain over 40,000 tree species (Slik et al., 2015).

3.3 Carbon market

As deforestation challenges the viability and future of the timber industry, the past decade has seen an increase in research interest on the potential of forests to store carbon. Conservationists have sought to persuade several stakeholders—including policy makers and politicians—that carbon markets open an opportunity to preserve remaining forest stands. Several countries have pursued the REDD+ agenda buoyed by commitments from some countries to finance carbon conservation through protection of natural forests. For instance, at the 2007 COP 13, in Bali, the Norwegian government announced a US$1 billion donation to Brazilian REDD+ initiatives and US$500 million towards the creation and implementation of national-based, REDD+ activities in Tanzania. In addition, the United Kingdom pledged US$200 million towards the Congo Basin Forest Fund, to aid forest conservation activities in Central Africa. In 2010 Norway signed a letter of intent with Indonesia to provide up to US$1 billion for similar mechanisms of carbon retention on the ground. The United States has provided more than US$1.5 billion in support of REDD+ and other sustainable landscape activities since 2010, with the same sourced from Finland and Australia (for more details on REDD+ projects see the Voluntary REDD+ Database at http://www.reddplusdatabase.org/).

The carbon market and protection of remaining tropical forests have the potential to inject billions of dollars into the economy of tropical countries, potentially catalysing the transition towards sustainable use of tropical forests. However, the disproportionate role of mega-trees in forest carbon stocks and water cycling makes them a key focus for these and other conservation initiatives.

4. Contemporary threats to tropical mega-trees

The majority of mega-tree species are locally rare and may also exhibit narrow geographic distributions, potentially making them naturally

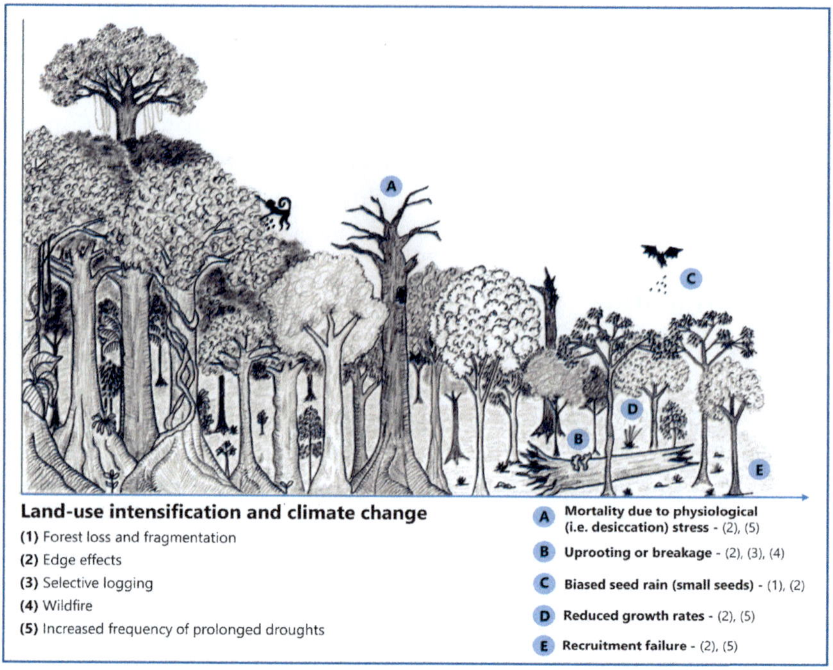

Fig. 4 The main threats to mega-tree species in human-modified landscapes. Numbers and letters indicate cause-effect relationships between driving processes (left) and consequences to mega-tree species (right), respectivelly. *Hand-drawn illustration by Jannah Oliveira.*

vulnerable to extinction (Hubbell et al., 2008; Newbold et al., 2018; Rabinowitz, 1981). To give but one example, at least 90 *Shorea* emergent dipterocarp tree species are endemic to Borneo (Ghazoul, 2016). In addition to such natural rarity, mega-tree species vulnerability can be summarized into three major threats: landscape modification (i.e. habitat loss, fragmentation and edge effects), selective logging and climate change (Fig. 4).

4.1 Landscape modification

Habitat loss represents by far the most important threat to tropical mega-trees, since a substantial fraction of species exhibit narrow geographic distributions in old-growth forests, which continue to be converted into human-modified landscapes (Lewis et al., 2015). For instance, tropical forest cover in Southeast Asia is reduced by an average net rate of 1.6 million ha year^{-1} (~0.6% year^{-1}; Stibig et al., 2014), and this figure roughly describes what other major blocks of tropical forests have experienced in recent decades. On average, tropical deforestation rates are about 1% per year (Achard et al., 2002; Hansen et al., 2013), but deforestation is not a random process. Historically, deforestation has been concentrated in lowland forests, particularly across agriculturally suitable areas, such as flat terrains on fertile soils. The deforestation trajectory experienced by the Atlantic Forest of South America, which has been reduced to ~12% of its original forest cover, illustrates this global trend. While lowland forests once accounted for >90% of the original Atlantic Forest cover, this forest type now represents less than 30% of the remaining habitat and a very limited proportion of the protected areas (Tabarelli et al., 2010). Yet mega-tree species are primarily restricted to lowland forests worldwide (Corlett and Primack, 2011).

Habitat loss also implies habitat fragmentation and the resulting expansion of forest edges, as old-growth forests are converted into human-modified landscapes comprising a collection of small forest remnants embedded within an open-habitat matrix, such as pastures and croplands (Fischer and Lindenmayer, 2007). Forest edges tend to be dominated by abiotic edge effects (sensu Murcia, 1995), including higher light and wind turbulence, increased incidence of wildfires and exposure to pesticides (Barlow et al., 2016; Laurance et al., 2018; Silvério et al., 2019). Forest edges thus usually consist of more open and desiccated habitats compared to forest core areas. Edge-affected habitats can typically penetrate ca. 300 m into the forest, and at times much further (2–3 km) where edge-related fires occur (Laurance et al., 2018). As land use intensifies, human-modified landscapes progressively retain edge-affected habitats rather than forest interior or forest stands free of edge effects (Lewis et al., 2015). For instance, over 90% of the remaining Brazilian Atlantic forest is within 1 km of an edge (Haddad et al., 2015). This includes many small, edge-dominated forest fragments and secondary forests following land abandonment.

Edge-affected habitats are usually unsuitable for mega-trees due to a myriad of processes. Increased wind turbulence induces uprooting, and adult mortality is often high due to physiological (i.e. desiccation) stress along forest edges (Laurance et al., 2000). In addition, the large stature achieved by tropical mega-trees makes them more prone to hydraulic failures, particularly in environments under extremely desiccated conditions (Brum et al., 2019), such as forest edges (Laurance et al., 2000). Moreover, mega-tree seedlings are usually intolerant to both high light habitats and competition with proliferating light-demanding pioneer tree species (Benítez-Malvido and Martínez-Ramos, 2003; Santo-Silva et al., 2013). Thus, small remnants and forest edges often lack mega-trees and the forest emergent layer (Laurance et al., 2006; Oliveira et al., 2008; Santos et al., 2008). This is consistent with the notion that edge effects drive remaining forest areas into intermediate if not early-successional stages, in which shade-tolerant mega-tree species are not expected to occur (Tabarelli et al., 2008). Additionally, edge-affected habitats are more prone to wildfires and available evidence suggests that mega-trees may be more fire-sensitive, although the underlying mechanism is still poorly understood (Barlow et al., 2003).

Most of the threats mentioned here relate to adult tree vulnerability. However, mega-tree species are potentially threatened by other processes, such as limited pollination and seed dispersal, as are many tropical tree species relying on sensitive pollinators, such as hummingbirds, large bees, and seed dispersal agents, such as large-bodied frugivores (Caughlin et al., 2015; Cordeiro and Howe, 2001; Neuschulz et al., 2016). Human-modified landscapes tend to become progressively 'empty' (Redford, 1992) and those dominated by small forest fragments often lack long-distance seed dispersal, which can be fatal to successful seedling recruitment in shade-tolerant, large-seeded mega-tree species (Costa et al., 2012; de Melo et al., 2006). To make matters worse, a substantial proportion of tropical mega-tree species bear large seeds, dispersed by medium- to large-bodied vertebrates, including primates, large-gaped birds and terrestrial mammals such as tapirs, large primates and large rodents (da Silva and Tabarelli, 2000; Muller-Landau and Hardesty, 2005; Peres et al., 2016). Such recruitment bottlenecks have been documented in particular cases, but we are still far from a comprehensive understanding on how limitation in biotic flows threaten mega-tree species persistence. In sum, human-modified landscapes dominated by edge-affected habitats are often unsuitable for most large canopy and emergent trees, as already shown for tropical landscapes in Amazonia and the Atlantic Forest (Joly et al., 2014; Laurance et al., 2018).

4.2 Timber extraction

Logging represents a serious threat to commercial tropical tree species because (1) selective logging usually targets trees $\geq$50 cm DBH, (2) tropical timber production is typically based on the exploitation of old-growth native forests rather than tree plantations, and (3) a substantial proportion of conventional logging operations are illegal, which implies 'high-grading' of old-growth stands, in which most large stems are harvested, thereby eliminating the most reproductive individuals (Cannon et al., 1998; Shima et al., 2018).

Annually, around 1 billion m^3 of wood from over 300 tropical tree species is commercially harvested, particularly from Southeast Asia and South America (ITC/ITTO, 2002; ITTO, 2014). Available estimates indicate that between 27 and 50 million m^3 of wood from commercial timber species is extracted every year across the Amazon alone, a large proportion of which is from protected areas (Asner et al., 2005). Accordingly, several mega-tree species traded in both national and international timber markets have been officially declared threatened with extinction, such as *Swietenia macrophylla*, *B. excelsa*, *D. excelsa* and *Cedrela* spp. (Steege et al., 2015). Historically, the timber market has operated as a boom-and-bust system and any region covered by tropical forest now includes mega-tree species, which were overexploited in the past to the point of population decline or collapse, often leading to regional-scale economic extinction (Richardson and Peres, 2016). This includes dozens of timber species that were once widely commercialized and are now largely unavailable due to population collapse. Given the lack of subsequent population recovery, secondary economic options are often adopted and include transitions to more damaging forms of land use, if not wholesale forest conversion (Shima et al., 2018).

4.3 Climate change

Climate change can directly affect mega-tree species, particularly where this results in increased frequency of severe droughts (Bennett et al., 2015; Brum et al., 2019; Phillips et al., 2010), which is the case in many tropical forest regions (Boisier et al., 2015; Dai, 2012). Global-scale analyses have documented higher rates of mortality among large tropical forest trees, potentially due to higher physiological vulnerability to water stress (Bennett et al., 2015; Phillips et al., 2010), particularly in soft-wooded species and taxa requiring moist habitats (Esquivel-Muelbert et al., 2019; Greenwood et al., 2017). Large crowns above the forest canopy are more exposed to elevated solar radiation and vapour pressure deficits, rendering conditions more challenging to maintain a positive carbon balance, as evaporative demands increase under reduced water availability. Moreover, there is close covariance between tree size, rooting depth and embolism-risk, so that most mega-trees operate near the limit of hydraulic safety, which explains their higher mortality rates during extreme drought events (Brum et al., 2019). Additionally, mega-trees exhibit reduced growth rates during droughts (Bennett et al., 2015). This is consistent with the prediction of lower forest productivity in terms of wood production in tropical forests experiencing drier and hotter conditions and/or severe droughts (Brando et al., 2008; Feeley et al., 2007). Declines in productivity associated with climate change may be compensated by increased CO_2 levels, but the impact of CO_2 fertilization on mega-trees is still poorly understood (but see Bartlett et al., 2019; Esquivel-Muelbert et al., 2019).

Indirectly, climate change can threaten mega-trees via the synergistic interaction between droughts, forest fires and tree mortality. This is particularly the case for severe droughts associated with strong El Niño Southern Oscillation (ENSO) events in Amazonia. Droughts induce desiccation and increased mortality among large canopy trees, which in turn opens up the canopy, increasing understorey dry-down rates and the amount of dead biomass. This can result in a feedback loop, connecting droughts and tree mortality through increased forest flammability (Holdsworth and Uhl, 1997; Nepstad et al., 2007). For instance, a complete phase-shift in structure and composition—from primary closed-canopy forest with high basal area to scrubby savannah-like vegetation—has been observed in twice- and thrice-burned forests in the Central Amazon (Barlow and Peres, 2008; Pellegrini, forthcoming). Positive synergisms between logging, ENSO-related droughts and increased mega-tree mortality have also been documented in Borneo (Slik, 2004). Other potential climate change-related impacts on mega-trees includes liana proliferation and increased tree mortality (Phillips et al., 2002). On the other hand, it has been argued that climate change and CO_2 fertilization may favour some mega-tree species (Esquivel-Muelbert et al., 2019). This does not mean, however, that such 'winners' are able to persist in human-modified landscapes.

5. Synthesis and way forward

Research on the complex and manifold ecological roles exerted by mega-tree species—on biodiversity persistence, provision of ecosystem services and livelihoods of forest-dependent peoples—is at best embryonic. As a thought-provoking exercise we could assume that the relative biomass sustained by mega-tree species represent a proxy for the amount of foliage, fruits, flowers, litter and microhabitats mega-trees offer to plant attendants and their networks of interactions, which in many cases amount to irreplaceable resources. This would imply that a large fraction of tropical biodiversity relies on mega-trees in one way or another (as proposed in the Fig. 2 framework), but this remains a working hypothesis yet to be examined. Despite such limited knowledge, it is clear that large tree species represent a substantial portion of the tropical tree flora and a large share of the forest phytomass, with a disproportionate contribution to biodiversity persistence of many directly or indirectly associated taxa (Table 1). Moreover, mega-trees have historically supported the tropical timber industry (Box 2, Table 2), and provide invaluable ecosystem services and goods for either local subsistence or trade (Figs. 3).

Protecting such a keystone species group is challenging, as threats to mega-tree species do not operate in isolation but result from a historical process that is rapidly gaining momentum globally: old-growth forest conversion towards human-modified landscapes as a 'perfect-storm' of threats including habitat loss, edge effects, logging, defaunation and wildfires. Climate change adds to this risk and represents a powerful overarching threat for several mega-tree species that until recently were considered to be 'safe' or of 'least concern' given the remoteness of their remnant populations.

This is an alarming prospect, since the most tried-and-tested approach to ensure tropical biodiversity conservation relies on protecting large blocks of old-growth forest, including the so-called biodiversity corridors and their derivations (Melo et al., 2013). Cumulative evidence reinforces the notion that old-growth forests represent irreplaceable habitats for countless large, forest-dependent tree species and that complementary strategies such as land-sharing (sensu Fischer et al., 2014) can be incorporated into conservation plans. However, large tree species can now succumb to mortality induced by severe droughts even in remote areas of old-growth forest (Bennett et al., 2015; Brum et al., 2019). Similar to several montane floras that have been increasingly squeezed to higher altitudes, there are few places left to hide for several mega-tree species with climate change impacts on tropical forests.

With this perspective, the future of tropical mega-tree flora depends on our ability to protect extensive tracts of old-growth forest and their vulnerability to climate change, particularly severe droughts. This also relies on regional-scale persistence of forest cover, to reduce edge effects and ensure key hydrological functions such as rainfall interception, storage and evapotranspiration, which can reduce the impact of droughts and prevent widespread forest fires (Lawrence and Vandecar, 2015). Imperatives to protect mega-trees can be used to strengthen global scale justification to spare/restore old-growth forests and mitigate climate change, and also to promote socioeconomic development via sustainable forestry. Both local and global demands could be negotiated by establishing multiple-use productive forest landscapes, with the added-value of forestry provision of globally relevant ecosystem services. Without this drive, tropical ecosystems will continue their trajectory towards degraded landscapes and thus demand massive investments in restoration if forest ecosystem services are to be recovered (Fig. 5). Benefits from productive forest landscapes include their high capacity for biodiversity persistence and provision of ecosystem services, maintenance of local livelihoods, high potential for economic development, and opportunities for NTFP-based markets (Fig. 5). This broad concept of productive forest landscapes is largely aligned with the Sustainable Development Goals (Griggs et al., 2013) and should be considered as a crucial tool for global sustainability.

On a positive note, local initiatives that promote productive forest landscapes are already underway (e.g. forest for people; Larson et al., 2010), but these are still limited in the tropics (García-Fernández et al., 2008). For a desirable scaling-up of these efforts, challenges are yet to be overcome through policies and research. Briefly, effective policy should support local governance, involving multiple stakeholders in the planning process, and integrating traditional practices into formal forestry regulations (Guariguata et al., 2012). These measures must be enhanced and rapidly scaled up to preclude agribusiness frontiers from colonizing the last tracts of old-growth forests and previously disturbed landscapes from experiencing further degradation (Lewis et al., 2015). Finally, productive forest landscapes must be made financially feasible through a combination of mechanisms and products provided by tropical forests. Instruments such as REDD+ and payments for ecosystem services could be explicitly integrated to support productive forest landscapes (Fig. 5). In terms of research, it is essential to recognize and properly value all services provided by mega-trees and their forest habitats, including potential NTFPs and a proper taxonomic treatment of commercial timber species to ensure sustainable exploitation. The knowledge basis required to scale up productive forest landscapes represents a knowledge gap and goes beyond the scope of this chapter.

Finally, the productive forest landscape approach poses economic, political, technological and social challenges, as sparing tropical forests still results in elevated opportunity costs that few stakeholders can afford to pay (Fisher et al., 2011; Kremen et al., 2000). Despite these challenges, the future of tropical forests, their mega-tree species, and local livelihoods are inextricably linked, and productive forest landscapes would likely bring benefits for future generations and tropical biodiversity.

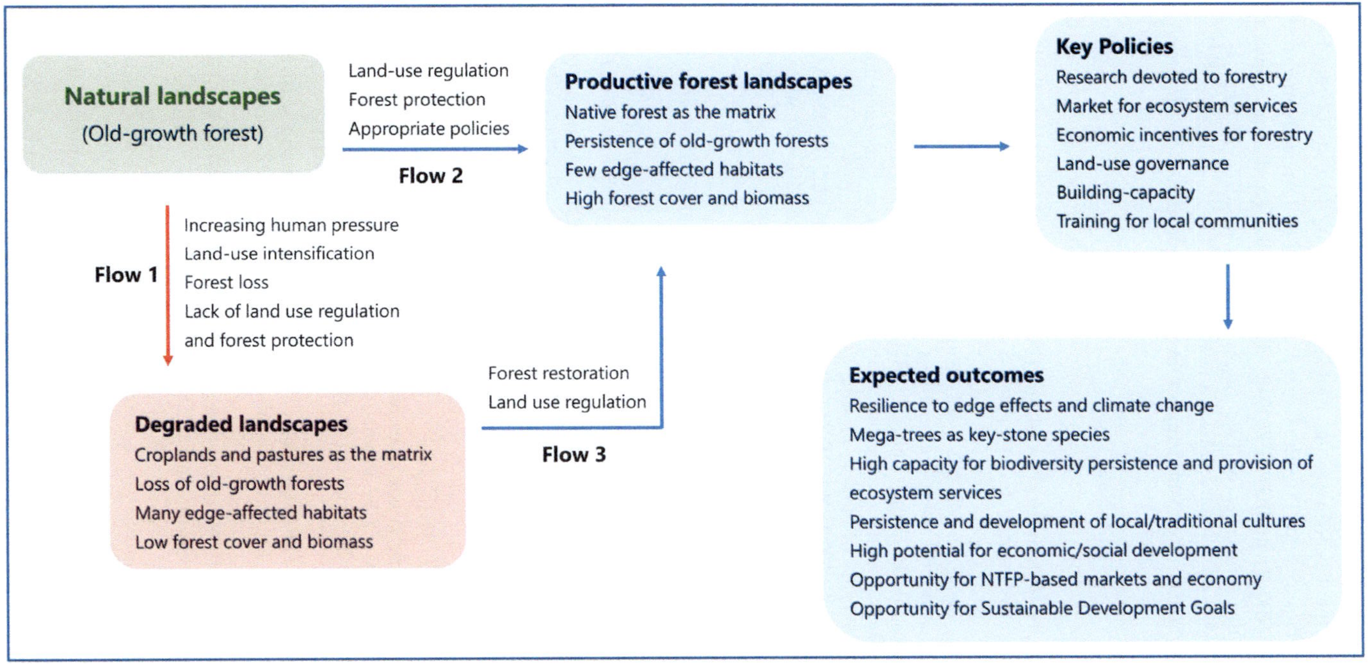

Fig. 5 Old-growth forest or natural landscapes frequently transition into degraded human-modified landscapes (Flow 1), which tend to support reduced forest cover, particularly of old-growth forest. Alternatively, natural landscapes could move towards productive forest landscapes either directly (Flow 2) or via restoration (Flow 3). Productive forest landscapes are structurally different from both natural and degraded habitats, as the dominant landscape matrix consists of sustainably exploited native forests. Moreover, these landscapes retain old-growth forest patches that ensure high levels of structural connectivity. Productive forest landscapes provide a wide spectrum of opportunities for biodiversity conservation, provision of globally important ecosystem services and sustainable socioeconomic development. However, scaling-up these landscapes demands an extensive set of public policies spanning a variety of topics. *Adapted from Melo, F.P.L., Arroyo-Rodríguez, V., Fahrig, L., Martínez-Ramos, M., Tabarelli, M., 2013. On the hope for biodiversity-friendly tropical landscapes. Trends Ecol. Evol. 28, 462–468.*

Acknowledgements

This review was funded by the Coordenação de Aperfeiçoamento de Pessoal de Nivel Superior—Brazil (CAPES)—Finance code 001, the Conselho Nacional de Desenvolvimento Científico e Tecnológico—CNPq (grant 403770/2012-2), and a Newton Fund Institutional Partnership award between the University of East Anglia and Universidade Federal de Pernambuco. B.X.P. was funded by a PhD scholarship from CNPq. I.R.L. and M.T. also thank CNPq for productivity grants. We thank Jannah Oliveira for hand-drawing all figures of this chapter.

References

Achard, F., Eva, H.D., Stibig, H.-J., Mayaux, P., Gallego, J., Richards, T., Malingreau, J.-P., 2002. Determination of deforestation rates of the world's humid tropical forests. Science 297, 999–1002.

Ali, A., Lin, S.-L., He, J.-K., Kong, F.-M., Yu, J.-H., Jiang, H.-S., 2019. Big-sized trees overrule remaining trees' attributes and species richness as determinants of aboveground biomass in tropical forests. Glob. Chang. Biol. 25, 2810–2824.

Alvarez-Cordero, E., 1996. Biology and Conservation of the Harpy Eagle in Venezuela and Panama. PhD Thesis, University of Florida, Florida, USA.

Asner, G.P., Asner, G.P., Knapp, D.E., Broadbent, E.N., Oliveira, P.J.C., Keller, M., Silva, J.N., 2005. Selective logging in the Brazilian Amazon. Science 310, 480–482.

Baker, T.R., Phillips, O.L., Malhi, Y., Almeida, S., Arroyo, L., Di Fiore, A., et al., 2004. Variation in wood density determines spatial patterns in Amazonian forest biomass. Glob. Chang. Biol. 10, 545–562.

Balick, M.J., Mendelsohn, R., 1992. Assessing the economic value of traditional medicines from tropical rain forests. Conserv. Biol. 6, 128–130.

Balzotti, C.S., Asner, G.P., Taylor, P.G., Cole, R., Osborne, B.B., Cleveland, C.C., Porder, S., Townsend, A.R., 2017. Topographic distributions of emergent trees in tropical forests of the Osa Peninsula, Costa Rica. Ecography 40, 829–839.

Barlow, J., Lennox, G.D., Ferreira, J., Berenguer, E., Lees, A.C., Nally, R.M., et al., 2016. Anthropogenic disturbance in tropical forests can double biodiversity loss from deforestation. Nature 535, 144–147.

Barlow, J., Peres, C.A., 2008. Fire-mediated dieback and a compositional cascade in Amazonian forests. Philos. Trans. R. Soc. B 363, 1787–1794.

Barlow, J., Peres, C.A., Lagan, B.O., Haugaasen, T., 2003. Large tree mortality and the decline of forest biomass following Amazonian wildfires. Ecol. Lett. 6, 6–8.

Bartlett, M.K., Detto, M., Pacala, S.W., 2019. Predicting shifts in the functional composition of tropical forests under increased drought and CO_2 from trade-offs among plant hydraulic traits. Ecol. Lett. 22, 67–77.

Bastin, J.-F., Rutishauser, E., Kellner, J.R., Saatchi, S., Pélissier, R., Hérault, B., et al., 2018. Pan-tropical prediction of forest structure from the largest trees. Glob. Ecol. Biogeogr. 27, 1366–1383.

Beck, H., Thebpanya, P., Filiaggi, M., 2010. Do Neotropical peccary species (Tayassuidae) function as ecosystem engineers for anurans? J. Trop. Ecol. 26, 407–414.

Beer, C., Reichstein, M., Tomelleri, E., Ciais, P., Jung, M., Carvalhais, N., et al., 2010. Terrestrial gross carbon dioxide uptake: global distribution and covariation with climate. Science 329, 834–838.

Benítez-Malvido, J., Martínez-Ramos, M., 2003. Impact of forest fragmentation on understory plant species richness in Amazonia. Conserv. Biol. 17, 389–400.

Bennett, A.C., Mcdowell, N.G., Allen, C.D., Anderson-Teixeira, K.J., 2015. Larger trees suffer most during drought in forests worldwide. Nat. Plants 1, 15139.

Benzing, D.H., 2008. Vascular Epiphytes: General Biology and Related Biota. Cambridge University Press, Cambridge.
Bernal, R.A., Valente, A., Pissarra, J., 2011. Wood identification of 18th century furniture: interpreting wood naming inventories. Int. J. Conserv. Sci. 2, 165–178.
Bhagwat, S.A., Rutte, C., 2006. Sacred groves: potential for biodiversity management. Front. Ecol. Environ. 4, 519–524.
Bildstein, K.L., Schelsky, W., Zalles, J., Ellis, S., 1998. Conservation status of tropical raptors. J. Raptor Res. 32, 3–18.
Boisier, J.P., Ciais, P., Ducharne, A., Guimberteau, M., 2015. Projected strengthening of Amazonian dry season by constrained climate model simulations. Nat. Clim. Chang. 2658.
Bonan, G.B., 2008. Forests and climate change: forcings, feedbacks, and the climate benefits of forests. Science 320, 1444–1449.
Boyle, W.A., Ganong, C.N., Clark, D.B., Hast, M.A., 2008. Density, distribution, and attributes of tree cavities in an old-growth tropical rain forest. Biotropica 40, 241–245.
Brando, P.M., Nepstad, D.C., Davidson, E.A., Trumbore, S.E., Ray, D., Camargo, P., 2008. Drought effects on litterfall, wood production and belowground carbon cycling in an Amazon forest: results of a throughfall reduction experiment. Philos. Trans. R. Soc. B 363, 1839–1848.
Brightsmith, D.J., 2005. Competition, predation and nest niche shifts among tropical cavity nesters: ecological evidence. J. Avian Biol. 36, 74–83.
Brown, I.F., Martinelli, L.A., Thomas, W.W., Moreira, M.Z., Cid Ferreira, C.A., Victoria, R.A., 1995. Uncertainty in the biomass of Amazonian forests: an example from Rondônia, Brazil. For. Ecol. Manag. 75, 175–189.
Brum, M., Vadeboncoeur, M.A., Ivanov, V., Asbjornsen, H., Saleska, S., Alves, L.F., Penha, D., Dias, J.D., Arag, L.E.O.C., Barros, F., Bittencourt, P., Pereira, L., Oliveira, R.S., 2019. Hydrological niche segregation defines forest structure and drought tolerance strategies in a seasonal Amazon forest. J. Ecol. 107, 318–333.
Cannon, C.H., Peart, D.R., Leighton, M., 1998. Tree species diversity in commercially logged Bornean rainforest. Science 281, 1366–1368.
Caughlin, T.T., Ferguson, J.M., Lichstein, J.W., Zuidema, P.A., Bunyavejchewin, S., Levey, D.J., 2015. Loss of animal seed dispersal increases extinction risk in a tropical tree species due to pervasive negative density dependence across life stages. Proc. R. Soc. B 282 (1798).
Chambers, J.Q., Higuchi, N., Schimel, J., 1998. Ancient trees in Amazonia. Nature 391, 135–136.
Charles, E., Basset, Y., 2005. Vertical stratification of leaf-beetle assemblages (Coleoptera: Chrysomelidae) in two forest types in Panama. J. Trop. Ecol. 21, 329–336.
Cheyne, S.M., Höing, A., Rinear, J., Sheeran, K., 2012. Sleeping site selection by agile gibbons: the influence of tree stability, fruit availability and predation risk. Folia Primatol. 83, 299–311.
Chmel, K., Riegert, J., Paul, L., Novotný, V., 2016. Vertical stratification of an avian community in New Guinean tropical rainforest. Popul. Ecol. 58, 535–547.
Clark, D.B., Clark, D.A., 1996. Abundance, growth and mortality of very large trees in Neotropical lowland rain forest. For. Ecol. Manag. 80, 235–244.
Clay, J.W., Clement, C.R., 1993. Selected Species and Strategies to Enhance Income Generation From Amazonian Forests. Food and Agriculture Organization of the United Nations (FAO), Rome.
Cockle, K.L., Martin, K., Drever, M.C., 2010. Supply of tree-holes limits nest density of cavity-nesting birds in primary and logged subtropical Atlantic forest. Biol. Conserv. 143, 2851–2857.
Cockle, K.L., Martin, K., Wesołowski, T., 2011a. Woodpeckers, decay, and the future of cavity-nesting vertebrate communities worldwide. Front. Ecol. Environ. 9, 377–382.

Cockle, K., Martin, K., Wiebe, K., 2011b. Selection of nest trees by cavity-nesting birds in the Neotropical Atlantic Forest. Biotropica 43, 228–236.
Cockle, K.L., Martin, K., Robledo, G., 2012. Linking fungi, trees, and hole-using birds in a Neotropical tree-cavity network: pathways of cavity production and implications for conservation. For. Ecol. Manag. 264, 210–219.
Cordeiro, N.J., Howe, H.F., 2001. Low recruitment of trees dispersed by animals in African forest fragments. Conserv. Biol. 15, 1733–1741.
Corlett, R.T., Primack, R.B., 2011. Tropical Rain Forests: An Ecological and Biogeographical Comparison, second ed. Wiley-Blackwell, Chichester.
Cornelius, C., Cockle, K., Politi, N., Berkunsky, I., Sandoval, L., Ojeda, V., Rivera, L., Hunter, M., Martin, K., 2008. Cavity-nesting birds in Neotropical forests: cavities as a potentially limiting resource. Ornitol. Neotrop. 19, 253–268.
Costa, J.B.P., Melo, F.P.L., Santos, B.A., Tabarelli, M., 2012. Reduced availability of large seeds constrains Atlantic forest regeneration. Acta Oecol. 39, 61–66.
Curran, L.M., Trigg, S.N., McDonald, A.K., Astiani, D., Hardiono, Y.M., Siregar, P., Caniago, I., Kasischke, E., 2004. Lowland forest loss in protected areas of Indonesian Borneo. Science 303, 1000–1003.
Cusack, C.M., 2011. The Sacred Tree: Ancient and Medieval Manifestations, first ed. Cambridge Publishing, Cambridge.
Dafni, A., 2007. Rituals, ceremonies and customs related to sacred trees with a special reference to the Middle East. J. Ethnobiol. Ethnomed. 3, 28.
Dai, A., 2012. Increasing drought under global warming in observations and models. Nat. Clim. Chang. 3, 52–58.
da Silva, J.M.C., Tabarelli, M., 2000. Tree species impoverishment and the future flora of the Atlantic forest of northeast Brazil. Nature 404, 72–74.
de Jong, W., Mendelsohn, R.O. 1992. Managing the non-timber forest products of Southeast Asia. Unpublished manuscript prepared for the World Bank, Southeast Asia Division.
de Melo, F.P.L., Dirzo, R., Tabarelli, M., 2006. Biased seed rain in forest edges: evidence from the Brazilian Atlantic forest. Biol. Conserv. 132, 50–60.
Diaz-Martin, Z., Swamy, V., Terborgh, J., Alvarez-loayza, P., Cornejo, F., 2014. Identifying keystone plant resources in an Amazonian forest using a long- term fruit-fall record Identifying keystone plant resources in an Amazonian forest using a long-term fruit-fall record. J. Trop. Ecol. 30, 291–301.
Dijk, J.F.W., 1999. An assessment of non-wood forest product resources for the development of sustainable commercial extraction. In: Vantomme, P., Sunderland, T., Clark, L. (Eds.), Non-Wood Forest Products of Central Africa: Current Research Issues and Prospects for Conservation and Development. FAO, Rome, pp. 37–49.
Ding, Y., Liu, G., Zang, R., Zhang, J., Lu, X., Huang, J., 2016. Distribution of vascular epiphytes along a tropical elevational gradient: disentangling abiotic and biotic determinants. Sci. Rep. 6, 19706.
Doughty, C.E., Goulden, M.L., 2008. Are tropical forests near a high temperature threshold? J. Geophys. Res. 113, G00B07.
Dunn, J.C., Cristóbal-Azkarate, J., Veà, J.J., 2009. Differences in diet and activity pattern between two groups of Alouatta palliata associated with the availability of big trees and fruit of top food taxa. Am. J. Primatol. 71, 654–662.
Ellinger, N., Schlatte, G., Jerome, N., Hödl, W., 2001. Habitat use and activity patterns of the Neotropical arboreal lizard Tropidurus (=Uracentron) azureus werneri (Tropiduridae). J. Herpetol. 35, 395–402.
Ellwood, M.D.F., Foster, W.A., 2004. Doubling the estimate of invertebrate biomass in a rainforest canopy. Nature 429, 549–551.
Eltz, T., Brühl, C.A., van der Kaars, S., Linsenmair, K.E., 2002. Determinants of stingless bee nest density in lowland dipterocarp forests of Sabah, Malaysia. Oecologia 131, 27–34.

Eltz, T., Bruhl, C.A., Imiyabirb, Z., Linsenmair, E., 2003. Nesting and nest trees of stingless bees (Apidae: Meliponini) in lowland dipterocarp forests in Sabah, Malaysia, with implications for forest management. For. Ecol. Manag. 172, 301–313.
Enquist, B.J., Norberg, J., Bonser, S.P., Violle, C., Webb, C.T., Henderson, A., Sloat, L.L., Savage, V.M., 2015. Scaling from traits to ecosystems: developing a general trait driver theory via integrating trait-based and metabolic scaling theories. Adv. Ecol. Res. 52, 249–318.
Erwin, T.L., 1983. Tropical forest canopies: the last biotic frontier. Bull. Entomol. Soc. Am. 29, 14–20.
Esquivel-Muelbert, A., Baker, T.R., Dexter, K.G., Lewis, S.L., Brienen, R.J.W., Feldpausch, T.R., et al., 2019. Compositional response of Amazon forests to climate change. Glob. Chang. Biol. 25, 39–56.
Evelyn, M.J., Stiles, D.A., 2003. Roosting requirements of two frugivorous bats (Sturnira lilium and Arbifeus intermedius) in fragmented Neotropical forest. Biotropica 35, 405–418.
Falster, D.S., Westoby, M., 2003. Plant height and evolutionary games. Trends Ecol. Evol. 18, 337–343.
FAO, 2018. The State of the World's Forests 2018—Forest Pathways to Sustainable Development. Rome.
Feeley, K.J., Wright, S.J., Supardi, M.N.N., Kassim, A.R., Davies, S.J., 2007. Decelerating growth in tropical forest trees. Ecol. Lett. 10, 461–469.
Fischer, J., Lindenmayer, D.B., 2007. Landscape modification and habitat fragmentation: a synthesis. Glob. Ecol. Biogeogr. 16, 265–280.
Fischer, J., Abson, D.J., Butsic, V., Chappell, M.J., Ekroos, J., Hanspach, J., Kuemmerle, T., Smith, H.G., Wehrden, H., 2014. Land sparing versus land sharing: moving forward. Conserv. Lett. 7, 149–157.
Fisher, B., Edwards, D.P., Giam, X., Wilcove, D.S., 2011. The high costs of conserving Southeast Asia's lowland rainforests. Front. Ecol. Environ. 9, 329–334.
Funasaki, M., Barroso, H.S., Fernandes, V.L.A., Menezes, I.S., 2016. Amazon rainforest cosmetics: chemical approach for quality control. Quim. Nova 39, 194–209.
Gadgil, M., 2018. Sacred groves. Sci. Am. 319, 48–57.
García-Fernández, C., Ruiz-Pérez, M., Wunder, S., 2008. Is multiple-use forest management widely implementable in the tropics? For. Ecol. Manag. 256, 1468–1476.
Gascon, C., 1991. Population- and community-level analyses of species occurrences of central Amazonian rainforest tadpoles. Ecology 72, 1731–1746.
Gentry, A.H., Hecht, M.K., Wallace, B., Prance, G.T., 1982. Patterns of Neotropical plant species diversity. In: Evolutionary Biology. Springer US, Boston, pp. 1–84.
Gérard, J., Guibal, D., Paradis, S., Cerre, J.-C., 2017. Tropical Timber Atlas. International Tropical Timber Organization (ITTO), Paris.
Gerhardt, R.P., 2004. Cavity nesting in raptors of Tikal National Park and vicinity, Petén, Guatemala. Ornitol. Neotrop. 15, 477–483.
Ghazoul, J., 2016. Dipterocarp Biology, Ecology, and Conservation. Oxford University Press, Oxford.
Gibbons, P., Lindenmayer, D., 2002. Tree Hollows and Wildlife Conservation in Australia. CSIRO Publishing.
Goldstein, G., Andrade, J.L., Meinzer, F.C., Holbrook, N.M., Cavelier, J., Jackson, P., Celis, A., 1998. Stem water storage and diurnal patterns of water use in tropical forest canopy trees. Plant Cell Environ. 21, 397–406.
Gonmadje, C., Picard, N., Gourlet-fleury, S., Maxime, R., Freycon, V., Sunderland, T., Mckey, D., Doumenge, C., 2017. Altitudinal filtering of large-tree species explains above-ground biomass variation in an Atlantic Central African rain forest. J. Trop. Ecol. 33 (2017), 143–154.
González, J.d.J.V., Vargas, F.H., 2011. Nesting density of Harpy Eagles in Darien with population size estimates for Panama. J. Raptor Res. 45, 199–211.

Gonzalez, D.C.G., Quiel, C.R., Zotz, G., Bader, M.Y., 2017. Species richness and biomass of epiphytic vegetation in a tropical montane forest in western Panama. Trop. Conserv. Sci. 10, 1–17.

Greenwood, S., Ruiz-Benito, P., Martínez-Vilalta, J., Lloret, F., Kitzberger, T., Allen, C.D., Fensham, R., Laughlin, D.C., Kattge, J., Bönisch, G., Kraft, N.J., Jump, A.S., 2017. Tree mortality across biomes is promoted by drought intensity, lower wood density and higher specific leaf area. Ecol. Lett. 20, 539–553.

Griggs, D., Stafford-Smith, M., Gaffney, O., Rockström, J., Öhman, M.C., Shyamsundar, P., Steffen, W., Glaser, G., Kanie, N., Noble, I., 2013. Sustainable development goals for people and planet. Nature 495, 305–307.

Grimes, A., Loomis, S., Jahnige, P., Burnham, M., Onthank, K., Alarcón, R., Cuenca, W.P., Martinez, C.C., David Neill, M.B., Bennett, B., Mendelsohn, R., 1994. Valuing the Rain Forest: the economic value of nontimber forest products in Ecuador. Ambio 23, 405–410.

Guariguata, M.R., Sist, P., Nasi, R., 2012. Multiple use management of tropical production forests: how can we move from concept to reality? For. Ecol. Manag. 263, 170–174.

Haberman, D.L., 2013. People Trees: Worship of Trees in Northern India. Oxford University Press, Oxford.

Haddad, N.M., Brudvig, L.A., Clobert, J., Davies, K.F., Gonzalez, A., Holt, R.D., et al., 2015. Habitat fragmentation and its lasting impact on Earth's ecosystems. Sci. Adv. 1, 1–9.

Hansen, M.C., Potapov, P.V., Moore, R., Hancher, M., Turubanova, S.A., Tyukavina, A., Thau, D., Stehman, S.V., Goetz, S.J., Loveland, T.R., Kommareddy, A., Egorov, A., Chini, L., Justice, C.O., Townshend, J.R.G., 2013. High-resolution global maps of 21st-century forest cover change. Science 342, 850–854.

Hardwick, S.R., Toumi, R., Pfeifer, M., Turner, E.C., Nilus, R., Ewers, R.M., 2015. The relationship between leaf area index and microclimate in tropical forest and oil palm plantation: forest disturbance drives changes in microclimate. Agric. For. Meteorol. 201, 187–195.

Herwitz, S.R., 1985. Interception storage capacities of tropical rainforest canopy trees. J. Hydrol. 77, 237–252.

Holdsworth, A.R., Uhl, C., 1997. Fire in Amazonian selectively logged rain forest and the potential for fire reduction. Ecol. Appl. 7, 713–725.

Houghton, R.A., Byers, B., Nassikas, A.A., 2015. A role for tropical forests in stabilizing atmospheric CO2. Nat. Clim. Chang. 5, 1022–1023.

Hubbell, S.P., He, F., Condit, R., Borda-de-Agua, L., Kellner, J., ter Steege, H., 2008. How many tree species are there in the Amazon and how many of them will go extinct? Proc. Natl. Acad. Sci. U. S. A. 105, 11498–11504.

IBGE, 2007. Produção da extração vegetal e da silvicultura. Instituto Brasileiro de Geografia e Estatistica, Rio de Janeiro.

IBGE, 2010. Produção da Extração Vegetal e da Silvicultura. Instituto Brasileiro de Geografia e Estatística, Rio de Janeiro.

Inoue, T., Nakamura, K., Salmah, S., Abbas, I., 1993. Population dynamics of animals in unpredictably-changing tropical environments. J. Biosci. 18, 425–455.

ITC/ITTO, 2002. Tropical Timber Products: Development of Further Processing in ITTO Producer Countries. International Trade Centre UNCTAD/WTO and International Tropical Timber Organization, Geneva.

ITTO, 2014. Biennial Review and Assessment of the World Timber Situation. International Tropical Timber Organization, Yokohama.

ITTO, 2018. Tropical Timber Market Report. International Tropical Timber Organization, Yokohama.

Jenny, J.P., 2012. Neotropical Birds of Prey: Biology and Ecology of a Forest Raptor Community. Cornell University Press, New York.

Joly, C.A., Metzger, J.P., Tabarelli, M., 2014. Experiences from the Brazilian Atlantic Forest: ecological findings and conservation initiatives. New Phytol. 204, 459–473.

Jones, C.G., Lawton, J.H., Shachak, M., 1994. Organisms as ecosystem engineers. Oikos 69, 373–386.

Jordan, C., Kline, J., 1977. Transpiration of trees in a tropical rainforest. J. Appl. Ecol. 14, 853–860.

Jucker, T., Hardwick, S.R., Both, S., Elias, D.M.O., Ewers, R.M., Milodowski, D.T., Swinfield, T., Coomes, D.A., 2018. Canopy structure and topography jointly constrain the microclimate of human-modified tropical landscapes. Glob. Chang. Biol. 24, 5243–5258.

Kajobe, R., 2015. Nesting biology of equatorial Afrotropical stingless bees (Apidae; Meliponini) in Bwindi Impenetrable National Park, Uganda. J. Apic. Res. 46, 245–255.

Kandari, L.S., Bisht, V.K., Bhardwaj, M., Thakur, A.K., 2014. Conservation and management of sacred groves, myths and beliefs of tribal communities: a case study from north-India. Environ. Syst. Res. 3, 16.

Katayama, A., Kume, T., Komatsu, H., Ohashi, M., Nakagawa, M., Yamashita, M., Otsuki, K., Suzuki, M., Kumagai, T., 2009. Effect of forest structure on the spatial variation in soil respiration in a Bornean tropical rainforest. Agric. For. Meteorol. 149, 1666–1673.

Kawecki, T.J., 1993. Age and size at maturity in a patchy environment: fitness maximization versus evolutionary stability. Oikos 66, 309–317.

Klimas, C.A., Kainer, K.A., Wadt, L.H.O., 2012. The economic value of sustainable seed and timber harvests of multi-use species: an example using Carapa guianensis. For. Ecol. Manag. 268, 81–91.

Kremen, C., Niles, J.O., Dalton, M.G., Daily, G.C., Ehrlich, P.R., Fay, J.P., Grewal, D., Guillery, R.P., 2000. Economic incentives for rain forest conservation across scales. Science 288, 1828–1832.

Kunz, T.H., Lumsden, L.F., 2003. Ecology of cavity and foliage roosting bats. In: Kunz, T.H., Fenton, M.B. (Eds.), Bat Ecology. University of Chicago press, Chicago.

Kunz, T.H., Braun de Torrez, E., Bauer, D., Lobova, T., Fleming, T.H., 2011. Ecosystem services provided by bats. Ann. N. Y. Acad. Sc. 1223, 1–38.

Laird, S., 1995. The Natural Management of Tropical Forests for Timber and Non-timber Products. Oxford Forestry Institute, University of Oxford, Oxford.

Larson, A.M., Barry, D., Dahal, G.R., Colfer, C.J.P., 2010. Forests for People Community Rights and Forest Tenure Reform. Center for International Forest Research (CIFOR), The Cromwell Press Group, London.

Laurance, W.F., Delamônica, P., Laurance, S.G., Vasconcelos, H.L., Lovejoy, T.E., 2000. Rainforest fragmentation kills big trees. Nature 404, 836.

Laurance, W.F., Nascimento, H.E.M., Laurance, S.G., Condit, R., Angelo, S.D., Andrade, A., 2004. Inferred longevity of Amazonian rainforest trees based on a long-term demographic study. For. Ecol. Manag. 190, 131–143.

Laurance, W.F., Nascimento, H.E.M., Laurance, S.G., Andrade, A.C., Fearnside, P.M., Ribeiro, J.E.L., Capretz, R.L., 2006. Rain forest fragmentation and the proliferation of successional trees. Ecology 87, 469–482.

Laurance, W.F., Sayer, J., Cassman, K.G., 2014. Agricultural expansion and its impacts on tropical nature. Trends Ecol. Evol. 29, 107–116.

Laurance, W.F., Camargo, J.L.C., Fearnside, P.M., Lovejoy, T.E., Williamson, G.B., Mesquita, R.C.G., Meyer, C.F.J., Bobrowiec, P.E.D., Laurance, S.G.W., 2018. An Amazonian rainforest and its fragments as a laboratory of global change. Biol. Rev. 93, 223–247.

Lawrence, W., Vandecar, K., 2015. Effects of tropical deforestation on climate and agriculture. Nat. Clim. Chang. 5, 27–36.
Leonhardt, S.D., Blüthgen, N., 2009. A sticky affair: resin collection by Bornean stingless bees. Biotropica 41, 730–736.
Levis, C., Costa, F.R.C., Bongers, F., Peña-Claros, M., Clement, C.R., Junqueira, A.B., et al., 2017. Persistent effects of pre-Columbian plant domestication on Amazonian forest composition. Science 355, 925–931.
Lewis, R.E., 1986. A rain-forest raptor in danger. Oryx 20, 170–175.
Lewis, S.L., Edwards, D.P., Galbraith, D., 2015. Increasing human dominance of tropical forests. Science 349, 827–832.
Lindenmayer, D.B., Laurance, W.F., 2016. The unique challenges of conserving large old trees. Trends Ecol. Evol. 31, 416–418.
Lindenmayer, D.B., Laurance, W.F., 2017. The ecology, distribution, conservation and management of large old trees. Biol. Rev. 92, 1434–1458.
Lohbeck, M., Poorter, L., Martinez-Ramos, M., Bongers, F., 2015. Biomass is the main driver of changes in ecosystem process rates during tropical forest succession. Ecology 96, 1242–1252.
Lutermann, H., Verburgt, L., Rendigs, A., 2010. Resting and nesting in a small mammal: sleeping sites as a limiting resource for female grey mouse lemurs. Anim. Behav. 79, 1211–1219.
Lutz, J.A., Furniss, T.J., Johnson, D.J., Davies, S.J., Allen, D., Alonso, A., et al., 2018. Global importance of large-diameter trees. Glob. Ecol. Biogeogr. 27, 849–864.
Malhi, Y., 2012. The productivity, metabolism and carbon cycle of tropical forest vegetation. J. Ecol. 100, 65–75.
Malhi, Y., Wood, D., Baker, T.R., Wright, J., Phillips, O.L., Cochrane, T., et al., 2006. The regional variation of aboveground live biomass in old-growth Amazonian forests. Glob. Chang. Biol. 12, 1107–1138.
Malhi, Y., Gardner, T.A., Goldsmith, G.R., Silman, M.R., Zelazowski, P., 2014. Annu. Rev. Env. Resour. 39, 125–159.
Malhi, Y., Jackson, T., Bentley, P.L., Lau, A., Shenkin, A., Herold, M., Calders, K., Bartholomeus, H., Disney, M., 2018. New perspectives on the ecology of tree structure and tree communities through terrestrial laser scanning. Interface Focus 8, 52.
Manning, A.D., Fischer, J., Lindenmayer, D.B., 2006. Scattered trees are keystone structures—implications for conservation. Biol. Conserv. 132, 311–321.
Martins, A.C.M., Willig, M.R., Presley, S.J., Marinho-Filho, J., 2017. Forest ecology and management effects of forest height and vertical complexity on abundance and biodiversity of bats in Amazonia. For. Ecol. Manag. 391, 427–435.
Melo, F.P.L., Arroyo-Rodríguez, V., Fahrig, L., Martínez-Ramos, M., Tabarelli, M., 2013. On the hope for biodiversity-friendly tropical landscapes. Trends Ecol. Evol. 28, 462–468.
Miranda, E.B., Menezes, J.F., Farias, C.C., Munn, C., Peres, C.A., 2019. Species distribution modeling reveals strongholds and potential reintroduction areas for the world's largest eagle. PLoS One 14, e0216323.
Momose, K., Yumoto, T., Nagamitsu, T., Kato, M., Nagamasu, H., Sakai, S., Harrison, R.D., Itioka, T., Hamid, A.A., Inoue, T., 1998. Pollination biology in a lowland dipterocarp forest in Sarawak, Malaysia. I. Characteristics of the plant-pollinator community in a lowland dipterocarp forest. Am. J. Bot. 85, 1477–1501.
Muller-Landau, H.C., Hardesty, D., 2005. Seed dispersal of woody plants in tropical forests: concepts, examples and future directions. In: Burslem, D.F.R.P., Pinard, M.A., Hartley, S.E. (Eds.), Biotic Interactions in the Tropics: Their Role in the Maintenance of Species Diversity. Cambridge University Press, Cambridge, pp. 267–309.

Murcia, C., 1995. Edge effects in fragmented forests: implications for conservation. Trends Ecol. Evol. 10, 58–62.
Nagamitsu, T., Inoue, T., 2005. Floral resource utilization by stingless bees (Apidae, Meliponini). In: Roubik, D.W., Sakaiand, S., Karim, H.A.A. (Eds.), Pollination Ecology and the Rain Forest: Sarawak Studies. Springer New York, New York, pp. 73–88.
Nakamura, A., Kitching, R.L., Cao, M., Creedy, T.J., Fayle, T.M., Freiberg, M., et al., 2017. Forests and their canopies: achievements and horizons in canopy science. Trends Ecol. Evol. 32, 438–451.
Nepstad, D., de Carvalho, C., Davidson, E., Jipp, P.H., Lefebvre, P.A., Negreiros, G.H., da Silva, E.D., Stone, T.A., Trumbore, S.E., Vieira, S., 1994. The role of deep roots in the hydrological and carbon cycles of Amazonian forests and pastures. Nature 372, 666–669.
Nepstad, D.C., Azevedo-Ramos, C., Lima, E., McGrath, D., Pereira, C., Merry, F., 2004. Managing the Amazon timber industry. Conserv. Biol. 18, 575–577.
Nepstad, D.C., Tohver, I.M., Ray, D., Moutinho, P., Cardinot, G., 2007. Mortality of large trees and lianas following experimental drought in an Amazon forest. Ecology 88, 2259–2269.
Neuschulz, E.L., Mueller, T., Schleuning, M., Böhning-Gaese, K., 2016. Pollination and seed dispersal are the most threatened processes of plant regeneration. Sci. Rep. 6, 29839.
Newbold, T., Hudson, L.N., Contu, S., Hill, S.L.L., Beck, J., Liu, Y., Meyer, C., Phillips, H.R.P., Scharlemann, J.P.W., Purvis, A., 2018. Widespread winners and narrow-ranged losers: land use homogenizes biodiversity in local assemblages worldwide. PLoS Biol. 16, e2006841.
Newton, I., 1994. The role of nest sites in limiting the numbers of hole-nesting birds: a review. Biol. Conserv. 70, 265–276.
Newton, P., Watkinson, A.R., Peres, C.A., 2012. Spatial, temporal, and economic constraints to the commercial extraction of a non-timber forest product: copaíba (*Copaifera* spp.) oleoresin in Amazonian reserves. Econ. Bot. 66, 165–177.
Nieder, J., Prosper, J., Michaloud, G., 2001. Epiphytes and their contribution to canopy diversity. Plant Ecol. 153, 51–63.
Niroula, G., Singh, N.B., 2015. Religion and conservation: a review of use and protection of sacred plants and animals in Nepal. J. Instr. Sci. Technol. 20, 61–66.
Oliveira, M.A., Santos, A.M.M., Tabarelli, M., 2008. Profound impoverishment of the large-tree stand in a hyper-fragmented landscape of the Atlantic forest. For. Ecol. Manag. 256, 1910–1917.
Pellegrini, A. The distribution and structure of long-term and large-scale fire manipulation experiments. Adv. Ecol. Res., forthcoming.
Peres, C.A., 1993. Structure and spatial organization of an Amazonian terra firme forest primate community. J. Trop. Ecol. 9, 259–276.
Peres, C.A., 2000. Identifying keystone plant resources in tropical forests: the case of gums from Parkia pods. J. Trop. Ecol. 16, 287–317.
Peres, C.A., Baider, C., Zuidema, P.A., Wadt, L.H.O., Kainer, K.A., Gomes-Silva, D.A.P., Salomão, R.P., Simões, L.L., Franciosi, E.R.N., Cornejo, Valverde F., Gribel, R., Shepard Jr., G.H., Kanashiro, M., Coventry, P., Yu, D.W., Watkinson, A.R., Freckleton, R.P., 2003. Demographic threats to the sustainability of Brazil nut exploitation. Science 302, 2112–2114.
Peres, C.A., Emilio, T., Schietti, J., Desmoulière, S.J., Levi, T., 2016. Dispersal limitation induces long-term biomass collapse in overhunted Amazonian forests. Proc. Natl. Acad. Sci. U. S. A. 113, 892–897.

Peres, C.A., Schiesari, L.C., Dias-Leme, C.L., 1997. Vertebrate predation of Brazil-nut, an agouti-dispersed Amazonian seed crop (Bertholletia excelsa, Lecythidaceae): a test of the escape hypothesis. J. Trop. Ecol. 13, 69–79.

Peters, C.M., Gentry, A.H., Mendelsohn, R.O., 1989. Valuation of an Amazonian rainforest. Nature 339, 655–656.

Phillips, O.L., Vásquez-Martínez, R., Arroyo, L., Baker, T.R., Killeen, T., Lewis, S.L., et al., 2002. Increasing dominance of large lianas in Amazonian forests. Nature 418, 770–774.

Phillips, O.L., van der Heijden, G., Lewis, S.L., López-González, G., Aragão, L.E.O.C., Lloyd, J., et al., 2010. Drought–mortality relationships for tropical forests. New Phytol. 187, 631–646.

Phoonjampa, R., Koenig, A., Borries, C., Gale, G.A., Savini, T., 2010. Selection of sleeping trees in pileated gibbons (Hylobates pileatus). Am. J. Primatol. 72, 617–625.

Poorter, L., van der Sande, M.T., Thompson, J., Arets, E.J.M.M., Alarcón, A., Álvarez-Sánchez, J., et al., 2015. Diversity enhances carbon storage in tropical forests. Glob. Ecol. Biogeogr. 24, 1314–1328.

Poorter, L., van der Sande, M.T., Arets, E.J.M.M., Ascarrunz, N., Enquist, B., Finegan, B., et al., 2017. Biodiversity and climate determine the functioning of Neotropical forests. Glob. Ecol. Biogeogr. 26, 1423–1434.

Rabinowitz, D., 1981. Seven forms of rarity. In: Synge, H. (Ed.), The Biological Aspects of Rare Plant Conservation. Wiley, pp. 205–217.

Redford, K.H., 1992. The empty forest: many large animals are already ecologically extinct in vast areas of Neotropical forest where the vegetation still appears intact. Bioscience 42, 412–422.

Richardson, V.A., Peres, C.A., 2016. Temporal decay in timber species composition and value in Amazonian logging concessions. PLoS One 11, e0159035.

Samejima, H., Marzuki, M., Nagamitsu, T., Nakasizuka, T., 2004. The effects of human disturbance on a stingless bee community in a tropical rainforest. Biol. Conserv. 120, 577–587.

Santos, A., Peres, C.A., Oliveira, M.A., Grillo, A., Alves-Costa, C.P., Tabarelli, M., 2008. Drastic erosion in functional attributes of tree assemblages in Atlantic forest fragments of northeastern Brazil. Biol. Conserv. 141, 249–260.

Santo-Silva, E.E., Almeida, W.R., Melo, F.P.L., Zickel, C.S., Tabarelli, M., 2013. The nature of seedling assemblages in a fragmented tropical landscape: implications for forest regeneration. Biotropica 45, 386–394.

Savage, J.M., 2002. The Amphibians and Reptiles of Costa Rica: A Herpetofauna Between Two Continents, Between Two Seas, first ed. University of Chicago Press, Chicago and London.

Scheffers, B.R., Evans, T.A., Williams, S.E., Edwards, D.P., 2014. Microhabitats in the tropics buffer temperature in a globally coherent manner. Biol. Lett. 10, 20140819.

Sergio, F., Caro, T., Brown, D., Clucas, B., Hunter, J., Ketchum, J., McHugh, K., Hiraldo, F., 2008. Top predators as conservation tools: ecological rationale, assumptions, and efficacy. Annu. Rev. Ecol. Evol. Syst. 39, 1–19.

Shanahan, M., Compton, S.G., 2001. Vertical stratification of figs and fig-eaters in a Bornean lowland rain forest: how is the canopy different? Plant Ecol. 153, 121–132.

Shima, K., Yamada, T., Okuda, T., Fletcher, C., Kassim, A.R., 2018. Dynamics of tree species diversity in unlogged and selectively logged Malaysian forests. Sci. Rep. 8, 1024.

Shiva, M.P., Jantan, I., 1998. Non-timber forest products from dipterocarps. In: Appanah, S., Turnbull, J.M. (Eds.), A Review of Dipterocarps: Taxonomy, Ecology and Silviculture. Center for International Forestry Research, Bogor, pp. 187–197.

Silvério, D.V., Brando, P.M., Bustamante, M.M.C., Putz, F.E., Marra, D.M., Levick, S.R., Trumbore, S.E., 2019. Fire, fragmentation, and windstorms: a recipe for tropical forest degradation. J. Ecol. 107, 656–667.

Simmons, N.B., Voss, R.S., 1998. The Mammals of Paracou, French Guiana: A Neotropical Lowland Rainforest Fauna. Part 1, Bats. Bulletin of the American Museum of Natural History, New York.

Siqueira-Filho, J.A., Tabarelli, M., 2006. Bromeliad species of the Atlantic forest of north-east Brazil: losses of critical populations of endemic species. Oryx 40, 218–224.

Sist, P., Mazzei, L., Blanc, L., Rutishauser, E., 2014. Large trees as key elements of carbon storage and dynamics after selective logging in the Eastern Amazon. For. Ecol. Manag. 318, 103–109.

Slik, J.W.F., 2004. El Niño droughts and their effects on tree species composition and diversity in tropical rain forests. Oecologia 141, 114–120.

Slik, J.W.F., Paoli, G., McGuire, K., Amaral, I., Barroso, J., Bastian, M., et al., 2013. Large trees drive forest aboveground biomass variation in moist lowland forests across the tropics. Glob. Ecol. Biogeogr. 22, 1261–1271.

Slik, J.W.F., Arroyo-rodríguez, V., Aiba, S., Alvarez-loayza, P., Alves, L.F., Ashton, P., et al., 2015. An estimate of the number of tropical tree species. Proc. Natl. Acad. Sci. U. S. A. 112, 7472–7477.

Steege, H., Pitman, N.C.A., Killeen, T.J., Laurance, W.F., Peres, C.A., Guevara, J.E., et al., 2015. Estimating the global conservation status of more than 15,000 Amazonian tree species. Sci. Adv. 1, e1500936.

Stephenson, N.L., Das, A.J., Condit, R., Russo, S.E., Baker, P.J., Beckman, N.G., et al., 2014. Rate of tree carbon accumulation increases continuously with tree size. Nature 507, 90–93.

Stibig, H.J., Achard, F., Carboni, S., Rasi, R., Miettinen, J., 2014. Change in tropical forest cover of Southeast Asia from 1990 to 2010. Biogeosciences 11, 247–258.

Stuntz, S., Ziegler, C., Simon, U., Zotz, G., 2002a. Diversity and structure of the arthropod fauna within three canopy epiphyte species in central Panama. J. Trop. Ecol. 18, 161–176.

Stuntz, S., Simon, U., Zotz, G., 2002b. Rainforest air-conditioning: the moderating influence of epiphytes on the microclimate in tropical tree crowns. Int. J. Biometeorol. 46, 53–59.

Sunderland, T.C.H., Obama, C., 1999. Preliminary market survey of the non-wood forest products of Equatorial Guinea. In: Sunderland, T.C.H., Clark, L.E., Vantomme, P. (Eds.), Non-wood Forest Products of Central Africa: Current Research Issues and Prospects for Conservation and Development. FAO, Rome, pp. 211–220.

Tabarelli, M., Lopes, A.V., Peres, C.A., 2008. Edge-effects drive tropical forest fragments towards an early-successional system. Biotropica 40, 657–661.

Tabarelli, M., Aguiar, A.V., Ribeiro, M.C., Metzger, J.P., Peres, C.A., 2010. Prospects for biodiversity conservation in the Atlantic Forest: lessons from aging human-modified landscapes. Biol. Conserv. 143, 2328–2340.

Terborgh, J., 1986a. Keystone plant resources in the tropical forests. In: Soule, M.E. (Ed.), Conservation Biology: The Science of Scarcity and Diversity. Sinauer, Sunderland, pp. 330–344.

Terborgh, J., 1986b. Community aspects of frugivory in tropical forests. In: Estrada, A., Fleming, T.H. (Eds.), Frugivores and Seed Dispersal, pp. 371–384.

Tews, J., Brose, U., Grimm, V., Tielbörger, K., Wichmann, M.C., Schwager, M., Jeltsch, F., 2004. Animal species diversity driven by habitat heterogeneity/diversity: the importance of keystone structures. J. Biogeogr. 31, 79–92.

Tripati, S., Sujatha, M., Rao, R.V., Rao, K.S., 2005. Use of timber in shipbuilding industry: identification and analysis of timber from shipwrecks off Goa coast, India. Curr. Sci. 89, 1022–1027.

van der Hoek, Y., Gaona, G.V., Martin, K., 2017. The diversity, distribution and conservation status of the tree- cavity-nesting birds of the world. Divers. Distrib. 23, 1120–1131.

van der Sande, M.T., Peña-Claros, M., Ascarrunz, N., Arets, E.J.M.M., Licona, J.C., Toledo, M., Poorter, L., 2017. Abiotic and biotic drivers of biomass change in a Neotropical forest. J. Ecol. 105, 1223–1234.

Vieilledent, G., Cornu, C., Sanchez, A.C., Pock-Tsy, J.-M.L., Danthu, P., 2013. Vulnerability of baobab species to climate change and effectiveness of the protected area network in Madagascar: towards new conservation priorities. Biol. Conserv. 166, 11–22.

Vieira, E., Monteiro-Filho, E.L.A., 2003. Vertical stratification of small mammals in the Atlantic rain forest of South-Eastern Brazil. J. Trop. Ecol. 19, 501–507.

Vitt, L.J., 1991. Ecology and life history of the scansorial arboreal lizard Plica plica (Iguanidae) in Amazonian Brazil. Can. J. Zool. 69, 504–511.

Voris, H.K., 1977. Comparison of herpetofaunal diversity in tree buttresses of evergreen tropical forests. Herpetologica 33, 375–380.

Wagner, K., Mendieta-leiva, G., Zotz, G., 2015. Host specificity in vascular epiphytes: a review of methodology, empirical evidence and potential mechanisms. AoB Plants 7, 1–25.

Waterman, P.G., 1994. Trees for drugs and other extractives: future prospects. In: Tropical Trees: Potential for Domestication, Rebuilding Genetic Resources. HMSO, London.

Wells, K.D., 2010. The Ecology and Behaviour of Amphibians, first ed. University of Chicago Press, Chicago.

Whitfield, S.M., Pierce, M.S., 2005. Tree buttress microhabitat use by a Neotropical leaf-litter herpetofauna. J. Herpetol. 39, 192–198.

Woods, C.L., Cardelús, C.L., Dewalt, S.J., 2015. Microhabitat associations of vascular epiphytes in a wet tropical forest canopy. J. Ecol. 103, 421–430.

Zidar, C., Elisens, W., 2009. Sacred giants: depiction of Bombacoideae on Maya ceramics in Mexico, Guatemala, and Belize. Econ. Bot. 63, 119–129.

Zotz, G., Vollrath, B., 2003. The epiphyte vegetation of the palm Socratea exorrhiza—correlations with tree size, tree age and bryophyte cover. J. Trop. Ecol. 19, 81–90.

CHAPTER EIGHT

Monitoring tropical insects in the 21st century

Greg P.A. Lamarre[a,b,g], Tom M. Fayle[a,c], Simon T. Segar[d], Benita C. Laird-Hopkins[a,b], Akihiro Nakamura[e], Daniel Souto-Vilarós[a,f], Shuntaro Watanabe[i,j], Yves Basset[a,b,g,h,*]

[a]Institute of Entomology, Biology Centre, Czech Academy of Sciences, Ceske Budejovice, Czech Republic
[b]Faculty of Sciences, University of South Bohemia, Ceske Budejovice, Czech Republic
[c]Institute of Tropical Biology and Conservation, Universiti Malaysia Sabah, Kota Kinabalu, Sabah, Malaysia
[d]Department of Crop and Environment Sciences, Harper Adams University, Newport, United Kingdom
[e]CAS Key Laboratory of Tropical Forest Ecology, Chinese Academy of Sciences, Yunnan, China
[f]Department of Ecology, Faculty of Science, Charles University Prague, Staré Město, Czech Republic
[g]ForestGEO, Smithsonian Tropical Research Institute, Balboa, Ancon, Panama
[h]Maestria de Entomologia, Universidad de Panama, Panama, Panama
[i]Field Science Education and Research Center, Kyoto University, Kyoto, Japan
[j]Faculty of Sciences, Kagoshima University, Kagoshima, Japan
*Corresponding author: e-mail address: bassety@si.edu

Contents

Abstract

Research has repeatedly shown that ongoing habitat loss and the increasing frequency of extreme climatic events have altered fundamental biological processes, threatening biodiversity and ecosystem functioning worldwide. However, the multitude of interacting factors underlying the impacts of these threats remain poorly understood in

Advances in Ecological Research, Volume 62
ISSN 0065-2504
https://doi.org/10.1016/bs.aecr.2020.01.004

tropical forests. This is unfortunate because the majority of terrestrial biodiversity resides in these habitats. The responses of organisms to environmental changes are often studied indirectly, for example, using discrete snapshots at different time periods at the same location. This is due to the challenges of compiling long-term data, especially for invertebrates. Evaluating how functionally important species assemblages will respond to anthropogenic threats requires continuous long-term monitoring. This is vital for arthropods because they are responsible for crucial ecosystem services and human food security. Here, we review existing arthropod monitoring schemes and consider some innovative avenues for future research that promise to improve monitoring of this important group in tropical systems.

1. Threats to insects in the Anthropocene

Recent reports suggest catastrophic declines in insect abundance with potentially dire implications for ecosystem functioning (Hallmann et al., 2017; Lister and Garcia, 2018; but see McGregor et al., 2019; Saunders et al., 2019; Willig et al., 2019). These declines mostly concern insect communities in temperate countries (Sánchez-Bayo and Wyckhuys, 2019), insect biomass in Europe and Puerto Rico (Dirzo et al., 2014; Hallmann et al., 2017; Lister and Garcia, 2018; Seibold et al., 2019), butterflies and moths in the United Kingdom (Long et al., 2016), insect pollinators worldwide (Goulson et al., 2015), or more generally the "associated biodiversity" that supports human food production through ecosystem services (FAO, 2014). In particular, pollinating insects have experienced significant declines for several decades in many parts of the world (Carvalheiro et al., 2013). This is a serious concern because pollination represents a critical ecosystem service (Garibaldi et al., 2013) and declines in pollinators have been directly linked with declines in the plants with which they interact (Potts et al., 2010). Most data indicating this decline originates from censuses of honeybee colonies and temperate bumblebees or from decreases in delivery of pollination services (Goulson et al., 2015). Case studies of wild pollinator population dynamics in tropical regions are rather infrequent and almost exclusively concern bees (Knoll, 2016).

The main threats to insect populations have already been identified. Habitat loss and fragmentation via agricultural expansion, land-use intensification and associated pollution (e.g. pesticides, light) are among the greatest threats to insects, causing population declines and shifts in community composition (Sánchez-Bayo and Wyckhuys, 2019). These threats are especially detrimental for endemic and specialized species that rely on specific host plants and/or habitats (Newbold et al., 2018). Intensive agriculture has also involved repeated or uncontrolled use of insecticides, such as neonicotinoids, which may have an impact on both target and non-target insect species (Forister et al., 2016). This insecticide use has been linked to population declines of economically important insects such as pollinators (Goulson et al., 2015; Sánchez-Bayo and Wyckhuys, 2019). Furthermore, the ongoing expansion of large-scale agriculture in the tropics may also negatively impact plant communities, shifting composition and altering the population dynamics of the insects feeding on them (Habel et al., 2019a). This is of particular importance, because most insects interact in some way with plants, and an estimated 85% of angiosperms, including most tropical tree species, are directly pollinated by insects (Ollerton et al., 2011). Most insect species are associated with living plants (Price, 2002) and are highly specialized in terms of food resources and microhabitats (Novotny and Basset, 2000). Hence, one of the greatest challenges facing modern agriculture is the development of low-impact, sustainable pest management and agriculture practices in order to promote more ecologically efficient food production (Laurance et al., 2014). In anthropogenically modified areas, ecologists have also shown that nocturnal pollination may potentially be disrupted by increasing light pollution, although this remains to be tested in tropical systems (but see MacGregor et al., 2015; Van Langevelde et al., 2017).

In addition to habitat degradation, climate change represents a considerable threat to insect communities (Chen et al., 2009; Deutsch et al., 2008). Globally, and for yet another year, 2018 was the warmest since records began (Cheng et al., 2019). In Central America, an increase in air temperature of 2–5 °C is expected by 2080 (Stocker et al., 2014). Extreme climatic events are also expected to occur with increasing frequency (Blunden and Arndt, 2017; IPCC, 2012, 2018; Ripple et al., 2020). While studies suggest that pronounced dieback may be triggered at 3–4 °C of global warming in Central American tropical forests (IPCC, 2018), there may catastrophic consequences for biomass productivity at just 1.5 °C increases (Lyra et al., 2017). These predicted changes may lead to local extinctions and/or alterations in the geographic distribution of species and hence, of local communities. Insects are particularly sensitive to climate change. This is in part due to their strong ecological links to plants; the slightest alteration to plant growth, abundance or phenology in a future disrupted climate can have severe consequences for associated insect communities (Pau et al., 2010, 2011). Individuals may either have to rapidly adapt or migrate in order to avoid extinction, with changes in insect populations and distributions ultimately leading to alterations in ecosystem functioning (Kaspari et al., 2015; Thomas et al., 2004). However, our knowledge of insect responses to climate change is still limited (Hegland et al., 2009; Memmott et al., 2007). What we know is almost exclusively drawn from temperate systems (Sánchez-Bayo and Wyckhuys, 2019), and mainly relates to Lepidoptera (Altermatt, 2010; Diamond et al., 2011; McGregor et al., 2019).

Invasive species, and more broadly, changes in the distribution of pests and pathogens may also threaten insect populations, resulting in changes in ecosystem processes (Bohan et al., 2017a; Simler-Williamson et al., 2019; Yamamura et al., 2006). Changes in the populations of natural enemies, for example, are likely to have adverse effects on the regulation of primary consumers such as insect pests. Natural enemies provide a crucial service and their loss can be irreversible (Salcido et al., 2020) with negative consequences for human food security (FAO, 2014). Both insect pests and insect-spread pathogens can also negatively impact tree growth, causing mortality, altering forest structure and composition and potentially shifting ecosystem processes (Simler-Williamson et al., 2019).

The multiple interacting threats resulting from anthropogenic activities have negative consequences not only for insects themselves but also for the interacting species that rely on them. However, the responses of insects to anthropogenic threats are, even for common species, far from well understood in tropical ecosystems. In turn, long-term, continuous monitoring is urgently needed in tropical regions in order to test whether the results from studies in temperate systems showing dramatic declines in insect populations apply globally (Wagner, 2019).

2. Why monitoring insects is crucial in tropical ecosystems

Tropical entomologists are in broad agreement that we have only quantified a small proportion of insect diversity (Stork, 2018). Understanding insect biodiversity in tropical ecosystems is of paramount importance for a number of reasons. First, insects represent the largest proportion of macroscopic terrestrial animal life, both in terms of species richness (Stork, 2018) and biomass (Pimentel and Andow, 1984). Second, the greatest diversity of both insects and the species with which they interact are observed in tropical rainforests (Basset et al., 2012). Third, insects play crucial roles in forest dynamics, through interactions such as herbivory, seed predation, pollination, parasitism/mutualism, decomposition and nutrient cycling (Strong et al., 1984). Fourth, the ecosystem services that wild insects provide are of significant economic value, worth an annual estimated $57 billion in the United States (Losey and Vaughan, 2006) and $12 billion in Brazil (Giannini et al., 2015). Fifth, tropical insects may be particularly sensitive to global climate changes for several reasons: (1) The low seasonality of tropical regions is thought to result in evolution of insect species with narrower thermal limits (Kaspari et al., 2015). For example, tropical montane stream insects have a narrower thermal tolerance than temperate species, suggesting they will be particularly vulnerable to climate change (Polato et al., 2018). Projected air temperatures in the tropics appear to be close to the thermal limits of many insects, beyond which muscular coordination can be disrupted or lost (Deutsch et al., 2008). (2) Tropical insects are thought to be sensitive to desiccation and hence to drought or variation in rainfall. However, a recent study showed that canopy ants in tropical forest, when experiencing extreme temperatures have a more permeable cuticle that increases evaporative water loss making them more resistant to desiccation (Bujan et al., 2016). (3) Climate change may produce phenological mismatches with insect resources, such as host plants or prey (Facey et al., 2014; Singer and Parmesan, 2010). Only species that can track climate change by adjusting their phenologies to their resources will be able to persist under a changing climate (Cleland et al., 2012). This has been demonstrated mostly for temperate species whose host plants/prey are often only available seasonally but, to our knowledge, phenological mismatches due to climate change have not been studied in tropical forests.

3. Overview of insect monitoring programs in the tropics

In temperate ecosystems numerous long-term monitoring studies have been conducted and historical insect records have been compiled (Habel et al., 2019a), both over continuous time scales and using repeated snapshots with short time gaps (Augenstein et al., 2012; Hallmann et al., 2017; Wenzel et al., 2006). While this has helped to understand the responses of arthropod communities to land-use change and climate change (Sánchez-Bayo and Wyckhuys, 2019; Seibold et al., 2019; Wagner, 2019), the scientific rigour of such studies has been questioned (Leather, 2018; Thomas et al., 2019; Willig et al., 2019). In the tropics, the strongest long-term time-series data of insect species from a single location come from fruit-feeding butterflies (Grøtan et al., 2012; Leidner et al., 2010; Srygley et al., 2010, 2014; Valtonen et al., 2013). These studies, performed in locations buffered from the effects of agricultural practices, do not suggest a decline in abundance, in contrast to those observations from temperate ecosystems. In fact, insect communities in tropical rainforests may actually be experiencing an initial positive effect of rising temperature (Basset and Lamarre, 2019). A study conducted in Panama on butterflies indicated that decreases in rainfall during El Niño events benefitted lianas. Consequently, lianas produced more leaves which led to an increase in the population sizes of the butterflies that feed on them (Srygley et al., 2010, 2014). However, a "concentration effect" may occur, where following the loss of suitable habitats elsewhere, populations concentrate into remaining favourable and less disturbed habitats (Spaniol et al., 2019).

Specialized interactions such as those between plants and their pollinators are particularly interesting study systems because phenological mismatch may occur (Vamosi et al., 2016). Roubik (2001) sampled euglossine orchid bees consistently between 1979 and 2000 in Panama in a relatively undisturbed forest. The assemblage showed no detectable change in diversity and no overall decline in abundance over the study period, but responses of individual species varied. Only 4 species declined, whereas 23 species remained unchanged and 9 species increased. These fluctuations appeared to be driven primarily by El Niño events, which produced a floral resource flush in the dry season, followed by an increase in bee populations (Roubik, 2001). A similar pattern has been found for nocturnal bees between 1977 and 1983 (Wolda and Roubik, 1986) with no directional changes in bee abundance but with fluctuations probably relating to seasonal flower abundance (see also Wolda, 1992). In Brazil, a study conducted on orchid bees for 7 years did not show any significant decline or any effect of climate anomalies on the assemblage (Knoll, 2016).

Other insect groups studied reveal varying patterns in the tropics. In French Guiana, Dejean and colleagues (2011) studied the responses of social wasps in plantations between 1997 and 2003. These authors found that recent climate fluctuations may have reduced the number of wasp colonies because of lowered wasp resistance to parasites and pathogens. Eleven years of monitoring tropical ants in an Ecuadorian rainforest revealed that overall community species richness remained relatively constant, showing a high temporal turnover of species but with no or only weak evidence of directional changes towards a new community (Donoso, 2017). In Borneo, neither abundance nor species richness of scarab beetle communities changed in undisturbed forests between 1994 and 1999 (Kishimoto-Yamada and Itioka, 2013). On Barro Colorado Island (BCI, Panama), even small differences in the mean annual air temperature and precipitation (Anderson-Teixeira et al., 2015) appear to affect the populations of certain species of saturniid moths (Basset et al., 2017) and reduviid assassin bugs (Lucas et al., 2016) with contrasting directional responses among taxa on BCI (Fig. 1(7)).

Clearly, we currently lack sufficient insect monitoring data to either refute or support claims of global insect decline with respect to tropical regions (Janzen and Hallwachs, 2019). Our lack of knowledge concerning the response of tropical insects facing climate change and other major threats is in part due to lack of funding for long-term projects of this kind. We are not the first to note the challenge of finding support for long-term projects within the constraints of standard funding cycles. Our knowledge of the past, present and future directions of tropical insect populations depend greatly on ongoing support for taxonomists, museums and biodiversity studies. Therefore, because comprehensive data for tropical insects are lacking, caution is needed regarding extrapolations of temperate results to global scales (Simmons et al., 2019; Willig et al., 2019).

4. Case study: The ForestGEO arthropod initiative

The ForestGEO Arthropod Initiative is an ongoing long-term monitoring scheme for tropical arthropods. The network of permanent forest dynamic plots monitored by the Forest Global Earth Observatory (ForestGEO) is an international collaborative network of scientists with 69 forest research sites across the Americas, Africa, Asia, Europe, and Oceania dedicated to advancing long-term studies in forest dynamics (Anderson-Teixeira et al., 2015). Within these forest plots each tree >1 cm DBH is tagged, identified and its DBH measured once every 5 years. This program provides a unique opportunity for long-term monitoring of insect populations and can capitalize on existing floristic, phenological and climatic data from these plots, thus simplifying efforts to study tropical insects and their interactions with plants (Godfray et al., 1999).

The ForestGEO Arthropod Initiative is currently monitoring focal arthropod groups at seven such permanent tropical plots. Here we present a case study of one of these sites, Barro Colorado Island (BCI) in the Panama Canal. The site receives an average annual rainfall of 2662 mm, with annual average daily maximum and minimum air temperatures of 31 and 23.6 °C, respectively. Since 1981 BCI has already experienced an increase of 0.36 °C

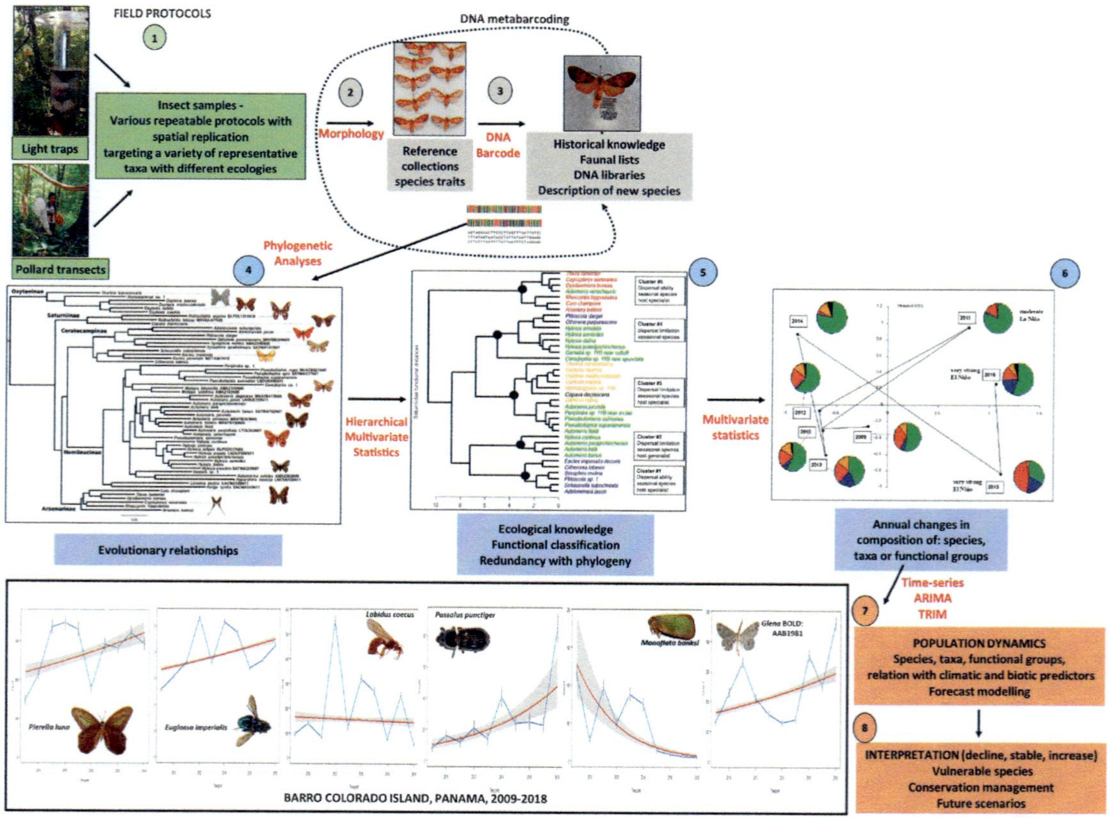

Fig. 1 Workflow of the ForestGEO Arthropod Initiative monitoring scheme for long-term monitoring of population dynamics for common, poorly-known and species-rich insect assemblages. *Modified from Basset, Y., Lamarre, G.P.A., Ratz, T., Segar, S.T., Decaëns, T., Rougerie, R., Miller, S.E., Perez, F., Bobadilla, R., Lopez, Y., Ramirez, J.A., Aiello, A., Barrios, H., 2017. The Saturniidae of Barro Colorado Island, Panama: a model taxon for studying the long-term effects of climate change? Ecol. Evol. 7 (23) 9991–10004, https://doi.org/10.1002/ece3.3515 and Basset and Lamarre (2019).*

in mean annual temperature and a 17.9% increase in mean annual precipitation (Anderson-Teixeira et al., 2015). The principal objective of the ForestGEO Arthropod Initiative is to detect long-term changes in the abundance and composition of focal assemblages of arthropods, driven primarily by climatic cycles and changes, as opposed to short-term stochastic changes (Basset et al., 2013, 2017). Initiated in 2009, the field protocol enables the quantification of population dynamics for 19 focal arthropod taxa including 10 Lepidoptera taxa, both diurnal (Hesperiidae, Riodinidae, Nymphalidae, Pieridae, Papilionidae, Lycaenidae) and nocturnal (Saturniidae, Geometridae, Arctiinae, Pyraloidea); 3 Coleoptera taxa (Dynastinae, Passalidae, Platypodinae); 2 Hemiptera taxa (Flatidae, Reduviidae); 2 Hymenoptera taxa (Euglossini, Halictidae) and 2 social insect taxa, ants and termites (including workers and alates; see Box 1 for a

BOX 1 Monitoring social insects in the tropics.

Social insects represent core ecological components of ecosystems, with high biomasses, particularly in the tropics. They drive ecosystem processes such as pollination (bees, wasps), predation (ants, wasps), soil turnover (termites, ants), seed dispersal (ants), and decomposition (termites). As with other tropical arthropods, they experience a range of anthropogenic threats, which are expected to alter the ecosystem processes that they mediate. Hence, long-term monitoring of social insect populations and communities is vital. However, monitoring of social insects involves some considerations that do not apply to the usual methods used for monitoring of solitary organisms. In order to facilitate inclusion of social insects into long-term monitoring programs for tropical arthropods, we discuss these considerations here.

Which considerations are specific to monitoring social insects?

Social insects operate from colony centres that move relatively infrequently. In this sense, social insects are similar to plants (Andersen, 1991), as they represent spatially stable colonies, each using resources in the local area and disseminating propagules to reproduce (winged reproductives for social insects, seeds for plants). These propagules then experience strong competition from nearby colonies while trying to initiate a new colony. Hence, the success of long-term monitoring programs for trees, for example the ForestGEO network (Anderson-Teixeira et al., 2015), should be an inspiration for long-term monitoring of social insects. For ants and termites, presence of a wingless worker means that a colony has successfully established in the local area, within close distance of the sampling point. As a caveat, the limits of mobility for entire colonies of social insects are poorly-known, and colony monitoring of species with high

Continued

BOX 1 Monitoring social insects in the tropics.—cont'd

colony mobility may not reflect colony mortality. For example, species nesting in ephemeral habitats, such as ants living in small dead twigs, appear to move more frequently than those in more permanent habitats (Byrne, 1994), and hence the latter should be targeted for long-term colony monitoring in the tropics. Furthermore, presence of colonies does not necessarily mean that there is successful reproduction occurring, and hence long-lived but non-reproductive colonies might be misinterpreted as representing healthy populations. This issue could be avoided by monitoring population structure (i.e. abundances of colonies at different life stages), rather than just recording the presence of individuals. Light trapping for reproductive individuals should also be included to help overcome this issue.

Sampling design and statistical analysis

Sampling methods used for social insects can either be based on whole colony sampling or sampling of individual workers. Intensive methods that involve searching for colonies can be considered to create a long-term scheme to parallel that used in permanent vegetation plots (Anderson-Teixeira et al., 2015). If this is not practical (e.g. cryptic nests that are difficult to survey in rainforests), then a scheme based on sampling individual workers should be considered (e.g. Winkler leaf litter sifting). However, individual workers from a single sample cannot be considered as independent data points, since they are all likely to come from a single colony. One method to overcome this is to use presence-absence data for analyses (Agosti et al., 2000). Such data can be considered to relate to colony densities if distances between sampling points are sufficient to prevent a single colony from foraging at multiple sampling points. Hence, spatial arrangement of sampling points should take into consideration foraging ranges of the group. Statistical methods appropriate for presence-absence data should then be used. These include, diversity indices such as species richness, indices for differences in composition between samples such as Sorensen's index (Fayle et al., 2010), incidence-based species richness estimators such as ICE or Chao 2 (Longino, 2000), Generalized linear models with binomial errors for presence/absence of individual species, or Poisson errors for species counts. Abundances of social insects may still be worth recording, since abundance and biomass are likely to reflect the degree to which they are able to mediate ecosystem processes. Abundance at baits can also give insights into behavioural dominance of different species (Drescher et al., 2011; Dworschak and Bluethgen, 2010). Finally, since the objective of monitoring (of communities) is to reliably capture as many species in the community as possible, optimisation studies can be carried out. These allow selection of particular combinations of sampling methods to ensure that the greatest number of species is collected given the resources available (Yusah et al., 2012).

discussion of considerations relating to monitoring social insects). All field protocols targeting the above-mentioned taxa are standard, reproducible entomological methods (Basset et al., 2015) that reflect real population changes rather than changes in insect activity (Basset et al., 2013). To date, the ForestGeo Arthropod Initiative at BCI has recorded more than half a million specimens representing over 2300 focal species. The initiative has accumulated 11 years of continuous data, including four seasonal replicates per year, so that time-series to date include more than 40 data points. In Fig. 1, we describe the general workflow of the monitoring components with time-series examples for common species (Basset and Lamarre, 2019).

Our first attempt at understanding climate-driven changes in tropical insects focused on saturniid moths of BCI, Panama (Basset et al., 2017; Box 2). Our data on species abundance and composition suggested that the year 2011 (a weak-to-moderate La Niña resulting in increased rainfall) was different from other years, with distinct peaks in populations for six of the 14 saturniid species studied (see also Box 2). The number of saturniid species with peak in abundance for each sampled year appeared to be related to the magnitude of these anomalous events (La Niña, El Niño), which was also reflected in changes in faunal composition.

These continuous long-term datasets are crucial for two reasons: (a) because continuous or frequent monitoring (as opposed to resurveys or snapshots at longer time periods; see Chen et al., 2009; Colwell et al., 2008; Donoso, 2017) is critical to measure annual change in insect populations, to evaluate the main drivers of arthropod responses to climate changes and anomalies, and to estimate the detection probability of vulnerable and resilient species (Basset et al., 2017; Newbold et al., 2018; Roubik, 2001) and (b) because tropical insects often exhibit short generation times, these data already potentially represent a significant number of generations (as low as eight generations per year for some insect pests; Nair, 2007). This contrasts with plant monitoring, where for some taxa monitoring schemes still cover much less than the time of a single generation.

The ForestGEO Arthropod initiative proposes a comprehensive framework that we hope will serve as a baseline study to be replicated at other monitoring sites. Because it would be unrealistic to monitor all taxa in tropical forests, we explore three different approaches to the selection of indicator taxa that may be highly sensitive to climate changes but are still representative of many insect species sharing taxonomical (Fig. 1(3)), evolutionary (Fig. 1(4)) or functional relationships (Fig. 1(5)). The workflow includes exploration of the temporal dynamics of tropical insect

BOX 2 Functional traits of tropical moths and climate change: an example.

Saturniidae (Lepidoptera) are an interesting model taxon in comparison to other insect groups since the adults do not feed, have short life spans, and, particularly for the heavy-bodied species, are unable to fly until their thoracic muscles have reached a temperature of 35 °C, achieved by muscular 'shivering' (Blest, 1960). For saturniids, rising air temperature means allocating less time to shivering and more time to finding mating partners and searching for suitable oviposition sites. Warming may represent a significant advantage for species with particularly short adult life spans (Janzen, 1984). For an 8-year period (2009–2016), we analysed population dynamics of saturniid moths on BCI and tested the importance of using functional classification in climate change studies (Basset et al., 2017). First, we generated a morphological and life-history traits-by-species matrix from collected specimens to compute a principal component analysis (PCA) prior to the construction of the functional hierarchical classification. Second, the different components (axes) of the PCA were used as combinations of traits to define a Euclidean functional space with reduced uncorrelated dimensions, where each species is represented by its PCA axis scores (Hawking and New, 2002). This approach is an efficient way to produce high quality functional spaces where the weight of each synthetic trait is relevant to the fraction of the total trait variance it explains (Basset et al., 2017; Maire et al., 2015). Then, we computed the complete functional classification using hierarchical *daisy* clustering methods based on functional clusters defined by the saturniid species scores on all axes (Fig. 1(5)). By exploring the functional space of these tropical moths, we showed that the species sharing life-history traits related to dispersal (body size), annual seasonality (phenology), and host plant specialization (plant use) can be clustered into five distinct functional groups (Fig. 1(5)). The use of the functional traits also revealed that tropical saturniids display contrasting functional attributes and that the functional groups exhibit different directional responses, possibly related to climate changes or extreme events (such as in the years 2011 and 2015). Species belonging to one distinct functional group, including large species with good dispersal ability, high host specialization and high seasonality, collectively showed a significant moderate increase over time. We speculate that this may form the basis for explaining the positive effect of rising air temperature on certain saturniid populations (Basset et al., 2017). Moreover, our results also emphasize the need to study below the species level and start exploring intraspecific responses in tropical insects that may provide clues about future adaptation and local responses. The use of functional traits and groups over taxonomic categories may also improve our ability to interpret complex results when confronted with diverse groups such as tropical insects. Furthermore, this simplification is potentially useful when dealing with a multitude of species, in order to present more succinct results to the general public and decision-makers.

communities (Fig. 1(6)) and the directional responses of contrasting trophic groups (related to their function in ecosystem processes) to climate changes and anomalies (Fig. 1(7)).

Finally, the initiative hopes to deliver an easily-interpretable synthesis of the future scenarios for tropical insect communities in order to encourage the rapid adoption of conservation measures for vulnerable species and their habitats (Basset and Lamarre, 2019; Fig. 1(8)). In the next section, we discuss some innovative avenues for future research that promise to improve arthropod monitoring in tropical systems.

5. Addressing present and future challenges in monitoring tropical insects

5.1 Inclusion of functional traits in monitoring programs

Traditional community ecology uses the identity of species to understand how changes in habitat condition influence community composition and structure. This approach, however, has its limitations. Critically, the biology of a given species is often unknown and, especially in tropical regions, many species remain undescribed (Basset et al., 2012; Stork et al., 2015). The use of functional traits has been well established in community ecology, because unlike species identities, functional traits are related to species performance and ecosystem processes (McGill et al., 2006; Weiher, 2011). Measurement of functional traits allows inferences to be made about a species' role in the ecosystem, without prior natural history knowledge. However, ecological traits related to a species' function in the ecosystem are numerous and the choice of which functional traits to include should be made based on the underlying mechanisms that constrain species performance and influence ecosystem processes (Weiher, 2011). Moretti et al. (2017) summarized 29 functional traits that can be measured for terrestrial invertebrates and are categorized into 5 types: morphology, feeding, life history, physiology and behaviour.

Functional traits are now increasingly being included in studies of species' response to climate change. Most morphological traits can be measured a posteriori using preserved specimens collected from previous surveys or monitoring. Other functional and life-history traits, such as feeding guild, clutch size, and life span, may be obtained from existing literature if the biology of species is well known, although such information is unlikely to be available for most tropical biota. However, most functional traits related to physiology (e.g. metabolic rate) and behaviour (e.g. activity time and locomotion speed) require live specimens. Hence such traits must be selected a priori before monitoring starts (e.g. increase in metabolic rate with increasing temperature, Dillon et al., 2010). Continuous traits are measured from a number of individuals of the same species and average values are used for analysis. Intraspecific variation is common but generally of smaller magnitude than interspecific variation (Weiher, 2011).

Intraspecific changes in functional traits should be monitored through time, accounting for other factors such as genotypic variation, phenotypic plasticity, acclimation and ontogeny (Weiher, 2011). Morphological traits such as body size (Fontanilla et al., 2019) and body colour (Dalrymple et al., 2017; Xing et al., 2018) can be easily measured. These traits are interpreted as phenotypic correlates of functional attributes (Weiher, 2011). For example, temperature generally decreases with increasing elevation, and ectotherms at higher elevations need to become efficient in gaining heat while minimizing heat loss. Bishop et al. (2016) investigated body colour and body size of ant assemblages along elevational transects and found that ants became darker (to maximize heat gain) and larger (to minimize heat loss) with increasing elevation. A recent study on butterflies also suggests that univoltine butterflies are less sensitive to climatic extremes than are multivoltine species (McDermott et al., 2017), emphasizing the importance of including life-history traits in studies related to climate change. Species that share similar functional attributes (not necessarily those that are closely phylogenetically related) may have similar responses to climate fluctuations and anomalies, and these responses may be easier to interpret than more traditional species-specific approaches (Didham et al., 1996; Valtonen et al., 2013). In temperate systems, functional categories have been used for Lepidoptera as relevant surrogates for life-history changes induced by climatic shifts (Pau et al., 2010). Within the context of climate change, it is therefore necessary to identify the key functional traits associated with mechanisms by which single species or groups of species influence ecosystem properties (Lavorel and Garnier, 2002). However, in the tropics, where the knowledge of life-history traits is limited, relevant species traits have often been limited to a few taxa, such as butterflies or ants (Diamond et al., 2011; Slade et al., 2013).

Therefore, compiling records of functional traits and life-history traits, sharing databases publicly and initiating large-scale collaborations among tropical, behavioural, evolutionary, ecological, and physiological entomologists is a fruitful future research direction.

5.2 Inclusion of tolerance to temperature and desiccation in monitoring programs

Ectotherms are particularly sensitive to changes in environmental temperatures and precipitation patterns as many of their physiological functions are influenced by temperature and water balance (Benoit and Denlinger, 2010; Deutsch et al., 2008; Easterling et al., 2000; Paaijmans et al., 2013). A growing body of research relates to determining the tolerance of ectotherm species, including insects, to changes in temperature and desiccation risk (Pinsky et al., 2019, Terblanche et al., 2011, 2017).

Changes in temperature can affect an insect's phenology, geographical distribution and performance (Deutsch et al., 2008). One common physiological measurement consists of experimentally quantifying the critical thermal maximum (CTmax) and minimum (CTmin), the temperatures at which an ectotherm loses motor control or locomotion (Bujan et al., 2016; Kaspari et al., 2015, 2016). This information is vital to understand the energy and the (future) cost allocated to thermoregulation among functionally important species in tropical forests. Work performed by García-Robledo et al. (2015) in Costa Rica focused on thermal tolerance in rolled-leaf beetles along a tropical elevational gradient. Leaving beetles in falcon tubes in a water bath to measure CTmax, these authors observed that species at low elevations are more tolerant to high temperature than those at mid and high elevations. A similar methodology was used to determine the difference in thermal tolerance between nocturnal and diurnal ants in Mexico (García-Robledo et al., 2018). The authors demonstrated that ants from tropical rainforests show lower tolerances to high temperatures than those from semi-desert and subtropical dry forests. Kaspari et al. (2015) showed that CTmax and desiccation resistance are inversely related in canopy ants, but positively correlated in understorey ants, that larger ants have broader thermal tolerances than smaller ants, and that canopy nesting species are less sensitive to temperature, supporting the thermal adaptation hypothesis (Bujan et al., 2016; Kaspari et al., 2015, 2016). Ideally, insect physiological traits for species that are part of long-term monitoring programs should be measured under manipulated air temperature and humidity based on projected climate scenarios (IPCC, 2018). For example, the lower and the upper critical temperatures (CTmax and CTmin) of larvae and adult insect species across a range of temperature and humidity conditions could be determined using thermo-incubators. In turn, these data may help to refine functional classification and the predictive time-series models and evaluate how species will cope with climate change in the future and their risks of extinction. For example, we would expect that species that are already close to their upper thermal limit would experience negative impacts on their populations as a result of increases in temperature due to climate change.

5.3 Long-term monitoring of interaction networks in tropical ecosystems

There is increasing understanding regarding the responses of interaction networks to both anthropogenic and non-anthropogenic environmental factors, for example changes with altitude (Maunsell et al., 2015; Morris et al., 2015; Plowman et al., 2017), latitude (Dalsgaard et al., 2011; Forister et al., 2015), habitat fragmentation (Emer et al., 2013), and habitat conversion (Tylianakis et al., 2007). Although this information is valuable, there remains no long-term program monitoring network structure changes in the tropics, despite calls for such projects (Kaiser-Bunbury and Bluthgen, 2015).

Because of the lack of long-term monitoring data, we must extrapolate from responses of networks to existing gradients of climate and anthropogenic habitat modification. This is vital because many aspects of ecosystem functioning are mediated through interactions, and hence network structure can affect ecosystem processes (Poisot et al., 2013). Furthermore, understanding network structure can give insights into ongoing ecosystem stability, because network structure can predict impacts of cascading extinctions (Dunne et al., 2009), for example, through identification of keystone species (Ebenman et al., 2004). Many existing studies relate to changes in host specificity along environmental gradients and can provide clues to climate change related network responses. These shifts in specificity can be driven to a great degree by the well-known large-scale changes in species richness, the most comprehensively studied of which being tropical-temperate latitudinal differences. For example, parasitoid networks do not vary with latitude, in any of a range of specificity metrics, once variation in host diversity is accounted for (Morris et al., 2014). A similar pattern of increased numbers of interactions with lower-level species by each higher-level species is observed for bird-plant frugivory networks, in which birds are less specialized in the tropics, because there are more available fruit species to feed on there (Dalsgaard et al., 2017). Such patterns are also observed in relation to altitude; for both herbivore-parasitoid networks (Maunsell et al., 2015), and mutualistic symbiotic ant-plant networks (Plowman et al., 2017) reduced host diversity results in reduction in generality with increasing altitude.

Hence, changes in the diversity of lower-level species in trophic networks are likely to be an important driver of overall network structure (Pellissier et al., 2013). For systems in which long-term monitoring indicates reductions in species richness, we would hypothesize that this will result in increases in the level of observed specialization.

However, all of the above-mentioned studies rely on snapshots of network patterns at single points in time, and hence long-term impacts of environmental changes on network structure are likely to be missed. An understanding of how interaction networks respond to different kind of threats (land-use change, including habitat fragmentation, and climate anomalies) and how these responses vary between network types (e.g. pollination, parasitoidism), is urgently needed if we are to mitigate against irreversible impacts.

5.4 Evolutionary drivers of network re-wiring in the face of climate changes

Species are not independent data points (Harvey and Pagel, 1991) and often share functional similarities with their relatives (Wiens et al., 2010). It is important to account for these phylogenetic patterns as part of monitoring schemes. Species level relationships are being increasingly well estimated through the use of sequence data and the construction of phylogenetic hypotheses. For example, grafting methods linking shallow sequencing (e.g. COI for insects) to multigene datasets increase connectivity for understudied tropical groups (Basset et al., 2017; Chesters, 2017). Armed with phylogenies, researchers can harness the predictive power of trait conservatism, ideally in combination with detailed functional measurements with which to qualify this assumption. Basic quantification of phylogenetic signal in species extinction risk over time, for example, can help to identify lineages pre-disposed to local extinction when monitoring data are comprehensive enough (Basset et al., 2015). Moreover, we can predict species' responses to climate and habitat change, tease out key traits or even map the response of local species to invasive ones (Pearse and Altermatt, 2013). This approach is critical as we race to understand the loss of species that may not even have a name yet (centinelian extinction), let alone an entry in a functional database. Extrapolating species level responses or their position in a network based on phylogenetic and trait data (Segar et al., 2017) also allows us to understand the eco-evolutionary drivers of network re-wiring in the face of global change. There are several predictive approaches currently available to researchers.

Phylogenetic Generalized Linear Mixed models (PGLMMs) (Ives and Helmus, 2011; Pearse et al., 2015) and Generalized Estimating Equations (Paradis and Claude, 2002) offer an extremely flexible framework for use in tropical arthropod monitoring schemes because they can both quantify the role of phylogenetic, functional and temporal correlations. Bayesian implementation of generalized mixed models is similarly promising and flexible (Hadfield, 2010). We anticipate that machine learning (e.g. phylogenetic 'Flexible Discriminant Analysis', phyloFDA) will provide exciting opportunities for fully predictive network and species level approaches for interrogating monitoring datasets (Motani and Schmitz, 2011; Pichler et al., 2019). Scripts for these analyses should be user friendly for local stakeholders and embedded in an accessible workflow. In combination with increasingly automated monitoring and molecular identification tools (Bohan et al., 2017a; Keck et al., 2018) a predictive phylogenetic framework will help to understand both historical responses and forecast future trends in the Anthropocene.

5.5 Developments of high-throughput DNA sequencing technologies

Beyond phylogenetic relationships between species, documenting changes over time has been a central goal of evolutionary biology, particularly when studying genetic responses of species to environmental change (Wandeler et al., 2007). Often, these studies depend on present day genetic snapshots with little room for manipulation beyond common garden or translocation experiments, or studies performed in controlled laboratory conditions (Flanagan et al., 2018). The use of natural history collections (Bi et al., 2013) or long-term monitoring programs (Silvertown et al., 2006) allows researchers to extend the temporal scale that these studies encompass. This enables them to identify periods of loss of genetic diversity, reductions in effective population sizes, changes in gene flow patterns and even measure rates of introgression from non-native introductions to native populations (reviewed in Wandeler et al., 2007).

High-throughput DNA sequencing technologies now allow for extending these studies to much larger scales, both in terms of genomic regions and numbers of samples, at rapidly decreasing costs and for non-model organisms. Even though many of these historical long-term collections were not done with genetic studies in mind, some sequencing techniques are able to retrieve meaningful data even from low quantities of degraded DNA (Staats et al., 2013). The continued developments in preservation of field samples, now with DNA studies in mind, has set the stage for genetic monitoring as a way of studying population changes, particularly during this period of extreme anthropogenic environmental change (Flanagan et al., 2018; Schwartz et al., 2007). For instance Dormontt et al. (2014), using historical and contemporary collections, were able to reconstruct the invasion history of *Senecio madagascariensis* in Australia since its introduction in 1918. There was also an increase in genetic diversity, likely resulting from a second introduction event sometime during the 20th century. This resolution would not be possible from surveying present day samples alone, highlighting the importance of historic collections and long-term monitoring programs (Dormontt et al., 2014). Similarly, the level of detail from high-throughput sequencing technologies continues to increase allowing us to characterize species interactions, as is the case with network reconstructions based on DNA metabarcoding (Gous et al., 2019).

Historic collections and those from long-term monitoring not only record the time and place of individual collection, but may also include information regarding their host plant association and surrounding habitats. For example, pollen carried by flower-visiting insects can be used to identify recently visited flowers, giving us an indication of plant community structure and visitation rates (see Fig. 2). Gous et al. (2019) used this approach by applying DNA metabarcoding to pollen extracted from South African bees sampled over a 93-year period. Although taxonomic resolution was generally low (we lack a plant universal barcode with comparable species level resolution to COI), the authors were able to identify the pollen carried by these bees to genus level and were even able to identify plant taxa that were not listed regional databases (Gous et al., 2019). Genetic methodological, technological and statistical techniques continue to develop for the study of taxonomical, functional and evolutionary responses of terrestrial biodiversity to climate-driven changes.

Overall, new molecular methods such as DNA metabarcoding are very promising for the elimination of taxonomic barriers allowing biological monitoring at low cost (Yu et al., 2012). Such methods are especially urgent for highly diverse tropical forests. Metabarcoding represents a powerful and less labour-intensive approach for the detection and monitoring of indicator species and functionally important groups (Basset et al., 2017; Beng et al., 2016). This approach aims to compare the occurrence of these groups over time to study their response to environmental changes and climatic anomalies. For example, adult moths associated with flowering plants can be sampled using light traps and have the potential to be routinely identified using

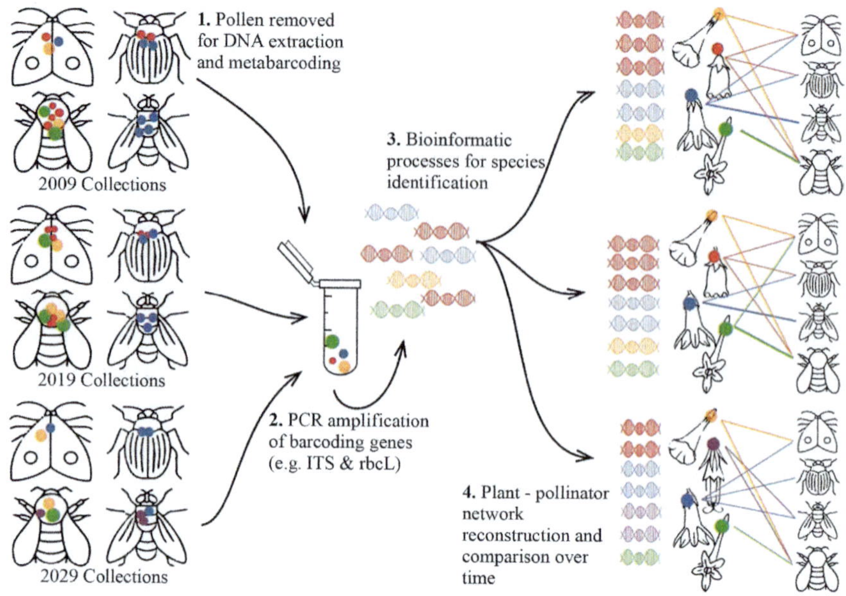

Fig. 2 Reconstructing pollination networks from monitoring and/or historic collections. Pollinating insects often carry a snapshot of the plant community they visited at the time of capture. Pollen DNA metabarcoding techniques allow researchers to extract pollen DNA from individuals and amplify standard barcoding genes. High-throughput sequencing can generate millions of short reads which can be sorted to high taxonomic resolution and used to reconstruct pollination networks from the sampled individuals. Depending on the timespan of these monitoring efforts, researchers can potentially identify floral resource use by pollinators over different sampling periods (i.e. different seasons) and even trace the re-wiring of these networks due to alien plant introductions or loss of native species.

new molecular methods (Ji et al., 2013). The ForestGEO Arthropod Initiative (see above) seeks to implement and replicate new monitoring protocols for tropical insects based on molecular techniques at other monitoring sites (see Tang et al., 2015) and to thus explore the utility of metabarcoding in long-term monitoring. However, one of the first challenges ahead, prior to extending the use of metabarcoding in further monitoring sites, is to obtain clean, complete DNA barcode libraries for these focal tropical taxa, because results from metabarcoding studies are highly dependent on database quality.

5.6 Machine learning

Image recognition and classification techniques have made advances that hold promise for monitoring of tropical arthropods. Image-based species identification technology has progressed as a result of the development of computer vision technology. Computer vision works by enabling computers to see, identify and process images in the same way that humans do (Szeliski, 2010) using machine learning. Computer vision machine learning approaches can be categorized as being both (1) manual feature extraction and (2) deep learning (Fig. 3). The manual feature extraction approach consists of two phases: feature extraction and classification. In the feature extraction stage, features of the classification target in the image are extracted based on pixel values. The classification stage is used to calculate the probabilities of each image belonging to a range of classes, in this case species, or higher-level taxa. Common classification methods include machine learning algorithms such as Support Vector Machines (SVM), Random Forest (RF) and Artificial Neuronal Network (ANN).

Feature extraction aims to extract the target object accurately from images. However, development of these algorithms is still challenging and often present problems specific to the kinds of objects being identified. In the past, manually deriving characteristic features was the main method used. Generally, object feature description methods can be categorized as using either global features or local features (Martineau et al., 2017). Early research on insect recognition used global features based on colour, texture or geometric invariants, such as colour histograms, geometric shape (eccentricity, perimeter, area, etc.), or other relatively simple features (Liu et al., 2016). However, because simple features are very sensitive to rotation, scale, translation, and viewpoint changes, this classification method did not work well on objects with large intraspecific variation or high interspecific similarity. Moreover, these methods are often labour intensive because they require manual image manipulation in order to identify key features. To address this issue, researchers began using local feature-based representation of insect images, which allows learning with much less direct user interaction (Cheng and Guyer, 2012; Solis-Sánchez et al., 2011; Venugoban and Ramanan, 2014). The most popular of these local feature-based methods employ custom features, which are feature extraction algorithms such as Local Binary Patterns (LBP; Ojala et al., 2002), the Scale Invariant Feature Transform (SIFT; Lowe, 2004), Histogram of Oriented Gradients (HOG; Dalal and Triggs, 2005). For example, a local feature-based insect identification scheme using the SIFT descriptor to classify five common pest species in orchard habitats achieved 88.4% accuracy (Wen et al., 2009). A hierarchical combination model based on local feature and global feature models was able to classify eight insect species with accuracy of 86.6% (Wen and Guyer, 2012). However, these approaches require careful choice of

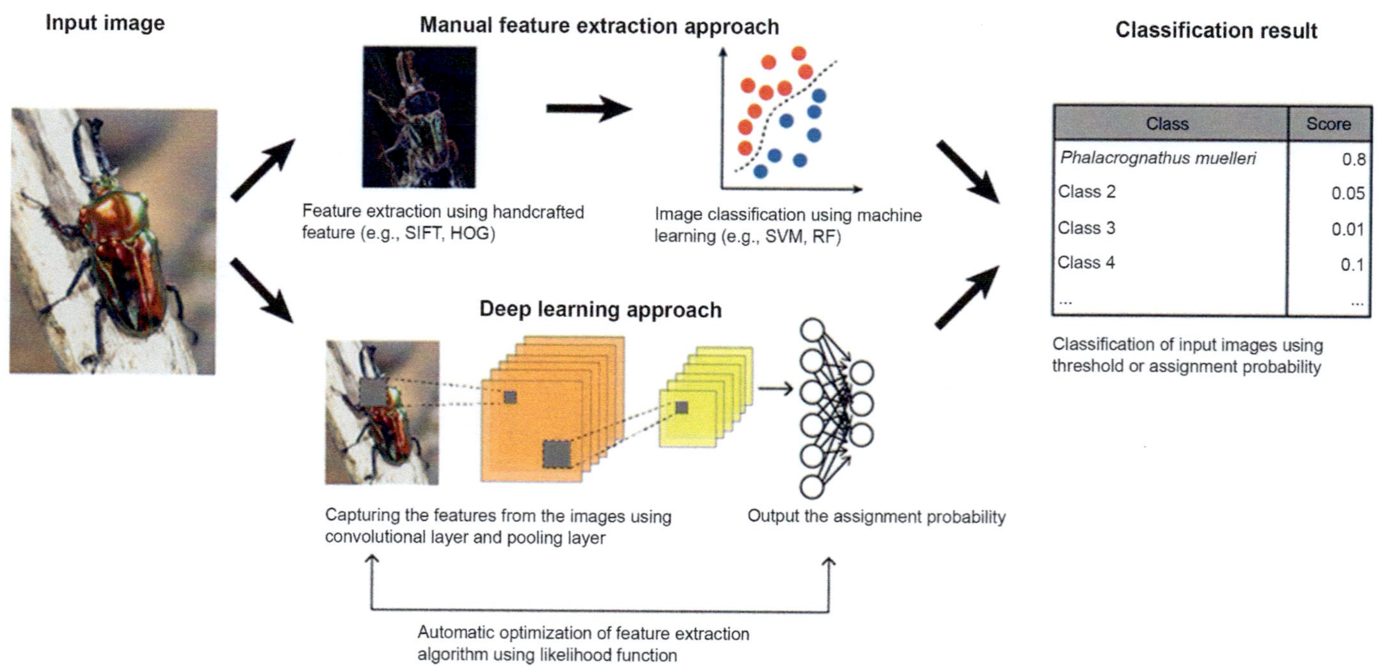

Fig. 3 Schematic diagram of computer vision machine learning pipeline for image-based species identification. In the manual feature extraction approach, image classification is performed by machine learning based on algorithms (e.g. handcrafted feature) designed by domain experts. On the other hand, in deep learning approach, convolutional neural networks automatically extract the feature and developing a robust classification model by learning a training data.

features, and a sophisticated design for the preprocess procedure (i.e. ways to aggregate them). Additionally, for many features it has not been possible to extract them manually in the past (Wäldchen and Mäder, 2018).

Recently, deep learning based approaches have allowed the construction of sophisticated recognition systems. Deep learning (DL) is a specific kind of machine learning and it builds upon ANN. DL can simultaneously optimize the feature extraction stage and classification stage through during the learning of training images. A strong advantage of DL is automatic feature learning; the method uses training data to categorize image content without a priori specification of image features (LeCun et al., 2015). Recently, DL has been employed in ecological studies (e.g. Carranza-Rojas et al., 2017; Norouzzadeh et al., 2018) and insect classification studies (e.g. Liu et al., 2016; Thenmozhi and Reddy, 2019). Liu et al. (2016) classified 12 species of paddy field pest insects with greater than 95% accuracy. Thenmozhi and Reddy (2019) combined DL and transfer learning for classification of 40 crop pest species and achieved over 90% accuracy. How these methods can be scaled-up to species-rich assemblage remain to be tested.

In addition to these developments in software, there are now massive repositories of ecological image data uploaded from field-based cameras or captured by citizen scientists available to implement public monitoring programs and for ecological analysis. For example, to date iNaturalist has archived 29 million images of 244,183 species taken by citizens and researchers (iNaturalist.org, 2019). Furthermore, some natural history museum collections have been digitized, which has resulted in large and historically relevant digital datasets. For example, the iDigBio portal currently provides more than 1.8 million georeferenced images of vascular plant specimens (Willis et al., 2017). Such large image datasets in combination with the recent advances in machine learning based-image recognition and classification technologies are bringing image-based species identification and its application for ecological monitoring closer to reality (Wäldchen and Mäder, 2018; Weinstein, 2017).

Machine learning can potentially identify the taxa of specimens prepared from ongoing monitoring programs. Such specimens are usually stored in boxes and therefore images can be obtained in bulk (i.e. capturing the image of all specimens in a specimen box). Alternatively, images of live specimens can be captured in the field by entomologists (e.g. capturing the images of insects on the white sheet of the light trap) or by the many scientists and other people around the world that carry a smartphone with a high-resolution camera. However, there remains the question as to what level of accuracy (correctly identifying specimens) and precision (taxonomic resolution) machine learning can achieve in monitoring programs. Martineau et al. (2017) surveyed 44 existing image-based insect classification publications and found that laboratory photography of insects is still the most widely used method, with positioning of insects being carried out manually in most cases. Martineau et al. (2017) also pointed out that there is still little work being done to develop insect identification systems that can be used in the field rather than in the laboratory. To evaluate the utility of image-based species identification techniques for insect monitoring, further research based on images taken in the field is necessary.

6. Conclusion

Current incentivization in modern agriculture encourages habitat loss, fragmentation and the use of herbicides and pesticides, all of which represent threats to insects. Extreme climate events, invasive species, pests, and pathogens are also threatening local insect populations and causing changes in ecosystem processes. However, species responses to individual anthropogenic threats in tropical regions are not yet well understood, mostly because we lack adequate monitoring schemes. Despite their ecological and economic importance, insects are still underrepresented in scientific research, leading to potential bias in our understanding of the functional ecology of ecosystems. The dramatic decrease in funding studies related to biodiversity science has also resulted in a lack of training, leading to a decrease in fundamental and applied knowledge in entomology. The public should be educated about the importance of insects (Lamarre et al., 2018). We need a bio-literate society that protects insects to ensure humanity's own survival (Basset and Lamarre, 2019).

This review highlights the fact that insects may be particularly sensitive to climate changes and so deserve more attention. Only continuous long-term monitoring will enable us to understand insect population dynamics in the face of future climate-driven changes. Functional traits can be used to delineate key functional arthropod taxa in tropical forests. Those focal insects (herbivores, pollinators, predators, social insects) need to be monitored in long-term programs using reproducible, standardized field protocols. This approach can then be proposed to stakeholders, cities and national parks, leading eventually to key insights into the conservation of vulnerable species and habitats. However, much work is needed to achieve this final goal. For example, we need to determine the tolerance of tropical ectotherms, including insects, to future changes in temperature and humidity and to know which species or functional groups are vulnerable to extinction. We also need to generate time-series across multiple sites and forecast models of insect assemblages to understand their future directional changes and to define early warning systems of the health of tropical forests. More work is also needed in distinguishing between directional changes and background variation in communities, and the nature of climate change events and their impacts on terrestrial ectotherms at various spatial scales (e.g. involving collaboration with climate scientists). More collaborative cross-disciplinary projects focused on protecting insect populations are vital to mitigate potential disruption or loss of the important ecosystem services they provide. Given current expectations of warming and an increased frequency of extreme weather events (such as the Amazonian and Australian fires of 2019 and 2020) these actions are urgent.

Exciting new avenues of research in interaction ecology and molecular biology can be explored in the context of climate change, once monitoring data is comprehensive enough. We will also be able to provide quantification of phylogenetic signals in species extinction risk over time and to identify tropical insect lineages pre-disposed to local extinction. New sequencing technologies also allow the identification of periods of loss of genetic diversity and reductions in effective population sizes. Metabarcoding is very promising for the elimination of some taxonomic barriers for tropical taxa of functional importance, allowing biological monitoring at low cost. Scientists are also exploring the use of machine learning and citizen science to potentially identify species from pinned or live specimens captured from monitoring projects. Finally we, the scientists, should empower insect research and teaching in tropical regions, which support the majority of terrestrial diversity. This can be done by reinforcing local capacity building in tropical locations where scientists are setting up projects, something that should be included in grant applications and perhaps included in legislation in these partner countries.

Acknowledgements

Over the years and in different countries, ForestGEO insect data were collected and processed efficiently by many research assistants, all collectively thanked here. G.P.A.L. is supported the Czech Science Foundation Junior Grant (GACR 19-15645Y) and an ERC (669609) during the writing of the article. A.N. is supported by National Natural Science Foundation of China General Program (31770472) and Chinese Academy of Sciences South-East Asian Biodiversity Research Institute Key Area Expansion Projects (Y4ZK111B01). S.T.S. acknowledges departmental support from Harper Adams

University. T.M.F. was supported by a Czech Science Foundation Standard Grant (19-14620S) (LifeWebs). B.L.-H. is supported by the Czech Science Foundation Junior Grant (GACR 19-15645Y) and the University of South Bohemia. Y.B. is supported by ForestGEO and SENACYT—Sistema Nacional de Investigación.

References

Agosti, D., Majer, J., Alonso, L.E., Schultz, T., 2000. Ants: Standard Methods for Measuring and Monitoring Biodiversity. Smithsonian Institution Press.

Altermatt, F., 2010. Climatic warming increases voltinism in European butterflies and moths. Proc. Biol. Sci. 277 (1685), 1281–1287.

Andersen, A.N., 1991. Parallels Between Ants and Plants—Implications for Community Ecology. Oxford University Press, Oxford.

Anderson-Teixeira, K.J., Davies, S.J., Bennett, A.C., Gonzalez-Akre, E.B., Muller-Landau, H.C., Joseph Wright, S., Abu Salim, K., Almeyda Zambrano, A.M., Alonso, A., Baltzer, J.L., Basset, Y., Bourg, N.A., Broadbent, E.N., Brockelman, W.Y., Bunyavejchewin, S., Burslem, D.F.R.P., Butt, N., Cao, M., Cardenas, D., Chuyong, G.B., Clay, K., Cordell, S., Dattaraja, H.S., Deng, X., Detto, M., Du, X., Duque, A., Erikson, D.L., Ewango, C.E.N., Fischer, G.A., Fletcher, C., Foster, R.B., Giardina, C.P., Gilbert, G.S., Gunatilleke, N., Gunatilleke, S., Hao, Z., Hargrove, W.W., Hart, T.B., Hau, B.C.H., He, F., Hoffman, F.M., Howe, R.W., Hubbell, S.P., Inman-Narahari, F.M., Jansen, P.A., Jiang, M., Johnson, D.J., Kanzaki, M., Kassim, A.R., Kenfack, D., Kibet, S., Kinnaird, M.F., Korte, L., Kral, K., Kumar, J., Larson, A.J., Li, Y., Li, X., Liu, S., Lum, S.K.Y., Lutz, J.A., Ma, K., Maddalena, D.M., Makana, J.-R., Malhi, Y., Marthews, T., Mat Serudin, R., McMahon, S.M., McShea, W.J., Memiaghe, H.R., Mi, X., Mizuno, T., Morecroft, M., Myers, J.A., Novotny, V., de Oliveira, A.A., Ong, P.S., Orwig, D.A., Ostertag, R., den Ouden, J., Parker, G.G., Phillips, R.P., Sack, L., Sainge, M.N., Sang, W., Sri-ngernyuang, K., Sukumar, R., Sun, I.F., Sungpalee, W., Suresh, H.S., Tan, S., Thomas, S.C., Thomas, D.W., Thompson, J., Turner, B.L., Uriarte, M., Valencia, R., Vallejo, M.I., Vicentini, A., Vrška, T., Wang, X., Wang, X., Weiblen, G., Wolf, A., Xu, H., Yap, S., Zimmerman, J., 2015. CTFS-ForestGEO: a worldwide network monitoring forests in an era of global change. Glob. Chang. Biol. 21, 528–549. https://doi.org/10.1111/gcb.12712.

Augenstein, B., Ulrich, W., Habel, J.C., 2012. Directional temporal shifts in community structure of butterflies and ground beetles in fragmented oligotrophic grasslands of Central Europe. Basic Appl. Ecol. 13, 715–724.

Basset, Y., Lamarre, G.P.A., 2019. Toward a world that values insects. Science 364 (6447), 1230–1231. https://doi.org/10.1126/science.aaw7071.

Basset, Y., Cizek, L., Cuénoud, P., Didham, R.K., Guilhaumon, F., Missa, O., Novotny, V., Ødegaard, F., Roslin, T., Schmidl, J., Tishechkin, A.K., Winchester, N.N., Roubik, D.W., Aberlenc, H.-P., Bail, J., Barrios, H., Bridle, J.R., Castaño-Meneses, G., Corbara, B., Curletti, G., Duarte da Rocha, W., De Bakker, D., Delabie, J.H.C., Dejean, A., Fagan, L.L., Floren, A., Kitching, R.L., Medianero, E., Miller, S.E., Gama de Oliveira, E., Orivel, J., Pollet, M., Rapp, M., Ribeiro, S.P., Roisin, Y., Schmidt, J.B., Sørensen, L., Leponce, M., 2012. Arthropod diversity in a tropical forest. Science 338, 1481–1484.

Basset, Y., Eastwood, R., Sam, L., Lohman, D.J., Novotny, V., Treuer, T., Osorio-Arenas, M.A., 2013. Cross-continental comparisons of butterfly assemblages in rainforests: implications for biological monitoring. Insect Conserv. Divers. 6, 223–233.

Basset, Y., Barrios, H., Segar, S., Srygley, R.B., Aiello, A., Warren, A.D., Delgado, F., Coronado, J., Lezcano, J., Arizala, S., Rivera, M., Perez, F., Bobadilla, R.,

Lopez, Y., Ramirez, J.A., 2015. The butterflies of Barro Colorado Island, Panama: local extinction since the 1930s. PLoS One 10, e0136623. https://doi.org/10.1371/journal.pone.0136623.

Basset, Y., Lamarre, G.P.A., Ratz, T., Segar, S.T., Decaëns, T., Rougerie, R., Miller, S.E., Perez, F., Bobadilla, R., Lopez, Y., Ramirez, J.A., Aiello, A., Barrios, H., 2017. The Saturniidae of Barro Colorado Island, Panama: a model taxon for studying the long-term effects of climate change? Ecol. Evol. 7 (23), 9991–10004. https://doi.org/10.1002/ece3.3515.

Beng, K., Tomlinson, K., Shen, X., et al., 2016. The utility of DNA metabarcoding for studying the response of arthropod diversity and composition to land-use change in the tropics. Sci. Rep. 6, 24965. https://doi.org/10.1038/srep24965.

Benoit, J.B., Denlinger, D.L., 2010. Meeting the challenges of on-host and off-host water balance in blood-feeding arthropods. J. Insect Physiol. 56 (10), 1366–1376.

Bi, K., Linderoth, T., Vanderpool, D., Good, J.M., Nielsen, R., Moritz, C., 2013. Unlocking the vault: next-generation museum population genomics. Mol. Ecol. 22 (24), 6018–6032. https://doi.org/10.1111/mec.12516.

Bishop, T., Robertson, M., Gibb, H., Van Rensburg, B., Braschler, B., Chown, S., Foord, S., Munyai, T., Okey, I., Tshivhandekano, P.G., Werenkraut, V., Parr, C., 2016. Ant assemblages have darker and larger members in cold environments. Glob. Ecol. Biogeogr. 25 (12), 1489–1499.

Blest, A.D., 1960. The evolution, ontogeny and quantitative control of the settling movements of some New World saturniid moths, with some comments on distance communication by honey-bees. Behaviour 16, 188–253.

Blunden, J., Arndt, D.S., 2017. State of the climate in 2016. Bull. Am. Meteorol. Soc. 98, Si–S277.

Bohan, D.A., Dumbrell, A.J., Massol, F., 2017a. Network of invasion: a synthesis of concepts. Academic Press, London, UK.

Bujan, J., Yanoviak, S.P., Kaspari, M., 2016. Desiccation resistance in tropical insects: causes and mechanisms underlying variability in a Panama ant community. Ecol. Evol. 6 (17), 6282–6291. https://doi.org/10.1002/ece3.2355.

Byrne, M.M., 1994. Ecology of twig-dwelling ants in a wet lowland tropical forest. Biotropica 26, 61–72.

Carranza-Rojas, J., Goeau, H., Bonnet, P., Mata-Montero, E., Joly, A., 2017. Going deeper in the automated identification of Herbarium specimens. BMC Evol. Biol. 17, 181.

Carvalheiro, L.G., et al., 2013. Species richness declines and biotic homogenisation have slowed down for NW-European pollinators and plants. Ecol. Lett. 16 (7), 870–878.

Chen, I.C., Shiu, H.-J., Benedick, S., Holloway, J.D., Chey, V.K., Barlow, H.S., Hill, J.K., Thomas, C.D., 2009. Elevation increases in moth assemblages over 42 years on a tropical mountain. PNAS 106 (5), 1479–1483. https://doi.org/10.1073/pnas.0809320106.

Cheng, L., Guyer, D., 2012. Image-based orchard insect automated identification and classification method. Comput. Electron. Agr. 89, 110–115.

Cheng, L., et al., 2019. 2018 continues record global ocean warming. Adv. Atmos. Sci. 36, 249–252. https://doi.org/10.1007/s00376-019-8276-x.

Chesters, D., 2017. Construction of a species-level tree of life for the insects and utility in taxonomic profiling. Syst. Biol. 66, 426–439. https://doi.org/10.1093/sysbio/syw099.

Cleland, E.E., Allen, J.M., Crimmins, T.M., Dunne, J.A., Pau, S., Travers, S.E., Zavaleta, E.S., Wolkovich, E.M., 2012. Phenological tracking enables positive species responses to climate change. Ecology 93, 1765–1771.

Colwell, R.K., Brehm, G., Cardelus, C.L., Gilman, A.C., Longino, J.T., 2008. Global warming, elevational range shifts, and lowland biotic attrition in the wet tropics. Science 322, 258–261.

Dalal, N., Triggs, B., 2005. Histograms of oriented gradients for human detection. Proc. IEEE Conf. Comput. Vis. Pattern Recognit. 1, 886–893.

Dalrymple, R., Flores-Moreno, H., Kemp, D., White, T., Laffan, S., Hemmings, F., D, H.T., Moles, A., 2017. Abiotic and biotic predictors of macroecological patterns in bird and butterfly coloration. Ecol. Monogr. 88, 204–224.

Dalsgaard, B., Magård, E., Fjeldså, J., González, A.M.M., Rahbek, C., Olesen, J.M., Ollerton, J., Alarcón, R., Araujo, A.C., Cotton, P.A., 2011. Specialization in plant-hummingbird networks is associated with species richness, contemporary precipitation and quaternary climate-change velocity. PLoS One 6, e25891.

Dalsgaard, B., Schleuning, M., Maruyama, P.K., Dehling, D.M., Sonne, J., Vizentin-Bugoni, J., Zanata, T.B., Fjeldså, J., Böhning-Gaese, K., Rahbek, C., 2017. Opposed latitudinal patterns of network-derived and dietary specialization in avian plant–frugivore interaction systems. Ecography 40 (12), 1395–1401.

Dejean, A., Céréghino, R., Carpenter, J.M., Corbara, B., Hérault, B., Rossi, V., Leponce, M., Orivel, J., Bonal, D., 2011. Climate change impact on neotropical social wasps. PLoS One 6, e27004.

Deutsch, C.A., et al., 2008. Impacts of climate warming on terrestrial ectotherms across latitude. Proc. Natl. Acad. Sci. U. S. A. 105 (18), 6668–6672. https://doi.org/10.1073/pnas.0709472105.

Diamond, S.E., Frame, A.M., Martin, R.A., Buckley, L.B., 2011. Species' traits predict phenological responses to climate change in butterflies. Ecology 92, 1005–1012.

Didham, R.K., Ghazoul, J., Stork, N.E., Davis, A.J., 1996. Insects in fragmented forests: a functional approach. Trends Ecol. Evol. 11, 255–260.

Dillon, M.E., Wang, G., Huey, R.B., 2010. Global metabolic impacts of recent climate warming. Nature 467, 704–706.

Dirzo, R., Young, H.S., Galetti, M., Ceballos, G., Isaac, N.J.B., Collen, B., 2014. Defaunation in the Anthropocene. Science 345, 401–406.

Donoso, D.A., 2017. Tropical ant communities are in long-term equilibrium. Ecol. Indic. 83, 515–523.

Dormontt, E.E., Gardner, M.G., Breed, M.F., Rodger, J.G., Prentis, P.J., Lowe, A.J., 2014. Genetic bottlenecks in time and space: reconstructing invasions from contemporary and historical collections. PLoS One 9, e106874. https://doi.org/10.1371/journal.pone.0106874.

Drescher, J., Feldhaar, H., Blüthgen, N., 2011. Interspecific aggression and resource monopolization of the invasive ant *Anoplolepis gracilipes* in Malaysian Borneo. Biotropica 43, 93–99.

Dunne, J.A., Williams, R.J., Martinez, N.D., 2009. Cascading extinctions and community collapse in model food webs, Phil. Trans. R. Soc. B 364, 1711–1723. https://doi.org/10.1098/rstb.2008.0219.

Dworschak, K., Bluethgen, N., 2010. Networks and dominance hierarchies: does interspecific aggression explain flower partitioning among stingless bees? Ecol. Entomol. 35, 216–225.

Easterling, D.R., et al., 2000. Climate extremes: observations, modeling, and impacts. Science 289, 2068–2075.

Ebenman, B., Law, R., Borrvall, C., 2004. Community viability analysis: the response of ecological communities to species loss. Ecology 85, 2591–2600.

Emer, C., Venticinque, E., Fonseca, C.R., 2013. Effects of dam-induced landscape fragmentation on amazonian ant–plant mutualistic networks. Conserv. Biol. 27, 763–773.

Facey, S.L., Ellsworth, D.S., Staley, J.T., Wright, D.J., Johnson, S.N., 2014. Upsetting the order: how climate and atmospheric change affects herbivore–enemy interactions. Curr. Opin. Insect Sci. 5, 66–74. https://doi.org/10.1016/j.cois.2014.09.

FAO, 2014. Agroecology for food security and nutrition. In: Proceedings of the FAO International Symposium, 18–19 September 2014. Biodiversity & Ecosystem Services in Agricultural Production Systems, Rome, Italy. http://www.fao.org/3/a-i4729e.pdf.

Fayle, T.M., Turner, E.C., Snaddon, J.L., Chey, V.K., Chung, A.Y.C., Eggleton, P., Foster, W.A., 2010. Oil palm expansion into rain forest greatly reduces ant biodiversity in canopy, epiphytes and leaf-litter. Basic Appl. Ecol. 11, 337–345, 1215. https://doi.org/10.1016/j.baae.2009.12.009.

Flanagan, S.P., Forester, B.R., Latch, E.K., Aitken, S.N., Hoban, S., 2018. Guidelines for planning genomic assessment and monitoring of locally adaptive variation to inform species conservation. Evol. Appl. 11 (7), 1035–1052. https://doi.org/10.1111/eva.12569.

Fontanilla, A.M., Nakamura, A., Xu, Z., Cao, M., Kitching, R.L., Tang, Y., Burwell, C.J., 2019. Taxonomic and functional ant diversity along tropical, subtropical, and subalpine elevational transects in Southwest China. Insects 10, 128.

Forister, M.L., Novotny, V., Panorska, A.K., Baje, L., Basset, Y., Butterill, P.T., Cizek, L., Coley, P.D., Dem, F., Diniz, I.R., Drozd, P., Fox, M., Glassmire, A.E., Hazen, R., Hrcek, J., Jahner, J.P., Kaman, O., Kozubowski, T.J., Kursar, T.A., Lewis, O.T., Lill, J., Marquis, R.J., Miller, S.E., Morais, H.C., Murakami, M., Nickel, H., Pardikes, N.A., Ricklefs, R.E., Singer, M.S., Smilanich, A.M., Stireman, J.O., Villamarín-Cortez, S., Vodka, S., Volf, M., Wagner, D.L., Walla, T., Weiblen, G.D., Dyer, L.A., 2015. The global distribution of diet breadth in insect herbivores. Proc. Natl. Acad. Sci. 112, 442–447. https://doi.org/10.1073/pnas.1423042112.

Forister, M.L., et al., 2016. Increasing neonicotinoid use and the declining butterfly fauna of lowland California. Biol. Lett. 12, 20160475.

García-Robledo, C., et al., 2015. Limited tolerance by insects to high temperatures across tropical elevational gradients and the implications of global warming for extinction. PNAS 113 (3), 680–685.

García-Robledo, C., et al., 2018. Lower thermal tolerance in nocturnal than in diurnal ants: a challenge for nocturnal ectotherms facing global warming. Ecol. Entomol. 43 (2), 162–167. https://doi.org/10.1111/een.12481.

Garibaldi, L.A., et al., 2013. Wild pollinators enhance fruit set of crops regardless of honey bees abundance. Science 339 (6127), 1608–1611.

Giannini, T.C., Cordeiro, G.D., Freitas, B.M., Saraiva, A.M., Imperatriz-Fonseca, V.L., 2015. The dependence of crops for pollinators and the economic value of pollination in Brazil. J. Econ. Entomol. 108, 849–857.

Godfray, H.C.J., Lewis, O.T., Memmott, J., 1999. Studying insect diversity in the tropics. Philos. Trans. R. Soc. Lond. A 354, 1811–1824.

Goulson, D., Nicholls, E., Botías, C., Rotheray, E.L., 2015. Bee declines driven by combined stress from parasites, pesticides, and lack of flowers. Science 347, 1255957.

Gous, A., Swanevelder, D.Z.H., Eardley, C.D., Willows-Munro, S., 2019. Plant–pollinator interactions over time: pollen metabarcoding from bees in a historic collection. Evol. Appl. 12 (2), 187–197. https://doi.org/10.1111/eva.12707.

Grøtan, V., Lande, R., Engen, S., Saether, B.-E., DeVries, P.J., 2012. Seasonal cycles of species diversity and similarity in a tropical butterfly community. J. Anim. Ecol. 81, 714–723.

Habel, J.C., Trusch, R., Schmitt, T., et al., 2019a. Long-term large-scale decline in relative abundances of butterfly and burnet moth species across south-western Germany. Sci. Rep. 9, 14921. https://doi.org/10.1038/s41598-019-51424-1.

Hadfield, J.D., 2010. MCMC methods for multi-response generalized linear mixed models: the MCMCglmm R package. J. Stat. Softw. 33, 1–22.

Hallmann, C.A., Sorg, M., Jongejans, E., Siepel, H., Hofland, N., Schwan, H., Stenmans, W., Müller, A., Sumser, H., Hörren, T., Goulson, D., Kroon, H., 2017.

More than 75 percent decline over 27 years in total flying insect biomass in protected areas. PLoS One 12, e0185809.
Harvey, P.H., Pagel, M.D., 1991. The Comparative Method in Evolutionary Biology. Oxford University Press, Oxford.
Hawking, J.H., New, T.R., 2002. Interpreting dragonfly diversity to aid in conservation assessment: lessons from the Odonata assemblage at Middle Creek, north-eastern Victoria, Australia. J. Insect Conserv. 6, 171–178.
Hegland, S.J., et al., 2009. How does climate warming affect plant-pollinator interactions? Ecol. Lett. 12, 184–195.
iNaturalist.org, 2019. iNaturalist Research-Grade Observations. Occurrence dataset, https://doi.org/10.15468/ab3s5x. accessed via GBIF.org on 2019-11-13.
IPCC, 2012. Summary for policymakers. In: Field, C.B., Barros, V., Stocker, T.F., Qin, D., Dokken, D.J., Ebi, K.L., … Midgley, P.M. (Eds.), Managing the Risks of Extreme Events and Disasters to Advance Climate Change Adaptation. A Special Report of Working Groups I and II of the Intergovernmental Panel on Climate Change. Cambridge University Press, pp. 1–19.
IPCC, 2018. Global warming of 1.5°C. In: Masson-Delmotte, V., Zhai, P., Pörtner, H.-O., Roberts, D., Skea, J., Shukla, P.R., Pirani, A., Moufouma-Okia, W., Péan, C., Pidcock, R., Connors, S., Matthews, J.B.R., Chen, Y., Zhou, X., Gomis, M.I., Lonnoy, E., Maycock, T., Tignor, M., Waterfield, T. (Eds.), An IPCC Special Report on the Impacts of Global Warming of 1.5°C Above Pre-Industrial Levels and Related Global Greenhouse Gas Emission Pathways, in the Context of Strengthening the Global Response to the Threat of Climate Change, Sustainable Development, and Efforts to Eradicate Poverty.
Ives, A.R., Helmus, M.R., 2011. Generalized linear mixed models for phylogenetic analyses of community structure. Ecol. Monogr. 81, 511–525.
Janzen, D.H., 1984. Two ways to be a tropical big moth: Santa Rosa saturniids and sphingids. Oxford Survey Evol. Biol. 1, 85–140.
Janzen, D.H., Hallwachs, W., 2019. Where might be many tropical insects? Biol. Conserv. 233, 102–108.
Ji, Y., Ashton, L., Pedley, S.M., et al., 2013. Reliable, verifiable and efficient monitoring of biodiversity via metabarcoding. Ecol. Lett. 16, 1245–1257.
Kaiser-Bunbury, C.N., Bluthgen, N., 2015. Integrating network ecology with applied conservation: a synthesis and guide to implementation. AoB Plants 7.
Kaspari, M., et al., 2015. Thermal adaptation generates a diversity of thermal limits in a rainforest ant community. Glob. Chang. Biol. 21 (3), 1092–1102. https://doi.org/10.1111/gcb.12750.
Kaspari, M., et al., 2016. Thermal adaptation and phosphorus shape thermal performance in an assemblage of rainforest ants. Ecology 97 (4), 1038–1047.
Keck, F., Vasselon, V., Rimet, F., Bouchez, A., Kahlert, M., 2018. Boosting DNA metabarcoding for biomonitoring with phylogenetic estimation of operational taxonomic units' ecological profiles. Mol. Ecol. Resour. 18, 1299–1309. https://doi.org/10.1111/1755-0998.12919.
Kishimoto-Yamada, K., Itioka, T., 2013. Seasonality in phytophagous scarabaeid (Melolonthinae and Rutelinae) abundances in an 'aseasonal' Bornean rainforest. Insect Conserv. Divers. 6, 179–188.
Knoll, F.R.N., 2016. Variation in the abundance of neotropical bees in an unpredictable seasonal environment. Neotrop. Entomol. 45, 129–138.
Lamarre, G.P.A., Juin, Y., Lapied, E., Le Gall, P., Nakamura, A., 2018. Using field-based entomological research to promote awareness about forest ecosystem conservation. Nat. Conserv. 29, 39–56.
Laurance, W.F., et al., 2014. Agricultural expansion and its impacts on tropical nature. Trends Ecol. Evol. 29, 107–116.

Lavorel, S., Garnier, E., 2002. Predicting changes in community composition and ecosystem functional from plant traits: revisiting the Holy Grail. Funct. Ecol. 16, 545–556.
Leather, S.R., 2018. "Ecological Armageddon"—more evidence for the drastic decline in insect numbers. Ann. Appl. Biol. 172 (1), 1–3.
LeCun, Y., Bengio, Y., Hinton, G., 2015. Deep learning. Nature 521 (7553), 436–444.
Leidner, A.K., Haddad, N.M., Lovejoy, T.E., 2010. Does tropical forest fragmentation increase long-term variability of butterfly communities? PLoS One 5, e9534.
Lister, B.C., Garcia, A., 2018. Climate-driven declines in arthropod abundance restructure a rainforest food web. PNAS 115, E10397.
Liu, Z., Gao, J., Yang, G., Zhang, H., He, Y., 2016. Localization and classification of paddy field pests using a saliency map and deep convolutional neural network. Sci. Rep. 6, 20410. https://doi.org/10.1038/srep20410.
Long, O.M.C., Warren, R., Price, J., Brereton, T.M., Botham, M.S., Franco, M.A., 2016. Sensitivity of UK butterflies to local climatic extremes: which life stages are most at risk? J. Anim. Ecol. 86, 108–116. https://doi.org/10.1111/1365-2656.12594.
Longino, J.T., 2000. What to do with the data. In: Agosti, D., Majer, J.D., Alonso, L.E., Schultz, T.R. (Eds.), Ants, Standard Methods for Measuring and Monitoring Biodiversity. Smithsonian Institution Press, Washington and London, pp. 186–203.
Losey, J.E., Vaughan, M., 2006. The economic value of ecological services provided by insects. Bioscience 56, 311–323.
Lowe, D.G., 2004. Distinctive image features from scale-invariant keypoints. Int. J. Comput. Vis. 60, 91–110.
Lucas, M., Forero, D., Basset, Y., 2016. Diversity and recent population trends of assassin bugs (Hemiptera: Reduviidae) on Barro Colorado Island, Panama. Insect Conserv. Divers. 9, 546–558.
Lyra, A., et al., 2017. Projections of climate change impacts on central America tropical rainforest. Clim. Change 141 (1), 93–105. https://doi.org/10.1007/s10584-016-1790-2.
MacGregor, C.J., Pocock, M.J.O., Fox, R., Evans, D.M., 2015. Pollination by nocturnal Lepidoptera, and the effects of light pollution: a review. Ecol. Entomol. 40, 187–198.
Maire, E., Grenouillet, G., Brosse, S., Villéger, S., 2015. How many dimensions are needed to accurately assess functional diversity? A pragmatic approach for assessing the quality of functional spaces. Glob. Ecol. Biogeogr. 24, 728–740.
Martineau, M., Conte, D., Raveaux, R., Arnault, I., Munier, D., Venturini, G., 2017. A survey on image-based insect classification. Pattern Recogn. 65, 273–284.
Maunsell, S.C., Kitching, R.L., Burwell, C.J., Morris, R.J., 2015. Changes in host–parasitoid food web structure with elevation. J. Anim. Ecol. 84, 353–363.
McDermott, L.O., Warren, R., Price, J., Brereton, T.M., Botham, M.S., Franco, A., 2017. Sensitivity of UK butterflies to local climatic extremes: which life stages are most at risk? J. Anim. Ecol. 86 (1), 108–116,pmid: 27796048.
McGill, B.J., Enquist, B.J., Weiher, E., Westoby, M., 2006. Rebuilding community ecology from functional traits. Trends Ecol. Evol. 21, 178–185.
McGregor, C., Williams, J.H., Bell, J.R., Thomas, C.D., 2019. Moth biomass increases and decreases over 50 years in Britain. Nat. Ecol. Evol. 3, 1645–1649.
Memmott, J., Craze, P.G., Waser, N.M., Price, M.V., 2007. Global warming and the disruption of plant–pollinator interactions. Ecol. Lett. 10, 710–717. https://doi.org/10.1111/j.1461-0248.2007.01061.x.
Moretti, M., Dias, A.T.C., de Bello, F., Altermatt, F., Chown, S.L., Azcárate, F.M., Bell, J.R., Fournier, B., Hedde, M., Hortal, J., Ibanez, S., Öckinger, E., Sousa, J.P., Ellers, J., Berg, M.P., 2017. Handbook of protocols for standardized measurement of terrestrial invertebrate functional traits. Funct. Ecol. 31, 558–567.

Morris, R.J., Gripenberg, S., Lewis, O.T., Roslin, T., 2014. Antagonistic interaction networks are structured independently of latitude and host guild. Ecol. Lett. 17, 340–349.
Morris, R.J., Sinclair, F.H., Burwell, C.J., 2015. Food web structure changes with elevation but not rainforest stratum. Ecography 38, 792–802.
Motani, R., Schmitz, L., 2011. Phylogenetic versus functional signals in the evolution of form-function relationships in terrestrial vision. Evolution 65, 2245–2257.
Nair, K.S.S., 2007. Tropical Insect Forest Pests. Ecology, Impact, and Management. Cambridge University Press.
Newbold, T., Lawrence, N.H., Contu, S., Hill, S.L.L., Beck, J., et al., 2018. Widespread winners and narrow-ranged losers: land use homogenizes biodiversity in local assemblages worldwide. PLoS Biol. 16 (12), e2006841.
Norouzzadeh, M.S., Nguyen, A., Kosmala, M., Swanson, A., Palmer, M.S., Packer, C., Clune, J., 2018. Automatically identifying, counting, and describing wild animals in camera-trap images with deep learning. Proc. Natl. Acad. Sci. U. S. A. 115, E5716–E5725.
Novotny, V., Basset, Y., 2000. Rare species in communities of tropical insect herbivores: pondering the mystery of singletons. Oikos 89, 564–572.
Ojala, T., Pietikäinen, M., Mäenpää, T., 2002. Multiresolution gray-scale and rotation invariant texture classification with local binary patterns. IEEE Trans. Pattern Anal. Mach. Intell. 24, 971–987.
Ollerton, J., Winfree, R., Tarrant, S., 2011. How many flowering plants are pollinated by animals? *Oikos* 120, 321–326. https://doi.org/10.1111/j.1600-0706.2010.18644.x.
Paaijmans, K.P., Heinig, R.L., Seliga, R.A., Blanford, J.I., Blanford, S., et al., 2013. Temperature variation makes ectotherms more sensitive to climate change. Glob. Chang. Biol. 19, 2373–2380.
Paradis, E., Claude, J., 2002. Analysis of comparative data using generalized estimating equations. J. Theor. Biol. 218, 175–185. https://doi.org/10.1006/jtbi.2002.3066.
Pau, S., Okin, G.S., Gillespie, T.W., 2010. Asynchronous response of tropical forest leaf phenology to seasonal and El Nino-driven drought. PLoS One 5 (6), e11325.
Pau, S., Wolkovich, E.M., Cook, B.I., Davies, T.J., Kraft, N.J.B., Bolmgren, K., … Cleland, E.E., 2011. Predicting phenology by integrating ecology, evolution and climate science. Glob. Chang. Biol. 17, 3633–3643.
Pearse, I.S., Altermatt, F., 2013. Predicting novel trophic interactions in a non-native world. Ecol. Lett. 16, 1088–1094. https://doi.org/10.1111/ele.12143.
Pearse, W.D., Cadotte, M.W., Cavender-Bares, J., Ives, A.R., Tucker, C.M., Walker, S.C., Helmus, M.R., 2015. pez: phylogenetics for the environmental sciences. Bioinformatics 31, 2888–2890. https://doi.org/10.1093/bioinformatics/btv277.
Pellissier, L., et al., 2013. Turnover of plant lineages shapes herbivore phylogenetic beta diversity along ecological gradients. Ecol. Lett. 16, 600–608.
Pichler, M., Boreux, V., Klein, A.-M., 2019. Schleuning M. and Hartig F., Machine learning algorithms to infer trait matching and predict species interactions in ecological networks. Methods Ecol Evol. **11**, 281–293. https://doi.org/10.1111/2041-210X.13329.
Pimentel, D., Andow, D.A., 1984. Pest management and pesticide impacts. Insect Sci. Appl. 5, 141–149.
Pinsky, M.L., Eikeset, A.M., McCauley, D.J., Payne, J.L., Sunday, J.M., 2019. Greater vulnerability to warming of marine versus terrestrial ectotherms. Nature 569, 108–111. https://doi.org/10.1038/s41586-019-1132-4.
Plowman, N.S., Hood, A.S., Moses, J., Redmond, C., Novotny, V., Klimes, P., Fayle, T.M., 2017. Network reorganization and breakdown of an ant–plant protection mutualism with elevation. Proc. R. Soc. B 284, 20162564, The Royal Society.
Poisot, T., Mouquet, N., Gravel, D., 2013. Trophic complementarity drives the biodiversity–ecosystem functioning relationship in food webs. Ecol. Lett. 16 (7), 853–861.

Polato, N.R., et al., 2018. Narrow thermal tolerance and low dispersal drive higher speciation in tropical mountains. Proc. Natl. Acad. Sci. U. S. A. 115 (49), 12471–12476. https://doi.org/10.1073/pnas.1809326115.

Potts, S.G., Biesmeijer, J.C., et al., 2010. Global pollinator declines: trends, impacts and drivers. Trends Ecol. Evol. 25, 345–353.

Price, P.W., 2002. Resource-driven terrestrial interaction webs. Ecol. Res. 17, 241–247.

Ripple, W.J., et al., 2020. World scientists' warning of a climate emergency. BioScience 70 (1), 8–12. biz088, https://doi.org/10.1093/biosci/biz088.

Roubik, D.W., 2001. Ups and downs in pollinator populations: when is there a decline? Conserv. Ecol. 5, 1–22.

Salcido, D.M., Forister, M.L., Garcia Lopez, H., Dyer, L.A., 2020. Loss of dominant caterpillar genera in a protected tropical forest. Sci Rep 10, 422. https://doi.org/10.1038/s41598-019-57226-9.

Sánchez-Bayo, F., Wyckhuys, K.A.G., 2019. Worldwide decline of the entomofauna: a review of its drivers. Biol. Conserv. 232, 8–27. https://doi.org/10.1016/j.biocon.2019.01.020.

Saunders, M.E., Janes, J.K., O'Hanlon, J.C., 2019. Understanding the Evidence Informing the Insect Apocalypse Myth. preprint, https://doi.org/10.32942/osf.io/2cqws.

Schwartz, M.K., Luikart, G., Waples, R.S., 2007. Genetic monitoring as a promising tool for conservation and management. Trends Ecol. Evol. 22 (1), 25–33. https://doi.org/10.1016/j.tree.2006.08.009.

Segar, S.T., Volf, M., Isua, B., Sisol, M., Redmond, C.M., Rosati, M.E., Gewa, B., Molem, K., Dahl, C., Holloway, J.D., Basset, Y., Miller, S.E., Weiblen, G.D., Salminen, J.-P., Novotny, V., 2017. Variably hungry caterpillars: predictive models and foliar chemistry suggest how to eat a rainforest. Proc. R. Soc. B Biol. Sci. 284, 20171803. https://doi.org/10.1098/rspb.2017.1803.

Seibold, S., Gossner, M.M., Simons, N.K., et al., 2019. Arthropod decline in grasslands and forests is associated with landscape-level drivers. Nature 574, 671–674. https://doi.org/10.1038/s41586-019-1684-3.

Silvertown, J., Poulton, P., Johnston, E., Edwards, G., Heard, M., Biss, P.M., 2006. The park grass experiment 1856–2006: its contribution to ecology. J. Ecol. 94, 801–814. https://doi.org/10.1111/j.1365-2745.2006.01145.x.

Simler-Williamson, A.B., Rizzo, D.M., Cobb, R.C., 2019. Interacting effects of global change on forest pest and pathogen dynamics. Annu. Rev. Ecol. Evol. Syst. 50, 381–403.

Simmons, B.I., Balmford, A., Bladon, A.J., et al., 2019. Worldwide insect declines: an important message, but interpret with caution. Ecol. Evol. 9, 3678–3680. https://doi.org/10.1002/ece3.5153.

Singer, M.C., Parmesan, C., 2010. Phenological asynchrony between herbivorous insects and their hosts: signal of climate change or pre-existing adaptive strategy? Philos. Trans. R. Soc., B 365 (1555), 3161–3176. https://doi.org/10.1098/rstb.2010.0144.

Slade, E.M., Merckx, T., Riutta, T., Bebber, D.P., Redhead, D., Riordan, P., Macdonald, D.W., 2013. Life-history traits and landscape characteristics predict macro-moth responses to forest fragmentation. Ecology 94, 1519–1530.

Solis-Sánchez, L.O., Castañeda-Miranda, R., Garcia-Escalante, J.J., Torres-Pacheco, I., Guevara-Gonzalez, R.G., Castañeda-Miranda, C.L., Alaniz-Lumbrerasa, P.D., 2011. Scale invariant feature approach for insect monitoring. Comput. Electron. Agric. 75, 92–99.

Spaniol, L.R., et al., 2019. Combining functional traits and phylogeny to disentangling Amazonian butterfly assemblages on anthropogenic gradients. Ecosphere 10 (8), e02837. https://doi.org/10.1002/ecs2.2837.

Srygley, R.B., Dudley, R., Oliveira, E.G., Aizprua, R., Pelaez, N.Z., Riveros, A.J., 2010. El Nino and dry season rainfall influence hostplant phenology and an annual butterfly migration from neotropical wet to dry forests. Glob. Chang. Biol. 16, 936–945. https://doi.org/10.1111/j.1365-2486.2009.01986.x.

Srygley, R.B., Dudley, R., Oliveira, E.G., Riveros, A.J., 2014. El Niño, host plant growth, and migratory butterfly abundance in a changing climate. Biotropica 46, 90–97.

Staats, M., Erkens, R.H.J., van de Vossenberg, B., Wieringa, J.J., Kraaijeveld, K., Stielow, B., … Bakker, F.T., 2013. Genomic treasure troves: complete genome sequencing of herbarium and insect museum specimens. PLoS One 8 (7), e69189. https://doi.org/10.1371/journal.pone.0069189.

Stocker, T., et al., 2014. Climate Change 2013: The Physical Science Basis. IPCC, Cambridge University Press, Cambridge.

Stork, N.E., 2018. How many species of insects and other terrestrial arthropods are there on Earth? Annu. Rev. Entomol. 63, 31–45.

Stork, N.E., et al., 2015. New approaches narrow global species estimates for beetles, insects, and terrestrial arthropods. Proc. Natl. Acad. Sci. U. S. A. 112 (24), 7519–7523. https://doi.org/10.1073/pnas.1502408112.

Strong, D.R., Lawton, J.H., Southwood, T.R.E., 1984. Insects on Plants: Community Patterns and Mechanisms. Blackwell, Oxford, p. 313. 1984.

Szeliski, R., 2010. Computer Vision: Algorithms and Applications. Springer, Berlin.

Tang, M., Hardman, C.J., Ji, Y., Meng, G., Liu, S., Tan, M., et al., 2015. Highthroughput monitoring of wild bee diversity and abundance via mitogenomics. Methods Ecol. Evol. 6, 1034–1043.

Terblanche, J.S., et al., 2011. Ecologically relevant measures of tolerance to potentially lethal temperatures. J. Exp. Biol. 214, 3713–3725.

Terblanche, J.S., Mitchell, K.A., Uys, W., Short, C., Boardman, L., 2017. Thermal limits to survival and activity in two life stages of false codling moth *Thaumatotibia leucotreta* (Lepidoptera, Tortricidae). Physiol. Entomol. 42, 379–388. https://doi.org/10.1111/phen.12210.

Thenmozhi, K., Reddy, U.S., 2019. Crop pest classification based on deep convolutional neural network and transfer learning. Comput. Electron. Agricult. 164, 104906.

Thomas, A.G., Telfer, M.G., Roy, D.B., Preston, C.D., Greenwood, J.J.D., Asher, J., Fox, R., Clarke, R.T., Lawton, J.H., 2004. Comparative losses of British butterflies, birds, and plants and global extinction crisis. Science 303, 1879–1881.

Thomas, C.D., Jones, T.H., Hartley, S.E., 2019. "Insectageddon": a call for more robust data and rigorous analyses. Glob. Chang. Biol. 25, 1891–1892.

Tylianakis, J.M., Tscharntke, T., Lewis, O.T., 2007. Habitat modification alters the structure of tropical host–parasitoid food webs. Nature 445, 202–205.

Valtonen, A., Molleman, F., Chapman, C.A., Carey, J.R., Ayres, M.P., Roininen, H., 2013. Tropical phenology: bi-annual rhythms and interannual variation in an Afrotropical butterfly assemblage. Ecosphere 4, 1–28.

Vamosi, J.C., Gong, Y.B., Adamowicz, S.J., Packer, L., 2016. Forecasting pollination declines through DNA barcoding: the potential contributions of macroecological and macroevolutionary scales of inquiry. New Phytol. 213, 11–18. https://doi.org/10.1111/nph.14356.

Van Langevelde, F., van Grunsven, R.H.A., Veenendaal, E.M., Fijen, T.P.M., 2017. Artificial night lighting inhibits feeding in moths. Biol. Lett. 13, 20160874.

Venugoban, K., Ramanan, A., 2014. Image classification of paddy field insect pests using gradient-based features. Int. J. Mach. Learn. Comput. 4 (1), 1–5.

Wagner, D.L., 2019. Insect declines in the Anthropocene. Annu. Rev. Entomol. 65, 457–480. https://doi.org/10.1146/annurev-ento011019-025151.

Wäldchen, J., Mäder, P., 2018. Machine learning for image based species identification. Methods Ecol. Evol. 9, 2216–2225.

Wandeler, P., Hoeck, P.E.A., Keller, L.F., 2007. Back to the future: museum specimens in population genetics. Trends Ecol. Evol. 22 (12), 634–642. https://doi.org/10.1016/j.tree.2007.08.017.

Weiher, E., 2011. A primer of trait and functional diversity. In: Magurran, A.E., McGill, B.J. (Eds.), Biological Diversity: Frontiers in Measurement and Assessment. Oxford University Press, Oxford, pp. 175–193.

Weinstein, B.G., 2017. A computer vision for animal ecology. J. Anim. Ecol. 87, 533–545.

Wen, C., Guyer, D., 2012. Image-based orchard insect automated identification and classification method. Comput. Electron. Agric. 89, 110–115.

Wen, C., Guyer, D.E., Li, W., 2009. Local feature-based identification and classification for orchard insects. Biosyst. Eng. 104, 299.

Wenzel, M., Schmitt, T., Weitzel, M., Seitz, A., 2006. The severe decline of butterflies on western German calcareous grasslands during the last 30 years: a conservation problem. Biol. Conserv. 128, 542–552.

Wiens, J.J., Ackerly, D.D., Allen, A.P., Anacker, B.L., Buckley, L.B., Cornell, H.V., Damschen, E.I., Jonathan Davies, T., Grytnes, J.A., Harrison, S.P., et al., 2010. Niche conservatism as an emerging principle in ecology and conservation biology. Ecol. Lett. 13, 1310–1324. https://doi.org/10.1111/j.1461-0248.2010.01515.x.

Willig, M.R., Woolbright, L., Presley, S.J., Schowalter, T.D., Waide, R.B., et al., 2019. Populations are not declining and food webs are not collapsing at the Luquillo Experimental Forest. PNAS 116, 12143–12144.

Willis, C.G., Ellwood, E.R., Primack, R.B., Davis, C.C., Pearson, K.D., Gallinat, A.S., Yost, J.M., Nelson, G., Mazer, S.J., Rossington, N.L., Sparks, T.H., Soltis, P.S., 2017. Old plants, new tricks: phenological research using herbarium specimens. Trends Ecol. Evol. 32, 531–546.

Wolda, H., 1992. Trends in abundance of tropical forest insects. Oecologia 89, 47–52.

Wolda, H., Roubik, D.W., 1986. Nocturnal bee abundance and seasonal bee activity in a Panamanian forest. Ecology 67, 426–433.

Xing, S., Bonebrake, T.C., Ashton, L.A., Kitching, R.L., Cao, M., Sun, Z., Ho, J.C., Nakamura, A., 2018. Colors of night: climate–morphology relationships of geometrid moths along spatial gradients in southwestern China. Oecologia 188, 537–546.

Yamamura, K., Yokozawa, M., Nishimori, M., Ueda, Y., Yokosuka, T., 2006. How to analyze long-term insect population dynamics under climate change: 50-year data of three insect pests in paddy fields. Popul. Ecol. 48 (1), 31–48.

Yu, D.W., et al., 2012. Biodiversity soup: Metabarcoding of arthropods for rapid biodiversity assessment and biomonitoring. Methods Ecol. Evol. 3 (4), 613–623. https://doi.org/10.1111/j.2041-210X.2012.00198.x.

Yusah, K.M., Fayle, T.M., Harris, G., Foster, W.A., 2012. Optimizing diversity assessment protocols for high canopy ants in tropical rain forest. Biotropica 44, 73–81.

Further reading

Basset, Y., Miller, S.E., Gripenberg, S., Ctvrtecka, R., Dahl, C., Leather, S.R., Didham, R.K., 2019. An entomocentric view of the Janzen-Connell hypothesis. Insect Conserv. Divers. 12, 1–8.

Bohan, D.A., Vacher, C., Tamaddoni-Nezhad, A., Raybould, A., Dumbrell, A.J., Woodward, G., 2017b. Next-generation global biomonitoring: large-scale, automated reconstruction of ecological networks. Trends Ecol. Evol. 32, 477–487. https://doi.org/10.1016/j.tree.2017.03.001.

Cagnolo, L., Valladares, G., Salvo, A., Cabido, M., Zak, M., 2009. Habitat fragmentation and species loss across three interacting trophic levels: effects of life-history and food-web traits. Conserv. Biol. 23, 1167–1175.

De Bello, F., Lavorel, S., Diaz, S., et al., 2010. Towards an assessment of multiple ecosystem processes and services via functional traits. Biodivers. Conserv. 19, 2873–3893.

Dyer, L., Singer, M., Lill, J., Stireman, J., Gentry, G., Marquis, R., Ricklefs, R., Greeney, H., Wagner, D., Morais, H., 2007. Host specificity of Lepidoptera in tropical and temperate forests. Nature 448, 696–699.

Fayle, T., Turner, E.C., Basset, Y., Ewers, R.M., Reynolds, G., Novotny, V., 2015. Whole-ecosystem experimental manipulations of tropical forests. Trends Ecol. Evol. 30 (6), 334–346.

Fayle, T., Wanji, C., Turner, E.C., Yusah, K.M., 2017. Living together in novel habitats: a review of land-use change impacts on mutualistic ant-plant symbioses in tropical forests. In: Ant-Plant Interactions: Impacts of Humans on Terrestrial Ecosystems. Cambridge University Press, pp. 52–72.

Habel, J.C., et al., 2019b. Mitigating the precipitous decline of terrestrial European insects: requirements for a new strategy. Biodivers. Conserv. 28, 1343–1360.

Newbold, T., Hudson, L.N., Hill, S.L., Contu, S., Lysenko, I., Senior, R.A., Börger, L., Bennett, D.J., Choimes, A., Collen, B., 2015. Global effects of land use on local terrestrial biodiversity. Nature 520, 45–50.

Novotny, V., Drozd, P., Miller, S.E., Kulfan, M., Janda, M., Basset, Y., Weiblen, G.D., 2006. Why are there so many species of herbivorous insects in tropical rainforests? Science 313, 1115–1118.

Parmesan, C., 2006. Ecological and evolutionary responses to recent climate change. Annu. Rev. Ecol. Evol. Syst. 37, 637–669.

Sala, O.E., Stuart Chapin III, F., Armesto, J.J., Berlow, E., Bloomfield, J., Dirzo, R., Huber-Sanwald, E., Huenneke, L.F., Jackson, R.B., Kinzig, A., Leemans, R., Lodge, D.M., Mooney, H.A., Mn, O., Poff, N.L., Sykes, M.T., Walker, B.H., Walker, M., Wall, D.H., 2000. Global biodiversity scenarios for the year 2100. Science 287, 1770–1774. https://doi.org/10.1126/science.287.5459.1770.

Swanson, A., Kosmala, M., Lintott, C., Simpson, R., Smith, A., Packer, C., 2015. Snapshot Serengeti, high-frequency annotated camera trap images of 40 mammalian species in an African savanna. Sci. Data 2, 150026. https://doi.org/10.1038/sdata.2015.26.

Terblanche, J.S., 2014. Physiological performance of field-released insects. Curr. Opin. Insect Sci. 4, 60–66.

Thomas, J.A., 2016. Butterfly communities under threat. Science 353, 216–218.

Tilman, D., Fargione, J., Wolff, B., D'Antonio, C., Dobson, A., Howarth, R., Schindler, D., Schlesinger, W.H., Simberloff, D., Swackhamer, D., 2001. Forecasting agriculturally driven global change. Science 292, 281–284.

Tylianakis, J.M., Didham, R.K., Bascompte, J., Wardle, D.A., 2008. Global change and species interactions in terrestrial ecosystems. Ecol. Lett. 11, 1351–1363. https://doi.org/10.1111/j.1461-0248.2008.01250.x.

Wilson, E.O., 2017. Biodiversity research requires more boots on the ground. Nat. Ecol. Evol. 1, 1590–1591.

CHAPTER NINE

Advances and prospects of environmental DNA in neotropical rainforests

Lucie Zinger[a,*], Julian Donald[b], Sébastien Brosse[b], Mailyn Adriana Gonzalez[c], Amaia Iribar[b], Céline Leroy[d,e], Jérôme Murienne[b], Jérôme Orivel[e], Heidy Schimann[e], Pierre Taberlet[f,g], Carla Martins Lopes[h]

[a]Ecole Normale Supérieure, PSL Research University, CNRS, Inserm, Institut de Biologie de l'Ecole Normale Supérieure (IBENS), Paris, France
[b]Evolution et Diversité Biologique (EDB), Université Toulouse 3 Paul Sabatier, CNRS, IRD, UMR 5174, Toulouse, France
[c]Programa Ciencias de la Biodiversidad, Instituto de Investigación de Recursos Biológicos Alexander von Humboldt. Bogotá, Colombia
[d]AMAP, Univ Montpellier, CIRAD, CNRS, INRAE, IRD, Montpellier, France
[e]INRAE, CNRS, AgroParisTech, CIRAD, Université de Guyane, Université des Antilles, UMR Ecologie des Forêts de Guyane (EcoFoG), Campus agronomique, Kourou, France
[f]Laboratoire d'Ecologie Alpine (LECA), CNRS, Université Grenoble Alpes, Grenoble, France
[g]UiT—The Arctic University of Norway, Tromsø Museum, Tromsø, Norway
[h]Departamento de Zoologia, Instituto de Biociências, Universidade Estadual Paulista (UNESP), Rio Claro, SP, Brazil
*Corresponding author: e-mail address: lucie@zinger.fr

Contents

Advances in Ecological Research, Volume 62
ISSN 0065-2504
https://doi.org/10.1016/bs.aecr.2020.01.001

Abstract

The rainforests of the Neotropics shelter a vast diversity of plant, animal and microscopic species that provide critical ecosystem goods and services for both local and worldwide populations. These environments face a major crisis due to increased deforestation, pollution, and climate change, emphasizing the need for more effective conservation efforts. The adequate monitoring of these ecosystems has proven a complex and time consuming endeavour, which depends on ever dwindling taxonomic expertise. To date, many species remain undiscovered, let alone described, with otherwise limited information regarding known species population distributions and densities. Overcoming these knowledge shortfalls and practical limitations is becoming increasingly possible through techniques based on environmental DNA (eDNA), i.e., DNA that can be obtained from environmental samples (e.g. tissues, soil, sediment, water, etc.). When coupled with high-throughput sequencing, these techniques now enable realistic, cost-effective, and standardisable biodiversity assessments. This opens up enormous opportunities for advancing our understanding of complex and species-rich tropical communities, but also in facilitating large-scale biomonitoring programs in the neotropics. In this review, we provide a brief introduction to eDNA methods, and an overview of their current and potential uses in both terrestrial and aquatic ecosystems of neotropical rainforests. We also discuss the limits and challenges of these methods for our understanding and monitoring of biodiversity, as well as future research and applied perspectives of these techniques in neotropical rainforests, and beyond.

1. Introduction

Faced with the current environmental crisis, there is an ever growing need to accurately assess existing policy and legislation which aims to protect ecosystems, such as the Paris Climate Agreement, the REDD+ framework, and the Aichi targets (Marques et al., 2014), as exemplified by the IPBES framework (Díaz et al., 2019). This is particularly true for the neotropical moist broadleaf forests, i.e., those occurring from southern Mexico and Florida to Argentina (Morrone, 2014; Olson et al., 2001). Of these forests, the rainforests occurring across Amazonia are the most substantial, covering 40% of the region, and representing the primary source of biodiversity across most taxa (Antonelli et al., 2018b; Jenkins et al., 2013; Olson et al., 2001).

The biodiversity of neotropical rainforests provides critical ecosystem goods and services for both local and worldwide populations (Chaplin-Kramer et al., 2019; Rice et al., 2018), but these are threatened by increasing human pressures. The region has experienced a 10-fold increase in population densities over the past few decades (Tritsch and Le Tourneau, 2016), coupled with a drastic increase in human activities such as deforestation, agricultural expansion, mining and infrastructure construction (e.g. roads, damns; Castello et al., 2013; Rice et al., 2018). These unsustainable land transformations considerably modify abiotic conditions across habitats, and lead to species extinctions, resulting in altered ecosystem functioning and service provision (FAO, 2019; Rice et al., 2018). In addition, current predictions for the Amazon basin suggest that climate change will translate to increased droughts, forest-to-savanna transitions, carbon stock losses, and an alteration of the hydrologic and biogeochemical cycles which currently structure this ecosystem (Davidson et al., 2012; Nepstad et al., 2008).

Assessing the fate of biodiversity with global change and the efficiency of management policies relies largely on the measurement of biological variation at genetic, population, community and ecosystem levels. Such measures, termed 'Essential Biodiversity Variables' (EBV; Pereira et al., 2013; Table 1) are most effective when they can be measured in a standardized way that can be employed at varying scales. Currently, these measurements are based on sampling and direct observation of individuals and their description as species by taxonomists. However, obtaining EBVs for neotropical forests is not straightforward. The majority of species occurring in the Neotropics are rare and often exhibit a high level of cryptic diversity (Antonelli et al., 2018a; ter Steege et al., 2013; Zizka et al., 2018), making them difficult to describe. The description of such hyperdiverse ecosystems thus relies on considerable taxonomic expertise, yet these skills are in decline (Paknia et al., 2015).

Such a shortfall inherently affects our understanding of species spatial distribution, abundance, evolutionary history, feeding and habitat preferences, as well as functional properties (Hortal et al., 2015). Even when species are identifiable, uncertainties surrounding their spatial distribution remain considerable for neotropical rainforests, since biodiversity assessments are often spatially restricted and biased towards a limited number of accessible areas. These issues pose major limitations to characterizing these ecosystems, to better anticipating their responses to global change, and ultimately to implementing effective policies of biodiversity conservation across the region.

Environmental DNA (eDNA) based methods (Fig. 1) are now considered as key tools to overcome the aforementioned challenges (Deiner et al., 2017; Taberlet et al., 2012a, 2018; Table 1), providing numerous advantages over classical inventory approaches. Firstly, DNA for taxonomic identification allows an objective analysis of sequence composition, as opposed to more subjective determination using specimen morphology. Secondly,

Table 1 Essential biodiversity variables (EBVs) and potential utility of eDNA-based methods to measure them in neotropical rainforests.

EBV class	EBV candidate	Utility of eDNA	Sections or references
Genetic composition	Co-ancestry	Fairly useful	Sections 2 and 3.4
	Allelic diversity and population genetic differentiation	Fairly useful	Sigsgaard et al. (2016)
	Breed and variety diversity	Unknown	NA
Species populations	Species distribution	Very useful	Sections 1–2, 3.1–3.2, and 4.3–4.4
	Population abundance	Poorly useful	Sections 2, 3.3–3.4
	Population structure by age/size class	Useless	NA
Species traits	Phenology	Fairly useful	Sections 3.1 and 4.1
	Morphology and Reproduction	Useless	NA
	Physiology and movement	Fairly useful	Sections 4.1–4.2
Community composition	Taxonomic diversity	Very useful	Sections 1–3
	Species interactions	Very useful	Sections 1 and 4.3
Ecosystem function	Net Iary or IIary productivity	Poorly useful	NA
	Nutrient retention	Useless	NA
	Disturbance regime	Fairly useful	Sections 3.1–3.2 and 4.5
Ecosystem structure	Habitat structure	Fairly useful	Sections 3.1–3.2 and 4.5
	Ecosystem extent and fragmentation	Fairly useful	Sections 3.1–3.2 and 4.5
	Ecosystem composition by functional type	Useful	Sections 4.1 and 4.3

EBVs are as defined by Pereira et al. (2013). Sections of this review or reference paper discussing such applications, or associated limitations are also indicated. NA: no documentation available yet. Usefulness levels are attributed depending on the biases of eDNA for each EBV candidate, the potential costs, as well as the extent to which eDNA information has to be complemented by other sources (e.g. species functional traits)

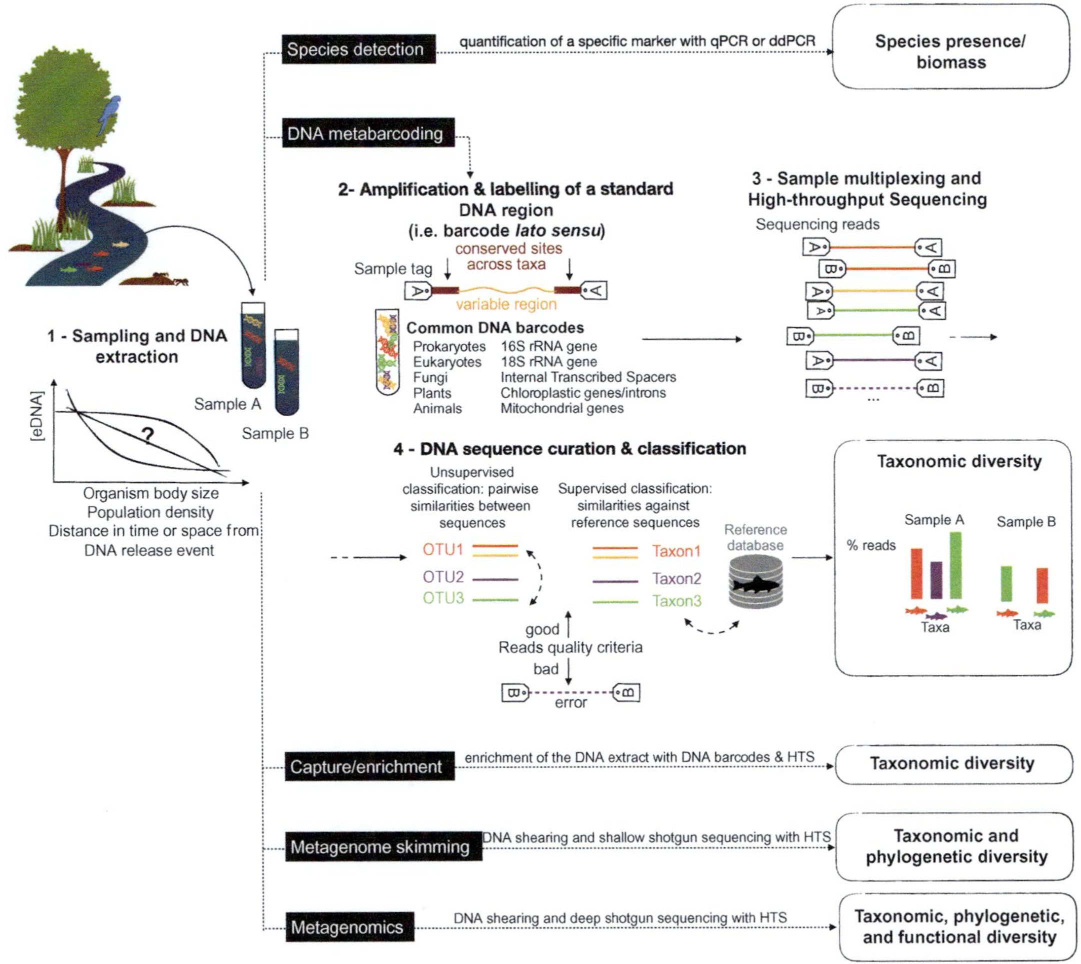

Fig. 1 See legend on next page.

the sampling of the DNA released in the environment by organisms, or environmental DNA (eDNA) is straightforward, due to its prevalence almost everywhere.

In its narrowest sense, eDNA corresponds to the mixture of DNA that can be found in any environmental matrix, whether consisting of soil, sediment or water. This DNA can belong to organisms that are present within the sample in an active or dormant stage (e.g. microbes, spores, pupae, or seeds). Alternatively, it can belong to organisms living in the sample vicinity, since organisms continuously expel DNA into the environment through excretion, secretion, decomposition, or sloughing of tissues. An environmental sample therefore contains a 'metagenome', i.e., a pool of complete or partial genomes from many different species. This metagenome is made up of DNA that can be intracellular or extracellular, dissolved or adsorbed on organic or mineral particles (Nagler et al., 2018).

In its broadest sense, eDNA also corresponds to the DNA that can be extracted from any biological material collected in natural systems, whether it corresponds to a single specimen or a whole community (e.g. bulk samples made of a mass trapping of arthropods or fish larvae). In both cases, the DNA recovered from such a sample does not only contain that of the specimens, but also encompasses the genes/genomes of the specimens symbionts, parasites, or more generally of their microbiota, as well as of their prey (Hacquard et al., 2015; Taberlet et al., 2018).

Fig. 1 Overview of the main eDNA-based methods with a focus on DNA metabarcoding applied to fish diversity assessment. The broad information that can be retrieved through each of these methods is depicted in white boxes. Step 1 corresponds to DNA sampling and extraction, which is common to all eDNA-based methods (black boxes). Each step of DNA metabarcoding is then described: Step 2 depicts the DNA amplification step and which DNA regions are generally used. It also shows how multiple samples can be sequenced in parallel: by adding a small sample-specific nucleotidic label in the 5′ region of each primer (here corresponding to sample A) prior to or after DNA amplification. Step 3 illustrates a multiplex of samples that has been sequenced in a single sequencing run. Between ca. 500–1000 samples can be multiplexed on Illumina sequencers depending on the sample diversity and sequencing technology. The sequencing step can be seen as a sampling process; the more diverse the pool of amplicons (i.e. containing different barcodes), the more sequencing reads are required to appropriately describe the sample diversity and composition. The dashed sequence in sample B illustrates a tag-jump event. Step 4 broadly summarizes the bioinformatic procedures used to curate/annotate the sequencing data and ultimately retrieve a site by OTU/species table.

Thus, the biodiversity retrieved from an eDNA sample is trans-kingdom and multitrophic. Combined with high-throughput sequencing (HTS), environmental DNA-based methods (Section 2; Fig. 1) now make large-scale and multi-taxa surveys possible from material that is easy to collect, requiring minimal taxonomic expertise. So far mostly used in temperate environments, such surveys could considerably speed up the acquisition of EBVs in general (Jetz et al., 2019), and in species-rich and challenging ecosystems such as neotropical rainforests (Table 1).

First, eDNA can provide information on the occurrence of invasive species (Takahara et al., 2013; Valentin et al., 2018), human and agricultural pathogens or pests (Bass et al., 2015; Harwood et al., 2014; Lievens et al., 2006), endangered species or populations (Harper et al., 2018; Tessler et al. 2018) and of wild species in general (Kirshtein et al., 2007; Scibetta et al., 2012). Likewise, it can be used to monitor species that indicate the health of ecosystems (i.e. bioindicators), in particular when these are microbes or invertebrates, of which identification requires advanced and often rare taxonomic skills (Mächler et al., 2014; Pawlowski et al., 2014), especially in tropical ecosystems (Bowles and Courtney, 2018; Rousseau et al., 2013, and references herein).

Second it can provide reliable information on the diversity and community composition of soil or aquatic microbes (e.g. Gilbert et al., 2012; Lauber et al., 2009; Zinger et al., 2011), as well as of invertebrates (Bista et al., 2017; Pansu et al., 2015; Zinger et al., 2019a), fish, amphibian, and mammalian communities (Boussarie et al., 2018; Schnell et al., 2018). eDNA can be further used as a standard impact assessment tool in both aquatic (Chariton et al., 2010; Li et al., 2018) and terrestrial ecosystems (e.g. Drenovsky et al., 2010), or as an evaluation tool for the success of restoration and conservation strategies (Bohmann et al., 2014; Perring et al., 2015). Finally, eDNA can provide information for multiple taxon at the same time (e.g. Li et al., 2018; Zinger et al., 2019a), and thus on biological interactions (Vacher et al., 2016). For example, using the eDNA retrieved from the faeces or gut content of a given species can reveal feeding habits (Pompanon et al., 2012), as well as host-microbiota and the occurrence of potential pathogens/parasites (Bass et al., 2015). This enables the study of full ecological networks across environmental or land disturbance gradients.

The objectives of this review are therefore (i) to provide a brief overview of eDNA-based methods, (ii) to assess their implementation to describe biodiversity in both terrestrial and aquatic ecosystems of neotropical rainforests, (iii) to highlight the limits and challenges of these methods for providing reliable assessments of EBVs in these environments, and (iv) to propose several avenues for future research in this field.

2. Overview of eDNA methods

The study of eDNA is made possible through the extraction of DNA from its environmental/biological matrix and its separation from any chemicals that can affect DNA amplification or sequencing reactions (e.g. humic substances, polyphenols, etc.). Once the DNA extract is obtained, four main methods are now routinely applied depending on the final objective (Fig. 1). They rely either on the amplification or enrichment of a target genomic region of the metagenome (i.e. species detection, DNA metabarcoding, or capture/enrichment), or on the direct—or 'shotgun'—sequencing of the metagenome (i.e. metagenome skimming or metagenomics). We briefly describe each of these approaches below and in particular emphasize DNA metabarcoding throughout this review, as this method is currently the most widely used in the field, in particular in neotropical rainforests. For more detail regarding the molecular and bioinformatics procedures involved, we refer the reader to dedicated literature (e.g. Bálint et al., 2016; de Bruijn, 2011; Deiner et al., 2017; Taberlet et al., 2018).

The 'species detection' approach consists of detecting/quantifying the amount of a DNA marker that is specific to a single or a small set of species. This approach is most relevant when one aims to detect a species with a high level of sensitivity, including low density populations or dormant/juvenile life forms. The DNA markers used for this approach must correspond to a highly polymorphic locus, enabling the design of primers which are highly species-specific. The approach currently preferred is a direct quantification of the number of copies of the target DNA marker through quantitative PCR (i.e. qPCR, sometimes referred to as real time PCR; Rees et al., 2014) or digital droplet PCR (ddPCR; Doi et al., 2015). These two quantitative methods can help to assess species population density or biomass in the studied area (e.g. Pilliod et al., 2013). This approach is relatively cheap, since it does not rely on sequencing, and is therefore more suitable for large-scale or temporal studies, although is limited to focusing on only one or a reduced set of species.

'DNA metabarcoding' (Taberlet et al., 2012b) is the most popular approach to study eDNA (see Fig. 1 for more detailed information). This approach has also been referred to in the literature as 'amplicon sequencing', 'ecometagenetics', 'metataxogenomics', but should not be confused with 'metagenomics', which we define below. As with species detection, DNA metabarcoding relies on the amplification of a target DNA region by PCR. However in this case, the DNA region targeted is used as a *barcode* to discriminate the species comprising the metagenome under study. A relatively large number of samples processed with DNA metabarcoding can be sequenced in a single HTS run (Fig. 1). The obtained sequencing reads are then processed bioinformatically to retrieve a list of species (or Operational Taxonomic Units, OTUs).

Enrichment capture on eDNA is very similar to DNA metabarcoding in that it consists of sequencing the same targeted regions. However, it differs in that the target DNA to be sequenced is not enriched through PCR amplification, but instead by capturing it with multiple, taxon-specific DNA probes bound to magnetic beads. This approach is often used for the analysis of ancient DNA of single species or simple species assemblages (e.g. Carpenter et al., 2013) and is increasingly used for the analysis of modern eDNA and complex communities (e.g. Shokralla et al., 2016; Wilcox et al., 2018), although the sensitivity and limitations of this approach are yet to be evaluated.

The last alternative relies on shotgun sequencing, i.e., random sequencing of DNA molecules from the environmental sample. 'Metagenomics' is the most direct and comprehensive DNA-based technique, and consists of sequencing as much of the metagenome as possible so as to retrieve organisms taxonomic identity, their phylogenetic relationships, as well as to their metabolic properties. However, it is also the most challenging approach. First, much of the information contained within metagenomes remains undescribed. Second, a metagenome contains a huge diversity of genes and noncoding regions, of which a tiny fraction are highly repeated (e.g. ribosomal RNA genes), and a majority of which are rare. Fully describing this complexity therefore requires substantial sampling, in this case sequencing effort, which today remains costly. Finally, most environmental samples are dominated by microbial DNA, which reduces the probability of detecting larger organisms. Consequently, metagenomics is for now mostly used in environmental microbiology (e.g. de Bruijn, 2011) or for ancient DNA analyses (Thomsen and Willerslev, 2015).

'Metagenome skimming' is a cheap version of metagenomics (Linard et al., 2015; Papadopoulou et al., 2015), albeit more expensive than methods targeting a particular DNA region. In this case, the metagenome is sampled at a shallow sequencing depth so as to sequence only highly repeated DNA regions, i.e., the ribosomal RNA gene regions and the organelle genomes for eukaryotes. These regions can then be partially or fully reconstructed, and thus used to identify the species present but also their phylogenetic relationships.

3. Current use and challenges of eDNA applications in neotropical rainforests

Current studies of biodiversity in neotropical rainforests that rely on eDNA-based methods mainly describe community composition and diversity changes along environmental or disturbance gradients in order to identify patterns in diversity and their drivers. These studies are reviewed below across ecosystems and focal organisms, and examined to determine what eDNA from different sources can reveal regarding ecological communities from neotropical rainforests and how sampling can be tailored to suit the ecological question. We will restrict our review to contemporary environments, as—to our knowledge—eDNA approaches per se have not been used yet in the neotropical rainforests for palaeoecological purposes. We refer interested readers to dedicated reviews on this particular application (Rawlence et al., 2014; Taberlet et al., 2018; Thomsen and Willerslev, 2015).

3.1 In terrestrial ecosystems

3.1.1 Microbial communities

These have been mostly analysed in the soil environment, with the study of eDNA from soil samples having a relatively long history in soil microbial ecology (Tiedje et al., 1999 for an early review). Available studies for neotropical rainforests have shown that soil prokaryotic and microeukaryotic communities vary across altitudes (Nottingham et al., 2018), soil conditions, forest types and tree species composition (Ritter et al., 2019; Vasco-Palacios et al., 2019). Numerous studies also report steep changes in composition with increased drought (Kivlin and Hawkes, 2016a; Pajares et al., 2018; Waring and Hawkes, 2015), deforestation and reconversion to different types of sylviculture (Carney et al., 2004; Kivlin and Hawkes, 2016a,b; Ndaw et al., 2009), arable farming (Mendes et al., 2015; Paula et al., 2014; Rodrigues et al., 2013; e.g. Franco et al., 2019), and even as a result of pre-columbian activities (Grossman et al., 2010; Kim et al., 2007; Navarrete et al., 2010). Likewise, soil microbial diversity differs between old-growth and secondary forests (Araújo et al., 2014; McGee et al.,

2019). All these studies exemplify the utility of soil eDNA for providing microbial-derived EBVs that are meaningful for monitoring the impact of climate change and land use practices.

3.1.2 Invertebrates

Soil micro- and macroinvertebrates (i.e. nematodes, earthworms, insects and springtails) have seldom been studied with eDNA from neotropical rainforest soil samples (Ritter et al., 2019; Wu et al., 2011; Zinger et al., 2016), and in such cases, rather as part of the whole soil eukaryote diversity, through the use of universal primers. A global-scale analysis suggests that neotropical rainforests are dominated by arthropods and enriched in soil annelids (Wu et al., 2011). Locally, soil micro- and mesofauna communities exhibit primarily random spatial patterns that are more pronounced for the mesofauna as compared to microscopic organisms, as shown at a forest site in French Guiana (Zinger et al., 2019a).

The large majority of studies of soil or aboveground invertebrates have so far rather relied on eDNA extracted from bulk samples and analysed through DNA metabarcoding, which is a fast alternative to time consuming sorting and identification of hundreds to thousands of specimens that are difficult to identify. Using this approach, Porazinska et al. (2012) were able to observe strong variation in soil nematodes communities across sites and habitats of Costa Rican rainforests. This approach has also enabled the description of aboveground terrestrial arthropods, such as sandflies occurring at several sites in French Guiana (Kocher et al., 2017c), or arthropods from a forest canopy in Honduras (Creedy et al., 2019). The latter study also tested the effect of animal size on species detection, with results suggesting such effects are not visible when sequencing depth is sufficient. Enrichment capture has also been used to analyse bulk samples of arthropods sampled with malaise traps in a forest of Costa Rica (Shokralla et al., 2016). This method was found to be more accurate in describing biodiversity than DNA metabarcoding on the same samples and classical observations.

3.1.3 Mammals

Using eDNA from bulk samples of faeces or hematophagous arthropods also seems particularly promising for sampling terrestrial vertebrate diversity as well. For example, DNA extracted from owl pellets in central Brazil provided meaningful information regarding the diversity of small mammals (Rocha et al., 2015). Likewise, vertebrate communities are better described by the DNA contained in blood feeding arthropods collected with Malaise traps and pitfall traps than with classical or camera trap-based inventories, as shown for forests in Panama and Brazil (Lynggaard et al., 2019; Rodgers et al., 2017). This approach further revealed variation in vertebrate community composition, consistent with a gradient of anthropogenic pressures in French Guiana, with a decline of diversity in the areas experiencing the highest pressures (Kocher et al., 2017b). Alternatively, water samples could also be used to study terrestrial mammals, since water bodies should accumulate and transport material from the whole catchment areas through erosion (Sales et al., 2019a).

3.1.4 Plants

Initial attempts to describe plant diversity with eDNA used bulk samples of dried, fine roots isolated by hand from soil cores that were collected following a grid or regular sampling scheme in the Barro Colorado Island in Panama (Barberán et al., 2015; Jones et al., 2011). With this approach, one soil core exhibited an average diversity of ca. four plant species and the DNA imprint of each tree individual was detectable from 1 to ca. 20 m from the stem. Similar figures can be retrieved by directly using soil as starting material, as shown in a lowland rainforest in French Guiana (Fig. 2A, see also Taberlet et al., 2018; Yoccoz et al., 2012). Thus, root and soil eDNA can offer new insights into plant root distribution in the soil and their functional implications. The aboveground plant community might be better assessed by targeting plant DNA markers on bulk samples of herbivorous arthropods, but to our knowledge, this approach has not been tested yet.

3.1.5 Constraints and limits

The above shows that organisms from terrestrial environments are either studied using environmental DNA extracted from soils, which are noticeable reservoirs of both intra- and extracellular DNA and mostly contain the signature of soil organisms, or using bulk samples of invertebrates. The former material is probably the easiest to sample from a practical point of view, and less biased/variable than different arthropod sampling techniques (Missa et al., 2009). However, one critical aspect when studying diversity using eDNA extracted from soil is the heterogeneous and complex nature of soil substrates themselves, in terms of physical, chemical and biological properties (Bardgett and van der Putten, 2014). This can be an issue when comparing contrasting environments. Typically, the amount of available extracellular DNA, useful for detecting non-microbial organisms, is

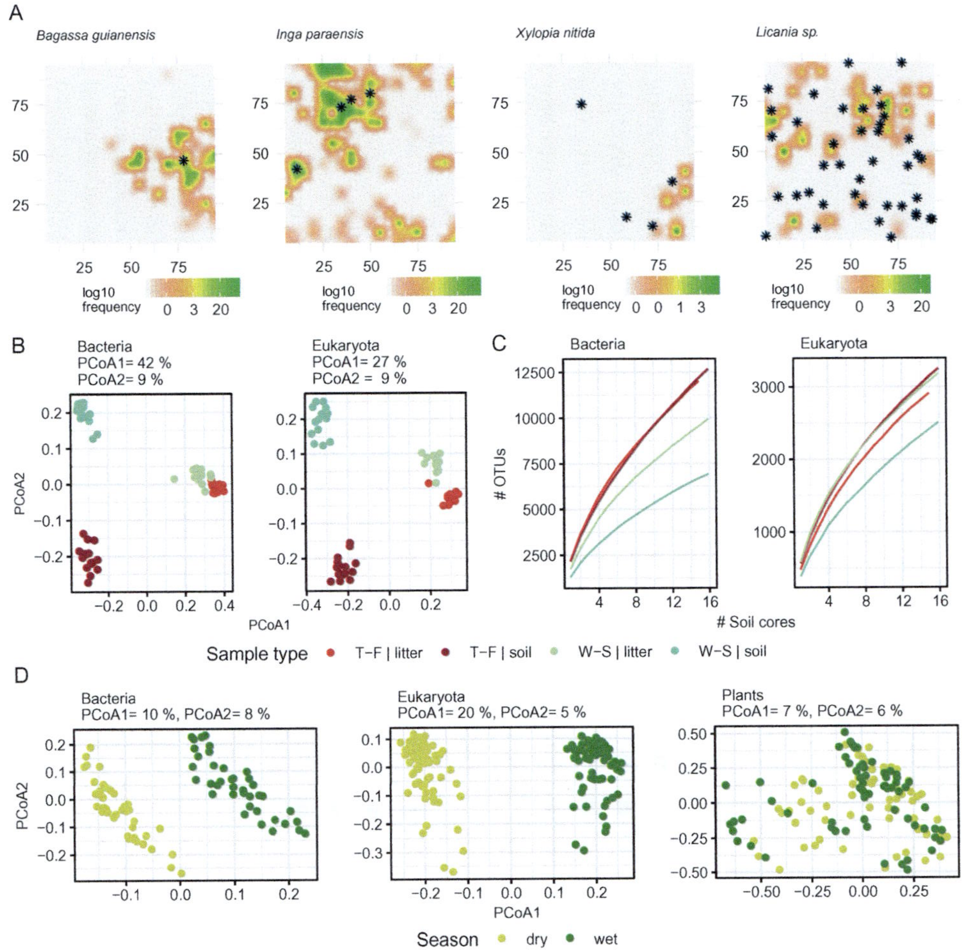

Fig. 2 See legend on next page.

strongly constrained by soil chemical properties. For example, DNA has a much stronger affinity to clay than sand (Levy-Booth et al., 2007), which thus could introduce bias to comparisons between white-sand vs. *terra firme* forest soils. This heterogeneity not only applies horizontally in space but also vertically, with clear differences in prokaryotic and microeukaryotic communities between the organo-mineral horizon and the litter layer from a taxonomic point of view (Fig. 2B), and most likely from a functional one (Basset et al., 2015; Fragoso and Lavelle, 1992; Ritter et al., 2019). This raises the question of if and how one should integrate this vertical dimension.

The same applies for how much soil samples should be collected across space, whatever the taxon targeted. For example, *terra firme* soils/litter and white-sand litter may require higher sampling effort than white-sand soils to estimate the plot-scale diversity, due to higher spatial heterogeneity (Fig. 2C). Alternatively, sampling effort could be reduced when comparing diversity or community turnover across conditions by for example building composite samples at different sampling points to capture local diversity while keeping down experimental costs. At the sample scale, extracting eDNA from volumes of material that are larger than those from most commercial soil DNA extraction kits (typically 250 μg) should best capture local diversity, which is now possible for ≥10 g of starting material, as shown for neotropical rainforest soils (e.g. Zinger et al., 2016). Thus the required

Fig. 2 Examples of soil/litter eDNA signals in 1 ha forest plots of French Guiana. (A) Comparison of the eDNA imprints of different tree species in soil samples collected every 5 m across a 1 ha plot in the Nouragues Reserve, and in the top 10 cm of the soil layer. The colour gradient represents the log 10 relative abundance of sequencing reads from each species. Black stars correspond to the location of tree stems with diameter at breast height ≥10 cm. The two left panels show signatures that are consistent with the locations of conspecific stems. The two right panels show inconsistent trends, where 'false absences' (i.e. absence of DNA when a stem is present) is likely due to deep rooting systems and 'false presences' to roots of small trees not included in the botanical inventory. It is unlikely that they correspond to pollen, seeds or litter, because such material should be present around the other conspecific stems. (B–C) DNA metabarcoding based analyses of bacterial and eukaryotic communities from soils and litter samples (ca. 10 g and 0.5 m^3 each, respectively) collected in 1 ha plots of a *terra firme* forest (Nouragues Reserve) and white-sand forest (Mana). The plots show differences (B) in community composition as measured with the Bray-Curtis index on hellinger-transformed data, summarized with a principal coordinate analysis and (C) in plot-scale diversity and spatial heterogeneity, as depicted with species accumulation curves. (D) Seasonal variations in bacterial, eukaryotic and plant community composition in the same plot as in (A) and retrieved with soil eDNA. The figure has been produced using the same indices and techniques as in (B).

sampling effort is likely to be highly system dependent, and further analyses across habitats will help better define sampling standards for neotropical rainforests when using soil as a starting material.

In any case, soil samples are unlikely to be the most relevant material for sampling the diversity of plants or aboveground animals at the plot scale because eDNA is poorly transported in soils and thus highly patchily distributed (Levy-Booth et al., 2007; Nagler et al., 2018). This patchiness most likely also results from the reduced DNA persistence in tropical rainforest soils due to high demands of the living biomass for phosphorus, which is otherwise highly limitating (Dalling et al., 2016), thus reducing the probability of detecting large organisms. Accordingly, experiments show that dead root DNA is almost totally degraded after 15 days (Bithell et al., 2014). Likewise, microbes and soil fauna communities exhibit marked seasonal and yearly dynamics (Fragoso and Lavelle, 1992; Kivlin and Hawkes, 2016a,b; Pajares et al., 2018), and so most likely does their DNA as compared to that of rooted plants, which continuously release DNA in soils (Fig. 2D).

3.2 In aquatic ecosystems

3.2.1 Microbial communities

Several studies using water eDNA have been conducted across different systems to study microbial communities. For example, Tessler et al. (2017) showed that bacterial communities from Brazilian floodplain lakes were highly distinct from other areas of the globe, while within Brazilian sites, the composition was overall fairly similar. Other studies suggest the opposite for microeukaryotic plankton: Brazilian rivers seem to exhibit marked spatial patterns with relatively high community turnover, even within the same location (Lentendu et al., 2019). These discrepancies raise the question as to whether they arise from biological differences between microeukaryotes and bacteria or from methodological inconsistencies, which emphasize the need for increased efforts in studying aquatic microbial communities in these ecosystems.

Tank bromeliads (Bromeliaceae) accumulate rainwater at the base of each leaf axil and thus represent freshwater islands in a terrestrial matrix. They harbour various aquatic organisms ranging from prokaryotes to macroinvertebrates (Benzing, 2000; Leroy et al., 2016). eDNA methods have provided insights into their community structure through either metagenomics (Rodriguez-Nuñez et al., 2018) or DNA metabarcoding (Louca et al., 2016, 2017), revealing bacterial communities that are substantially different from freshwater lake sediments and soil, but remarkably similar in functional structure due to an adaptation to oxygen-limited conditions.

3.2.2 Invertebrates

The use of water/sediment eDNA for targeting aquatic invertebrates (aquatic insects, crustaceans) has, to our knowledge, not yet been applied to neotropical rainforest ecosystems. A recent study has shown its usefulness for assessing macroinvertebrates community composition in the tropical freshwaters of Singapore (Lim et al., 2016), suggesting that such an approach could be relevant to neotropical rainforest ecosystems. As for terrestrial environments, the use of bulk samples for aquatic systems is emerging, such as with the study of Talaga et al. (2017), which details the development of DNA reference libraries for Guianese mosquito larvae to distinguish species from bulk samples of freshwater invertebrates. Still, eDNA studies of freshwater invertebrates in neotropical rainforests are currently limited by knowledge deficits related to their taxonomy and ecology and a lack of previously implemented studies. Although several macroinvertebrate indices enabling the biological evaluation of freshwater ecosystems are available (e.g. Couceiro et al., 2012; Dedieu et al., 2016), these are seldom used because to our knowledge, there is currently no environmental law or regulation relying on these in this ecoregion. One exception in that respect is French Guiana, which must comply with the European Water Framework Directive.

3.2.3 Fishes

The potential of water/sediment eDNA has received comparatively much more attention for studying fish communities. This has been particularly stimulated by the strong limitations of traditional sampling methods, which provide biased estimates and/or cause substantial fish mortalities. Indeed, gill nets provide only partial inventories, and ichthyocides such as rotenone, which were widely employed in the past, are increasingly banned. Electric fishing, which is often a good sampling alternative in other environments can be inefficient in neotropical streams because of the very low conductivity of the water (Allard et al., 2014). Hence, fish eDNA has rapidly emerged as the most promising non-invasive alternative to traditional sampling for small streams, rivers, lakes and the sea. Cilleros et al. (2019) compared eDNA and traditional sampling (nets and ichthyocides) both in small streams and rivers across French Guiana. Not only did they find that species assemblages were congruent between eDNA and traditional records, but also that eDNA results were more efficient in distinguishing the fauna from different river drainages. eDNA also enables the study of fish communities at cryptic life stages, i.e., the ichthyoplankton. Nobile et al. (2019) used DNA metabarcoding on mock communities built from fish eggs and larvae in the Grande River in Brazil, and obtained an average detection rate higher than 95%, and a relatively good estimate of larvae abundances. Likewise, capture enrichment on bulk samples for catfish larvae from the Peruvian Amazon provided a good description of the community in terms of both species and abundance (Maggia et al., 2017; Mariac et al., 2018).

3.2.4 Vertebrates

Several studies have focussed on vertebrates inhabiting aquatic environments for at least a part of their life, such as amphibians. Comparing traditional visual and audio survey techniques with DNA metabarcoding of water samples showed that eDNA accurately reflects the conclusions of the other methods while cutting the length of fieldwork required studying for frog communities in freshwater streams in the Brazilian Atlantic forest (Lopes et al., 2017; Sasso et al., 2017). Likewise, a comparison of cost models suggests that eDNA-based surveys are a cost-efficient alternative to traditional surveys in amphibian species-rich areas such as in the neotropical forest-savannah ecotones of Bolivia (Bálint et al., 2018). All these studies further show that eDNA methods circumvent biases of traditional approaches linked with species abundance and life history traits. Indeed, they not only allow for the detection of species closely associated with streams, but also of frog species at cryptic life stages (e.g. tadpoles or eggs). These are often missed by traditional surveys, but detectable with eDNA since they release DNA into the environment irrespective of their life stage. Likewise, eDNA is also able to detect endangered species in a non-destructive way, such as for the bromeliad inhabiting Trinidad golden tree frog (Brozio et al., 2017). Beyond amphibians, Sales et al. (2019a) also detected eDNA from both aquatic and terrestrial mammals when sampling water in the Amazon's mainstream and tributaries, in addition to a river of the Brazilian Atlantic forest. Comparing these results with camera trapping data confirmed the congruence between the methods (Sales et al., 2019a). Interestingly, some of the species detected using eDNA from water samples belong to strictly terrestrial species such as bats or anteaters, which can be explained by the fact that water conveys DNA from terrestrial to aquatic ecosystems. However, further studies are needed to validate this protocol for capturing terrestrial vertebrate diversity.

3.2.5 Constraints and limits

The above shows that eDNA for studying aquatic ecosystems can be extracted from either water or sediment samples, or bulk samples. For bulk samples, the trapping system is likely to be an important factor, as for traditional observations. For water or sediment samples, the interpretation of eDNA data from these two substrates remains unclear. Apart from microbial communities that highly differ between these two environments due to contrasted oxygen nutrient availability (Thompson et al., 2017; Zinger et al., 2011), the discrepancy between the results obtained from water and sediments when targeting larger organisms has been highlighted by several studies. Some studies have shown that fish eDNA concentration in sediments is higher and detectable over longer timescales than in water (Sales et al., 2019b; Turner et al., 2015). However, other studies found that sediments were less effective than water samples, e.g., allowing the recovery of only 10% of the fish species in an oligotrophic lake in Mexico (Valdez-Moreno et al., 2019). Water remains to date the most commonly used substrate for eDNA studies in neotropical rainforests due to its ease of collection. Sampling of eDNA is mainly conducted using filtration that is either directly performed in the field or subsequently in the laboratory (Lopes et al., 2017; Sales et al., 2019b).

For both water and sediment samples, the concentration of eDNA in the environmental matrix strongly determines how much material should be collected to appropriately sample freshwater diversity. For example, Cantera et al. (2019) sampled up to 340 L of water in streams and rivers in French Guiana to study the impact of sampling effort on fish detection. They showed that with a total filtration of 68 L, 91% of fish diversity could be detected in streams, and 74% in rivers. These results resonate with those obtained by Lopes et al. (2017), showing that filtering larger quantities of water (from 20 to 60 L) increases the detection probability for amphibian species and thus covering local amphibian diversity in the Brazilian Atlantic forest. Nevertheless, according to Cantera et al. (2019), filtering 34 L of water is sufficient for the recovery of 64% of the local fish fauna in Guianese streams and rivers, with a strong redundancy between eDNA replicates. Such a limited sampling effort seems hence sufficient to distinguish fish communities between sites and between ecosystem types (i.e. streams vs. rivers).

The concentration of eDNA in freshwaters is a function of the local living biomass, but also of the transport and degradation rate of eDNA in freshwater ecosystems, which depends on environmental conditions

(Barnes and Turner, 2016; Barnes et al., 2014). These processes require further investigation in both waters and sediments of neotropical rainforest ecosystems, in order to best define the sampling effort required to conduct reliable eDNA studies in these areas. It is now well established that low pH conditions, high oxygen demand and primary production, and high temperatures all accelerate the degradation of aquatic eDNA (Barnes et al., 2014; Strickler et al., 2015), which is likely to strongly vary across neotropical rainforest rivers and streams. As such, a study found an unexpected higher mammal species richness in the Brazilian Atlantic forest compared to the Amazon (Sales et al., 2019a), which is suspected to arise from a higher degradation rate of eDNA due to the low pH of the Amazon waters (pH $\leq$4).

Another important point, strongly linked to the degradation rate, is the transportation of eDNA with water flow. Studies are ongoing on this aspect in neotropical rivers, but Cantera et al. (2019) report that fish species detected from a stream site were no longer detected in eDNA samples collected in a river site located 300 m downstream from the confluence with the river. This suggests either a rapid degradation and hence a relatively short distance of eDNA transportation in neotropical waters, or more generally a high dilution downstream, which should make eDNA detection more difficult at sites distant from where it has been released. Finally, precipitations and stream size should also define local eDNA concentrations. For example, Sales et al. (2019b) reported noticeable compositional differences between samples collected from the same location following a 3-week interval. While this might be due to real variation in species composition, it is also possible that variation in water volume linked to increased precipitation at the time of sampling affected species recovery.

3.3 Common field, wet, and dry lab biases

Besides the clade- and environment-specific considerations mentioned above, the processing of eDNA data typically consists of a series of methodological steps (Fig. 1) that are all subject to various biases (Dickie et al., 2018; Zinger, et al., 2019b). We will briefly outline some of them and their associated solution when crucial for applications in neotropical rainforests ecosystems, as these issues are extensively addressed elsewhere (e.g. Alberdi et al., 2019; Deiner et al., 2017; Taberlet et al., 2018). This discussion will be mostly focused on DNA metabarcoding, as it is the approach the most widely used in eDNA research.

At the sampling step, the extent of the sampling area, sampling point locations, number of biological replicates, sample conditioning and transport, etc. are all important points to critically consider to avoid compromising the results (reviewed in Dickie et al., 2018; Taberlet et al., 2018) and will inherently depend on the particularities of the ecosystem and taxon under study (see above). Appropriate sample conditioning is also critical in tropical climates, in which microbial growth and DNA degradation is faster and more likely to occur during sample transport. Sample cooling in ice can considerably slow down DNA degradation and microbial growth, but this is seldom logistically feasible when working in remote and warm sites. To circumvent this limitation, DNA extraction can be done directly in the field with specified protocols requiring minimal infrastructure (e.g. Zinger et al., 2016; see Taberlet et al., 2018 for a detailed protocol). Alternatively, the sample can be desiccated with *silica gel* for soils or sediments, or more generally conserved with preservation buffers. These are typically used for aquatic eDNA samples, either for conserving water filters on which eDNA has been captured (Cilleros et al., 2019) or for direct addition to water samples, although preservation buffers seems less effective than sample cooling for eDNA recovery and taxon detection (Sales et al., 2019b).

After collection, the molecular processing of samples also has a variety of biases that can reduce the detection or distort the abundance of the taxa retrieved, an important limit for species population EBVs (Table 1). DNA extraction methods are not equally efficient in extracting and purifying DNA, due to variable success of cell lysis for microbes, and more generally to strong variations in the chemical composition of the starting material, with some being noticeable PCR inhibitors (e.g. humic acids). The methods employed for the extraction of DNA should be tailored to the starting material and question, or it may miss or overrepresent certain taxa. Once DNA is extracted, PCR amplification should be done with primers whose specificity-to- and generality-within the clade of interest should have been verified following a thorough literature review, preliminary tests, or the use of in silico PCR softwares (e.g. Elbrecht and Leese, 2017; Ficetola et al., 2010). Use of inappropriate primers will both strongly bias the retrieved taxa abundances and in some cases, their detection altogether.

Both PCR amplification and sequencing can also generate artefactual DNA fragments/sequences, especially when the target DNA is rare (reviewed in Taberlet et al., 2018). These artefacts are generally in low abundance and very similar to genuine fragments (e.g. only one or a few different nucleotides). They are hence difficult to identify and can artificially inflate taxonomic diversity estimates, this attribute being a community composition EBV candidate (Table 1). Nevertheless, such errors can be reduced by clustering DNA sequences at a certain sequence identity level using supervised or unsupervised approaches (Fig. 1). However, it should be noted that the bioinformatics tools used, as well as their associated parameters (e.g. clustering methods and thresholds, sequences distance indices) are not all equally efficient in reducing this artefactual variability, and can even fail to detect genuine biological variability (Bálint et al., 2016; Coissac et al., 2012; Deiner et al., 2017; Taberlet et al., 2018; Zinger and Philippe, 2016). The same applies when using supervised approaches, as the taxonomic assignment quality of a sequence/OTU inherently relies on the completeness and accuracy of the reference databases. For example, using an incomplete reference database, i.e., without conspecific sequences, can lead to an increase of 20% of erroneous taxonomic assignments as compared to the use of a complete one, as shown for Amazonian mammals (Kocher et al., 2017a).

Diversity estimates can also be inflated through the presence of genuine DNA fragments that are not initially present in the sample. The most obvious source of such a problem is exogenous contamination, which can occur not only at the sampling step, but also at the extraction, PCR, and sequencing steps because labs and reagents all contain a number of contaminants (Salter et al., 2014). Beside this problem, the multiplexing of samples within a single sequencing library or sequencing lane also produces apparent cross-sample contamination. The exact underlying mechanisms remain not well understood, but DNA fragments that are multiplexed seem to exchange the small tags used to identify their sample of origin (Fig. 1), a bias often referred to as 'tag-switches', 'tag-jumps', or 'cross-talks' (e.g. Esling et al., 2015; Schnell et al., 2015). Although this bias produces contaminants at usually low abundances, it can have strong consequences if downstream analysis relies on presence/absence and occurrences.

Given the different artefacts mentioned above, the reader should now be aware that the inclusion of negative and positive controls at the sampling, extraction, amplification and sequencing steps as well as technical replicates is critically important to ensure not only data reliability but also to optimize the processing and curation procedures of the obtained sequences through bioinformatics pipelines. The problem of false positives can be reduced by using PCR-independent methods, such as metagenomics/metagenome skimming, or capture enrichment. However both approaches still require substantial developments and cost reductions to be applicable in large-scale studies. In addition, these approaches are not error-free. They still include tag-jumps or sequencing errors (Taberlet et al., 2018; Wilcox et al., 2018) that remain difficult to detect and filter out.

Artefactual signals can have dramatic effects on estimates and patterns of alpha, and to a lesser extent beta diversity (Calderón-Sanou et al., 2019), as well as on model parameters inference such as for Hubbell's neutral model (Sommeria-Klein et al., 2016). Since these artefacts are generally low in frequency, end-users should also be careful when focusing on rare taxa. This corresponds to the majority of species in neotropical rainforest ecosystems (Antonelli et al., 2018a; ter Steege et al., 2013; Zizka et al., 2018), which suggests that it is unlikely that current eDNA-based approaches provide reliable estimates of species richness, i.e., the number of species being present in the ecosystem studied. Nevertheless, these approaches can still provide meaningful information on alpha or beta diversity patterns by using diversity indices penalizing low-abundance OTUs or taxa such as those based on Hill numbers, which includes well known indices such as Shannon or Simpson diversity (Chao et al., 2014). These have been shown to provide more reliable ecological inferences (Calderón-Sanou et al., 2019), and should be favoured over other indices where singletons (e.g. Chao, ACE, Fisher's alpha indices) or rare species have a strong weight (e.g. inferences based on species abundance distribution or on presence-absence data). Nevertheless, new occupancy models able to detect both false negative and false positives are currently emerging (Ficetola et al., 2016; Guillera-Arroita et al., 2017), and their inclusion in current data curation procedures will certainly allow overcoming the above-mentioned limitations.

3.4 Biological interpretation of eDNA

Beyond the methodological considerations raised above, eDNA has specific intrinsic properties which must be considered when interpreting derived results. Even if eDNA data resembles a traditional species abundance table, the abundances correspond to sequencing read counts and species correspond to species, genera, or to OTUs defined at a given level of sequence similarity. This difference can have strong implications for the type of EBV that eDNA can actually measure (Table 1), as well as on ecological inferences depending on the question addressed and types of inference tools used, in particular when they involve theoretical frameworks and models that rely heavily on species and abundances (e.g. niche or neutral models, species abundance distributions).

A first uncertainty is on the extent to which sequences or OTUs can be used as a proxy for species. In most eDNA studies, species or OTUs are defined by using a threshold of 97% of sequence similarity. This threshold has been historically defined for full-length barcode genes (e.g. Hebert et al., 2003; Schoch et al., 2012; Stackebrandt and Goebel, 1994). However, current eDNA studies target small regions within these barcodes (Fig. 1) in order to comply with both the sequencing limits of current HTS instruments and, when applicable, with the fragmented nature of extracellular DNA. This constraint inherently comes with a loss of taxonomic resolution, which may have consequences for subsequent ecological inferences. The 'Amplicon-' or 'Exact Sequence Variant' concept (ASV or ESV, Callahan et al., 2017) has been recently proposed to, amongst other reasons, circumvent this problem, yet this remains sensitive to some molecular artefacts. Sometimes interpreted as intraspecific variability, which can be a desirable output of eDNA (Table 1), ASVs may also yield ecological signals that differ from what one should expect when considering species. Finally, eDNA markers do not have the same taxonomic resolution across clades. For example, the fungal Internal Transcribed Spacers, or the metazoan cytochrome oxidase subunit I (COI) can exhibit some intraspecific variability for certain groups, and only genus to family level variability for others (Schoch et al., 2012). Phylogenetic-based approaches can to a certain extent deal with these limitations. However, while these can be employed with metagenomics or metagenome skimming data (Andújar et al., 2015; Papadopoulou et al., 2015), the short and hypervariable nature of most classical DNA markers used for DNA metabarcoding do not enable making robust phylogenetic inferences, which limits the use of such data to retrieve co-ancestry relationships (Table 1). For such data, the phylogenetic diversity should be retrieved through phylogenetic placement methods, provided that a robust backbone phylogenetic tree is available (e.g. Czech et al., 2019; Matsen et al., 2010), which remains challenging for neotropical taxa (see Section 4.1.).

The other uncertainty of eDNA data relates to the meaning of sequencing reads counts. As mentioned in Section 3.3, a DNA extract is subjected to a suite of molecular manipulations that can distort the original distribution of DNA fragment abundances. Adding spiked DNA of known composition and concentrations in environmental samples could allow for the retrieval of absolute values of eDNA molecules (e.g. Smets et al., 2015; Thomas et al., 2016). However, while the abundance (relative or not) of eDNA molecules has been found to correlate with organism biomass in simple experimental set ups (e.g. Nobile et al., 2019) or when quantifying single species *in natura* with qPCR (reviewed in Taberlet et al., 2018), several factors can alter this relationship, and hence, assessment of population abundance (Table 1). First, eDNA persistence and transport in the environment makes it difficult to know whether this biomass is local and contemporary. This is likely to be especially true for soils or sediments as compared to water, the latter being more exposed to high temperature and UV radiations, which favour DNA degradation (Barnes and Turner, 2016; Nagler et al., 2018). Even if this bias is limited, relating eDNA abundance to population abundance per se remains challenging. Indeed, the number of DNA marker gene copies depends on the taxon, on the tissues from which eDNA is released, the biomass/size of the organisms, but also its life stage (Maruyama et al., 2014). To our knowledge, there is no tool which can retrieve individual counts from sequencing reads or eDNA molecules at the scale of the biological community. These uncertainties have often led researchers to prefer presence-absence metrics over abundance-based ones. However, unless the representativeness of the data curation procedure can be proven, we advocate against such reasoning due to the high error rate of PCR and sequencing based approaches (see Section 3.3.).

Given the above-mentioned differences in the intrinsic nature of eDNA data as compared to traditional species abundance tables, this raises the question of whether one can draw ecological inferences with classical tools. Typically, it remains largely uncertain whether inferring community diversity and related characteristics from eDNA-based species abundance distribution or using process models involving explicitly species and individuals is a correct approach. For example, adaptation of Hubbell's model to account for body size or biomass could be more appropriate (O'Dwyer et al., 2009; Sommeria-Klein et al., 2016). There is hence a need for development of related tools and theories in ecology that would better comply with the nature of eDNA data.

4. Future directions and perspectives

The past decade has seen enormous advances in the development and extension of eDNA-based approaches, as well as a large number of potential applications in various environments, including neotropical rainforests. However, these applications remain largely underused in this part of the world when compared with other far less diverse regions (this paper; Mulatu et al., 2017; Belle et al., 2019). This is because countries harbouring lower diversity are in general more developed economically: infrastructure for molecular-based research is accessible, with associated personnel now relatively well trained for eDNA data generation and analysis. On the other hand, the Nagoya Protocol on Access and Benefit Sharing restricts the access of genetic resources to the country where the sample has been collected, protecting local countries, which are often less economically developed, from unethical practices by collectors outside of and within the scientific community. We argue that current efforts to develop eDNA-based research in neotropical countries should be encouraged and strengthened through international collaborations between researchers from Neotropical countries and researchers from countries that have already overcome issues relating to methodological application, technical infrastructure and skill acquisition. Such efforts will enable the acquisition of EBVs related to taxonomic diversity, but also beyond to provide information such as species distributions or biotic interactions (Table 1), as well as associated underlying processes. In this final section of the review, we will explore how eDNA can be better used to improve research methods and their subsequent applications, and in doing so ultimately contribute to improving conservation programs and management strategies for these hyperdiverse ecosystems.

4.1 Making better sense of eDNA data with better reference databases

A key limit to current eDNA studies in neotropical rainforests is the provision of relatively poor taxonomic information. This drawback arises in part from the limitations of eDNA-based methods mentioned above, but is further exacerbated when dealing with neotropical taxa in that they are largely underrepresented in current DNA reference databases, and/or they have an unresolved taxonomy. This is particularly true for microeukaryotes, for which a significant proportion of OTU and sequencing reads remain unassigned to a taxon, even at the phylum level (Ritter et al., 2019; Zinger et al., 2019a). The deficit in DNA references also applies to less cryptic organisms. For example, only 58% of the São Paulo tree flora has genetic records in international DNA reference databases (de Lima et al., 2018). While eDNA does facilitate the identification of challenging taxa at gross taxonomic levels, it is therefore unlikely to provide a satisfactory solution for resolving the Linnean shortfall and provide on its own information on EBVs related to species evolutionary history and functional traits. We hence argue that the future of eDNA remains inherently intertwined with the continued efforts of taxonomists and naturalists to sample, identify and store physical specimens in order to complement DNA reference databases, but also to describe their morphology, evolutionary history, functional traits, and to solve taxonomic problems (Dormontt et al., 2018; Pinheiro et al., 2019; Sheth and Thaker, 2017).

Augmenting the completeness of DNA reference databases is crucial not only to facilitate the assignment of unknown sequences. It is also essential to ensure, or verify the plausibility of the retrieved signal, which can be extremely noisy as discussed above. However, one of the difficulties in improving DNA reference databases is the current lack of consensus when choosing the DNA regions to be used across studies. Indeed, these may differ from the ones used in curated reference databases linked to voucher specimens such as the BOLD system for animals and plants (Ratnasingham and Hebert, 2007) or databases dedicated to the ribosomal clusters for microorganisms (e.g. UNITE, Abarenkov et al., 2010; SILVA, Quast et al., 2013), which only contain gold standard barcoding genes (i.e. COI for animals, *rbc*L or *mat*K for plants, and ITS for fungi). This is because gold standard barcodes are not necessarily compatible with all applications of eDNA, which often require DNA primers that target broad taxonomic groups and DNA markers that are short to suit existing sequencing technologies or the degraded state of eDNA. Conserved priming sites across broad taxonomic groups are often absent within these gold standard barcodes, an issue highlighted for animals (Deagle et al., 2014) and plants (Hollingsworth et al., 2011). As a consequence, existing primer sets targeting classical barcode subregions are often biased towards certain taxa or on the contrary lack of specificity because they contain too many degenerate bases (Collins et al., 2019; Deagle et al., 2014). Alternative DNA markers fulfilling these conditions are often located in mitochondrial or chloroplastic introns or ribosomal genes (Fig. 1) which are better conserved. However these regions also often exhibit lower taxonomic resolution and are much less referenced in DNA databases. When choosing a DNA marker, the end-user must hence usually compromise between more precise taxonomic information versus unbiased sampling of biodiversity. These considerations go beyond the scope of this review and we refer interested users to dedicated literature on the subject (Deagle et al., 2014; Hollingsworth et al., 2011; Taberlet et al., 2018).

As stated above, the choice of a given DNA marker strongly relies on the biological question to be addressed, the starting material used and because current reference databases have large deficits in neotropical organisms. Therefore, we encourage the construction of custom reference databases for the targeted DNA region from local taxa that are likely to be detected with the eDNA analysis, as done for example in studies using the mt 12S rRNA gene of neotropical mammals (Kocher et al., 2017a) and of Guianese fishes (Cilleros et al., 2019), or for the ITS1 region for the Basidiomycota of French Guiana (Jaouen et al., 2019). Although often considered as a costly endeavour, it can be achieved at relatively low expense (as low as ca. 5 $USD/specimen) by using freshly collected specimens, or herbarium/museum collections (e.g. Dormontt et al., 2018) and by multiplexing thousands specimens in a single HTS run. Another promising alternative that will alleviate the lack of standard DNA markers across studies lies in the building of 'marker-free' DNA reference databases. This is now possible with genome skimming (Dodsworth, 2015), which is similar to metagenome skimming but relies on a single specimen. This approach produces sequences usable for both gold standard and other barcodes as it generates sequences of the complete organelle genomes and full nuclear ribosomal regions (Coissac et al., 2016). Although this remains relatively expensive (as low as ca. 100 $USD/specimen), it is likely to become more affordable with continued decreases in sequencing costs.

In addition to compiling DNA information across species, reference databases could complement taxonomic data with ecophysiological characteristics, such as foliar, root or seed traits for plants, and morphological characteristics such as body size for animals. Such information would be extremely valuable, allowing eDNA studies to go beyond the simple description of taxonomic and phylogenetic diversity of the studied system (Table 1). For example, inferring taxon function or gross ecological traits from eDNA data is now possible for bacteria and fungi through databases that compile both metabolic, life history traits, or broad lifestyle types (e.g. PiCrust, Langille et al., 2013; FUNGuild, Nguyen et al., 2016). To our knowledge, such tools are currently not directly available for macroorganisms, although several databases compiling taxonomic and functional information in a number of groups have been developed (e.g. FishBase, Froese and Pauly, 2019; TRY, Kattge et al., 2019; Atlantic Bird Traits, Rodrigues et al., 2019; or the Global Ants Database, Gibb et al., 2019). Their coupling with DNA reference databases would certainly help advance the field of eDNA studies to include more process-based approaches.

4.2 Toward eDNA-based occurrence portals for the neotropics?

The greatest strength of eDNA-based approaches is their relative ease of implementation for both long-term and large-scale monitoring of complex communities. Even if these data are not necessarily well resolved at the species level, they still constitute invaluable occurrence data and thereby provide more information on species distributions, another EBV (Table 1), that is currently largely lacking for neotropical rainforest taxa (Antonelli et al., 2018a).

To date, eDNA data and metadata reporting the location, time and exact protocol of the sampling are disseminated individually using study specific web repositories, as in data papers (e.g. Murienne et al., 2019) or more general repositories (e.g. Dryad, https://datadryad.org, Zenodo, https://zenodo.org; or the Short Read Archives from GenBank, https://www.ncbi.nlm.nih.gov/sra). However, the construction of dedicated portals compiling eDNA-based taxa occurrence can now be envisioned for all neotropical rainforests and beyond following the examples of the occurrence portal GBIF (https://www.gbif.org), the BOLD system (http://www.boldsystem.org) which integrates DNA data with occurrence, or the EMP (http://www.earthmicrobiome.org) which compiles occurrence and diversity of microbial taxa across the globe. The success of such an endeavour depends on the adequate standardization of data, a challenge given that ecological signal from eDNA data is influenced by the technique used, the DNA region targeted, and the protocols of molecular biology and bioinformatics chosen. While defining standards for such purposes will certainly facilitate the integration of data across studies, it is also likely that this will be difficult to apply to all desired situations, which may ultimately undermine scientific advances. Several alternatives have been proposed to circumvent this issue. The first is to adopt sequence taxonomy classification as a standard unit (Ramirez et al., 2018). As highlighted above, such an approach heavily depends on taxonomic expertise and enriched DNA reference databases to make the best use of eDNA data. The second is the implementation of 'eDNA biobanking', i.e., the development of storage facilities for eDNA samples that could be reused with different technologies (Jarman et al., 2018).

Although less precise than traditionally collected occurrence data, which are limited in other ways, Sections 3.1.2 demonstrate how eDNA-based studies can unveil the abiotic determinants of neotropical diversity. Increasing eDNA-based taxonomic inventories across environmental gradients will provide insights into taxa environmental/physiological tolerances/preferences (Table 1), information which remains scarce in neotropical rainforests. From a more applied perspective, increasing eDNA sampling across land use gradients will enable the identification of indicator taxa for environmental impacts or umbrella taxa that are specific to this ecoregion. However, this application currently remains limited by the difficulty in retrieving population size information from eDNA as discussed above. Without significant developments for this particular aspect, eDNA-based approaches will likely remain of limited utility when assessing the conservation status of neotropical taxa.

4.3 Shedding new light on biotic interactions

The increasing use of eDNA will also certainly fill the current gap of knowledge on species interactions (Table 1) by improving the description of complex and multitrophic communities for both well studied taxa and more elusive organisms. Such assessments are urgently needed at a time where environmental changes already cause direct species loss and cascading extinction via bottom-up or top-down effects, especially in tropical ecosystems, including neotropical rainforests, where biotic interactions are often expected to be highly specific (Barnes et al., 2017).

It is now possible to analyse the diet of a particular species by collecting faeces, gut contents or even the DNA traces herbivores or pollinators leave on plants (Thomsen and Sigsgaard, 2019). These applications are routinely used in temperate ecosystems (Alberdi et al., 2019; Bohmann et al., 2014; Taberlet et al., 2018). By contrast, only few diet studies have been performed on neotropical organisms, i.e., on tapirs from French Guiana (Hibert et al., 2013), on white-face capuchins from Costa Rica (Mallott et al., 2018), on neotropical vampire bats (Bohmann et al., 2018) and rodents (Lopes et al., 2015), and on particular arthropods (Kocher et al., 2017b; Paula et al., 2016; Rodgers et al., 2017). New protocols of diet assessment based on faeces or gut contents are now available and optimized to reduce host DNA concentration in DNA extracts (e.g. Krehenwinkel et al. 2017). Such improvements considerably reduce the costs associated with molecular treatments and sequencing and hence allow for the implementation of large-scale studies of full food-webs composed of understudied and hyperdiverse taxa. This will certainly enable improved characterization of trophic niche and breadth for many neotropical taxa, thereby improving documentation of feeding behaviour in relation to species functional traits and competitive interactions.

Likewise, eDNA can be used to unravel plant-pollinator networks. Pollinators yield substantial amounts of pollen on their bodies, and conversely the surfaces of leaves and flower petals also harbour traces of DNA belonging to visiting pollinators. This material can be used to build reliable plant-pollinator insect interactions, as shown in temperate ecosystems (Pornon et al., 2016; Thomsen and Sigsgaard, 2019). The applicability of the methods has, to our knowledge, not yet been tested in neotropical rainforests and remains to be critically assessed due to the particular climatic conditions, much greater richness, and also the greater amount of vertebrate pollinators in these ecosystems, which can be more challenging to sample than arthropods.

Similarly, improved understanding of host-microbiota interactions can have important implications for threatened species conservation (West et al., 2019). This can be done by studying microbial communities occurring at the surface or within larger organisms in a more comprehensive way than before. So far, existing studies have principally aimed to describe microbial communities and, in some cases, their assembly mechanisms. This has been done mostly for leaf or root endophytes in trees (Bonfim et al., 2016; Donald et al., 2020; Kembel et al., 2014; Schroeder et al., 2019), palms (Donald et al., 2019), grasses (Higgins et al., 2014) or fern species (Del Olmo-Ruiz and Elizabeth Arnold, 2017) and for the microbiota of frogs to assess its potential role in the resistance to the chytrid fungus *Batrachochytrium dendrobatidis* (Catenazzi et al., 2018; Hughey et al., 2017). To our knowledge the microbiota associated with neotropical mammals has been only assessed for the endangered Andean Bear (Borbón-García et al., 2017), and the same holds true for the microbiota of invertebrates, which has been so far mostly studied on emblematic arthropods such as ants (Pringle and Moreau, 2017; Sapountzis et al., 2015). Although few studies have shown experimentally that the plant microbiota can promote the growth and survival of seedlings (Christian et al., 2017; Leroy et al., 2019), much remains to be done to understand the functional contribution of the microbiota to host health, and how this can affect community level distribution or diversity patterns (e.g. Janzen-Connell effects accounting for the whole microbiota).

The approaches discussed above mostly enable reconstructing bi- or tri-partite networks, but future applications are likely to span the whole ecological network to advance our understanding of the resistance and resilience of biological communities to disturbance. Indeed, eDNA can provide co-occurrence data for multiple taxonomic, functional and trophic groups retrieved from soil, sediments or water. While these co-occurrences do not represent biological interactions per se, these can assist in the discovery of a large variety of interactions at larger temporal/spatial scales, provided that these inferences are evaluated with a priori knowledge of the system or statistical tools (Vacher et al., 2016).

4.4 Epidemiology and healthcare

Neotropical rainforests ecosystems harbours many emerging infectious diseases, and use of eDNA for monitoring their agents or vectors has enormous implications for human health. Most parasites and pathogens are usually only detected when aggregating on or in their hosts and without eDNA, their detection remains challenging in the environment (Bass et al., 2015). Recent results from Sengupta et al. (2019) indicate that free living larval aquatic phases of *Schistosoma* can be detected with eDNA from water samples, opening an avenue to the control of this neglected tropical disease affecting >250 million people worldwide, mainly in Africa, but with human infestations in several regions of South and Central America. Although using eDNA as diagnostic evidence for pathogens or parasites requires extensive validation before it is used in notification procedures or detection programs (Bass et al., 2015), developments of such methods in the region would considerably improve the monitoring and fight against agents of tropical diseases.

A number of human diseases require a vector, typically an insect, to transmit the pathogen and surveillance programs usually rely on monitoring potential vector populations. Such a task can prove daunting given a single night of trapping using a standard CDC trap (Center for Disease and Control) could yield thousands of mosquitoes/sandflies that need to be identified to species level. eDNA-based approaches greatly reduce the time and costs related to these identifications (Kocher et al., 2017c; Talaga et al., 2017), and could be used for routine monitoring of vector species and help in the control of vector-borne diseases.

Classical epidemiological monitoring programs largely focused on pathogens or their vectors, yet it is increasingly recognized that the prediction of transmission risk should include a better understanding of the ecosystem as a whole. This is particularly true in a context of biodiversity erosion and habitat degradation, which could be connected to the emergence of diseases as a result of trophic food-web modifications. For example, deforestation has been suggested to lead to the emergence of diseases such as malaria (Vittor et al., 2009) or Buruli ulcer (Morris et al., 2016), through reductions in diversity and modifications to the species composition of aquatic food-webs. Because eDNA-based methods can provide not only rapid information on pathogens and vectors, but also a broad characterization of the whole ecological network, we believe they will strongly modify our approach to epidemiology and understanding of disease emergence in the next few years.

4.5 Conservation and impact assessments in neotropical rainforests and beyond

Managing ecosystems and biodiversity requires efficient detection of the species of interest, but also standard, cost- and time-effective protocols that can be implemented repeatedly across large spatial scales and through time, with low, or limited impact on organisms. Such protocols are currently not available for monitoring neotropical rainforests and, more generally, neotropical ecosystems. This review shows that eDNA-based methods fulfil these criteria while enabling characterization of the taxonomic composition of multiple trophic communities, and could even constitute proxies of other EBVs. These methods complement remote sensing tools since eDNA provides information at a much finer taxonomic resolution, thereby better complying with some of the Aichi Targets that focus on endangered and invasive species (Bush et al., 2017; Marques et al., 2014). Their use could hence greatly facilitate the establishment of Rapid Biodiversity Assessment programs.

eDNA-based rapid biodiversity assessments hold great potential for the evaluation of environmental impacts, in particular for the ever increasing unsustainable use of land in neotropical rainforests, as exemplified with soil organisms (e.g. Franco et al., 2019). Likewise, eDNA-based methods will be able to help evaluate the success of different restoration and conservation strategies (Fernandes et al., 2018). However, the use of eDNA for informing management and political decisions will inherently require the development of quick and standardized sampling protocols that work across varying environment types and can be easily applied by practitioners. Beyond standardization, which we show here to be a challenging issue, such an application implies the development of biotic integrity indices that are easily transferable to stakeholders, resource managers, and policy makers, and eDNA research is still in its infancy on this particular matter (Cordier et al., 2019). Nevertheless, we are confident that these limits can and will be overcome in the near future.

Aside from rainforests, the Neotropics holds large areas of other biomes that face threats that are not necessarily the same as for tropical rainforests but whose diversity remains poorly described with both traditional and eDNA methods (Antonelli et al., 2018a). For example, eDNA could be particularly relevant to describe and monitor white-sand ecosystems which harbour a unique flora and fauna (Fine and Baraloto, 2016), but which are currently threatened by increases in cattle ranching, deforestation for firewood or mining for sand (Ferreira et al., 2013). Likewise, it could be used for savanna and dry forest conservation, habitats which currently experience greater pressures than other neotropical biomes, typically as deforestation, localized human disturbance and increasing drought frequency and intensity (González-M et al., 2018; Strassburg et al., 2017). For example, the revision of Brazil's Forest Code in 2012, the Cerrado (Brazilian savanna) indirectly encouraged Brazilian agribusiness to invest in this biome (Soares-Filho et al., 2014; Strassburg et al., 2017). Estuaries, including mangrove forests, also represent neglected and threatened habitats in the Neotropics, while harbouring rich communities and serving as a nursery for many fish and crustaceans (Mumby et al., 2004). In these environments, the turbidity and strong water currents make species inventories difficult, a limit that could be circumvented with eDNA (Belle et al. 2019). A last example is the Pantanal biome, a savanna wetland which hosts a unique diversity, supports essential ecosystem services, and is currently under strong human pressure (Alho, 2008). Descriptive and monitoring studies using eDNA analysis in these neglected yet important ecosystems would therefore help to better characterize their diversity and how they respond to various pressures.

From a more basic perspective, the possibility to implement comprehensive, large-scale and long-term biodiversity observatories will certainly help to gain insights into the origin and maintenance of neotropical biodiversity, and its singularity in many ecosystems. Reconstructing past ecosystems from ancient DNA (Thomsen and Willerslev, 2015) would be extremely valuable in such a case, and would further improve our understanding of the long-term dynamics of neotropical ecosystems, and hence better predict their future. However, it remains unclear whether eDNA can persist in the long term in tropical ecosystems and further studies are required in this area. Nevertheless, long-term dynamics can be assessed through monitoring initiatives along transitions between different biomes. For example, savannas and dry forests constitute transitory or alternative stable states of rainforests in response to global changes (Dexter et al., 2018; Nepstad et al., 2008), and monitoring these sites through eDNA should provide useful information on their dynamics, enabling the identification of early warning markers of major ecological transitions. Acquisition of such data will prove valuable for anticipating the status of these environments and prioritizing corresponding conservation or restoration actions to mitigate such transitions.

Acknowledgements

Some of the research presented in this paper received funding from the French Agence Nationale de la Recherche (ANR) (METABAR: ANR-11-BSV7-0020; GlobNets: ANR-16-CE02-0009; DEBIT: ANR-17-CE02-0007-01), from 'Investissement d'Avenir' grants managed by the ANR (CEBA: ANR-10-LABX-25-01; TULIP: ANR-10-LABX-0041), and from São Paulo Research Foundation (FAPESP #2013/50741-7).

References

Abarenkov, K., et al., 2010. The UNITE database for molecular identification of fungi—recent updates and future perspectives. New Phytol. 186 (2), 281–285.

Alberdi, A., et al., 2019. Promises and pitfalls of using high-throughput sequencing for diet analysis. Mol. Ecol. Resour. 19 (2), 327–348.

Alho, C.J.R., 2008. Biodiversity of the Pantanal: response to seasonal flooding regime and to environmental degradation. Braz. J. Biol. 68 (Suppl. 4), 957–966.

Allard, L., et al., 2014. Electrofishing efficiency in low conductivity neotropical streams: towards a non-destructive fish sampling method. Fish. Manag. Ecol. 21 (3), 234–243.

Andújar, C., et al., 2015. Phylogenetic community ecology of soil biodiversity using mitochondrial metagenomics. Mol. Ecol. 24 (14), 3603–3617.

Antonelli, A., Zizka, A., et al., 2018a. Amazonia is the primary source of Neotropical biodiversity. Proc. Natl. Acad. Sci. U. S. A. 115 (23), 6034–6039.

Antonelli, A., Ariza, M., et al., 2018b. Conceptual and empirical advances in Neotropical biodiversity research. PeerJ 6, e5644.

Araújo, A.S.F., et al., 2014. Soil bacterial diversity in degraded and restored lands of Northeast Brazil. Antonie Van Leeuwenhoek 106 (5), 891–899.

Bálint, M., et al., 2016. Millions of reads, thousands of taxa: microbial community structure and associations analyzed via marker genes. FEMS Microbiol. Rev. 40 (5), 686–700.

Bálint, M., et al., 2018. Accuracy, limitations and cost efficiency of eDNA-based community survey in tropical frogs. Mol. Ecol. Resour. 18 (6), 1415–1426.

Barberán, A., et al., 2015. Relating belowground microbial composition to the taxonomic, phylogenetic, and functional trait distributions of trees in a tropical forest. Ecol. Lett. 18 (12), 1397–1405.

Bardgett, R.D., van der Putten, W.H., 2014. Belowground biodiversity and ecosystem functioning. Nature 515 (7528), 505–511.

Barnes, M.A., Turner, C.R., 2016. The ecology of environmental DNA and implications for conservation genetics. Conserv. Genet. 17, 1–17.

Barnes, M.A., et al., 2014. Environmental conditions influence eDNA persistence in aquatic systems. Environ. Sci. Technol. 48 (3), 1819–1827.

Barnes, A.D., et al., 2017. Direct and cascading impacts of tropical land-use change on multitrophic biodiversity. Nat. Ecol. Evol. 1 (10), 1511–1519.

Bass, D., et al., 2015. Diverse applications of environmental DNA methods in parasitology. Trends Parasitol. 31 (10), 499–513.

Basset, Y., et al., 2015. Arthropod distribution in a tropical rainforest: tackling a four dimensional puzzle. PLoS One 10 (12), e0144110.

Belle, C.C., Stoeckle, B.C., Geist, J., 2019. Taxonomic and geographical representation of freshwater environmental DNA research in aquatic conservation. Aquat. Conserv. Mar. Freshwat. Ecosyst. 29 (11), 1996–2009.

Benzing, D.H., 2000. Bromeliaceae: Profile of an Adaptive Radiation. Cambridge University Press.

Bista, I., et al., 2017. Annual time-series analysis of aqueous eDNA reveals ecologically relevant dynamics of lake ecosystem biodiversity. Nat. Commun. 8, 14087.

Bithell, S.L., et al., 2014. DNA analysis of soil extracts can be used to investigate fine root depth distribution of trees. AoB Plants 7, lu091.
Bohmann, K., et al., 2014. Environmental DNA for wildlife biology and biodiversity monitoring. Trends Ecol. Evol. 29, 358–367.
Bohmann, K., et al., 2018. Using DNA metabarcoding for simultaneous inference of common vampire bat diet and population structure. Mol. Ecol. Resour. 18 (5), 1050–1063.
Bonfim, J.A., et al., 2016. Dark septate endophytic fungi of native plants along an altitudinal gradient in the Brazilian Atlantic forest. Fungal Ecol. 20, 202–210.
Borbón-García, A., et al., 2017. Captivity shapes the gut microbiota of Andean bears: insights into health surveillance. Front. Microbiol. 8, 1316.
Boussarie, G., et al., 2018. Environmental DNA illuminates the dark diversity of sharks. Sci. Adv. 4 (5), eaap9661.
Bowles, D.E., Courtney, G.W., 2018. Advances in aquatic insect systematics and biodiversity in the Neotropics: introduction. Aquat. Insects 39 (2-3), 89–93.
Brozio, S., et al., 2017. Development and application of an eDNA method to detect the critically endangered trinidad golden tree frog (*Phytotriades auratus*) in bromeliad phytotelmata. PLoS One 12 (2), e0170619.
Bush, A., et al., 2017. Connecting Earth observation to high-throughput biodiversity data. Nat. Ecol. Evol. 1 (7), 176.
Calderón-Sanou, I., et al., 2019. From environmental DNA sequences to ecological conclusions: how strong is the influence of methodological choices. J. Biogeogr. 47 (1), 193–206.
Callahan, B.J., McMurdie, P.J., Holmes, S.P., 2017. Exact sequence variants should replace operational taxonomic units in marker-gene data analysis. ISME J. 11 (12), 2639–2643.
Cantera, I., et al., 2019. Optimizing environmental DNA sampling effort for fish inventories in tropical streams and rivers. Sci. Rep. 9 (1), 3085.
Carney, K.M., Matson, P.A., Bohannan, B.J.M., 2004. Diversity and composition of tropical soil nitrifiers across a plant diversity gradient and among land-use types. Ecol. Lett. 7 (8), 684–694.
Carpenter, M.L., et al., 2013. Pulling out the 1%: whole-genome capture for the targeted enrichment of ancient DNA sequencing libraries. Am. J. Hum. Genet. 93, 852–864.
Castello, L., et al., 2013. The vulnerability of Amazon freshwater ecosystems. Conserv. Lett. 6 (4), 217–229.
Catenazzi, A., et al., 2018. Widespread elevational occurrence of antifungal bacteria in Andean amphibians decimated by disease: a complex role for skin symbionts in defense against Chytridiomycosis. Front. Microbiol. 9, 465.
Chao, A., Chiu, C.-H., Jost, L., 2014. Unifying species diversity, phylogenetic diversity, functional diversity, and related similarity and differentiation measures through Hill numbers. Annu. Rev. Ecol. Evol. Syst. 45, 297–324.
Chaplin-Kramer, R., et al., 2019. Global modeling of nature's contributions to people. Science 366 (6462), 255–258.
Chariton, A.A., et al., 2010. Ecological assessment of estuarine sediments by pyrosequencing eukaryotic ribosomal DNA. Front. Ecol. Environ. 8 (5), 233–238.
Christian, N., et al., 2017. Exposure to the leaf litter microbiome of healthy adults protects seedlings from pathogen damage. Proc. R. Soc. B Biol. Sci. 284 (1858), 20170641.
Cilleros, K., et al., 2019. Unlocking biodiversity and conservation studies in high-diversity environments using environmental DNA (eDNA): a test with Guianese freshwater fishes. Mol. Ecol. Resour. 19 (1), 27–46.
Coissac, E., Riaz, T., Puillandre, N., 2012. Bioinformatic challenges for DNA metabarcoding of plants and animals. Mol. Ecol. 21, 1834–1847.
Coissac, E., et al., 2016. From barcodes to genomes: extending the concept of DNA barcoding. Mol. Ecol. 25 (7), 1423–1428.

Collins, R.A., et al., 2019. Non-specific amplification compromises environmental DNA metabarcoding with COI. Methods Ecol. Evol. 10 (11), 1985–2001.
Cordier, T., et al., 2019. Embracing environmental genomics and machine learning for routine biomonitoring. Trends Microbiol. 27 (5), 387–397.
Couceiro, S.R.M., et al., 2012. A macroinvertebrate multimetric index to evaluate the biological condition of streams in the Central Amazon region of Brazil. Ecol. Indic. 18, 118–125.
Creedy, T.J., Ng, W.S., Vogler, A.P., 2019. Toward accurate species-level metabarcoding of arthropod communities from the tropical forest canopy. Ecol. Evol. 9 (6), 3105–3116.
Czech, L., Barbera, P., Stamatakis, A., 2019. Methods for automatic reference trees and multilevel phylogenetic placement. Bioinformatics 35 (7), 1151–1158.
Dalling, J.W., et al., 2016. Nutrient availability in tropical rain forests: the paradigm of phosphorus limitation. In: Tropical Tree Physiology. Springer, Cham, pp. 261–273.
Davidson, E.A., et al., 2012. The Amazon basin in transition. Nature 481 (7381), 321–328.
de Bruijn, F.J. (Ed.), 2011. Handbook of Molecular Microbial Ecology I: Metagenomics and Complementary Approaches. John Wiley & Sons.
de Lima, R.A.F., et al., 2018. Can plant DNA barcoding be implemented in species-rich tropical regions? A perspective from São Paulo State, Brazil. Genet. Mol. Biol. 41 (3), 661–670.
Deagle, B.E., et al., 2014. DNA metabarcoding and the cytochrome c oxidase subunit I marker: not a perfect match. Biol. Lett. 10 (9), 20140562.
Dedieu, N., et al., 2016. A multimetric macroinvertebrate index for the implementation of the European Water Framework Directive in French Guiana, East Amazonia. River Res. Appl. 32 (3), 501–515.
Deiner, K., et al., 2017. Environmental DNA metabarcoding: transforming how we survey animal and plant communities. Mol. Ecol. 26 (21), 5872–5895.
Del Olmo-Ruiz, M., Elizabeth Arnold, A., 2017. Community structure of fern-affiliated endophytes in three neotropical forests. J. Trop. Ecol. 33 (1), 60–73.
Dexter, K.G., et al., 2018. Inserting tropical dry forests into the discussion on biome transitions in the tropics. Front. Ecol. Evol. 6, 104.
Díaz, S., et al., 2019. Pervasive human-driven decline of life on Earth points to the need for transformative change. Science 366 (6471), eaax3100.
Dickie, I.A., Boyer, S., Buckley, H.L., 2018. Towards robust and repeatable sampling methods in eDNA-based studies. Mol. Ecol. 18 (5), 940–952.
Dodsworth, S., 2015. Genome skimming for next-generation biodiversity analysis. Trends Plant Sci. 20 (9), 525–527.
Doi, H., et al., 2015. Droplet digital polymerase chain reaction (PCR) outperforms real-time PCR in the detection of environmental DNA from an invasive fish species. Environ. Sci. Technol. 49 (9), 5601–5608.
Donald, J., et al., 2019. Tropical palm endophytes exhibit low competitive structuring when assessed using co-occurrence and antipathogen activity analysis. Front. Forests Glob. Chang. 2, 86.
Donald, J., et al., 2020. A test of community assembly rules using foliar endophytes from a tropical forest canopy. J. Ecol. Available at: https://doi.org/10.1111/1365-2745.13344.
Dormontt, E.E., et al., 2018. Advancing DNA barcoding and metabarcoding applications for plants requires systematic analysis of herbarium collections-an Australian perspective. Front. Ecol. Evol. 6, 134.
Drenovsky, R.E., et al., 2010. Land use and climatic factors structure regional patterns in soil microbial communities. Glob. Ecol. Biogeogr. 19 (1), 27–39.
Elbrecht, V., Leese, F., 2017. PrimerMiner: an r package for development and in silico validation of DNA metabarcoding primers. Methods Ecol. Evol. 8 (5), 622–626.

Esling, P., Lejzerowicz, F., Pawlowski, J., 2015. Accurate multiplexing and filtering for high-throughput amplicon-sequencing. Nucleic Acids Res. 43 (5), 2513–2524.

FAO, 2019. The state of the world's biodiversity for food and agriculture. In: Bélanger, J., Pilling, D. (Eds.), FAO Commission on Genetic Resources for Food and Agriculture Assessments, Rome, pp. 572.

Fernandes, K., et al., 2018. DNA metabarcoding-a new approach to fauna monitoring in mine site restoration. Restor. Ecol. 26 (6), 1098–1107.

Ferreira, L.V., et al., 2013. A extração ilegal de areia como causa do desaparecimento de campinas e campinaranas no Estado do Pará, Brasil. Pesquisas (Botânica) 64, 157–173.

Ficetola, G.F., et al., 2010. An in silico approach for the evaluation of DNA barcodes. BMC Genomics 11, 434.

Ficetola, G.F., Taberlet, P., Coissac, E., 2016. How to limit false positives in environmental DNA and metabarcoding? Mol. Ecol. Resour. 16, 604–607.

Fine, P.V.A., Baraloto, C., 2016. Habitat endemism in white-sand forests: insights into the mechanisms of lineage diversification and community assembly of the neotropical flora. Biotropica 48 (1), 24–33.

Fragoso, C., Lavelle, P., 1992. Earthworm communities of tropical rain forests. Soil Biol. Biochem. 24 (12), 1397–1408.

Franco, A.L.C., et al., 2019. Amazonian deforestation and soil biodiversity. Conserv. Biol. 33 (3), 590–600.

Froese, R., Pauly, D., 2019. FishBase. World Wide Web electronic publication. Available at: https://www.fishbase.org.

Gibb, H., et al., 2019. The Global Ants Database. Available at: http://globalants.org/.

Gilbert, J.A., et al., 2012. Defining seasonal marine microbial community dynamics. ISME J. 6 (2), 298–308.

González-M, R., et al., 2018. Disentangling the environmental heterogeneity, floristic distinctiveness and current threats of tropical dry forests in Colombia. Environ. Res. Lett. 13 (4), 045007.

Grossman, J.M., et al., 2010. Amazonian anthrosols support similar microbial communities that differ distinctly from those extant in adjacent, unmodified soils of the same mineralogy. Microb. Ecol. 60 (1), 192–205.

Guillera-Arroita, G., et al., 2017. Dealing with false-positive and false-negative errors about species occurrence at multiple levels. Methods Ecol. Evol. 8 (9), 1081–1091.

Hacquard, S., et al., 2015. Microbiota and host nutrition across plant and animal kingdoms. Cell Host Microbe 17 (5), 603–616.

Harper, L.R., et al., 2018. Needle in a haystack? A comparison of eDNA metabarcoding and targeted qPCR for detection of the great crested newt (Triturus cristatus). Ecol. Evol. 8 (12), 6330–6341.

Harwood, V.J., et al., 2014. Microbial source tracking markers for detection of fecal contamination in environmental waters: relationships between pathogens and human health outcomes. FEMS Microbiol. Rev. 38 (1), 1–40.

Hebert, P.D.N., et al., 2003. Biological identification through DNA barcodes. Proc. R. Soc. Lond. B Biol. Sci. 270, 313–321.

Hibert, F., et al., 2013. Unveiling the diet of elusive rainforest herbivores in next generation sequencing era? The tapir as a case study. PLoS One 8, e60799.

Higgins, K.L., et al., 2014. Communities of fungal endophytes in tropical forest grasses: highly diverse host- and habitat generalists characterized by strong spatial structure. Fungal Ecol. 8, 1–11.

Hollingsworth, P.M., Graham, S.W., Little, D.P., 2011. Choosing and using a plant DNA barcode. PLoS One 6 (5), e19254.

Hortal, J., et al., 2015. Seven shortfalls that beset large-scale knowledge of biodiversity. Annu. Rev. Ecol. Evol. Syst. 46 (1), 523–549.

Hughey, M.C., et al., 2017. Skin bacterial microbiome of a generalist Puerto Rican frog varies along elevation and land use gradients. PeerJ 5, e3688.
Jaouen, G., et al., 2019. Fungi of French Guiana gathered in a taxonomic, environmental and molecular dataset. Sci. Data 6 (1), 206.
Jarman, S.N., Berry, O., Bunce, M., 2018. The value of environmental DNA biobanking for long-term biomonitoring. Nat. Ecol. Evol. 2 (8), 1192–1193.
Jenkins, C.N., Pimm, S.L., Joppa, L.N., 2013. Global patterns of terrestrial vertebrate diversity and conservation. Proc. Natl. Acad. Sci. U. S. A. 110 (28), E2602–E2610.
Jetz, W., et al., 2019. Essential biodiversity variables for mapping and monitoring species populations. Nat. Ecol. Evol. 3 (4), 539–551.
Jones, F.A., et al., 2011. The roots of diversity: below ground species richness and rooting distributions in a tropical forest revealed by DNA barcodes and inverse modeling. PLoS One 6, e24506.
Kattge, J., et al., 2019. TRY plant trait database—enhanced coverage and open access. Glob. Chang. Biol. 26, 119–188. Available at: https://doi.org/10.1111/gcb.14904.
Kembel, S.W., et al., 2014. Relationships between phyllosphere bacterial communities and plant functional traits in a neotropical forest. Proc. Natl. Acad. Sci. U. S. A. 111 (38), 13715–13720.
Kim, J.-S., et al., 2007. Bacterial diversity of terra preta and pristine forest soil from the Western Amazon. Soil Biol. Biochem. 39 (2), 684–690.
Kirshtein, J.D., et al., 2007. Quantitative PCR detection of *Batrachochytrium dendrobatidis* DNA from sediments and water. Dis. Aquat. Organ. 77 (1), 11–15.
Kivlin, S.N., Hawkes, C.V., 2016a. Temporal and Spatial Variation of Soil Bacteria Richness, Composition, and Function in a Neotropical Rainforest. PLoS One 11 (7), e0159131.
Kivlin, S.N., Hawkes, C.V., 2016b. Tree species, spatial heterogeneity, and seasonality drive soil fungal abundance, richness, and composition in Neotropical rainforests. Environ. Microbiol. 18 (12), 4662–4673.
Kocher, A., de Thoisy, B., Catzeflis, F., et al., 2017a. Evaluation of short mitochondrial metabarcodes for the identification of Amazonian mammals. Methods Ecol. Evol. 8 (10), 1276–1283.
Kocher, A., de Thoisy, B., Catzeflis, F., et al., 2017b. iDNA screening: disease vectors as vertebrate samplers. Mol. Ecol. 26 (22), 6478–6486.
Kocher, A., Gantier, J.-C., et al., 2017c. Vector soup: high-throughput identification of Neotropical phlebotomine sand flies using metabarcoding. Mol. Ecol. Resour. 17 (2), 172–182.
Krehenwinkel, H., et al., 2017. A cost-efficient and simple protocol to enrich prey DNA from extractions of predatory arthropods for large-scale gut content analysis by Illumina sequencing. Methods Ecol. Evol. 8 (1), 126–134.
Langille, M.G.I., et al., 2013. Predictive functional profiling of microbial communities using 16S rRNA marker gene sequences. Nat. Biotechnol. 31 (9), 814–821.
Lauber, C.L., et al., 2009. Pyrosequencing-based assessment of soil pH as a predictor of soil bacterial community structure at the continental scale. Appl. Environ. Microbiol. 75, 5111–5120.
Lentendu, G., et al., 2019. Protist biodiversity and biogeography in lakes from four Brazilian river—floodplain systems. J. Eukaryot. Microbiol. 66 (4), 592–599.
Leroy, C., et al., 2016. The contribution of microorganisms and metazoans to mineral nutrition in bromeliads. J. Plant Ecol. 9 (3), 241–255.
Leroy, C., et al., 2019. How significant are endophytic fungi in bromeliad seeds and seedlings? Effects on germination, survival and performance of two epiphytic plant species. Fungal Ecol. 39, 296–306.

Levy-Booth, D.J., et al., 2007. Cycling of extracellular DNA in the soil environment. Soil Biol. Biochem. 39 (12), 2977–2991.
Li, F., et al., 2018. Application of environmental DNA metabarcoding for predicting anthropogenic pollution in rivers. Environ. Sci. Technol. 52 (20), 11708–11719.
Lievens, B., et al., 2006. Real-time PCR for detection and quantification of fungal and oomycete tomato pathogens in plant and soil samples. Plant Sci. 171 (1), 155–165.
Lim, N.K.M., et al., 2016. Next-generation freshwater bioassessment: eDNA metabarcoding with a conserved metazoan primer reveals species-rich and reservoir-specific communities. R. Soc. Open Sci. 3 (11), 160635.
Linard, B., et al., 2015. Metagenome skimming of insect specimen pools: potential for comparative genomics. Genome Biol. Evol. 7 (6), 1474–1489.
Lopes, C.M., et al., 2015. DNA metabarcoding diet analysis for species with parapatric vs sympatric distribution: a case study on subterranean rodents. Heredity 114, 525–536.
Lopes, C.M., et al., 2017. eDNA metabarcoding: a promising method for anuran surveys in highly diverse tropical forests. Mol. Ecol. Resour. 17 (5), 904–914.
Louca, S., et al., 2016. High taxonomic variability despite stable functional structure across microbial communities. Nat. Ecol. Evol. 1 (1), 15.
Louca, S., et al., 2017. Functional structure of the bromeliad tank microbiome is strongly shaped by local geochemical conditions. Environ. Microbiol. 19 (8), 3132–3151.
Lynggaard, C., et al., 2019. Vertebrate diversity revealed by metabarcoding of bulk arthropod samples from tropical forests. Environ. DNA 56, 1637.
Mächler, E., et al., 2014. Utility of environmental DNA for monitoring rare and indicator macroinvertebrate species. Freshwat. Sci. 33 (4), 1174–1183.
Maggia, M.E., et al., 2017. DNA metabarcoding of amazonian ichthyoplankton swarms. PLoS One 12 (1), e0170009.
Mallott, E.K., Garber, P.A., Malhi, R.S., 2018. trnL outperforms rbcL as a DNA metabarcoding marker when compared with the observed plant component of the diet of wild white-faced capuchins (*Cebus capucinus*, Primates). PLoS One 13 (6), e0199556.
Mariac, C., et al., 2018. Metabarcoding by capture using a single COI probe (MCSP) to identify and quantify fish species in ichthyoplankton swarms. PLoS One 13 (9), e0202976.
Marques, A., et al., 2014. A framework to identify enabling and urgent actions for the 2020 Aichi Targets. Basic Appl. Ecol. 15 (8), 633–638.
Maruyama, A., et al., 2014. The release rate of environmental DNA from juvenile and adult fish. PLoS One 9, e114639.
Matsen, F.A., Kodner, R.B., Armbrust, E.V., 2010. pplacer: linear time maximum-likelihood and Bayesian phylogenetic placement of sequences onto a fixed reference tree. BMC Bioinf. 11, 538.
McGee, K.M., et al., 2019. Determinants of soil bacterial and fungal community composition toward carbon-use efficiency across primary and secondary forests in a Costa Rican conservation area. Microb. Ecol. 77 (1), 148–167.
Mendes, L.W., et al., 2015. Land-use system shapes soil bacterial communities in Southeastern Amazon region. Appl. Soil Ecol. 95, 151–160.
Missa, O., et al., 2009. Monitoring arthropods in a tropical landscape: relative effects of sampling methods and habitat types on trap catches. J. Insect Conserv. 13 (1), 103.
Morris, A.L., et al., 2016. Deforestation-driven food-web collapse linked to emerging tropical infectious disease, Mycobacterium ulcerans. Sci. Adv. 2 (12), e1600387.
Morrone, J.J., 2014. Biogeographical regionalisation of the Neotropical region. Zootaxa 3782, 1–110.
Mulatu, K.A., et al., 2017. Biodiversity monitoring in changing tropical forests: a review of approaches and new opportunities. Remote Sens. (Basel) 9 (10), 1059.

Mumby, P.J., et al., 2004. Mangroves enhance the biomass of coral reef fish communities in the Caribbean. Nature 427 (6974), 533–536.
Murienne, J., et al., 2019. Aquatic eDNA for monitoring French Guiana biodiversity. Biodivers. Data J. 7, e37518.
Nagler, M., et al., 2018. Extracellular DNA in natural environments: features, relevance and applications. Appl. Microbiol. Biotechnol. 102 (15), 6343–6356.
Navarrete, A.A., et al., 2010. A molecular survey of the diversity of microbial communities in different Amazonian agricultural model systems. Diversity 2 (5), 787–809.
Ndaw, S.M., et al., 2009. Relationships between bacterial diversity, microbial biomass, and litter quality in soils under different plant covers in northern Rio de Janeiro State, Brazil. Can. J. Microbiol. 55 (9), 1089–1095.
Nepstad, D.C., et al., 2008. Interactions among Amazon land use, forests and climate: prospects for a near-term forest tipping point. Philos. Trans. R. Soc. B 363 (1498), 1737–1746.
Nguyen, N.H., et al., 2016. FUNGuild: an open annotation tool for parsing fungal community datasets by ecological guild. Fungal Ecol. 20, 241–248.
Nobile, A.B., et al., 2019. DNA metabarcoding of Neotropical ichthyoplankton: enabling high accuracy with lower cost. Metabarcod. Metagenom. 3, e35060.
Nottingham, A.T., et al., 2018. Microbes follow Humboldt: temperature drives plant and soil microbial diversity patterns from the Amazon to the Andes. Ecology 99 (11), 2455–2466.
O'Dwyer, J.P., et al., 2009. An integrative framework for stochastic, size-structured community assembly. Proc. Natl. Acad. Sci. U. S. A. 106 (15), 6170–6175.
Olson, D.M., et al., 2001. Terrestrial ecoregions of the world: a new map of life on earth. Bioscience 51 (11), 933.
Pajares, S., et al., 2018. Environmental controls on soil microbial communities in a seasonally dry tropical forest. Appl. Environ. Microbiol. 84 (17), e00342–18.
Paknia, O., Rajaei, S., H. & Koch, A., 2015. Lack of well-maintained natural history collections and taxonomists in megadiverse developing countries hampers global biodiversity exploration. Org. Divers. Evol. 15 (3), 619–629.
Pansu, J., et al., 2015. Landscape-scale distribution patterns of earthworms inferred from soil DNA. Soil Biol. Biochem. 83 (0), 100–105.
Papadopoulou, A., Taberlet, P., Zinger, L., 2015. Metagenome skimming for phylogenetic community ecology: a new era in biodiversity research. Mol. Ecol. 24 (14), 3515–3517.
Paula, F.S., et al., 2014. Land use change alters functional gene diversity, composition and abundance in Amazon forest soil microbial communities. Mol. Ecol. 23 (12), 2988–2999.
Paula, D.P., et al., 2016. Uncovering trophic interactions in arthropod predators through DNA shotgun-sequencing of gut contents. PLoS One 11 (9), e0161841.
Pawlowski, J., et al., 2014. Environmental monitoring through protist next-generation sequencing metabarcoding: assessing the impact of fish farming on benthic foraminifera communities. Mol. Ecol. Resour. 14, 1129–1140.
Pereira, H.M., et al., 2013. Ecology. essential biodiversity variables. Science 339 (6117), 277–278.
Perring, M.P., et al., 2015. Advances in restoration ecology: rising to the challenges of the coming decades. Ecosphere 6 (8), 1–25.
Pilliod, D.S., et al., 2013. Estimating occupancy and abundance of stream amphibians using environmental DNA from filtered water samples. Can. J. Fish. Aquat. Sci. 70, 1123–1130.
Pinheiro, H.T., et al., 2019. Will DNA barcoding meet taxonomic needs? Science 365 (6456), 873–874.
Pompanon, F., et al., 2012. Who is eating what: diet assessment using next generation sequencing. Mol. Ecol. 21, 1931–1950.

Porazinska, D.L., et al., 2012. Nematode spatial and ecological patterns from tropical and temperate rainforests. PLoS One 7 (9), e44641.
Pornon, A., et al., 2016. Using metabarcoding to reveal and quantify plant-pollinator interactions. Sci. Rep. 6, 27282.
Pringle, E.G., Moreau, C.S., 2017. Community analysis of microbial sharing and specialization in a Costa Rican ant–plant–hemipteran symbiosis. Proc. R. Soc. B Biol. Sci. 284 (1850), 20162770.
Quast, C., et al., 2013. The SILVA ribosomal RNA gene database project: improved data processing and web-based tools. Nucleic Acids Res. 41 (Database issue), D590–D596.
Ramirez, K.S., et al., 2018. Detecting macroecological patterns in bacterial communities across independent studies of global soils. Nat. Microbiol. 3 (2), 189–196.
Ratnasingham, S., Hebert, P.D.N., 2007. bold: the barcode of life data system (http://www.barcodinglife.org). Mol. Ecol. Notes 7 (3), 355–364.
Rawlence, N.J., et al., 2014. Using palaeoenvironmental DNA to reconstruct past environments: progress and prospects. J. Quat. Sci. 29, 610–626.
Rees, H.C., et al., 2014. The detection of aquatic animal species using environmental DNA—a review of eDNA as a survey tool in ecology. J. Appl. Ecol. 51 (5), 1450–1459.
Rice, J., et al., 2018. Summary for policymakers of the regional assessment report on biodiversity and ecosystem services for the Americas of the Intergovernmental Science-Policy Platform on Biodiversity and Ecosystem Services. In: IPBES book. Available at: https://www.fs.usda.gov/treesearch/pubs/56826.
Ritter, C.D., et al., 2019. Locality or habitat? Exploring predictors of biodiversity in Amazonia. Ecography 42 (2), 321–333.
Rocha, R.G., et al., 2015. DNA from owl pellet bones uncovers hidden biodiversity. Syst. Biodivers. 13 (4), 403–412.
Rodgers, T.W., Xu, C.C.Y., Giacalone, J., 2017. Carrion fly-derived DNA metabarcoding is an effective tool for mammal surveys: evidence from a known tropical mammal community. Mol. Ecol. 17 (6), e133–e145.
Rodrigues, J.L.M., et al., 2013. Conversion of the Amazon rainforest to agriculture results in biotic homogenization of soil bacterial communities. Proc. Natl. Acad. Sci. U. S. A. 110 (3), 988–993.
Rodrigues, R.C., et al., 2019. ATLANTIC BIRD TRAITS: a data set of bird morphological traits from the Atlantic forests of South America. Ecology 100 (6), e02647.
Rodriguez-Nuñez, K.M., Rullan-Cardec, J.M., Rios-Velazquez, C., 2018. The metagenome of bromeliads phytotelma in Puerto Rico. Data Brief 16, 19–22.
Rousseau, L., et al., 2013. Soil macrofauna as indicators of soil quality and land use impacts in smallholder agroecosystems of western Nicaragua. Ecol. Indic. 27, 71–82.
Sales, N.G., et al., 2019a. Assessing the Potential of Environmental DNA Metabarcoding for Monitoring Neotropical Mammals: A Case Study in the Amazon and Atlantic Forest, Brazil. bioRxiv, p. 750414. Available at: https://www.biorxiv.org/content/biorxiv/early/2019/08/31/750414.full.pdf.
Sales, N.G., et al., 2019b. Influence of preservation methods, sample medium and sampling time on eDNA recovery in a neotropical river. Environ. DNA 1 (2), 119–130.
Salter, S.J., et al., 2014. Reagent and laboratory contamination can critically impact sequence-based microbiome analyses. BMC Biol. 12, 87.
Sapountzis, P., et al., 2015. Acromyrmex leaf-cutting ants have simple gut microbiota with nitrogen-fixing potential. Appl. Environ. Microbiol. 81 (16), 5527–5537.
Sasso, T., et al., 2017. Environmental DNA characterization of amphibian communities in the Brazilian Atlantic forest: potential application for conservation of a rich and threatened fauna. Biol. Conserv. 215, 225–232.
Schnell, I.B., Bohmann, K., Gilbert, M.T.P., 2015. Tag jumps illuminated—reducing sequence-to-sample misidentifications in metabarcoding studies. Mol. Ecol. Resour. 15 (6), 1289–1303.

Schnell, I.B., et al., 2018. Debugging diversity—a pan-continental exploration of the potential of terrestrial blood-feeding leeches as a vertebrate monitoring tool. Mol. Ecol. Resour. 18 (6), 1282–1298.

Schoch, C.L., et al., 2012. Nuclear ribosomal internal transcribed spacer (ITS) region as a universal DNA barcode marker for Fungi. Proc. Natl. Acad. Sci. U. S. A. 109 (16), 6241–6246.

Schroeder, J.W., et al., 2019. Host plant phylogeny and abundance predict root-associated fungal community composition and diversity of mutualists and pathogens. R. Shefferson, ed J. Ecol. 107 (4), 1557–1566.

Scibetta, S., et al., 2012. A molecular method to assess Phytophthora diversity in environmental samples. J. Microbiol. Methods 88 (3), 356–368.

Sengupta, M.E., et al., 2019. Environmental DNA for improved detection and environmental surveillance of schistosomiasis. Proc. Natl. Acad. Sci. U. S. A. 116 (18), 8931–8940.

Sheth, B.P., Thaker, V.S., 2017. DNA barcoding and traditional taxonomy: an integrated approach for biodiversity conservation. Genome 60 (7), 618–628.

Shokralla, S., et al., 2016. Environmental DNA barcode sequence capture: targeted, PCR-free sequence capture for biodiversity analysis from bulk environmental samples. bioRxiv. Available at: https://doi.org/10.1101/087437.

Sigsgaard, E.E., et al., 2016. Population characteristics of a large whale shark aggregation inferred from seawater environmental DNA. Nat. Ecol. Evol. 1, 0004.

Smets, W., et al., 2015. A method for simultaneous measurement of soil bacterial abundances and community composition via 16S rRNA gene sequencing. PeerJ Prepr. 3, e1622, e1622.

Soares-Filho, B., et al., 2014. Land use. Cracking Brazil's forest code. Science 344 (6182), 363–364.

Sommeria-Klein, G., et al., 2016. Inferring neutral biodiversity parameters using environmental DNA data sets. Sci. Rep. 6, 35644.

Stackebrandt, E., Goebel, B.M., 1994. Taxonomic note: a place for DNA-DNA reassociation and 16S rRNA sequence analysis in the present species definition in bacteriology. Int. J. Syst. Evol. Microbiol. 44 (4), 846–849.

Strassburg, B.B.N., et al., 2017. Moment of truth for the Cerrado hotspot. Nat. Ecol. Evol. 1 (4), 99.

Strickler, K.M., Fremier, A.K., Goldberg, C.S., 2015. Quantifying effects of UV-B, temperature, and pH on eDNA degradation in aquatic microcosms. Biol. Conserv. 183, 85–92.

Taberlet, P., Coissac, E., Hajibabaei, M., et al., 2012a. Environmental DNA. Mol. Ecol. 21 (8), 1789–1793.

Taberlet, P., Coissac, E., Pompanon, F., et al., 2012b. Towards next-generation biodiversity assessment using DNA metabarcoding. Mol. Ecol. 21 (8), 2045–2050.

Taberlet, P., et al., 2018. Environmental DNA: For Biodiversity Research and Monitoring. Oxford University Press.

Takahara, T., Minamoto, T., Doi, H., 2013. Using environmental DNA to estimate the distribution of an invasive fish species in ponds. PLoS One 8, e56584.

Talaga, S., et al., 2017. DNA reference libraries of French Guianese mosquitoes for barcoding and metabarcoding. PLoS One 12 (6), e0176993.

ter Steege, H., et al., 2013. Hyperdominance in the Amazonian tree flora. Science 342 (6156), 1243092.

Tessler, M., et al., 2017. A global eDNA comparison of freshwater bacterioplankton assemblages focusing on large-river floodplain lakes of Brazil. Microb. Ecol. 73 (1), 61–74.

Tessler, M., et al., 2018. Bloodlines: mammals, leeches, and conservation in southern Asia. Syst. Biodivers. 16 (5), 488–496.

Thomas, A.C., et al., 2016. Quantitative DNA metabarcoding: improved estimates of species proportional biomass using correction factors derived from control material. Mol. Ecol. Resour. 16 (3), 714–726.

Thompson, L.R., et al., 2017. A communal catalogue reveals Earth's multiscale microbial diversity. Nature 551 (7681), 457–463.
Thomsen, P.F., Sigsgaard, E.E., 2019. Environmental DNA metabarcoding of wild flowers reveals diverse communities of terrestrial arthropods. Ecol. Evol. 9 (4), 1665–1679.
Thomsen, P.F., Willerslev, E., 2015. Environmental DNA—an emerging tool in conservation for monitoring past and present biodiversity. Biol. Conserv. 183, 4–18.
Tiedje, J.M., et al., 1999. Opening the black box of soil microbial diversity. Appl. Soil Ecol. 13 (2), 109–122.
Tritsch, I., Le Tourneau, F.-M., 2016. Population densities and deforestation in the Brazilian Amazon: new insights on the current human settlement patterns. Appl. Geogr. 76, 163–172.
Turner, C.R., Uy, K.L., Everhart, R.C., 2015. Fish environmental DNA is more concentrated in aquatic sediments than surface water. Biol. Conserv. 183, 93–102.
Vacher, C., et al., 2016. Learning ecological networks from next-generation sequencing data. Adv. Ecol. Res. 54, 1–39.
Valdez-Moreno, M., et al., 2019. Using eDNA to biomonitor the fish community in a tropical oligotrophic lake. PLoS One 14 (4), e0215505.
Valentin, R.E., et al., 2018. Early detection of invasive exotic insect infestations using eDNA from crop surfaces. Front. Ecol. Environ. 16 (5), 265–270.
Vasco-Palacios, A.M., et al., 2019. Carbon content and pH as important drivers of fungal community structure in three Amazon forests. Plant and Soil. Available at: https://doi.org/10.1007/s11104-019-04218-3.
Vittor, A.Y., et al., 2009. Linking deforestation to malaria in the Amazon: characterization of the breeding habitat of the principal malaria vector, Anopheles darlingi. Am. J. Trop. Med. Hyg. 81 (1), 5–12.
Waring, B.G., Hawkes, C.V., 2015. Short-term precipitation exclusion alters microbial responses to soil moisture in a wet tropical forest. Microb. Ecol. 69 (4), 843–854.
West, A.G., et al., 2019. The microbiome in threatened species conservation. Biol. Conserv. 229, 85–98.
Wilcox, T.M., et al., 2018. Capture enrichment of aquatic environmental DNA: a first proof of concept. Mol. Ecol. Resour. 18 (6), 1392–1401.
Wu, T.H., et al., 2011. Molecular study of worldwide distribution and diversity of soil animals. Proc. Natl. Acad. Sci. U. S. A. 108, 17720–17725.
Yoccoz, N.G., et al., 2012. DNA from soil mirrors plant functional and structural diversity. Mol. Ecol. 21, 3647–3655.
Zinger, L., Philippe, H., 2016. Coalescing molecular evolution and DNA barcoding. Mol. Ecol. 25 (9), 1908–1910.
Zinger, L., et al., 2011. Global patterns of bacterial beta-diversity in seafloor and seawater ecosystems. PLoS One 6, e24570.
Zinger, L., et al., 2016. Extracellular DNA extraction is a fast, cheap and reliable alternative for multi-taxa surveys based on soil DNA. Soil Biol. Biochem. 96, 16–19.
Zinger, L., Taberlet, P., et al., 2019a. Body size determines soil community assembly in a tropical forest. Mol. Ecol. 28 (3), 528–543.
Zinger, L., Bonin, A., et al., 2019b. DNA metabarcoding—need for robust experimental designs to draw sound ecological conclusions. Mol. Ecol. 28 (8), 1857–1862.
Zizka, A., et al., 2018. Finding needles in the haystack: where to look for rare species in the American tropics. Ecography 41 (2), 321–330.

CHAPTER TEN

Informing marine spatial planning decisions with environmental DNA

Alessia Bani[a,†], Maarten De Brauwer[b,†], Simon Creer[c], Alex J. Dumbrell[a], Gino Limmon[d], Jamaluddin Jompa[e], Sophie von der Heyden[f,‡], Maria Beger[b,g,*,‡]

[a]School of Life Sciences, University of Essex, Essex, United Kingdom
[b]School of Biology, Faculty of Biological Sciences, University of Leeds, Leeds, United Kingdom
[c]School of Natural Sciences, Bangor University, Bangor, Gwynedd, United Kingdom
[d]Maritime and Marine Science Center of Excellence, Pattimura University, Ambon, Indonesia
[e]Graduate School, Hasanuddin University, Makassar, Indonesia
[f]Evolutionary Genomics Group, Department of Botany and Zoology, University of Stellenbosch, Stellenbosch, South Africa
[g]Centre for Biodiversity and Conservation Science, School of Biological Sciences, University of Queensland, Brisbane, QLD, Australia
*Corresponding author: e-mail address: M.Beger@leeds.ac.uk

Contents

Abstract

Marine management areas provide a key tool for efforts towards sustainable development, reconciling socio-economic goals with those for biodiversity conservation. Decisions about where and when to establish spatial management areas in the oceans are currently hampered by the uncertainties of incomplete, or overly general, information about biodiversity. The analysis of environmental DNA (eDNA) provides a

† Joint first author.
‡ Joint last author.

Advances in Ecological Research, Volume 62
ISSN 0065-2504
https://doi.org/10.1016/bs.aecr.2020.01.011

potentially powerful tool to overcome this lack of data in the future. Here we present directions to develop robust approaches to integrate eDNA and spatial planning processes, aiming to provide guidance to underpin tool development.

The potential of eDNA use in conservation is widely recognised, although direct applications almost exclusively focus on detection of invasive or threatened species and not spatial management decisions. The implementation of broader interaction between the fields of conservation science and eDNA analysis could create substantial benefits to biodiversity conservation and management. In particular, eDNA analysis can provide information on biodiversity over spatial-temporal scales that are currently prohibitive in spatial planning studies.

Here, we provide an overview of how eDNA is currently used in conservation practice, in addition to understanding its limitations and benefits within the context of spatial planning. With the goal to harness rapid technological developments in both molecular and conservation sciences, we provide a horizon scan of the future of eDNA analysis and its application to inform biodiversity conservation in a rapidly changing world.

1. Introduction

The management of biodiversity in tropical marine ecosystems relies to a large extent on spatial management actions, such as marine protected areas (Tittensor et al., 2019; Weeks et al., 2014), no-take zones (Russ et al., 2015), and fisheries management zones (McClanahan and Hicks, 2011). Spatial planning of these management actions plays a key role in implementing objective-driven transparent prioritisations (Carvalho et al., 2017; Wilson et al., 2007), often as part of national commitments to sustainable development goals and Convention of Biological Diversity (CBD) agreements. However, spatial planning for biodiversity conservation often fails to represent biodiversity patterns and associated bio-physical processes, because up-to-date spatial databases comprising the required data are often insufficient and lack spatial, temporal, and taxonomic coverage, particularly in resource-poor tropical developing countries. The representation of biodiversity in such spatial plans often relies on very broad proxies such as ecosystem extent (Beger et al., 2010; Green et al., 2009; Makino et al., 2015), habitat type (Boon and Beger, 2016; Grantham et al., 2013; Makino et al., 2015), or bioregions (Fernandes et al., 2005) and only rarely are there attempts to capture ecosystem condition (Vercammen et al., 2019). Given the global-scale declines observed in tropical marine ecosystems (Hughes et al., 2017; Unsworth et al., 2016), up-to-date and high-resolution biodiversity data play a pivotal role in implementing rapid management responses to the climate crisis. One potential solution to these challenges lies with the emergence of environmental DNA (eDNA) technologies, defined as "genetic material obtained directly from environmental samples (soil, sediment, water, etc.) without any obvious signs of biological source material" (Thomsen and Willerslev, 2015) (Box 1).

Spatial planning evaluates the trade-offs between protecting biodiversity features and ensuring socio-economic sustainability in a transparent, quantitative, and repeatable manner (Carvalho et al., 2017; Kukkala and

BOX 1 A basic guide to eDNA studies.

The analysis of microbial life using eDNA approaches has been commonplace for over 20 years, but their use to detect macro-organisms to investigate large scale ecological processes is more recent (Ficetola et al., 2007). Recently, eDNA methods have gained increasing attention as a possible alternative to survey rare or cryptic species, or to replace lethal or invasive survey techniques (Barnes and Turner, 2016; Jeunen et al., 2019). Environmental samples used in eDNA studies collect a mixture of DNA fragments originating from the various organisms present in that environment, regardless of whether these organisms are visible or morphologically identifiable in the source material. This pool of DNA is then extracted with commercial kits or other well-established protocols. It is then often necessary to amplify the amount of DNA present via Polymerase Chain Reaction technology (PCR, see Glossary), before identifying which taxa it originates from (although methods are increasingly moving away from amplification-based approaches). While widely used for microorganisms, metagenomics (see Glossary) approaches are not routinely used in eDNA studies, with, to our knowledge, only one study that has attempted this method (Stat et al., 2017). The main limitation for metagenomics is the very limited percentage of macro-organism DNA sequences that can be amplified (Stat et al., 2017).

eDNA studies can be broadly classified into studies focussing on community composition or those that target specific organisms or even for population-level studies (Porter and Hajibabaei, 2018a). When targeting individual species, species-specific primers (employed during PCR) ensure that only the target species is amplified, with researchers utilising quantitative (or more recently digital) PCR (see Glossary) to provide estimates of biomass or cellular abundance (Porter and Hajibabaei, 2018a). This approach is common in the detection of invasive species, such as American bullfrogs (*Rana catesbeiana*) in France (Ficetola et al., 2007), the Asian Carp (*Hypophthalmichthys* sp.) in the USA (Bohmann et al., 2014), and Asian date mussels (Arcuatula senhousia) (Holman et al., 2019), alongside detecting threatened species; for example, the great crested

Continued

BOX 1 A basic guide to eDNA studies.—cont'd

newt (*Triturus cristatus*) in the United Kingdom, Endangered skates (*Zearaja maugeana*) in Australia, or nearly extinct freshwater fish (*Misgurnus fossilis*) of conservation concern in Denmark (Biggs et al., 2015; Sigsgaard et al., 2015; Weltz et al., 2017). Currently, eDNA tools are being tested and specialised companies are already offering related services (see for example https://www.naturemetrics.co.uk/wildlife-services/gcn-edna/) particularly in single-species management of invasive or threatened species.

For studies aiming to record all species present in a sample, High Throughput Sequencing (see Glossary HTS) in the form of metabarcoding (see Glossary) is currently the most commonly applied approach. Metabarcoding is the taxonomic identification of multiple species extracted from eDNA samples (Deiner et al., 2017a). Metabarcoding primers employed during PCR aim to capture broad taxonomic groups, for example, amplifying all the eukaryotes present, or targeting specific groups such as fish or crustaceans. These primers anneal to complementary sequence in the mixed pool of DNA fragments and only amplify copies of a selected genomic region that contains enough sequence information to facilitating species identification. These amplicons are then sequenced using one of the available HTS technologies (Illumina for example uses nucleotide labelled with different fluorochromes that are read by a laser while Nanopore using differences in membrane potential, see Van Dijk et al. (2018) for a review on HTS). The DNA sequence reads, generated via HTS, are then analysed to determine the species composition of the original sample. A series of bioinformatics tools are employed to generate ecologically-relevant data for biomonitoring and/or spatial planning. Typically, sequences are clustered in groups based on a predefined similarity threshold (OTUs; see Glossary) or left ungrouped to capture total genetic variation (ASVs; see Glossary). OTUs and/or ASVs are then matched to sequences of known taxonomic identity held in large databases, which then completes the identification of the species present.

The analysis of eDNA provides data that leverages the ability to monitor species composition and distribution in a quicker and often easier way than more traditional approaches (Bista et al., 2017; Bohan et al., 2017; Cristescu and Hebert, 2018). However, the probability of species detection via eDNA approaches, differs in many and often unknown ways; field and laboratory methods (McGee et al., 2019), sampling depths (Eilers et al., 2012, DiBattista et al., 2019), environmental substrate (Holman et al., 2019), and the chemical, physical, oceanographic and biological factors that influence eDNA degradation (Rees et al., 2014), all introduce potential biases. For example, water samples are more homogenous than sediment samples that can contain significant small-scale heterogeneities (Koziol et al., 2019). The distance eDNA travels from a source is highly variable, ranging from 50 m (Jerde et al., 2016) to >200 m (Jane et al., 2015; Pont et al., 2018). Heterogeneous sources of macrobial

BOX 1 A basic guide to eDNA studies.—cont'd

eDNA consist of different particles (mucus, skin, faeces etc.) that are transported, settled on the benthos, and resuspended in a complex and stochastic manner (Jerde et al., 2016; Shogren et al., 2017) and that may degrade differently according to their size (Jo et al., 2017; Wei et al., 2018) but see the contrary in Bista et al. (2018), Bylemans et al. (2018a). Field sampling in tropical marine environments relies on many different strategies, including surface water (Cilleros et al., 2019), benthic water (Boussarie et al., 2018), and sediment samples (DiBattista et al., 2019), mirroring non-standardised sampling in the eDNA and metabarcoding fields (McGee et al., 2019). In addition, variation in laboratory procedures may influence the comparability of results across studies (Berry et al., 2019; Kelly et al., 2019; McGee et al., 2019) however the biggest variability is how the data are produced and analysed.

Illumina technology is currently most commonly used, but other options are available and comparisons between different sequencing techniques can be challenging (Porter and Hajibabaei, 2018b). One of the most debated steps in the bioinformatics workflow is sequence clustering. Researchers can cluster their sequences in OTU (see Glossary) based on a similarity threshold or treat them as ASV (see Glossary) without clustering (Deiner et al., 2017a). Incorrect clustering can have strong effects on alpha diversity indices introducing overestimation or under estimation within the community (Pawlowski et al., 2018). Taxonomy can be assigned using a variety of programs that are based on different approaches including BLAST, MG-RAST or RDP which can lead to different outputs (Deiner et al., 2017a). The use of different databases can similarly lead to different annotations and potential errors. Contrary to BOLD, the NCBI database is not curated but recently has been demonstrated to be reliable for eDNA analysis (Leray et al., 2019), especially because error in taxonomic assignments are easier to correct as more information on biogeography are available for macro-organisms (Deiner et al. 2017a).

We suggest the following reviews/articles for a better understanding of the above discussed possibilities and challenges that eDNA studies face. For general application of eDNA, we recommend Rees et al. (2014) and Cristescu and Hebert (2018), for better understanding of PCR and bioinformatics application on eDNA we strongly suggest Deiner et al. (2017a) and Kelly et al. (2019).

Moilanen, 2013; Margules and Pressey, 2000; Wilson et al., 2007). Such planning supports conservation management in efficiently reaching specific objectives by integrating multiple ecological and socio-economic variables to decide which areas should receive different conservation treatments. Importantly, effective spatial management decisions rely on carefully developed conservation objectives that are quantifiable and SMART (Specific Measurable Achievable Relevant Time-bound (Maxwell et al., 2015). Key principles include connectivity (e.g. considering metapopulation dynamics and larval dispersal, (Makino et al., 2015)), adequacy (e.g. protecting enough of each species/habitat, and sites that support communities in good condition (Magris et al., 2014)), representation of local biodiversity, and efficiency (e.g. minimise impact on users and other costs (Wilson et al., 2009)). Therefore, uncertainties associated with the distribution of biodiversity features and supporting processes, i.e., failure to properly implement the first three principles, can lead to management decisions that are imprecise, select unsuitable areas, or create conflict between users (Game et al., 2014).

The logistical and cost restrictions associated with underwater surveys of biodiversity, even in shallow marine habitats, the lack of taxonomic expertise, and the lack of environmental predictors operating over appropriate spatiotemporal scales for species distribution modelling (e.g. contrasting $250\,m^2$ survey areas vs $100\,km^2$ analysis areas in common databases such as BioOracle, (Assis et al., 2018)), all currently severely hamper spatial management decisions. While there has been some progress towards integrating genetic information into conservation prioritisation (Beger et al., 2014; Nielsen et al., 2017), this has yet to be extended to include eDNA data. eDNA could offer solutions to many of the challenges related to scale or taxonomic expertise in collecting data needed for spatial planning (Box 1). For example, eDNA could give wider geographical coverage and span broader taxonomic ranges of biodiversity than is currently possible to record (Deiner et al., 2017a, 2017b; Stat et al., 2017).

Arguably one of the most exciting recent tool developments in ecology is the increasing adaptation of the analysis of eDNA to answer ecosystem-level questions, moving onwards from its origins in environmental microbiology and ancient DNA studies (Clark et al., 2018). The vast potential of eDNA analysis has led to an explosion of novel, and previously unimaginable advances in the fields of ecology (e.g. Berry et al., 2019; Deiner et al., 2017a; De Vere et al., 2017), palaeontology (e.g. Sønstebø et al., 2010; Willerslev et al., 2014), and conservation (e.g. Cilleros et al., 2019; Weltz et al., 2017). eDNA has a remarkably broad spatiotemporal application spanning a diversity of environmental matrices (soil, air, aquatic systems), resolution (from haplotypes that allow for the analysis of population level structure to entire communities) and time-frames (hours to millennia)

(Ruppert et al., 2019). As such, eDNA is becoming a tool of choice for the monitoring of biodiversity, including in tropical marine environments (see for example Carvalho et al., 2019; DiBattista et al., 2019; Nichols and Marko, 2019; Uthicke et al., 2018; Stat et al., 2017) (Fig. 1). However, tropical eDNA studies mostly come from a few regions such as the Red Sea (Carvalho et al., 2019; DiBattista et al., 2017) and Australia (e.g. Stat et al., 2017), leaving other regions such as Southeast Asia, Africa, or South/Central America almost unexplored (Jerde et al., 2019). The systematic, large-scale use of eDNA and spatial planning could mean an incredible step forward in addressing this bias, where less developed, high biodiversity regions that are most threatened by human impacts are least understood and protected.

Much of the current eDNA literature has focused on tool development and addressing basic ecological questions. So far, few studies have interpreted how eDNA may be used to reveal spatiotemporal patterns and processes that could support spatial conservation decisions. In addition, inconsistencies in generating eDNA data and their analysis still limit the direct comparison and

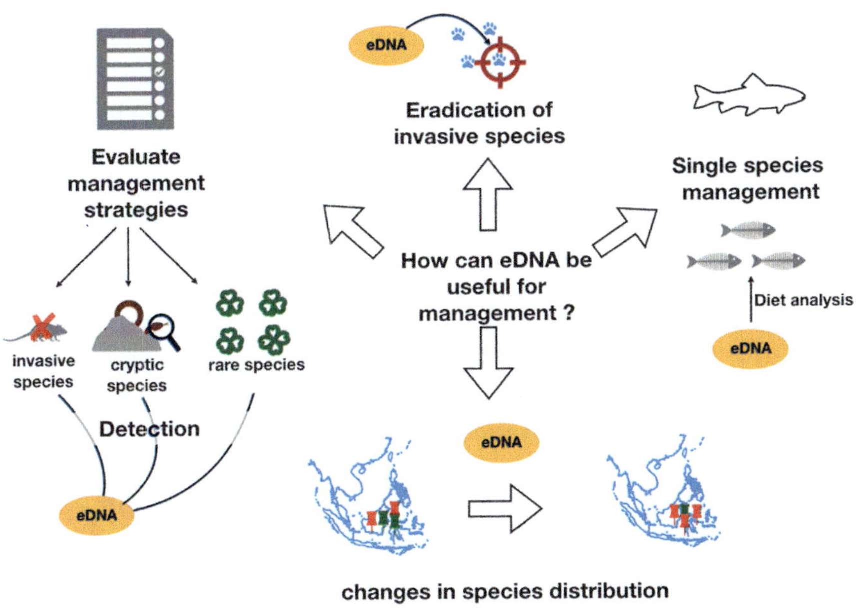

Fig. 1 Application of eDNA in conservation science. eDNA has been utilised in species detection (including cryptic, threatened, rare and invasive species), in diet analysis, or to survey biodiversity in complex environments. Each of these is applicable within different management contexts (Table 1).

integration of different eDNA datasets that would allow for comparative analyses across larger spatial areas (McGee et al., 2019). Hence, streamlining potential future developments of eDNA tools and their application in ecology to underpin and support conservation decisions and processes presents a major opportunity in molecular and conservation sciences.

Here we assess how eDNA could be applied to spatial planning and suggest best-practice guidelines of eDNA in conservation management. Further, we propose the necessary research developments for eDNA to effectively contribute to the management and conservation of biodiversity that is applicable not only to tropical marine ecosystems, but also in terrestrial, freshwater and marine conservation efforts. We provide an overview of how eDNA is currently used in conservation practice, in addition to reviewing its limitations and benefits. We then complete a horizon scan of the future of eDNA as a tool to inform biodiversity conservation.

2. Current use of eDNA in natural resource management and conservation

eDNA data is often implied to be of high value for conservation initiatives (e.g. Ruppert et al., 2019; Stat et al., 2017), but very few authors define how this would work in practice, beyond suggesting improved detection rates, or extend to the population level (Rees et al., 2014; Ruppert et al., 2019; Stat et al., 2017). eDNA has been used to detect invasive species such as bullfrogs, pythons, fish, and mussels (Balasingham et al., 2017; Ficetola et al., 2007; Holman et al., 2019; Hunter et al., 2015; Klymus et al., 2017) (Fig. 1). Biomonitoring of cryptic or threatened species such as sawfishes, orang-utans, and crayfish that are challenging to monitor using conventional methods have benefitted from eDNA detection (Ikeda et al., 2016; Ishige et al., 2017; Simpfendorfer et al., 2016). Stewart et al. (2017) used eDNA to record the spatiotemporal distribution of the Yangtze finless porpoise (*Neophocaena phocaenoides*) to assess the effectiveness of protected areas in China. In addition, recent eDNA work has detected pathogens that might threaten rare populations of amphibians or crayfish (Kamoroff and Goldberg, 2017; Wu et al., 2018) (Fig. 1).

eDNA studies in marine environments have included sediment, water column samples at different depths, plankton tows and stomach content. Different source materials harbour different communities and can provide complementary information (Holman et al., 2019). For example, sediments harbour more information about cryptic and benthic organisms

(Pearman et al., 2018) but limited information on the fish assemblage that can be retrieved from water column samples at the same site (Koziol et al., 2019). Gut contents of predators can provide important information on their feeding behaviour or the distribution of prey (Correia et al., 2017). eDNA has also been successfully applied in monitoring the diet requirements of several species including sea lions (*Neophoca cinerea*) (Berry et al., 2017), or little penguins (*Eudyptula minor*) (Deagle et al., 2010), with studies of gut contents of parasites leading to a better understanding of the population structure of their hosts (Meekan et al., 2017).

In practice, however, few studies have used eDNA for species detection over large spatial or temporal scales in highly diverse environments such as the tropics (Cilleros et al., 2019; Stat et al., 2019). Long-term studies have tracked the effects of climate change on plankton biodiversity (Berry et al., 2019) and described shark diversity over large spatial scales (Boussarie et al., 2018). Temporal studies have tracked seasonality in coastal fish communities (Sigsgaard et al., 2017b) or in lake communities (Bista et al., 2017). The potential applications of eDNA analyses in threatened species management and conservation science might seem endless (Fig. 1), but eDNA approaches have yet to be used for community analyses over large spatial scales or as part of land-based or marine spatial planning.

3. Strategically integrating eDNA analysis and spatial planning

eDNA data have the potential to revolutionise access to biodiversity information throughout the spatial planning process (Fig. 2). Yet to truly understand this potential, we need to assess current and potential technological developments in both eDNA analysis and conservation science. An important component in applying eDNA to meeting many conservation objectives, is how to treat point-based eDNA data for mapping across continuous land- and sea-scapes. Even with the reduced expense that eDNA could bring, fine-scale sampling across large spatial areas is likely to still have high fiscal, labour, and computational costs for the foreseeable future. Therefore, innovative ways of mapping eDNA will be needed to meet conservation objectives. eDNA analyses result in Operational Taxonomic Units (OTUs) that serve as a proxy for species (see Glossary). The resulting sampling site $\times$ OTU matrix will still not cover the full spatial extent required for spatial planning analysis, therefore necessitating either interpolation between sampling sites (Beger et al., 2014; Nielsen et al., 2017), or

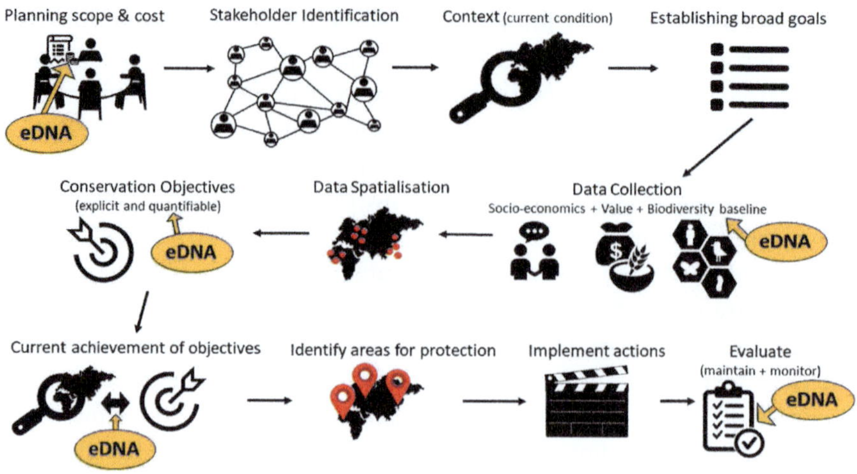

Fig. 2 Flowchart of the spatial planning process, with potential uses of eDNA at relevant stages. *Planning*: the benefits, limitations, feasibility, and cost of eDNA approaches should be carefully considered at the planning stage; *Data collection*: eDNA could be used to provide biodiversity baseline data; *Conservation objectives*: quantifiable diversity objectives could be set by using metrics generated from eDNA data; *Current achievement of objectives*: data collected with eDNA can be compared to objectives; *Evaluation*: eDNA data can be used to monitor progress on objectives. *Modified from Pressey, R.L., Bottrill, M.C., 2008. Opportunism, threats, and the evolution of systematic conservation planning.* Conserv. Biol. *22 (5), 1340–1345.*

predictive statistical modelling of OTU distributions (Fig. 2). Such species distribution modelling is accomplished by relating occurrence, presence/absence, or abundance data to biophysical and socio-economic predictor variables (Broennimann et al., 2012; Elith and Leathwick, 2009; Guisan et al., 2013). Translation from eDNA data to abundance data is currently still problematic. While some studies have found positive correlations between the abundance of organisms and the quantity of eDNA molecules (Takahara et al., 2012), other studies have shown that eDNA quantities are dependent upon several factors including age, development stage, and environmental factors (Jo et al., 2019; Maruyama et al., 2014; Robson et al., 2016).

Given the potentially small spatial resolution of eDNA data and the uncertainty associated with OTU assignment, new modelling methods and small-scale biophysical parameter databases will be required. Thereafter, OTUs could be used as a proxy for species in spatial planning, and potentially be used in three main ways when defining the spatial planning objectives (Fig. 2): (1) setting conservation targets for the amount of the distribution of each OTU to be protected; (2) generating broad multivariate community types for multiple taxa and setting conservation targets for these; and (3) setting (different) targets for OTUs with different distributional patterns, such as patchy vs consistently distributed. Information derived from eDNA analyses then can help evaluate current achievements of these objectives in a gap analysis (Fig. 2, Vimal et al., 2011). With repeat sampling, changes in OTU composition and relative abundance will be able to provide information on spatiotemporal community variation (Berry et al., 2019; Bista et al., 2017) and should expand our knowledge on intra and interspecific connectivity as well as population diversity especially with the developing of organelles sequencing (Adams et al., 2019). In addition, through concerted sequencing efforts, growing barcode and genomic reference databases should facilitate species-level identification of OTUs. This may provide further insight into the prediction of the distribution of species and the use of these data in spatial planning, specifically; whether the ecological or functional traits of species can be used to further refine species distribution models.

Finally, eDNA analysis can serve as a monitoring tool, where the achievement of objectives is assessed against resampled sites over a longer term (Fig. 2), and where changes in management may be adopted when required as part of an adaptive management framework (Williams and Brown, 2016). The relative ease of collecting water samples for eDNA analysis from tropical marine environments will not only allow an increase in geographical scope, but also detect a higher number of species that could be monitored simultaneously, theoretically covering the entire ecosystems' diversity. A weak point of eDNA analysis is the lack of appropriately curated and extensively populated barcode reference databases, leading to incomplete taxonomic assignment, which may be particularly problematic for mega-diverse tropical marine ecosystems. Misclassification of OTUs or ASVs is linked also to the relative short length of the amplified fragments due the degree of degradation of eDNA molecules that could reduce the possibility to discriminate between closely related organisms (Porter and Hajibabaei, 2018b).

Therefore, eDNA integration into spatial planning may require setting conservation objectives and implementing management actions for taxa (e.g. OTUs) at high taxonomic resolution (e.g. family level), while developing better databases (Table 2) (Porter and Hajibabaei, 2018b). Currently, spatial planning for known species applies mostly to cetaceans, reptiles and fishes, for which DNA reference databases are relatively well populated. Large global databases exist for mitochondrial (mtDNA) COI markers (Porter and Hajibabaei, 2018a), but most eDNA studies on freshwater and marine fish diversity utilise mtDNA 12S rRNA markers (Collins et al., 2019), for which reference databases are significantly smaller, as COI markers amplify phytoplankton too, thereby reducing the number of reads available for fish diversity (see Box 1 and Glossary for details on marker regions used in eDNA work). Focussing on generating reference databases across the tree of life, in particular to adequately represent the groups with key functional roles, is pivotal for the integration of eDNA results into spatial planning (Porter and Hajibabaei, 2018b). As more eDNA data are generated, similar efforts should be applied to expand the taxonomic resolution of all the main taxonomic or functional groups of organisms.

4. Opportunities and challenges for eDNA analysis to inform spatial planning for conservation and management

Conservation organisations and government management agencies recognise the opportunities provided by eDNA. The unparalleled scope to detect Tree of Life assemblages of entire ecosystems with eDNA opens up multiple pathways to achieve more ambitious environmental management objectives. However, specifically within the context of spatial management, the benefits and limitations of eDNA need to be made clear, so that patterns of diversity recovered from eDNA can be interpreted appropriately. Currently, the lack of a consistent framework hampers the translation of eDNA data into spatial prioritisation plans, although eDNA data could be used to meet a wide variety of conservation objectives (Table 1). Major eDNA applications, from single species to whole community studies, include: (1) detection of low abundance species, (2) shortening of the time required to produce data, (3) cost reduction and (4) non-invasive or non-destructive sampling (Table 1).

Increasingly, it is clear that not all species can be protected or saved from extinction in the face of ubiquitous anthropogenic impacts and limited conservation funds (Beyer et al. 2018; Pressey and Bottrill, 2008). Instead, environmental management needs to focus on maintaining functional integrity (D'Agata et al., 2016). eDNA research may provide new opportunities to define ecosystem functionality (Cordier et al., 2017), and thus to develop approaches to protect these functions in a more targeted way (Sutherland et al., 2009, 2018). Rather than measuring a small subset of species

Table 1 Potential spatial planning objectives used in spatial planning and how the use of eDNA could influence future practice for tropical marine systems, references provided where available.

Conservation objective	Conservation action	Data needed	Current challenges	eDNA opportunities	Examples of eDNA use
Identify and eradicate invasive species	Robinson et al. (2019)	Presence/ absence of invasive species	Low detection probability	Increased detection probability + decreased cost	Holman et al. (2019)
Manage rare/ threatened species	Stewart et al. (2017)	Presence/ absence of species	Low detection probability	Increased detection probability + decreased cost	Stewart et al. (2017)
Represent cryptic species	DiBattista et al. (2017)	Cryptic species detection	Low detection probability	Increased detection probability + decreased cost	Holman et al. (2019)
Baseline biodiversity assessment	Stat et al. (2017)	Species presence data from entire assemblage	Expensive to collect fully representative samples	Can be collected in single sample	Stat et al., 2017 Cilleros et al., 2019
Protect ecosystem functions of poorly studied taxa	Costello et al. (2015)	Little known taxa	High taxonomic expertise required	Reference database return species / OTUs for most taxa	Not available yet
Manage ecological networks	Tulloch et al. (2018)	Species abundance data from entire assemblage	Intensive data collection + unlikely to cover entire assemblage	Full assemblage collected + reduced sampling effort	Not available yet
Manage trophic functioning	Mills et al. (2011)	Invertebrate data	Low detection probability + intensive data collection	Increased detection probability + Tree of Life assemblage collection possible	Not available yet

Continued

Table 1 Potential spatial planning objectives used in spatial planning and how the use of eDNA could influence future practice for tropical marine systems, references provided where available.—cont'd

Conservation objective	Conservation action	Data needed	Current challenges	eDNA opportunities	Examples of eDNA use
Protect ecosystem functions of microorganism (e.g. macronutrient cycling)	Louca et al. (2016)	Microbial functional gene	Target studies on specific gene	Single sample can provide information on microbial functionality as well as baseline biodiversity data	Delmont et al. (2018)
Find priority areas for connected protected area networks	Beger et al. (2014), Nielsen et al. (2017)	Population genetics of target species	Invasive + intensive sampling effort	Non-invasive + reduced sampling effort	Sigsgaard et al. (2017a)
Avoid protecting exclusion areas	Daigle et al. (2020)	In-depth knowledge of ecosystem function + presence/ absence data target species	Not possible or not cost-effective	Increased detection probability + decreased cost	Not available yet
Monitoring: Assess trends in species populations	McDonald-Madden et al. (2010)	Abundance data	Low detection probability + labour intensive to get accurate abundance of target species	Not deployed yet, easier to standardise through time	Bista et al. (2017), Sigsgaard et al. (2017b)
Monitoring: Assess sex and/or age of populations	Stewart et al. (2018)	In-depth knowledge of the species	Invasive + intensive sampling effort	Not deployed yet	Not available yet
Monitoring: Maintain large scale (global) sampling of population trends	Thompson et al. (2016)	Biodiversity data large geographic scale	Extensive, standardised sampling effort + expensive	Decreased sampling cost + easier to standardise	Gilbert et al. (2014)

(e.g. large predators, benthic cover), eDNA sampling can detect a large proportion of micro- and macrobiota within each replicate and test how OTU interactions change between samples to ultimately discover which taxa really drive ecosystem functions (Makiola et al., 2020). For example, eDNA could identify key generalist taxa that use a wide array of environments under varying impacts, and such species could be highly resilient species of value to conservation and restoration management actions. A better understanding of how species respond to varying impacts will allow managers to detect warning signs of changing ecosystems and adapt the spatial management in response.

With the potential of eDNA to target the full species assemblage, analytical methods such as ecological network analysis are likely to gain importance (Table 1) (Evans et al., 2016). Network analyses can be of use in highlighting the sensitive groups that should be targeted in management or biomonitoring (Derocles et al., 2018). Co-occurrence network analysis could reveal the way in which conservation actions (or lack thereof) affect the biodiversity of entire ecosystems (Tulloch et al., 2018). Networks could also shed light on hitherto unknown mechanisms; rare species might be influenced by other taxa that are overlooked in many of the presently used survey methods. Trophic functioning of protected versus unprotected areas could be examined in new ways as eDNA offers the potential to accurately quantify producers or invertebrates at the base of the food chain as easily as the large fishes, which currently receive the bulk of research attention (Cinner et al., 2016, 2018; Edgar et al., 2014; Martin et al., 2017; Mora et al., 2011). Stomach content analysis using eDNA has also been shown to be more accurate than visual analysis and can provide much needed data on trophic interactions (Jo et al., 2016). The higher taxonomic resolution that can be achieved using eDNA compared to conventional methods can further open up the way for far more specific spatial planning than is currently possible (Table 1).

Representing and maintaining genetic connectivity between different populations can form an important conservation objective in spatial planning (Carvalho et al., 2017). Such analyses of population genetics will increasingly benefit from eDNA data collection and advances in sequencing technology as multiple individuals (Sigsgaard et al., 2017a) and/or multiple taxa haplotypes (see Glossary) (Adams et al., 2019) can be studied at the same time. Inferring information on population genetics will become easier in the nearer future with expected cost reductions in organelle sequencing (full mitochondria or nuclear markers, see Deiner et al., 2017b) and more comprehensive reference databases. Such advances are fundamental for monitoring and will provide information to spatial planners and managers to act rapidly and to establish best practices in ecosystem management and biodiversity conservation (Boehm et al., 2017).

5. Guidelines: The practicalities of spatial planning with eDNA

As with any other method, the key to obtaining accurate data relevant to conservation objectives is a well thought out experimental design and appropriate sampling method(s). While questions about the spatial coverage of eDNA are not yet fully resolved, enough is known to guide sampling design. The range at which the signature of eDNA can be detected varies from 50 m to several kilometres depending on the environment and conditions, which should be reflected in the aims and design of a study (e.g. Deiner and Altermatt, 2014; Jeunen et al., 2019). In the marine environment, eDNA can discriminate between different marine habitats, even at small spatial scales (Jeunen et al., 2019). In rivers, however, eDNA can potentially travel much further, up to 12 km, depending on river flow rates and DNA degradation time (Deiner and Altermatt, 2014; Jane et al., 2015; Pont et al., 2018). Therefore, eDNA sampling design for spatial planning depends on conservation objectives and planning area. For example, if sampling effort aims to provide information on specific habitats, it is essential to get as close as possible to the source of the eDNA which best describes that habitat. If the goal is, however, to sample a large catchment or conduct initial, general biodiversity surveys, downstream sampling could be more advisable. Moreover, the number of samples to collect (i.e. replication) needs careful consideration under both aforementioned scenarios. Based on previous studies, more replicates with smaller volumes are preferable to a larger volume with less replication (Dickie et al., 2018; Shaw et al., 2017), and high replication ($n \gg 3$) is necessary to increase the detection rates of rarer taxa (Mächler et al., 2018; Rees et al., 2014).

While the relatively rapid degeneration of DNA makes eDNA sampling an ideal method to obtain estimates of recent diversity in a given area, this in turn brings uncertainties to designing eDNA protocols for spatial planning. Sampling might have to be repeated at multiple times throughout the year or target particular seasons (e.g. spawning events, migrations, monsoon vs dry season, etc.) depending on the goal of the sampling program (De Souza et al., 2016) and conservation objectives and spatial planning time frames. Thus, a thorough ecological understanding of the study system remains essential when designing monitoring plans (see for example Bylemans et al., 2018b). This extends to the choice of source material. eDNA in sediments can provide precise information on different taxonomic levels, but data are less precise for obtaining information over short temporal scales. In addition, samples taken from sediments and the water column can harbour different eDNA signals (Holman et al., 2019), potentially as these capture biodiversity over longer and shorter temporal scales respectively. Conversely, sampling water could be a good solution for representing relative short time frames for a wide range of species (Collins et al., 2018).

A crucial aspect to effective tropical marine resource management is also integrating uncertainties and habitat status into spatial planning (Vercammen et al., 2019), particularly for occupancy uncertainty of species or habitats. eDNA studies should therefore aim to provide such information to be relevant to conservation management. Quantifying chances of false negatives or false positives is currently not common practice in eDNA studies (but see Hunter et al., 2015); future inclusion of these metrics would greatly benefit spatial planning. The primary risks of false positives are through contamination during sampling or extraction stages, or by wrongly assigning OTUs to a certain taxonomic species. It is essential that care is taken to avoid contamination where possible (for example, by using closed filters to capture eDNA), particularly in challenging fieldwork conditions or when engaging citizen scientists (Biggs et al., 2015; Julian et al., 2019). Increased DNA degradation rates under different conditions could increase chances of false negatives and have unintended consequences on conservation measures, such as removal of protective measures (Chadès et al., 2008).

Finally, current eDNA methods might not be suitable for all organisms of interest (e.g. Walker et al., 2017), or existing primer sets might not be able to detect the presence of phylogenetically distinct taxa such as Syngnathidae (Nester et al. in press). A combination of multiple primers (e.g. a "universal" 18S primer set, combined with one or more targeted primers) could therefore result in more robust data on community composition and increased detection of rare species (Berry et al., 2019; Deiner et al., 2017a) without incurring prohibitive additional costs. Technological advances such as metagenomic or whole organelle sequencing will continue to improve eDNA methodology (Porter and Hajibabaei, 2018b). Metagenomic sequencing has the potential to remove primer and PCR biases present in metabarcoding, however, its current use in eDNA analysis remains limited due to difficulties in assigning sequences to macro-organisms (Stat et al., 2017). It is therefore advised to test the methods proposed for use with novel combinations of rare or endangered species of interest to ensure efficiency.

6. The future of eDNA in conservation planning science

Technological innovations can provide a step-change in conservation science and practice in the face of escalating global biodiversity declines, but such technology needs to be developed in the context of well-defined conservation problems and applications (Iacona et al., 2019). Similarly, to enhance the relevance of new directions in eDNA approaches for spatial planning, it is necessary to carefully examine and standardise important features in the methodology that relate to spatial and taxonomic comprehensiveness (Table 2), ecological relevance, and SMART (Maxwell et al., 2015) implementation of spatial planning. Spatially comprehensive and real-time biodiversity data generation will be where eDNA analysis is likely to excel, with the potential to combine eDNA with autonomous sampling and machine learning to create global monitoring networks (Bohan et al., 2017). For example, Australia and New Zealand are considering using eDNA to support biological surveillance (Cristescu and Hebert, 2018). Such systems could link with app-based spatial planning (Daigle et al., 2020) or planning web-platforms (MARXAN, www.marxan.org, SEASKETCH https://www.seasketch.org/) that could apply pipelines to correctly apply eDNA data. Such technology supported spatial planning also has great potential in achieving community buy-in (Game et al., 2014). Automated eDNA recording and monitoring systems could underpin new developments in dynamic ocean planning, where eDNA data could inform where, how and what is managed on short timeframes such as a week or biweekly (Dunn et al., 2016; Hobday et al., 2011; Lewison et al., 2015). Such improvements will lead to a reduced number of field scientists and bigger studies across larger spatial and temporal scales. Thus, quantifying and reporting the impacts on ecosystem and their relative services that are largely affected by climate changes, resource overexploitation or pollution at an unprecedented resolution.

As the use of eDNA in spatial planning becomes more common, scientists from disparate fields will require simple guidelines (Fig. 3). As the use of eDNA in spatial planning would benefit from remote sampling and app-based application, sharing of data and especially of metadata would become fundamental. Nowadays only the raw sequences are requested to be stored in public database, accessible to everyone, while metadata that are equally important for large regional/global comparisons are often missing or incomplete. The use of Otlet-style data and sample sharing (https://otlet.io/) should become a good practice step in experimental setup with clear and well documented metadata upload systems (Fig. 3).

Table 2 Horizon scanning table of research needs to optimise eDNA integration in spatial planning, references provided where available.

Factor	Challenge	Future	(Review) References
Abiotic and biotic degradation	Very limited information on tropical environments (e.g. Temperature, pH, UV intensity)	Quantify decay rate-biophysical conditions relationships for different settings	Lopes et al. (2017), Sales et al. (2019), Valdez-Moreno et al. (2019)
Sampling and analysis method efficiency	Differences in sampling method (substrate, volume, extraction, primers) influence results.	Quantify differences associated with different sampling methods	McGee et al. (2019), Holman et al. (2019)
	Sample size and replication varies across studies	Develop hierarchical standardised protocols for multi-purpose eDNA analyses.	Dickie et al. (2018), Mächler et al. (2018), Rees et al. (2014)
	Use of negative controls (i.e. filtration (field) and extraction blank) only.	Include positive controls (i.e. a mixture of known DNA from different species and at different concentrations)	Zinger et al. (2019), Evans et al. (2017)
Bioinformatics and statistical analysis	Data analysis methods are variable, i.e., different pipelines and algorithms	Develop a robust bioinformatic pipeline that could be used across different eDNA experimental set ups	Eckert et al. (2018), Deiner et al. (2017a)
Reference databases are lacking	Lack of curated databases for taxonomic assignment More reference data required on different markers than COI Databases lack information such as geographic position or environmental variables Rate of false positive and false negative observations not captured	Better population of databases entries and improved curation, with barcodes for identified taxa, in parallel with better bioinformatics algorithms Development of occupancy models to quantify error rates (ongoing, but more depth needed) Occupancy models should take in account the hierarchical nature of the experiment and include false positives	Porter and Hajibabaei (2018a) Ficetola et al. (2015), Ficetola et al. (2016) Davis et al. (2018), Doi et al. (2019), Strickland and Roberts (2019)

Continued

Table 2 Horizon scanning table of research needs to optimise eDNA integration in spatial planning, references provided where available.—cont'd

Factor	Challenge	Future	(Review) References
Method boundaries	Turning read abundance into estimated biomass	Develop, test, and document potential biomass/ abundance eDNA methods	Contradictory results based on few species studied (Robson et al., 2016, Sassoubre et al., 2016)
Data access and compatibility	Data storage not centralised, metadata not available	Standardised metadata recording should be established and metadata appropriately deposited. Centralise eDNA data in open access databases (with solid funding) to increase use by wider community	
Unifying spatial planning with eDNA data framework and guidelines	Very little overlap in expertise between molecular ecologists and conservation scientists	Better and bigger databases that could be used for spatial planning	Beger et al. (2014), Nielsen et al. (2017), Von der Heyden et al. (2014)

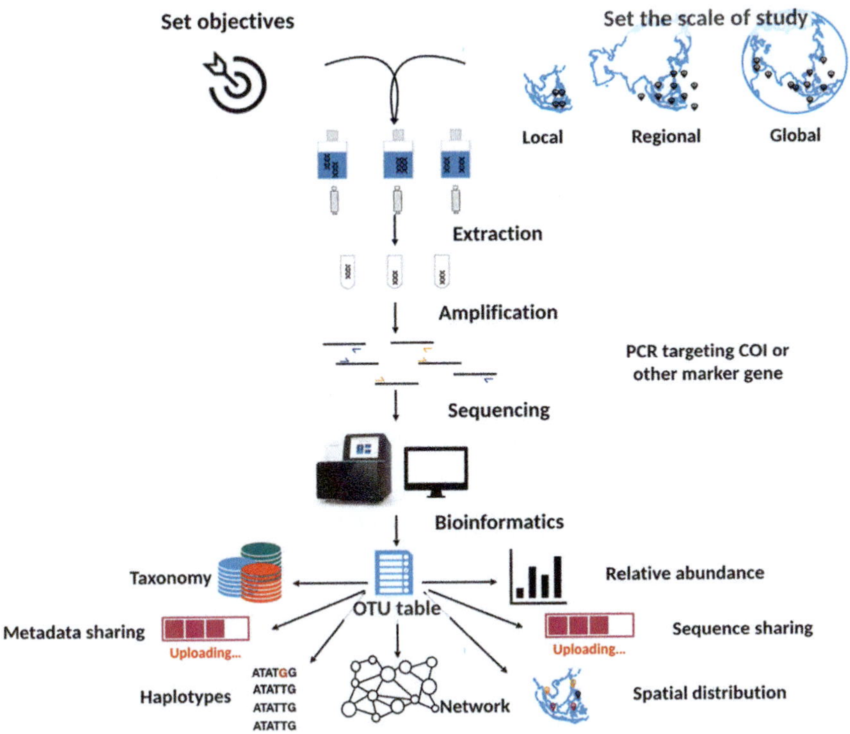

Fig. 3 Schematic guidelines for future eDNA projects that provide biodiversity data for spatial planning.

7. Conclusions

The natural world is currently facing multiple interacting threats on an unprecedented scale that will considerably impact how human communities connect with natural resources. Adequate resource and conservation management of tropical marine ecosystems based on real-time, large-scale biodiversity is more important than ever, yet collecting and analysing such data remains challenging. It is beyond doubt that eDNA can and will play an increasingly large role in environmental research and it will likely increase the scope of future spatial planning. This review has outlined current and future possibilities on how to do this, as well as provide information on how to integrate eDNA in planning and how to avoid the most common mistakes. The rapidly developing applications of eDNA might seem daunting to non-experts, particularly since the technique is still very recent. While research gaps and methodological uncertainties exist, the method is ready to be tested for integration in spatial planning on a large scale. Effective marine spatial planning decisions depend on accurate and timely knowledge of the system to be managed. eDNA provides a step-change in how we think about the availability of biodiversity data, and has the potential to completely redefine the spatio-temporal context of how ecological systems are managed.

Glossary

Molecular terminology

12S rRNA Mitochondrial gene that is used for taxonomic assignment especially for fishes.

Barcode of Life Data (BOLD) Public database of COI gene sequenced across all the Tree of Life.

Cytochrome oxidase I (COI) Alternatively known as COX1 or CO1, it is a mitochondrial gene that encodes the main subunit of the cytochrome c complex. It is widely used to barcode eukaryotes. The reference database for this gene is known as BOLD database.

Environmental DNA (eDNA) DNA directly extracted from environmental samples (soil, sediment, water, etc.) without any knowledge of the original organism.

Haplotypes A group of alleles that are inherited together from a single parent, for example mitochondrial haplotypes. Haplogroup are haplotypes that shared a common ancestor with a single nucleotide polymorphism mutation.

High Throughput Sequencing (HTS) A technique able to determine the nucleotide composition of millions of nucleic acid sequences. Different types of sequencing are now available and include Illumina, PacBio or NanoPore. Every sequencing method uses different strategies to generate the nucleic acid sequence, for example Illumina uses fluorescent nucleotides while NanoPore uses current change when DNA strand passes through a membrane protein. For more information please see Van Dijk et al. (2018).

Metabarcoding Taxonomic identification of millions of sequences in one experiment generated by PCR amplification on eDNA samples. This is possible using one of the HTS techniques.

Metagenomics Different from metabarcoding, metagenomics analyses do not require PCR amplification prior to sequencing. During the process, all DNA molecules are amplified together, which limits the error connected with PCR amplification.

Operational Taxonomic Unit (OTU) Sequences (reads) obtained from HTS are grouped together, to minimise the influence of PCR and sequencing error, based on threshold dissimilarity (usually 3%). OTUs clusters are generated in programs such as VSEARCH or USEARCH. They are NOT species but an approximation to species. Clustering together multiple reads will inevitably reduce the information on nucleotide variations within that OTU.

PCR (Polymerase Chain Reaction) A molecular technique that allows the exponential amplification of a target fragment/region of DNA from a mixture of DNA fragments. The desired fragment to amplify is recognised from the other fragments in the mixture by specific primers (small single strand oligonucleotides) complementary to the desired sequence. The process is based on sequential cycles of heating and cooling at specific temperature. In the first step, the double strand DNA molecules are separated into single strands by high temperatures. In the second step, temperature is lowered, and primers bind to the complementary sequences of the targeted regions of DNA. In the third step, temperature is increased to the working optimum for the polymerase enzyme.

The enzyme adds nucleotides to assemble the complementary sequence of the target DNA. During a PCR reaction, the three steps are repeated several times (between 25 and 30 cycles) and for each cycle the quantity of amplified DNA increases exponentially.

Quantitative PCR (qPCR) Quantitative PCR is a variant of PCR. The main difference between the two is that qPCR is able to quantify how many fragments of DNA are amplified during each step in the reaction, leading to quantitative data.

Amplicon Sequence Variance (ASV) ASV does not include grouping reads based on dissimilarity but retains all the reads that are generated by HTS after a denoising step (removing of sequencing errors and chimera). Single nucleotide variation sequences are maintained in the dataset allowing the discrimination between different haplotypes within the same species. Programs that generate ASVs are DADA2 or Deblur. For a more detailed description of the difference between OTU and ASV see Callahan et al. (2017).

Conservation terminology

Convention on Biological Diversity (CBD) International convention signed by 168 countries which aims to conserve biological diversity, promote sustainable use of biological diversity and ensure the fair and equitable sharing of the benefits arising out of the utilisation of genetic resources.

Data spatialisation Transformation of site-specific environmental, biodiversity, and socio-economic data into spatially-explicit map-based representations, typically achieved through distribution modelling or interpolation.

Gap analysis Method used to identify problems or gaps that are likely to decrease the efficiency of protected area managements. Gaps can range from exclusion of species or habitats, to missing ecological processes, or problems in the management process itself.

Marine spatial planning Framework to decide where to implement different management and conservation actions by evaluating the trade-offs between protecting biodiversity features and ensuring socio-economic sustainability in a transparent, quantitative, and repeatable manner.

Objectives Quantitative specification of management goals for a certain ecosystem, habitat type, or species. Objectives can be ecological, social, or economic, but should be detailed and quantifiable.

SMART (spatial planning) A conservation approach where objectives are Specific (clearly defined), Measurable (specific on what will be measured and how), Achievable (realistic in light of existing ecological and social conditions), Relevant (complementary to project goals), and Time-bound (clear timeline).

Acknowledgements

The authors would like to acknowledge funding from the NERC Wallacea (UK) and Ristekdikti (Indonesia) joint programme to AB, MDB, JJ, GL, AD and MB. MB also has received funding from the European Union's Horizon 2020 research and innovation programme under the Marie Skłodowska-Curie grant agreement No TRIM-DLV-747102. SvdH received funding through the National Research Foundation of South Africa, Competitive Support for Rated Researchers grant number 105842. SvdH and SC also received funding through the Royal Society Newton Mobility grant NI170129. We also thank two anonymous reviewers or their constructive comments.

References

Adams, C.I., Knapp, M., Gemmell, N.J., Jeunen, G.-J., Bunce, M., Lamare, M.D., Taylor, H.R., 2019. Beyond biodiversity: can environmental DNA (eDNA) cut it as a population genetics tool? Genes 10, 192.

Assis, J., Tyberghein, L., Bosch, S., Verbruggen, H., Serrão, E.A., De Clerck, O., 2018. Bio-ORACLE v2. 0: extending marine data layers for bioclimatic modelling. Glob. Ecol. Biogeogr. 27, 277–284.

Balasingham, K.D., Walter, R.P., Heath, D.D., 2017. Residual eDNA detection sensitivity assessed by quantitative real-time PCR in a river ecosystem. Mol. Ecol. Resour. 17, 523–532.

Barnes, M.A., Turner, C.R., 2016. The ecology of environmental DNA and implications for conservation genetics. Conserv. Genet. 17, 1–17.

Beger, M., Simon, L., Game, E., Ball, I., Treml, E., Watts, M., Possingham, H.P., 2010. Incorporating functional ecological connectivity into spatial decision making for conservation. Conserv. Lett. 3, 359–368.

Beger, M., Selkoe, K.A., Treml, E., Barber, P.H., von der Heyden, S., Crandall, E.D., Toonen, R.J., Riginos, C., 2014. Evolving coral reef conservation with genetic information. Bull. Mar. Sci. 90, 159–185.

Berry, T.E., Osterrieder, S.K., Murray, D.C., Coghlan, M.L., Richardson, A.J., Grealy, A.K., Stat, M., Bejder, L., Bunce, M., 2017. DNA metabarcoding for diet analysis and biodiversity: a case study using the endangered Australian sea lion (Neophoca cinerea). Ecol. Evol. 7, 5435–5453.

Berry, T.E., Saunders, B.J., Coghlan, M.L., Stat, M., Jarman, S., Richardson, A.J., Davies, C.H., Berry, O., Harvey, E.S., Bunce, M., 2019. Marine environmental DNA biomonitoring reveals seasonal patterns in biodiversity and identifies ecosystem responses to anomalous climatic events. PLoS Genet. 15, e1007943.

Beyer, H., Kennedy, E., Beger, M., Chen, A., Cinner, J., Darling, E.S., Eakin, M., Gates, R., Heron, S., Knowlton, N., Obura, D., Palumbi, S., Possingham, H.P., Puotinen, M., Runting, R., Skirving, W., Spalding, M., Wilson, K., Wood, S., Veron, J., Hoegh-Guldberg, O., 2018. Risk-sensitive planning for conserving coral reefs under rapid climate change. Conserv. Lett, e12587.

Biggs, J., Ewald, N., Valentini, A., Gaboriaud, C., Dejean, T., Griffiths, R.A., Foster, J., Wilkinson, J.W., Arnell, A., Brotherton, P., 2015. Using eDNA to develop a national citizen science-based monitoring programme for the great crested newt (Triturus cristatus). Biol. Conserv. 183, 19–28.

Bista, I., Carvalho, G.R., Walsh, K., Seymour, M., Hajibabaei, M., Lallias, D., Christmas, M., Creer, S., 2017. Annual time-series analysis of aqueous eDNA reveals ecologically relevant dynamics of lake ecosystem biodiversity. Nat. Commun. 8, 14087.

Bista, I., Carvalho, G.R., Tang, M., Walsh, K., Zhou, X., Hajibabaei, M., Shokralla, S., Seymour, M., Bradley, D., Liu, S., 2018. Performance of amplicon and shotgun sequencing for accurate biomass estimation in invertebrate community samples. Mol. Ecol. Resour. 18, 1020–1034.

Boehm, A.B., Ismail, N.S., Sassoubre, L.M., Andruszkiewicz, E.A., 2017. Oceans in peril: grand challenges in applied water quality research for the 21st century. Environ. Eng. Sci. 34, 3–15.

Bohan, D.A., Vacher, C., Tamaddoni-Nezhad, A., Raybould, A., Dumbrell, A.J., Woodward, G., 2017. Next-generation global biomonitoring: large-scale, automated reconstruction of ecological networks. Trends Ecol. Evol. 32, 477–487.

Bohmann, K., Evans, A., Gilbert, M.T.P., Carvalho, G.R., Creer, S., Knapp, M., Douglas, W.Y., De Bruyn, M., 2014. Environmental DNA for wildlife biology and biodiversity monitoring. Trends Ecol. Evol. 29, 358–367.

Boon, P.Y., Beger, M., 2016. The effect of contrasting threat mitigation objectives on spatial conservation priorities. Mar. Policy 68, 23–29.

Boussarie, G., Bakker, J., Wangensteen, O.S., Mariani, S., Bonnin, L., Juhel, J.-B., Kiszka, J.J., Kulbicki, M., Manel, S., Robbins, W.D., 2018. Environmental DNA illuminates the dark diversity of sharks. Sci. Adv. 4, eaap9661.

Broennimann, O., Fitzpatrick, M.C., Pearman, P.B., Petitpierre, B., Pellissier, L., Yoccoz, N.G., Thuiller, W., Fortin, M.J., Randin, C., Zimmermann, N.E., 2012. Measuring ecological niche overlap from occurrence and spatial environmental data. Glob. Ecol. Biogeogr. 21, 481–497.

Bylemans, J., Furlan, E.M., Gleeson, D.M., Hardy, C.M., Duncan, R.P., 2018a. Does size matter? An experimental evaluation of the relative abundance and decay rates of aquatic environmental DNA. Environ. Sci. Technol. 52, 6408–6416.

Bylemans, J., Gleeson, D.M., Lintermans, M., Hardy, C.M., Beitzel, M., Gilligan, D.M., Furlan, E.M., 2018b. Monitoring riverine fish communities through eDNA metabarcoding: determining optimal sampling strategies along an altitudinal and biodiversity gradient. Metabarcoding Metagenom. 2, e30457.

Callahan, B.J., McMurdie, P.J., Holmes, S.P., 2017. Exact sequence variants should replace operational taxonomic units in marker-gene data analysis. ISME J. 11, 2639–2643.

Carvalho, S.B., Velo-Antón, G., Tarroso, P., Portela, A.P., Barata, M., Carranza, S., Moritz, C., Possingham, H.P., 2017. Spatial conservation prioritization of biodiversity spanning the evolutionary continuum. Nat. Ecol. Evol. 1, 0151.

Carvalho, S., Aylagas, E., Villalobos, R., Kattan, Y., Berumen, M., Pearman, J.K., 2019. Beyond the visual: using metabarcoding to characterize the hidden reef cryptobiome. Proc. R. Soc. B 286, 20182697.

Chadès, I., McDonald-Madden, E., McCarthy, M.A., Wintle, B., Linkie, M., Possingham, H.P., 2008. When to stop managing or surveying cryptic threatened species. Proc. Natl. Acad. Sci. U S A 105, 13936–13940.

Cilleros, K., Valentini, A., Allard, L., Dejean, T., Etienne, R., Grenouillet, G., Iribar, A., Taberlet, P., Vigouroux, R., Brosse, S., 2019. Unlocking biodiversity and conservation studies in high-diversity environments using environmental DNA (eDNA): a test with Guianese freshwater fishes. Mol. Ecol. Resour. 19, 27–46.

Cinner, J.E., Huchery, C., Macneil, M.A., Graham, N.A.J., McClanahan, T.R., Maina, J., Maire, E., Kittinger, J.N., Hicks, C.C., Mora, C., Allison, E.H., D'Agata, S., Hoey, A., Feary, D.A., Crowder, L., Williams, I.D., Kulbicki, M., Vigliola, L., Wantiez, L., Edgar, G., Stuart-Smith, R.D., Sandin, S.A., Green, A.L., Hardt, M.J., Beger, M., Friedlander, A., Campbell, S.J., Holmes, K.E., Wilson, S.K., Brokovich, E., Brooks, A.J., Cruz-Motta, J.J., Booth, D.J., Chabanet, P., Gough, C., Tupper, M., Ferse, S.C.A., Sumaila, U.R., Mouillot, D., 2016. Bright spots among the world's coral reefs. Nature 535, 416–419.

Cinner, J.E., Maire, E., Huchery, C., Macneil, M.A., Graham, N.A.J., Mora, C., McClanahan, T.R., Barnes, M.L., Kittinger, J.N., Hicks, C.C., D'Agata, S., Hoey, A., Gurney, G.G., Feary, D.A., Williams, I.D., Kulbicki, M., Vigliola, L., Wantiez, L., Edgar, G., Stuart-Smith, R.D., Sandin, S.A., Green, A.L., Hardt, M.J., Beger, M., Friedlander, A., Wilson, S.K., Brokovich, E., Brooks, A.J., Cruz-Motta, J.J., Booth, D.J., Chabanet, P., Gough, C., Tupper, M., Ferse, S.C.A., Sumaila, U.R., Pardede, S., Mouillot, D., 2018. The gravity of human impacts mediates coral reef conservation gains. Proc. Natl. Acad. Sci. U S A 115, E6116–E6125.

Clark, D.R., Ferguson, R.M., Harris, D.N., Matthews Nicholass, K.J., Prentice, H.J., Randall, K.C., Randell, L., Warren, S.L., Dumbrell, A.J., 2018. Streams of data from drops of water: 21st century molecular microbial ecology. Wiley Interdiscip. Rev. Water 5, e1280.

Collins, R.A., Wangensteen, O.S., O'Gorman, E.J., Mariani, S., Sims, D.W., Genner, M.J., 2018. Persistence of environmental DNA in marine systems. Commun. Biol. 1, 185.

Collins, R.A., Bakker, J., Wangensteen, O.S., Soto, A.Z., Corrigan, L., Sims, D.W., Genner, M.J., Mariani, S., 2019. Non-specific amplification compromises environmental DNA metabarcoding with COI. Methods Ecol. Evol. 10, 1985–2001.

Cordier, T., Esling, P., Lejzerowicz, F., Visco, J., Ouadahi, A., Martins, C., Cedhagen, T., Pawlowski, J., 2017. Predicting the ecological quality status of marine environments from eDNA metabarcoding data using supervised machine learning. Environ. Sci. Technol. 51, 9118–9126.

Correia, E., Granadeiro, J.P., Regalla, A., Dias, E., Almeida, A., Catry, P., 2017. Predatory pelagic fishes of the Bijagós Archipelago (Guinea-Bissau) show high overlap in diets dominated by sardinella. Afr. J. Mar. Sci. 39, 389–396.

Costello, M.J., Vanhoorne, B., Appeltans, W., 2015. Conservation of biodiversity through taxonomy, data publication, and collaborative infrastructures. Conserv. Biol. 29 (4), 1094–1099.

Cristescu, M.E., Hebert, P.D., 2018. Uses and misuses of environmental DNA in biodiversity science and conservation. Annu. Rev. Ecol. Evol. Syst. 49, 209–230.

D'Agata, S., Vigliola, L., Graham, N.A., Wantiez, L., Parravicini, V., Villéger, S., Mou-Tham, G., Frolla, P., Friedlander, A.M., Kulbicki, M., 2016. Unexpected high vulnerability of functions in wilderness areas: evidence from coral reef fishes. Proc. R. Soc. B Biol. Sci. 283, 20160128.

Daigle, R.M., Metaxas, A., Balbar, A.C., McGowan, J., Treml, E.A., Kuempel, C.D., Possingham, H.P., Beger, M., 2020. Operationalizing ecological connectivity in spatial conservation planning with Marxan Connect. Methods Ecol. Evol. https://besjournals.onlinelibrary.wiley.com/doi/abs/10.1111/2041-210X.13349.

Davis, A.J., Williams, K.E., Snow, N.P., Pepin, K.M., Piaggio, A.J., 2018. Accounting for observation processes across multiple levels of uncertainty improves inference of species distributions and guides adaptive sampling of environmental DNA. Ecol. Evol. 8, 10879–10892.

De Souza, L.S., Godwin, J.C., Renshaw, M.A., Larson, E., 2016. Environmental DNA (eDNA) detection probability is influenced by seasonal activity of organisms. PLoS One 11, e0165273.

De Vere, N., Jones, L.E., Gilmore, T., Moscrop, J., Lowe, A., Smith, D., Hegarty, M.J., Creer, S., Ford, C.R., 2017. Using DNA metabarcoding to investigate honey bee foraging reveals limited flower use despite high floral availability. Sci. Rep. 7, 42838.

Deagle, B.E., Chiaradia, A., McInnes, J., Jarman, S.N., 2010. Pyrosequencing faecal DNA to determine diet of little penguins: is what goes in what comes out? Conserv. Genet. 11, 2039–2048.

Deiner, K., Altermatt, F., 2014. Transport distance of invertebrate environmental DNA in a natural river. PloS One 9, e88786.

Deiner, K., Bik, H.M., Mächler, E., Seymour, M., Lacoursière-Roussel, A., Altermatt, F., Creer, S., Bista, I., Lodge, D.M., De Vere, N., 2017a. Environmental DNA metabarcoding: transforming how we survey animal and plant communities. Mol. Ecol. 26, 5872–5895.

Deiner, K., Renshaw, M.A., Li, Y., Olds, B.P., Lodge, D.M., Pfrender, M.E., 2017b. Long-range PCR allows sequencing of mitochondrial genomes from environmental DNA. Methods Ecol. Evol. 8, 1888–1898.

Delmont, T.O., Quince, C., Shaiber, A., Esen, Ö.C., Lee, S.T.M., Rappé, M.S., McLellan, S.L., Lücker, S., Eren, A.M., 2018. Nitrogen-fixing populations of Planctomycetes and Proteobacteria are abundant in surface ocean metagenomes. Nat. Microbiol. 3, 804–813.

Derocles, S.A., Bohan, D.A., Dumbrell, A.J., Kitson, J.J., Massol, F., Pauvert, C., Plantegenest, M., Vacher, C., EVANS, D.M., 2018. Biomonitoring for the 21st century: integrating next-generation sequencing into ecological network analysis. In: Advances in Ecological Research. Elsevier.

DiBattista, J.D., Coker, D.J., Sinclair-Taylor, T.H., Stat, M., Berumen, M.L., Bunce, M., 2017. Assessing the utility of eDNA as a tool to survey reef-fish communities in the Red Sea. Coral Reefs 36, 1245–1252.

DiBattista, J.D., Reimer, J.D., Stat, M., Masucci, G.D., Biondi, P., De Brauwer, M., Bunce, M., 2019. Digging for DNA at depth: rapid universal metabarcoding surveys (RUMS) as a tool to detect coral reef biodiversity across a depth gradient. PeerJ 7, e6379.

Dickie, I.A., Boyer, S., Buckley, H.L., Duncan, R.P., Gardner, P.P., Hogg, I.D., Holdaway, R.J., LEAR, G., Makiola, A., Morales, S.E., 2018. Towards robust and repeatable sampling methods in eDNA-based studies. Mol. Ecol. Resour. 18, 940–952.

Doi, H., Fukaya, K., Oka, S.-I., Sato, K., Kondoh, M., Miya, M., 2019. Evaluation of detection probabilities at the water-filtering and initial PCR steps in environmental DNA metabarcoding using a multispecies site occupancy model. Sci. Rep. 9, 3581.

Dunn, D.C., Maxwell, S.M., Boustany, A.M., Halpin, P.N., 2016. Dynamic ocean management increases the efficiency and efficacy of fisheries management. Proc. Natl. Acad. Sci. U. S. A 113, 668–673.

Eckert, I.M., Littlefair, J.E., Zhang, G.K., Chain, F.J., Crease, T.J., Cristescu, M.E., 2018. Bioinformatics for biomonitoring: species detection and diversity estimates across next-generation sequencing platforms. Next Generation Biomonitoring, PT 2, Adv. Ecol. Res. 59, 1–32.

Edgar, G.J., Stuart-Smith, R.D., Willis, T.J., Kininmonth, S., Baker, S.C., Banks, S., Barrett, N.S., Becerro, M.A., Bernard, A.T.F., Berkhout, J., Buxton, C.D., Campbell, S.J., Cooper, A.T., Davey, M., Edgar, S.C., Forsterra, G., Galvan, D.E., Irigoyen, A.J., Kushner, D.J., Moura, R., Parnell, P.E., Shears, N.T., Soler, G., Strain, E.M.A., Thomson, R.J., 2014. Global conservation outcomes depend on marine protected areas with five key features. Nature 506, 216–220.

Eilers, K.G., Debenport, S., Anderson, S., Fierer, N., 2012. Digging deeper to find unique microbial communities: the strong effect of depth on the structure of bacterial and archaeal communities in soil. Soil Biol. Biochem. 50, 58–65.

Elith, J., Leathwick, J., 2009. The contribution of species distribution modelling to conservation prioritization. In: Moilanen, A., Wilson, K.A., Possingham, H.P. (Eds.), Spatial conservation prioritisation: Quantitative methods and computational tools. Oxford University Press, Oxford, UK, pp. 70–93. Chapter 6.

Evans, D.M., Kitson, J.J., Lunt, D.H., Straw, N.A., Pocock, M.J., 2016. Merging DNA metabarcoding and ecological network analysis to understand and build resilient terrestrial ecosystems. Funct. Ecol. 30, 1904–1916.

Evans, N.T., Li, Y., Renshaw, M.A., Olds, B.P., Deiner, K., Turner, C.R., Jerde, C.L., Lodge, D.M., Lamberti, G.A., Pfrender, M.E., 2017. Fish community assessment with eDNA metabarcoding: effects of sampling design and bioinformatic filtering. Can. J. Fish. Aquat. Sci. 74, 1362–1374.

Fernandes, L., Day, J., Lewis, A., Slegers, S., Kerrigan, B., Breen, D., Cameron, D., Jago, B., Hall, J., Lowe, D., Innes, J., Tanzer, J., Chadwick, V., Thompson, L., Gorman, K., Simmons, M., Barnett, B., Sampson, K., De'Ath, G., Mapstone, B., Marsh, H., Possingham, H., Ball, I., Ward, T., Dobbs, K., Aumend, J., Slater, D., Stapleton, K., 2005. Establishing representative no-take areas in the great barrier reef: large-scale implementation of theory on marine protected areas. Conserv. Biol. 19, 1733–1744.

Ficetola, G.F., Thuiller, W., MIAUD, C., 2007. Prediction and validation of the potential global distribution of a problematic alien invasive species—the American bullfrog. Divers. Distrib. 13, 476–485.

Ficetola, G.F., Pansu, J., Bonin, A., Coissac, E., Giguet-Covex, C., De Barba, M., Gielly, L., Lopes, C.M., Boyer, F., Pompanon, F., 2015. Replication levels, false presences and the estimation of the presence/absence from eDNA metabarcoding data. Mol. Ecol. Resour. 15, 543–556.

Ficetola, G.F., Taberlet, P., Coissac, E., 2016. How to limit false positives in environmental DNA and metabarcoding? Mol. Ecol. Resour. 16, 604–607.

Game, E.T., Meijaard, E., Sheil, D., McDonald-Madden, E., 2014. Conservation in a wicked complex world; challenges and solutions. Conserv. Lett. 7, 271–277.

Gilbert, J.A., Jansson, J.K., Knight, R., 2014. The earth microbiome project: successes and aspirations. BMC Biol. 12, 69.

Grantham, H.S., Agostini, V.N., Wilson, J., Mangubhai, S., Hidayat, N., Muljadi, A., Muhajir, Rotinsulu, C., Mongdong, M., Beck, M.W., Possingham, H.P., 2013. A comparison of zoning analyses to inform the planning of a marine protected area network in Raja Ampat, Indonesia. Mar. Policy 38, 184–194.

Green, A., Smith, S.E., Lipsett-Moore, G., Groves, C., Peterson, N., Sheppard, S., Lokani, P., Hamilton, R., Almany, J., Aitsi, J., Bualia, L., 2009. Designing a resilient network of marine protected areas for Kimbe Bay, Papua New Guinea. Oryx 43, 1–11.

Guisan, A., Tingley, R., Baumgartner, J.B., Naujokaitis-Lewis, I., Sutcliffe, P.R., Tulloch, A.I., Regan, T.J., Brotons, L., McDonald-Madden, E., Mantyka-Pringle, C., 2013. Predicting species distributions for conservation decisions. Ecol. Lett. 16, 1424–1435.

Hobday, A.J., Hartog, J.R., Spillman, C.M., Alves, O., 2011. Seasonal forecasting of tuna habitat for dynamic spatial management. Can. J. Fish. Aquat. Sci. 68, 898–911.

Holman, L.E., De Bruyn, M., Creer, S., Carvalho, G., Robidart, J., Rius, M., 2019. Detection of introduced and resident marine species using environmental DNA metabarcoding of sediment and water. Sci. Rep. 9, 1–10.

Hughes, T.P., Barnes, M.L., Bellwood, D.R., Cinner, J.E., Cumming, G.S., Jackson, J.B.C., Kleypas, J., Van De Leemput, I.A., Lough, J.M., Morrison, T.H., Palumbi, S.R., Van Nes, E.H., Scheffer, M., 2017. Coral reefs in the Anthropocene. Nature 546, 82–90.

Hunter, M.E., Oyler-McCance, S.J., Dorazio, R.M., Fike, J.A., Smith, B.J., Hunter, C.T., Reed, R.N., Hart, K.M., 2015. Environmental DNA (eDNA) sampling improves occurrence and detection estimates of invasive Burmese pythons. PloS One 10, e0121655.

Iacona, G., Ramachandra, A., McGowan, J., Davies, A., Joppa, L., Koh, L.P., Fegraus, E., Game, E., Guillera-Arroita, G., Harcourt, R., 2019. Identifying technology solutions to bring conservation into the innovation era. Front. Ecol. Environ. 17, 591–598.

Ikeda, K., Doi, H., Tanaka, K., Kawai, T., Negishi, J.N., 2016. Using environmental DNA to detect an endangered crayfish Cambaroides japonicus in streams. Conserv. Genet. Resour. 8, 231–234.

Ishige, T., Miya, M., Ushio, M., Sado, T., Ushioda, M., Maebashi, K., Yonechi, R., Lagan, P., Matsubayashi, H., 2017. Tropical-forest mammals as detected by environmental DNA at natural saltlicks in Borneo. Biol. Conserv. 210, 281–285.

Jane, S.F., Wilcox, T.M., McKelvey, K.S., Young, M.K., Schwartz, M.K., Lowe, W.H., Letcher, B.H., Whiteley, A.R., 2015. Distance, flow and PCR inhibition: e DNA dynamics in two headwater streams. Mol. Ecol. Resour. 15, 216–227.

Jerde, C.L., Olds, B.P., Shogren, A.J., Andruszkiewicz, E.A., Mahon, A.R., Bolster, D., Tank, J.L., 2016. Influence of stream bottom substrate on retention and transport of vertebrate environmental DNA. Environ. Sci. Technol. 50, 8770–8779.

Jerde, C.L., Wilson, E.A., Dressler, T.L., 2019. Measuring global fish species richness with eDNA metabarcoding. Mol. Ecol. Resour. 19, 19–22.

Jeunen, G.J., Knapp, M., Spencer, H.G., Lamare, M.D., Taylor, H.R., Stat, M., Bunce, M., Gemmell, N.J., 2019. Environmental DNA (eDNA) metabarcoding reveals strong discrimination among diverse marine habitats connected by water movement. Mol. Ecol. Resour. 19, 426–438.

Jo, H., Ventura, M., Vidal, N., Gim, J.S., Buchaca, T., Barmuta, L.A., Jeppesen, E., Joo, G.J., 2016. Discovering hidden biodiversity: the use of complementary monitoring of fish diet based on DNA barcoding in freshwater ecosystems. Ecol. Evol. 6, 219–232.

Jo, T., Murakami, H., Masuda, R., Sakata, M.K., Yamamoto, S., Minamoto, T., 2017. Rapid degradation of longer DNA fragments enables the improved estimation of distribution and biomass using environmental DNA. Mol. Ecol. Resour. 17, e25–e33.

Jo, T., Murakami, H., Yamamoto, S., Masuda, R., Minamoto, T., 2019. Effect of water temperature and fish biomass on environmental DNA shedding, degradation, and size distribution. Ecol. Evol. 9, 1135–1146.

Julian, J.T., Glenney, G.W., Rees, C., 2019. Evaluating observer bias and seasonal detection rates in amphibian pathogen eDNA collections by citizen scientists. Dis. Aquat. Organ. 134, 15–24.

Kamoroff, C., Goldberg, C.S., 2017. Using environmental DNA for early detection of amphibian chytrid fungus Batrachochytrium dendrobatidis prior to a ranid die-off. Dis. Aquat. Organ. 127, 75–79.

Kelly, R.P., Shelton, A.O., Gallego, R., 2019. Understanding pcR processes to draw meaningful conclusions from environmental DnA studies. Sci. Rep. 9, 1–14.

Klymus, K.E., Marshall, N.T., Stepien, C.A., 2017. Environmental DNA (eDNA) metabarcoding assays to detect invasive invertebrate species in the Great Lakes. PLoS One 12, e0177643.

Koziol, A., Stat, M., Simpson, T., Jarman, S., Dibattista, J.D., Harvey, E.S., Marnane, M., McDonald, J., Bunce, M., 2019. Environmental DNA metabarcoding studies are critically affected by substrate selection. Mol. Ecol. Resour. 19, 366–376.

Kukkala, A.S., Moilanen, A., 2013. Core concepts of spatial prioritisation in systematic conservation planning. Biol. Rev. 88, 443–464.

Leray, M., Knowlton, N., Ho, S.-L., Nguyen, B.N., Machida, R.J., 2019. GenBank is a reliable resource for 21st century biodiversity research. Proc. Natl. Acad. Sci. U. S. A. 116, 22651–22656.

Lewison, R., Hobday, A.J., Maxwell, S., Hazen, E., Hartog, J.R., Dunn, D.C., Briscoe, D., Fossette, S., O'Keefe, C.E., Barnes, M., 2015. Dynamic ocean management: identifying the critical ingredients of dynamic approaches to ocean resource management. Bioscience 65, 486–498.

Lopes, C.M., Sasso, T., Valentini, A., Dejean, T., Martins, M., Zamudio, K.R., Haddad, C.F., 2017. eDNA metabarcoding: a promising method for anuran surveys in highly diverse tropical forests. Mol. Ecol. Resour. 17, 904–914.

Louca, S., Parfrey, L.W., Doebeli, M., 2016. Decoupling function and taxonomy in the global ocean microbiome. Science 353 (6305), 1272–1277.

Mächler, E., Osathanunkul, M., Altermatt, F., 2018. Shedding light on eDNA: neither natural levels of UV radiation nor the presence of a filter feeder affect eDNA-based detection of aquatic organisms. PLoS One 13, e0195529.

Magris, R.A., Pressey, R.L., Weeks, R., Ban, N.C., 2014. Integrating connectivity and climate change into marine conservation planning. Biol. Conserv. 170, 207–221.

Makino, A., Klein, C.J., Possingham, H.P., Yamano, H., Yara, Y., Ariga, T., Matsuhasi, K., Beger, M., 2015. The effect of applying alternate IPCC climate scenarios to marine reserve design for range changing species. Conserv. Lett. 8, 320–328.

Makiola, A., Compson, Z.G., Baird, D.J., Barnes, M.A., Boerlijst, S.P., Bouchez, A., Brennan, G., Bush, A., Canard, E., Cordier, T., Creer, S., Curry, R.A., David, P., Dumbrell, A.J., Gravel, D., Hajibabaei, M., Hayden, B., Van Der Hoorn, B., Jarne, P., Jones, J.I., Karimi, B., Keck, F., Kelly, M., Knot, I.E., Krol, L., Massol, F., Monk, W.A., Murphy, J., Pawlowski, J., Poisot, T., Porter, T.M., Randall, K.C., Ransome, E., Ravigné, V., Raybould, A., Robin, S., Schrama, M., Schatz, B., Tamaddoni-Nezhad, A., Trimbos, K.B., Vacher, C., Vasselon, V., Wood, S., Woodward, G., Bohan, D.A., 2020. Key questions for next-generation biomonitoring. Front. Environ. Sci. 7.

Margules, C.R., Pressey, R.L., 2000. Systematic conservation planning. Nature 405, 243.

Martin, T.S.H., Connolly, R.M., Olds, A.D., Ceccarelli, D.M., Fenner, D.E., Schlacher, T.A., Beger, M., 2017. Subsistence fishing on Pacific atolls can maintain near-pristine fish communities. ICES J. Mar. Sci. 74, fsx043.

Maruyama, A., Nakamura, K., Yamanaka, H., Kondoh, M., Minamoto, T., 2014. The release rate of environmental DNA from juvenile and adult fish. PLoS One 9, e114639.

Maxwell, S.L., Milner-Gulland, E.J., Jones, J.P.G., Knight, A.T., Bunnefeld, N., Nuno, A., Bal, P., Earle, S., Watson, J.E.M., Rhodes, J.R., 2015. Being smart about SMART environmental targets. Science 347, 1075–1076.

McClanahan, T.R., Hicks, C.C., 2011. Changes in life history and ecological characteristics of coral reef fish catch composition with increasing fishery management. Fish. Manag. Ecol. 18, 50–60.

McDonald-Madden, E., Baxter, P.W., Fuller, R.A., Martin, T.G., Game, E.T., Montambault, J., Possingham, H.P., 2010. Monitoring does not always count. Trends Ecol. Evol. 25, 547–550.

McGee, K.M., Robinson, C., Hajibabaei, M., 2019. Gaps in DNA-based biomonitoring across the globe. Front. Ecol. Evol. 7, 337.

Meekan, M., Austin, C.M., Tan, M.H., Wei, N.-W.V., Miller, A., Pierce, S.J., Rowat, D., Stevens, G., Davies, T.K., Ponzo, A., 2017. iDNA at sea: recovery of whale shark (Rhincodon typus) mitochondrial DNA sequences from the whale shark copepod (Pandarus rhincodonicus) confirms global population structure. Front. Mar. Sci. 4, 420.

Mills, M., Jupiter, S.D., Pressey, R.L., Ban, N.C., Comley, J., 2011. Incorporating effectiveness of community-based management in a national marine gap analysis for Fiji. Conserv. Biol. 25, 1155–1164.

Mora, C., Aburto-Oropeza, O., Ayala Bocos, A., Ayotte, P.M., Banks, S., Bauman, A.G., Beger, M., Bessudo, S., Booth, D.J., Brokovich, E., Brooks, A., Chabanet, P., Cinner, J., Cortés, J., Cruz-Motta, J.J., Cupul Magaña, A., Demartini, E.E., Edgar, G.J., Feary, D.A., Ferse, S.C.A., Friedlander, A.M., Gaston, K.J., Gough, C., Graham, N.A.J., Green, A., Guzman, H.M., Hardt, M., Kulbicki, M., Letourneur, Y., López Pérez, A., Loreau, M., Loya, Y., Martinez, C., Mascareñas-Osorio, I., Morove, T., Nadon, M.-O., Nakamura, Y., Paredes, G., Polunin, N.V.C., Pratchett, M.S., Reyes Bonilla, H., Rivera, F., Sala, E., Sandin, S., Soler, G., Stuart-Smith, R., Tessier, E., Tittensor, D.P., Tupper, M., Usseglio, P., Vigliola, L., Wantiez, L., Williams, I.D., Wilson, S.K., Zapata, F.A., 2011. Global human footprint on the linkage between biodiversity and ecosystem functioning in reef fishes. PLoS Biol. 9, e1000606. https://doi.org/10.1371/journal.pbio.1000606.

Nester G., De Brauwer M., Kozial A., West K., DiBattista J., White N., Power M., Heydenrych M., Harvey E.S. and Bunce M., Development and evaluation of fish eDNA metabarcoding assays facilitates the detection of cryptic seahorse taxa (family: Syngnathidae). Environmental DNA, in press.

Nichols, P.K., Marko, P.B., 2019. Rapid assessment of coral cover from environmental DNA in Hawai'i. Environ. DNA 1, 40–53.

Nielsen, E.S., Beger, M., Henriques, R., Selkoe, K.A., Von der Heyden, S., 2017. Multispecies genetic objectives in spatial conservation planning. Conserv. Biol. 31, 872–882.

Pawlowski, J., Kelly-Quinn, M., Altermatt, F., Apothéloz-Perret-Gentil, L., Beja, P., Boggero, A., Borja, A., Bouchez, A., Cordier, T., Domaizon, I., 2018. The future of biotic indices in the ecogenomic era: integrating (e) DNA metabarcoding in biological assessment of aquatic ecosystems. Sci. Total Environ. 637, 1295–1310.

Pearman, J.K., Leray, M., Villalobos, R., Machida, R., Berumen, M.L., Knowlton, N., Carvalho, S., 2018. Cross-shelf investigation of coral reef cryptic benthic organisms reveals diversity patterns of the hidden majority. Sci. Rep. 8, 8090.

Pont, D., Rocle, M., Valentini, A., Civade, R., Jean, P., Maire, A., Roset, N., Schabuss, M., Zornig, H., Dejean, T., 2018. Environmental DNA reveals quantitative patterns of fish biodiversity in large rivers despite its downstream transportation. Sci. Rep. 8, 10361.
Porter, T.M., Hajibabaei, M., 2018a. Over 2.5 million COI sequences in GenBank and growing. PloS One 13, e0200177.
Porter, T.M., Hajibabaei, M., 2018b. Scaling up: a guide to high-throughput genomic approaches for biodiversity analysis. Mol. Ecol. 27, 313–338.
Pressey, R.L., Bottrill, M.C., 2008. Opportunism, threats, and the evolution of systematic conservation planning. Conserv. Biol. 22 (5), 1340–1345.
Rees, H.C., Maddison, B.C., Middleditch, D.J., Patmore, J.R., Gough, K.C., 2014. The detection of aquatic animal species using environmental DNA–a review of eDNA as a survey tool in ecology. J. Appl. Ecol. 51, 1450–1459.
Robinson, C.V., Garcia de Leaniz, C., Rolla, M., Consuegra, S., 2019. Monitoring the eradication of the highly invasive topmouth gudgeon (Pseudorasbora parva) using a novel eDNA assay. Environ. DNA 1, 74–85.
Robson, H.L., Noble, T.H., Saunders, R.J., Robson, S.K., Burrows, D.W., Jerry, D.R., 2016. Fine-tuning for the tropics: application of eDNA technology for invasive fish detection in tropical freshwater ecosystems. Mol. Ecol. Resour. 16, 922–932.
Ruppert, K.M., Kline, R.J., Rahman, M.S., 2019. Past, present, and future perspectives of environmental DNA (eDNA) metabarcoding: a systematic review in methods, monitoring, and applications of global eDNA. Glob. Ecol. Conserv. 17, e00547.
Russ, G.R., Miller, K.I., Rizzari, J.R., Alcala, A.C., 2015. Long-term no-take marine reserve and benthic habitat effects on coral reef fishes. Mar. Ecol. Prog. Ser. 529, 233–248.
Sales, N.G., Wangensteen, O.S., Carvalho, D.C., Mariani, S., 2019. Influence of preservation methods, sample medium and sampling time on eDNA recovery in a neotropical river. Environ. DNA 1, 119–130.
Sassoubre, L.M., Yamahara, K.M., Gardner, L.D., Block, B.A., Boehm, A.B., 2016. Quantification of environmental DNA (eDNA) shedding and decay rates for three marine fish. Environ. Sci. Technol. 50, 10456–10464.
Shaw, J.L., Weyrich, L., Cooper, A., 2017. Using environmental (e) DNA sequencing for aquatic biodiversity surveys: a beginner's guide. Mar. Freshw. Res. 68, 20–33.
Shogren, A.J., Tank, J.L., Andruszkiewicz, E., Olds, B., Mahon, A.R., Jerde, C.L., Bolster, D., 2017. Controls on eDNA movement in streams: transport, retention, and resuspension. Sci. Rep. 7, 5065.
Sigsgaard, E.E., Carl, H., Møller, P.R., Thomsen, P.F., 2015. Monitoring the near-extinct European weather loach in Denmark based on environmental DNA from water samples. Biol. Conserv. 183, 46–52.
Sigsgaard, E.E., Nielsen, I.B., Bach, S.S., Lorenzen, E.D., Robinson, D.P., Knudsen, S.W., Pedersen, M.W., Al Jaidah, M., Orlando, L., Willerslev, E., 2017a. Population characteristics of a large whale shark aggregation inferred from seawater environmental DNA. Nat. Ecol. Evol. 1, 0004.
Sigsgaard, E.E., Nielsen, I.B., Carl, H., Krag, M.A., Knudsen, S.W., Xing, Y., Holm-Hansen, T.H., Møller, P.R., Thomsen, P.F., 2017b. Seawater environmental DNA reflects seasonality of a coastal fish community. Mar. Biol. 164, 128.
Simpfendorfer, C.A., Kyne, P.M., Noble, T.H., Goldsbury, J., Basiita, R.K., Lindsay, R., Shields, A., Perry, C., Jerry, D.R., 2016. Environmental DNA detects critically endangered largetooth sawfish in the wild. Endanger. Species Res. 30, 109–116.
Sønstebø, J., Gielly, L., Brysting, A., Elven, R., Edwards, M., Haile, J., Willerslev, E., Coissac, E., RIOUX, D., SANNIER, J., 2010. Using next-generation sequencing for molecular reconstruction of past Arctic vegetation and climate. Mol. Ecol. Resour. 10, 1009–1018.

Stat, M., Huggett, M.J., Bernasconi, R., Dibattista, J.D., Berry, T.E., Newman, S.J., Harvey, E.S., Bunce, M., 2017. Ecosystem biomonitoring with eDNA: metabarcoding across the tree of life in a tropical marine environment. Sci. Rep. 7, 12240.

Stat, M., John, J., Dibattista, J.D., Newman, S.J., Bunce, M., Harvey, E.S., 2019. Combined use of eDNA metabarcoding and video surveillance for the assessment of fish biodiversity. Conserv. Biol. 33, 196–205.

Stewart, K., Ma, H., Zheng, J., Zhao, J., 2017. Using environmental DNA to assess population-wide spatiotemporal reserve use. Conserv. Biol. 31, 1173–1182.

Stewart, J., Hegarty, A.M., Young, C., Fowler, A.M., 2018. Sex-specific differences in growth, mortality and migration support population resilience in the heavily exploited migratory marine teleost Mugil cephalus (Linnaeus 1758). Mar. Freshw. Res. 69 (3), 385–394.

Strickland, G.J., Roberts, J.H., 2019. Utility of eDNA and occupancy models for monitoring an endangered fish across diverse riverine habitats. Hydrobiologia 826, 129–144.

Sutherland, W., Adams, W., Aronson, R., Aveling, R., Blackburn, T., Broad, S., Ceballos, G., Cote, I., Cowling, R., Da Fonseca, G., 2009. One hundred questions of importance to the conservation of global biological diversity. Conserv. Biol. 23, 557–567.

Sutherland, W.J., Butchart, S.H., Connor, B., Culshaw, C., Dicks, L.V., Dinsdale, J., Doran, H., Entwistle, A.C., Fleishman, E., Gibbons, D.W., 2018. A 2018 horizon scan of emerging issues for global conservation and biological diversity. Trends Ecol. Evol. 33, 47–58.

Takahara, T., Minamoto, T., Yamanaka, H., Doi, H., Kawabata, Z.I., 2012. Estimation of fish biomass using environmental DNA. PloS One 7, e35868.

Thompson, M.S., Bankier, C., Bell, T., Dumbrell, A.J., Gray, C., Ledger, M.E., Lehmann, K., McKew, B.A., Sayer, C.D., Shelley, F., 2016. Gene-to-ecosystem impacts of a catastrophic pesticide spill: testing a multilevel bioassessment approach in a river ecosystem. Freshw. Biol. 61, 2037–2050.

Thomsen, P.F., Willerslev, E., 2015. Environmental DNA–an emerging tool in conservation for monitoring past and present biodiversity. Biol. Conserv. 183, 4–18.

Tittensor, D.P., Beger, M., Boerder, K., Boyce, D.G., Cavanagh, R.D., Cosandey-Godin, A., Crespo, G.O., Dunn, D.C., Ghiffary, W., Grant, S.M., Hannah, L., Halpin, P.N., Harfoot, M., Heaslip, S.G., Jeffery, N.W., Kingston, N., Lotze, H.K., McGowan, J., McLeod, E., McOwen, C.J., O'Leary, B.C., Schiller, L., Stanley, R.R.E., Westhead, M., Wilson, K.L., Worm, B., 2019. Integrating climate adaptation and biodiversity conservation in the global ocean. Science Advances 5, eaay9969.

Tulloch, A.I., Chadès, I., Lindenmayer, D.B., 2018. Species co-occurrence analysis predicts management outcomes for multiple threats. Nat. Ecol. Evol. 2, 465.

Unsworth, R.K.F., Jones, B.L., Cullen-Unsworth, L.C., 2016. Seagrass meadows are threatened by expected loss of peatlands in Indonesia. Glob. Chang. Biol. 22, 2957–2958.

Uthicke, S., Lamare, M., Doyle, J.R., 2018. eDNA detection of corallivorous seastar (Acanthaster cf. solaris) outbreaks on the great barrier reef using digital droplet PCR. Coral Reefs 37, 1229–1239.

Valdez-Moreno, M., Ivanova, N.V., Elias-Gutierrez, M., Pedersen, S.L., Bessonov, K., Hebert, P.D., 2019. Using eDNA to biomonitor the fish community in a tropical oligotrophic lake. PloS One 14, e0215505.

Van Dijk, E.L., Jaszczyszyn, Y., Naquin, D., Thermes, C., 2018. The third revolution in sequencing technology. Trends Genet. 34, 666–681.

Vercammen, A., McGowan, J.A., Knight, A.T., Pardede, S., Muttaqin, E., Harris, J., Ahmadia, G.N., Estradivari, Dallison, T., Selig, E.R., Beger, M., 2019. Evaluating the impact of accounting for coral cover in large-scale marine conservation prioritisations. Divers. Distrib. 25, 1564–1574.

Vimal, R., Rodrigues, A.S., Mathevet, R., Thompson, J.D., 2011. The sensitivity of gap analysis to conservation targets. Biodivers. Conserv. 20, 531–543.

Von der Heyden, S., Beger, M., Toonen, R.J., Van Herwerden, L., Juinio-Meñez, M.A., Ravago-Gotanco, R., Fauvelot, C., Bernardi, G., 2014. The application of genetics to marine management and conservation: examples from the Indo-Pacific. Bull. Mar. Sci. 90, 123–158.

Walker, D.M., Leys, J.E., Dunham, K.E., Oliver, J.C., Schiller, E.E., Stephenson, K.S., Kimrey, J.T., Wooten, J., Rogers, M.W., 2017. Methodological considerations for detection of terrestrial small-body salamander eDNA and implications for biodiversity conservation. Mol. Ecol. Resour. 17, 1223–1230.

Weeks, R., Alino, P.M., Atkinson, S., Beldia, P., Binson, A., Campos, W.L., Djohani, R., Green, A.L., Hamilton, R., Horigue, V., Jumin, R., Kalim, K., Kasasiah, A., Kereseka, J., Klein, C., Laroya, L., Magupin, S., Masike, B., Mohan, C., Pinto, R.M.D., Vave-Karamui, A., Villanoy, C., Welly, M., White, A.T., 2014. Developing marine protected area networks in the coral triangle: good practices for expanding the coral triangle marine protected area system. Coast. Manag. 42, 183–205.

Wei, N., Nakajima, F., Tobino, T., 2018. A microcosm study of surface sediment environmental DNA: decay observation, abundance estimation, and fragment length comparison. Environ. Sci. Technol. 52, 12428–12435.

Weltz, K., Lyle, J.M., Ovenden, J., Morgan, J.A., Moreno, D.A., Semmens, J.M., 2017. Application of environmental DNA to detect an endangered marine skate species in the wild. PloS One 12, e0178124.

Willerslev, E., Davison, J., Moora, M., Zobel, M., Coissac, E., Edwards, M.E., Lorenzen, E.D., Vestergård, M., Gussarova, G., Haile, J., 2014. Fifty thousand years of Arctic vegetation and megafaunal diet. Nature 506, 47.

Williams, B.K., Brown, E.D., 2016. Technical challenges in the application of adaptive management. Biol. Conserv. 195, 255–263.

Wilson, K.A., Underwood, E.C., Morrison, S.A., Klausmeyer, K.R., Murdoch, W.W., Reyers, B., Wardell-Johnson, G., Marquet, P.A., Rundel, P.W., McBride, M.F., Pressey, R.L., Bode, M., Hoekstra, J.M., Andelman, S., Looker, M., Rondinini, C., Kareiva, P., Shaw, M.R., Possingham, H.P., 2007. Conserving biodiversity efficiently: what to do, where, and when. PLoS Biol. 5, e223.

Wilson, K.A., Carwardine, J., Possingham, H.P., 2009. Setting conservation priorities. Ann. N. Y. Acad. Sci. 1162, 237–264.

Wu, D., Struwe, W.B., Harvey, D.J., Ferguson, M.A.J., Robinson, C.V., 2018. N-glycan microheterogeneity regulates interactions of plasma proteins. Proc. Natl. Acad. Sci. U. S. A. 115, 8763–8768.

Zinger, L., Bonin, A., Alsos, I.G., Bálint, M., Bik, H., Boyer, F., Chariton, A.A., Creer, S., Coissac, E., Deagle, B.E., 2019. DNA metabarcoding—need for robust experimental designs to draw sound ecological conclusions. Mol. Ecol. 28, 1857–1862.

Further reading

Djurhuus, A., Port, J., Closek, C.J., Yamahara, K.M., Romero-Maraccini, O., Walz, K.R., Goldsmith, D.B., Michisaki, R., Breitbart, M., Boehm, A.B., 2017. Evaluation of filtration and DNA extraction methods for environmental DNA biodiversity assessments across multiple trophic levels. Front. Mar. Sci. 4, 314.

Moushomi, R., Wilgar, G., Carvalho, G., Creer, S., Seymour, M., 2019. Environmental DNA size sorting and degradation experiment indicates the state of Daphnia magna mitochondrial and nuclear eDNA is subcellular. Sci. Rep. 9, 1–9.

CHAPTER ELEVEN

Copulas and their potential for ecology

Shyamolina Ghosh[a], Lawrence W. Sheppard[a], Mark T. Holder[b], Terrance D. Loecke[c], Philip C. Reid[d,e], James D. Bever[a], Daniel C. Reuman[a,f,*]

[a]Department of Ecology and Evolutionary Biology and Kansas Biological Survey, University of Kansas, Lawrence, KS, United States
[b]Department of Ecology and Evolutionary Biology and Biodiversity Institute, University of Kansas, Lawrence, KS, United States
[c]Environmental Studies Program and Kansas Biological Survey, University of Kansas, Lawrence, KS, United States
[d]School of Biological & Marine Sciences, University of Plymouth, Drake Circus, Plymouth, United Kingdom
[e]Continuous Plankton Recorder Survey, The Marine Biological Association, The Laboratory, Citadel Hill, Plymouth, United Kingdom
[f]Laboratory of Populations, Rockefeller University, New York, NY, United States
[*]Corresponding author: e-mail address: reuman@ku.edu

Contents

Abstract

All branches of ecology study relationships among and between environmental and biological variables. However, standard approaches to studying such relationships, based on correlation and regression, provide only some of the complex information contained in the relationships. Other statistical approaches exist that provide a complete description of relationships between variables, based on the concept of the

Advances in Ecological Research, Volume 62
ISSN 0065-2504
https://doi.org/10.1016/bs.aecr.2020.01.003

copula; they are applied in finance, neuroscience and elsewhere, but rarely in ecology. We explore the concepts that underpin copulas and the potential for those concepts to improve our understanding of ecology. We find that informative copula structure in dependencies between variables is common across all the environmental, species-trait, phenological, population, community, and ecosystem functioning datasets we considered. Many datasets exhibited asymmetric tail associations, whereby two variables were more strongly related in their left compared to right tails, or vice versa. We describe mechanisms by which observed copula structure and tail associations can arise in ecological data, including a Moran-like effect whereby dependence structures are inherited from environmental variables; and asymmetric or nonlinear influences of environments on ecological variables, such as under Liebig's law of the minimum. We also describe consequences of copula structure for ecological phenomena, including impacts on extinction risk, Taylor's law, and the temporal stability of ecosystem services. By documenting the importance of a complete description of dependence between variables, advancing conceptual frameworks, and demonstrating a powerful approach, we encourage widespread use of copulas in ecology, which we believe can benefit the discipline.

1. Introduction

All branches of ecology study relationships among biological variables and relationships between environmental and biological variables. However, commonly used correlation and regression approaches to studying such relationships are limited, and provide only a partial description of the relationship. For instance, datasets showing different relationships may have the same correlation coefficient (Fig. 1A and B). The variables of Fig. 1A (respectively, Fig. 1B) are principally related in the left (respectively, right) portions of their distributions, an asymmetric pattern of association that can have ecological significance, as discussed below, but that is not captured by correlation. Although correlation is only one way to study relationships between variables, other common metrics also provide only partial information.

Well-developed approaches do exist, however, and are applied widely in other fields, that provide a description of the association between two or more variables that is mathematically provably complete (Joe, 2014; Mai and Scherer, 2017; Nelsen, 2006). The approaches are based on the concept of the *copula*. Copula approaches separate information in a bivariate random variable into two parts: the information in the marginal distributions (which says nothing about the association between the variables), and the remaining information (which is solely about the association). We here

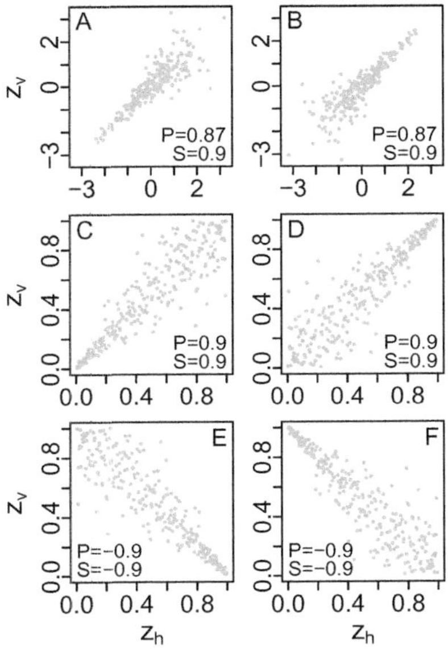

Fig. 1 (A)–(D) Bivariate datasets showing diverse relationships between variables. Some of these datasets nevertheless have the same Pearson (P) or Spearman (S) correlation coefficients. (C) and (D) show normalized ranks of (A) and (B), respectively. (E) and (F) Reflections of (C) and (D), respectively, about a vertical axis. Subscripts h and v in the axis labels stand for "horizontal" and "vertical".

introduce some simple concepts, based on ranks, that relate to copulas and give a conceptual flavour of copula ideas to be introduced formally in the "Background" section below (Section 2). Given a sample (x_i, y_i), $i=1, 2, \ldots, n$ (e.g., Fig. 1A and B), information about the structure of the association between x and y can be separated from information contained in the marginal distributions by considering the plot of u_i vs v_i, where u_i is the rank of x_i in the set of x_j $(j=1,2,\ldots,n)$, divided by $n+1$; and v_i is defined in the same way but using the y_i. The rank of the smallest element in a set is 1. The u_i and v_i are called *normalized ranks* of the x_i and y_i; they relate to the empirical cumulative distribution functions of the x_i and y_i, respectively. We henceforth abbreviate "cumulative distribution function" as "cdf". See, for instance, Fig. 1C and D, which show the normalized ranks of A-B. Ranking makes the marginal distributions of the component datasets uniform, isolating the dependence structure. Dependence structure and marginals can then be studied separately. Normalized rank plots such as Fig. 1C and D relate to copulas in a way to be described (Section 2). Note that Pearson correlation, even though it is the most commonly used measure of association, is modified by monotonic transformation of the component variables (Appendix S1 and Fig. S1 in Supplementary material in the online version at https://doi.org/10.1016/bs.aecr.2020.01.003), and therefore reflects not only dependence information, but also information on the marginals (Genest and Favre, 2007). Spearman and Kendall correlations, being rank-based, are not influenced by monotonic transformations of the variables. They provide information solely on the dependence between those variables.

A main benefit of a copula approach is that it can detect associations in the tails of distributions, and asymmetries of such associations. Tail association (introduced formally in Section 2) is association between extreme values of variables. If smaller values of two positively associated variables are more strongly associated than are larger values, the variables are said to have stronger lower- or left-tail association than right- or upper-tail association (Fig. 1C); and vice versa if larger values are more strongly associated (Fig. 1D). Datasets of the same Spearman or Kendall correlation can have a range of tail associations (Fig. S2 in Supplementary material in the online version at https://doi.org/10.1016/bs.aecr.2020.01.003), which can be quantified (see sections on Background and methods for Q1 below, Sections 2 and 4). Although copulas can be used to model the entire complex dependence structure of data, we have found tail associations to be a useful component of that structure, so tail association is a main focus of this paper. We give four examples, below, which illustrate some of the ecological meaning of tail associations.

The goal of this paper is to explore the potential for applications of copulas in ecology and to estimate to what extent ecological understanding may benefit from using copulas. We provide an introduction to copula concepts in Section 2. Readers can find additional abbreviated (Anderson et al., 2018; Genest and Favre, 2007) and comprehensive (Joe, 2014; Mai and Scherer, 2017; Nelsen, 2006) introductions elsewhere. Copulas have already been used effectively in a few studies to illuminate ecological phenomena (Anderson et al., 2018; de Valpine et al., 2014; Popovic et al., 2019), but such usage is still rare. Whereas Anderson et al. (2018) provide a specific method, based on copulas, for important problems of the analysis of multivariate population count data, we instead advocate more broadly for the application of copulas across ecology. Our results suggest that environmental, ecological and evolutionary processes may commonly generate complex dependence structures, including asymmetric tail associations, and that greater use of copulas can illuminate underlying processes. We believe copula approaches are among the tools all ecologists should be considering for analysis of their data in the 21st century.

One way we may expect, a priori, a study of tail associations and copula dependence structures to better reflect ecological relationships relates to Liebig's law of the minimum. Liebig's law is the idea that growth is controlled not by total resources but by the resource which is scarcest relative to organism needs. If, for instance, the growth of a plant depends on soil nitrogen, N, and other factors, a plot of growth rates vs soil N may look more like Fig. 1A or C than like Fig. 1B or D, i.e., the two variables may show left-tail association: N controls plant growth, producing a clear relationship, only when it is limiting. This aspect of the association would be visible in a rank-based plot, but may not be revealed by correlation.

A second reason why we may expect detailed understanding of dependence structures to benefit ecological research is that prior work demonstrated, using copulas, complex structure in the spatial dependence of environmental variables (Goswami et al., 2018; Li et al., 2013; Serinaldi, 2008; She and Xia, 2018). An environmental variable measured through time in two locations may show strong tail associations between the locations if intense meteorological events are also widespread, as seems frequently to be the case: extreme values are associated with intense events, so happen in both places at the same time, whereas moderate values of the environmental variable may be associated with local phenomena, which differ between the locations. Spatial dependence in an environmental variable tends to beget spatial dependence in fluctuations of populations or other ecosystem variables (this is called *spatial synchrony*) influenced by the environmental variable. This is called the *Moran effect*. If asymmetric tail associations, or other complex dependence structure, is transmitted from environmental to ecological variables, then we would expect complex dependence structure and tail associations to be a common feature of the spatial synchrony of population, community, biogeochemical and other environmentally influenced ecological variables. Synchrony attracts major interest in ecology (Liebhold et al., 2004; Sheppard et al., 2016; Walter et al., 2017).

A third reason why we may expect an understanding of complex dependence structures and tail associations to benefit ecology is that such an approach may help illuminate causal mechanisms between variables. If two species, Sp1 and Sp2, are strong competitors, abundances of the two species across quadrats should be negatively related, as in Fig. 1E and F. If Sp1 is plotted on the horizontal axis and Sp2 on the vertical, then

Fig. 1E may suggest Sp1 is the dominant competitor: when Sp1 is abundant, Sp2 is necessarily rare because it is suppressed; whereas when Sp1 is rare, Sp2 may be abundant, or may also be relatively rare due to limiting factors other than Sp1. Alternatively, Fig. 1F may suggest Sp2 is the dominant competitor. Other causal hypotheses may produce similar dependence structure, and it is usually impossible to obtain complete information on causal pathways from analyses which are fundamentally based on associations. However, examination of tail associations may be combined with biological reasoning to rule out some causal hypotheses which could not previously be eliminated. Popovic et al. (2019) used copulas to illuminate causal relationships between species.

A fourth reason why we may expect an understanding of tail associations and complex dependence structures to be useful for ecology has to do with spatially aggregated or averaged quantities. Many ecological variables of applied importance depend on the spatial average or sum of local quantities. For instance, regional methane and CO_2 fluxes are the sum of local fluxes; and the total economic value of a fishery is the sum of local catches. We will explore how tail associations between fluctuations of local quantities can influence fluctuations of the spatial mean or sum, and how this may influence higher organizational levels in ecology and human concerns. To illustrate this idea we cite Li et al. (2013), who demonstrated that the overall reliability of wind-generated electricity depends sensitively on details of the dependence between wind speeds at multiple generator sites. Spatially aggregated ecological variables may be subject to a similar effect. For instance, if populations of a pest species in different locations are all positively associated and are also more strongly related to each other in their right tails, then local outbreaks will tend to occur together, creating regional epidemics. Stronger left-tail associations in a pest species, even if overall correlation were the same, would have more benign effects.

We approached our overall goal of exploring the potential of copula approaches for ecology by addressing the following specific questions. (Q1) Do datasets in ecology have dependence structure distinct from that of a multivariate Gaussian/normal distribution (here called non-normal copula structure)? Do positively associated ecological variables show tail associations distinct from those of a normal distribution, and in particular do they show asymmetric tail associations? Normal copula structure is assumed by standard approaches that use multivariate normal distributions or distributions obtained by transforming the marginals of a normal distribution. So Q1 asks whether ecological data contain dependency information distinct from a standard or default case. (Q2) If the answer to Q1 is "yes", then what are some possible causes/mechanisms of non-normal copula structure and asymmetric tail associations in ecology? And (Q3), what are the consequences of non-normal copula structure and asymmetric tail associations for ecological understanding and applications? After Section 2, "Background on copulas", we address Q1 (Sections 3–5 of the paper) by analysing several datasets, including environmental, species-trait, phenological, population, community, and ecosystem functioning data, selected to span multiple sub-fields and organizational levels of ecology. We address Q2 (Sections 6 and 7) using simple models. We address Q3 (Sections 8 and 9) using both data and models. Section 10 is the Discussion. Multiple analyses are brought to bear on each question, and these are summarized diagrammatically in Fig. 2, which can serve as a post hoc guide to the paper. Copula approaches have been used to great effect in neuroscience (Onken et al., 2009), bioinformatics (Kim et al., 2008), medical research (Emura and Chen, 2016), direct study of environmental variables (Goswami et al., 2018; Li et al., 2013; Serinaldi, 2008; She and Xia, 2018), and finance (Li, 2000), and have also been used effectively, but rarely so far, in ecology (Anderson et al., 2018; de Valpine et al., 2014; Popovic et al., 2019). We argue that benefits of wider use of copula descriptions of dependence in ecology will be substantial.

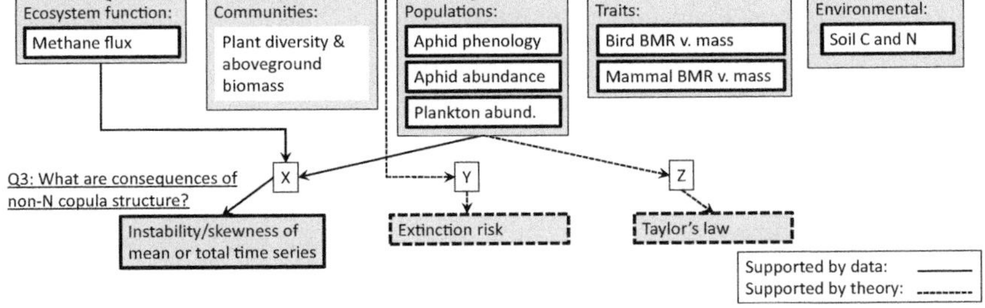

Fig. 2 Summary and guide for analyses presented in the text. Middle boxes correspond to ecological datasets for which Q1 was examined (in Sections 3–5, see also Table 1). Upper boxes correspond to potential causes of non-normal (non-N) copula structure that were examined (in Sections 6 and 7 of the text). Lower boxes correspond to potential consequences we examined of non-normal copula structure for ecological understanding and for applications (in Sections 8 and 9 of the text). Arrow labels (A–D for analyses pertaining to causes, Q2; and X–Z for analyses pertaining to consequences, Q3) link to locations in the text.

2. Background on copulas

We give a brief introduction to copulas, focusing, for simplicity, on bivariate copulas and on concepts needed for the rest of the paper. We assume familiarity with the basic language of probability distributions and random variables. See Nelsen (2006), Genest and Favre (2007), Joe (2014), and Mai and Scherer (2017) for general background on copulas, and see also Anderson et al. (2018), part of which is a short introduction to copulas for ecologists. A bivariate *copula* can be defined as a bivariate cdf with both margins uniform on $(0, 1)$ (Joe (2014), p. 7). This will be the cdf of a bivariate random variable with uniform marginals, and the terminology *copula* is sometimes also applied to this random vector. A foundational theorem of Sklar (1959) (see also Mai and Scherer (2017), p. 16) says that if F is the cdf of a random vector (X, Y), with margins F_X and F_Y, then there exists a copula C such that for all (x, y) in the Euclidean plane, $F(x, y) = C(F_X(x), F_Y(y))$. The theorem also says C is unique if F_X and F_Y are continuous. Thus C couples the bivariate cdf, F, with the cdfs of the marginals. Finally, Sklar's theorem says that if D is any bivariate copula and G_A and G_B are univariate cdfs, then $D(G_A(x), G_B(y))$ is a valid cdf of some bivariate random variable. The applications of this study fall within the case in which univariate marginal distributions have continuous, strictly monotonic cdfs, and this case is simpler. So we henceforth make such an assumption. Table S1 in Supplementary material in the online version at https://doi.org/10.1016/bs.aecr.2020.01.003 provides a summary of notation.

Sklar's theorem implies that any random vector can be constructed from a unique copula and marginal cdfs, and, furthermore, any copula and any univariate cdfs give rise to a random vector; so Sklar's theorem provides a correspondence between random vectors and pairs consisting of a copula and two univariate cdfs, thereby making it possible to study a random vector by studying these two constituents. Marginals contain no information about the dependence structure of a random vector, so the copula contains all such information—it is a complete description of the dependence between variables. The univariate cdfs associated with a random vector (X, Y) are its margins, F_X and F_Y, and the associated copula is the copula, C, guaranteed by Sklar's theorem (and guaranteed unique, thanks to the assumptions of continuity and monotonicity made above). In fact, and making Sklar's correspondence more concrete, we show in Appendix S2 in Supplementary material in the online version at https://doi.org/10.1016/bs.aecr.2020.01.003 that C is the bivariate distribution function of the random variable $(F_X(X), F_Y(Y))$. Conversely, given a copula D and univariate cdfs G_A and G_B, Sklar's theorem guarantees that $D(G_A(x), G_B(y))$ is a cdf, so the random variable with this cdf is the random variable corresponding to D, G_A and G_B under Sklar's correspondence. Again making Sklar's correspondence concrete, we show in Appendix S2 in Supplementary material in the online version at https://doi.org/10.1016/bs.aecr.2020.01.003 that this random variable is $(G_A^{-1}(U), G_B^{-1}(V))$,where G_A^{-1} and G_B^{-1} are the inverses of G_A and G_B and (U, V) is the random variable with cdf D (which has uniform marginals). If we conflate a copula with the random vector of which the copula is the cdf, then Sklar's correspondence is between random variables, (U, V), with uniform marginals and random variables, (X, Y), with arbitrary continuous, strictly monotonic marginals. The correspondence is simply via the application of the univariate cdfs, or their inverses: $U = F_X(X)$, $V = F_Y(Y)$, $X = F_X^{-1}(U)$, $Y = F_Y^{-1}(V)$.

Thus Sklar's theorem makes it possible to construct bivariate distributions in two separate steps: by specifying the marginal distributions and by specifying the dependence structure. Construction of several bivariate distributions is carried out in this way in Fig. 3. The contrast between Fig. 3C and D illustrates in particular how a bivariate distribution can be changed, while retaining the same marginals, by changing the copula. Fig. 3C is a bivariate normal distribution, but Fig. 3D clearly is not, although its marginals are the same. Fig. 3D shows stronger association of the two variables in the left than in the right tails. Sklar's theorem also makes it possible to study and model dependence separately from marginal distributions. For instance, in a pedagogical example, Anderson et al. (2018) modelled the dependence between abundance data for two fish species, red moki and black angelfish, by first modelling the marginal distributions and then modelling the dependence using Gaussian copulas (their Section 2). See the Discussion for a few words on multivariate copulas, which can be developed in the same way as above (Anderson et al., 2018; Joe, 2014; Mai and Scherer, 2017; Nelsen, 2006).

We introduce some standard copula families, as examples, and also because we will fit these families to data. The bivariate normal copula family, which consists of copulas of bivariate normal distributions, is a one-parameter family. The parameter, p, corresponds to the correlation of the related bivariate normal distribution, and controls the degree of association between the variables. A normal copula with $p = 0.7$ was already

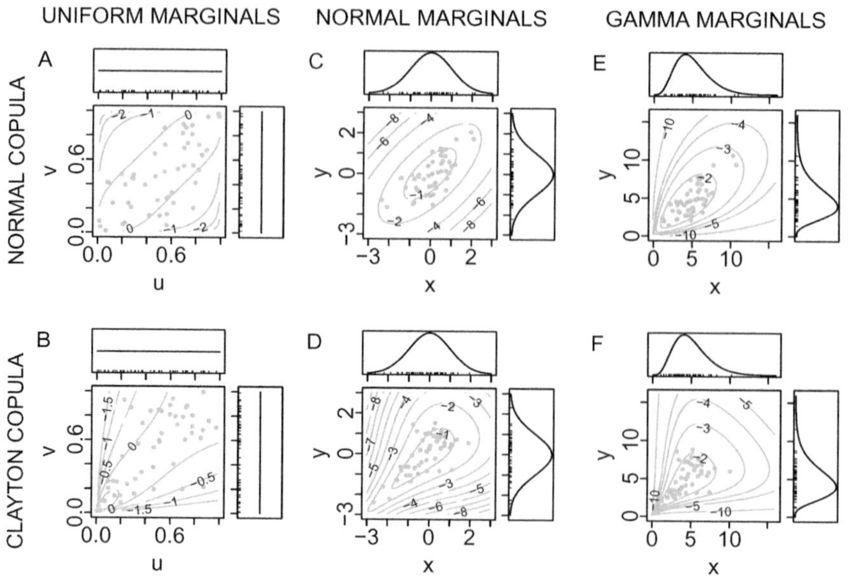

Fig. 3 Normal (A) and Clayton (B) copulas were combined with standard normal marginals (C, D) and gamma marginals (E, F) via Sklar's theorem to produce bivariate distributions. Note that a normal copula is distinct from a normal marginal distribution: a normal copula is the copula of a bivariate normal distribution. See Section 2 for more information on normal, Clayton, and other copulas. Each copula was used with both sets of marginals and each set of marginals was used with both copulas, to demonstrate that: both copula and marginals contribute fundamentally to the resulting distribution, the copula contributing the information on the dependence between the variables; and that one can select the copula and marginals independently. Bivariate distributions, including the copulas, are depicted via their log-scale probability density functions (pdfs), and marginals are depicted via their linear-scale pdfs. Grey dots are 50 random samples from each distribution. The parameter 0.7 was used for the normal copula, the parameter 2 for the Clayton copula, and shape and scale parameters 5 and 1 for the gamma marginals. Variables *u* and *v* were used for copulas and *x* and *y* for distributions created by pairing a copula with (non-uniform) marginals.

introduced (Fig. 3A). The formula for the copula is $\Phi_{2,p}(\Phi_1^{-1}(u), \Phi_1^{-1}(v))$, where Φ_1^{-1} is the inverse of the cdf of a (univariate) standard normal and $\Phi_{2,p}$ is the cdf of a bivariate normal distribution with mean $(0, 0)$ and covariance matrix having 1s on the diagonal and p in the off diagonal positions. Formulas for all the copula families we use were provided by Brechmann and Schepsmeier (2013). However, formulas for copulas are often not as directly informative about copula properties as probability density function (pdf) pictures; Fig. 4 has example pdfs corresponding to bivariate normal copulas for various values of p. Note that the pdfs are symmetric around

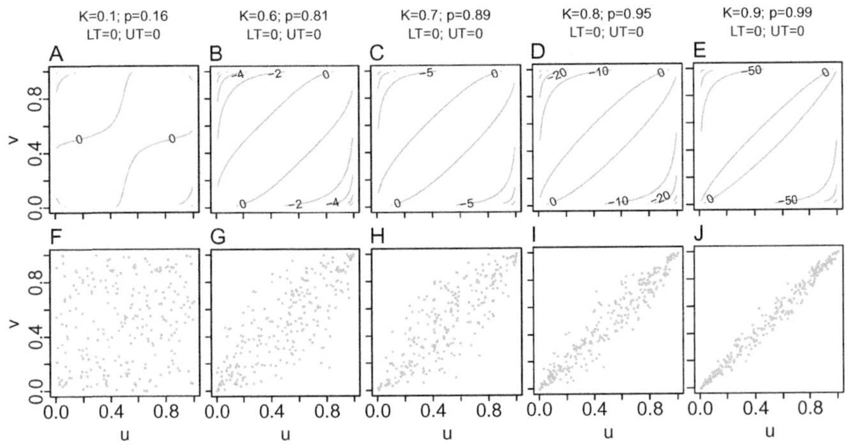

Fig. 4 Log-transformed pdfs (A–E) and samples (F–J) from example normal copulas. *K* is Kendall correlation; *p* is the value of the parameter for the normal family (it is a one-parameter family); and LT and UT are the measures of lower- and upper-tail dependence, respectively (for the definitions of these, see the section on Background on copulas, Section 2). The parameter range for the family is $p \in [-1, 1]$, and lower- and upper-tail dependence are always 0.

the diagonal line $v = -u + 1$ (Fig. 4), so normal copulas have symmetric associations between the two variables in the left and right tails. The Clayton copula family, of which we already pictured an example (Fig. 3B), is another one-parameter family. In contrast to normal copulas, Clayton copulas have stronger left- than right-tail association. The formula is $[\max(u^{-p} + v^{-p} - 1, 0)]^{-1/p}$, for parameter p; though this again provides less direct intuition than do example pdfs (Fig. 5). Higher values of the parameter, p produce copulas with higher Kendall or Spearman correlation. The survival Clayton family is the symmetric opposite of the Clayton family, showing stronger right- than left-tail association (Fig. S3 in Supplementary material in the online version at https://doi.org/10.1016/bs.aecr.2020.01.003). The BB1 copula family is a two-parameter family, thereby providing more flexibility: for some parameters BB1 copulas have stronger left- than right-tail association, and for others the reverse (Fig. 6). See Nelsen (2006), Joe (2014) and Brechmann and Schepsmeier (2013) for further information on these and other families.

Having introduced tail association conceptually and referred to it in examples, we now need a precise definition of a measure of tail association. We use the term *tail association* to describe the general idea of the strength of association between two variables in the tails of their distributions. One way

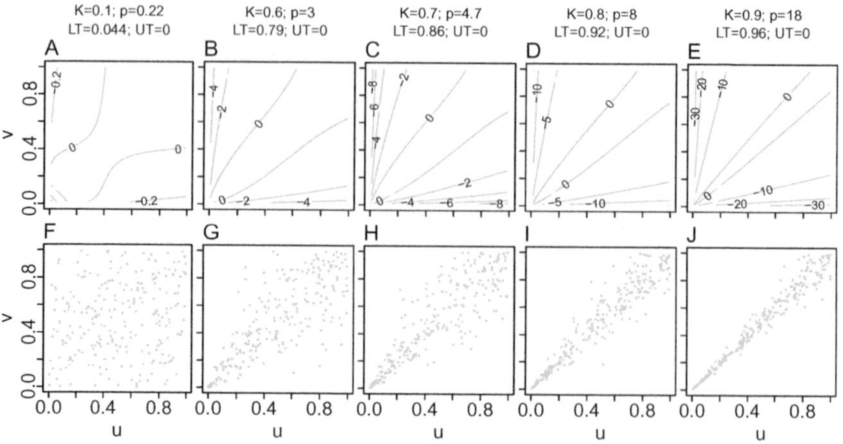

Fig. 5 Log-transformed pdfs (A–E) and samples (F–J) from example Clayton copulas. *K* is Kendall correlation; *p* is the value of the parameter for the Clayton family (it is a one-parameter family); and LT and UT are the measures of lower- and upper-tail dependence, respectively (for the definitions of these, see the section on Background on copulas, Section 2). The parameter range for the family is $p \in (0, \infty)$, lower-tail dependence is $2^{-1/p}$ and upper-tail dependence is 0.

this is measured is called *tail dependence* (Joe (2014), Section 2.13; Nelsen (2006), Section 5.4), defined here. Given a random vector (X, Y) with margins F_X, F_Y and defining $U = F_X(X)$, $V = F_Y(Y)$, the upper-tail dependence of X and Y is defined as $\lambda_U = \lim_{u \to 1^-} P\left[Y > F_Y^{-1}(u) | X > F_X^{-1}(u)\right]$. This equals $\lim_{u \to 1^-} P[U > u | V > u]$, which in turn equals $\lim_{u \to 1^-} P[U > u, V > u] / P[V > u] = \lim_{u \to 1^-} P[U > u, V > u]/(1 - u)$, which shows upper-tail dependence is defined symmetrically in the two variables. All the variables we consider were positively associated when they were significantly associated, so we think of positively associated variables when conceptualizing the definitions here. Lower tail dependence is defined analogously, as $\lambda_L = \lim_{u \to 0^+} P\left[Y \le F_Y^{-1}(u) | X \le F_X^{-1}(u)\right]$. Tail dependence is a property of (U, V), so depends only on the copula and not on the marginals combined with it. It can be shown (Nelsen (2006), Section 5.4) that $\lambda_L = \lambda_U = 0$ for the normal copula, $\lambda_L = 2^{-1/p}$ and $\lambda_U = 0$ for a Clayton copula with parameter $p > 0$, and $\lambda_L = 0$ and $\lambda_U = 2^{-1/p}$ for a survival Clayton copula with parameter $p > 0$. In Section 4 we will introduce ways tail dependence can be applied to data, by first fitting copulas to the data. We will also introduce nonparametric measures of tail association.

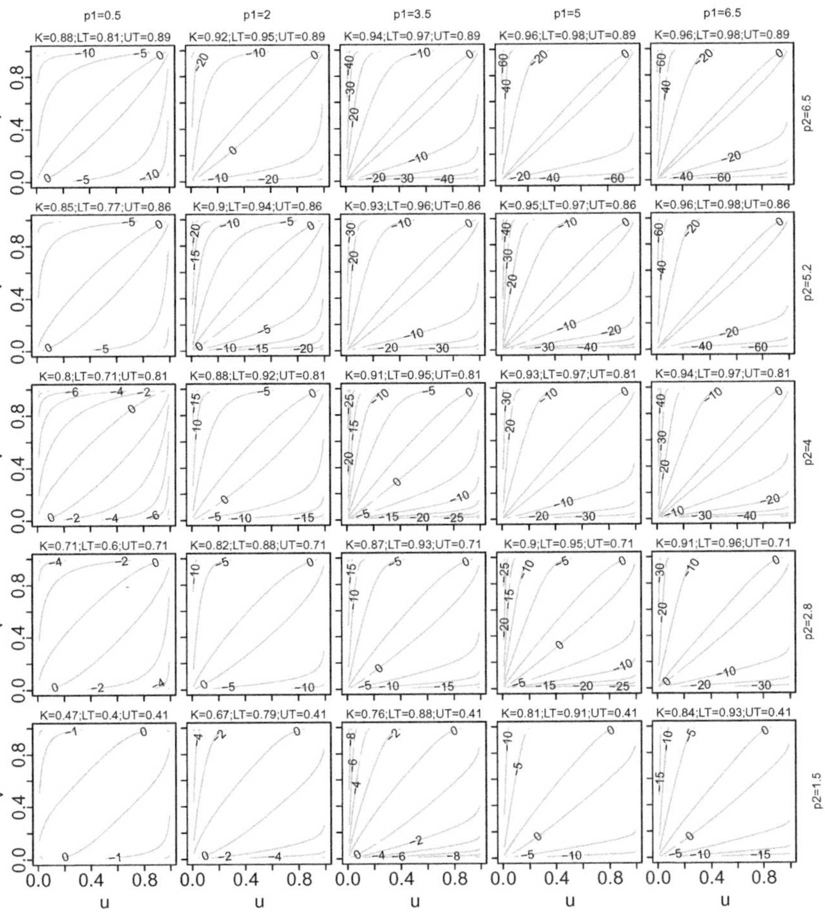

Fig. 6 Log-transformed pdfs for example BB1 copulas. *K* is Kendall correlation; p_1 and p_2 denote the two parameters of the family (it is a two-parameter family); and LT and UT are our measures of lower- and upper-tail dependence, respectively (for the definitions of these, see the section on Background on copulas, Section 2). The parameter ranges for the family are $p_1 \in (0, \infty)$ and $p_2 \in [1, \infty)$, lower-tail dependence is $2^{-1/(p_1 p_2)}$ and upper-tail dependence is $2 - 2^{1/p_2}$.

Plots using normalized ranks were proposed in the Introduction as being conceptually similar to copulas; we explain the connection. As stated above, the copula associated with a random variable (X, Y) is the cdf of the random variable $(F_X(X), F_Y(Y))$. Given a sample (x_i, y_i), $i = 1, \ldots, n$ from (X, Y), let $\widehat{F}_X$ and $\widehat{F}_Y$ be the empirical cdfs associated with the x_i and y_i, respectively. These are step functions of x and y, respectively, that start at 0 for low values of x and y and jump by $1/n$ at each of the data points. Therefore $\widehat{F}_X(x_j)$ is the rank of x_j in the set $\{x_i : i = 1, \ldots, n\}$, divided by n, and analogously $\widehat{F}_Y(y_j)$ is the rank of y_j in the set $\{y_i : i = 1, \ldots, n\}$, divided by n. For large n, $\widehat{F}_X(x_i)$ and $\widehat{F}_Y(y_i)$ approximate the cdfs F_X and F_Y, respectively. But $\widehat{F}_X(x_i)$ equals the normalized rank of x_i times $(n+1)/n$, and likewise $\widehat{F}_Y(y_i)$ equals the normalized rank of y_i times $(n+1)/n$. So the normalized rank in turn approximates the empirical cdf and therefore the cdf. Thus the normalized rank pairs (u_i, v_i) for $i = 1, \ldots, n$ can be regarded as approximate samples from the random variable $(F_X(X), F_Y(Y))$, which is the random variable associated with the copula of (X, Y). The scatterplot of v_i vs u_i can be used to infer aspects of copula structure (see Section 4 below).

3. Data

The datasets we used included environmental, species-trait, phenological, population, community, and ecosystem functioning data (Table 1), selected to span multiple fields and levels of organization within ecology. Copula structure between atmospheric weather variables such as rainfall or wind speed, as measured in multiple locations through time, has been examined previously in the meteorological literature (e.g., Li et al., 2013; Serinaldi, 2008). We therefore examined environmental variables from the soil instead, using the Rapid Carbon Assessment database (RaCA; Wills et al., 2014). The database comprises measurements of soil organic carbon and total soil nitrogen (megagrams C or N per hectare of soil surface) from 5907 locations across the coterminous United States (Fig. S4 in Supplementary material in the online version at https://doi.org/10.1016/bs.aecr.2020.01.003 for locations). Species-trait data were average species basal metabolic rate (BMR, KJ per hour) and body mass (grams) for 533 species of birds (McNab, 2009) and 638 species of mammals (McNab, 2008). These data have been much studied, but to our knowledge the copula structure of the association has not been examined. Species-trait data such as these reflect the coevolution of the two traits. Phenological data were aphid first-flight dates from 10 locations (Fig. S5 in Supplementary material in the online version at https://doi.org/10.1016/bs.aecr.2020.01.003) across the United Kingdom (UK) for 20 aphid species (Table S2 in Supplementary material in the online version at https://doi.org/10.1016/bs.aecr.2020.01.003), for the 35 years 1976 to 2010. These time series were computed from the Rothamsted Insect Survey (RIS) suction-trap dataset (Bell et al., 2015; Harrington, 2014). The first of our two population-level datasets was also derived from the RIS suction-trap data, and comprised total counts of aphids trapped for the same locations, species, and years. Our second population dataset comprised average annual plankton abundance estimates for 14 locations (Fig. S6 in Supplementary material in the online version at https://doi.org/10.1016/bs.aecr.2020.01.003) in the North Sea and British seas for 22 taxa (Table S3 in Supplementary material in the online version at https://doi.org/10.1016/bs.aecr.2020.01.003) for the 56 years 1958–2013. The locations are 2° by 2° grid cells. These data were computed from the Continuous Plankton Recorder survey of the UK Marine Biological Association. Community-level data, obtained from the Cedar Creek Ecosystem Science Reserve, were plant aboveground biomass (Tilman, 2018a) and Shannon's diversity index (computed from plant species percent cover data (Tilman, 2018b)) for 168 plots (Tilman, 2018c), each 9 m by 9 m, as sampled in the years 1996–2000 and 2007, each year analysed separately (Tilman et al., 2001, 2006). Finally, ecosystem functioning data were methane (CH_4) fluxes between the soil or water surface and the troposphere, measured at 13 locations at daily to weekly intervals from September 2015 to September 2016 at the Great Miami Wetland mitigation bank in Trotwood, Ohio (Holland et al., 1999; Jarecke et al., 2016; Smyth et al., 2019). Each included location was measured on at least 50 dates. See Appendix S3 in Supplementary material in the online version at https://doi.org/10.1016/bs.aecr.2020.01.003 for additional data details.

Our environmental, species-trait, and community datasets happen to be "bivariate" datasets in the sense that they comprise two quantities measured at different locations or for different species (Table 1). Our phenology, population, and ecosystem functioning datasets are "multivariate" in that they comprise, for each taxon, measurements through time at multiple locations; copula structure was studied for each location pair.

4. Concepts and methods for Q1

Recall that Q1 has three parts: Do datasets in ecology have non-normal copula structure? Do positively associated ecological variables show tail associations distinct from those of a normal distribution? And in particular do they show asymmetric tail associations? We addressed Q1 first via a model selection procedure in which several families of copulas were fitted to our ecological datasets and fits were compared via the Akaike and Bayesian Information Criteria (AIC and BIC). One of the fitted copulas was the normal copula, making possible comparisons of the degree to which the normal vs other copula families were good descriptions of data.

Table 1 Summary table for the data we used.

Data	Category	No. of measurements we used	References
Soil C and N	Environmental	5907 locations	US Rapid Carbon Assessment database (RaCA), Wills et al. (2014)
Bird body masses and BMR	Species-trait	533 birds	McNab (2009)
Mammal body masses and BMR	Species-trait	638 mammals	McNab (2008)
Green spruce and other aphid species abundances	**Population**	**10 locations with at least 30-yr. timeseries for each of 20 sp.**	**Rothamsted Insect Survey**
Leaf-curling plum and other aphid first flight dates	**Phenological**	**10 locations with at least 30-yr. timeseries for each of 20 sp.**	**Rothamsted Insect Survey**
***Ceratium furca* and other plankton taxa abundances**	**Population**	**14 locations with at least 45-yr. timeseries for each of 22 taxa**	**Continuous Plankton Recorder Survey**
Plant diversity and aboveground biomass	Community level	168 plots	Cedar Creek Ecosystem Science Reserve, biodiversity experiment e120
Methane-flux	**Ecosystem functioning**	**13 locations with at least 50 dates of data**	**Great Miami Wetland Mitigation Bank, Smyth et al. (2019)**

Bold entries are the multivariate data, the rest are bivariate datasets (see Section 3). Basal metabolic rate = BMR.

For bivariate datasets (x_i, y_i) for $i=1, \ldots, n$, model selection involved several steps. First, we produced normalized ranks u_i and v_i as in the Introduction. Second, we tested the independence of the u_i and v_i using the statistic $\sqrt{(9n(n-1))/(4n+10)}|\tau|$, where τ is Kendall's tau for the data. Genest and Favre (2007) argue that this statistic is approximately standard normally distributed. We used the implementation of this test in BiCopIndTest in the VineCopula package in R. We tested for independence because our model selection algorithms were ineffective if data could not be distinguished from independent data, since many copula families include the independent copula. If independence could be rejected (0.05 level), model selection proceeded. Third, we fit 16 bivariate copula families (see below) to the normalized ranks via maximum likelihood. The approach of fitting copula families to normalized ranks was recommended by Genest et al. (1995) and Shih and Louis (1995). Their estimator of copula parameters, which we use, is consistent, asymptotically normal, and fully efficient at independence (Genest et al., 1995). Genest and Favre (2007) recommend carrying out inferences of dependence structures (which was our goal here) using normalized ranks. We used the fitting implementation given in BiCopEst in VineCopula. Fourth, we obtained AIC and BIC values and accompanying model weights AIC_w and BIC_w (Burnham and Anderson, 2003) for each fitted copula. BiCopEst also provided lower- and upper-tail dependence of the best-fitting member of each family. AIC_w values were used to get model-averaged lower- and upper-tail dependence values using standard model averaging formulas (Burnham and Anderson, 2003); likewise for BIC.

Thus for each bivariate dataset, the end products of our procedure were threefold, corresponding to the three parts of Q1 listed in the Introduction: (A) the AIC, BIC, AIC_w and BIC_w values for each of our 16 copula families (see below for list), including the normal family, providing an inference as to whether the normal copula or an alternative was better supported by data; (B) lower- and upper-tail dependence measures for each fitted copula and model-averaged tail dependence measures, providing information on whether, and to what extent, tail dependence differed from the tail dependence of a normal copula (i.e., 0); and (C) the difference of lower- and upper-tail dependence measures for each fitted copula and model averages of those quantities, providing information on whether tail dependence was asymmetric.

Model selection methods give the *relative* support of several models, but do not indicate whether any of the models are an objectively good fit. To test that, we tested the goodness of fit of our AIC-best copula family using a bootstrapping procedure of Wang and Wells (2000) and Genest et al. (2006), implemented as BiCopGofTest in VineCopula. The procedure performed one test based on a Cramer-von Mises statistic and another based on a Kolmogorov-Smirnov statistic. To keep computation times reasonable, a run using 100 bootstraps was performed. If the p-value from either test was <0.2, tests were re-run with 1000 bootstraps.

We fit 16 bivariate copula families, exhibiting a variety of lower- and upper-tail dependence characteristics, with bivariate datasets. The purpose of using a large collection of families was to include a variety of alternative dependence structures to have a robust model selection and multimodel inference procedure. For that purpose, it is not important for the reader to understand the details of these copulas, and, additionally, these copulas have been described in detail elsewhere (Brechmann and Schepsmeier, 2013; Joe, 1997). So we only briefly identify each family, say a few words about its tail dependence, and provide pictorial descriptions (Figs. 4–6; S3 and S7–S18 in Supplementary material in the online version at https://doi.org/10.1016/bs.aecr.2020.01.003). The pictorial descriptions can also be used to aid quick comparisons of copula families when alternative families are being considered for future applications. We used several families that can exhibit positive lower-tail dependence (of strength depending on parameters) and zero upper-tail dependence: the Clayton, survival Gumbel, survival Joe and survival BB6 families. We used several families that can exhibit positive upper-tail dependence (strength depends on parameters) and zero lower-tail dependence: the survival Clayton, Gumbel, Joe and BB6 families. We used families that show zero upper- and lower-tail dependence: the normal and Frank families. These families have pdfs symmetric about the line $v = -u + 1$. We also used several families that can show both upper- and lower-tail dependence, in relative amounts depending on parameter values: the BB1, survival BB1, BB7 and survival BB7 families. We also used the BB8 family, which shows zero lower-tail dependence, and zero upper-tail dependence except for a boundary case for the parameters. And we used the survival BB8 copula, which shows zero upper-tail dependence, and zero lower-tail dependence except for a boundary case. "Survival" families are rotations of the copula with the similar name by 180 degrees. See Joe (1997) and Brechmann and Schepsmeier (2013) for details on all families. We used the implementations provided in the VineCopula package for the R programming language. See Figs. 4–6; S3 and S7–S18 in Supplementary material in the online version at https://doi.org/10.1016/bs.aecr.2020.01.003 for visual depictions of the pdfs of these copulas and how pdfs and tail dependence are influenced by the parameters. See Table 2 for a summary.

For multivariate datasets, we performed the bivariate analysis for all possible pairwise combinations of distinct locations. We carried out pairwise bivariate analyses instead of trying to fit a multivariate copula, for simplicity and because that approach was sufficient to answer our research questions; but see the Discussion for a few words on multivariate copulas. We present the number of pairs for which a non-normal copula was the AIC-best copula, and we characterize AIC differences between the normal and AIC-best copulas across location pairs. We also computed the model-averaged lower- and upper-tail statistics, and differences between these, for each pair of locations, and we characterize the distributions of these values across location pairs.

In addition to our model selection approach, we also used a nonparametric approach, to provide greater confidence in our answers to Q1. We used three statistics which quantify the extent to which the normalized ranks u_i and v_i are related in any part of their distributions. We here describe the statistics, with additional details in Appendix S4 in Supplementary material in the online version at https://doi.org/10.1016/bs.aecr.2020.01.003. The statistics are defined with positively associated variables in mind. All our variables were positively associated when they were significantly associated. Given two bounds $0 \leq l_b < u_b \leq 1$, define the lines $u+v=2l_b$ and $u+v=2u_b$, which intersect the unit square (Fig. 7). Our statistics quantify the association between u_i and v_i in the region bounded by these lines. Using $l_b=0$ and $u_b \leq 0.5$ gives information about association in the left parts of the distributions of u and v, and using $u_b=1$ and $l_b \geq 0.5$ gives information about association in the right parts. The first statistic, a partial Spearman correlation,

$$\mathrm{cor}_{l_b,u_b}(u,v) = \frac{\sum (u_i - \mathrm{mean}(u))(v_i - \mathrm{mean}(v))}{(n-1)\sqrt{\mathrm{var}(u)\mathrm{var}(v)}}, \tag{1}$$

is the portion of the Spearman correlation of u_i and v_i that is attributable to the points between the bounds. Here sample means and variances are computed using all n data points, but the sum is over only the indices i for which $u_i+v_i>2l_b$ and $u_i+v_i<2u_b$. Larger values of the partial Spearman

Table 2 Summary table for the 16 copula families we used (see Section 2).

Copula family	No. of parameters	Tails with stronger association	Reference
Normal	1	Neither (symmetric)	Fig. 4
Clayton	1	Lower	Fig. 5
Survival Clayton	1	Upper	Fig. S3 in Supplementary material
Gumbel	1	Upper	Fig. S7 in Supplementary material
Survival Gumbel	1	Lower	Fig. S8 in Supplementary material
Joe	1	Upper	Fig. S9 in Supplementary material
Survival Joe	1	Lower	Fig. S10 in Supplementary material
Frank	1	Neither (symmetric)	Fig. S11 in Supplementary material
BB1	2	Either (depending on params)	Fig. 6
Survival BB1	2	Either (depending on params)	Fig. S12 in Supplementary material
BB6	2	Upper	Fig. S13 in Supplementary material
Survival BB6	2	Lower	Fig. S14 in Supplementary material
BB7	2	Either (depending on params)	Fig. S15 in Supplementary material
Survival BB7	2	Either (depending on params)	Fig. S16 in Supplementary material
BB8	2	Upper or symmetric (depending on params)	Fig. S17 in Supplementary material
Survival BB8	2	Lower or symmetric (depending on params)	Fig. S18 in Supplementary material

For the parameters we consider, all copula families model positive association between variables.

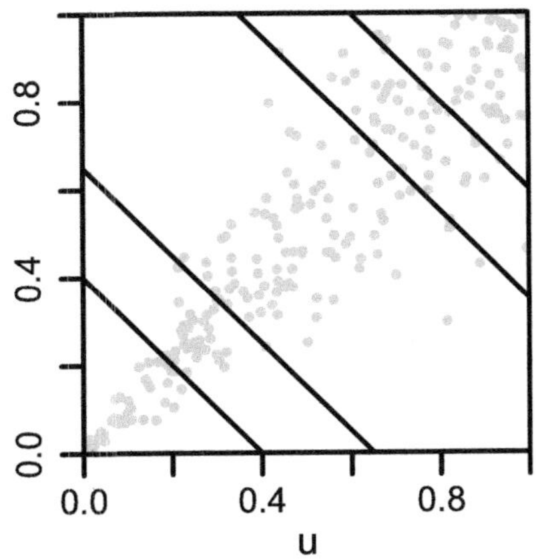

Fig. 7 The partial Spearman correlation, $cor_{l_b,u_b}(u,v)$, within a band can be computed for any band (Section 4) to describe how the strength of association between *u* and *v* varies from one part of the two distributions to another, as can the statistics $P_{l_b,u_b}(u,v)$ and $D^2_{l_b,u_b}(u,v)$. Diagonal lines show two bands, the data in the left/lower band showing stronger association than those in the right/upper band.

correlation indicate stronger positive association. The sum of $cor_{0,0.5}(u,v)$ and $cor_{0.5,1}(u,v)$ (or some other choice of $cor_{l_{b_k},u_{b_k}}(u,v)$ for bounds l_{b_k}, u_{b_k} that partition the interval (0, 1)) equals the standard Spearman correlation as long as no points fall exactly on the bounds. We also defined a statistic P_{l_b,u_b} (Appendix S4 in Supplementary material in the online version at https://doi.org/10.1016/bs.aecr.2020.01.003), which has a similar interpretation to cor_{l_b,u_b}. Our third statistic, $D^2_{l_b,u_b}$, is the average squared distance between points satisfying $u_i+v_i>2l_b$ and $u_i+v_i<2u_b$ and the line $v=u$. Unlike cor_{l_b,u_b} and P_{l_b,u_b}, for which large values indicate strong association between the bounds, small values of $D^2_{l_b,u_b}$ indicate strong association. These statistics are not estimators of the tail dependence quantities defined previously, but rather are conceptually similar measures of associations in the tail portions of the distributions when appropriate values of l_b and u_b are selected.

For large datasets (large *n*), we used l_b and u_b close together without incurring undue sampling variation in our statistics, and we considered multiple bands (l_b, u_b) to understand how association varies across different parts of the distributions. But for datasets with smaller *n* we considered only $l_b=0$, $u_b=0.5$ and $l_b=0.5$, $u_b=1$, abbreviating $cor_l=cor_{0,0.5}$ (*l* is for "lower") and $cor_u=cor_{0.5,1}$ (*u* is for "upper"). Likewise $P_l=P_{0,0.5}$, $P_u=P_{0.5,1}$, $D^2_l=D^2_{0,0.5}$, $D^2_u=D^2_{0.5,1}$. To test for asymmetry of association in upper and lower portions of distributions, we used differences cor_l-cor_u,

$P_l - P_u$ and $D_u^2 - D_l^2$ for smaller datasets (note the opposite order in the last of these); and $\text{cor}_{0,u_b} - \text{cor}_{1-u_b,1}$, $P_{0,u_b} - P_{1-u_b,1}$ and $D_{1-u_b,1}^2 - D_{0,u_b}^2$ with u_b close to 0 for large datasets. We tested our statistics in Appendix S5 (see also Figs. S19 and S20) in Supplementary material in the online version at https://doi.org/10.1016/bs.aecr.2020.01.003.

For bivariate datasets, we compared values of cor_{l_b,u_b}, P_{l_b,u_b}, D_{l_b,u_b}^2, $\text{cor}_{0,u_b} - \text{cor}_{1-u_b,1}$, $P_{0,u_b} - P_{1-u_b,1}$ and $D_{1-u_b,1}^2 - D_{0,u_b}^2$ to distributions of the same statistics computed on surrogate datasets that were produced from the empirical data by randomizing it in a special way to have no tail dependence (Appendix S6 in Supplementary material in the online version at https://doi.org/10.1016/bs.aecr.2020.01.003). The surrogate/randomized datasets had exactly the same marginal distributions as the empirical data and had very similar Kendall or Spearman correlation (the surrogate algorithm had two versions, one for preserving each correlation coefficient), but had normal copula structure. Thus our comparisons tested the null hypothesis that our statistics took values on the empirical data no different from what would have been expected if the copula structure of the data were normal, but the data were otherwise statistically unchanged. The comparison of one of the statistics cor_{l_b,u_b}, P_{l_b,u_b}, D_{l_b,u_b}^2, as computed on the empirical data, to the distribution of its values computed on surrogate datasets provides a test of whether association between the variables in the part of the distributions specified by l_b and u_b is different from what would be expected from a null hypothesis of normal copula structure. Thus the comparison addresses the first two parts of Q1 for bivariate datasets. Significant deviations correspond to deviations from normal copula structure. In particular, deviations using $l_b = 0$ and u_b small (say, 0.1) correspond to lower-tail associations different from that of a normal copula; likewise, using $l_b = 0.9$ and $u_b = 1$ tests for upper-tail associations different from a normal copula. The comparison of one of the statistics $\text{cor}_{0,u_b} - \text{cor}_{1-u_b,1}$, $P_{0,u_b} - P_{1-u_b,1}$ and $D_{1-u_b,1}^2 - D_{0,u_b}^2$, as computed on the empirical data, to the distribution of its values on surrogate datasets provides a test of asymmetry of tail associations. Thus this comparison helps address the third part of Q1.

For multivariate datasets, we calculated Spearman and Kendall correlations and cor_l, cor_u, P_l, P_u, D_l^2, D_u^2, $\text{cor}_l - \text{cor}_u$, $P_l - P_u$, and $D_u^2 - D_l^2$ for all pairs of sampling locations, and we then computed means across the pairs for each statistic. We used a spatial resampling scheme (Appendix S7 in Supplementary material in the online version at https://doi.org/10.1016/bs.aecr.2020.01.003) to calculate confidence intervals of these means.

The scheme is identical to that proposed by Bjørnstad and Falck (2001) and also used by Walter et al. (2017). Code and data for this project are archived at www.github.com/sghosh89/BIVAN.

5. Results for Q1: Ecological datasets have non-normal copula structure and asymmetric tail associations

Copula structures were non-normal and showed asymmetric tail associations for most, but not all datasets we examined, answering Q1 in the affirmative. To make results easier to absorb, we present results first for an example bivariate dataset, then in summary for all bivariate datasets, then for an example multivariate dataset, then in summary for all multivariate datasets.

For the soil C and N data (Section 3, Table 1, Fig. 8A), variables were nonindependent ($p=0$, to within the precision available from BiCopIndTest in the VineCopula package). Nonindependence is also visually apparent. The Kendall correlation was 0.6. We fitted our 16 copulas to the normalized ranks (Fig. 8E) and computed AIC and BIC weights and corresponding lower- and upper-tail statistics for the fitted distributions (Table 3). BB1 was the minimum-AIC copula, with AIC −6708.25, whereas the normal copula had much higher AIC, −6333.16, answering the first part of Q1 for these data: the data have a dependence structure

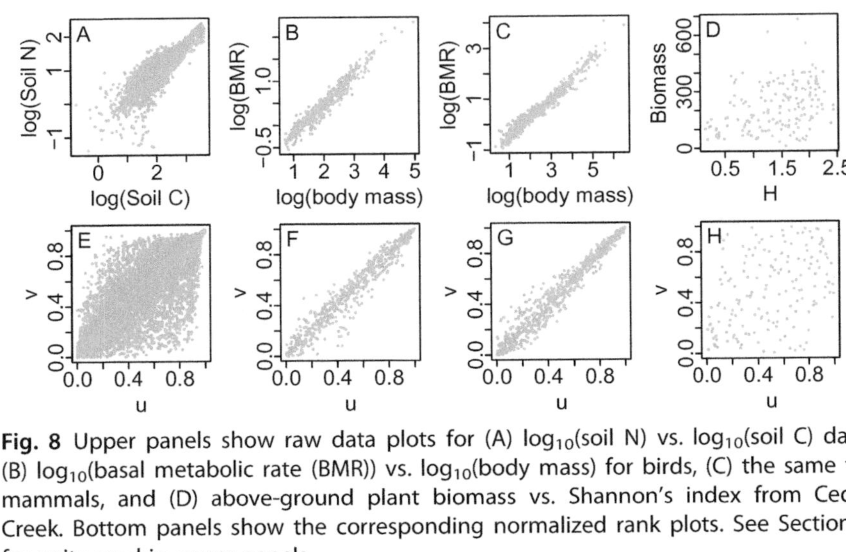

Fig. 8 Upper panels show raw data plots for (A) $\log_{10}$(soil N) vs. $\log_{10}$(soil C) data, (B) $\log_{10}$(basal metabolic rate (BMR)) vs. $\log_{10}$(body mass) for birds, (C) the same for mammals, and (D) above-ground plant biomass vs. Shannon's index from Cedar Creek. Bottom panels show the corresponding normalized rank plots. See Section 3 for units used in upper panels.

Table 3 Model fitting results for soil C and N dataset using 16 copula families: normal (N), Frank (F), Clayton (C), survival Clayton (SC), Gumbel (G), survival Gumbel (SG), Joe (J), survival Joe (SJ), BB1, survival BB1 (SBB1), BB6, survival BB6 (SBB6), BB7, survival BB7 (SBB7), BB8, and survival BB8 (SBB8).

Copula	AIC	AIC_w	BIC	BIC_w	LT	UT
BB1	−6708.25	0.82	−6694.88	0.82	0.4344	0.6151
SBB1	−6705.27	0.18	−6691.90	0.18	0.5348	0.6157
SBB7	−6688.98	0.00	−6675.61	0.00	0.6243	0.6839
BB7	−6684.60	0.00	−6671.24	0.00	0.6034	0.6829
G	−6504.34	0.00	−6497.65	0.00	0.0000	0.6789
BB6	−6501.94	0.00	−6488.57	0.00	0.0000	0.6790
N	−6333.16	0.00	−6326.47	0.00	0.0000	0.0000
SG	−6061.93	0.00	−6055.25	0.00	0.6652	0.0000
SBB6	−6059.17	0.00	−6045.80	0.00	0.6653	0.0000
SC	−5602.48	0.00	−5595.79	0.00	0.0000	0.7315
F	−5599.82	0.00	−5593.14	0.00	0.0000	0.0000
BB8	−5566.65	0.00	−5553.28	0.00	0.0000	0.0000
J	−5522.63	0.00	−5515.95	0.00	0.0000	0.7425
SBB8	−5506.94	0.00	−5493.57	0.00	0.0000	0.0000
C	−4866.90	0.00	−4860.22	0.00	0.7018	0.0000
SJ	−4793.86	0.00	−4787.18	0.00	0.7170	0.0000

Table rows are sorted by AIC. AIC_w, AIC weight; BIC_w, BIC weight; LT, the lower-tail dependence statistic for the indicated copula family with fitted parameters; UT, the same for upper-tail dependence.

markedly distinct from a normal copula. Recall that AIC differences of 2 or 3 are considered meaningful and differences of 8 or greater are definitive, so that data provide essentially no support for the higher-AIC model in that case (Burnham and Anderson, 2003). Model-averaged lower- and upper-tail dependence statistics were 0.453 and 0.615, respectively (AIC weights were used for averaging). These values are distinct from what a normal copula would give, namely, 0, helping to answer the second part of Q1 for these data. The values also differed substantially from each other, helping answer the third part of Q1. The numbers reflect stronger upper- than lower-tail dependence, and this is also visible in the extreme upper-right and lower-left corners of the copula plot (Fig. 8E). Apparently C and N are more strongly associated with each other in soils for which they are both plentiful than in soils for which they are both scarce, and C and N are more strongly associated in both tails than would be modelled using a normal copula model or methods that assume such a model.

However, our model selection procedures do not reveal whether tail dependence parameters are *significantly* different from 0 and from each other. Furthermore, we caution that, while our model selection results do convincingly show non-normal copula structure, model-averaged tail-dependence statistics may have been biased because even the lowest-AIC copula (BB1) was a poor fit ($P=0$ and 0.001 for the Cramer-von Mises and Kolmogorov-Smirnov goodness-of-fit tests, respectively, to within the precision available from BiCopGofTest). Our nonparametric results, detailed next, provide information about significance of tail associations and asymmetries of tail associations.

The values of cor_{l_b,u_b}, P_{l_b,u_b} and $D^2_{l_b,u_b}$ for the soil C and N data were compared to distributions of their values on 1000 Kendall- or Spearman-preserving normal surrogates, separately in two comparisons for each of the ranges $(l_b, u_b)=(0, 0.1), \ldots, (0.9, 1)$ (Fig. 9). Results confirmed that tail associations of data are stronger in both the lower and upper tails than would be expected under a normal-copula null hypothesis (Fig. 9B–F). The values of the statistics $\mathrm{cor}_{0,0.1}-\mathrm{cor}_{0.9,1}$, $P_{0,0.1}-P_{0.9,1}$ and $D^2_{0.9,1}-D^2_{0,0.1}$ for the soil C and N data were also compared to distributions of their values on surrogates (Table 4). Results confirmed that upper-tail association was significantly stronger than lower-tail association; i.e., C and N values are more related when high than when low. These results are consistent with the model selection results, but go beyond them by providing information about significance. Thus our results provide an affirmative answer to Q1 for the soil C and N data; this answer is represented in Fig. 2 as a solid outline around "soil C and N" in the right-most box in the middle row.

Table 5 summarizes Q1 results for the bivariate datasets, with some details in the Appendices in Supplementary material in the online version at https://doi.org/10.1016/bs.aecr.2020.01.003. The two variables were significantly related for all datasets (Table 5, row 1). A non-normal copula always emerged as the lowest-AIC copula (Table 5, row 2). The normal copula was a poor fit compared to the lowest-AIC copula (Table 5, rows 3–4), except for Cedar Creek, for which the AIC difference between these fits was marginal. Thus we answer the first part of Q1 in the affirmative for 3 of 4 of our bivariate datasets. Often, either lower- or upper-tail

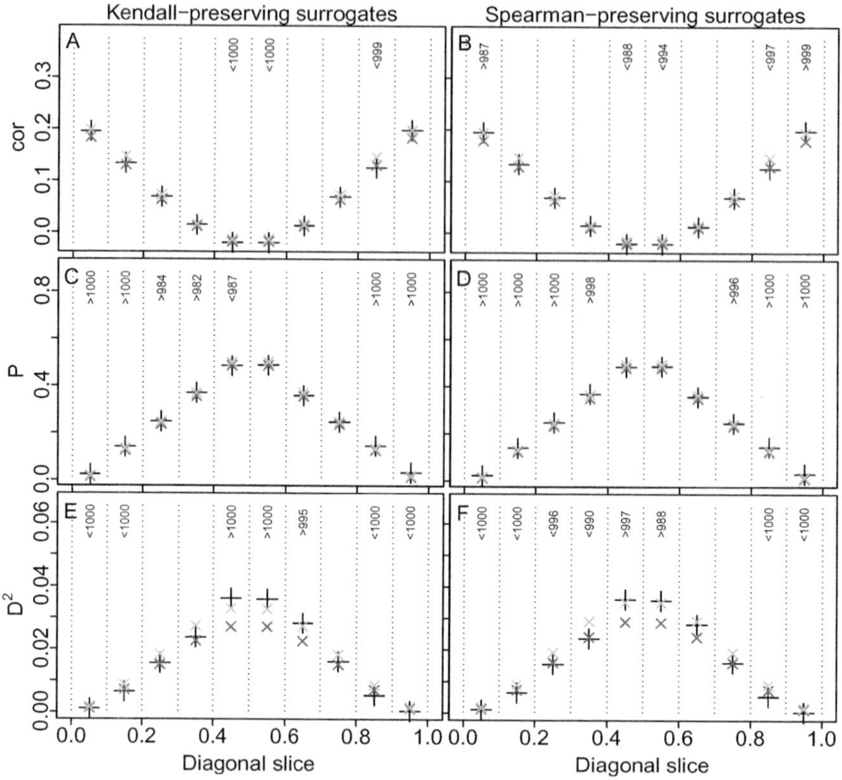

Fig. 9 Nonparametric tests for tail association and other deviations from normal copula structure, for soil C and N data. As described in the main text, the values of the statistics cor_{l_b,u_b}, P_{l_b,u_b} and $D^2_{l_b,u_b}$ for real data (crosses) were compared to distributions of their values on 1000 Kendall- or Spearman-preserving normal surrogates of the data (dark and light grey ×'s show 0.025 and 0.975 quantiles), separately in two comparisons for each of the ranges $(l_b, u_b) = (0, 0.1), \ldots, (0.9, 1)$. Whenever the cross was outside the range given by the ×'s, text at the top of panels indicates the number of surrogate values the real-data value was greater than or less than. For instance, a value $>N$ (respectively, $<N$) means the value of the statistic on real data was greater than (respectively, less than) its value on N surrogates. When the statistic was greater than 975 or less than 975 surrogate values, it indicates significance (95% confidence level). When cor or P values (respectively, D^2 values) were greater than surrogates, it means association in that part of the distributions was stronger than (respectively, weaker than) expected from a normal-copula null hypothesis.

dependence statistics differed substantially from 0 (Table 5, rows 7–8), and/or these statistics differed from each other (Table 5, row 9), helping to answer the second and third parts of Q1. Though again the model selection results do not provide information on significance of these

Table 4 Soil C and N results for nonparametric tests of whether asymmetry of tail association was significant, compared to a normal-copula null hypothesis.

Statistic	Kendall	Spearman
$\text{cor}_{FS} - \text{cor}_{LS}$	<695	<696
$P_{FS} - P_{LS}$	<1000	<1000
$D^2_{LS} - D^2_{FS}$	<1000	<1000

The values of the listed statistics ($\text{cor}_{FS} - \text{cor}_{LS} = \text{cor}_{0,0.1} - \text{cor}_{0.9,1}$, $P_{FS} - P_{LS} = P_{0,0.1} - P_{0.9,1}$, $D^2_{LS} - D^2_{FS} = D^2_{0.9,1} - D^2_{0,0.1}$; here FS stands for "first slice" and LS for "last slice") for real data were compared to their values for 1000 Kendall- or Spearman-preserving normal surrogates, in separate comparisons. A table entry $<X$ indicates the value of the given statistic on the data was less than its value on X of the surrogates, so entries of the form $<X$ for X equal to 975 or above indicate that upper-tail dependence was significantly stronger than lower-tail dependence. Results were significant only for the P and D^2 statistics.

differences, and are subject to the caveat that, for some datasets, even the lowest-AIC copula was not an objectively good fit (Table 5, rows 5 and 6), nonparametric results (Table 5, row 10) showed that tail associations deviated significantly from what would be expected from a normal-copula null model, and were also significantly asymmetric, except for the Cedar Creek data. See Figs. S21–S23 in Supplementary material in the online version at https://doi.org/10.1016/bs.aecr.2020.01.003 for analogues to Fig. 9 for the bivariate datasets other than the soil C and N data, see Tables S4–S6 in Supplementary material in the online version at https://doi.org/10.1016/bs.aecr.2020.01.003 for analogues to Table 3, and see Table S7 in Supplementary material in the online version at https://doi.org/10.1016/bs.aecr.2020.01.003 for analogues to Table 4. The first three datasets had stronger upper- than lower-tail dependence, and the last dataset had the reverse, though nonsignificantly. Apparently species body masses and metabolic rates, for both birds and mammals, are more strongly associated with each other when both these quantities are large than when they are small. Our generally affirmative (except for Cedar Creek), empirically based answer to Q1 for the bivariate datasets is represented in Fig. 2 as solid outlines around the names of the soil C and N and bird and mammal body mass vs metabolic rate datasets, in the middle row of boxes. We did the same analyses with Cedar Creek data from other available years, 1996–1999 and 2007. For 1996 and 2007, independence of biomass and Shannon's index could not be rejected; for 1997–1999, results were similar to 2000.

We present green spruce aphid abundance (Section 3; Table 1) results as an example multivariate analysis. Independence was rejected for each

Table 5 Summary of Q1 results for bivariate datasets.

	Soil C and N	Bird masses and BMR	Mammal masses and BMR	Cedar Creek data
1. p-value, independence test	0	0	0	0
2. Minimum-AIC copula	BB1	G	G	F
3. AIC for best copula	−6708.25	−1447.36	−1854.31	−22.68
4. AIC for normal copula	−6333.16	−1336.77	−1678.76	−20.42
5. p-value, Cramer-von Mises goodness of fit test	0	0.067	0.002	0.39
6. p-value, Kolmogorov-Smirnov goodness of fit test	0.001	0.279	0	0.67
7. Model averaged lower-tail dependence	0.4529	0	0	0.0551
8. Model averaged upper-tail dependence	0.6152	0.877	0.893	0.0034
9. Model averaged lower- minus upper-tail dependence	−0.1623	−0.877	−0.893	0.0517
10. Placement of $\text{cor}_{FS} - \text{cor}_{LS}$ in surrogate distribution (see caption)	<696	<999	<993	>897

The p-values (rows 5–6) are for the minimum-AIC copula. Model averaging used for rows 7–9 was based on AIC weights. Row 10 shows values as in Table 4, upper right table entry. Although the result shown in row 10 for the soil C and N data is nonsignificant, see Table 4 for significant results using the P and D^2 statistics. The first 3 datasets use $(l_b, u_b) = (0, 0.1)$ for the first slice (FS) and $(0.9, 1)$ for the last slice (LS), whereas $(0, 0.25)$ and $(0.75, 1)$ were used for Cedar Creek due to fewer data for that system. Results for Cedar Creek for the year 2000 are shown. See Table 3 for copula family abbreviations.

pair of the 10 sampling locations. Best-fitting (lowest-AIC) copulas were non-normal for the large majority of location pairs (72 of 90 pairs), and the AIC of the normal copula minus the minimum AIC, averaged across location pairs, was 2.732. Whereas 2.732 would be only a marginally meaningful AIC difference for a single location pair, it is more meaningful as an average across many pairs. Relatedly, and illustrating the concept here, the chances of getting 72 or more non-normal location pairs if the chances were equal of getting a normal vs a non-normal result (and taking into account that the location pair (A, B) will necessarily produce the same result as the pair (B, A)) is low, 3.287×10^{-5}. The average AIC difference 2.732 was much less than the AIC difference between −6708.25 and −6333.16 that was obtained for the soil C and N data at least in part because the soil

C and N data were much more numerous (Table 1). These results help answer the first part of Q1: non-normal copula structure appears to be meaningfully common in these data, even if not universal. Goodness of fit tests in every case failed to reject the hypothesis that the AIC-best copula family was also an objectively adequate description of the data; i.e., the collection of copula families we used was sufficiently broad to characterize these data. Model-averaged lower- and upper-tail dependence statistics had 2.5^{th} and 97.5^{th} quantiles (across location pairs) (0.123, 0.721) and (0.002, 0.41), respectively, thereby commonly differing from what a normal copula would give (i.e., 0), and helping to answer the second part of Q1: these data have greater tail dependence (lower and upper) than would be expected from a normal-copula null hypothesis. We note however that our model selection procedures again do not reveal whether tail-dependence parameters are *significantly* different from 0, and we refer the reader to nonparametric results below for that information. Model-averaged lower- minus upper-tail dependence statistics were positive for all but a few location pairs (86 out of 90). The chances of getting 86 or more positive values, here, under a null hypothesis of equal chances for positive and negative values (and again accounting for the fact that location pairs (A, B) and (B, A) will show the same result) was again low, 2.944×10^{-11}. Thus the spatial synchrony of rarity in the green spruce aphid is stronger than the spatial synchrony of outbreaks. This answers the third part of Q1.

Nonparametric statistics verified that tail associations were asymmetric for the green spruce aphid abundance data. Mean values across all pairs of locations of the Spearman and Kendall correlations and the statistics cor_l, cor_u, P_l, P_u, D_l^2, and D_u^2 were positive and confidence intervals excluded 0 (Table 6). Mean values of the statistics $\text{cor}_l - \text{cor}_u$, $P_l - P_u$, and $D_u^2 - D_l^2$ also were always positive and had confidence intervals that excluded 0 (Table 6).

Table 7 summarizes Q1 results for multivariate data. Results supported the conclusions that non-normal copula structure, non-normal tail dependence, and asymmetric tail associations were common, though not universal, answering Q1 in the affirmative for these data. Most location pairs were nonindependent (Table 7, row 2). The large majority of nonindependent location pairs had best-fitting copulas that were not the normal copula (Table 7, row 3), and AIC values of best-fitting copulas were, on average across location pairs, between 2.714 (*Ceratium furca* abundance data) and 5.764 (leaf-curling plum aphid first-flight data) lower than AIC values for the normal copula (Table 7, row 4). Best-fitting copulas were nearly always

Table 6 Average values of statistics across all locations pairs and confidence intervals based on spatial resampling for green spruce aphid abundance data.

	2.5^{th} quantile	Mean	97.5^{th} quantile
Spearman	0.490	0.580	0.668
Kendall	0.351	0.428	0.505
cor_l	0.291	0.335	0.375
cor_u	0.195	0.250	0.307
P_l	0.097	0.116	0.135
P_u	0.063	0.084	0.106
D_l^2	0.019	0.026	0.033
D_u^2	0.030	0.039	0.047
$cor_l - cor_u$	0.034	0.085	0.128
$P_l - P_u$	0.015	0.032	0.048
$D_u^2 - D_l^2$	0.005	0.013	0.020

Table 7 Summary of Q1 results for multivariate datasets.

	Green spruce aphid abundance	Leaf-curling plum aphid first flight	*Ceratium furca* abundance	Methane flux
1. Location pairs (excluding self comparisons)	90	90	182	156
2. Number of nonindependent pairs	90	90	162	102
3. Number of pairs with non-normal copula as best fit	72	82	134	96
4. Average of AIC differences (normal AIC minus min AIC) across location pairs	2.732	5.764	2.714	4.482
5. Number of well-fitting location pairs	90	83	160	102
6. 2.5^{th} and 97.5^{th} quantiles of model-avg. lower-tail dependence	(0.123, 0.721)	(0, 0.152)	(0.019, 0.629)	(0.004, 0.461)

Table 7 Summary of Q1 results for multivariate datasets.—cont'd

	Green spruce aphid abundance	Leaf-curling plum aphid first flight	*Ceratium furca* abundance	Methane flux
7. 2.5th and 97.5th quantiles of model-avg. upper-tail dependence	(0.002, 0.41)	(0.175, 0.798)	(0, 0.505)	(0, 0.592)
8. Percent pairs showing stronger lower tail dependence	95.56	0	79.01	72.55
9. Mean with 2.5th and 97.5th quantiles of $cor_l - cor_u$ across all location pairs	0.085, (0.034, 0.128)	−0.15, (−0.207, −0.096)	0.077, (0.031, 0.118)	0.007, (−0.039, 0.05)

Rows 3–8 of the table were computed for the nonindependent pairs (row 2) only. A well-fitting location pair (row 5) was one for which the best-fitting copula had p-values >0.01 for both the Cramer-von Mises and Kolmogorov-Smirnov goodness of fit tests.

considered an adequate fit (Table 7, row 5). Some of the datasets (green spruce aphid abundance, *Ceratium furca* abundance, methane-flux) showed stronger lower- than upper-tail dependence (Table 7, rows 6–8), whereas leaf-curling plum aphid first flight data showed the reverse. Thus the spatial synchrony of rarity was stronger than that of outbreaks for green spruce aphid and *Ceratium furca* abundance, for instance, whereas leaf-curling aphid first flights were more spatially related to each other when these flights happened late than when they happened early. All of the values in Table 7, rows 8 deviated highly significantly from what would have been expected under a null hypothesis of equal chances for positive and negative values.

Asymmetry results were generally verified by nonparametric approaches, with the exception of the methane-flux data. For instance, the 95% confidence intervals of the mean over pairs of sampling locations of the statistic $cor_l - cor_u$ were $(-0.207, -0.096)$ for the leaf-curling plum aphid first flight data (Table 7, row 9), indicating greater upper-tail dependence, and consistent with the results of Table 7, row 8. For the *Ceratium furca* abundance data, confidence intervals were (0.031, 0.118), indicating greater lower-tail dependence, and again consistent with the results of Table 7, row 8. For the methane-flux data, confidence intervals were (−0.039, 0.05); the asymmetry in tail dependence revealed by the model selection results for the methane-flux data was apparently not strong enough to also be revealed by the nonparametric analyses. Analogues to Table 6 are in Tables S8–S10 in Supplementary material in the online version at https://doi.org/10.1016/bs.aecr.2020.01.003. Our generally affirmative, empirically based answer to Q1 for the multivariate datasets is represented in Fig. 2 as solid outlines around the names of those datasets, in the middle row of boxes. We also carried out the same analyses for abundance and first-flight data for the 18 aphid species for which we had data other than the green spruce and leaf-curling plum aphids, as well as for the 21 plankton taxa for which we had abundance data other than *Ceratium furca* (results not shown). Results supported the conclusion that non-normal copula structure and tail dependence, and asymmetry of tail associations, are common.

6. Concepts and methods for Q2

Recall that Q2 is: what are some possible causes/mechanisms of non-normal copula structure and asymmetric tail associations in ecology? Having demonstrated that non-normal copula structure and asymmetric tail associations are common, we addressed Q2 by exploring, using models, three possible mechanisms for these phenomena. Our models presented are initial explorations, only, of whether the proposed mechanisms may explain observed patterns. As such, simple models were used. Comprehensive explorations of model parameter space and alternative model formulations were left for future work.

The first mechanism relates to the ideas in the Introduction about Liebig's law of the minimum, and to nonlinear influences of environmental variables on ecological variables. If an environmental variable influences an ecological variable disproportionately in one of its tails, we explored whether the ecological variable could then exhibit asymmetric tail associations across space. Let $E_i(t)$ be an ecological variable measured at location i $(i=1,2)$ at time t. Assume the dynamics $E_i(t+1)=bE_i(t)+g(\varepsilon_i(t))+a\delta_i(t)$, where the $\delta_i(t)$ are standard-normally distributed and independent across time and locations, $a=0.2$, $b=0.1, -0.1, 0.5, -0.5$ in different simulations, and the $(\varepsilon_1(t), \varepsilon_2(t))$ were drawn, independently through time, from a bivariate normal distribution with $\text{var}(\varepsilon_i)=1$ and $\text{cov}(\varepsilon_1, \varepsilon_2)=0.8$. Thus basic ecological dynamics follow the very simple AR(1) formulation, influenced by a "regional" environmental factor which is correlated across locations (ε) and by a local factor (δ). The function g describes how $\varepsilon_i(t)$ influences $E_i(t)$.

We used g equal to g_1 or g_2 in different simulations: $g_1(\varepsilon)$ equals ε if $\varepsilon < 0$ and 0 otherwise; $g_2(\varepsilon)$ equals ε if $\varepsilon > 0$ and 0 otherwise. Thus g_1 represents environmental effects that negatively impact populations, but only below the threshold $\varepsilon = 0$; and g_2 represents effects that positively impact populations, but only above $\varepsilon = 0$. The values of b provide a modicum of exploration of whether ecological dynamics may also influence how complex dependencies between the $E_i(t)$ arise: negative b corresponds to overcompensating dynamics and positive b to undercompensating dynamics; larger $|b|$ means slower return to equilibrium after a disturbance.

For each b and g_i, we simulated the model for 25,000 time steps and retained the $E_i(t)$ for the final 2500 time steps. We applied our nonparametric statistic $\text{cor}_{0,0.1} - \text{cor}_{0.9,1}$ and our Spearman- or Kendall-preserving normal-copula surrogate comparison methods (see Section 4) to these outputs to discover if the model could produce asymmetric tail associations (and therefore non-normal copula structure) between the $E_i(t)$. Because $(\varepsilon_1, \varepsilon_2)$ and (δ_1, δ_2) have normal copula structure (they were drawn from bivariate normal distributions), the Moran mechanism analysed below does not operate here.

Our second mechanism is an extension of the well-known Moran effect, and was summarized conceptually in the Introduction. We consider a linear model, as well as two parameterizations of a nonlinear population model which includes density dependence of population growth. For the linear model, let $E_i(t)$ again be an ecological variable, $i = 1, 2$. We use AR(1) dynamics, $E_i(t+1) = \beta E_i(t) + \sqrt{1-\beta^2}\varepsilon_i(t)$, with $\beta = 0.5$. The environmental noises $\varepsilon_i(t)$ were standard-normal random variables that were independent for distinct times, t, but exhibited different kinds of dependence across locations in different simulations (see below). The variable E is general. It could represent deviations of a population density from a carrying capacity, deviations of total plant community biomass from an average value, flux of a biogeochemical variable such as methane, or other quantities. The nonlinear population model was a stochastic, multi-habitat-patch version of the Ricker model, $P_i(t+1) = P_i(t)\exp[r(1 - P_i(t)/K) + \sigma\varepsilon_i(t)]$ for $i = 1, 2$, using $r = 0.5$, $K = 100$, $\varepsilon_i(t)$ as above, and $\sigma = 0.1$ or $\sigma = 1$ in different simulations. When $\sigma = 0.1$, population dynamics stay close to the carrying capacity, K, where the model equation can be well approximated by a linear equation, and the nonlinearities of the model therefore have limited influence on dynamics. When $\sigma = 1$, model dynamics are strongly nonlinear because the stochastic component of the model causes populations to stray far from the carrying capacity. We refer to these as the weak-noise and strong-noise cases, though the importance of the noise here is that, when strong, it brings the nonlinearities of the model into play.

For each of the three model setups above (the linear model and the nonlinear model with weak and strong stochasticity), for the Clayton and survival Clayton copula families, for each $\tau = 0.1, 0.2, \ldots, 0.9$, and for each of 50 replicate simulations, we generated 5000 noise pairs $(\varepsilon_1(t), \varepsilon_2(t))$ from the bivariate random variable with standard-normal marginals and with the given copula family and the given Kendall correlation τ. We then used this noise to drive the model, and retained both the noise and population values for the final 500 time steps. For each simulation, the following statistics were then computed for noises and populations: Pearson, Spearman and Kendall correlations, cor_l, cor_u, P_l, P_u, D_l^2, D_u^2, $\text{cor}_l - \text{cor}_u$, $P_l - P_u$, and $D_u^2 - D_l^2$. Values were plotted against τ for noises and populations. If the hypothesis from the Introduction was reasonable that characteristics of the copula structure of spatial dependence in an ecological variable may be inherited from characteristics of spatial dependence in an environmental variable through a Moran-like effect, then plots should be similar for populations and noises.

The next mechanism we investigated is evolutionary, and pertains to bivariate trait data across species, e.g., our bird and mammal data. This mechanism is a hypothetical explanation for the bias toward right-tail association observed in those data (Fig. 8F and G). The hypothesis is that asymmetric tail association occurs in evolutionary changes in bivariate characters, and gives rise to asymmetric tail association between the two character values across extant species. We simulated bivariate character evolution on an estimate of the phylogeny, taken from Genoud et al. (2018), of 817 mammal species. The root character state and change across each branch were randomly chosen from matrices of one million independent draws from bivariate distributions showing one of five distinct types of copula structure: (1) extreme or (2) moderate left-tail dependence, (3) symmetric tail dependence, or (4) moderate or (5) extreme right-tail dependence (Appendix S8 in Supplementary material in the online version at https://doi.org/10.1016/bs.aecr.2020.01.003). All distributions had standard-normal marginals and Spearman correlation 0.875 between components, so our simulations assess the impact of copula structure only. For each of the five copulas, mammalian character evolution was simulated 100 times. For each simulation, symmetry of tail associations of the two characters across phylogeny tips was assessed using our nonparametric statistics. We hypothesized that cases 1 and 2 above would yield stronger left- than right-tail associations in tip characters, and cases 4 and 5 would yield the reverse. The simulator was written in Python and used version 4.4 of the DendroPy package (Sukumaran and Holder, 2010).

7. Results for Q2: Moran effects and asymmetric dependencies produce non-normal copula structure

Our model with asymmetric environmental effects produced outputs with visually apparent asymmetry of tail associations between the ecological variables E_i in the two locations, to an extent that depended on b. For $b=0.1$ (Fig. 10) and $b=-0.1$ (Fig. S24A and B in Supplementary material in the online version at https://doi.org/10.1016/bs.aecr.2020.01.003), for both g_1 and g_2, asymmetry of tail association was strong; for $b=\pm 0.5$, asymmetry was weaker but still apparent (Fig. S24C–F in Supplementary material). Lower-tail (respectively, upper-tail) spatial associations in the effects of noise (g_1, respectively, g_2) produced lower-tail (respectively, upper-tail) spatial associations in the ecological variable, E_i. Results using our statistic $\text{cor}_{0,0.1} - \text{cor}_{0.9,1}$ strongly reflected the visually apparent asymmetry (Table 8). Thus asymmetry of environmental effects is a mechanism that may be partly responsible for non-normal copula structure and asymmetric tail associations across space in ecological variables: when populations or other ecological quantities are influenced

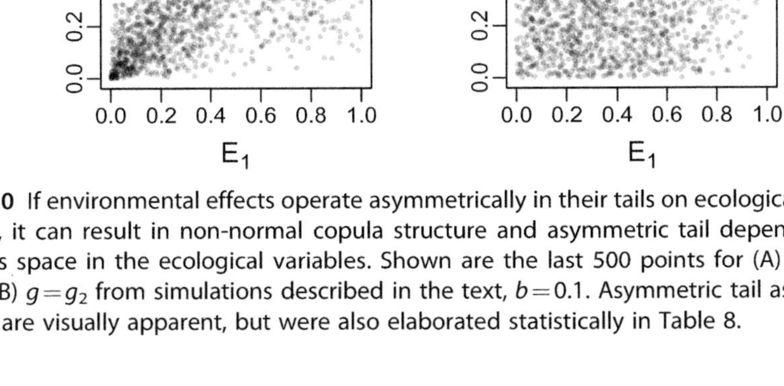

Fig. 10 If environmental effects operate asymmetrically in their tails on ecological variables, it can result in non-normal copula structure and asymmetric tail dependence across space in the ecological variables. Shown are the last 500 points for (A) $g=g_1$ and (B) $g=g_2$ from simulations described in the text, $b=0.1$. Asymmetric tail associations are visually apparent, but were also elaborated statistically in Table 8.

Table 8 Asymmetric sensitivity model, nonparametric statistics results.

b	**g**	**Kendall**	**Spearman**
0.1	g_1	>1000	>1000
0.1	g_2	<1000	<1000
−0.1	g_1	>1000	>1000
−0.1	g_2	<1000	<1000
0.5	g_1	>1000	>1000
0.5	g_2	<1000	<1000
−0.5	g_1	>1000	>1000
−0.5	g_2	<1000	<1000

Outputs of simulations from the first model detailed in Section 6 (i.e., Figs. 10 and S24 in Supplementary material in the online version at https://doi.org/10.1016/bs.aecr.2020.01.003) were subjected to nonparametric statistical analyses described in Section 4. The statistic $\text{cor}_{FS} - \text{cor}_{LS}$ was computed for the output of the model with the indicated b and g, and the value was compared to 1000 values of the same statistic computed on surrogate time series randomized to have normal copula structure (see Section 4). A table entry $<X$ (respectively, $>X$) indicates the value of the given statistic on the data was less (respectively, more) than its value on X of the surrogates, so entries of the form $<X$ (respectively, $>X$) for X equal to 975 or above indicate that upper-tail (respectively, lower-tail) dependence was significantly stronger than lower-tail (respectively, upper-tail) associations compared to a normal-copula null hypothesis. "FS" stands for "first slice" and refers to the bounds 0 to 0.1; "LS" stands for "last slice" and refers to the bounds 0.9 to 1.

principally by low (respectively, high) values of an environmental variable, one may expect left-tail (respectively, right-tail) associations across space for the ecological variable. This result is represented in Fig. 2 as the solid box around "Nonlinear environmental effects, Liebig's law" and the arrows labelled "A". It is explained in the Discussion why the box and some of the arrows are solid, instead of dashed, although the results here are theoretical.

The hypothesis that characteristics of the copula structure of spatial dependence in an ecological variable can be inherited from an environmental variable was found to be reasonable, because it held for our models—the Moran effect seems to extend to copula structure and tail associations—however, similarities between environmental- and ecological-variable copula structure were reduced when dynamics were strongly nonlinear. For simulations using AR(1) models, our correlation and tail asymmetry statistics were always similar for noise and populations (Fig. 11 for the Clayton copula, Fig. S25 in Supplementary material in the online version at https://doi.org/10.1016/bs.aecr.2020.01.003 for the survival Clayton copula).

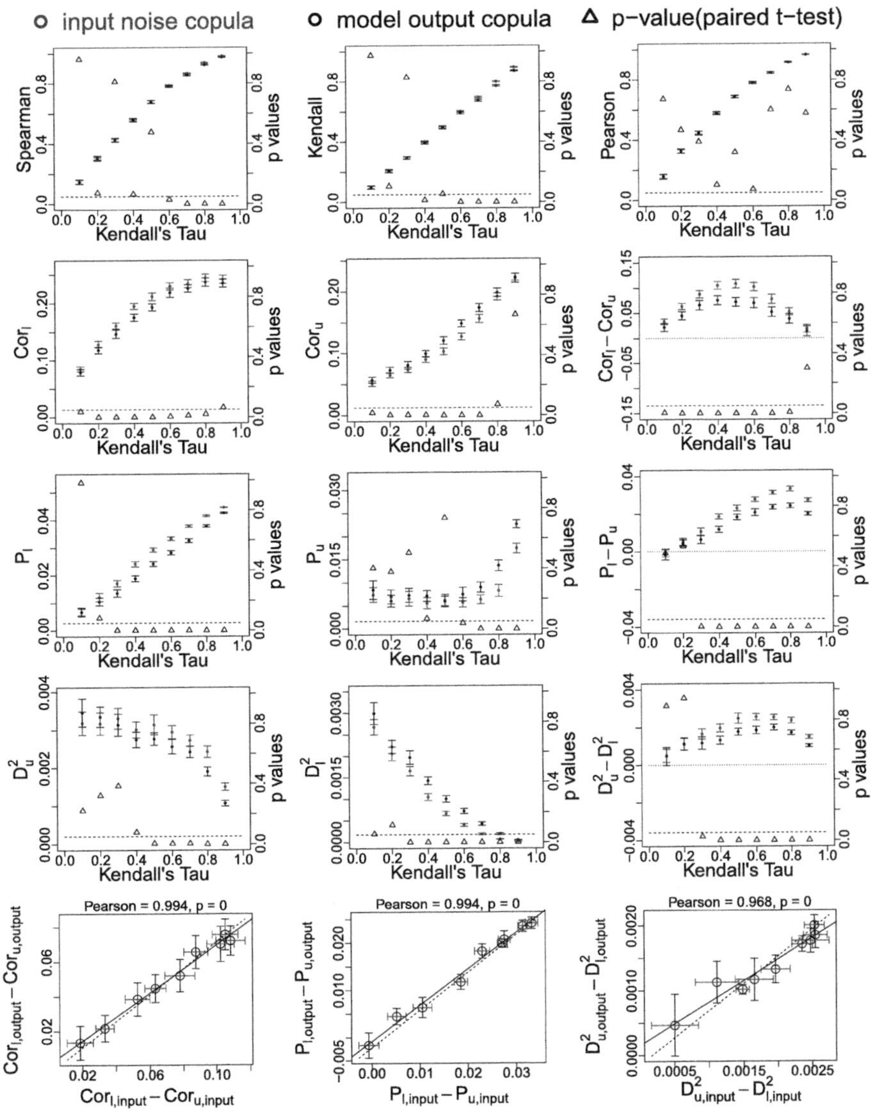

Fig. 11 Comparison of correlation and tail association statistics between environmental noise inputs and ecological-variable outputs, AR(1) model, Clayton copula. Grey and black points give means over 50 replicate simulations of the listed statistics, grey points are for noise inputs and black points are for model outputs. Error bars are standard errors. *p*-values (triangles, right axis) are for a paired *t*-test of the null hypothesis that the distributions have the same mean. For bottom panels, headers give Pearson correlations for the points. The regression line through the points (solid) was similar to the 1–1 line (dashed). See Section 6 for details.

Though there were significant differences for many statistics and simulations, these were small compared to the overall tendency for larger (respectively, smaller) values of our statistics, as applied to noise, to be paired with larger (respectively, smaller) values of the same statistics for model outputs. For our nonlinear model with weak noise, values of the statistics were again quite similar for noise and model outputs (Fig. S26 for Clayton and Fig. S27 for survival Clayton in Supplementary material in the online version at https://doi.org/10.1016/bs.aecr.2020.01.003). Though there were again significant differences for many statistics and simulations, these were again small relative to overall variation of values of the statistics. Since many ecological models are nonlinear, this result provides the reasonable expectation for a Moran-effect-like correspondence between noise and model-output dependence structure across space for typical ecological dynamics, as long as environmental noise is small enough that dynamics stay relatively close to the model equilibrium. Theoretical results that hold for "weak noise" in this sense are common in ecology. For strong noise and using our nonlinear model, our correlation and tail-association-asymmetry statistics, generally speaking, were approximately similar between noise and model outputs; however, similarity was reduced, and for a few simulations, asymmetry statistics had opposite signs for noise and model outputs (Figs. S28 and S29 in Supplementary material in the online version at https://doi.org/10.1016/bs.aecr.2020.01.003). For instance, using a Clayton copula with a large Kendall τ, $\text{cor}_l - \text{cor}_u$ was slightly positive for noise, but slightly negative for population outputs (Fig. S28 in Supplementary material in the online version at https://doi.org/10.1016/bs.aecr.2020.01.003).

We repeated our analyses using the nonlinear model with $r=1.3$. The deterministic one-habitat-patch Ricker model exhibits a monotonic approach to a stable equilibrium at K when $r<1$ (undercompensating dynamics, e.g., the value $r=0.5$ used initially), but exhibits an oscillatory approach when $r>1$ (overcompensating dynamics, e.g., $r=1.3$). For weak noise, similarities were again dominant between values of our statistics on noise and population time series. For strong noise, however, discrepancies were often glaring. Apparently noise of standard deviation 1 interacted especially strongly with model nonlinearities when the model was in its overcompensatory regime. We repeated all analyses using the Gumbel, survival Gumbel, Joe, and survival Joe copulas, with substantially similar main conclusions (not shown).

Thus it is reasonable to hypothesize that a Moran-effect-like mechanism may produce non-normal copula structure and asymmetric tail associations across space in ecological variables. This result is represented in Fig. 2 as the dashed box around "Moran effects" and the arrows labelled "B".

For our evolutionary model, the hypothesis was correct, for the models we employed, that asymmetric tail association in evolutionary changes can produce similarly asymmetric tail association between characters across phylogeny tips. Once character evolution was simulated 100 times for each of the dependence structures we considered for evolutionary changes, we had 817 bivariate characters for each of 500 simulations (see Section 6). We computed our nonparametric asymmetry statistics $\mathrm{cor}_{0,0.2} - \mathrm{cor}_{0.8,1}$, $P_{0,0.2} - P_{0.8,1}$, and $D^2_{0.8,1} - D^2_{0,0.2}$ for each simulation output and produced a histogram for each statistic and for each dependence structure (Fig. 12). Results showed that asymmetries of tail associations in evolutionary changes were associated with similar asymmetries of tail associations in extant characters. Thus we cannot reject the possibility that this is a proximate mechanism behind observed asymmetric tail associations in our bird and mammal

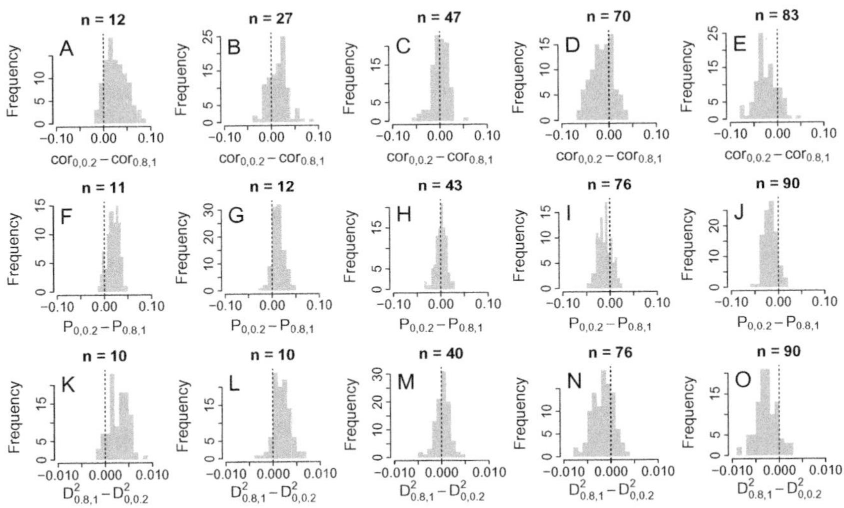

Fig. 12 Three measures of asymmetry of tail associations ($\mathrm{cor}_{0,0.2} - \mathrm{cor}_{0.8,1}$, A–E; $P_{0,0.2} - P_{0.8,1}$, F–J; $D^2_{0.8,1} - D^2_{0,0.2}$, K–O) between two characters, across extant species, were computed for each of 100 simulations of mammalian character evolution for each of five types of tail dependence between evolutionary changes in the characters (extreme left-tail dependence, A, F, K; moderate left-tail dependence, B, G, L; symmetry of tail dependence, C, H, M; moderate right-tail dependence, D, I, N; and extreme right-tail dependence, E, J, O). See Section 6 for details. The number above each panel indicates how many statistics out of 100 were less than or equal to 0. Values substantially lower (respectively, higher) than 50 indicate greater left (respectively, right) tail-dependence between characters.

data. This result is represented in Fig. 2 as the dashed box around "If character evolution has copula structure" and the arrow labelled "D". This topic is revisited in the Discussion, as is the box "Asymmetric species interactions" and the arrow labelled "C" in Fig. 2.

8. Concepts and methods for Q3

Recall that Q3 is: what are the consequences of non-normal copula structure and asymmetric tail associations for ecological understanding and applications? We addressed Q3 by exploring, using both data and models, three hypothesized consequences. As for Q2, the models we used are intended as initial explorations, only. Therefore only simple models are used, and comprehensive explorations of the sensitivity of results to model structure and parameters are left for future work.

The hypothesis was presented in the Introduction that the distribution (through time) of a spatially averaged quantity should be influenced by dependencies between the local quantities being averaged, including their copula structure and tail associations. To make this hypothesis more precise, suppose an ecological variable $E_i(t)$ is measured at locations $i=1, \ldots, N$ and times $t=1, \ldots, T$, and the spatial mean $\sum_i E_i(t)/N$ is of interest. The $E_i(t)$ could be, for instance, local abundances of a pest or exploited species, or local fluxes of a greenhouse gas. If E_i and E_j are associated primarily in their right tails for most location pairs i and j, then exceptionally large values tend to occur at the same time in most locations. We hypothesize that this can increase the skewness of the distribution of the spatial mean. Similarly, left-tail associations between local variables should decrease skewness. Strong positive skewness of the spatial-mean time series corresponds to "spikiness" of that time series, i.e., occasional very large values, which corresponds to instability through time. The spatial-mean time series and its skewness may be quantities of principal importance for pest or resource abundance, for which extreme values (spikes) in the spatial mean may have large effects.

We tested the above hypothesis using our multivariate datasets (Table 1). For each dataset we calculated the spatial mean time series, and then the skewness through time of that mean. Then, for each dataset, we compared the value obtained to a distribution of values of the same quantity for each of 10,000 surrogate datasets. Surrogate datasets were produced by randomizing the empirical data in a special way to have the copula structure of a multivariate-normal distribution, but to retain exactly the same distributions of values for each sampling location as the original data (Appendix S9 in Supplementary material in the online version at https://doi.org/10.1016/bs.aecr.2020.01.003). Surrogates also had very similar Spearman correlations between pairs of sampling locations as the data. Our comparisons therefore tested the null hypothesis that the skewness of the spatial mean took values on the empirical data no different from what one would expect if the copula structure of the data were multivariate normal (i.e., the same copula as a multivariate normal distribution), but the data were otherwise statistically unchanged. Significant differences indicate that non-normal copula structure in the data contributed to the skewness of the spatial mean time series, i.e., to its instability and "spikiness" through time.

For green spruce aphid abundance data, *C. furca* abundance data, and methane-flux data, because these datasets exhibited stronger lower- than upper-tail associations (Table 7), we compared empirical and surrogate skewness values via a one-tailed test in the left tail: the p-value was the fraction of surrogate skewnesses less than the skewness for the empirical data. The test examines whether stronger lower-tail associations between local time series caused the spatial average to have significantly lower skewness than would have been expected with symmetric tail associations. For leaf-curling plum aphid data, because that dataset exhibited stronger upper- than lower-tail associations (Table 7), we did the analogous one-tailed test in the right tail. The test examines whether upper-tail associations caused the spatial average to have significantly higher skewness than would have been expected with symmetric tail associations.

We also examined a hypothesis that asymmetric tail associations across space of an environmental variable can influence the extinction risk of a metapopulation. We hypothesized that environmental noises exhibiting greater left-tail associations across habitat patches would cause higher metapopulation extinction risks because then very bad years for the component populations occur simultaneously in many patches, reducing rescue effects. Here we assume, for simplicity, that low values of the environmental variable are "bad" for the populations and high values are "good". We tested the reasonableness of this hypothesis using a metapopulation extension of the Lewontin-Cohen model, $\vec{P}(t+1) = D\lambda(t)\vec{P}(t)$, where the ith component of the length-N vector $\vec{P}(t)$ represents population density in the ith habitat patch at time t. The $N \times N$ matrix $\lambda(t)$ was diagonal with i^{th} diagonal entry $\exp(r+\varepsilon_i(t))$. Here r is a growth rate; we used $r=0$. The $\varepsilon_i(t)$ represent environmental noises. They were standard-normally distributed, were independent through time, and showed the same spatial correlations for every simulation, but were made to exhibit stronger right- or left-tail associations between patches in different simulations (Appendix S10 in Supplementary material in the online version at https://doi.org/10.1016/bs.aecr.2020.01.003). The $N \times N$ matrix D was a dispersal matrix modelling local or global dispersal at rate d, in different simulations (Appendix S10 in Supplementary material in the online version at https://doi.org/10.1016/bs.aecr.2020.01.003). After each step, if the density in a patch was <1, it was set to 0. We simulated the model 10,000 times for each combination of parameters, starting from $p_0 = 50$ in each patch, and calculated extinction risk after 25 time steps.

Finally, and pursuing ideas from Reuman et al. (2017), we tested whether the copula structure of the dependences between population variables measured in different locations has consequences for the spatial version of Taylor's law. Taylor's law is a commonly observed and widely studied (Cohen et al., 2013; Reuman et al., 2017; Taylor, 1961; Taylor et al., 1988; Xu et al., 2016) empirical pattern that relates the variances of groups, g, of population measurements to the means of the groups via a power law, $v_g = a \times m_g^b$, or equivalently, $\log(v_g) = \log(a) + b \times \log(m_g)$. Here b is called the Taylor's law *exponent* or *slope*, and $\log(a)$ is the Taylor's law intercept. There are several versions of Taylor's law. For spatial Taylor's law, given population density or abundance data $x_i(t)$ measured in locations $i = 1, \ldots, N$ and times $t = 1, \ldots, T$, a group g consists of all the measurements $x_i(t)$, $i = 1, \ldots, N$ made in different locations at the same time. So means m_g and variances v_g are computed across space. We refer to the matrix with $x_i(t)$ in the tth row and the ith column as a *population matrix*. We consider that Taylor's law holds true for a dataset if the $\log(v_g)$ vs $\log(m_g)$ scatter plot for the dataset shows a linear, homoskedastic pattern. Taylor's law has been verified empirically for hundreds of taxa and has been applied in numerous fields including fisheries management, estimation of species persistence times, and agriculture (Cohen and Xu, 2015; Reuman et al., 2017), so the question of whether copula structure of spatial dependence influences Taylor's law may have applied significance.

To explore the influence of copula structure and tail associations on spatial Taylor's law, we carried out a series of simulations that generated sets of population matrices that had distinct copula structure between locations but that were otherwise statistically similar. For $k = 1, \ldots, 1000$ and X representing a Clayton, normal, Frank, or a survival Clayton copula family, we generated $N \times n$ population matrices $m^{(X,k)}$ with the following properties: (1) for any given values of j and k, the values $\{m_{t,j}^{(X,k)} : t = 1, \ldots, N\}$ were exactly the same, as unordered sets, for all X, i.e., the same actual population values were used for location j in simulation k, regardless of X; (2) for any given k, and j_1 and j_2 such that $j_1 \neq j_2$, the Spearman correlations (computed through time) $\text{cor}_t(m_{t,j_1}^{(X,k)}, m_{t,j_2}^{(X,k)})$ were the same, to within sampling variation, i.e., correlations through time of populations in two locations were the same, up to sampling variation, regardless of X; (3) the copula for the dependence between $m_{t,j_1}^{(X,k)}$ and $m_{t,j_2}^{(X,k)}$ was from the family X. We used $N = 50$ and $n = 25$. Details of how these matrices were generated are in Appendices S11 and S12 in Supplementary material in the online version at https://doi.org/10.1016/bs.aecr.2020.01.003. Thus for each k, the population matrices $m^{(X,k)}$ for X taking the values Clayton, normal, Frank, and survival Clayton were statistically similar except for the copula structure between locations, which was X. Thus comparing how the $\log(v_g)$ vs $\log(m_g)$ relationship may manifest differently for $m^{(X,k)}$ for different values of X constitutes a test of the influence of copula structure on Taylor's law. We selected X equal to the Clayton, normal, Frank, and survival Clayton copulas to explore a range of tail association patterns.

A variety of Taylor's law statistics were computed for each population matrix $m^{(X,k)}$. First, the $N = 50$ spatial means, m_g, and variances, v_g, were computed, and $\log(v_g)$ vs $\log(m_g)$ plots were considered. Linearity of the $\log(v_g)$ vs. $\log(m_g)$ relationship was tested for each simulation by comparing the linear regression through these 50 points to a quadratic alternative via an F-test, producing a p-value result. If these p-values were uniformly distributed across the unit interval for the 1000 replicate simulations which were generated for X, it supported the linearity assumption of Taylor's law for X; whereas if they were clustered toward smaller values the test tended to reject that assumption. We also tested, for each simulation, the assumption of Taylor's law that the $\log(v_g)$ vs. $\log(m_g)$ plot was homoskedastic: we regressed the absolute residuals of the linear regression of $\log(v_g)$ vs. $\log(m_g)$ against the predictions of that regression. A significant p-value result of this test indicates heteroskedasticity. For each simulation we also computed the root mean squared error of $\log(v_g)$ vs. $\log(m_g)$ data from the linear regression of $\log(v_g)$ vs. $\log(m_g)$,. as well as the intercept and slope of the linear regression. Finally we recorded the quadratic coefficient of the regression of $\log(v_g)$ against $\log(m_g)$ and $(\log(m_g))^2$, and the mean curvature of the quadratic regression equation across the values $\log(m_g)$. The distributions of all these statistics across the 1000 replicate simulations were compared for X the Clayton, normal, Frank, and survival Clayton copulas, to determine if copula structure influences Taylor's law.

9. Results for Q3: Tail dependence influences skewness of the spatial average, extinction risk, and Taylor's law

Empirical results were consistent with the hypothesis that the skewness of a spatial-average time series is influenced by tail associations between the local quantities being averaged. For datasets that exhibited highly asymmetric tail associations in earlier analyses (green spruce aphid abundance data had stronger lower-tail association, and, respectively, leaf-curling plum aphid first flight data had stronger upper-tail association, Table 7), skewness of the spatial average was less than (respectively, greater than) a significant fraction of surrogate skewness values (Fig. 13A and B). For datasets with moderately stronger lower than upper-tail dependence (*Ceratium furca* abundance and methane data), skewness of the spatial average showed a non-significant or marginally significant tendency toward being less than surrogate skewnesses (Fig. 13C and D). Thus copula structure and asymmetric tail associations are important for spatially averaged quantities and can influence the temporal stability of those quantities. This result is represented in Fig. 2 as the solid box around "Instability/skewness of mean or total time series" and the solid arrows labelled "X".

Consistent with our extinction risk hypothesis, left-tail-associated environmental fluctuations increased metapopulation extinction risk for the spatial Lewontin-Cohen model, for $N=5$ and $N=25$ habitat patches, and for local and global dispersal (Fig. 14). This result is represented in Fig. 2 as the dashed box around "Extinction risk" and the dashed arrow labelled "Y".

Also consistent with hypothesis, copula structure had a substantial effect on Taylor's law for the models we considered. Taylor's law was strongly influenced, and was often even invalidated in its assumptions of linearity and homoskedasticity, by non-normal copula structure. For normal copula structure (i.e., for simulations that gave normal-copula dependence between populations in different locations), p-values for the linearity and homoskedasticity tests were roughly uniformly distributed across replicate simulations and Taylor's law appeared visually to be a reasonable approximation of the $\log(v_g)$ vs $\log(m_g)$ relationship (Fig. 15B, E and F). Furthermore, quadratic coefficients and curvature values were close to 0 (Fig. 15J and K), and root mean squared errors from the linear regression

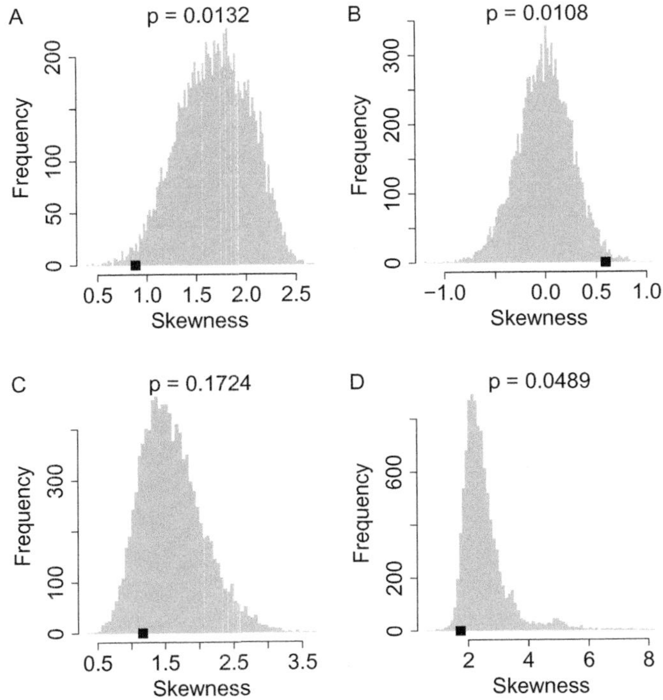

Fig. 13 Skewness of spatially averaged green spruce aphid abundance (A), leaf-curling plum aphid first flight dates (B), *Ceratium furca* abundance (C), and methane-flux (D) time series compared to a multivariate normal-copula null hypothesis. Black dots are empirical skewnesses; see text and Appendix S9 in Supplementary material in the online version at https://doi.org/10.1016/bs.aecr.2020.01.003 for details of the null hypothesis. Results show a tendency for skewness of the spatial average to be affected as hypothesized by asymmetric tail associations.

were relatively small (Fig. 15G). But linearity or homoskedasticity were violated more frequently for non-normal copula structure (Fig. 15A, C, D–F); quadratic coefficients and curvatures were frequently non-zero (Fig. 15J and K); and root mean squared errors from the linear regression were much higher (Fig. 15G). Slopes and intercepts of the linear regression were also strongly affected by copula structure (Fig. 15H and I), though some of the effect here was because linear regressions do not always adequately represent the $\log(v_g)$ vs $\log(m_g)$ relationship when copula structure was not normal. Thus our results substantiated the hypothesis, at least for the models we used, that Taylor's law can be influenced by copula structure and asymmetric tail associations. This is represented in Fig. 2 by the arrow labelled "Z".

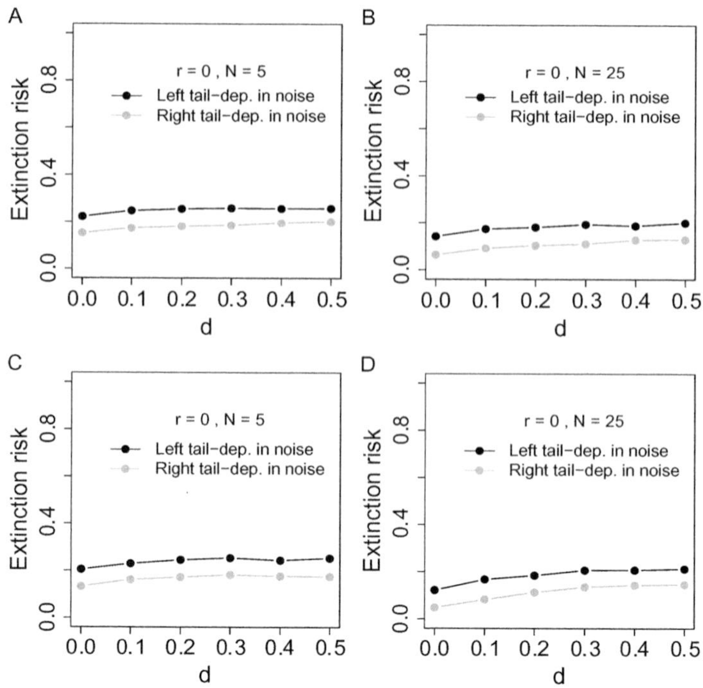

Fig. 14 Extinction risk for the metapopulation extension of the Lewontin-Cohen model, after 25 time steps, was higher for environmental noise with stronger left-tail association across space. Dispersal was local (A, B) or global (C, D), with dispersal rate *d* (Appendix S10 in Supplementary material in the online version at https://doi.org/10.1016/bs.aecr.2020.01.003). Simulations used *N* patches for $N=5$ (A, C) and $N=25$ (B, D).

10. Discussion

We showed that non-normal copula structure and asymmetric tail associations are common across multiple sub-disciplines in ecology, although these facets of data are only occasionally accounted for (Anderson et al., 2018; de Valpine et al., 2014; Popovic et al., 2019). We hypothesized mechanisms that may cause non-normal copula structures and asymmetric tail associations; we discuss below how commonly some of our mechanisms may operate. We also demonstrated important consequences of non-normal copula structure and asymmetric tail associations for ecology. For instance, the skewness of a spatial-average time series is influenced by asymmetric tail associations between its constituent time series: predominantly right-tail-associated local time series can lead to "spiky" spatially averaged

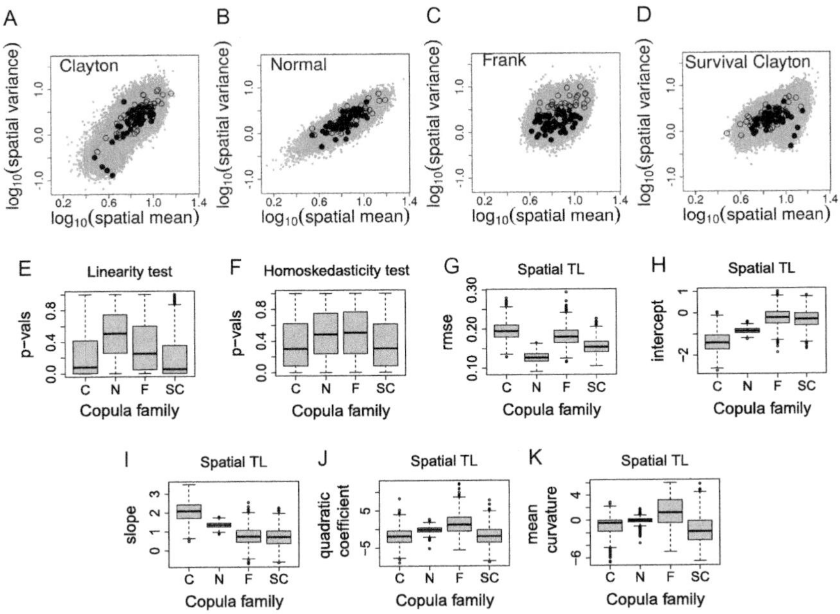

Fig. 15 Spatial Taylor's law results. Log(spatial variance) vs. log(spatial mean) relationships for 1000 simulations over 50 time steps for 25 locations using Clayton (A), normal (B), Frank (C), and survival Clayton (D) copula structures between locations. Results from all 1000 simulations were plotted on the same axes for each panel, but two example simulations are shown in solid and hollow black circles to help assess to what extent variation on the plots was between simulations or within simulations. For each simulation, we tested the linearity (E) and homoskedasticity (F) of the log(variance) vs. log(mean) plot for that simulation, and quantified the root mean squared error of points from the linear regression line (G). We also calculated the linear regression intercept (H) and slope (I), the quadratic term of a quadratic regression through the points on the log(variance) vs. log(mean) plot (J), and the curvature of that quadratic regression (K). Distributions of values across all 1000 simulations are displayed. Spearman-preserving surrogates were used, though results using Kendall-preserving surrogates were similar. See Section 8 and Appendices S11 and S12 in Supplementary material in the online version at https://doi.org/10.1016/bs.aecr.2020.01.003 for details.

time series, with large outbreaks; and predominantly left-tail-associated local time series can lead to spatially averaged time series showing accentuated "crashes". Thus tail associations could have implications for pests and exploited species. Extinction risk and Taylor's law can also be altered by tail association patterns across space. In our view, our results make it reasonable to suggest that a more comprehensive understanding of many ecological phenomena may be possible if a complete copula characterization of associations between variables is employed. Copula statistics are well developed (Joe, 2014; Mai and Scherer, 2017; Nelsen, 2006), and have been introduced in accessible formats (Anderson et al., 2018; Genest and Favre, 2007). Furthermore, open-source computer implementations exist (e.g., the copula and VineCopula packages in R). Ecologists can apply these tools immediately. We created several interrelated randomization procedures (Appendices S6, S9, S12 in Supplementary material in the online version at https://doi.org/10.1016/bs.aecr.2020.01.003) that built upon existing copula methods.

The approaches we demonstrated should apply equally well to data from tropical or temperate ecosystems. There is no reason to expect that non-normal copula structure and asymmetric tail associations should be special properties of datasets from temperate regions. The mechanisms we proposed of non-normal copula structure and asymmetric tail associations seem equally likely to apply anywhere. For instance, the Moran effect, which underpins one of our proposed mechanisms (Fig. 2, B), is a standard mechanism that occurs whenever environmental variables influence populations. And Liebig's law and nonlinear environmental influences on ecosystems, which underpin another one of our mechanisms (Fig. 2, A), are widely demonstrated phenomena.

Our first proposed causal mechanism (Fig. 2, A) may well operate commonly, for two reasons. First, Liebig's law and the idea of limiting nutrients are dominant paradigms in ecology, and many studies have documented nonlinear or threshold influences of environmental variables on ecological quantities. Second, fluctuations in environmental variables through time are very commonly correlated across space. Because these factors, which are the essential ingredients of the mechanism, are common, it is reasonable to hypothesize that the mechanism may operate commonly and may be a dominant cause of asymmetric tail associations and non-normal copula structure of ecological dependencies across space. We provide further support for the mechanism in our discussion of green spruce aphids and winter temperature below.

There are also reasons to hypothesize that our Moran mechanism (Fig. 2, B) may operate commonly: Moran effects are common (Defriez and Reuman, 2017a, 2017b; Liebhold et al., 2004; Sheppard et al., 2016), and non-normal copula structures and asymmetric tail associations are often found in environmental variables. If intense meteorological events are also widespread, then environmental variables associated with these events should take extreme values simultaneously across large spatial areas, producing tail associations in measurements made through time at different locations. Non-extreme values may instead be associated with local phenomena, and therefore may be less correlated across large areas. Serinaldi (2008) examined the spatial dependence of rainfall in Central Italy. Gumbel or Student 2-copulas were candidates for modelling dependence, and neither of these is a normal copula. A long-term study (1950–2014) in the Loess Plateau of China (She and Xia, 2018) showed that a Gumbel copula effectively modelled the spatial dependence of drought variables. The Gumbel copula has asymmetric tail association. Bivariate copula analysis was also used in forecasting the co-occurrence of extreme events (flood or drought) over the North Sikkim Himalayas using spatial datasets (Goswami et al., 2018).

We suggested in the Introduction that asymmetric competitive relationships between species could yield asymmetric tail associations between abundance measurements for the species. This is another theoretical mechanism for non-normal copula structure and asymmetric tail associations, represented in Fig. 2 by the box "Asymmetric species interactions" and the arrow labelled "C". It could be tested by analysing copulas of abundances of competing species, sampled across space or time. We note that, whereas all the datasets we studied here have been positively associated when they were significantly associated, for negatively associated variables such as abundances of competing species, the definitions of left- and right-tail association no longer apply, strictly speaking: the left tail of one distribution corresponds to the right tail of the other. One must be careful with terminology, but it is still possible to study asymmetries of association.

Our simulations of character evolution suggest the hypothesis that changes through evolutionary time in bird and mammal BMR and body size may exhibit greater right- than left-tail association, contrary to standard normality assumptions of character evolution models. This is a hypothesis only, because the greater right-tail association shown in Fig. 8F and G could have come about in another, unknown way rather than via the mechanism we suggested which implicated asymmetric tail associations in evolutionary change. Our simulations show that asymmetric tail associations in evolutionary changes are sufficient, but may not be necessary, to produce the observed asymmetric tail associations in characters of extant species. For instance (see below, and Appendix S13 and Fig. S30 in Supplementary material in the online version at https://doi.org/10.1016/bs.aecr.2020.01.003), systematically missing data can also produce tail dependence and may have influenced results for the

BMR-body mass datasets. Even if the hypothesized evolutionary mechanism (Fig. 2, D) is correct, our results only replace one question, i.e., why do we see greater right- than left-tail association between BMR and body mass, with another, i.e., why might we see greater right- than left-tail association in evolutionary changes in these traits?

Performing statistical tests for associations between continuous traits across different species was a primary motivating example of phylogenetic comparative methods. For example, Felsenstein's method of phylogenetically independent contrasts (Felsenstein, 1985) is currently the second most-cited paper in the history of The American Naturalist journal (Huey et al., 2019). Frustratingly, the field still lacks a dependable procedure for dealing with branch lengths, which are a crucial input to the method. Ideally, the branch lengths used to correct for phylogenetic effects would represent the expected amounts of change for the characters that are being analysed. Because researchers almost never have a reliable method for providing such branch lengths, most researchers rely on ultrametric trees—those for which the duration of the branch in time can be treated as a proxy for the branch length. Frequently these branch lengths are transformed to assess sensitivity to different assumptions about the degree of phylogenetic inertia displayed by the traits under study (Harmon, 2018; Ives, 2018). Even if one were able to simply use a time-based set of branch lengths, assigning dates to nodes in phylogenies is difficult. DNA sequence data can provide estimates of branch lengths, but these estimates are dependent on the adequacy of models which correct sequences for multiple substitutions occurring at the same location. Biases in estimating the evolutionary distance can affect downstream analyses (Phillips, 2009). Additionally, changes in the rate of molecular evolution make the estimation of dates difficult (Heath and Moore, 2014) even when branch lengths in time are accurately estimated.

Without reliable branch length estimates, it is difficult to interpret the significance of the magnitude of changes in traits across a tree. Developing tests of association based on copula structure may possibly lead the way to more robust methods for studying associations when we lack defensible estimates of branch lengths. We note that our phylogenetic analyses here consisted merely of simulations to assess whether interesting copula structure in a simple evolutionary process could leave a detectable signal on the trait data for extant species. Substantial work remains to be done before we have a copula-based method for analysing data on a phylogenetic tree. But the nonparametric nature of approaches based solely on data ranks, as many of our methods are, may be a promising avenue for avoiding inaccuracies that arise via the highly structured model assumptions implicit in the method of phylogenetically independent contrasts and related methods.

Additionally, character evolution was simulated using one random draw from the relevant matrix per phylogeny branch. This was principally because branch lengths are often hugely uncertain. If branch lengths were well known, alternative simulation strategies may include selecting one draw from the matrix of evolutionary changes per unit time, or selecting one draw for the whole branch but rescaling the variances of the selected character changes according to the length of the branch. These choices amount to the same thing for normal copula structure, but not otherwise. Modelling choices such as these may have influenced our results. Additional research seems warranted testing the realism of our hypothesized mechanism and simulations.

Relationships between BMR and body mass relate to a trade-off between mass-specific BMR (BMR per unit body mass) and body mass itself. Copulas probably interrelate with life-history trade-off theory in additional ways beyond what we demonstrated. For instance, it is well known that energy allocation to a life function, F (e.g., reproduction) will reduce the energy that can be allocated to other functions, G_1, G_2, G_3 (e.g., growth, predation avoidance). This is the principle of allocation. But F can trade off against any or all of the G_i. Therefore, for large F, approaching absolute limitations, there may be a strong association between F and G_1, for instance. For small F, there may be little association because resources not allocated to F can instead be allocated to any combination of the G_i. This constitutes asymmetric tail association between F and G_1. Winemiller and Rose (1992) described a three-way trade-off in fishes between age of reproductive maturity, juvenile survivorship, and fecundity. The trade-off should, in theory, produce a tight association between age of maturity and fecundity for fishes with low age of maturity, but little such association for later-maturing fishes because those species may invest the resources not invested in maturing quickly into either fecundity or juvenile survival. These ideas suggest that copulas may be useful for studying multidimensional life-history trade-offs. But applications will require careful attention to the possible consequences of biased sampling: if the degree of completeness of a dataset is associated with one or more of the characters, then statistical artefacts can bias conclusions (Appendix S13 and Fig. S30 in Supplementary material in the online version at https://doi.org/10.1016/bs.aecr.2020.01.003). Multidimensional copulas may also be the appropriate copula approach to studying multidimensional trade-offs. We used only bivariate analyses in this study solely for simplicity. But statistical theory on multidimensional copulas is also well developed (Joe, 2014; Mai and Scherer, 2017; Nelsen, 2006), and open-source computer implementations exist for multivariate as well as bivariate copula methods (e.g., the VineCopula package for R). Such approaches may be a useful next step in life-history theory.

Additional mechanisms of non-normal copula structure and asymmetric tail associations probably also operate. For instance, measurement error may modify copula structure. Our models investigating potential causes of copula structure were intentionally simple in other respects, too, not including factors such as delayed density dependence, dispersal, population stage structure, trophic interactions, etc.; and we did not comprehensively explore parameter space for our models. We re-emphasize that fuller explorations, in future work, of some of our models and of variant models may be informative. We hope by enumerating a few potential mechanisms of copula structure we will inspire additional research on the potentially numerous mechanisms that may operate in diverse datasets, and their relative importance under different circumstances.

We also elaborated potential consequences of copula structure for ecological phenomena and understanding. Our results showed that the skewness of the spatial average of local time series is influenced by their tail associations. But the same logic should also apply to any collection of time series, whether associated with locations in space or not. Another potential application is time series of abundances of all species from a single community, e.g., all plants in a quadrat surveyed repeatedly over time. A large literature has focused on synchrony vs compensatory dynamics between such time series, and the influence of interspecific relationships on the variability of community properties such as total biomass (e.g., Doak et al. (1998), Tilman (1999), Tilman et al. (2006), Gonzalez and Loreau (2009)). Typically, variability of community biomass is measured with the coefficient of variation, but skewness may also be of interest because it can help characterize "spikiness" through time. Future work on copula structure of interspecific relationships in communities and its implications for community variability seems likely to be valuable.

Although we demonstrated that tail associations between environmental variables can influence extinction risk, substantial work remains to determine the importance of this effect. First, we used a non-density-dependent model. Do similar results pertain when density dependence is involved? Second, we considered metapopulation extinction risk, but the large field of population viability analysis (PVA) via stochastic matrix modelling (Caswell, 2000; Morris and Doak, 2002) uses a framework in which a single population's vital rates (e.g., life-stage-specific fecundity and survival rates) are considered to vary stochastically through time due to environmental variation. Do relationships between different vital rates exhibit asymmetric tail associations, and do tail associations influence extinction risk in this context? Finally, is the copula structure of environmental variables or vital rates changing through time, and, if so, how do such changes influence extinction risks? Climate change is known to amplify the factors that lead to extreme weather events (Hansen et al., 2012) and hence may alter spatial tail associations for weather variables.

Our hypotheses and results cover the presence, causes, and consequences of non-normal copula structure and asymmetric tail associations in ecological systems (Fig. 2), but this was done using a variety of datasets and models. We here take a closer look at the green spruce aphid, because it simultaneously illustrates causes and consequences within one system. Green spruce aphid abundance, as measured in the data we use, is strongly positively associated with the temperature of the previous winter (Sheppard et al. (2016), their Supplementary Fig. 6). Winter temperature for year t was here taken to be an average for December of year $t-1$ through March of year t, was available for the locations of aphid sampling, and was preprocessed in the same way as Sheppard et al. (2016). For each of our 10 sampling locations, we therefore examined the copula of winter temperature and aphid abundance time series for the location, finding stronger left- than right-tail associations in 7 of the 8 locations for which independence of winter temperature and aphid abundance could be rejected, according to the data specific to the location (Table 9). Apparently winter temperature has an asymmetric influence on aphid abundance in that cold winters generally produce low abundances but warm winters often do not yield higher abundances than moderate winters. One of our hypothesized mechanisms (Fig. 2, A), which our modelling results supported (Section 7), therefore suggests that spatial dependence between green spruce aphid counts in different locations should show stronger left- than right-tail associations. This is exactly what was observed (Table 7), providing empirical evidence supporting the mechanism (this is why the box around "Nonlinear environmental effects, Liebig's law" and some of the arrows labelled "A" are solid

Table 9 Summary results for analysis of copulas for winter temperature vs green spruce aphid abundance for each of the 10 sampling locations, using the same model selection and nonparametric methods detailed in Section 4.

Site	Best-fit copula	Best-fit AIC	Normal AIC	*p*, CvM	*p*, KS	Model-avg. LT	Model-avg. UT	Model-avg. LT minus UT	$cor_l - cor_u$	$P_l - P_u$	$D_u^2 - D_l^2$
2	C	−10.35	−8.19	0.860	0.830	0.373	0.019	0.354	0.152	0.046	0.024
3	SJ	−18.79	−14.85	0.520	0.750	0.529	0.003	0.526	0.165	0.079	0.027
4	C	−25.48	−17.39	0.520	0.230	0.703	0.001	0.702	0.206	0.080	0.030
5	C	−9.37	−7.26	0.640	0.340	0.393	0.022	0.371	0.187	0.073	0.034
6	F	−22.13	−18.95	0.810	0.410	0.262	0.212	0.050	−0.038	−0.024	−0.006
7	C	−21.56	−19.96	0.320	0.480	0.564	0.028	0.536	0.055	0.023	0.007
8	SJ	−7.33	−3.72	0.131	0.079	0.427	0.012	0.415	0.147	0.047	0.016
9	NA	NA	NA	NA	NA	NA	NA	NA	NA	NA	NA
10	SJ	−8.21	−4.91	0.222	0.149	0.446	0.010	0.436	0.123	0.047	0.015
11	NA	NA	NA	NA	NA	NA	NA	NA	NA	NA	NA

The column labelled *p*, CvM is the *p*-value result for the goodness of fit test using the Cramer-von Mises statistic. The column labelled *p*, KS is the *p*-value result for the goodness of fit test using the Kolmogorov-Smirnov statistic. Entries with an NA are because model selection and other statistics were only employed for copulas for which independence was rejected (5% significance level).

instead of dashed in Fig. 2). The consequences of such tail dependence for the skewness of spatially averaged aphid counts was described previously (Fig. 13A). Thus the asymmetric influence of winter temperature ultimately causes spatially averaged aphid abundance time series to have lower skewness (i.e., less spikiness, and greater stability through time) than they would otherwise. It seems likely that asymmetric influence of winter temperature on populations may be a common phenomenon, so effects such as we have documented for green spruce aphid may be common.

We carried out most of our analyses on data ranks, u_i and v_i, for good reasons mentioned in Sections 1, 2 and 4 and reviewed here; but under some circumstances it may also be appropriate to use techniques which parallel our nonparametric statistics but that use unranked data, x_i, y_i. As mentioned previously, Genest and Favre (2007) and others recommend carrying out inferences about dependence structures (which was our goal) using normalized ranks, stating explicitly that "statistical inference concerning dependence structures should always be based on ranks". The reason for this is that measures of dependence which use unranked data conflate information about the marginal distributions of variables with information about the association between variables. For instance, suppose the true population densities $p_{1,t}$ and $p_{2,t}$ of two species of fish in a lake in year t are unknown, but are assessed using catch per unit effort (CPUE), a standard approach in fisheries science (Zale et al., 2013). The CPUE measurements, which we denote $c_{1,t}$ and $c_{2,t}$, may be differently correlated through time than the true densities would be, if they were known, if the function f relating $p_{i,t}$ and $c_{i,t}$ is nonlinear and if Pearson correlation is used. The function f modifies the marginal distributions, which modifies measures such as Pearson correlation that conflate dependence and marginal information. Rank-based measures such as Spearman or Kendall correlation, or any of our copula approaches, will be unaffected by nonlinear monotonic functions such as f. The same difficulty will pertain in any case for which a measurement of an ecological quantity is a nonlinear index of the true quantity of interest. Nevertheless, if one is certain that measurements are linearly related to the true quantities being measured, statistics based on unranked data may be more appropriate under some circumstances, in part because ecological mechanisms which produce relationships between variables are influenced not only by ranks but also by the relative-size information in the unranked data. Judgements of what statistics are appropriate depend, of course, on the purpose of the analysis. The goal of this study was to study dependence in isolation, so we used ranks. If one is interested in tail associations using unranked data, a partial Pearson correlation can be developed straightforwardly, by replacing u and v in Eq. 1 by x and y, but still computing the sum in that equation over data points with ranks u_i and v_i constrained by the bounds $u_i + v_i > 2l_b$ and $u_i + v_i < 2u_b$. This would result in a different approach to tail associations that would conflate dependence and marginal information but that may also have its own utility for some applications.

Non-copula methods may be useful under some circumstances for exploring asymmetries of tail association, but copula approaches have theoretical and practical advantages. It was suggested to us that asymmetries of tail association can also be explored using a regression model with heteroskedastic error term, $y = mx + b + \varepsilon(x)$. Whereas this approach seems capable of yielding insight for some datasets, it implicitly assumes a particular causal relationship between x and y, an untenable assumption for many applications. Furthermore, no equivalent of Sklar's theorem exists under this approach, so information on marginal distributions cannot be separated from association information, and the two types of information will both influence regression results. Sklar's theorem means that the copula associated with two variables contains all and only the information about the association. Such statements of mathematical completeness will not be available for other methods.

Data requirements for copula methods will vary depending on multiple factors, but our results show that requirements are not beyond what is common in ecology. With copula methods as with any statistical analysis, the data required to detect an effect or a phenomenon depends on the strength of the effect. For instance, detection of mild asymmetry of tail association will require more data than detection of strong asymmetry. Different copula or copula-related methods will also require different amounts of data. For instance, our nonparametric methods such as the partial Spearman correlation should be effective with datasets only slightly bigger than typical guidelines for standard correlation methods; whereas our model-selection approaches may be most effective with more data. Some of our datasets were large, but others were much smaller, e.g., aphid datasets comprised 30 points, and our analyses still provided robust results. Thus many copula methods will be suitable for datasets of sizes that are common in ecology. Partial Spearman correlation seems particularly widely applicable.

Copula methods are numerous, and go well beyond the cases we have considered in this introductory paper. We hope our work helps inspire applications of copulas in ecology both of and beyond the specific tools we employed. For instance, multivariate copulas are useful for studying relationships between multiple interacting variables, and have been used in finance to model tail risk for portfolio optimization problems (Czado, 2019; Joe and Kurowicka, 2011). Analogous ecological applications exist that may be amenable to the same approach, e.g., ecosystem functioning variables such as total primary productivity or carbon flux are the sum of the contributions of multiple species in the same way that the value of an investment portfolio is the sum of the values of its constituent assets. Ecologists may likewise be interested in managing the risk that an ecosystem functioning variable will take an extreme value. As another example, we assumed for this paper that univariate marginal distributions have continuous, strictly monotonic cdfs. This was for simplicity and because the simpler approach was sufficient for our research questions. But count data in ecology have been analysed with approaches not making this assumption (Anderson et al., 2018).

Acknowledgements

We thank Joel E. Cohen for making us aware of copulas in the first place. We thank the many contributors to the large datasets we used; D. Stevens and P. Verrier for data extraction; and David Tilman, Lauren Hallet, Jonathan Walter, Thomas Anderson, Lei Zhao, Andrew Rypel, and editors and anonymous reviewers for helpful suggestions. We thank James Bell of the Rothamsted Insect Survey (RIS). The RIS, a UK Capability, is funded by the Biotechnology and Biological Sciences Research Council under the Core Capability Grant BBS/E/C/000J0200. SG, LWS and DCR were partly funded by US National Science Foundation grants 1714195 and 1442595, the James S McDonnell Foundation, and the California Delta Science Program.

References

Anderson, M., et al., 2018. A pathway for multivariate analysis of ecological communities using copulas. Ecol. Evol. 9, 3276–3294.

Bell, J.R., et al., 2015. Long-term phenological trends, species accumulation rates, aphid traits and climate: five decades of change in migrating aphids. J. Anim. Ecol. 84, 21–34.

Bjørnstad, O., Falck, W., 2001. Nonparametric spatial covariance functions: estimation and testing. Environ. Ecol. Stat. 8, 53–70.

Brechmann, E., Schepsmeier, U., 2013. Modelling dependence with C- and D-vine copulas: the R package CDVine. J. Stat. Softw. 52, 1–27.

Burnham, K.P., Anderson, D.R., 2003. Model Selection and Multimodel Inference: A Practical Information-Theoretic Approach. Springer Science & Business Media.

Caswell, H., 2000. Matrix Population Models: Construction, Analysis, and Interpretation. Sinauer Associates, Sunderland, MA.

Cohen, J., Xu, M., 2015. Random sampling of skewed distributions implies Taylor's power law of fluctuation scaling. Proc. Natl. Acad. Sci. U. S. A. 112, 7749–7754.

Cohen, J.E., et al., 2013. Taylor's law applies to spatial variation in a human population. Genus 69 (1), 25–60.

Czado, C., 2019. Analyzing Dependent Data With Vine Copulas: A Practical Guide With R. Springer.

de Valpine, P., et al., 2014. The importance of individual development variation in stage-structured population models. Ecol. Lett. 17, 1026–1038.
Defriez, E., Reuman, D., 2017a. A global geography of synchrony for marine phytoplankton. Glob. Ecol. Biogeogr. 26, 867–877.
Defriez, E., Reuman, D., 2017b. A global geography of synchrony for terrestrial vegetation. Glob. Ecol. Biogeogr. 26, 878–888.
Doak, D., et al., 1998. The statistical inevitability of stability-diversity relationships in community ecology. Am. Nat. 151, 264–276.
Emura, T., Chen, Y.-H., 2016. Gene selection for survival data under dependent censoring: a copula-based approach. Stat. Methods Med. Res. 25, 2840–2857.
Felsenstein, J., 1985. Phylogenies and the comparative method. Am. Nat. 125, 1–15.
Genest, C., Favre, A.-C., 2007. Everything you always wanted to know about copula modeling but were afraid to ask. J. Hydrol. Eng. 12, 347–368.
Genest, C., et al., 1995. A semiparametric estimation procedure of dependence parameters in multivariate families of distributions. Biometrika 82, 543–552.
Genest, C., et al., 2006. Goodness-of-fit procedures for copula models based on the probability integral transformation. Scand. J. Stat. 33, 337–366.
Genoud, M., et al., 2018. Comparative analyses of basal rate of metabolism in mammals: data selection does matter. Biol. Rev. 93, 404–438.
Gonzalez, A., Loreau, M., 2009. The causes and consequences of compensatory dynamics in ecological communities. Annu. Rev. Ecol. Evol. Syst. 40, 393–414.
Goswami, U.P., et al., 2018. Copula-based probabilistic characterization of precipitation extremes over North Sikkim Himalaya. Atmos. Res. 212, 273–284.
Hansen, J., et al., 2012. Perception of climate change. Proc. Natl. Acad. Sci. U. S. A. 109, E2415–E2423.
Harmon, L., 2018. Phylogenetic Comparative Methods: Learning From Trees. EcoEvoRxiv.
Harrington, R., 2014. The Rothamsted Insect Survey strikes gold. Antenna 38, 159–166.
Heath, T., Moore, B., 2014. Bayesian inference of species divergence times. In: Bayesian Phylogenetics: Methods, Algorithms, and Applications. Chapman & Hall/CRC Press, pp. 277–318.
Holland, E.A., et al., 1999. Soil CO_2, N_2O, and CH_4 exchange. In: Robertson, G.P., et al., (Eds.) Standard Soil Methods for Long-Term Ecological Research. Oxford University Press, pp. 185–201.
Huey, R.B., et al., 2019. Revisiting a key innovation in evolutionary biology: Felsenstein's "phylogenies and the comparative method", Am. Nat. 193, 755–772.
Ives, A., 2018. Mixed and Phylogenetic Models: A Conceptual Introduction to Correlated Data. Leanpub.
Jarecke, K.M., et al., 2016. Coupled soil oxygen and greenhouse gas dynamics under variable hydrology. Soil Biol. Biochem. 95, 164–172.
Joe, H., 1997. Multivariate Models and Multivariate Dependence Concepts. Chapman; Hall/CRC.
Joe, H., 2014. Dependence Modeling With Copulas. Chapman & Hall/CRC Monographs on Statistics & Applied Probability, Taylor & Francis.
Joe, H., Kurowicka, D., 2011. Dependence Modelling: Vine Copula Handbook. World Scientific.
Kim, J.-M., et al., 2008. A copula method for modeling directional dependence of genes. BMC Bioinf. 9, 225.
Li, D., 2000. On default correlation: a copula function approach. J. Fixed Income 9, 43–54.
Li, Y., et al., 2013. Copula-ARMA model for multivariate wind speed and its applications in reliability assessment of generating systems. J. Electr. Eng. Technol. 8, 421–427.
Liebhold, A., et al., 2004. Spatial synchrony in population dynamics. Annu. Rev. Ecol. Evol. Syst. 35, 467–490.

Mai, J.-F., Scherer, M., 2017. Simulating Copulas: Stochastic Models, Sampling Algorithms, and Applications, second ed. Imperial College Press, London.
McNab, B.K., 2008. An analysis of the factors that influence the level and scaling of mammalian BMR. Comp. Biochem. Physiol. A Mol. Integr. Physiol. 151, 5–28.
McNab, B.K., 2009. Ecological factors affect the level and scaling of avian BMR. Comp. Biochem. Physiol. A Mol. Integr. Physiol. 152, 22–45.
Morris, W., Doak, D., 2002. Quantitative Conservation Biology: Theory and Practice of Population Viability Analysis. Sinauer Associates, Sunderland, MA.
Nelsen, R.B., 2006. An Introduction to Copulas, second ed. Springer.
Onken, A., et al., 2009. Analyzing short-term noise dependencies of spike-counts in macaque prefrontal cortex using copulas and the flashlight transformation. PLoS Comput. Biol. 11, e1000577.
Phillips, M.J., 2009. Branch-length estimation bias misleads molecular dating for a vertebrate mitochondrial phylogeny. Gene 441, 132–140.
Popovic, G., et al., 2019. Untangling direct species associations from indirect mediator species effects with graphical models. Methods Ecol. Evol. 10, 1571–1583.
Reuman, D.C., et al., 2017. Synchrony affects Taylor's law in theory and data. Proc. Natl. Acad. Sci. U. S. A. 114, 6788–6793.
Serinaldi, F., 2008. Analysis of inter-gauge dependence by Kendall's τ_K, upper tail dependence coefficient, and 2-copulas with application to rainfall fields. Stoch. Environ. Res. Risk Assess. 22, 671–688.
She, D., Xia, J., 2018. Copulas-based drought characteristics analysis and risk assessment across the Loess Plateau of China. Water Resour. Manag. 32, 547–564.
Sheppard, L.W., et al., 2016. Changes in large-scale climate alter spatial synchrony of aphid pests. Nat. Clim. Chang. 6, 610.
Shih, J., Louis, T., 1995. Inferences on the association parameter in copula models for bivariate survival data. Biometrics 51, 1384–1399.
Sklar, A., 1959. Fonctions de répartition à n dimensions et leurs marges. 8, Publications de l'institute de statistique de l'Université de Paris, pp. 229–231.
Smyth, A.R., et al., 2019. Using high-frequency soil oxygen sensors to predict greenhouse gas emissions from wetlands. Soil Biol. Biochem. 128, 182–192.
Sukumaran, J., Holder, M.T., 2010. DendroPy: a Python library for phylogenetic computing. Bioinformatics 26, 1569–1571.
Taylor, L., 1961. Aggregation, variance and the mean. Nature 189, 732–735.
Taylor, L., et al., 1988. Specificity of the spatial power-law exponent in ecology and agriculture. Nature 332, 721.
Tilman, D., 1999. The ecological consequences of changes in biodiversity: a search for general principles. Ecology 80, 1455–1474.
Tilman, D., 2018a. Plant aboveground biomass data: Biodiversity II: Effects of Plant Biodiversity on Population and Ecosystem Processes. Environmental Data Initiative. Available at https://doi.org/10.6073/pasta/ab7b3604fb36e6875535fee68bd500a5.
Tilman, D., 2018b. Plant species percent cover data: Biodiversity II: Effects of Plant Biodiversity on Population and Ecosystem Processes. Environmental Data Initiative. Available at https://doi.org/10.6073/pasta/ae57e6e9d87dd5fd204422eeec37b442.
Tilman, D., 2018c. Plot coordinates for the E120 Big Biodiversity Field: Biodiversity II: Effects of plant Biodiversity on Population and Ecosystem Processes. Environmental Data Initiative. Available at https://doi.org/10.6073/pasta/940934edb437799bba2ca0a996cf072c.
Tilman, D., et al., 2001. Diversity and productivity in a long-term grassland experiment. Science 294, 843–845.
Tilman, D., et al., 2006. Biodiversity and ecosystem stability in a decade-long grassland experiment. Nature 441, 629.

Walter, J., et al., 2017. The geography of spatial synchrony. Ecol. Lett. 20, 801–814.
Wang, W., Wells, M., 2000. Model selection and semiparametric inference for bivariate failure-time data. J. Am. Stat. Assoc. 95, 62–72.
Wills, S., et al., 2014. Overview of the U.S. rapid carbon assessment project: sampling design, initial summary and uncertainty estimates. In: Hartemink, A.E., McSweeney, K. (Eds.), Soil Carbon. Springer International Publishing, Cham, pp. 95–104.
Winemiller, K., Rose, K., 1992. Patterns of life-history diversification in North American fishes: implications for population regulation. Can. J. Fish. Aquat. Sci. 49, 2196–2218.
Xu, M., et al., 2016. Taylor's power law and fixed-precision sampling: application to abundance of fish sampled by gillnets in an African lake. Can. J. Fish. Aquat. Sci. 74, 87–100.
Zale, A., et al., 2013. Fisheries Techniques, 3rd Edition. American Fisheries Society.

CPI Antony Rowe
Eastbourne, UK
March 26, 2020

Acknowledgments

At the University of Windsor, I am grateful to Dr Michael Sui, vice president of research, and to Dr Robert Orr, acting dean of arts and social sciences, for generously supporting the publication of this work. Additional support came from the university's Centre for Research in Reasoning, Argumentation and Rhetoric, which took the project under its wing. Invaluable research assistance was given by Annette L. Demers at the university's Paul Martin Law Library.

Thanks are also due to Professors Maurice Charland, Thomas Flanagan, Desmond Morton, and Jennifer Reid for their early support of this project. Indeed, all the contributors have been exceedingly supportive in putting this book together. Our initial submission was helped by good suggestions from three anonymous reviewers for McGill-Queen's University Press and was brought to completion by the thoughtful and exacting copyediting of Robert Lewis. Sheila Flavel composed a very thorough index, Kerry Sloan helped with some of the more difficult passages in Riel's speeches, and Heng Wee Tan designed the book's perspicacious cover. I thank these people for lending their talents and goodwill to our project. But most of all, I am thankful to Jacqueline Mason, our editor at the press, who assiduously steered the project through treacherous shoals.

Closer to home I am grateful for discussions with friends Christopher Tindale, Carmela Patrias, and Wayne Thorpe, who gave both encouragement and advice in just proportions. Closest of all to home is Jane McLeod, whom I thank for never faltering.

HVH
Windsor, Ontario
December 2013

RIEL'S DEFENCE

Introduction

HANS V. HANSEN

The focus of this collection of essays is two speeches given by the Métis leader Louis Riel at his trial for high treason. Nearly all Canadians have heard about Riel, and most people with an interest in Canadian history know about the speeches, but few have paused to consider them in detail. They are the last public statements of defence by a man who has an ineradicable place in Canadian history, who combined his personal vision with a national vision, and who left it with us under extraordinary circumstances.

Riel's speeches interest us partly because of their autobiographical character: they give us a self-portrait of an intelligent, ambitious, charismatic, and conflicted man, the leader of the Métis in the years when they began to assert their identity as a people and make their land claims in the North-West Territories. They also provide a first-person account of events by one of the principal actors who influenced the course of history as it unfolded in the 1870s and 1880s. In these addresses by Riel – meant both for those present in the court and those beyond – we find insights about his constitutional thought, arguments about the basis of the Métis land claims, a record of grievances against the federal government, an outline of a plan for immigration in the North-West, and a personal plea for justice, all themes that are with us still. The tragedy of Riel's life, and the drama of the speeches, is heightened by his courage in standing very much alone against the power of the Crown – and his own lawyers – both before sentence was passed and after, and in movingly arguing his own case and that of the Métis.

Riel's trial was and continues to be a source of legal controversy, and he mentioned some of the reasons why in his first speech. However, the focus of this collection of essays is not the trial per se, although it does comprise the narrow and immediate context of the speeches. Riel believed that the troubles that led to the North-West Rebellion in 1885, and subsequently to his trial, could be traced back to the Red River Uprising in 1869–70 (II: 11).[1] Accordingly, the first part of this introduction outlines the background of issues and events that weighed heavily with Riel and that found their way into the speeches. The second part of the introduction gives a brief overview of the essays written especially for this volume, essays that aim to interpret, analyze, and explore the themes of Riel's two addresses, the first to the jury and the second to the court.

BACKGROUND

The Course of Events

After Confederation in 1867, Canada set about arranging to acquire the huge land area known as Rupert's Land, granted to the Hudson's Bay Company by the British Crown in 1670. Especially of interest was the area that lay north of the 49th parallel and west of the Great Lakes, known as the North-West Territories. Several reasons motivated the Canadian government to acquire the region, including (1) a need for more agricultural land, (2) the desire to connect with British Columbia on the West Coast, and (3) the pre-emption of any move by the United States to take possession of the land. The possibility of Canada admitting the North-West Territories was mentioned in the British North American Act of 1867.[2] Canada bought the land from the Hudson's Bay Company for £300,000, and Britain transferred administration of the region to Canada on 1 December 1869. In preparation for taking possession of the region, and in anticipation of the arrival of settlers from Ontario, Ottawa sent an advance party to the Red River region to survey the land.

At the time, there were about 11,000 people living in the Red River region, most of them Métis.[3] The term "Métis" has been used to designate descendants of French and First Nations people as well as of English and First Nations people. For the most part, however, it is used to refer to those with French and Native parentage.[4] This is the community into which Louis Riel had been born in 1844 and the group that he has been thought to primarily represent. These Métis, who had adopted the Roman Catholic religion, spoke Native languages as well as French. They lived by hunting buffalo, transporting goods across the Prairies, and making pemmican for the fur-trading companies. The biggest concentration of Métis was along the Red and Assiniboine Rivers, where they occupied river lots. Although in most cases they had no legal deeds to the land on which they lived, they nevertheless believed they had a right to it.[5]

The Métis were uneasy about the transfer of the North-West Territories from the Hudson's Bay Company to Canada. They feared they might lose their landholdings, and they were also wary that they would be swamped by the influx of Protestant settlers who would come to dominate the region politically and culturally. When the surveyors arrived, the Métis resisted (II: 11). This led to a series of events beginning in November 1869 now known as the Red River Uprising – led by the young Louis Riel – and culminating in the entry of a small part of present-day Manitoba into Confederation as a province in the following year (II: 15).

Although the Manitoba Act of 1870 gave control over natural resources to Ottawa, it nevertheless appeared to give the Métis much of what they had wanted, including land, language, education, and religious rights.[6] Nevertheless, there was a significant emigration from the Red River area to settlements along the South Saskatchewan River throughout the 1870s. One reason for this was that some Métis desired to continue the traditional way of life of following the buffalo, another was that it took a long time to settle the land claims in Manitoba, and a third was that the Métis were being swamped by the influx of English-speaking Protestant settlers from the east and, as anticipated, found their community was being marginalized.[7] Riel maintained that since the Métis no longer felt secure in Manitoba, they had subsequently sold their land for much less than it was worth and had then moved west (II: 26).

There were already Métis living along the Saskatchewan River when their numbers were augmented by the arrival of the Métis immigrants from Red River. Throughout the 1870s and early 1880s, many petitions and requests had been sent to Ottawa by different groups in the

North-West Territories, but they mostly went unanswered or received the reply that the government would deal with the requests "in due course." Because communication between the Métis and the Dominion government was poor, many people were left uncertain of their land rights, and accordingly the Métis – both French-speaking and English-speaking, as well as the white settlers[8] – were frustrated and grew anxious. This led them to seek out Riel, who was then a school teacher in Montana, and invite him to come to the North-West in order to help with their political cause (I: 13). Consequently, Riel returned to Canada in July 1884 and took up residence among the Métis along the South Saskatchewan River. In December of that year, a petition written by Riel and others was sent to Ottawa asking for, among other things, additional sustenance for the First Nations, land grants for the Métis, and patents for those of them who already occupied land. Whatever steps the federal government may have been taking to address these grievances, it was perceived as being indifferent to the demands of the petitioners,[9] and Riel and the Métis formed a provisional government in February 1885, hoping to compel Ottawa to negotiate with the residents of the Saskatchewan region as they had done in Manitoba fifteen years before.[10] This time, however, the federal government was better prepared to resist the Métis's demands. Using the newly constructed Canadian Pacific Railroad, it sent a military force into the West to put down the rebellion. The Métis's military efforts, guided by the skilled buffalo-hunt leader Gabriel Dumont, were able to hold off the Canadian forces in a number of smaller skirmishes at Duck Lake and Fish Creek in March and April, but without the military support of the English Métis and the First Nations, they were eventually defeated at the Battle of Batoche in May 1885. Riel was soon arrested and taken to Regina, where later that summer he was put on trial for high treason.[11]

Issues Related to Riel

Against his wishes, Riel's team of lawyers pursued the defence that he was insane at the time of the North-West Rebellion and therefore not criminally responsible for his actions. Riel disowned this defence, realizing that it undermined the legitimacy of his cause (I: 16, 25; II: 4–7). Although he admitted that between the Red River Uprising in 1869–70 and the North-West Rebellion in 1885 he had spent time in asylums in Quebec, he maintained he had been put there without reason (I: 15).

Questions of Riel's sanity are connected to his avowal that he was a prophet and was given a mission from God. He refers to this explicitly in the first speech (I: 10–14, 34; see also II: 43). In support of the claim that he had a mission, Riel made the further claim that he had prophetic powers (I: 26–8). Based on testimony by others, expert witnesses inferred that Riel must have been insane at the time of the rebellion. Some of the symptoms of Riel's mental illness (identified as megalomania) were his alleged egotism and strong reaction to being contradicted. Hence there are several references in the speeches accounting for his egotism (I: 6, 18; II: 26) and giving context to the way he behaved when people disagreed with him (I: 16; II: 7, 33, 35).

Although the Red River Uprising was not, on the whole, violent, it did lead to the execution of one of the Orangemen from Ontario (II: 42), an act that incensed Ontarians at the time and continued to plague Riel for the rest of his life. During the negotiations that led to Manitoba's entry into Confederation, the delegates from Red River believed, as did local clergy and the lieutenant governor of Manitoba, that Riel and his comrades who had taken control of the Red River settlement in the winter of 1869–70 would be pardoned for their actions. But the pardon for Riel never came. In fact, the Ontario government offered a $5,000 reward for the arrest of Riel (II: 31). This meant that he could be prosecuted if he was captured, and it kept him from taking his seat in the House of Commons, a seat to which Manitobans had twice elected him. In lieu of a pardon, the federal government decided it was better to remove Riel from political influence in the North-West and eventually struck a deal whereby he would voluntarily go into exile for at least five years in return for $5,000 (I: 11; II: 23, 24, 27, 28, 30, 31, 36).[12]

In addition to having to endure exile, Riel had other grievances against the federal government. He believed that he should have been paid for his contribution in marshalling the Métis to resist the Fenian invasion of Manitoba in 1871 (II: 30), as well as for the two months he spent as leader of the Manitoba government in the period right after the province entered Confederation (II: 28, 34). Furthermore, he believed he was entitled to be reimbursed for being excluded from the Métis land grant (II: 28). In the months leading up to the North-West Rebellion in 1885, there was an attempt by the federal government, through intermediaries, to reach an agreement with Riel that would give him financial compensation on the condition that he abandon his political activity in Saskatchewan. Testimony at the trial[13] indicated that a sum of $35,000 was discussed (II: 29, 42), giving the Crown cause to depict Riel as being motivated solely by self-interest in leading the North-West Rebellion.[14] Riel did not hide from the accusation that part of his motivation for his actions in the North-West Rebellion was personal and financial, but he insisted that his own interest was always secondary to his pursuit of the welfare of the Métis and others settled in the North-West (I: 6).[15]

A large part of Riel's second speech focuses on what he considered the principle for allocating land to the Métis during Canada's western expansion. The Manitoba Act had set aside 1.4 million acres for the Métis, and since the area of Manitoba at the time it entered Confederation was 9.5 million acres, this was about one-seventh of the land. Riel believed that this ratio was a precedent that should be followed in dividing up the rest of the land in the North-West Territories so that the Métis would have one-seventh of that land too (II: 16, 20). He went further and proposed a plan for how the West could be populated by giving sevenths to other cultural or national groups, coming either from the United States or Europe. Among the groups he mentioned that would join the Métis and First Nations were the Poles, Germans, Italians, Scandinavians, Irish, Belgians, and Jews (II: 17). Riel also intimated that if Canada did not deal with him fairly, several of these nationalities would enter the North-West and, by force, help him with his political program in return for their seventh of the land (II: 17–19).

Legal Issues Relevant to the Trial

Riel's trial was and remains controversial. The 1885 North-West Rebellion took place in a part of Canada that is now the province of Saskatchewan but was then a part of the North-West Territories. The legal system in the territories was not as fully developed as it was in the provinces, being a mixture of surviving Hudson's Bay Company regulations, British laws valid in Canada, and statutes and laws newly passed by the Dominion government in Ottawa.[16] This uncertainty was then and continues to be a source of debate about what Riel should have been charged with and about where his trial should have taken place. It also makes possible the view that the federal government chose legal avenues that would favour its case against Riel.[17]

The Crown decided to prosecute Riel under the Statute of Treasons passed in the reign of Edward III in 1352 – long before there was a Canada and long before there were Métis. If found guilty under this law, the death penalty was inevitable. Riel was the only one of the insurgents from the North-West Rebellion of 1885 to be charged under this law; other Métis were charged with the lesser crime of felony treason and received relatively short prison sentences. Eight First Nations men were executed for their part in the killing of white settlers at Frog Lake in March 1885, but they had been charged with murder, not treason.[18]

The question of venue for the trial is a central point of controversy because of its several consequences. Riel had wanted to be tried before the Supreme Court of Canada sitting in Quebec, a legal as well as a political impossibility. The more pressing issue is whether he should have been tried in the province of Manitoba instead of the North-West Territories. A trial in the territories was a disadvantage to Riel since there the law required only a jury of six men (as it turned out, six non-Métis men) and a stipendiary judge (a government-appointed judge who was potentially vulnerable to political interference). If the trial had been held in Manitoba, as seems to have been the original intent, Riel's chances of acquittal would have been better since there would then have been a twelve-man jury, some of whom would have been French-speaking Métis, and a politically independent judge. After some initial indecision on the government's part, it was determined that the trial should be held in the North-West Territories, where the alleged offence had occurred. Toward the conclusion of his first speech, Riel reminded the jury of the contentious nature of the trial, remarking both on the judge's status and on the questionable size of the jury (I: 38).

These are the foremost issues and themes in the two speeches. Most of the first address is given over to reacting to the charges made and testimony given during the trial, whereas the second speech reviews the history of events, both those personal to Riel and those affecting the inhabitants of the North-West Territories. Despite his extraordinary oratorical efforts, Riel was found guilty of high treason and hanged in November 1885.

OVERVIEW OF THE ESSAYS

The authors of the essays in this collection come from a variety of academic backgrounds: philosophy, law, history, political science, argumentation studies, religion, and communication. Accordingly, they approach the study of Riel's speeches from different perspectives and find different aspects of the speeches to be of interest. Still, each of the essays may be seen as belonging to one of three groups. The first two essays, by Desmond Morton and Nicole O'Byrne respectively, give us valuable historical background for appreciation of the issues in Riel's speeches. The next four essays, by Thomas Flanagan, Christopher Tindale, myself, and Kerry Sloan, are specifically concerned with the matter, the analysis, and the evaluation of the speeches. The last five essays, by Paul Groarke, Benjamin Authers, Jennifer Reid, Maurice Charland, and Louis Groarke, consider aspects of the speeches from particular vantage points as they take up legal, literary, political, philosophical, and comparative aspects of the texts.

It is largely to George F.G. Stanley and Desmond Morton that we owe the renewed interest in Riel and his career that arose in the second half of the twentieth century. Whereas Stanley's books *The Birth of Western Canada* and *Louis Riel*[19] allow us to see Riel-the-rebel in a sympathetic light, Morton's edition of the trial transcript, *The Queen v Louis Riel*, makes Riel's two remarkable speeches, as well as the troubling details of the trial, widely available. In his essay in this volume, Morton revisits his earlier interest in Riel and the military campaign in the North-West, commenting on the question of Riel's citizenship, his mental state, the fairness of his trial, Ottawa's strategy, contemporary judgments of Riel, and the impact of his speech. Morton's perceptive essay provides a broad background for the various themes explored in the subsequent chapters.

O'Byrne's essay gives us insights into the extent of Riel's knowledge of British constitutional law. "Responsible government" is referred to at least a half-dozen times in the first speech and twice more in the second one, and the concept of "rights," or "public liberties," is invoked about two-dozen times in the second speech. Riel was well aware of the difference between a province and a territory and of the significance of having control over resources. Whereas the original four provinces – Nova Scotia, New Brunswick, Quebec, and Ontario – entered Confederation with full control of their natural resources,

Manitoba, without similar jurisdiction, was a constitutional anomaly. Not only was this a legal difference, but it also had consequences that encumbered the social and practical growth of Manitoba – a situation Riel wanted to amend and thus prevent in the development of the North-West Territories. O'Byrne's chapter is especially significant for a reading of Riel's second speech, his address to the court, wherein he recounts the history of the Red River Uprising and maintains that the troubles in the North-West, fifteen years later, had their genesis in the faults of the "Manitoba treaty."

In his essay, Flanagan considers why Riel's speeches were unsuccessful in both their short- and long-term effects. This is curious because, as Flanagan shows, Riel's background and natural abilities suggest that he could have made much better, more persuasive speeches. One factor that weakened the first speech, thinks Flanagan, is that Riel crowded too many different themes into it. This made it lack coherence and diminished the effect of his best arguments.

Flanagan examines Riel's speeches in light of the three Aristotelian artistic means of persuasion – *ethos*, *pathos*, and *logos* – and observes a significant quantitative difference in their presence in the two speeches. In the first speech, it is *ethos* that is the predominant means of persuasion, whereas in the second speech, it is *logos*. Riel's attempt to establish his *ethos* in the first speech is compromised by his tying it too closely to his religious mission, namely his belief that he was the prophet of the New World; and his arguments, or *logos*, in the second speech, which trace the political failings of the Canadian government and its mistreatment of Riel, are offset by the speaker's two lengthy explanations of his highly unorthodox immigration plan for the North-West.

In the following chapter, Tindale classifies Riel's speech to the jury as falling within the genre of defence speeches and, accordingly, as exhibiting some of the characteristics of that genre. He also identifies some of the familiar rhetorical tropes employed by Riel, such as drawing attention to something by saying it will not be mentioned, as well as "turning the tables." But the rhetorical strategy that interests Tindale the most is that of allusion, in which audience members are influenced by making them aware of something by means of imitation. Riel's speech, it is proposed, alludes to the speech that Socrates gave in his defence. As documented in Louis Groarke's chapter, the similarities in the rhetorical situations of Riel and Socrates are indeed extraordinary. Tindale adds to this that the rhetorical strategies pursued by the two speakers have much in common as well.

As mentioned, commentators have observed that Riel's speeches are poorly organized, which has put in doubt their value as objects of study. However, Tindale finds that Riel responded remarkably well to the exigency of the courtroom in which he found himself. Not only did he give arguments in his own defence, but he also dealt extemporaneously with his lawyers and with testimony given by witnesses. In conclusion, Tindale finds that Riel did "a masterful job of using the materials available to him" and that his speech is one of the best in the genre of defence speeches.

In the next essay, which also expresses a concern about the order of Riel's first speech, I set out to distinguish the narrative order from the logical order of the argumentation. To describe the narrative order, I divide the speech into nine parts (in addition to the opening and conclusion), each devoted to an issue Riel wanted to address. I suggest an alternative narrative ordering of the same parts that might have made for a more effective speech. The logical order is quite different. It organizes in an evidentiary way the arguments Riel gave, such that it can be seen which arguments are in support of other arguments in relation to the main conclusion Riel was urging, namely that he be acquitted. Three main lines of argument are discerned: one is that Riel's actions were justified, another is that he was not insane, and a third is that there was something amiss about the legal proceedings. In identifying the components in each of these lines of argumentation, and their relation to each other, a picture of the speech's overall logical order emerges.

In her essay, Sloan explores the complexities of Riel's second speech – the speech he gave after the jury had found him guilty of treason. She considers it to be a fugue that interweaves different thematic strands: it integrates the statement of Métis rights with a unique immigration policy; it depicts the troubles and misfortunes of Riel's life as a micro version of the trials and grievances of the Métis; and, very subtly, it harmonizes Indigenous religious thought with Riel's political views.

Sloan resists the idea that Riel's religious and political thought was wholly Western, as some have maintained. She finds antecedents to Riel's unique views in the Iroquois League of Nations, as well as in the Anishinabek prophecies, which spoke of different cultures having shared responsibility for the land. We are then led to see Riel's vision for the West as one that combines elements of European and Native ideas. Riel believed that the Métis society should be the model for immigration to the North-West. European nationalities would be invited to come there, and through intermarriage a new and greater nation would arise, organized under Métis principles that allowed for the existence of multiple distinct cultures in one confederation. In this way, the region would eventually become wholly Métis, and the lands lost to the onset of Canadian western expansion would be regained. Sloan sees this as significant in that the Métis, like Riel, had become exiles in their own land, and it was through immigration, organized in accordance with Métis principles, that they could return home since the land from which they had become estranged would now be peopled by them.

In his essay, Paul Groarke takes up the question of identifying the legal grounds of the defence in Riel's first speech. He observes that the jury's being all white and all Protestant meant that it would be sympathetic to the government's case, which, he thinks, led Riel's lawyers to opt for a defence strategy based on insanity. But Riel wanted to be tried on the substantive issues in the case, and Groarke is of the opinion that Riel's lawyers failed in their duty to their client in not following his wishes. The lawyers went further: they refused to let Riel speak during the examination of witnesses, thus in effect making his "legal representation contingent on the fact that he did not defend himself."

When Riel did finally get to speak, his first goal was to persuade the jurors of his sanity, and the second was to defend himself on the merits of the case. Groarke takes Riel's own legal defence to rest on two arguments. The first was that the North-West was not legally part of Canada because the region had been annexed without the consent of the inhabitants; consequently, there could be no "rebellion," only a military opposition. The Canadian government's record of negligence with regard to the region, both materially and politically, reinforces this view. The second part of Riel's defence was that he and the Métis were acting in self-defence, something they had a right and a duty to do under natural law, or the doctrine of lawful rebellion, a basic principle upon which Riel's legal defence ultimately should have rested. Groarke concludes that since Riel's case went to the jury as an insanity defence, his substantive legal arguments have never received due consideration.

In the next essay, Authers focuses on the questions of responsibility and insanity that are so central to Riel's trial and his speeches. The legal background is the M'Naghten case of 1843, which brought together four concepts: punishment, responsibility, sanity, and the capacity to tell right from wrong. The parties to Riel's trial were well acquainted with the M'Naghten Rules. The prosecution argued that Riel was sane and therefore responsible, making him eligible for punishment; his defence lawyers tried to show that Riel was insane and therefore not responsible, meaning that he could not be punished for his actions. Authers steps outside the confines of the legal discourse of Riel's time and considers how the question of responsibility was treated in the popular literature of the day. He looks especially at John Mackie's *The Rising of the Red Man* (1902), which depicts Riel not only as a madman but also as an imposter. Based on either characterization, Riel and his cause are depicted as having no legitimacy.

Riel rejected his own lawyers' insanity defence, according to Authers, because it was inconsistent with the validity of his political actions. To be found insane is to be put outside the community of moral agents; it is humiliating, and puts one's status as a person in doubt. Accordingly, throughout his speech, Riel tried to gain control of how his mental state was depicted by insisting that he was a responsible agent. He argued that whereas the federal government had been irresponsible in not dealing with the inhabitants of the North-West in a timely and fair manner, Riel had taken on the responsibility of speaking for those people and identifying their rights. In this way, Riel presented himself as a responsible agent, politically and mentally, and depicted the Dominion government as being neither.

In her essay, Reid distinguishes between three levels of debatable issues in Riel's trial. There were the issues of whether Riel was sane and of whether he should be found guilty of treason; at a higher level, there was the question of whether, within the Canadian legal system at the time, the court at Regina had the authority to try Riel; and at a still higher level, there were philosophical questions concerning the ultimate legitimacy of states. It is this last issue that intrigues Reid and that she explores in Riel's two speeches to uncover his critique of Canada and his thoughts on the legitimacy of the concept of state sovereignty.

Riel would not accept that the expansion westward should be an extension of only Anglo-Saxon Canada; he made it clear in his second speech that his thinking was that the North-West should be settled by a variety of nationalities that would share the land equitably with the First Nations and the Métis. Furthermore, in his insistence that the dealings with the westerners must accord with the rights of nations, Riel showed that he thought of the people of the North-West as constituting a series of nations, albeit smaller than Canada; nevertheless, the rights of small nations are as valid as those of bigger, more powerful nations. Not only the rights of nations, but also, in Riel's view, the British Constitution and laws, prescribed a standard of justice by which Canada had to abide in dealing with the North-West. Thus Riel held that there were restrictions, over and above Canadian law, that could regulate the disagreements between Canada and the inhabitants of the North-West, restrictions that Canada had failed to heed. Finally, Riel did not think that a state could be self-legitimizing, as Canada purports to be, but that its legitimacy must come from without. It was at this point that Riel's religious thought came into play since, as he argued, only God could start a nation. By bringing these aspects of Riel's thought into focus, Reid gives us a view of Riel's unique geopolitical thought and provides us with a compelling hypothesis to explain his second speech, his address to the court.

Charland's essay explores Riel's trial and speeches using two sets of concepts from Jean-François Lyotard. The first set is that of *litige* and *différend*. A *litige* is a kind of disagreement where both parties share a common idiom and where there is a recognized adjudicator for resolving differences; in a *différend*, in contrast, the regulation of the conflict is in the language of the accuser, so the other party is divested of the means to argue and thus becomes a victim. The second set of concepts is the *polis* and the *pagus*. The *polis* is the city or the state where law and the canons of reason are recognized; the *pagus* is the region outside the polis, where there is no determinable audience and where law is yet to be made.

Charland's analysis sees Riel's situation in 1885 vis-à-vis Canada as being that of a *différend*: the court in Regina could not recognize his arguments inside the framework of a trial for high treason. Since Riel's position was that he was neither insane nor a traitor, to overcome the impossible *différend* situation, he had to construct a larger framework in which he could be heard, one in which the competing interests of Canada and the people of the North-West could both have standing. Thus Charland interprets Riel as taking a rhetorical stance that had him speaking to the jury from the *pagus*. He had to persuade the jury to see things his way, to take his perspective, and, so to speak, to join him in the region where the law of the *polis* itself could be judged. For him to do this, his character had to be sufficiently credible, and it is to this factor that the ultimate evaluation of Riel's speech can be traced. Charland's analysis of the address to the jury reveals the near impossibility of Riel's rhetorical situation and gives us a vivid analysis of the factors that circumscribed it.

The final essay in this collection, by Louis Groarke, undertakes to make a comparison of the trials, character, and speeches of Socrates and Louis Riel, using this as yet another way to gain a fresh perspective on Riel. Among many of the similarities between the two men is that both were on trial for crimes against the state, that both made long speeches, and that both were martyrs for their cause. Both had the same strong and unshakeable conviction that they were right, both justified their activities by claiming to be on missions, and both followed signs that they took to come from gods or God. There are many other points of similarity between Socrates and Riel that Groarke draws to our attention, but there are also important differences between them. One difference is that Socrates was primarily a philosopher, engaged in a contemplative life, whereas Riel was more of a man of action, being a political organizer and leader. Moreover, Socrates was an insider, a man whose Athenian citizenship was never in doubt and who preferred death to exile; in contrast, Riel was an outsider who had suffered exile and who struggled to find a place for his people and himself inside Canada. Groarke's comparison of Socrates and Riel is both a way of paying homage to Riel and, by drawing out similarities and differences, a way of showing us the extraordinary difficulty of his situation.

In subjecting Riel's two trial addresses to analysis, comparison, and criticism, it is our intent to pay tribute to their enduring relevance and importance. The different disciplinary perspectives that are brought to bear bring into focus different aspects of the speeches and the issues that surround them. In this way, the essays complement each other, and when combined, they provide us with a broader and deeper understanding and appreciation of Riel's speeches, which in turn allows us to realize more fully his abilities, his character, and his vision.

NOTES

1 References to Riel's speeches are to paragraphs as they are numbered in the version of the texts printed in this volume. For example, "(II: 11)" indicates the second speech, paragraph 11. Inserted in the text of the speeches are the page numbers of the trial transcript as reprinted in Desmond Morton, ed., *The Queen v Louis Riel: Canada's Greatest State Trial* (Toronto: University of Toronto Press, 1974).

2 Article 146.

3 There were about 6,000 French-speaking Métis, 4,000 English Métis (or "Native English"), and 1,000 Canadians in Red River in 1869. See D.N. Sprague, *Canada and the Métis, 1869–1985* (Waterloo, ON: Wilfrid Laurier University Press, 1988), 45.

4 On the origin of the Métis, see J.M. Bumsted, *Louis Riel v. Canada: The Making of a Rebel* (Winnipeg: Great Plains, 2001), ch. 1; and Bob Beal and Rod Macleod, *Prairie Fire: The North-West Rebellion* (Edmonton: Hurtig, 1984), ch. 1. Interestingly, Riel himself does not use the term "Métis" in either of the speeches studied in this volume. Instead, he refers to his people as "half-breeds," thereby distinguishing them from Aboriginal people and European settlers. The term "Métis" does not occur in either of the Bills of Rights associated with the Red River Uprising (December 1869 and February 1970) or in the St Laurent petition of December 1884, but it is used in the Revolutionary Bill of Rights issued from Batoche in March 1885.

5 See Sprague, *Canada and the Métis*, chs 2–4; and Thomas Flanagan, *Riel and the Rebellion: 1885 Reconsidered*, 2nd ed. (Toronto: University of Toronto Press, 2000), ch. 2.

6 See the Manitoba Act of 1870, articles 22 (ss. 1, 2), 23, 31, and 32.

7 See Bumsted, *Louis Riel v. Canada*, ch. 11; George F.G. Stanley, *The Birth of Western Canada: A History of the Riel Rebellions* (1936; reprint, Toronto: University of Toronto Press, 1960), ch. 12; and Flanagan, *Riel and the Rebellion*, ch. 1.

8 Stanley, *Birth of Western Canada*, ch. 12.

9 Flanagan, *Riel and the Rebellion*, 108.

10 Ibid., ch. 4.

11 See Beal and Macleod, *Prairie Fire*, ch. 14; and Maggie Siggins, *Riel: A Life of Revolution* (Toronto: HarperCollins, 1994), ch. 18.

12 On the amnesty question, see Bumsted, *Louis Riel v. Canada*, chs 8–10.

13 Morton, ed., *Queen v Louis Riel*, 194–5.

14 Ibid., 74–5.

15 For a discussion of the indemnity issue, see Flanagan, *Riel and the Rebellion*, ch. 5; and Siggins, *Riel*, 360–2.

16 See, for example, the Temporary Government of Rupert's Land Act of 1869, article 5.

17 For contrasting views of the fairness of the trial, see Flanagan, *Riel and the Rebellion*, ch. 6; and J.M. Bumsted, "Another Look at the Riel Trial for Treason," in *Canadian State Trials III: Political Trials and Security Measures, 1840–1914*, ed. Barry Wright and Susan Binnie, 411–50 (Toronto: University of Toronto Press, 2009).

18 See Blair Stonechild and Bill Waiser, *Loyal till Death: Indians and the North-West Rebellion* (Markham, ON: Fifth House, 1997), 221.

19 Stanley, *Birth of Western Canada*; and George F.G. Stanley, *Louis Riel* (Toronto: McGraw-Hill Ryerson, 1963).

Preface to the Texts of Louis Riel's Addresses to the Jury and the Court

HANS V. HANSEN

Access to the two speeches that Louis Riel gave at his trial can be difficult. The first speech, addressed to the jury, is the best known and most often reprinted. The second speech, addressed to the court after the verdict and before sentencing, is longer, not as well known, and seldom reprinted. A recent collection, *Great Canadian Speeches* (2004), edited by Dennis Gruending, regrettably contains only about half of the 7,500-word speech to the jury. A greater part of the same speech was reprinted in *Sundog Highway* (2000), edited by Larry Warwaruk, but it is also incomplete. Both of these publications of the speech, as well as the one in *Great Canadian Speeches: Words That Shaped a Nation* (2008), edited by Brian Busby, take their text from what has been the most widely available source of the speech, the complete trial transcript in *The Queen v Louis Riel* (1974), edited by Desmond Morton. Unfortunately, Morton's book is now out of print; moreover, it contains a number of typographical errors that have also found their way into an electronic version of Riel's speech to the jury, published on the Famous Trials website (2004), edited by Douglas Linder.

Riel's speeches as found in the Morton edition are a reproduction of the original *Epitome of Parliamentary Documents in Connection with the North-West Rebellion, 1885* (1886) (hereafter "the Epitome edition"). This Epitome edition is not easily available, and the same may be said of another version of Riel's speeches published as *The Queen vs. Louis Riel* (1886) (hereafter "the Queen's Printer edition"). The Queen's Printer edition should not be confused with Morton's 1974 edition, published almost ninety years later, although they have nearly identical titles.

The most carefully prepared version of Riel's speeches is to be found in *The Collected Writings of Louis Riel*, volume 3, *1884–1885* (1985), edited by Thomas Flanagan. It is based on the Epitome edition, preferred by Flanagan for two reasons: the speeches have been better edited and hence read better, and they are also the basis of the Morton edition, which is "more familiar to modern readers."[1] Unlike all the other editions of the speeches mentioned, Flanagan's includes informative notes that are valuable to Riel scholars. However, scholarly editions of collected works are expensive, and even those of important political figures such as Riel are for the most part found only on the shelves of archives and university libraries. Thus the version of Riel's speeches prepared by Flanagan is not easily available to the general public. Besides, the audience for a scholarly edition of collected works is considerably smaller than the wider audience that reads history: high school and university students, amateur historians, politicians, speech writers, and members of the general public. It is for this reason that I have added notes to both speeches, identifying the people and events to which Riel refers. It is hoped that these notes will help those who are not familiar with Canada's post-Confederation history to understand the speeches more easily.

The different texts of Riel's speeches vary in a number of ways, most of them relatively insignificant. Flanagan observes, "There are more than a hundred variant readings, mostly trivial disagreements as to whether Riel said such things as 'this' or 'the,' 'have' or 'had,' and 'land' or 'lands.' Under the circumstances of printing a stenographic version of a speech given by a man speaking a foreign language, many editorial decisions had to be made. There is no way to discover whether, for example, Riel meant to say 'have' but pronounced it in such a way that the stenographer transcribed 'had.'"[2] It is not surprising, then, that the two versions of Riel's speeches published in the year following his trial, the Epitome and the Queen's Printer editions, vary significantly in how they divide Riel's spoken words into paragraphs. Apparently, in spite of their presence at the trial, those who recorded Riel's words did not heed the extraverbal presentational clues about paragraphing, if there were any. This, I think, is an issue as important as whether Riel actually said "this" or "the" in a particular sentence. It is important because the way that a stream of spoken words is paragraphed attributes structure to the thought of the speaker. Since in Riel's trial the question of his sanity was of great importance, the structure of his speeches, including its division into paragraphs, might be taken as an indicator of his mental health. Thus, to be fair to Riel, and in the absence of any indication of how *he* might have wanted his speeches represented in print, the paragraphing should avoid lumping together different trains of thought in the same paragraph; and similarly a number of contiguous sentences expressing a single argument or a continuation of the same thought should not be separated into separate paragraphs. In other words, charity in reconstruction requires us to avoid, as much as possible, letting Riel's speeches look like the ramblings of a disorganized and incoherent speaker, if that appearance can be mitigated without affecting the content of his speeches or the order of his thoughts. Accordingly, I have not touched the order of Riel's thoughts, but I have made evident changes in the paragraphing of the speeches as well as some word corrections. The first speech is only slightly affected by these alterations, but the second speech, which is about 50 per cent longer than the first one, has undergone extensive reformatting in that a text consisting of five long complicated paragraphs has been turned into forty-five shorter ones. This paragraphing, in both speeches, I believe, will make the contents of the speeches more easily discernible to readers.

In addition, the following changes have been made:

- Paragraphs have been numbered to facilitate discussion of particular passages in the speeches. (The page numbers to the Morton edition are inserted in the text of the speeches. As well, an "Index of Comparative Versions" is included below, allowing readers to locate the same passage in the Morton and Flanagan editions.)
- The two speeches have been divided into eleven and nine parts respectively. The divisions are based on considerations of content.
- Some of the errors in the text that were due to misunderstandings of the court recorders have been corrected. This pertains to the misspellings of proper names. Typos have also been corrected. All such changes are recorded in the notes or are indicated by the use of square brackets.
- Inconsistencies in spelling and capitalization have been corrected without note, as have minor inconsistencies in the punctuation of abbreviations and in the use of quotation marks.
- In several instances where Riel appears to have been replying to testimony given earlier at the trial, the relevant passage is located in the Morton edition of the transcript.

Finally, in the interest of economy, the following abbreviations are used in the notes accompanying the speeches:

DCBO. *Dictionary of Canadian Biography*. http://www.biographi.ca/index-e.html (accessed 12 September 2013).

CWLR. *The Collected Writings of Louis Riel/Les Ecrits complets de Louis Riel*. Vol. 3, *1884–1885*. Ed. Thomas Flanagan. Edmonton: University of Alberta Press, 1985.

Q V LR. *The Queen v Louis Riel: Canada's Greatest State Trial*. Ed. Desmond Morton. Toronto: University of Toronto Press, 1974.

VERSIONS OF RIEL'S SPEECHES

Epitome of Parliamentary Documents in Connection with the North-West Rebellion, 1885, 191–9, 213–25. Ottawa: Maclean, Roger, and Company, 1886.

"The Prisoner's Address." In *The Queen vs. Louis Riel, Accused and Convicted of the Crime of High Treason*, 147–54, 155–66. Ottawa: Queen's Printer, 1886. (Accessed through Early Canadiana Online.)

The Queen v Louis Riel: Canada's Greatest State Trial, 311–25, 350–71. Ed. Desmond Morton. Toronto: University of Toronto Press, 1974.

"Address to the Jury" and "Address to the Court." In *The Collected Writings of Louis Riel/Les Ecrits complets de Louis Riel*, vol. 3, *1884–1885*, ed. Thomas Flanagan, 523–39, 540–62. Edmonton: University of Alberta Press, 1985.

"Excerpts from the Testimony of Louis Riel, 1885." In *Sundog Highway*, ed. Larry Warwaruk, 79–90. Regina: Coteau Books, 2000.

"Louis Riel: Address to the Jury." In *Great Canadian Speeches*, ed. Dennis Gruending, 61–4. Markham, ON: Fitzhenry and Whiteside, 2004.

Louis Riel Trial, 1885, by Douglas Linder. Famous Trials website, 2004. http://law2.umkc.edu/faculty/projects/ftrials/riel/riel.html (accessed 12 September 2013).

"Louis Riel: His Final Statement, the *Queen versus Louis Riel*." In *Great Canadian Speeches: Words That Shaped a Nation*, ed. Brian Busby, 32–43. London: Arcturus, 2008.

INDEX OF COMPARATIVE VERSIONS

The facing charts enable cross-referencing between the versions of Riel's speeches in this volume (by paragraph number), *The Queen v*

Speech 1: Address to the Jury			Speech 2: Address to the Court		
This volume (para. no.)	Q V LR *(page no.)*	CWLR *(line no.)*	*This volume (para. no.)*	Q V LR *(page no.)*	CWLR *(line no.)*
1	311	1	1	350	1
2	311	10	2	350	3
3	312	22	3	350	5
4	312	35	4	350	7
5	312	45	5	350	17
6	312	53	6	351	24
7	313	61	7	351	27
8	313	79	8	351	39
9	314	93	9	351	52
10	314	98	10	352	67
11	314	109	11	352	74
12	315	127	12	353	97
13	315	141	13	353	107
14	315	162	14	354	139
15	316	173	15	354	158
16	316	186	16	355	172
17	316	194	17	355	195
18	317	213	18	356	236
19	317	222	19	357	256
20	317	230	20	357	273
21	318	241	21	358	289
22	318	259	22	358	311
23	319	282	23	359	324
24	319	288	24	359	348
25	319	301	25	360	363
26	319	306	26	360	396
27	320	325	27	361	413
28	320	340	28	361	433
29	321	356	29	362	445
30	321	362	30	362	460
31	321	367	31	363	475
32	321	371	32	363	488
33	321	375	33	363	501
34	321	385	34	364	531
35	322	405	35	364	545
36	322	415	36	365	561
37	322	418	37	365	572
38	323	430	38	365	580
39	323	440	39	366	604
40	323	445	40	367	628
41	323	461	41	368	666
42	324	470	42	368	700
43	324	478	43	369	731
44	324	488	44	370	742
45	324	491	45	370	775
46	325	510			

Louis Riel, edited by Desmond Morton (by page number), and *The Collected Writings of Louis Riel*, volume 3, *1884–1885*, edited by Thomas Flanagan (by line number).

NOTES

1 Thomas Flanagan, ed., *The Collected Writings of Louis Riel/Les ecrits complets de Louis Riel*, vol. 3, *1884–1885* (Edmonton: University of Alberta Press, 1985), 537n1, 560n1.

2 Ibid., 537n1.

Louis Riel's Address to the Jury, 31 July 1885

PREPARED BY HANS V. HANSEN

[p. 311] HIS HONOR:[1] Prisoner, have you any remarks to make to the jury? If so, now is your time to speak.

MR. LEMIEUX:[2] May it please your Honors.[3] At a former stage of the trial you will remember that the prisoner wished to cross-examine the witnesses, we objected at the time, thinking that it was better for the interest of the prisoner that we should do so.[4] The prisoner at this stage is entitled to make any statement he likes to the jury and he has been so warned by your Honor, but I must declare before the court that we must not be considered responsible for any declaration he may make.

HIS HONOR: Certainly, but he is entitled, and I am bound to tell him so.

I OPENING REMARKS

[1] PRISONER: Your Honors, gentlemen of the jury: It would be easy for me to-day to play insanity, because the circumstances are such as to excite any man, and under the natural excitement of what is taking place to-day (I cannot speak English very well, but am trying to do so, because most of those here speak English), under the excitement which my trial causes me would justify me not to appear as usual, but with my mind out of its ordinary condition. I hope with the help of God I will maintain calmness and decorum as suits this honourable court, this honourable jury.

[2] You have seen by the papers in the hands of the Crown that I am naturally inclined to think of God at the beginning of my actions. I wish if you – I do it you won't take it as a mark of insanity, that you won't take it as part of a play of insanity.

> Oh, my God, help me through Thy grace and the divine influence of Jesus Christ. Oh, my God, bless me, bless this honourable court, bless this honourable jury, bless my good lawyers who have come 700 leagues to try to save my [p. 312] life, bless also the lawyers for the Crown, because they have done, I am sure, what they thought their duty. They have shown me fairness which at first I did not expect from them. Oh, my God, bless all those who are around me through the grace and influence of Jesus Christ our Saviour, change the curiosity of those who are paying attention to me, change that curiosity into sympathy with me.

[3] The day of my birth I was helpless and my mother took care of me although she was not able to do it alone, there was someone to help her to take care of me and I lived. To-day, although a man, I am as helpless before this court in the Dominion of Canada and in this world as I was helpless on the knees of my mother the day of my birth. The North-West is also my mother; it is my mother country. And although my mother country is sick and confined in a certain way, there are some from Lower Canada[5] who came to help her to take care of me during her sickness, and I am sure that my mother country will not kill me more than my mother did forty years ago when I came into the world, because a mother is always a mother, and even if I have my faults, if she can see I am true she will be full of love for me.

2 REASONS FOR POLITICAL ACTIONS

[4] When I came into the North-West in July, the first of July, 1884, I found the Indians suffering. I found the half-breeds eating the rotten pork of the Hudson Bay Company and getting sick and weak every day. Although a half-breed, and having no pretension to help the whites, I also paid attention to them. I saw they were deprived of responsible government, I saw that they were deprived of their public liberties. I remembered that half-breed meant white and Indian, and while I paid attention to the suffering Indians and the half-breeds I

remembered that the greatest part of my heart and blood was white and I have directed my attention to help the Indians, to help the half-breeds and to help the whites to the best of my ability.

[5] We have made petitions, I have made petitions with others to the Canadian Government asking to relieve the condition of this country. We have taken time; we have tried to unite all classes, even if I may speak, all parties. Those who have been in close communication with me know I have suffered, that I have waited for months to bring some of the people of the Saskatchewan to an understanding of certain important points in our petition to the Canadian Government and I have done my duty. I believe I have done my duty.

3 PERSONAL GRIEVANCES

[6] It has [p. 313] been said in this box that I have been egotistic. Perhaps I am egotistic. A man cannot be individuality without paying attention to himself. He cannot generalize himself, though he may be general. I have done all I could to make good petitions with others, and we have sent them to the Canadian Government, and when the Canadian Government did answer, through the Under Secretary of State, to the Secretary of the Joint Committee of the Saskatchewan, then I began to speak of myself, not before; so my particular interests passed after the public interests.

[7] A good deal has been said about the settlement and division of lands; a good deal has been said about that. I do not think my dignity to-day here would allow me to mention the foreign policy, but if I was to explain to you or if I had been allowed to make the questions to witnesses, those questions would have appeared in an altogether different light before the court and jury. I do not say that my lawyers did not put the right questions. The observations I had the honour to make to the court the day before yesterday were good, they were absent of the situation, they did not know all the small circumstances as I did. I could mention a point, but that point was leading to so many that I could not have been all the time suggesting. By it I don't wish it understood that I do not appreciate the good works of my lawyers, but if I were to go into all the details of what has taken place, I think I could safely show you that what Captain Young[6] said that I

am aiming all the time at practical results was true, and I could have proved it. During my life I have aimed at practical results. I have writings, and after my death I hope that my spirit will bring practical results.

[8] The learned lawyers for the Crown have produced all the papers and scribbling that was under their hands. I thank them for not having brought out those papers which are so particular to myself, though as soon as they saw what they were they should not have looked at them. I have written not books but many things. All my papers were taken. I destined the papers to be published, if they were worth publishing, after my death. I told Parenteau,[7] one of the prisoners, to put all my books under ground. He did not do it. At that time they acknowledged my orders, that is why I say so. He did not put my books away in time and I am not sorry. I say I thank the learned lawyers for the Crown for having reserved so many things; and if, by the almighty power of God, I go free from this trial, I have such confidence in British fairness that all my papers will be [p. 314] returned [to] me, at least the originals, and if copies are wanted I will be willing to give them.

4 RIEL'S MISSION

[9] No one can say that the North-West was not suffering last year, particularly the Saskatchewan; for the other parts of the North-West I cannot say so much. But what I have done, and risked, and to which I have exposed myself, rested certainly on the conviction, I had to do, was called upon to do something for my country.

[10] It is true, gentlemen, I believed for years I had a mission, and when I speak of a mission you will understand me not as trying to play the role of insane before the grand jury so as to have a verdict of acquittal upon that ground. I believe that I have a mission, I believe I had a mission at this very time. What encourages me to speak to you with more confidence in all the imperfections of my English way of speaking, it is that I have yet and still that mission, and with the help of God, who is in this box with me, and He is on the side of my lawyers, even with the honourable court, the Crown and the jury, to help me, and to prove by the extraordinary help that there is a Provi-

dence to-day in my trial, as there was a Providence in the battles of the Saskatchewan.

[11] I have not assumed to myself that I had a mission. I was working in Manitoba first, and I did all I could to get free institutions for Manitoba; they have those institutions to-day in Manitoba, and they try to improve them, while myself, who obtained them, I am forgotten as if I was dead. But after I had obtained, with the help of others, a constitution for Manitoba, when the Government at Ottawa was not willing to inaugurate it at the proper time, I have worked till the inauguration should take place, and that is why I have been banished for five years. I had to rest five years, I was willing to do it. I protested, I said: "Oh, my God, I offer You all my existence for that cause, and please to make of my weakness an instrument to help men in my country." And seeing my intentions, the late Archbishop Bourget[8] said: "Riel has no narrow views, he is a man to accomplish great things," and he wrote that letter of which I hope that the Crown has at least a copy. And in another letter, when I became what doctors believed to be insane, Bishop Bourget wrote again and said: "Be ye blessed by God and man and take patience in your evils." Am I not taking patience? Will I be blessed by man as I have been by God? [p. 315]

[12] I say that I have been blessed by God, and I hope that you will not take that as a presumptuous assertion. It has been a great success for me to come through all the dangers I have in that fifteen years. If I have not succeeded in wearing a fine coat myself I have at the same time the great consolation of seeing that God has maintained my view; that He has maintained my health sufficiently to go through the world, and that he has kept me from bullets, when bullets marked my hat. I am blessed by God. It is this trial that is going to show that I am going to be blessed by man during my existence, the benedictions are a guarantee that I was not wrong when by circumstances I was taken away from my adopted land to my native land. When I see British people sitting in the court to try me, remembering that the English people are proud of that word "fair-play,"[9] I am confident that I will be blesssed by God and by man also.

[13] Not only Bishop Bourget spoke to me in that way, but Father Jean Baptiste Primeau,[10] the priest of Worcester, who was my director of

conscience, said to me: "Riel, God has put an object into your hands, the cause of the triumph of religion in the world, take care, you will succeed when most believe you have lost." I have got those words in my heart, those words of J.B. Primeau and the late Archbishop Bourget. But last year, while I was yet in Montana, and while I was passing before the Catholic church, the priest, the Reverend Father Frederick Ebeville,[11] curate of the Church of the Immaculate Conception, at Benton, said to me: "I am glad to see you; is your family here?" I said: "Yes." He said: "Go and bring them to the altar, I want to bless you before you go away." And with Gabriel Dumont[12] and my family we all went on our knees at the altar, the priest put on his surplice and he took holy water and was going to bless us, I said: "Will you allow me to pronounce a prayer while you bless me?" He said: "Yes, I want to know what it is." I told him the prayer. It is speaking to God: "My Father, bless me according to the views of Thy Providence which are bountiful and without measure." He said to me: "You can say that prayer while I bless you." Well, he blessed me and I pronounced that prayer for myself, for my wife, for my children, and for Gabriel Dumont.

[14] When the glorious General Middleton[13] fired on us during three days, and on our families, and when shells went and bullets went as thick as mosquitoes in the hot days of summer, when I saw my [p. 316] children, my wife, myself and Gabriel Dumont were escaping, I said that nothing but the blessing without measure of Father Frederick Ebeville could save me, and that can save me to-day from these charges. The benediction promised to me surrounded me all the time in the Saskatchewan, and since it seems to me that I have seen it. Captain Deane,[14] Corporal Prickert, and the corporal of the guard who have been appointed over me have been so gentle while the papers were raging against me shows that nothing but the benediction of God could give me the favour I have had in remaining so respected among these men.

5 QUESTIONS ABOUT SANITY AND CHARACTER

[15] To-day, when I saw the glorious General Middleton bearing testimony that he thought I was not insane,[15] and when Captain Young[16] proved that I am not insane, I felt that God was blessing me, and blot-

ting away from my name the blot resting upon my reputation on account of having been in the lunatic asylum of my good friend Dr. Roy.[17] I have been in an asylum, but I thank the lawyers for the Crown who destroyed the testimony of my good friend Dr. Roy, because I have always believed that I was put in the asylum without reason. To-day my pretension is guaranteed, and that is a blessing too in that way. I have also been in the lunatic asylum at Longue Pointe, and I wonder that my friend Dr. Lachapelle,[18] who took care of me charitably, and Dr. Howard[19] are not here. I was there perhaps under my own name.

[16] Even if I was going to be sentenced by you, gentlemen of the jury, I have this satisfaction if I die – that if I die I will not be reputed by all men as insane, as a lunatic. A good deal has been said by the two reverend fathers, André[20] and Fourmond.[21] I cannot call them my friends, but they made no false testimony. I know that a long time ago they believed me more or less insane. Father Fourmond said that I would pass from great passion to great calmness. That shows great control under contradiction, and according to my opinion and with the help of God I have that control.

[17] Mr. Charles Nolin,[22] when he went into the box, did not say that he was sworn with me in all the affairs that I did. Far from taking them as insane affairs, he was in them under the cover of an oath with four of us. He did not say that in the box. My word is perhaps not testimony, but if he was asked in the box to say if there was an oath taken he could not deny it, and he would have to name the four men, and he would have to name himself. When he speaks of resigning a contract in my favour,[23] I did not ask it, the Government would not give it to [p. 317] me; besides, he was engaged in a movement against the Government, and to take a contract from the Government was certainly a weakness upon his part, and I told him not to compromise his cause, and I told him to withdraw instead of going ahead till we saw if we were going to be listened to at all. He wanted me to make a bargain and renounce my American citizenship. I told him that it was a matter of more strength that I should be an American citizen, not that I want to make any ground of it, but as it took place naturally and as the fact existed I wanted to take advantage of it as such. I told him: "It is of advantage for you that you should have me an American citizen.

I have no bargain to make with you about my American papers, no bargain on such a matter as that."

[18] Mr. Charles Nolin speaks of my ambition, and other witnesses also. There are men among the prisoners who know that last year Mr. Renez and Mr. Joseph Fourget[24] came to the Saskatchewan and said that I could have a place in the council if I wanted it, and that it was a good chance for the half-breeds of the Saskatchewan. If I had been so anxious for position I would have grasped at this place, but I did not, and Mr. Nolin has some knowledge of that. I speak of those things to defend my character, as it has been said that I am egotistical.

6 USE OF ARMS WAS IN SELF-DEFENCE

[19] The agitation in the North-West Territories would have been constitutional, and would certainly be constitutional to-day if, in my opinion, we had not been attacked. Perhaps the Crown has not been able to find out the particulars, that we were attacked, but as we were on the scene it was easy to understand. When we sent petitions to the Government, they used to answer us by sending police, and when the rumours were increasing every day that Riel had been shot here or there, or that Riel was going to be shot by such and such a man, the police would not pay any attention to it.

[20] I am glad that I have mentioned the police, because of the testimony that has been given in the box during the examination of many of the witnesses. If I had been allowed to put questions to the witnesses, I would have asked them when it was I said a single word against a single policeman or a single officer. I have respected the policemen, and I do to-day, and I have respected the officers of the police; the paper that I sent to Major Crozier[25] is a proof [of] it: "We respect you, Major." There are papers which the Crown has in its hands, and which show that demoralization exists among the police, if you will allow me to say it in the court, as I have said it in writing. [p. 318]

7 DEFENCE OF CHARACTER

[21] Your Honors, gentlemen of the jury: If I was a man of to-day perhaps it would be presumptuous to speak in that way, but the truth is

good to say, and it is said in a proper manner, and it is without any presumption. It is not because I have been libelled for fifteen years that I do not believe myself something. I know that through the grace of God I am the founder of Manitoba. I know that though I have no open road for my influence, I have big influence, concentrated as a big amount of vapour in an engine. I believe by what I suffered for fifteen years, by what I have done for Manitoba and the people of the North-West, that my words are worth something. If I give offence, I do not speak to insult. Yes, you are the pioneers of civilization, the whites are the pioneers of civilization, but they bring among the Indians demoralization. Do not be offended, ladies, do not be offended, here are the men who can cure that evil; and if at times I have been strong against my true friends and fathers, the reverend priests of the Saskatchewan, it is because my convictions are strong. There have been witnesses to show that immediately after great passion I could come back to the great respect I have for them.

[22] One of the witnesses here, George Ness, I think, said that I spoke of Archbishop Taché,[26] and told him that he was a thief.[27] If I had had the opportunity I proposed I would have questioned him as to what I said, so that you would understand me. I have known Archbishop Taché as a great benefactor, I have seen him surrounded by his great property, the property of a widow, whose road was passing near. He bought the land around, and took that way to try and get her property at a cheap price. I read in the Gospel: "Ye Pharisees with your long prayers devour the widows."[28] And as Archbishop Taché is my great benefactor, as he is my father, I would say because he has done me an immense deal of good, and because there was no one who had the courage to tell him, I did, because I love him, because I acknowledge all he has done for me; as to Bishop Grandin, it was on the same grounds. I have other instances of Bishop Taché, and the witness could have said that the Reverend Father Moulin: "When you speak of such persons as Archbishop Taché, you ought to say that he made a mistake, not that he committed robbery." I say that we have been patient a long time, and when we see that mild words only serve as covers for great ones to do wrong, it is time when we are justified in saying that robbery is robbery everywhere, and the guilty [p. 319] ones are bound by the force of public opinion to take notice of it. The one who has the courage to speak out in that way, instead of being an outrageous

man, becomes in fact a benefactor to those men themselves, and to society.

[23] When we got to the Church of St. Anthony on the 18th,[29] there was a witness who said, I think George Ness, that I said to Father Moulin, "You are a Protestant."[30] According to my theory I was not going to speak in that way, but I said that we were protesting against the Canadian Government, and that he was protesting against us, and that we were two protestants in our different ways.

[24] As to religion, what is my belief? What is my insanity about that? My insanity, your Honors, gentlemen of the jury, is that I wish to leave Rome aside, inasmuch as it is the cause of division between Catholics and Protestants. I did not wish to force my views, because in Batoche to the half-breeds that followed me I used the word, *carte blanche*. If I have any influence in the new world it is to help in that way and even if it takes 200 years to become practical, then after my death that will bring out practical results, and then my children's children will shake hands with the Protestants of the new world in a friendly manner. I do not wish these evils which exist in Europe to be continued, as much as I can influence it, among the half-breeds. I do not wish that to be repeated in America. That work is not the work of some days or some years, it is the work of hundreds of years.

[25] My condition is helpless, so helpless that my good lawyers, and they have done it by conviction (Mr. Fitzpatrick in his beautiful speech has proved he believed I was insane) my condition seems to be so helpless that they have recourse to try and prove insanity to try and save me in that way. If I am insane, of course I don't know it, it is a property of insanity to be unable to know it.

8 PROPHETIC ABILITIES

[26] But what is the kind of mission that I have? Practical results. It is said that I had myself acknowledged as a prophet by the half-breeds. The half-breeds have some intelligence. Captain Young who has been so polite and gentle during the time I was under his care, said that what was done at Batoche,[31] from a military point of view was nice, that the line of defence was nice, that showed some intelligence. It

is not to be supposed that the half-breeds acknowledged me as a prophet if they had not seen that I could see something into the future. If I am blessed without measure I can see something into the [p. 320] future. We all see into the future more or less. As what kind of a prophet would I come, would it be a prophet who would all the time have a stick in his hand, and threatening, a prophet of evil? If the half-breeds had acknowledged me as a prophet, if on the other side priests come and say that I am polite, if there are general officers, good men, come into this box and prove that I am polite, prove that I am decent in my manner, in combining all together you have a decent prophet. An insane man cannot withhold his insanity, if I am insane my heart will tell what is in me.

[27] Last night while I was taking exercise the spirit who guides and assists me and consoles me, told me that to-morrow somebody will come *t'aider*, five English and one French word, *t'aider*, that is, "to help you." I am consoled by that. While I was recurring to my God, to our God, I said, but woe to me if you do not help me, and these words came to me in the morning, in the morning someone will come *t'aider*, that is to-day. I said that to my two guards and, you can go for the two guards. I told them that if the spirit that directs me is the spirit of truth it is to-day that I expect help. This morning the good doctor who has care of me came to me and said you will speak to-day before the court. I thought I would not be allowed to speak; those words were given to me to tell me that I would have liberty to speak. There was one French word in it, it meant I believe that there was to be some French influence in it, but the most part English. It is true that my good lawyers from the Province of Quebec have given me good advice.

[28] Mr. Nolin came into the box and said that Mr. Riel said that he had a noise in his bowels and that I told him that it meant something.[32] I wish that he had said what I said, what I wrote on the paper of which he speaks, perhaps he can yet be put in the box. I said to Nolin, "Do you hear?" "Yes," I said, "there will be trouble in the North-West," and was it so or not? Has there been no trouble in the North-West? Besides Nolin knows that among his nationality, which is mine, he knows that the half-breeds as hunters can foretell many things, perhaps some of you have a special knowledge of it. I have seen half-

breeds who say, "My hand is shaking, this part of my hand is shaking; you will see such a thing to-day," and it happens. Others will say, "I feel the flesh on my leg move in such a way, it is a sign of such a thing," and it happens. There are men who know that I speak right. If the witness spoke of that fact which he mentioned, to show that I [p. 321] was insane he did not remember that perhaps on that point he is insane himself, because the half-breed by the movement of the hand, sometimes of his shoulders, sometimes his legs, can have certain knowledge of what will happen.

9 VARIOUS REMARKS

[29] To bring Sir John[33] to my feet, if it was well reported it would appear far more reasonable than it has been made to appear. Mr. Blake[34] the leader of the Opposition is trying to bring Sir John to his feet in one way. He never had as much at stake as I had: although the Province of Ontario is great, it is not as great as the North-West.

[30] I am glad that the Crown has[35] proved that I am the leader of the half-breeds in the North-West. I will perhaps be one day acknowledged as more than a leader of the half-breeds, and if I am I will have an opportunity of being acknowledged as a leader of good in this great country.

[31] One of the witnesses said that I intended to give Upper Canada to the Irish. If he had no mystery he would have seen that Upper Canada could not be given to the Irish without being given to England; he rested only upon his imagination.

[32] There is another thing about the partition of lands into sevenths. I do not know if I am prepared to speak of it here because it would become public information, there is so much at stake that if I explained that theory Canada would not very long remain in quiet.

[33] Captain Deane has seen my papers, I have sent them somewhere but he has seen them, and after seeing them he came there and said that I was an intelligent man, and pretty shrewd. I have written these documents and they are in the hands of those whom I trust. I do not want to make them public during my trial; what I have made public during the sixty days we were in arms at Batoche. There have been

three different times when the council decided to send men to the States to notify the nationalities to come to our assistance, but these three delegations waited for my orders and have not started.[36] Why? Because I had an object.

[34] The half-breeds also knew that I told them that they would be punished, that I did not say it of my own responsibility, but that I said it in the same way as I have told them other things. It was said to me that the nation would be punished. Why? Because she had consented to leave Rome too quick. What was the meaning of that? There was a discussion about "too quick"; they said that they should do it at once. "Too quick" does not mean "too soon," if we say yes, it [p. 322] shows no consideration to the man. If God wants something, and if we say yes, that is not the way to answer him. He wants the conscience to say: "Yes, oh my God, I do Thy will"; and because the half-breeds quickly separated from Rome, in such a quick manner, it was disagreeable to God and they were punished, and I told them it would happen; fifty of those who are there can prove it. But, you will say, "You did not put yourself as a prophet?" The 19th century is to be treated in certain ways, and it is probably for that reason I have found the word "*exovede*,"[37] I prefer to be called one of the flock; I am no more than you are, I am simply one of the flock, equal to the rest. If it is any satisfaction to the doctors to know what kind of insanity I have, if they are going to call my pretensions insanity, I say humbly, through the grace of God, I believe I am the prophet of the new world.

[35] I wish you to believe that I am not trying to play insanity; there is in the manner, in the standing of a man, the proof that he is sincere, not playing. You will say, what have you got to say? I have to attend to practical results. Is it practical that you be acknowledged as a prophet? It is practical to say it. I think that if the half-breeds have acknowledged me, as a community, to be a prophet, I have reason to believe that it is beginning to become practical. I do not wish, for my satisfaction, the name of prophet, generally that title is accompanied with such a burden, that if there is satisfaction for your vanity, there is a check to it.

[36] To set myself up as Pope, no, no.[38] I said I believed that Bishop Bourget had succeeded in spirit and in truth. Why? Because while Rome did not pay attention to us, he, as a bishop, paid attention to us.

10 CRITICISM OF COURT AND GOVERNMENT

[37] You have given me your attention, your Honors; you have given me your attention, gentlemen of the jury, and this great audience, I see that if I go any further on that point I will lose the favour you have granted me up to this time, and as I am aiming all the time at practical results, I will stop here, master of myself,[39] through the help of God. I have only a few more words to say, your Honors. Gentlemen of the jury, my reputation, my liberty, my life, are at your discretion.[40] So confident am I, that I have not the slightest anxiety, not even the slightest doubt, as to your verdict. The calmness of my mind concerning the favourable decision which I expect, does not come from any unjustifiable presumption upon my part. I simply [p. 323] trust, that through God's help, you will balance everything in a conscientious manner, and that, having heard what I had to say, that you will acquit me.

[38] I do respect you, although you are only half a jury, but your number of six does not prevent you from being just and conscientious; your number of six does not prevent me giving you my confidence, which I would grant to another six men. Your Honor, because you appointed these men, do not believe that I disrespect you. It is not by your own choice; you were authorized by those above you, by the authorities in the North-West; you have acted according to your duty, and while it is, in our view, against the guarantees of liberty, I trust the Providence of God will bring out good of what you have done conscientiously.

[39] Although this court has been in existence for the last fifteen years, I thought I had a right to be tried in another court. I do not disrespect this court. I do respect it, and what is called by my learned and good lawyers, the incompetency of the court[41] must not be called in disrespect, because I have all respect.

[40] The only things I would like to call your attention to before you retire to deliberate are:

1st That the House of Commons, Senate and Ministers of the Dominion, and [those] who make laws for this land and govern it, are no representation whatever of the people of the North-West.

2nd That the North-West Council generated by the Federal Government has the great defect of its parent.

3rd The number of members elected for the council by the people make it only a sham representative legislature and no representative government at all.

British civilization which rules to-day the world, and the British constitution has defined such government as this is which rules the North-West Territories as [an] irresponsible government,[42] which plainly means that there is no responsibility, and by all the science which has been shown here yesterday you are compelled to admit if there is no responsibility, it is insane.

[41] Good sense combined with scientific theories lead to the same conclusion. By the testimony laid before you during my trial witnesses on both sides made it certain that petition after petition had been sent to the Federal Government, and so irresponsible is that Government to the North-West that in the course of several years besides doing nothing to satisfy the people of this great land, it has [p. 324] even hardly been able to answer once or to give a single response. That fact would indicate an absolute lack of responsibility, and therefore insanity complicated with paralysis.

[42] The Ministers of an insane and irresponsible Government and its little one – the North-West Council – made up their minds to answer my petitions by surrounding me slyly and by attempting to jump upon me suddenly and upon my people in the Saskatchewan. Happily when they appeared and showed their teeth to devour, I was ready: that is what is called my crime of high treason, and to which they hold me to-day. Oh, my good jurors, in the name of Jesus Christ, the only one who can save and help me, they have tried to tear me to pieces.

II LAST REMARKS

[43] If you take the plea of the defence that I am not responsible for my acts, acquit me completely since I have been quarrelling with an insane and irresponsible Government. If you pronounce in favour of the Crown, which contends that I am responsible, acquit me all the same. You are perfectly justified in declaring that having my reason

and sound mind, I have acted reasonably and in self-defence, while the Government, my accuser, being irresponsible, and consequently insane, cannot but have acted wrong, and if high treason there is, it must be on its side and not on my part.

HIS HONOR: Are you done?

[44] PRISONER: Not yet, if you have the kindness to permit me your attention for a while.

HIS HONOR: Well, proceed.

[45] RIEL: For fifteen years I have been neglecting myself. Even one of the most hard witnesses on me said that with all my vanity, I never was particular to my clothing; yes, because I never had much to buy any clothing. The Rev. Father André has often had the kindness to feed my family with a sack of flour, and Father Fourmond. My wife and children are without means, while I am working more than any representative in the North-West. Although I am simply a guest of this country – a guest of the half-breeds of the Saskatchewan – although as a simple guest, I worked to better the condition of the people of the Saskatchewan at the risk of my life, to better the condition of the people of the North-West, I have never had any pay. It has always been my hope to have a fair living one day. It will be for you to pronounce – if you say I was [p. 325] right, you can conscientiously acquit me, as I hope through the help of God you will. You will console those who have been fifteen years around me only partaking in my sufferings. What you will do in justice to me, in justice to my family, in justice to my friends, in justice to the North-West, will be rendered a hundred times to you in this world, and to use a sacred expression, life everlasting in the other.

[46] I thank your Honor for the favour you have granted me in speaking; I thank you for the attention you have given me, gentlemen of the jury, and I thank those who have had the kindness to encourage my imperfect way of speaking the English language by your good attention. I put my speech under the protection of my God, my Saviour, He is the only one who can make it effective. It is possible it should become effective, as it is proposed to good men, to good people, and to good ladies also.

NOTES

1 Judge Hugh Richardson (b. London, England, 1826; d. Ottawa, Ontario, 1913) was a stipendiary judge of the North-West Territories and a member of the North-West Council (*DCBO*).

2 François-Xavier Lemieux (b. Levis, Quebec, 1851; d. Quebec City, 1933) was one of Riel's lawyers (Assemblée nationale de Québec website).

3 The plural "your Honors" was used because there was also an associate justice of the trial, Henry Le Jeune.

4 See *Q v LR*, 205–15. Riel's lawyers refused to let him put questions to witnesses.

5 From 1791 to 1841, what is now Quebec was called Lower Canada, and what is now Ontario was called Upper Canada. "Upper" and "Lower" indicate relative positions in the St Lawrence Valley.

6 Captain George Holmes Young was the soldier who guarded Riel from the day of his capture, 15 May 1885, until he was imprisoned in Regina, 23 May 1885.

7 Pierre Parenteau was a member of the provisional government at Batoche in 1885.

8 Ignace Bourget (b. Lower Canada, 1799; d. Montreal, June 1885, while Riel was awaiting trial) was bishop of Montreal from 1840 (*DCBO*).

9 Charles Fitzpatrick, addressing the jury immediately before Riel's speech, had spoken of a "British jury," an "English jury," and "fair play" (*Q v LR*, 286).

10 In this paragraph, I have twice substituted "Primeau" for "Bruno." The court stenographers got the name wrong (*CWLR*, 538n14). Riel stayed briefly with Primeau in Worcester, Massachusetts, in December 1875.

11 The priest's name was Eberschweiler. This time it is Riel who got the name wrong (*CWLR*, 538n15).

12 Gabriel Dumont (b. Red River settlement, 1837; d. Bellevue, Saskatchewan, 1906) was a military leader of the Métis (*DCBO*).

13 General Frederick Dobson Middleton (b. Belfast, Ireland, 1825; d. London, England, 1898) was a commander of Canada's militia from 1884 to 1890; he led the forces that defeated the North-West uprising in 1885 (*DCBO*).

14 Richard Burton Deane was Riel's jailer in Regina.

15 Middleton was examined by Christopher Robinson on 31 July 1885: "*Q*. During all your intercourse with him did you see anything whatever to indicate any suspicion of unsoundness of mind in him? *A*. No. I cannot say I did – on the contrary. *Q*. Did it occur to you there was any reason to imagine the man was not perfectly sound in mind? *A*. No, I should say on the

contrary he was a man of rather acute intellect. He seemed quite able to hold his own upon any argument or topic we happened to touch upon" (*Q v LR*, 281).

16 Young was examined by Christopher Robinson on 31 July 1885: "*Q*. ... did you observe anything to arouse a suspicion or indicate that he [Riel] was of unsound mind? *A*. None at all, certainly not" (*Q v LR*, 277).

17 Riel was in an asylum at Longue Pointe, Quebec, from March 1876 until transferred to an asylum at Beauport, Quebec, in May 1877, from which he was released in January 1878. See George F.G. Stanley, *Louis Riel* (Toronto: McGraw-Hill Ryerson, 1963), 225. Dr François Roy was the medical superintendent of the asylum at Beauport and a witness for the defence.

18 Emmanuel-Persillier Lachapelle (b. north of Montreal, 1845; d. Minnesota, 1918) was a Montreal schoolmate of Riel's who became a prominent physician in Quebec. He certified Riel as insane in 1876 (*DCBO*).

19 Dr Henry Howard (b. Nenagh, Ireland, 1815; d. Montreal, 1887) worked at the asylum at Longue Pointe from 1875 to 1887 (*DCBO*).

20 Alexis André (b. France, 1832; d. Calgary, 1893) was a Catholic priest. He was sympathetic to the Métis but did not fully cooperate with Riel during the Saskatchewan uprising. He was the priest who attended Riel from the time of his conviction until his execution (*DCBO*).

21 Vital Fourmond was an Oblat father. He had been in the Saskatchewan area since 1875 (*Q v LR*, 239). *Q v LR* has "Fourmand"; all occurrences have been changed to "Fourmond."

22 Charles Nolin (b. St Boniface, Manitoba, 1837; d. near Battleford, Saskatchewan, 1907) was also part of the Red River Uprising. "In exchange for his freedom at the end of the [1885] hostilities Nolin agreed to become one of the Crown's chief witnesses against Riel. His testimony was particularly vindictive and ultimately isolated him from the rest of the community, which branded him a 'turncoat'" (*DCBO*). Nolin was Riel's cousin (*Q v LR*, 200).

23 Nolin was examined by the Crown: "In the beginning of January the Government asked for tenders to construct a telegraph line between Edmonton and Duck Lake. I tendered for it. *Q*. You withdrew your tender? *A*. Yes. *Q*. Why? *A*. ... the prisoner [Riel] came with Dumont and asked me to withdraw my contract in his favour, because the Government had not given him any answer to his claim for $35,000" (*Q v LR*, 195–6).

24 The court recorders did not get the names right. The people Riel referred to were Amédée-Emmanuel Forget (b. Lower Canada, 1847; d. Ottawa, 1923), a clerk of the North-West Council (*DCBO*), and Octave Régnier, the local school teacher. See Thomas Flanagan, *Louis 'David' Riel: 'Prophet of the New*

World," 2nd ed. (Toronto: University of Toronto Press, 1996), 144; and *CWLR*, 539n25.

25 Lief Newry Fitzroy Crozier (b. Ireland, 1846; d. Oklahoma, 1901) was a superintendent of the North-West Mounted Police and the commander of the detachment at Battleford (*DCBO*).

26 Alexandre-Antonin Taché (b. Rivière-de-Loup, Lower Canada, 1823; d. St Boniface, Manitoba, 1894) became bishop of St Boniface, Manitoba, in 1851 and archbishop in 1871 (*DCBO*).

27 George Ness was examined by George W. Burbidge on 28 July 1885: "*Q*. At that time, did he not have a difficulty with Father Moulin? *A*. He accused Bishop Taché and Bishop Grandin of being thieves and rogues" (*Q v LR*, 151).

28 "Woe unto you, scribes and Pharisees, hypocrites! for ye devour widows' houses, and for a pretence make long prayer: therefore, ye shall receive the greater damnation" (Matthew 23: 14).

29 March 1885.

30 Ness was examined by the Crown: "*Q*. Did he say anything about taking possession of the church at the same time? *A*. Yes. Riel says: I will take possession of the church. Father Moulin says: I protest your touching the church. Riel says: Look at him; he's a Protestant" (*Q v LR*, 145).

31 The Battle of Batoche, 9–12 May 1885, resulted in the defeat of the Métis.

32 Nolin, examined by the defence, is paraphrased: "one evening they were spending the night together at his house, and there was a noise in his [Riel's] bowels, and the prisoner asked him if he heard that, and the witness [Nolin] said yes, and then the prisoner told him that that was his liver, and that he had inspirations that worked through every part of his body" (*Q v LR*, 203).

33 Sir John A. Macdonald (b. Glascow, Scotland, 1815; d. Ottawa, 1891) was leader of the Conservative Party and Canada's first prime minister, serving from 1867 to 1873 and from 1878 to 1891. He was thus the prime minister both during the Red River Uprising led by Riel in 1869–70 and during the North-West Rebellion led by Riel in 1885 (*DCBO*).

34 Edward Blake (b. west of London, Upper Canada, 1833; d. Toronto, 1912) was the leader of the Liberal Party (the Opposition) from 1882 to 1887 (*DCBO*).

35 "has" replaces "have."

36 Period replaces semicolon.

37 According to the witness Thomas Jackson, Riel coined the word "*exovede*," saying it meant "one of the flock" (*Q v LR*, 176).

38 Jackson had testified that he thought Riel would set himself up as pope in Canada: "I think very likely he intended himself to take the position" (*Q v LR*, 180).

39 Dr Roy had testified that he did not believe that Riel was "in a condition to be the master of his acts" (*Q v LR*, 246).

40 This sentence is identical to the first sentence of Riel's "Notes for the Address to the Jury" (*CWLR*, 155–6). Much of the material in his notes finds its way into paragraphs 38 to 43.

41 Riel's lawyers had challenged the jurisdiction of the court in the North-West Territories to try someone on the charge of treason and had argued that the trial should be in Upper Canada or British Columbia (*Q v LR*, 8–44).

42 Riel was alluding to the fact that the Government of the North-West Territories was not a responsible government, as were those of the Canadian provinces. "Responsible government" is here a technical term meaning that the ministers of a government are responsible to the people's elected representatives.

Louis Riel's Address to the Court, 1 August 1885

PREPARED BY HANS V. HANSEN

I REACTION TO VERDICT

[p. 350] [1] PRISONER: Can I speak now?

MR. JUSTICE RICHARDSON: Oh, yes.

[2] PRISONER: Your Honors, gentlemen of the jury —

MR. JUSTICE RICHARDSON: There is no jury now, they are discharged.

[3] PRISONER: Well, they have passed away before me.

MR. JUSTICE RICHARDSON: Yes, they have passed away.

[4] PRISONER: But at the same time I consider them yet still there, still in their seat. The court has done the work for me, and although at first appearance it seems to be against me, I am so confident in the ideas which I have had the honour to express yesterday, that I think it is for good, and not for my loss. Up to this moment I have been considered by a certain party as insane, by another party as a criminal, by another party as a man with whom it was doubtful whether to have any intercourse. So there was hostility, and there was contempt, and there was avoidance. To-day, by the verdict of the court, one of those three situations has disappeared.

[5] I suppose that after having been condemned, I will cease to be called a fool, and for me, it is a great advantage. I consider it as a [p. 351] great advantage. If I have a mission – I say "if," for the sake of those who doubt, but for my part it means "since," since I have a mission, I cannot fulfil my mission as long as I am looked upon as an insane being – human being, as the moment I begin to ascend[1] that scale I begin to succeed.

[6] You have asked me, your Honors, if I have anything to say why my sentence should not be passed. Yes, it is on that point particularly my attention is directed.

[7] Before saying anything about it, I wish to take notice that if there has ever been any contradiction in my life, it is at this moment, and do I appear excited? Am I very irritable? Can I control myself? And it is just on religion and on politics, and I am contradicted at this moment on politics, and the smile that comes to my face is not an act of my will so much as it comes naturally from the satisfaction that I proved that I experienced seeing one of my difficulties disappearing. Should I be executed – at least if I were going to be executed – I would not be executed as an insane man. It would be a great consolation for my mother, for my wife, for my children, for my brothers, for my relatives, even for my protectors, for my countrymen. I thank the gentlemen who were composing the jury for having recommended me to the clemency of the court.

[8] When I expressed the great hopes that I have just expressed to you, I don't express it without grounds. My hopes are reasonable, and since they are recommended, since the recommendation of the jury to the Crown is for clemency, it would be easy for me, your Honor, to make an incendiary protest and take the three reasons which have been reasonably put forward by my good lawyers and learned lawyers about the jury, about their selection, about the one who selected them, and about the competency of the court; but why should I do it since the court has undertaken to prove that I am a reasonable man? Must not I take advantage of the situation to show that they are right, and that I am reasonable? And yesterday, when I said, by repeating the evidence which had been given against me, when I said in conclusion that you had a decent prophet, I have just to-day the great opportunity of proving it is so.

[9] Besides clearing me of the stain of[2] insanity, clearing my career of the stain of insanity, I think the verdict that has been given against me is a proof that I am more than ordinary myself, but that the circumstances and the help which is given to me is more than ordinary, are more than ordinary, [p. 352] and although I consider myself only as others, yet by the will of God, by His Providence, by the circumstances which have surrounded me for fifteen years, I think that I have been called on to do something which, at least in the North-West, nobody has done yet. And in some way I think, that, to a certain number of people, the verdict against me to-day is a proof that maybe I am a prophet, maybe Riel is a prophet, he suffered enough for it. Now, I have been hunted as an elk for fifteen years. David was[3] seventeen, I think I will have to be about two years still. If the misfortunes that I have had to go through were to be as long as those of old David, I would have two years still, but I hope it will come sooner.

[10] I have two reasons why I would ask that my sentence should not be passed upon me, against me.[4] You will excuse me, you know my difficulty in speaking English, and have had no time to prepare, your Honor, and even had I prepared anything, it would have been imperfect enough, and I have not prepared, and I wish you would excuse what I have to say, the way which I will be able to perhaps express it.

2 TROUBLES TRACED TO RED RIVER

[11] The troubles of the Saskatchewan are not to be taken as an isolated fact. They are the result of fifteen years' war. The head of that difficulty lies in the difficulty of Red River. The troubles of Red River were called the troubles of the North-West, and I would like to know if the troubles of Saskatchewan have not the name of being the troubles of the North-West. So the troubles of 1869 being the troubles of the North-West, and the troubles of 1885 being still the troubles of the North-West, the suggestion comes naturally to the mind of the observer if it is a continuation. The troubles of the North-West in 1885 are the continuation of the troubles in 1869, or if they are two troubles entirely different – I say they are not. Canada – no, I ought not to say Canada, because it was a certain number of individuals, perhaps 700 or 800, that can have passed for Canada, but they came to the Red River, and they wanted to take possession of the country without con-

sulting the people. True, it was the half-breed people. There were a certain number of white pioneers among the population, but the great majority were half-breeds. We took up arms against the invaders of the east without knowing them. They were so far apart of us, on the other side of the lakes,[5] that it cannot be said that we had any hatred against them. We did not know them. They came without notification, they came boldly. We said, who are they?[6] They said, we are the possessors of the [p. 353] country. Well, knowing that it was not true, we done against those parties coming from the east, what we used to do against the Indians from the south and west, when they would invade us.

[12] Public opinion in the States helped us a great deal. I don't mean to say that it is need to obtain justice on this side of the line that the States should interfere, but, at that time, as there was no telegraph communication between the eastern Provinces and the North-West, no railroad, and as the natural way of going to Canada was through the United States, naturally all the rumours, all the news, had to pass by the States, and on their passage they had to meet the remarks and observations of the American people. The American people were favourable to us. Besides, the Opposition in Canada done the same thing, and said to the Government: "Well, why did you go into the North-West without consulting the people?"[7]

[13] We took up arms, as I stated, and we made hundreds of prisoners, and we negotiated. A treaty was made. That treaty was made by a delegation of both parties. Whether you consider that organization of the Red River people at that time a provisional government, or not, the fact is that they were recognized as a body tribal, if you like to call it, as a social body with whom the Canadian Government treated. Did they treat with them as they treated with Indians? It will be for them to say, but they didn't. Since Sir John A. Macdonald and the late Sir George Cartier[8] were delegated by the Dominion Government to meet our delegates, delegates who had been appointed by me, the president – that is the name that was given to me by the council, the president of that council – that our delegates had been invited three times, first by Donald A. Smith,[9] a member of the Privy Council at that time, second by the Rev. Mr. Thibault[10] (the late Rev. Mr. Thibault), third by Archbishop Taché,[11] who had been called from Rome for the purpose of pacifying the North-West, when those three

delegates had invited us to send delegates, we thought that it was safe to send delegates, and I appointed the Rev. Father Ritchot,[12] now curate of St. Norbert, in Manitoba, I appointed the late Judge Black,[13] who died in Scotland, I appointed Alfred H. Scott,[14] he is dead also, and those three delegates started, with our bill of rights of twenty conditions,[15] to go and put it before the Canadian Government, and when our delegates came to Ottawa, the Government wanted to treat them as Indians I suppose. Father Ritchot said: "If you don't give me, in writing, my acknowledgment as a delegate, I will go back, and you [p. 354] will go with your bayonets to the North-West. Acknowledge my status! I am invited, and I come."[16] And what was the answer? Our delegates had been invited three times, how were they received in Canada? They were arrested – to show exactly what is the right of nations. They were arrested, they had a formal trial, but the fact remains that they were arrested, and the protest of the Rev. Father Ritchot is still in the document.

[14] However, there was a treaty. Sir John A. Macdonald was delegated, the late Sir George Cartier was delegated to treat with the people, with those three delegates. Now, how were they acknowledged? Were they acknowledged as the delegates of Riel? Oh, no, they were acknowledged as the delegates of the North-West. The late Mr. Howe,[17] in his acknowledgment of the delegates, and in notifying those who had been delegated by the Canadian Government to treat with them, told them that they were acknowledged as the delegates of the North-West. Then it was the cause of the North-West that they represented. It is acknowledged by the Canadian Government by that very same fact that fifteen years ago the treaty of which I am speaking was a treaty of the North-West, of the delegates of the North-West, and if, by trying to say that it was the delegates of the North-West, they wanted to avoid the fact that I was no being at all, the whole world knows that it is not so; they cannot avoid me. And Sir John A. Macdonald himself, in the report of the committee of enquiry about those very same troubles – the committee sat in 1874 – Sir John A. Macdonald said, I think,[18] we acknowledge Riel in his status of a governor.

[15] What was the treaty? Was it an Indian affair? If it had been an Indian affair Manitoba would not have been as it is, would not be as it is. We have the Manitoba Act.[19] There was an agreement between

the two delegations[20] how the whole North-West interest would be considered and how the Canadian Government would treat with the North-West. And then, having settled on the matters of principle, those very principles, the agreement was made that those very principles would be inaugurated in Manitoba first. There was a province erected with responsible government; the lands, they were kept by the Dominion. As the half-breed people were the majority of Manitoba, as at their stage of civilization they were not supposed to be able to administer their lands, we thought that at that time it was a reasonable concession to let them go, not [p. 355] because we were willing to let them go, but because it seemed impracticable to have the administration of the lands.[21]

[16] Still, one of the conditions of the treaty was that the people of the North-West wanted the administration of their lands.[22] The half-breeds had a million, and the land grant of 1,400,000 acres owned about 9,500,000, if I mistake not, which [is] about one-seventh of the lands of Manitoba. You will see the origin of my insanity and of my foreign policy. One-seventh of the land was granted to the people, to the half-breeds of Manitoba – English and French, Protestant and Catholic; there was no distinction whatever. But in the sub-division, in the allotment of those lands between the half-breeds of Manitoba, it came that they had 240 acres of land. Now, the Canadian Government say that we will give to the half-breeds of the North-West 240 acres. If I was insane I would say yes, but as I have had, thank God, all the time the consciousness that I had a certain degree of reason, I have made up my mind to make use of it, and to say that one-seventh of the lands in Manitoba, as to the inauguration of a principle in the North-West, had to bring to the half-breeds of the North-West at least as soon as possible the guarantee for the future that a seventh of the lands will also be given to them; and seeing and yourself understanding how it is difficult for a small population, as the half-breed population, to have their voices heard, I said what belongs to us ought to be ours. Our right to the North-West is acknowledged, our co-proprietorship with the Indians is acknowledged, since one-seventh of the land is given to us, but we have not the means to be heard.

3 IMMIGRATION AND "THE SEVENTH"

[17] What will we do? I said to some of my friends if there is no other way we will make the people who have no country understand that we have a country here which we have ceded on condition. We want a seventh of the lands, and if the bargain is not kept, it is null and void, and we have no right to retreat again. And if we cannot have our seventh of the lands from Canada, we will ask the people of the States, the Italians, to come and help us as emigrants.[23] The Irish, I will count them. Now, it is my turn; I thank you.[24] I count them, and I will show you if I made an insane enumeration of the parties. I said we will invite the Italians of the States, the Irish of the States, the Bavarians of the States, the Poles of the States, the Belgians of the States, and if they come and help us here to have the seventh, we will give them each a seventh;[25] and to show that we are not fanatics, that we are not partisans, that we do not wish only for [p. 356] the Catholic[s] but that we have a consideration for those who are not Catholics,[26] I said we will invite the Danes, we will invite the Swedes who are numerous in the States, and the Norwegians, to come around, and as there are Indians and half-breeds in British Columbia, and as British Columbia is a part of the immense North-West, we said not only for ourselves, but speaking of our children, we will make the proposition that if they help us to have our seventh on the two sides of the Rocky Mountains, they will each have a seventh, and if the Jews will help us, on condition that they acknowledge Jesus Christ as the son of God and the only Saviour of human kind, and if they will help us with their money,[27] we will give them a seventh; and I said also, if the principle of giving a seventh of the lands in the North-West – if the principle of giving a seventh of the lands in the North-West to the half-breeds is good, it ought to be good in the east also, and I said if it is not possible that our views should be heard, we will meet as American citizens. I will invite the Germans of the States, and I will say if you ever have an opportunity of crossing the line east, do it, and help the Indians and the half-breeds of the east to have a revenue equivalent to about one-seventh. And what would be the reward of the Germans? The reward of the Germans would be, if they were successful, to take part of the country and make a new German-Indian world somewhere in British North America; but that is the last resort, and if I had not had a verdict of guilty against me, I would have never said it. Yesterday it is just those

things that I have just avoided to say; when I said I have a reason to not mention them, and when I said, as one of the witnesses said, that my proclamation was in Pembina,[28] I think I am right, because of this trial you see that my pretension is, that I can speak a little of future events.

[18] My trial has brought out the question of the seventh, and although no one has explained the things as I do now, still there is enough said about the seventh of the land and that the division of the lands into seven, seven nationalities, while it ought to have been said between ten nationalities, that by telegraph to-day my proclamation is in Pembina, truly, and the States have my idea; they have my idea. The Fenian element, without any tangible object, have crossed the lines several times for only the sake of what they called revenge, but now that Riel, whose name is somewhat prominent for fifteen years, is known to be in his trouble for life and death, for himself and his nationality, now that my trial gives me a [p. 357] certain increase of the celebrity, now that those questions are appearing now before the public, that there is a land league in the States, that that very same element which possesses Fenianism is still there, and quiet, because they have no plan, because they have no idea around which [to] gather their numbers, and when they catch at it do they think that they will smile? And Gabriel Dumont on the other side of the line, is that Gabriel Dumont inactive? I believe not. He is trying to save me from this box. This is no threat. I have written it. I have written a document of that kind and put it in the hands of Captain Deane three weeks ago. This is not an inspiration of the moment.

[19] I have the right to thank God for the prevision[29] of what happens to-day, but there is another means. I don't wish that means, these means. I don't wish them to call people from the States on this side of the line. No, I wish it only if there is no other possibility, if there is no other resort, of course that is my wish. The last remedy, although it may be extreme, is always a remedy, and it is worth something to try it. But if there is justice, as I still hope, oh, dear, it seems to me I have become insane to hope still. I have seen so many men in my position and where are they? But Lepine[30] has had his scaffold also in Manitoba, and he was not executed. Why? Because he was recommended to the clemency of the court. The idea of the seventh, I have two hands,

and I have two sides to my head, and I have two countries. I am an American citizen and I have two countries, and I am taken here as a British subject. I don't abandon my idea of the seventh. I say because the other is an extreme and extremity, I don't wish for it until extremities have come, and I have, coming to extremities just now, but there are some hopes yet for me, my heart is full of hope; but my friends, I suppose that many of them think that I am gone.

[20] If Canada is just with me, if Canada respects my life, my life, my liberty and my reputation, they will give me all that they have taken away from me, and as I said yesterday, that immense influence which my acts are gathering for the last fifteen years, and which, as the power of steam contained in an engine[31] will have its sway, then what will it do? It will do that Riel will go perhaps to the Dominion ministry, and there instead of calling the parties in the States, he will by means, constitutional means of the country, invite the same parties from Europe as emigration, but let it be well understood that as my right has been acknowledged as a co-proprietor of the soil with the Indians, I want [p. 358] to assert that right. It is constitutionally acknowledged in the Manitoba Act by the 31st clause of that Act, and it does not say to extinguish the Indian title. It says two words, "extinguishing," and "1,400,000 acres of land,"[32] two words and as each child of the half-breeds got one-seventh, naturally I am at least entitled to the same. It is why I spoke of the seventh for Indians, not of the lands but of the revenue as it increases.

[21] But somebody will say, on what grounds do you ask one-seventh of the lands? Do you own the lands? In England, in France, the French and the English have lands, the first was in England, they were the owners of the soil and they transmitted [it] to generations. Now, by the soil they have had their start as a nation. Who starts nations? The very one who creates them, God. God is the master of the universe, our planet is His land, and the nation and the tribes are members of His family, and as a good father, He gives a portion of His lands to that nation, to that tribe, to everyone, that is His heritage, that is His share of the inheritance, of the people, or nation or tribe. Now, here is a nation strong as it may be, it has its inheritance from God. When they have crowded their country because they had no room to stay any more at home, it does not give them the right to come and take the

share of all tribes besides them. When they come they ought to say, well, my little sister, the Cree tribe, you have a great territory, but that territory has been given to you as our own land, it has been given to our fathers in England or in France and of course you cannot exist without having that spot of land. This is the principle:[33] God cannot create a tribe without locating it. We are not birds. We have to walk on the ground, and that ground is encircled of many things, which besides its own value, increases its value in another manner, and when we cultivate it we still increase that value.

[22] Well, on what principle can it be that the Canadian Government has given one-seventh to the half-breeds of Manitoba? I say it must be on this ground, civilization has the means of improving life that Indians or half-breeds have not. So when they come in our savage country, in our uncultivated land, they come and help us with their civilization, but we helped them with our lands, so the question comes: Your land, you Cree or you half-breed, your land is worth to-day one-seventh of what it will be when civilization will have opened it? Your country unopened is worth to you only one-seventh of what it will be when opened. I think it is a fair share to acknowledge the genius of [p. 359] civilization to such an extent as to give, when I have seven pair of socks, six, to keep one. They made the treaty with us. As they made the treaty, I say they have to observe it, and did they observe the treaty? No.

4 THE QUESTION OF AMNESTY

[23] There was a question of amnesty then, and when the treaty was made one of the questions was that before the Canadian Government would send a governor into Manitoba an Imperial amnesty should be proclaimed so as to blot out all the difficulties of the past. Instead of proclaiming a general amnesty before the arrival of the governor which took place on the 2nd of September 1870, the amnesty was proclaimed the 25th April, 1875, so I suffered for five years unprotected, besides I was expelled from the House twice.[34] I was they say outlawed, but, as I was busy as a member of the east, and had a trial in the west, I could not be in two places, and they say that I was outlawed, but no notification was sent to my house of any proceedings of the court. They say that I was outlawed and when the amnesty came five

years after the time that it should have come, I was banished for five years and Lepine deprived of his political rights forever. Why? Because he had given political rights to Manitoba? Is that all? No. Did the amnesty come from the Imperial Government? Not at all. It came from our sister colony in the east, and mind you, to make a miracle of it I said the one being great, and Riel being small, I will go on the other side and I am banished. It is a wonder I did not take and go to Mexico. Naturally I went to the States, amnesty was given by the Secretary of State at Ottawa, the party who treated with us. That is no amnesty. It is an insult to me. It has always been an insult to me. I said in Manitoba two years ago that it was an insult and I considered it as such, but are there proofs that amnesty had been promised? Yes, many.

[24] Archbishop Taché the delegate who has been called, the prelate who has been called from Rome to come and pacify the North-West received a commission to make, to accomplish that pacification, and in general terms was written his commission, and when he came into the North-West before I sent delegates, he said, "I will give you my word of honour as a delegate, that there will be an Imperial amnesty, not because I can promise it on my own responsibility, but because it has been guaranteed to me by the representative of the Crown and the Ministers themselves, the Ministers of the Crown,"[35] and instead of the Imperial amnesty came the amnesty of which I spoke and besides, an amnesty came five years too late, and which took the trouble of banishing me five years more. [p. 360]

MR. JUSTICE RICHARDSON: Is that all?

PRISONER: No, excuse me, I feel weak and if I stop at times, I wish you would be kind enough to —

5 MANITOBA ACT NOT FULFILLED

[25] But the last clause of the Manitoba Act speaks also a little of the North-West, speaks that temporary government will be put into the North-West until a certain time, not more than five years, and, gentlemen, the temporary government, how long has it lasted now? How long has it existed now? For fifteen years, and it will be temporary yet.

It is against the Manitoba [Act], it is against the treaty of the North-West that this North-West Council should continue to be in existence, and against the spirit of the understanding. Have I anything to say against the gentlemen who compose the North-West Council? Not at all, not more than I had yesterday to say against the jury and to say against the officials of this court, whom I respect all, but I speak of the institutions. No; I speak of the institutions in the North-West. The Manitoba treaty has not been fulfilled, neither in regard to me, neither in regard to Lepine. Besides the population of the half-breeds who have [been] found in the troubles of the North-West in Manitoba in 1870, and who have been found in the troubles of the North-West, what right have they to be there? Have they not received their 240 acres? I suppose that the half-breeds in Manitoba in 1870 did not fight for 240 acres of land, but it is to be understood that there were two societies who treated together; one was small, but in its smallness it had its rights. The other was great, but by its greatness it had no greater rights than the rights of the small, because the rights is the same for everyone, and when they began by treating the leaders of the small community as bandits, as outlaws, leaving them without protection, they disorganized that community. The right of nations wanted that the treaty of Manitoba should be fulfilled towards the little community of Red River in the same condition that they were when they treated. That is the right of nations, and when the treaty would have been fulfilled towards that small community in the same state as when it was when she treated, then the obligations would have been fulfilled and the half-breeds might have gone to the North-West, the Saskatchewan, and have no right to call for any other things for themselves, although they had a right to help their neighbours if they thought that they were in a bad fix, because charity is always charity.

[26] Now I say that the people of Manitoba have not been satisfied, nor the leaders, not the people, [p. 361] because during those five years, which elapsed between 1870 and 1875, there were laws made and those laws they embraced the people, the half-breed people, and because they hadn't their rights, because the leaders were always threatened in their existence, the people themselves did not feel any security and they sold their lands, because they thought they would never get first that seventh of the lands. They sold their lands because they saw they had no protection and they went west.[36] What have they

received in receiving the 240 acres? They have received 240 acres of land, and as a matter of fact I can prove that by circumstances many – one-half of them – sold for half of the price, $50 or $40, $60 or $25, and to show the state in which they had been kept those who came from the Red River and the half-breeds of Red River who were in the Red River trouble of 1870, appeared to be a wonder of egotism and of unreasonableness because they appeared to be in the troubles of 1885, which are the continuation of the troubles of the Red River.

6 RIEL FORCED INTO EXILE

[27] The amnesty has not been given by the right parties. Amnesty has not been given to Lepine, one of the leaders, who was then, as Dumont is to-day and myself. I was allowed to come back into the country after ten years; after I would be completely deprived of the chances which I had in 1870 to do something for my people and myself,[37] for emigration, so as to cut down my influence forever. It is why I did not come at that time, and thought I would never come to the country. Did I take my American paper, put my papers of American naturalization during my five years' banishment? No, I did not want to give to the States a citizen of banishment, but when my banishment had expired, when an officer at Battleford – somewhere on this side of the line, in Benton – invited me to come to the North-West I said: "No, I will go to an American court, I will declare my intention, now that I am free to go back, and choose another land."[38] It sored my heart. It sored my heart to say that kind of adieu to my mother, to my brothers, to my sisters, to my friends, to my countrymen, my native land, but I felt that in coming back to this country I could not re-enter it without protesting against all the injustice which I had been suffering, and in doing it I was renewing a struggle which I had not been able to continue as a sound man, as I thought I was, I thought it better to begin a career on the other side of the line.

[28] In Manitoba is that all about the amnesty? No.[39] My share of the 1,400,000 acres of land, have I received it? No, I have not received it. My friends, my [p. 362] mother have applied to have it. No, I could not. Everyone else could apply for theirs. Father, Mother, would apply for their sons and that was all right, but for my mother[40] to apply for me it was not, I did not get it. Last year there was a proof. Here, in the

box, not long ago when I asked [for] an indemnity, I was refused. Was that indemnity based on a fancy? I wanted my lands in Manitoba to be paid. Besides, when they treatied, the treaty was completed on the 31st May 1870, it was agreed to the 24th June, and Sir George Cartier had said, "Let Riel govern the country until the troops get there," and from the 24th June till the 23rd August I governed the country in fact, and what was the reward for it?

[29] When the glorious General Wolseley[41] came he rewarded me in saying Riel's banditti has taken flight, and he wanted to come during the night, at midnight, so as to have a chance to raise a row in Fort Garry and to have a glory to call for in the morning, but heaven was against him then. It rained so much that he could not get there during the night, and he had to come at 10 o'clock next morning. He entered one door of Fort Garry while I left [by] the other.[42] I kept in sight of him. I was small. I did not want to be in his road. But, as I know he had good eyes I say I will keep at a distance, where I can be seen, and if he wants to have me, he will come. A general knows where his enemy is, ought to know, and I kept about 300 yards ahead of him. While he was saying that Riel's banditti had taken flight, Riel was very near. That has been my reward. When I speak of an indemnity of $35,000, to call for something to complete the $100,000, I don't believe that I am exaggerating, your Honor.[43]

7 SERVICE AND COMPENSATION

[30] In 1871 the Fenians[44] came in Pembina. Major Irvine,[45] one of the witnesses, I was introduced to him, and when I brought to the governor 250 men, Governor Archibald[46] was then anxious to have my help because he knew that we were the door of Manitoba, and he said as the question of amnesty came he said, "If Riel comes forward we will protect him. 'Pour la circonstance actuelle,'[47] we will protect him."[48] As long as we need him, we will protect him, but as soon as we don't want him, as soon as we don't need him, we want him to fall back in the same position he is to-day, and that answer had been brought because it had been represented that while I would be helping the Government the parties would be trying to shoot me in the back. "Pour la circonstance actuelle," they said, "I will protect him."[49] What reward have I had by that? The first reward that I had was that that

took place in the first days of October 1871, before the year was ended. [p. 363]

[31] Of course they gave the chance to Riel to come out. A rebel had a chance to be loyal then. My friends, my glorious friend in Upper Canada, now the leader of the Opposition, Mr. Blake,[50] said, "We must prevent Mr. Riel from arriving."[51] When he was Minister in Upper Canada he issued a proclamation of $5,000 for those who would arrest Riel. That was my reward, my dowry, but the Canadian Government, what reward would they give me? In the next year there was going to be an election – 1872. If Riel remains in the country for the elections, it will be trouble, and he has a right to speak. We have made a treaty with him, we do not fulfil it; we promise him amnesty; he is outlawed; we take his country and he has no room even to sleep. He comes to our help. He governs the country during two months and the reward is that he is a bandit. He comes to the help of the Government with 250 men and the reward is $5,000 for his head.

[32] It was at that time that I took the name of David and did[52] I take [it] myself? The honourable Judge of the court at Manitoba, Mr. Dubuc,[53] to-day is the one who gave me the name of David. When I had to hide myself in the woods and when he wanted to write me that he should write under the name which would not be known, so that my letter could come to me, and I may say that in that way it is a legal name. From that point of view even, and I put in a parenthesis, why I have a right, I think as a souvenir of my friend in Upper Canada who caused the circumstance, who bought me that name, to make nothing special about it, and besides, when the King of Judea was speaking of the public services of David didn't he refer us to refer to him in that way? Yes, he did, and as something similar I thought it was only proper that I should take the name of David, but it was suggested to me in a mighty manner, and I could not avoid it.

[33] The Canadian Government said, "Well, Riel will be in the elections here, and he will have the right with all those grievances to speak, and he will embarrass the Government,"[54] so they call[ed] on my greater protector, Archbishop Taché, I don't know what; but in the month of Feburary 1872, Archbishop Taché came to me and said, "The authorities in Lower Canada want you to go on the other side of the

line until the crisis is passed."[55] "Well," I said, "if the crisis is concerning me only, it would be my interest to go there, but I am in a crisis which is the crisis of people of the country, and as it concerns the public besides me I will speak to the public as the public are speaking to me."[56] But the Archbishop gave me so good reason[s] that although I could not yield to those reasons, I came to a conclusion [p. 364] with him, and I said, "My Lord, you have titles to my[57] acknowledgment which shall never be blotted out of my heart, and although my judgment in this matter altogether differs from yours, I don't consider my judgment above yours and what seems to me reasonable might be more reasonable, although I think my course of action reasonable, perhaps yours is more reasonable."[58] I said, "If you command me, as my Archbishop, to go, and take on your shoulders the responsibility of leaving my people in the crisis I will go, but let it be known that it is not my word, that I do it to please you, and yet after you command me to do it – to show that in politics when I am contradicted I can give way,"[59] and they offered me £10 a month to stay on the other side of the line. I said, "To be in gaol I have a chance here in Manitoba, and I want something."[60] They asked me how much I wanted, and I said, "How long do you want to me to stay away?"[61] "Well," he said, "perhaps a year."[62] I tell you beforehand that I want to be here during the elections; that is what I asserted. I want to be here during the elections and it was agreed that they would give £800; £400 to Lepine, £400 to me, £300 to me personally, £300 for Lepine, £100 for my family, £100 for Lepine's family. That makes £800.

[34] How was it agreed that I should receive that money? I said to his lordship that the Canadian Government owe[s] me money, they libel me, and even on the question of libel, they do it so clearly that it does not need any trial to come to judgment. They have a judgment and will they make use of it? They owe me something for my reputation that they abuse every day. Besides I have done work and they never paid me for it. I will take that money as an account of what they will have to pay me one day. It was agreed in that manner and the money was given me in the chapel of St. Vital in the presence of Mr. Dubuc, judge now, and when I did not know at that time where the money came, surely came from, and when the little sack of £300 of gold was handed to me there on the table, I said to his lordship, my Lord, if the one who wants me to go away was here, and if I had to treat him as

he is trying to treat me, the little sack of gold ought to go through his head. That was my last protest.

[35] At that time, but before the election, public opinion was so excited against the one who had taken the responsibility of advising my leaving, that he called me back, and during the elections I was present, it was thirteen[63] years ago to-day. I am rewarded for what I have done through those thirteen[64] years. Sir George Cartier in 1872, just in that summer was beaten in [p. 365] Montreal. I speak of him not as a man of party, I speak of him as a Canadian, as [a] public man he was beaten by Mr. Jetté, of Montreal, by 1,200 majority, and they came to me. My election was sure in Provencher. I had fifteen or twenty men against me, and they came to me. Riel, do you want to resign your seat? I have not it yet. Oh, well you are to get it. Allow George Cartier to be elected here, and I said yes, to show that if I had at the time any inclination to become insane when I was contradicted in politics. But Lower Canada has more than paid me for the little consideration, great as my consideration, but that little mark I considered it a little mark of consideration, a little mark of great consideration for them.

[36] The people of Manitoba hadn't their government inaugurated at that time, they had a sham government. It was to be elected.[65] It was to be inaugurated after 1871. After the 1st January 1871, but we went on in 1874 and it was not inaugurated. As long as Riel was there, with his popularity, if the proper institutions had been inaugurated, Riel would have come in the House, the Provincial House, and of course it was considered to be a damage so as to keep me back. They did not give the people their rights, when it was constitutionally agreed they should have done. I struggled not only for myself but I struggled for the rights, for the inauguration of the principles of responsible and constitutional government in Manitoba. That was conceded about the time I was banished.

[37] While I was in the States was I happy? Yes, I was very happy to find a refuge, but I have met men who have come to me several times and say, "here, look out, here is a man on the other side of the line, and he is trying to take a revenge at you, when you water your horses,"[66] because they have left stains as much as possible on my name. I could not even water my horses on the Missouri without being guarded

against those who wanted my life. And it is an irony for me that I should be called David.

8 IMMIGRATION AGAIN

[38] Last year when I was invited, instead of coming to this country, I could with the plan that has appeared to me, I could have communicated with the Fenian organizations, I could have sent my books; I did not do it; and as a proof of it, while I had no means at all to communicate with my brother, you will see in Manitoba, letter to my brother Joseph where I speak of my books, that I could get any amount of money for that book if I wished it to be published, but that I thought that there was a better chance on this side of the line. And what chance is it? What I said, constitutionally speaking, if Riel [p. 366] succeeded that he should one day, as a public man, invite emigration from the different parts of the different countries of the world, and because this North-West is acknowledged to be partly his own, as a half-breed of this population, to make bargains for this North-West here with the Canadian Government, in such a way so that when the English population has had a full and reasonable share of this land, other nationalities, with whom we are in sympathy, should have also their share of it. When we gave the lands in Manitoba for one-seventh, we did not explain, we gave it to the Canadian Government, but in giving it to the Canadian Government it does not mean that we gave it with all the respect that I have for the English population, the Anglo-Saxon race, we did not give it only for the Anglo-Saxon race. There is the Irish in the east and the French in the west, and their proportion in the Canadian Government ought to receive a reasonable proportion of this land which is bought here; and it is hardly the same to give to some French-Canadians in the North-West and none at all to the Irish.

[39] I don't speak here to call the sympathies, because I am sentenced; I speak sound sense. I follow the line of natural and reasonable sympathies, but behind my thought, perhaps, you would be inclined to believe that it is a way formed to try to work against the English – no, I don't. I believe that the English constitution is an institution which has been perfected for the nations of the world, and while I speak of having in future, if not during my lifetime, after it, of having different

nationalities in the North-West here, my hope that they will succeed is, that they will have it amongst them, the great Anglo-Saxon race. As among the nations of Europe 2,000 years ago, the Roman people were the leading race, and were teaching the other nations good government; that is my opinion of the Anglo-Saxon race. I am not insane enough to regret[67] the great glory of the Anglo-Saxon race God has given to that race, and when God gives something to somebody it is for a good purpose, and because God gave glory to England, it is because He wanted the Anglo-Saxon race to work for His own glory, and I suppose it is not finished yet; they will continue – the Roman empire at the time of the decay[68] existed 400 years – still the king. The Anglo-Saxon, the British empire, if it has come to its highest point of glory, it may be called the king; but it is so great that it will take many hundred years, and fully as many as 400 years to lose[69] its prestige, and during that time I hope that this great North-West, with [p. 367] British influence will, by the emigration of which I speak, [provide] good government.

[40] But will I show insanity in hoping that that plan will be fulfilled? I will speak of the wish of my heart, I have been in what is called, asserted to be wrong to-day; I have been proved to be the leader. I hope that before long that very same thing which is said wrong will be known as good, and then I will remain the leader of it; and as the leader of what I am doing. I say my heart will never abandon the idea of having a new Ireland in the North-West by constitutional means, inviting the Irish of the other side of the sea to come and have a share here; a new Poland in the North-West by the same way, a new Bavaria in the same way, a new Italy in the same way, and on the other side in Manitoba, and since Manitoba has been erected it has been increased since 1870 at least by 9,500,000 acres of land; now it is 96,000,000, say there is 86,000,000 about, acres of land to which the half-breeds' title has not been extinguished, a seventh gives 12,000,000 of those lands and I want the French-Canadians to come and help us there to-day, to-morrow – I don't know when. I am called here to answer for my life, to have time that I should make my testimony, and on the other side of the mountains, there are Indians and as I have said half-breeds, and there is a beautiful island, Vancouver, and I think the Belgians will be happy there, and the Jews who are looking for a country for 1,800 years, the knowledge of which the nations have not been able

to attain yet. While they are rich and the lords of finance, perhaps, will they hear my voice one day, and on the other side of the mountains, while the waves of the Pacific will chant sweet music for them to console their hearts for the mourning of 1,800 years; perhaps will they say is the one thought of us in the whole Cree world, and if they help us there on the other side between the great Pacific and the great Rockies to have a share? The Jews from the States? No; what I wish is the natural course of emigration, that is what I want; my thoughts are for peace. During the sixty days that I have been in Batoche, I told you yesterday that there were three delegations appointed by the *exovede* to send on the other side for help, but there I did not see the safety that I was looking for, not that I distrust my countrymen, but such a great revolution will bring immense disasters, and I don't want during my life to bring disasters except those which I am bound to bring to defend my own life, and to avoid to take away from my country, disasters which threaten me [p. 368] and my friends and those who have confidence in me, and I don't abandon my ancestors either,[70] the acknowledgment that I have from my ancestors.

[41] My ancestors were among those that came from Scandinavia and the British Isles, 1,000 years ago. Some of them went to Limerick and were called Reilson,[71] and then they crossed into Canada and they were called Riel; so in me there is Scandinavian, and well rooted; there is the Irish, and there is the French, and there is some Indian blood. The Scandinavians, if possible, they will have a share, it is my plan, it is one of the illusions of my insanity, if I am insane, that they should have on the other side of these mountains, a new Norway, a new Denmark, and a new Sweden, so that those who spoke of the lands of the great North-West to be divided into seven, forgot that it was in ten. The French in Manitoba, the Bavarians, the Italians and the Polands – the Poles and the Irish in the North-West, and then five on the other side too. I have written those things since I am in gaol, those things have passed through the hands of Captain Deane, they are in the hands of the Lieutenant Governor, and something of it has reached Sir John, I think, I don't know.[72] I did not hide my thoughts, I went through the channel of natural emigration, of peaceful emigration, through the channel of constitutional means, to start the idea, and if possible to inaugurate it, but if I can't do it during my life I leave the ideas to be fulfilled in the future, and if it is not possible, you are reasonable men

and you know that the interests that I propose are of an immense interest, and if it is not, if the peaceful channel of emigration is not open to those races into the North-West, they are in such numbers in the States that when you expect it least, they will perhaps try to come on your borders and to look at the land, whether it is worth paying it a visit or not. That is the seventh of lands, that is about the seventh of lands. So you see that by the very nature of evidence that had been given here when the witnesses speak of a seventh of the land, that very same question originates from 1870, from the troubles of Red River which brought a treaty where the seventh of the lands took its existence, and I say if this court tries me for what has taken place in the North-West they are trying me for something which was in existence before then. This court was not in existence when the difficulties of which we speak now in Saskatchewan, began; it is the difficulties of 1869, and what I say is, I wish that I have a trial.

9 CONCLUDING REMARKS

[42] My wish is this, your Honor, that a commission be [p. 369] appointed by the proper authorities, but amongst the proper authorities of course I count on English authorities, that is the first proper authorities; that a commission be appointed; that that commission examines into this question, or if they are appointed to try me, if a special tribunal is appointed to try me, that I am tried first on this question:

- Has Riel rebelled in 1869?
- Second question. Was Riel a murderer of Thomas Scott,[73] when Thomas Scott was executed?
- Third question. When Riel received the money from Archbishop Taché, reported to be the money of Sir John, was it corruption money?
- Fourth. When Riel seized, with the council of Red River, on the property of the Hudson's[74] Bay Company, did he commit pillage?
- Fifth. When Riel was expelled from the House as a fugitive of justice in 1874, was he a fugitive of justice?

As at that time I had through the member for Hochelaga, now in Canada, and through Dr. Fiset[75] had communications with the Government, but another time, through the member of Hochelaga, Mr.

Alphonse Desjardins,[76] I had asked from the Minister of Justice an interview on the 4th of March, and that interview was refused to me. In the month of April [1874] I was expelled from the House. Lepine was arrested in 1873, and I was not; not because they did not want to take me. And while I was in the woods waiting for my election Sir John sent parties to me offering $35,000 if I would leave the country for three years, and if that was not enough to say what I wanted, and that I might take a trip over the water, besides over the world. At the time I refused it. This is not the first time that the $35,000 comes up, and if at that time I refused it was it not reasonable for me that I should think it a sound souvenir to Sir John? Am I insulting? No, I do not insult. You don't mean to insult me when you declare me guilty; you act according to your convictions, I do also according to mine. I speak true. I say they should try me on the question, whether I rebelled on the Saskatchewan in 1885.

[43] There is another question. I want to have one trial; I wish to have a trial that will cover the space of fifteen years, on which public opinion is not satisfied.[77] I have heard, without meaning any offence, when I spoke of one of the articles I mentioned, some gentlemen behind me saying, "Yes, he was a murderer." You see what remarks. It shows there is something not told. If told by law it would not be said. I wish to have my trial, as I am tried for nothing; and as I am tried for my career, I wish my career should be tried; not the last part of it. [p. 370] On the other side I am declared to be guilty of high treason, and I give myself as a prophet of the new world.

[44] I wish that while a commission sits on one side a commission of doctors should also sit and examine fully whether I am sane, whether I am a prophet or not; not insanity, because it is disposed of, but whether I am a deceiver and impostor. I have said to my good lawyers, I have written things which were said to me last night and which have taken place to-day; I said that before the court opened. Last night the spirit that guides and assists me told me the court will make an effort – your Honor, allow me to speak of your charge, which appeared to me to go on one side – the court made an effort, and I think that word is justified. At the same time there was another thing said to me; a commission will sit; there will be a commission. I did not hear yet that a commission is to take place. I ask for it. You will see if I am an

impostor thereby. The doctors will say when I speak of these things whether I am deceiving. If they say I am deceiving, I am not an impostor by will. I may be declared insane because I seek an idea which drives me to something right. I tell you in all what I say in most things I do, I do according to what is told to me. In Batoche many[78] things which I said have already happened. It was said to me not far from here and that is why I never wanted to send the half-breeds far, I wanted to keep them, and it was said to me I will not begin to work before 12 o'clock, and when the first battle opened I was taking my dinner at Duck Lake. When the battle began it was a little after 12 o'clock. I will not begin to work before 12 o'clock, and what had happened? And it was said to me if you do not meet the troops on such a road, you will have to meet them at the foot of a hill, and the half-breeds facing it. It is said my papers have been published. If they have been published, examine what took place, and you will see we had to meet General Middleton at the foot of the hill. It was also told me that men would stay in the "belle prairie," and the spirit spoke of those who would remain on the "belle prairie," and there were men who remained on the "belle prairie." And he admits it was looked upon as something very correct in the line of military art, it was not come from me or Dumont, it was the spirit that guides me.

[45] I have two reasons why I wish the sentence of the court should not be passed upon me. The first, I wish my trial should take place as I said, whether that wish is practical or not, I bow respectfully to [p. 371] the court. I ask that a commission of doctors examine me. As I am declared guilty I would like to leave my name, as far as conscience is concerned, all right. If a commission of doctors sits and if they examine me, they can see if I was sincere or not. I will give them the whole history, and I think while I am declared guilty of high treason it is only right I should be granted the advantages of giving my proofs whether I am sincere, that I am sincere. Now, I am judged a sane man, the cause of my guilt is that I am an impostor, that would be the consequence. I wish a commission to sit and examine me. There have been witnesses around me for ten years, about the time they have declared me insane, and they will show if there is in me the character of an impostor. If they declare me insane, if I have been astray, I have been astray not as an impostor, but according to my conscience. Your Honor that is all what I have to say.

NOTES

1 "ascend" for "ascent."
2 "of" for "on."
3 "was" for "has been."
4 The two reasons are given in the last paragraph of the speech.
5 The Great Lakes.
6 "?" replaces comma.
7 Quotation marks inserted.
8 George Étienne Cartier (b. Lower Canada, 1814; d. London, England, 1873) was a prominent member of the federal government and considered one of the Fathers of Confederation (DCBO).
9 Donald Alexander Smith (b. Scotland, 1820; d. London, England, 1914) was an officer of the Hudson's Bay Company, a businessman, a member of the House of Commons from 1871 to 1880 and 1887 to 1896, and a philanthropist (DCBO).
10 Jean-Baptiste Thibault (b. Lower Canada, 1810; d. Quebec, 1879) was a delegate from the Canadian government to the Red River settlement (DCBO).
11 Alexandre-Antonin Taché (b. Rivière-du-Loup, Lower Canada, 1823; d. St Boniface, Manitoba, 1894) became bishop of St Boniface, Manitoba, in 1851 and archbishop in 1871 (DCBO).
12 Noël-Joseph Ritchot (b. Lower Canada, 1825; d. Manitoba, 1895) was a delegate from the provisional government in Red River to the Canadian government in 1870 (DCBO).
13 John Black (b. Scotland, 1817; d. Scotland, 1879) was a delegate from the provisional government in Red River to the Canadian government in 1870 (DCBO).
14 Alfred Henry Scott (b. ca. 1840; d. St Boniface, Manitoba, 1872) was a delegate from the provisional government in Red River to the Canadian government in 1870 (DCBO).
15 This bill, which listed twenty rights, was drawn up by the executive of the provisional government in March 1870 and sent with Red River delegates to Ottawa. See George F.G. Stanley, *The Birth of Western Canada: A History of the Riel Rebellions* (1936; reprint, Toronto: University of Toronto Press, 1960), 110–13; or Maggie Siggins, *Riel: A Life of Revolution* (Toronto: HarperCollins, 1994), 449–51.
16 In lieu of "the North-West, acknowledged my status, I am invited, and I come."
17 Joseph Howe (b. Halifax, 1804; d. Halifax, 1873) was the secretary of state for the provinces responsible for bringing Manitoba into Confederation (DCBO).

18 Comma inserted.
19 The Manitoba Act was passed by Parliament on 12 May 1870 and came into effect on 15 July 1870.
20 "delegations" replaces "delegates."
21 Riel refers to the fact that Manitoba, unlike the other provinces, was not given control of its natural resources.
22 "One of the conditions of the treaty was that the people of the North-West wanted the administration of their lands" replaces "One of the conditions was that the lands were that the people of the North-West wanted the administration of their lands."
23 "emigrant" occurs throughout this speech where "immigrant" appears to have been meant.
24 This sentence may be in response to a reaction from the audience.
25 The idea of "the seventh" and division of lands among nationalities was mentioned by several witnesses (*Q v LR*, 78, 160, 179, 203, 222, 237).
26 Comma replaces period.
27 This statement was met with laughter (*CWLR*, 545, line 218).
28 Pembina is a small town in North Dakota, just south of the Manitoba border. It was a Métis settlement on the route from Winnipeg to St Paul, Minnesota. See testimony of J.H. Willoughby (*Q v LR*, 78).
29 "prevision" for "provision."
30 Ambroise-Dydime Lépine (b. St Boniface, Manitoba, 1840; d. St Boniface, Manitoba, 1923), who participated in the 1869–70 uprising in Manitoba, was tried for the murder of Thomas Scott, found guilty, and sentenced to hang, but the sentence was commuted to a two-year prison term (*DCBO*).
31 A similar image occurs in the first speech, paragraph 21.
32 Quotation marks inserted.
33 Colon inserted.
34 Riel was twice elected from the federal riding of Provencher in Manitoba to the House of Commons, first in October 1873 and again in February 1874. The House voted to disqualify him from taking his seat in February 1875.
35 Quotation marks inserted.
36 "west" for "east."
37 Comma inserted.
38 Quotation marks inserted.
39 Period inserted.
40 "mother" replaces "honor."
41 Garnet Joseph Wolseley (b. Dublin, 1833; d. London, England, 1913) was the commander of the Canadian military expedition to Red River in 1870.

42 This statement was met with laughter (*CWLR*, 551, line 452).
43 Charles Nolin testified that Riel wanted an indemnity from the Canadian government of $35,000 (*Q v LR*, 194–5).
44 The Fenians were an Irish revolutionary group opposed to British rule in Ireland. American members attempted to invade Canada a number of times. Riel refers to October 1871.
45 Acheson Gosford Irvine (b. Quebec, 1837; d. Quebec, 1916) was a military man and civil servant. See Canada's Penitentiary Museum website, http://www.penitentiarymuseum.ca/default/index.cfm/history/cedarhedge1 (accessed 9 October 2013).
46 Adams George Archibald (b. Nova Scotia, 1814; d. Nova Scotia, 1892) was the first lieutenant governor of Manitoba, serving from 1870 to 1872 (*DCBO*).
47 "for the present circumstances."
48 Quotation marks inserted.
49 Quotation marks inserted.
50 Edward Blake (b. west of London, Upper Canada, 1833; d. Toronto, 1912) was the leader of the federal Liberal Party (the Opposition) from 1882 to 1887 (*DCBO*).
51 Quotation marks inserted.
52 "did" for "didn't."
53 Joseph Dubuc (b. Lower Canada, 1840; d. Los Angeles, 1914) was a supporter of Métis aspirations (*DCBO*).
54 Quotation marks inserted.
55 Quotation marks inserted.
56 Quotation marks inserted and sentence ended.
57 "my" for "me."
58 Quotation marks inserted.
59 Quotation marks inserted.
60 Quotation marks inserted.
61 Quotation marks inserted.
62 Quotation marks inserted and sentence ended.
63 "thirteen" for "three."
64 "thirteen" for "three."
65 "elected" for "erected."
66 Quotation marks inserted.
67 "regret" for "regard" (*CWLR*, 555, line 616).
68 "decay" for "decade" (*CWLR*, 556, line 622).
69 "lose" for "loose."

70 "ancestors either," replaces "ancestors, either."
71 Likely Riel intended "Rielson."
72 This statement was met with laughter (*CWLR*, 557, line 681).
73 Thomas Scott (b. Northern Ireland, 1842; d. Red River, 1870) was executed by the provisional government in Red River, of which Riel was the leader (*DCBO*).
74 "Hudson's" for "Hudson."
75 Jean-Baptiste Romuald Fiset (b. St Cuthbert, Canada East, 1843; d. Rimouski, Quebec, 1917) was a Liberal member of Parliament sympathetic to Riel (Wikipedia website).
76 Charles-Alphonse Desjardins (b. Terrebonne, Lower Canada, 1841; d. Terrebonne, Quebec, 1912) was a member of Parliament from 1874 to 1896 and a friend and supporter of Riel (*DCBO*).
77 Omitted: "I have heard, without meaning any offence."
78 "many" for "any."

I

The Queen vs Louis Riel: The Historical Context

DESMOND MORTON

I am now at the end of a career that some colleagues and probably rather more students consider to have been a bit too long. My work on Louis Riel was published at the beginning of that career, for the most part between 1972 and 1974, probably more to defend an ancestor, Colonel W.D. Otter of Cut Knife Hill, than to avoid perishing from post-thesis drought.[1] The intervening years have not passed without some learning. Among other details, I have discovered how passionately our ancestors followed every detail of the great trials of their day. Newspapers even looked away from politics to report, *in extenso*, what the butler saw or what the nursemaid overheard her mistress say as she bent over the body. Of all Victorian-era trials, was any more avidly followed and debated by Canadians than the *Queen v. Louis Riel*?[2] Even an intervening century has hardly reduced the fervour of Riel's allies and of his critics.[3] And for good reason: the trial saw some of Canada's most famous courtroom performers do life-or-death battle in a stifling Regina courtroom. The *Queen v. Louis Riel* combined politics and the law with the passion of a client who had faint hope of outliving his conviction and who faced the utter humiliation of being judged insane.

Riel's two years as a mental patient, first at Longue Pointe and then in Quebec's notorious Beauport Asylum, had taught him that there were worse fates than death – although the Crown's inference that he had been a self-seeking imposter rivalled the humiliation of being declared insane. Being insulted could not compare to what Riel recalled as the "animal existence" of an inmate of an insane asylum in the 1880s.[4]

A FEAST FOR LEGAL CONNOISSEURS

The Regina trial provided a feast for a nation of judicial connoisseurs. Although the setting was a stifling, prefabricated former police office of the territorial government, the Crown was led by Christopher Robinson, a worthy son of that pillar of Canadian High Toryism John Beverly Robinson, the prosecutor in Canada's other famous series of treason trials, the Ancaster Assizes of 1813, and by Britton Bath Osler, a Liberal, linked as a brother to Canada's best known physician. Penniless and confined to the North-West Mounted Police (NWMP) barracks in Regina, Riel had neither means nor opportunity to provide for his own defence. However, the friends of his Montreal youth and Quebec political allies of his cause mobilized the funds to finance a notable team from the Quebec Bar: François-Xavier Lemieux, Charles Fitzpatrick, and James Greenshields. In the years to come, when Wilfrid Laurier and the Liberal Party displaced the Conservatives from power, Lemieux would become chief justice of Quebec, Fitzpatrick would become Laurier's minister of justice and a future chief justice of the Supreme Court of Canada, and Greenshields, a McGill University gold medallist, would become chief justice of the Superior Court of Quebec. Called to the bar in 1877, he would soon compile a brilliant series of acquittals in Montreal murder cases.[5]

As events evolved, Riel might have felt better served if his brilliant lawyers had stayed home. As he complained from the dock, while government lawyers were trying to prove him guilty of a capital offence, his own counsel were trying to prove him insane enough to spend the rest of his existence in a lunatic asylum. If his friends had succeeded, is it conceivable that anyone more than a century and a quarter later would still take much interest in the case of the *Queen v. Louis Riel*, much less erect the massive monuments on Legislature grounds that now honour his memory and his role in Manitoba and Saskatchewan history? Would historical re-enactors find more than token audiences for replays of an insane man's arguments in his own defence?

RIEL'S HISTORICAL TRANSFORMATION

Over the many years since his execution, Riel's reputation has slowly been transformed. In 1885 both Liberal and Conservative newspapers across English-speaking Canada revelled in the verdict and sentence.[6] Times have changed. Perhaps a collective sense of guilt over the non-fulfilment of the Métis land claims embodied in the Manitoba Act of 1870, and over the tragic experiences of Canada's First Nations, has earned their advocate, Louis Riel, a new and much more sympathetic hearing from twenty-first-century Canadians. Both in his Depression-era reinterpretation of the history of western Canada and in his biography of Riel, George F.G. Stanley offered English-speaking Canadians a far more sympathetic understanding of the Métis leader than they would ever find in the provincially approved history textbooks of mid-twentieth-century Canada.[7] In recent years, Stanley's version of Riel has largely been reinforced by the research of J.M. Bumsted and of Bob Beal and Rod MacLeod and by D.N. Sprague's *Canada and the Métis, 1869-1885*, not to mention the proselytizing prose of Maggie Siggins.[8]

Canadians in 1885 lived in a harsher, more judgmental environment than their twenty-first-century heirs. Death is no longer a penalty in Canadian criminal law, whatever public opinion may apparently wish. Even in 1885, the more savage the penalty, the more likely a jury to balk at conviction. That factor was probably at work in Regina by the end of July. Even before the reign of Queen Victoria began in 1837, the antique penalty for high treason, laid out in the Statute of Treasons of 1352,[9] had been abandoned. "Drawing and quartering" (chopping apart a traitor's corpse) had passed into abeyance, allowing the villain merely to be "hanged by the neck until dead."[10]

DID RIEL'S AMERICAN CITIZENSHIP MATTER?

With an eye to rhetoric, it is worth considering the medieval language of the charge read out by the clerk of the court, Dixie Watson, to which Riel and his counsel had to make answer:

> Louis Riel, then living within the dominion of Canada and under the protection of our Sovereign Lady the Queen, not regarding the duty of his allegiance nor having the fear of God in his heart,

> but being moved and seduced by the instigation of the devil as a false traitor against our said Lady the Queen, and wholly withdrawing the allegiance, fidelity and obedience which he should, and of right ought to bear towards Our Lady the Queen ... together with divers other false traitors ... most wickedly, maliciously and traitorously did levy and make war against our said Lady the Queen ... and did then maliciously and traitorously attempt and endeavour by force and arms to subvert and destroy the constitution and government of this realm, as by law established.[11]

An exile from Canada, Riel had applied for and been granted United States citizenship. It might have been expected that Washington would intervene on his behalf. We know that American friends he had made during his sojourn in Montana expected such intervention. In fact, precedent and William Blackstone's legal teaching made Riel's foreign citizenship irrelevant. As Alan Taylor reminds us in his history of the War of 1812, Americans held boldly to the principle that national allegiance was a matter of personal choice.[12] The British at the time insisted that citizenship was determined solely by birthplace. British-born naval and military deserters captured in American uniforms and ships companies were traitors and subject to the penalties of treason. More important for Riel, British law upheld the principle of "natural allegiance," compelling anyone resident in a country to respect its laws as part of a "debt of gratitude" owed by those who enjoyed the "peace, order and good government" of that country, regardless of citizenship. By coming to Batoche in 1884, Riel had subjected himself unequivocally to the laws of Canada and the British Empire.

If Riel was required by the rule of natural allegiance to face trial in Canada, did it have to be in Regina? Relying on an imperial statute promulgated in 1821, Riel's lawyers initially insisted that a capital offence would have to be tried either in an older, established province such as Ontario or in British Columbia, a province since 1871. Nor was it legitimate to try a capital offence without a prior grand jury hearing. Christopher Robinson demolished the argument by citing subsequent imperial statutes that gave Ottawa jurisdiction and a Canadian statute of 1880 that permitted capital offences to be tried in the North-West Territories.[13] Although that might have placed the venue at Prince Albert or Battleford, the newly named capital of the territories was a perfectly legal choice. Indeed, Judge Hugh Richardson had proved the point when, a year before, his sentence of hanging for a convicted murderer named Conner or Kohner was upheld on appeal to Manitoba's Court of Queen's Bench. In fact, Kohner had already used the gibbet on which Riel's life would end.

Well before 1885, the Canadian Parliament had decided that if it did not choose to load up the venerable blunderbuss of Edward III's Statute of Treasons, it could turn to the lesser crime of treason-felony. Canada had its own Act for the Better Security of the Crown and Government of 1868,[14] a milder version of Edward III's statute since a death sentence was not mandatory. This was the law under which other prominent Métis and Natives would ultimately be charged for their role in the 1885 rebellion, and most would be punished with a few years of imprisonment in a federal penitentiary.[15] Another treason law could have been fitted to Riel as a United States citizen. During the Fenian Raids, Parliament had adopted and adapted the Fenian Act of 1866,[16] making it treasonable for a citizen of a nation at peace with Canada to make war on Canada. This could have been applied to Louis Riel, but, like Canada's own treason-felony statute, it avoided the automatic death penalty Sir John A. Macdonald was determinted to exact.[17]

CANADA'S LEGAL STRATEGY IN 1885

In the wake of the 1885 rebellion, the federal government's legal strategy was to be as lenient as possible to the Métis and Native participants since Ottawa wanted to argue that they had been duped by a cunning and thoroughly sane Louis Riel. At the same time, the Crown would use the full rigour of the criminal law against Native offenders convicted of murder in the Frog Lake killings and other fatal encounters during the winter and spring of 1885. In this policy, Ottawa found an enthusiastic agent in Judge Charles A. Rouleau, the stipendiary magistrate based at Battleford, whose house and library had been burned during the Cree invasion of Battleford and who, during the siege, had fled south in terror of his own and his family's life. Rouleau set about the trial and conviction of accused Natives with more than a hint of vengefulness, and even a cursory view of his proceedings in Battleford reveals Natives accused without representation or even translation into their own language and a series of lengthy prison terms that frequently proved fatal to their Native victims.[18]

Ottawa's strategy for dealing with Riel was equally unforgiving, but the political risk of executing someone who was again becoming a folk hero in Quebec might be eased if the government could locate some co-conspirators among the Ontarion-born English-speaking settlers in Prince Albert and neighbouring communities. The decision by the Canadian Pacific Railway to adopt a southern route through Regina and Swift Current had devalued the property of those who had expected the railway to encourage homesteading and land sales in or near Saskatchewan's Park Belt. Many of the aggrieved had welcomed Riel's return as much as the Métis.[19]

Recognizing the longstanding hostility between Métis and Sioux in Manitoba in 1870, Riel had kept potential Native supporters at a distance. By 1885 he seems to have concluded that his policy had been unwise. Once at Batoche, he generally ignored his possible allies at Prince Albert and sought support primarily among Native bands whose lands bordered on the Métis settlements and whose grievances had been growing steadily. These tactics were not necessarily successful since Native leaders had no good reason to have confidence in Métis success. However, hard times and broken promises had divided many bands; dissatisfied members erected warrior's lodges and looked forward to profiting from any conflict between whites and Métis.[20]

No sooner had Riel despatched his envoys to Native bands than white settler support cooled. The terror of an Indian war seized all but a very few settlers. It also alienated any possible sympathy from official Washington. Unlike in 1870, when Riel had seemed a possible agent of annexation, American authorities offered no resistance and even some encouragement to Canadian efforts to despatch supplies via Chicago and St Paul. Senior officers like General Frederick Middleton, cases of arms and ammunition, and even some small militia units travelled to Winnipeg by way of the United States. Ottawa's policy in the North-West was designed to show who was in charge north of the 49th parallel, but there was no perceptible resistance from Washington. Only the financial crisis of the Canadian Pacific, and considerable exaggeration of the capacity of its unfinished rail line to deliver troops, horses, and guns to the Northwest, persuaded Ottawa to move its expeditionary force on its side of the international border and rescue its major investment from bankruptcy.

MATCHING RIEL WITH WHITE REBELS

Although Robinson and Osler attempted to fulfil the government's plan to find co-accused among the white settlers of Prince Albert, both lawyers soon concluded that it was futile. Posterity has not challenged their judgment.[21] Only W.H. Jackson, the secretary of the Settlers' League, an excitable young Ontarian who volunteered to become Riel's secretary, who converted to Catholicism, and who ended, like others in Batoche, as one of the prisoners confined to the local rectory, could plausibly be considered a co-conspirator. But Riel, not Jackson, was and remains the key leader in the Métis rebellion of 1885. After a one-day trial, Jackson was pronounced insane and sent to the Selkirk Asylum in Manitoba, where he soon escaped to the United States. He spent the balance of his life as Honoré Jaxon, a harmless and penniless labour organizer and socialist militant in Chicago and New York.[22] Far from appeasing Quebec, Jackson's treatment – contrasted to Riel's conviction and execution – only seemed to prove the gross favouritism Ontario Protestants enjoyed wherever the federal government was in control. Charges against other anglophones were systematically dropped for lack of evidence or because they co-operated as Crown witnesses against Riel.

Language was another issue at Riel's trial. His judge, jury, and prosecutors spoke only English. In consequence, for the most part, so too did Riel and his legal counsel. Could the judge, jury, and Crown counsel understand testimony in French? The court was provided with an interpreter, but the trial transcript, from which we must assess most issues of argumentation and rhetoric, was composed virtually entirely in English. Given the nature and technology of court reporting in Canada in the 1880s, it would be surprising if some passages were not composed or at least interpreted in part by the court stenographer. Journalists covering the trial reported that Riel sometimes spoke passionately and rapidly. At other points, his voice slowed to give his words greater impact. A newspaper account of the trial claimed that Riel, in his address to the full court, spoke for two hours;[23] an unhurried reading of the official transcript of his speech lasts a little over an hour. Although the heat in the courtroom undoubtedly made minutes drag, was anything omitted by the stenographer? There is now no evident way to find out.

RIEL'S MENTAL STATE

A core question for both Riel and posterity was whether the Métis leader was insane or, as the Crown insisted, a self-promoting imposter, arousing a civil war to lever a further bribe out of the federal government? "My name is Riel," he had allegedly told Lawrence Clarke, a Hudson's Bay Company factor and representative for Lorne, the Prince Albert district in the territorial assembly, "and I want Material."[24] Riel's lifelong fondness for puns gives the story a coincidental verisimilitude, and the sum that might have persuaded him to disappear back to Montana has usually been set at $35,000. Clarke had his own interest. Riel's presence was an argument for stationing more NWMP in his district so that he and other merchants could profit from their business.[25]

Riel had left Montana without any material guarantees, summoned by people who hoped he could repeat his 1870 feat of transforming a colony into an authentic, if somewhat expensive, province of Canada called Manitoba, while securing Métis land claims. He was legitimately concerned with the welfare of his wife and that their two, and soon to be three, children would be left in utter destitution. As a prisoner in the police barracks in Regina, Riel was heavily guarded against both his own potential escape and any plots to rescue or to murder him. His jailer, Captain Richard Burton Deane, the adjutant of the NWMP, loaded Riel with a heavy ball and chain whenever he was allowed to exercise in the barrack yard. As some compensation, Deane allowed his prisoner to use the NWMP commissioner's office as a workroom. Soon, Riel was busy drafting letters, composing religious poetry, and struggling to find ways to support his family. Would Deane allow him to be photographed, perhaps with his ball and chain? The prints could be sold at a price to admirers, critics, and the merely curious. Deane refused. Riel also wrote to the prime minister, seeking a grand state trial before the Supreme Court of Canada. In return, he pledged to do everything he could to protect the Macdonald government and to embarrass the Liberal Party, which he blamed for his persecution and penniless exile in the United States. Riel's reward for his political service would be his restoration to Manitoba politics and what he termed "a fair indemnity." Mindful of his jailer, Riel also reported that Captain Deane might make an ideal lieutenant governor for Manitoba when he became premier. Needless to add, Macdonald did not make Riel's letters public.[26]

DID RIEL RECEIVE A FAIR TRIAL?

After Riel was brought to his camp as a prisoner, General Middleton's first instructions from his minister, Sir Adolphe Caron, was to despatch him to Winnipeg, the biggest city in the Canadian West and part of an established province. Caron's boss, Sir John A. Macdonald, countermanded the order. Under the law in 1885, Riel had to be tried in the North-West Territories, whose capital was now Regina. It might also be a concern that Manitoba law promised the accused a twelve-member jury and, if the accused so chose, a jury evenly divided between English speakers and francophones. The likelihood of a unanimous conviction by such a jury was remote, although the change of location seems to have reflected the law, not opportunism.[27]

Telegrams promptly redirected Riel and his escort straight south to Regina, a town known recently as Pile of Bones Creek until it became the improvised home of both the NWMP and the territorial capital. In the cells at the police barracks, Riel would be safe from both rescue and assassination. (Both threats could realistically be contemplated.) When Riel's lawyers reached Regina, they and their sponsors were dismayed at the prospect of spending weeks and months in a tiny, hot, and primitive prairie town. On Riel's behalf, they demanded a month's delay so that they could collect witnesses, documents, and evidence. Judge Hugh Richardson, one of only three stipendiary magistrates in all of the North-West Territories, was British-born, a small-town Ontario lawyer and militia officer from Woodstock, briefly a Liberal-appointed civil servant, and now the life-and-death judge of Louis Riel.[28] The verdict, under territorial law, would be pronounced by a jury of only six men chosen from a panel of thirty-six. Riel's lawyers ignored their client's demands for trial by Canada's Supreme Court, but they did demand a change of venue for the trial to an older province where their friend's fate could be settled by twelve "good men and true." Next they insisted that witnesses like Gabriel Dumont as well as several Catholic missionaries, not to mention several senior civil servants from Ottawa, should be made available for cross-examination. Richardson welcomed such witnesses, but he had to confess that he could not guarantee Dumont and other prominent rebels would be secure from arrest and prosecution. Fathers Alexis André and Vital Fourmond could come in safety since they had been Riel's prisoners at Batoche and would be immune from prosecution. The senior civil servants could remain at their desks in Ottawa.[29]

CONTEMPORARY JUDGMENTS

If Captain Deane believed that Riel's letter to the prime minister was sufficient to prove Riel's madness, Macdonald would not have been pleased. Evidence of insanity could save Riel's life and aggravate the roaring chorus of voters who wanted him dead. In 1870 and again in 1885, the Métis leader had imposed the cost and embarrassment of major military expeditions as the price of preserving the young Dominion *a mari usque ad mare*. In 1870 Riel had justified the execution of Thomas Scott as the price of forcing Canada to respect his authority over the Red River. For Sir John A. Macdonald, Riel's life had become the price of keeping Canada united across its vast and restless western plains. In the earlier crisis, Macdonald had purchased Riel's disappearance with a discretionary payment from his government's secret service fund. Paid to disappear, Riel had returned to Batoche in 1884 to fish in troubled waters. An earlier return to friends in Quebec ended in a two-year stay at the Beauport Asylum when his behaviour of roaring like a bull and ripping up his clothes and bed sheets had exhausted the patience of his friends and relatives.[30] A somewhat calmer Riel had found work as a school teacher at a Catholic mission in Montana. As seen from Ottawa, Riel seemed entirely capable of manipulating Métis and Natives into a murderous rebellion. Devout Catholic missionaries might regard his heretical claims to be the prophet of the New World as a form of insanity, but few Protestants were remotely interested in Riel's religious aberrations. In the end, Fathers Alexis André, Louis Cochin, and Vital Fourmond could claim that they had brought Riel back to orthodox Roman Catholicism in return for the promise of a Christian burial.[31]

In the summer of 1884, Gabriel Dumont and other leading Métis had come south from Batoche to persuade the hero of their 1870 struggle to bring his classical and legal education back to their service as spokesman for the Métis claim to one-seventh of the arable land of the North-west Territories. It appears that Riel's Métis hosts were unaware of the religious visions that had fed the aggressive and seemingly unbalanced behaviour that had landed Riel in two successive Quebec asylums and that would reappear at Batoche when he sought to explain and to fulfil his divine mission as the prophet of the New World. In Ottawa, Macdonald saw Riel as a man who had betrayed the prime minister's generosity and shattered his tolerance. Riel might arouse a popular following in Quebec, but Macdonald put his faith in the power of the Catholic Church to suppress support for a man who had imprisoned Catholic missionaries in their own rectory and who had reviled Archbishop Alexandre-Antonin Taché of St Boniface as a common thief for his land dealings. Uproar in Quebec would soon die away. In contrast, Orange Ontario's hatred of the murderer of Thomas Scott was unassuaged after fifteen years. Their spokesmen in Cabinet and caucus reminded Macdonald that their voters would never forgive him if Louis Riel escaped the gallows or even ended his days in an insane asylum.

A FINAL EXAMINATION

Was Riel really insane? A present-centred perspective may become irrelevant or confusing. Psychiatry was hardly an exact science in 1884, whatever may be claimed for it now. Riel's jury heard medical testimony and opinions that had clearly failed, despite the best efforts of Riel's counsel, to convince them that the Métis leader's insanity met the test of the oft-mentioned, seldom-read M'Naghten Rules.[32] The two soldiers who had come to know Riel in the days and weeks after his surrender, General Middleton and Captain George Holmes Young, reported him to be sane and articulate. They were unlikely to have raised religious or political topics.

Long after the trial and its verdict and sentence had provoked outrage in Quebec, Macdonald conceded one of the demands Riel had made in his final address to the court. A commission was appointed to examine his mental state. Two doctors from eastern Canada, Michael Lavell, warden of Kingston Penitentiary, and François-Xavier Valade of Ottawa, who earned part of his income as a food inspector for the federal government, would join surgeon Andrew Jukes of the NWMP in Regina.[33] Their job was to assess Riel's current mental state as his date of execution approached.

Since the trial verdict had been upheld in appeals all the way from Manitoba to the Judicial Committee of the Privy Council in London,

Macdonald explained to Lavell, "You cannot therefore go beyond that verdict, and your enquiry will be limited to the simple question of whether he, at the time of your report, is sufficiently a reasonable and accountable being to know right from wrong."[34] Jukes, the police surgeon, had already made up his mind. He compared Riel to Immanuel Swedenborg, an eighteenth-century religious mystic whom Jukes had found persuasive. By that comparison, Riel was sane.

Lavell, an obstetrician, was impressed by Riel's concern for his own children, an allegiance that had recaptured priority in Riel's mind over the Métis leader's concern for his place in history. Like Jukes, Lavell was drawn to Riel as a man who had delusions about religion and politics but otherwise was normal and even, intellectually, impressive.

Valade, the sole French Canadian with any direct role in Riel's fate, was the most critical of the three experts. Valade was more upset than Lavell or Jukes about Riel's religious heresies and hallucinations, but like his colleagues, he was attracted by the man's personality. "I confess," he wrote to Macdonald, "I should be well pleased if justice and popular clamour could be satisfied without depriving this man of life." In contrast to Lavell's report, Valade concluded that Riel "is not an accountable being, that he is unable to distinguish between right and wrong on political and religious subjects, which I consider well marked typical forms of a kind of insanity under which he undoubtedly suffers, but on the other hand, I believe him to be quite sensible and can distinguish right from wrong."[35] Governor Edgar Dewdney despatched the Lavell and Valade opinions to Macdonald, concluding that the two experts had reached the same conclusion with only minor semantic differences.[36]

In his major speech at the trial, Riel had given priority to establishing his own sanity. Forced by Judge Richardson to choose between retaining his counsel or being left to organize his own defence, Riel had kept his lawyers and forfeited any right to cross-examine witnesses. That forced him to postpone any rebuttal of their testimony until he could address the jury in the last moments of his trial.

Riel had opened his speech to the jury with an apology for speaking English: "I cannot speak English very well, but am trying to do so, because most of those here speak English."[37] He next prayed to God, adding his hope that the jury would not "take it as part of a play of insanity."[38] Earlier, Riel's estranged cousin, Charles Nolin, had testified that Riel's visions took a physical form, covering his face with sweat and making his limbs palpitate. In his speech, Riel explained that such symptoms were common among those he referred to consistently as "half-breeds." His people, he explained, had much intuition about how tomorrow's hunt or tonight's meeting would turn out, and physical symptoms – twinges, sweating, shaking – were natural and normal experiences. His voices had told him on the night before his own day of testimony on 31 July that "somebody will come *t'aider*." That he could look forward to help in both languages, he claimed, had proved its veracity. The message itself had been confirmed for Riel on his way to court on the following day when police surgeon Jukes had quietly informed him that Judge Richardson would give him a chance to speak for himself, despite the opposition of his own Quebec lawyers.[39]

THE IMPACT OF RIEL'S SPEECH TO THE JURY

Anyone who reads Riel's resulting speech, delivered to a packed and sweltering courtroom, can see that it served Macdonald's interests better than Riel's. Certainly, Riel did not sound like a mad man, nor was he nearly as coherent about Métis or settler grievances as he would be on the following day, when Judge Richardson invited him to respond to the jury's guilty verdict. Riel's biographer, George Stanley, suggests that fear of being found insane, as his lawyers hoped, led Riel to improvise a speech but that, in his confusion and desperation, Riel had disorganized the fragmentary notes he had prepared before his trial.[40] When he used his notes in his postverdict speech, it was too late to influence anyone but the frigidly hostile Judge Richardson. Possibly, Riel's reminder of the state of his penniless family had evoked sympathy that led the jury to weaken their guilty verdict with a plea for mercy. Just as probably, this plea was included as a necessary accommodation of a recalcitrant jurist when unanimity was required for conviction. One of the Crown's lawyers, Britton Bath Osler, was later scathing about both the judge and the jurors, but prosecutors are not selected for their broad and generous opinions of the accused or his sympathizers.

A century and a quarter after Riel's trial, most Canadians have forgotten or never cared to know much about Louis Riel, but the majority of informed opinion would probably prefer to claim him as an innocent, if ineffective, advocate of a deeply aggrieved Métis nation and of their First Nations neighbours. What kind of war did he levy on the Queen when NWMP superintendent Lief Crozier rode to Duck Lake with a hundred armed police and special constables? In a more secular age, we may be tempted to regard Riel's visions and beliefs as evidence of insanity more than his assumptions of power over his own people. If Riel described his early benefactor Archbishop Taché as a thief for robbing a widow of her land, modern memories of sexual abuse by clergymen might draw public sympathy for Riel's claim that soft words encouraged the powerful to ignore their crimes, whereas harsh words could encourage a beneficial honesty.[41]

"The past," claimed the British novelist L.P. Hartley, "is a foreign country. They do things differently there." Canada in 1885 was not modern-day Canada. Failure to execute Riel, warned Manitoba premier John Norquay, would cost Sir John A. Macdonald's party any hope of re-election in 1887. In 2014 Macdonald's ingenious and remorseless prosecution of Riel seems an error in political judgment wholly comparable to Riel's execution of Thomas Scott in 1870. How could Riel know that his victim, Scott, would become a martyr for the Orange Order?

Contemporary history offered Sir John A. Macdonald an instructive warning about the danger of hasty and political executions. In 1881 Daniel Guiteau, a lawyer with serious symptoms of insanity, assassinated President James Garfield. In the heat of public outrage, Guiteau was swiftly tried, condemned, and executed. Outrage soon cooled, to be replaced by growing condemnation for the way an insane man had been hurried to judicial murder. By 1885, as J.M. Bumsted reminds us, the Guiteau case was widely regarded as a travesty of justice.[42] As close bystanders of American legal processes, Canadians might have taken note.[43]

Like his successors, from Wilfrid Laurier to Stephen Harper, Sir John A. Macdonald might also have remembered that it was no accident that in 1883 Quebecers had adopted "Je me souviens" as their provincial and national motto. Riel's execution made him part of the contested collective memory of both Canada and Quebec.

NOTES

1 In research for my doctoral dissertation, I had discovered an unpublished parliamentary return of messages between the defence minister, Sir Adolphe Caron, and others engaged in the 1885 campaign; see Desmond Morton and Reginald H. Roy, eds, *Telegrams of the North-West Campaign, 1885* (Toronto: Champlain Society, 1972), which includes most of the messages and was published with much assisstance from Barbara Wilson of the Public Archives. This led to *The Last War Drum: The North-West Campaign of 1885* (Toronto: Hakkert, 1972), published for the Canadian War Museum. Finally, I wrote *The Canadian General: Sir William Otter* (Toronto: Hakkert, 1972).

2 I published an introduction to and transcript of the Riel trial for the Canadian Social History series, *The Queen v Louis Riel* (Toronto: University of Toronto Press, 1974).

3 Among the many recent accounts of Louis Riel and his fate, one may cite the serious analytical work on the fate of Thomas Scott collected in J.M. Bumsted, *Thomas Scott's Body and Other Essays on Early Manitoba History* (Winnipeg: University of Manitoba Press, 2000); Bob Beal and Rod MacLeod, *Prairie Fire: The 1885 North-West Rebellion* (Edmonton: Hurtig, 1984), with its particularly fine account of the trials of Métis and Native accused; and D.N. Sprague, *Canada and the Métis, 1869–1885* (Waterloo, ON: Wilfrid Laurier University Press, 1988). Thomas Flanagan takes an unfashionably critical view of Riel in *Riel and the Rebellion: 1885 Reconsidered* (Saskatoon: Western Producer Prairie Books, 1983), but he shows more sympathy as editor and commentator of *The Diaries of Louis Riel* (Edmonton: Hurtig, 1976), among other works published by the Riel Project under the supervision of George F.G. Stanley. A more recent and valuable view of the Métis leader appears in Lewis H. Thomas's biography of Riel in the *Dictionary of Canadian Biography*, http://www.biographi.ca/en/bio/riel_louis_1844_85_11E.html (accessed 10 October 2013). Another sympathetic work is Maggie Siggins, *Riel: A Life of Revolution* (Toronto: HarperCollins, 1994).

4 See Riel's comment in midtrial, in Morton, ed., *Queen v Louis Riel*, 212.

5 On prosecution and defence counsel, see *Macmillan Dictionary of Canadian Biography* (Toronto: Macmillan, 1978), 262 on Fitzpatrick, 315 on Greenshields, 456 on Lemieux, 633-4 on Osler, and 713 on Robinson.

6 Toronto's *Globe* led other Liberal papers in adding the suspicion that the Conservatives would now allow Riel to be spirited away, and such a plot may well have been planned by a wealthy Regina Métis, Pascal Bonneau. It came to nothing. See Beal and MacLeod, *Prairie Fire*, 303–4.

7 George F.G. Stanley, *The Birth of Western Canada: A History of the Riel Rebellions* (Toronto: Longmans Green and Company, 1936), reversed many of the conventional interpretations of Louis Riel in the academic history of Canada. Stanley reinforced his position with the biography *Louis Riel* (Toronto: McGraw-Hill Ryerson, 1963), which was reprinted five times by 1972. See also note 3.

8 See note 3 for the contributions of these researchers.

9 15 Edw. 3, stat. 5, cap. 2.

10 On the versions of treason available to Canadian courts, see Beal and MacLeod, *Prairie Fire*, 294–6.

11 Charges against Riel were laid on "The information and complaint" of Alexander David Stewart, the chief of police in Hamilton, Ontario, and, as it happened, a brother-in-law of Colonel W.D. Otter. It was read by Dixie Watson, clerk of the court. See Morton, ed., *Queen v Louis Riel*, 3–7. The text of the charge is drawn from Stanley, *Louis Riel*, 345.

12 See Alan Taylor, *The Civil War of 1813: American Citizens, British Subjects, Irish Rebels and Indian Allies* (New York: Alfred A. Knopf, 2010), 3–5, passim.

13 See also Beal and MacLeod, *Prairie Fire*, 296.

14 31 Vict., cap. 69.

15 At the opening of the proceedings on 18 July 1885, Britton Bath Osler for the Crown explained the legitimacy of the trial process, including the absence of a grand jury and the small size of the trial jury. See Morton, ed., *Queen v Louis Riel*, 64–8.

16 Fenian Act of 1866, or An Act to Continue for a Limited Time the Several Acts Therein Mentioned, SC 1868 (31 Vict.), c. 29.

17 J.M. Bumsted, *Louis Riel v. Canada: The Making of a Rebel* (Winnipeg: Great Plains, 2001), 307–9. Passing through Winnipeg, on his way to organize the evidence against Riel, police chief Stewart chatted with a reporter for the *Winnipeg Sun*: "He intended to get all the evidence he could against Riel, and by any means, for ...," said the chief with a sly laugh, "I guess the idea is to hang him." Cited in Beal and MacLeod, *Prairie Fire*, 296.

18 Stanley, *Birth of Western Canada*, 194–242. The best account of the trials of the Métis, white, and Native accused is in Beal and MacLeod, *Prairie Fire*, ch. 17.

19 Morton, *Last War Drum*, 16–24. Among the aggrieved were Lieutenant Colonel A.T.H. Williams, the member of Parliament for East Durham, who had invested heavily in Park Belt land but who had put together a militia battalion for service in the North-West. His anger against Riel helped to push him into leading the attack on 12 May that ended the Canadian siege

of Batoche. He died later in the expedition, having made no secret of his contempt for General Middleton's cautious leadership. Williams's Tory successor held East Durham in a by-election but warned Macdonald, "God help him next time if he don't hang Riel." See ibid., 159; and H.A. Ward to Macdonald, 26 August 1885, Library and Archives Canada (LAC), Macdonald Papers, vol. 108, 43418.

20 Older Native chiefs like Big Bear, One Arrow, and Poundmaker opposed younger, more radical elements in their bands who formed warrior lodges and, despite their many grievances, showed strong misgivings about trusting their futures to Louis Riel. At Cut Knife Hill, Poundmaker had intervened to allow the Canadian troops to withdraw. It won him life-long respect from Otter and a prison sentence from Judge Rouleau. The Cree chief's photo remained part of Otter's office until he died. Big Bear had interceded with his warriors to protect the widows of men they killed. He, too, went to prison. Both men did not long outlive their captivity.

21 The trial of Tom Scott, a white political leader of the Métis, ended in acquittal. See Beal and MacLeod *Prairie Fire*, 316–22.

22 On Honoré Jaxon's subsequent career, see Beal and MacLeod, *Prairie Fire*, 339–40.

23 See Riel's final address to the court in Morton, ed., *Queen v Louis Riel*, 311–25.

24 Riel's desired "material" was a $35,000 down payment on the $100,000 indemnity he believed he had negotiated for all Métis families in Manitoba as compensation for giving up their land for the future province. See Flanagan, *Riel and the Rebellion*, ch. 5.

25 On Clarke, see Sprague, *Canada and the Métis*, 162–3; on the material grievances of the Métis, see ibid., 157–66.

26 Riel to Sir John A. Macdonald, 26 July 1885, LAC, Macdonald Papers, vol. 107, 43192-207. Although Riel had described Captain Deane as fit to become lieutenant governor in Manitoba, "ce noble officier" concluded that Riel's communications with the prime minister proved only that he was "cracked." See R.B. Deane, *Mounted Police Life in Canada: A Record of Thirty One Years' Service* (London and Toronto: Cassell and Co., 1916), 190. On the legal background of any plea of insanity, see Beal and MacLeod, *Prairie Fire*, 299–301.

27 On the hurried change of destination for Riel, see Morton and Roy, eds, *Telegrams of the North-West Campaign*, 288, 302, 308, 309, 310. On motivation, see Beal and MacLeod, *Prairie Fire*, 293–4.

28 On Judge Richardson, see *Macmillan Dictionary of Canadian Biography*, 701.

29 On the trial venue, see Flanagan, *Riel and the Rebellion*, 122–3.
30 On Riel's condition in 1876–78, see Stanley, *Louis Riel*, 224–30.
31 Morton, ed., *Queen v Louis Riel*, xxvi.
32 Flanagan, *Riel and the Rebellion*, 129–31.
33 On the medical commission, see Lewis H. Thomas's biography of Riel in the *Dictionary of Canadian Biography*, cited at note 3. Beal and MacLeod, *Prairie Fire*, 336–7, attribute the interpretation of the Valade and Lavell reports to Governor Dewdney.
34 Macdonald to Lavell, 31 October 1885, Public Archives of Canada, Macdonald Papers, vol. 106, 42638.
35 Valade to Macdonald, 8 November 1885, Public Archives of Canada, Macdonald Papers, vol. 106, 42650-1.
36 Thomas Flanagan's harshest criticism of Macdonald arose from the "Old Chieftain's" acceptance of Dewdney's misrepresentation of Valade's report on Riel. Valade's horror of Riel's heresies battled with his unexpected discovery of Riel's intelligence and charm. Father André, the charismatic Oblate who persuaded Riel to submit to church doctrine, also convinced Valade not to become an accessory in Riel's execution. Flanagan, *Riel and the Rebellion*, 143, states that Macdonald and his colleagues "resorted to forgery to conceal, or at least disguise Valade's original opinions." See also Morton, ed., *Queen v Louis Riel*, xxv–xxvi; and Beal and MacLeod, *Prairie Fire*, 337.
37 Morton, ed., *Queen v Louis Riel*, 311.
38 Ibid., 311–12.
39 Ibid., 320.
40 Stanley, *Louis Riel*, 358.
41 Morton, ed., *Queen v Louis Riel*, 318–19.
42 Bumsted, *Louis Riel v. Canada*, 283.
43 I owe the Guiteau analogy to Beal and MacLeod, *Prairie Fire*, 304–5.

2

"Through the Grace of God I Am the Founder of Manitoba": Louis Riel's Constitutional Thought

NICOLE C. O'BYRNE

The agitation in the North-West Territories would have been constitutional, and would certainly be constitutional to-day if, in my opinion, we had not been attacked.[1]

There is, perhaps, no figure in Canadian history more enigmatic than Louis Riel. For more than a hundred years, his legacy has been the subject of intense debate.[2] In this volume, several leading scholars analyze Riel's speeches at his trial in order to gain further insight into the inner workings of this complex man. In this chapter, I explore a relatively neglected area of study: Riel's understanding and reliance upon British constitutional law. During his trial, Riel's speeches were peppered with references to concepts such as responsible government, public liberties, and representation.[3] Understanding these constitutional ideas is key to understanding Riel's dispute with the Canadian government over Manitoba and the North-West. In the context of the trial, Riel's constitutional references were often cursory, and he did not explain his usage or understanding. However, Riel's understanding of the foundational tenets of British constitutionalism can be discerned if one explores his role as leader of the provisional government at the Red River settlement in 1869. Riel was a key figure during the negotiations leading up to Manitoba's entry into Confederation. In subsequent years, he maintained that he was a founding father of Manitoba.[4] As a British subject, Riel relied on the rights he believed were inherent in this status. And, as a leader at the Red River settlement, which in 1869 still existed as a separate British colony, he proceeded on the assumption that the settlers would collectively be able to rely on their rights as British subjects as well. This understanding helps to illuminate the nature of his dispute with the Canadian state and permeates the statements he made at his trial.

To place Riel's constitutional thinking into its historical context, one must begin by exploring some of the larger themes of nineteenth-century Canada. One of the most profound ideas was that of responsible and representative self-government in the provinces that would become Canada. The evolution of this idea was inextricably linked with the control and ownership of the public domain. As responsible self-government developed over the course of the nineteenth century, local governments were granted the authority to tax and control their public lands for the benefit of the local population. As early as 1839, Lord Durham emphasized the importance of the public domain in his report on the political situation in the Canadas. Writing in the context of the aftermath of the 1837 rebellions in Upper and Lower Canada, where one of the major grievances was the setting aside of large tracts of land as clergy reserves,[5] Durham wrote that the disposal of public lands is "an operation of Government, which has a paramount influence over the happiness of individuals, and the progress of society towards wealth and greatness ... upon the manner in which this business in conducted, it may almost be said that everything else depends."[6] Durham recognized that control of public domain lands was key to the establishment of responsible self-government: "the disposal of public lands in a new country has more influence on the prosperity of the people than any other branch of Government."[7]

Durham's recognition of the importance of the public domain to the functioning of governance prompted him to recommend that the imperial Parliament retain control of the natural resources in the colonies of British North America. This would ensure that policies in accordance with the aims of the imperial Parliament would be implemented in a uniform fashion without interference by local governments. The imperial government, however, rejected this recommendation, and the provinces of Upper and Lower Canada were granted control of their natural resources.[8] The development of responsible government and the local administration of the public domain by ministers who were directly responsible to an elected assembly became an integral part of British colonial policy in the nineteenth century.[9] The British North America Act of 1867[10] incorporated the principle of local ownership of the public domain into the Canadian Constitution. Section 109 reads, "All Lands, Mines, Minerals, and Royalties belonging to the several Provinces of Canada, Nova Scotia, and New Brunswick at the Union, and all sums then due payable for such Lands, Mines, Minerals, or Royalties, shall belong to the several Provinces of Ontario, Quebec, Nova Scotia, and New Brunswick in which the same are situate or arise, subject to any Trusts existing in respect thereof, and to any Interest other than that of the Province in the same." The only notable exceptions to this constitutional practice would be the three provinces of western Canada created by the Dominion Parliament of Canada: Manitoba in 1870 and Saskatchewan and Alberta in 1905.

The background to the anomalous constitutional position of the three prairie provinces begins with the actions of a Stuart monarch. In 1670 Charles II incorporated Prince Rupert and a group of investors as the "Governor and Company of Adventurers of England tradeing into Hudson's Bay."[11] The Charter of the Hudson's Bay Company of 1670 transferred proprietary interest and governmental authority in the Hudson Bay watershed, better known as Rupert's Land, to the Hudson's Bay Company (HBC). The HBC maintained its monopoly over the region for two centuries. However, by the mid-nineteenth century, Upper Canadians began looking toward Rupert's Land and the "Northwestern Territory"[12] in terms of potential expansion. As the nineteenth century progressed, land shortages were becoming an increasing problem, and people were beginning to recognize the need for an economic hinterland. The expansionist movement that developed in Upper Canada has been described in the following terms: "It was as though Canadians confined within a political and economic prison, saw suddenly opened before their eyes an avenue of escape to the West."[13] Confederation became the instrument of this movement toward western expansion.

In the mid-nineteenth century, George Brown, editor of Toronto's *Globe* newspaper, advocated vociferously for western expansion. He claimed that "[Canada] is fully entitled to possess whatever parts of the Great British American territory she can safely occupy."[14] In this period, the Toronto Board of Trade advocated expansion into the western territories and politicians debated the idea.[15] In 1856 an anonymous author, "Huron," directly challenged the HBC's rights to the region: "I desire to see Canada for the Canadians and not exclusively for a selfish community of traders, utter strangers of our country; whose only anxiety is to draw all the wealth they can from it, without contributing to its advantage even one farthing."[16] In another article, Huron declared that the HBC charter was null and void and that it was in "the interests of Canada to require that this giant monopoly be swept out of existence."[17] In subsequent years, western expansionism became a cause célèbre as the the *Globe*, the Toronto Board of Trade, the Reform Party, the Conservative Party, and business interests all made continuous pleas for a Canadian takeover of the North-West.[18] Part of this effort included a boosterism campaign designed to change attitudes toward the North-West. Prior to this period, the North-West had a reputation as an impenetrable semi-Arctic wilderness. These attitudes gradually shifted in response to continuous reports of the region's vast agricultural potential.[19]

The expansionist movement active in the 1850s and 1860s helped to persuade the government of the United Canadas to pursue a policy of annexation of other territories in British North America.[20] In the negotiations leading up to Confederation, the drafters included a provision specifically dealing with the admission of new colonies or provinces from the lands of the North-West. Section 146 reads,

> It shall be lawful for the Queen, by and with the Advice of Her Majesty's Most Honourable Privy Council, on Addresses from the Houses of Parliament of Canada, and from the Houses of the respective Legislatures of the Colonies or Provinces of Newfoundland, Prince Edward Island, and British Columbia, to admit those Colonies or Provinces, or any of them, into the Union, and on Address from the Houses of Parliament of Canada to admit Rupert's Land and the North-western Territory, or either of them, into the Union, on such Terms and Conditions in each Case as are in the Addresses expressed and as the Queen thinks fit to approve, subject to the Provisions of this Act; and the Provisions of any Order in Council in that Behalf shall have effect as if they had been enacted by the Parliament of the United Kingdom of Great Britain and Ireland.[21]

Thus the acquisition of the North-western Territory and Rupert's Land took place within the context of the creation of the Canadian federation as contemplated by the British North America Act of 1867.[22] The western lands were intended to become an integral part of Canada. With this purpose in mind, delegates from the Canadian government, William McDougall and George-Étienne Cartier, went to London in 1868 in order to negotiate the terms for transfer of Rupert's Land and the North-western Territory.[23]

The Canadian delegates arrived to find Lord Granville, secretary of state for the colonies, anxious to broker a deal between the HBC and Canada whereby Britain would be divested of all interests in Rupert's Land and the North-western Territory.[24] In this period, the imperial government considered the colonies to be a financial liability and sought ways to decrease its colonial holdings. The only obstacle to the transfer was the HBC's claim to financial compensation for the surrender of its charter rights.[25] The imperial government had been advised by its law officers that the HBC could not be compelled to relinquish its rights without compensation.[26] The HBC was not adverse to the settlement policy advocated by the Canadian government; however, the company was determined to receive an adequate return for its shareholders in compensation for the loss of its monopoly in British North America.[27] The Canadian delegates were reluctant to commit the Canadian government to paying for the relinquishment of charter rights when the imperial government had relinquished the charters of other corporations, such as the East India Company, without compensation.[28] The negotiations ground to a halt until Lord Granville threatened to turn the whole matter over to deliberation by the Judicial Committee of the Privy Council. Neither the HBC nor the Canadian government wanted to become involved in a costly and time-consuming judicial reference.[29] A deal was eventually struck, and the transfer of sovereignty was accomplished by the passage of the Rupert's Land Act of 1868,[30] which authorized the imperial Crown to accept the surrender of the HBC's charter rights. The agreement also provided that a deed of surrender would be drawn up when the HBC received £300,000 from the Canadian government and provision for a one-twentieth part of the land in any future township settled within the fertile belt. In order to facilitate this arrangement, the government of British prime minister William Gladstone promised to extend the Canadian government a loan guarantee for the £300,000.[31] The Canadian Parliament then passed An Act for the Temporary Government of Rupert's Land and the North-western Territory When United with Canada of 1869[32] to provide for temporary governance of the region.

Sir John A. Macdonald wanted to acquire Rupert's Land and the North-western Territory on one stated condition: "The land could not be handed over to them [the inhabitants], it was of the greatest importance to the Dominion to have possession of it, for the Pacific Railway must be built by means of the land through which it had to pass."[33] The lands were to be used to finance railway development and subsequent immigration schemes in order to create a transcontinental Dominion of Canada. It was for these purposes that the lands were held and administered by the Dominion government from 1870 to 1930.[34] Canadian administration of these lands solved a number of problems: it opened up new trade possibilities for eastern Canada, it eased the land crisis, it secured a route for the Pacific Railway, and it countered American expansionism.[35] The retention of the public domain allowed the Dominion to set unified policies over the entire area. The absence of local or provincial governments allowed the federal government to administer the lands to suit its commercial and economic interests.[36] The federal government wanted to develop the North-West as an economic hinterland for the more densely populated parts of the Dominion. The resources were retained by the central government, and revenue from fees, land sales, rentals, leases, and royalties was directed to the government in Ottawa. If Canada were a unitary state, this would have been unremarkable since every province or territory would be little more than a municipal corporation.[37] Macdonald may have preferred a unitary state for administrative efficiency; however, the residents of the North-West had other ideas.[38]

The Red River Resistance[39] was an unexpected and unforeseen complication for the Macdonald government. Preoccupation with the economic aspects of the National Policy caused the Canadian government to be careless about the actual transfer of the territory.[40] The details of the transfer were arranged by imperial authorities, the Canadian government, and the HBC as though the territory were terra nullius. The negotiations for the land transfer proceeded as though the transaction concerned only real estate and not the rights of the people who lived in the area.[41] The settlers of the North-West were not consulted about the impending transfer. This arrogance, or neglect, fostered a climate of anxiety and ill feelings toward the Canadian government.[42] Blinded by the desire to acquire territory, the Canadian government ignored the degree of political autonomy that existed in the settled areas of the North-West – the most prominent being the settlement at Red River. The settlers in this area owned the land by occupancy and were worried that they would lose their land rights in the transfer.[43] Alexander Begg, an observer of the events at Red River, recorded, "Madame rumour has full sway for she has it all her own way – nothing official has transpired to enlighten us – everything is conjecture and conjecture is the worst thing to be abroad in a country like this."[44] Wild rumours spread throughout the settlement in 1869, and the settlers at Red River were not certain what would happen.

Led by the charismatic Louis Riel, the Métis and other settlers expressed their discontent about what they considered to be a complete disregard of their rights as British subjects.[45] The provisional government set up by the residents of Red River forced the federal government to negotiate terms of entry for the Province of Manitoba into Confederation.[46] At a constitutional convention, held at Fort Garry from 25 January to 10 February 1870, a List of Rights was drawn up as a basis for negotiations with the government at Ottawa. The List of Rights contained a variety of demands: no liability for the £300,000 paid to the HBC; no direct taxation, except for municipal purposes; confirmation of land ownership; representation in Parliament; and local control of the public domain within a circle whose radius was to be the distance between Fort Garry and the international boundary.[47] The members of the convention spent several days debating whether the residents of Red River should be admitted to the Union as a territory or a province. In the notes of the convention published in the *New Nation* on 11 February 1870, Louis Riel put forth his argument:

> One important consideration which we must bear in mind, is, that as a Territory we escape a great deal of the heavy responsibility that may weigh on us as a Province. Of course it would be very flattering to our feelings to have all the standing and dignity of a Province. The exclusive powers to Provinces are considerable, and in themselves satisfactory, if we found them applicable to our case. (Mr. Riel then read the Confederation Act to show the powers conferred on Provinces.) He alluded specifically to article 5 [s.

> 92(5) of the British North America Act of 1867] which provides that the management and sale of the public lands belonging to the Provinces and of the timber and wood thereon, is vested in the Province. This, he alluded to, as one of the most important as far as we are concerned. In looking at the advantages and disadvantages of the provincial and territorial systems, we have to consider fully the responsibility of our undertaking. I do not say positively that it is for our own good to go in as a province; but I think that the position of the Province might suit us better than that of a Territory.[48]

It was Riel who informed the convention of the constitutional differences between provinces and territories under the British North America Act of 1867. The importance of this distinction should not be underestimated. Riel argued for provincial rather than territorial status because he had formed the opinion that the Métis of Red River would have more autonomy within a federal state if they entered Confederation as a province among equals as provided in section 92(5) of the act.[49] As a province among equals, Manitoba would have control and administration of the natural resources. This would have guaranteed the province jurisdictional authority to direct the use of the revenue derived from the lands and resources. From a constitutional point of view, Riel must have understood that control of the public domain would better protect the political and cultural interests of the Métis.

After much debate, the convention decided to pursue provincial status. In April 1870 a three-person delegation appointed by Riel was sent to Ottawa to discuss the entry of Manitoba into Confederation. Abbé Noël-Joseph Ritchot, Judge John Black, and Alfred Scott arrived with a List of Rights that demanded protection for the cultural and linguistic rights of the people of the Red River settlement. The eleventh item in the List of Rights demanded "[t]hat the Local Legislature of this province have full control over all the lands of the North-West."[50] According to W.L. Morton, "[Riel's] aim was to make such terms with Canada as would enable the people of the North West to control its local government in the early days of settlement, and as would allow them to possess themselves, as individuals and as a people, enough of the lands of the North West to survive as a people, and to benefit by the enhancement of the wealth of the North

West that settlement would cause."[51] In response to these demands, Cartier and Macdonald insisted that Canada had spent £300,000 to secure the charter rights from the HBC and that they anticipated spending more money to extinguish Indian title in the North-West by the negotiation of treaties. These expenses were presented to the delegates from Red River as justification for the Dominion government's continued retention of the natural resources.[52] The List of Rights, especially the demand for the public domain, did not please Macdonald and Cartier, who had planned to use the land to pay the indemnity to the HBC, to fund railway grants, and to provide for a free homestead policy.[53]

There has been much scholarly debate about what occurred during the negotiations between the delegates from the Red River settlement and the representatives from the federal government. It has been speculated that Abbé Ritchot believed that the cultural and linguistic interests of the Métis would be well protected by land grants secured for the children of Métis in the form of Métis scrip, protection for denominational schools, guarantee of land titles, and official bilingual status.[54] Abbé Ritchot recorded the following in his journal about the negotiations:

> Mr. Black finds it just that the Dominion should have control of the lands, he finds extravagant the pretensions of the inhabitants of the North West to claim the lands as theirs. I reply and prove that not only is it not extravagant but just and reasonable. Sir George supports me, Sir John is of the same opinion, but they reply that to reach a settlement it is necessary to make some concessions ... Then the ministers asked us what we wished to do in the matter of lands. Reply, the control of those lands as requested in our instructions. Impossible, said the ministers. We could by no means let go control of the lands at least unless we had compensation or conditions which for the populations actually there would be the equivalent of the control of the lands of their province.[55]

Abbé Ritchot conceded his demand that the natural resources be transferred to the province.

The major problem was that the goal of the negotiations was to reach an amicable settlement. The representatives of both sides were only vaguely aware of the importance of the financial settlement provisions. There was no guide for the new province's fiscal needs, and there was no existing government framework or provincial budget that could be used as a guideline for the actual financial needs of the new province. The fault rests largely with the federal government's representatives, who should have thought it necessary to formulate a comprehensive financial arrangement for the province instead of giving merely what Manitoba's negotiators were willing to accept.[56] In the context of bargaining, no one looked beyond the exigencies of the immediate situation. No provisions were made to ensure the province had enough revenue to deal with the expenses that would inevitably arise due to immigration and settlement as a result of the National Policy. The federal government granted provincial status to Manitoba and based the subsidy schedule on the needs of the older, more established provinces.[57]

The lack of financial farsightedness by all parties involved in the negotiations led to a situation whereby provincial status was granted in the Manitoba Act of 1870[58] with no provision for control by the province over its public domain. Section 30 of the act provides, "All ungranted or waste lands in the Province shall be, from and after the date of the said transfer, vested in the Crown, and administered by the Government of Canada *for the purposes of the Dominion*, subject to, and except and so far as the same may be affected by, the conditions and stipulations contained in the agreement for the surrender of Rupert's Land by the Hudson's Bay Company to Her Majesty" (emphasis added). The phrase "for the purposes of the Dominion" incorporated the National Policy directly into the Constitution of Manitoba. Historian Chester Martin describes the position of Manitoba at its inception: "A premature province – a scattered population of 11,000 people caught in the eddy of a premature but imperative national policy – was thrust, by the operation of a single statute, from primitive Hudson's Bay paternalism into the full responsibilities of self-government."[59] Manitoba's financial and constitutional position stood in marked contrast to the position of British Columbia and Prince Edward Island upon their entry into Confederation. In 1871 British Columbia entered Confederation will full control over its public domain. Prince Edward Island entered Confederation with no public lands left because they had been nearly entirely alienated by the Crown a century earlier. In the absence of this source of revenue, the federal government granted a subsidy in lieu in the amount of $45,000 per annum. Additionally, the federal government made a loan of $800,000 available to the province in order to repurchase land. On repurchase, this land became part of the public domain of Prince Edward Island over which the province had full control.[60] The constitutional position of Manitoba was an anomaly within the British Empire.

Echoing Riel's concerns, the Official Opposition sharply criticized the Manitoba Bill during debate in the House of Commons. Alexander Mackenzie, leader of the Opposition, characterized the bill as so "ludicrous ... that it only put one in mind of some of the incidents in Gulliver's Travels."[61] In defense of the bill, Macdonald pointed out that it represented a pragmatic solution to the realities facing the Dominion government. He pointed out that the demands for provincial status by the people of Red River had been acceded to and that the region would be granted self-government through the adoption of the Manitoba Act of 1870.[62] However, even Macdonald had doubts about the constitutionality of section 30 of the act. He recognized that it represented a departure from British constitutional practice with respect to the control of the public domain. Macdonald's papers contain a legal opinion on the validity of both section 30 of the Manitoba Act of 1870 and An Act for the Temporary Government of Rupert's Land and the North-western Territory when United with Canada of 1869.[63] The authors considered the temporary act to be constitutional because it did not provide for the creation of a province in the region. They did, however, express doubts with regard to the Manitoba Act because the retention of Crown lands was contrary to section 92(5) of the British North America Act of 1867. They recommended that the federal government request confirmation of the section in an imperial statute. The imperial Parliament, upon receiving the request from the Canadian government, passed the British North America Act of 1871.[64] The act empowered the Canadian Parliament to establish new provinces and to formulate the terms of their constitutions.[65] The Canadian Parliament could create provinces that had different constitutional powers than the original four provinces of Confederation; however, it could not unilaterally alter the constitutions of the new provinces once they had been created.[66] Thus the constitutional inequality with respect to natural resources imbued in section 30 of the Manitoba Act was confirmed by imperial statute and was enshrined in the Canadian Constitution.

As foreseen by Riel, the constitutional inequality and lack of revenue from control of natural resources caused problems for the Province of Manitoba. As early as 1872, provincial premiers began making annual pilgrimages to Ottawa to request "better terms." The annual subsidy provided by the federal government in the Manitoba Act fell far short of the amount of money needed by the new provincial government.[67] The province requested more money to deal with rising costs associated with the flow of immigration into the region. The lack of revenue from the public domain meant that the province had no means of increasing its revenue except through grants from the federal government. The province had little money to spend on the schools, bridges, and roads needed due to increased immigration into the area.[68] The federal government had retained the natural resources in order to co-ordinate the building of railways and homestead lands. And in 1905 the federal government retained control of the natural resources of Alberta and Saskatchewan as well when these provinces were carved out of the North-West.[69]

As one of the founders of Manitoba, Riel's interpretation of the British North America Act of 1867 is historically significant. His arguments, formulated during his tenure as leader of the provisional government at Red River, were unattributed by later generations of prairie politicians who relied on the logic. Nevertheless, for over fifty years, prairie politicians reiterated Riel's essential arguments regarding the constitutional differences between provinces and territories by arguing that unless a province had administrative control over its natural resources, it was simply a territory disguised as a province. Until the signing of the British North America Act of 1930,[70] nearly every provincial premier of Saskatchewan, Alberta, and Manitoba would echo Riel's demands for control of the land and resources as a means of generating provincial revenue in order to create better public policy for the citizens. As the leader of the provisional government at Red River, Louis Riel understood that control over the public domain was an essential element of responsible government. The loss of this was at the core of his dispute with the Canadian government and a central issue at his trial for treason. Largely lost in the sensationalism of the trial was Riel's concern with securing meaningful self-government for the settlers of Manitoba and the North-West – a right that every British subject throughout the empire possessed.

NOTES

The author would like to thank Hamar Foster and Douglas Owram for their thoughtful suggestions on various drafts of this chapter.

1 Both the title of this chapter and the epigraph are from Louis Riel's final statement to the jury, 31 July 1885. Reprinted in Desmond Morton, ed., *The Queen v Louis Riel: Canada's Greatest State Trial* (Toronto: University of Toronto Press, 1974), 318.

2 See, for example, Douglas Owram, "The Myth of Louis Riel," *Canadian Historical Review* 63 (1982): 315–36.

3 Riel explicitly used the terms "responsible" and "irresponsible" in reference to government, as reprinted in Morton, ed., *Queen v Louis Riel*, at 312, 323, 324, 354, and 365; and he referred to legal or moral rights at 312, 323, 353–5, 357–8, 360–1, and 363.

4 See, for example, Morton, ed., *Queen v Louis Riel*, 318. Louis Riel is now often referred to as a "founding father" of Manitoba. Louis Riel Day is celebrated in Manitoba on the third Monday in February.

5 The Constitution Act of 1791, or the Clergy Endowments (Canada) Act of 1791, 31 Geo. 3, c. 31, reserved one-seventh of Crown lands for the support of Protestant clergy. During the rebellions, this policy was challenged. See Alan Wilson, *The Clergy Reserves of Upper Canada: A Canadian Mortmain* (Toronto: University of Toronto Press, 1968).

6 See Gerald M. Craig, ed., *Lord Durham's Report: An Abridgement of "Report on the Affairs of British North America"* (Montreal and Kingston: McGill-Queen's University Press, 2007), 107.

7 Ibid., 118.

8 C. Cecil Lingard, *Territorial Government in Canada: The Autonomy Question in the Old North-West Territories* (Toronto: University of Toronto Press, 1946), 22.

9 Arthur Berriedale Keith, *Responsible Government in the Dominions* (Oxford: Clarendon, 1912), vol. 2, 1048; George G. James, "Constitutional and Political Aspects of Federal Control of Natural Resources in the Prairie Provinces, 1870–1930" (MA thesis, University of Saskatchewan, 1975), 7.

10 Renamed the Constitution Act of 1867, 30 & 31 Vict., c. 3 (UK), reprinted in RSC 1985, app. 2, no. 5.

11 See Bernard W. Funston and Eugene Meehan, eds, *Canadian Constitutional Documents Consolidated*, 2nd ed. (Toronto: Thomson Carswell, 2007), 63.

12 "North-western Territory" refers to the lands licensed to the HBC by the Crown prior to 1870. "North-West Territories" refers to the lands outside the

boundaries of the Province of Manitoba created in 1870. "North-West" is a generic term used to refer to the entire region.

13 R.G. Riddell, "A Cycle in the Development of the Canadian West," *Canadian Historical Review* 21 (1940): 268.

14 *Globe*, 10 December 1856.

15 Donald Swainson, "Canada Annexes the West: Colonial Status Confirmed," in *The Prairie West: Historical Readings*, ed. R. Douglas Francis and Howard Palmer (Edmonton: Pica Pica, 1985), 125.

16 *Globe*, 18 October 1856.

17 *Globe*, 31 October 1856.

18 Jim Mochoruk, *Formidable Heritage: Manitoba's North and the Cost of Development, 1870 to 1930* (Winnipeg: University of Manitoba, 2004), 12.

19 Douglas Owram, *Promise of Eden: The Canadian Expansionist Movement and the Idea of the West, 1856–1900* (Toronto: University of Toronto Press, 1980), 3.

20 Ibid., 26.

21 Reprinted in Funston and Meehan, eds, *Canadian Constitutional Documents*, 169.

22 Swainson, "Canada Annexes the West," 129.

23 R.S. Longley, "Cartier and McDougall, Canadian Emissaries to London, 1868–69," *Canadian Historical Review* 26 (1945): 25.

24 Harold A. Kevin McQuinn, "Great Britain and the Red River: An Examination of Imperial Involvement in the Transfer of the North-West Territory to Canada, and in the Red River Rebellion" (MA thesis, University of New Brunswick, 1975), 45.

25 David M.L. Farr, *The Colonial Office and Canada* (Toronto: University of Toronto Press, 1955), 75.

26 Longley, "Cartier and McDougall," 34.

27 McQuinn, "Great Britain and the Red River," 30.

28 Rudolph Robert, *Chartered Companies: Their Role in the Development of Overseas Trade* (London: G. Bell and Sons, 1969), 115.

29 McQuinn, "Great Britain and the Red River," 48.

30 31 & 32 Vict., c. 105 (Imp.).

31 Farr, *Colonial Office*, 76.

32 SC 1869, c. 3.

33 Great Britain, Colonial Office, *Correspondence Relative to the Recent Disturbances in the Red River Settlement* (London: Printed by W. Clowes for H.M.S.O., 1870), 143. See also Enclosure #2, "Report of the Canadian House of Commons Debates," *Ottawa Times*, 2 May 1870.

34 See Donald V. Smiley, "Canada and the Quest for a National Policy," *Canadian Journal of Political Science* 8 (1975): 40–62.
35 Mochoruk, *Formidable Heritage*, 12.
36 Vernon C. Fowke, *Canadian Agricultural Policy: The Historical Pattern* (Toronto: University of Toronto Press, 1947), 157.
37 Mochoruk, *Formidable Heritage*, 105. See also Donald G. Creighton, *John A. Macdonald* (Toronto: University of Toronto Press, 1998).
38 These economic development policies would eventually be referred to as the National Policy. See Richard Gwyn, *Nation Maker – Sir John A. Macdonald: His Life, Our Times*, vol. 2, *1867–1891* (Toronto: Random House, 2011), 277.
39 In recent literature, Red River "Resistance" has replaced Red River "Rebellion."
40 R.G. Riddell, "A Cycle in the Development of the Canadian West," *Canadian Historical Review* 21 (1940): 268.
41 Lewis Herbert Thomas, *The Struggle for Responsible Government in the North-West Territories, 1870–97* (Toronto: University of Toronto Press, 1956), 27.
42 Michael Dorland and Maurice Charland, *Law, Rhetoric, and Irony in the Formation of Canadian Civil Culture* (Toronto: University of Toronto Press, 2002), 159.
43 W.L. Morton, *Manitoba: The Birth of a Province* (Altona, MB: D.W. Friesen and Sons, 1965), xv.
44 W.L. Morton, ed., *Alexander Begg's Red River Journal and Other Papers Relative to the Red River Resistance of 1869–1870* (Toronto: Champlain Society, 1956), 152.
45 McQuinn, "Great Britain and the Red River," 50.
46 See Thomas Flanagan, *Riel and the Rebellion: 1885 Reconsidered*, 2nd ed. (Toronto: University of Toronto Press, 2000); Stanley A. Puchniak, "Riel's Red River Government: A Legitimate Government, 1869–70" (MA thesis, University of Ottawa, 1931); Alexander Begg, *The Creation of Manitoba, or A History of the Red River Troubles* (Toronto: A.H. Hovey, 1871).
47 Thomas, *Struggle*, 39.
48 *New Nation*, 11 February 1870.
49 Section 92(5) of the British North America Act of 1867 reads, "The management and sale of the public lands belonging to the province, and of the timber and wood thereon."
50 Begg, *Creation*, 327.
51 Morton, *Manitoba*, xvi.

52 Douglas N. Sprague, *Canada and the Métis, 1869–1885* (Waterloo, ON: Wilfrid Laurier University Press, 1988), 57.
53 Mochoruk, *Formidable Heritage*, 107.
54 Ibid., 108. Métis scrip was a form of currency issued to the Métis by the federal government that could be used to purchase Crown land.
55 Abbé Ritchot's journal entry for 17 April 1870 is quoted in Morton, *Manitoba*, 140.
56 James A. Maxwell, *Federal Subsidies to the Provincial Governments in Canada* (Cambridge, MA: Harvard University Press, 1937), 35.
57 Ibid., 37.
58 SC 1870, c. 3, s. 31.
59 Chester Martin, *"Dominion Lands" Policy* (Toronto: McClelland and Stewart, 1973), 206.
60 Chester Martin, *The Natural Resources Question: The Historical Basis of Provincial Claims* (Winnipeg: Phillip Purcell, 1920), 70.
61 Canada, *House of Commons Debates*, 2 May 1870, 1305–6.
62 Morton, *Manitoba*, xxvi.
63 Granville to Lisgar, 23 December 1869, Library and Archives Canada, Macdonald Papers, vol. 101/1, 40372.
64 Also called An Act Respecting the Establishment of Provinces in the Dominion of Canada of 1871, 34 & 35 Vict., c. 28 (UK).
65 Paul Gérin-Lajoie, *Constitutional Amendment in Canada* (Toronto: University of Toronto Press, 1950), 53.
66 Gerard V. LaForest, *Natural Resources and Public Property under the Canadian Constitution* (Toronto: University of Toronto Press, 1969), 29.
67 Maxwell, *Federal Subsidies*, 77.
68 See Mochoruk, *Formidable Heritage*, 110–35, for a detailed analysis of Manitoba's fight for "better terms."
69 Maxwell, *Federal Subsidies*, 37.
70 See "Schedule" to the British North America Act of 1930, renamed the Constitution Act of 1930, 20 & 21 Geo. 5, c. 26 (UK), reprinted in RSC 1985, app. 2, no. 26.

3

Louis Riel's Trial Speeches

THOMAS FLANAGAN

THE PROBLEM

Louis Riel is a highly recognized figure in Canadian history, but the two long speeches he gave at his treason trial in Regina are barely known, being seldom anthologized for reference or study by students. Some authors think they were great speeches. Maggie Siggins, for example, calls his first oration "a superb speech, one of the classics of Canadian history."[1] I argue, however, that both speeches were ineffective – long, rambling, disorganized, sometimes almost unintelligible, even with footnotes to explain all the obscure references to the people and events of Riel's lifetime. They may have had some emotional impact on a few members of the audience and jury,[2] but they did not succeed in their short-term objectives: to persuade the jury to acquit him (the first speech) and, after conviction, to persuade the court to recommend clemency (the second speech). Nor, because they were so hard to understand, were they successful in Riel's longer-term objective of communicating to the world his rationale for leading the Métis in the North-West Rebellion; he did a better job of that in other things he wrote while awaiting execution.[3]

Of course, there are a few memorable lines. Probably the best known is the complicated pun that Riel made near the end of his first speech on irresponsibility and insanity. His lawyers were arguing that, because of insanity, he was not responsible for his actions, whereas he accused the federal government of not being responsible (in the Parliamentary sense) because it was not elected by the people of the North-West. Hence, "I have been quarrelling with an insane and irresponsible Government."[4] The pun is clever but perhaps too contrived to be really effective.

My favourite line is from Riel's second speech: "This is the principle. God cannot create a tribe without locating it. We are not birds. We have to walk on the ground."[5] I used this quotation in my most recent book on Aboriginal property rights because it expresses territoriality as an intrinsic aspect of the human condition.[6] I am also moved by the pathos of Riel's statement to the jury: "The day of my birth I was helpless and my mother took care of me although she was not able to do it alone, there was someone to help to take care of me and I lived. To-day, although a man I am as helpless before this court, in the Dominion of Canada and in this world, as I was helpless on the knees of my mother the day of my birth."[7] But such moving lines are few and far between.

It seems surprising that Riel would have given such bad speeches when his life was on the line. Why didn't the prospect of hanging concentrate Riel's mind on the oratory of which he was capable?

Although expelled from the Collège de Montréal shortly before he would have finished his baccalaureate, Riel received a rigorous classical education there, with emphasis on Greek, Latin, and French literature. He was a charismatic leader and an inspiring public speaker who had twice roused his Métis people to resistance against the Canadian government. Beyond his demonstrated speaking ability, he was also an accomplished author. His correspondence and published pamphlets are written in a direct, vigorous, and sometimes rousing style. He was also a wordsmith who started writing poetry when he was a student and continued throughout his life. He had a perfect mastery of French metre and could create moving imagery as well as clever rhymes.

Of course, he was handicapped by speaking English in his trial speeches. As he told the jurors, "I cannot speak English very well, but am trying to do so, because most of those here speak English."[8] But his English was not that bad. He grew up in Red River, where French was the more common language, but English was also widely used in dealings with the Hudson's Bay Company and with the English halfbreeds. He studied English at the Collège de Montréal and in fact won a prize there for translation from English to French.[9] Montreal was as much English as French at the time, and Riel often visited his uncle John Lee, an English-speaking Irishman, in the suburb of Mile End.

He subsequently spent long stretches of time in the United States – 1866–68, 1873–75, and 1878–84 – and became an American citizen in 1883. When he was invited in June 1884 to come to the Saskatchewan Valley, he was living in Montana and supporting himself as a schoolmaster, teaching entirely in English. One only has to peruse his extensive English correspondence to see that his knowledge of the language, although not perfect, was thoroughly functional. Admittedly, his attempts at poetry were unintentionally comical because his English was not quite idiomatic, but he had a large vocabulary and a mastery of English grammar. Overall, he could express himself at least as well as future Canadian prime minister Jean Chrétien, whose political career was certainly not hampered by his accented English.

Riel's trial speeches give the impression of being almost extemporaneous; did they in fact suffer from a lack of preparation? He himself said that he had no time to prepare his second speech, delivered immediately after being found guilty by the jury.[10] That much is understandable, but it would be odd if he had not thoroughly prepared his first speech, for throughout his life he carefully drafted and redrafted all sorts of texts – those meant for public consumption,[11] as well as family correspondence and private revelations. He lived according to the aphorism that "there is no good writing, only good rewriting." And he had plenty of time to prepare that first speech, from his arrival in Regina on 23 May to the commencement of his trial on 28 July 1885. Yet we have only one brief fragment of a draft, which corresponds more or less to the conclusion of the speech, with its famous pun about irresponsible government and insanity.[12] He obviously had thought that out in advance.

We do not have the manuscript of this partial draft; what survives is part of a longer typescript of some of Riel's prison papers, which ended up in the Archives of the Archdiocese of St Boniface. It is not clear how this happened because Riel's jailers attempted to keep his papers; maybe he secretly gave some to the missionary priests who attended him, who may then have spirited them away and transferred them to their superior, Archbishop Alexandre-Antonin Taché. But we have no idea why only a typed copy of those papers can now be found and what happened to the originals. Amid all the uncertainty, it is impossible to say how much preparation Riel might have done for his first speech. As the old legal adage goes, absence of evidence is not evidence of absence. But one can say that the second one was understandably improvised; and even if the first one was prepared, it does not give that impression, except for the conclusion.

THE RHETORICAL SITUATION

If the explanation for Riel's incoherence at his trial is neither a lack of rhetorical ability, nor the necessity of speaking English, nor a lack of opportunity to prepare, what is it? The documentary evidence suggests that he wanted to say too much about too many different ideas in his mind. He sketched these out in a remarkable series of letters written to various people in June and July, which in a sense amounted to the real preparation for his trial speeches. But the preparation was ineffective because it was centrifugal rather than centripetal. As he experimented with different lines of argument in letters to various people, his thinking spiralled outward in multiple directions rather than converging inward upon a focused argument that might have had some impact at trial.

That outward spiral effect is a pity because he actually began with a more concise presentation in mind. As he was being conveyed to Regina, he made quite a lucid case to the Reverend Charles Bruce Pitblado, a Presbyterian military chaplain. Below are some excerpts:

> [Pitblado]: Why did you fight?
>
> [Riel]: The people compelled me to do so. I told them last winter that our bill of rights had done its work. They should let me go back to Montana. They would get all that was just in due time. They said I could not go away. The matter was stirred up and I must fight it out. I was not the ringleader. They obliged me to lead them in the fight. But the matter will no doubt [be] tried in a court of law when the truth will come out.
>
> [Pitblado]: How would you answer supposing you were charged with treason?
>
> [Riel]: I would say (1) that whilst I was a subject of the English government I was loyal to the constitution. If I, with others, was instrumental in securing the recognition of the rights of the half-breeds in Manitoba, my conduct was condoned, if not justified, by

the treaty which the government made with me [the Manitoba Act of 1870]. (2) The government recognized me as Governor of Manitoba for two months – from 24th of June till 24th of August I governed Manitoba. When Wolseley entered he proclaimed me a bandit – not a traitor. He never arrested me. When Governor Archibald came he shook hands with me, so did Colonel Irvine, and I served faithfully in trying to put down the Fenian invasion. The Government recognized my services. I resigned my seat in Provencher to a member of the Government [George-Étienne Cartier]. I have shown the Government many favours. They have never conferred a single favor on me. I have been true to the half-breeds; I have not rebelled against the Government. Besides, I am a citizen of the United States – I have my papers all right. A citizen of the United States can scarcely be a traitor to the Dominion.

[Riel]: We did not rebel. The matter is not a rebellion. We have never yet been treated with about our rights. The half-breeds of Manitoba have entered into a treaty. That treaty stipulates that the arrangement to be made with the half-breeds of the territories, would be similar to those made in Manitoba. No treaty has been made with us. We never transferred our rights, and before they are taken from us we wish to have a treaty made, and we think we have a right to expect that the conditions of that treaty will be similar to those made in the treaty which settled the half-breeds of Manitoba. No, my friend, this is not rebellion. We simply defend ourselves.[13]

If Riel had made such a tightly argued presentation to the jury, he would not necessarily have been acquitted, but he would certainly have given a better speech, one that might have influenced public opinion in the direction of clemency. But it was not to be.

On 16 June 1885 Riel wrote to his old school friend Romuald Fiset and two of his lawyers, François-Xavier Lemieux and Charles Fitzpatrick, sketching out a fairly tight case based upon self-defence. According to Riel, the Métis had taken hostages without consulting him. Then when Major Lief Crozier of the North-West Mounted Police advanced upon Batoche with 120 men and opened fire upon the Métis at Duck Lake, "I told our people to defend themselves in the name of God."[14] Riel's version of the facts would have been open to challenge, but the self-defence argument might have done him some good at trial.

On 24 June he broadened the scope of his defence, writing to Edgar Dewdney, lieutenant governor of the North-West Territories, and Captain Richard Burton Deane, in charge of the Regina jail, that he wanted to defend his entire political career, from 1869 to the present. He was sure, he said, that the evidence would exculpate the Conservatives and implicate the Liberals of Ontario. "With the help of God, my victory over the party of the Honorable [Edward] Blake and [Alexander] Mackenzie will be complete. The triumph of the Conservative Party over its opponents will be great."[15]

Riel elaborated still further in a letter to Deane, Dewdney, and Prime Minister John A. Macdonald on 6 July and in another letter to Macdonald on 16 July. He warned that Gabriel Dumont, having taken refuge in the United States, might rouse various American ethnic groups for an invasion of Canada to fulfil Riel's plan of creating new Métis nations in Canada through intermarriage with various European nationalities. But there was a way to preserve political harmony. Based on Macdonald's original idea of "legislative union" for the foundation of Canada, there could also be an "imperial union" of Great Britain and the United States. The leaders of this new colossus should "forget about Rome" and choose "someone known for his profound wisdom, for his virtue, for his knowledge and vast intelligence" to "promote the interests of the most perfect possible harmony among the different Christian sects of their Colossal Dominion."[16] Riel probably had Archbishop Taché, not himself, in mind for this elevated distinction, for he added that he would like to become premier of Manitoba, from which position he could negotiate "Better Terms" for Manitoba within Confederation ("Better Terms" was a major rallying cry in Manitoba politics in the 1880s).[17]

On 21 July Riel wrote to James W. Taylor, United States consul in Winnipeg, asking the American government to intervene because he had become an American citizen.[18] Three days later, he wrote to his erstwhile patron, Alexandre-Antonin Taché, the archbishop of St Boniface. This letter of 24 July 1885 is one of Riel's most important statements about his prophetic "mission."[19] He laid out the whole story, going all the way back to his childhood, of how he had become a prophet and how the blessings of church leaders such as Taché himself had led to the direct inspiration of the Holy Spirit. However, he modified his heretical revelations to make them at least a little more acceptable to Taché and the church. He proposed now, instead of a rupture with Rome, that the Pope should appoint Taché "Pontifex Major totius Novi Mundi" – "Greater Pontiff of the Entire New World."[20] With that title, Taché would become the head of "l'église catholique, apostolique et vitale du Noveau-Monde" – a renewed version of Roman Catholicism based on the divine revelations vouchsafed to Riel.[21]

Taking all these letters together, we see a number of themes that Riel contemplated using in his defence:

- that he had not urged the Métis to rebel, that they had been attacked by the Mounted Police, and that his role was only to lead them in defending themselves;
- that federal politicians, especially the Liberals, had treated him and the Métis unfairly for fifteen years, neglecting their interests and repeatedly breaking promises;
- that he was an American citizen;
- that he had monumental plans for changing North America, both political (the "imperial union" of Great Britain and the United States) and religious ("l'église catholique, apostolique et vitale du Noveau-Monde");
- that he was a divinely inspired prophet, nourished in the bosom of the Roman Catholic Church and chosen by God to bring the church to unprecedented heights of glory in the New World.

All these themes were important to Riel, and in the end he could not choose among them. Much to the detriment of coherence, he would try to expound them all in his two speeches in Regina.

His desire to say so much was consistent with his oft-repeated demand to have his case heard by the justices of the Supreme Court of Canada sitting in the province of Quebec. The first written record of this demand appears in a few lines of verse composed on 9 June:

Without any prejudice
To any human office:
    O Jesus! Do grant
    Me a fair trial;
    As fair as I want,
    Right in Montreal.[22]

Riel expressed his desire more formally in a telegram of 12 or 13 June to his old schoolmate Romuald Fiset, one of the main organizers of his defence fund: "Please, let committee and lawyers obtain from Government that my case be tried before Judges of Supreme Court and in lower Canada. There only can have all my important witnesses and costs would be smaller."[23] Similar requests appear repeatedly in the prison correspondence cited above. Riel rationalized his demand in terms of getting witnesses to appear, but of course the Supreme Court would also have been the perfect venue for a political trial, in which Riel could emphasize his cause and his career rather than focus on the details of the charges against him. In such a proceeding, Riel could indeed have developed all the themes mentioned in his prison correspondence.

His request, however, was a nonstarter. The Supreme Court of Canada is a court of appeal, not a court of original jurisdiction. Its legal mandate is not to conduct trials but to hear appeals of trial verdicts. There was simply no legal way for Riel's demand to be granted. Government officials did debate whether Riel should be tried in Regina or Winnipeg or somewhere else, but they never for a second considered putting Riel before the Supreme Court of Canada.[24] That meant that Riel could not stage a grandiose political trial but would have to be content with an ordinary criminal trial examining mundane questions of fact and criminal intent. Such a setting was not a favourable venue for defending his career and expounding his divine mission.

As if that were not challenge enough, Riel also felt obliged to use his speeches to challenge the defence argument made by his attorneys – that he was not responsible for his actions at the time of the rebellion and therefore not guilty by reason of insanity. Riel was not suicidal, but he preferred a guilty verdict and execution over being declared legally insane, for such a verdict would fatally undermine his claim to be a divinely inspired prophet. He had, therefore, to argue at cross purposes with his attorneys. He tried at one point to intervene during the trial to contradict something said by Charles Nolin, a hostile witness; but Judge Hugh Richardson told Riel he had to either discharge his lawyers or be quiet and let them argue the case.[25] Riel consented under protest: "As they [Riel's lawyers] have determined to go on, I will assert that, while I wish to retain them, I cannot abandon my dignity. Here I have to defend myself against the accusations of high treason, or I have to consent to the animal life of an asylum. I don't care about animal life if I am not allowed to carry with it the moral existence of an intellectual being."[26]

Riel, in other words, wanted to argue on two fronts. He wanted to defend himself not only against the charge of high treason but also against the label of insanity that his own lawyers were trying to put on him. The need to argue against both prosecution and defence at the same time created an impossible rhetorical dilemma. In theory, Riel could have resolved the dilemma by discharging his lawyers and hiring others willing to argue his case the way he wanted, but he had no money to do so. He was destitute, and his lawyers had been recruited by friends in Quebec who raised money to pay their expenses. If Riel wanted their help – and since the American government had refused to intervene to help an American citizen, he had no other way of obtaining the legal representation he needed – he had to let these lawyers manage his case.

A RHETORICAL ANALYSIS

In his *Rhetoric*, Aristotle describes three modes of persuasion that have become known as *ethos*, *pathos*, and *logos*:

> Of the modes of persuasion furnished by the spoken word there are three kinds. The first kind depends on the personal character of the speaker [*ethos*]; the second on putting the audience into a certain frame of mind [*pathos*]; the third on the proof, or apparent proof, provided by the words of the speech itself [*logos*]. Persuasion is achieved by the speaker's personal character when the speech is so spoken as to make us think him more credible ... Secondly, persuasion may come through the hearers, when the speech stirs their emotions ... Thirdly, persuasion is effected through the speech itself when we have proved a truth or an apparent truth by means of the persuasive arguments suitable to the case in question.[27]

Let us follow the Aristotelian framework by examining how Riel resorted to *ethos*, *pathos*, and *logos* in his trial speeches.

Using a simple quantitative analysis, the contents of Riel's first speech, consisting of 517 lines in the *Collected Writings* edition,[28] can be categorized as follows:

Ethos (Riel's prophetic "mission" and alleged insanity)	53%
Pathos (sympathy for Riel's plight)	10%
Logos (justification for Riel's actions)	23%
Other	14%

These statistics mean that Riel spent more than half of his time during his speech to the jury outlining his character, denying that he was insane, and affirming that he was a divinely inspired prophet.

The breakdown is dramatically different for the second speech, in which Riel addressed the court after being convicted of high treason. The second speech was more than 50 per cent longer than the first, 791 lines as compared to 517. It also devoted much more time to *pathos* and *logos* and less to *ethos*:

Ethos (Riel's prophetic "mission" and alleged insanity)	14%
Pathos (sympathy for Riel's plight)	34%
Logos (justification for Riel's actions)	49%
Other	3%

This did make some sense in the context of his trial. Sympathy for Riel's plight and justification for his actions could conceivably have led to clemency, if not from the judge (whose judicial discretion was severely limited) then from the Cabinet or Crown exercising the royal prerogative of mercy.[29]

Of course, a simple tally of lines spent on different topics does not fully capture the spirit of Riel's speeches. Another problem was his tendency to switch frequently from one topic to another, not only from *ethos* to *pathos* to *logos* but also from one subargument to another within each category. Although the enumeration is admittedly subjective, I counted a dozen such transitions in each speech. The net effect must have been to bewilder the audience.

Ethos

Riel spent a lot of time in his first speech trying to establish his credibility as a divinely inspired prophet. He first broached this subject with reference to his "mission": "It is true, gentlemen, I believed for years I had a mission, and when I speak of a mission you understand me not as trying to play the role insane before the grand jury so as to have a verdict of acquittal upon that ground. I believe that I have a mission."[30] He tried to establish the veracity of his mission by quoting a series of things that francophone Catholic priests and bishops had said to him over his lifetime – probably not the most persuasive way of approaching a jury made up of anglophone Protestants. But he also tried to make up for that by emphasizing his desire "to leave Rome aside" and "bring out practical results" by promoting a new religion in the New World, in which "my children's children will shake hands with the Protestants of the new world in a friendly manner."[31] Later, he added that his mission was to be a prophet: "I say humbly, through the grace of God, I believe I am the prophet of the new world."[32]

Presenting himself as a prophet with a divinely endowed mission also required Riel to combat the label of insanity that his lawyers were trying to use in his defence. "Even if I was going to be sentenced by you, gentlemen of the jury, I have the satisfaction if I die – that if I die I will not be reputed by all men as insane, as a lunatic."[33] He then went over bits and pieces of testimony from various witnesses, attempting to show that the things he was reported to have said and done had been misinterpreted or had made sense in context. Of course, in substantiating his mission in that way, he was doing the prosecution's work; but we have already seen that defending his mission was more important to him than defending his life. Not surprisingly, then, he actually welcomed the jury's guilty verdict when he spoke to the court in his second speech. "And in some way I think, that, to a certain number of people, the verdict against me to-day is a proof that maybe I am a prophet, maybe Riel is a prophet, he suffered enough for it."[34]

In a conventional analysis, Riel's attempt to establish credibility by claiming the status of divinely inspired prophet makes little sense. Since it required him to attack the case that his attorneys had built, it undermined his best chance of being acquitted. And in the larger court of public opinion in late-nineteenth-century North America, talking about his prophetic mission was likely to convince the public that he really was crazy – perhaps not in the legal sense of not being responsible for his actions but in the ordinary sense of not having anything worthwhile to say about public affairs, of being a monomaniac trapped in a private world of imagined revelations.

Pathos

To the limited extent that Riel evoked compassion in his first speech, it was the emotion of pity for helplessness. He depicted himself as an infant on his mother's lap, "and I am sure that my mother country will not kill me more than my mother did forty years ago when I came into the world, because a mother is always a mother, and even if I have my faults if she can see I am true she will be full of love for me."[35] Later on he repeated, "My condition is helpless, so helpless that my good lawyers, and they have done it by conviction ... my condition seems to be so helpless that they have recourse to try and prove insanity to try and save me in that way."[36] Riel's followers were also to be pitied: "When I came into the North-West in July, the first of July, 1884, I found the Indians suffering. I found the half-breeds eating the rotten pork of the Hudson Bay Company and getting sick and weak every day."[37]

In his second speech, Riel changed tack a bit. Instead of emphasizing his helplessness, he went into great detail about how badly the government had treated him over the years, most of all by denying him an amnesty for the execution of Thomas Scott and other acts committed during the 1869–70 uprising in Manitoba. This was the personal part of the second speech's narrative of Métis grievances over how the government had deprived them of their land claims. I have classified the personal as *pathos* and the political as *logos*, although the boundary between the two can be debated. In any case, he was trying to evoke outrage and compassion for a victim of injustice.

Logos

Riel's attempt in his first speech to justify his actions was extraordinarily brief. He referred in a couple of paragraphs to Métis grievances, without being specific about what they were, and to his attempts to submit a petition to the government. Then he claimed that the government had attacked the Métis: "The agitation in the North-West Territories would have been constitutional, and could certainly be constitutional to-day if, in my opinion, we had not been attacked."[38] Then at the end he brought up his famous pun about "insane and irresponsible" government: "British civilization which rules to-day the world, and the British constitution has defined such government as this is which rules the North-West Territories as irresponsible government, which plainly means that there is no responsibility, and by all the science which has been shown here yesterday you are compelled to admit if there is no responsibility, it is insane."[39] The pun was amusing but not persuasive because Riel had not adduced enough evidence to show that the Canadian government had not fulfilled its legal obligations to the people of the North-West Territories.

Curiously, he did present that sort of evidence at greater length in his second speech, when it was now too late to ward off a verdict of guilty. He rehearsed in considerable detail the government's alleged failure to respect Métis land rights in the North-West, as well as its failure to keep its promises to him. But Riel undermined this potentially powerful indictment of the government's record over the previous fifteen years by talking at great length about his visionary plan for encouraging immigration to the North-West and thereby creating a Métis mosaic as the immigrants intermarried with the Native inhabitants. This plan was connected in his mind with the fact that the Métis of Manitoba had received about a seventh of the lands of that province through section 31 of the Manitoba Act of 1870, which set aside 1.4 million acres for "the children of the half-breed heads of families."[40] But it all must have sounded like a meaningless jumble to the audience:

> Our right to the North-West is acknowledged, our co-proprietorship with the Indians is acknowledged, since one-seventh of the land is given us, but we have not the means to be heard. What will we do? I have said to some of my friends if there is no other way we will make the people who have no country understand that we have a country here which we have ceded on condition. We want a seventh of the lands, and if the bargain is not kept, it is null and void, and we have no right to retreat again. And if we cannot have our seventh of the lands from Canada, we will ask the people of the States, the Italians, to come and help us as emigrants. The Irish, I will count them. Now, it is my turn; I thank you. I count them, and I will show you if I made an insane enumeration of the parties. I said we will invite the Italians of the States, the Irish of the States, the Bavarians of the States, the Poles of the States, the Belgians of the States, and if they come and help us here to have the seventh, we will give them each a seventh; and to show that

> we are not fanatics, that we are not partisans, that we do not wish only for the Catholic, but that we have a consideration for those who are not Catholics, I said we will invite the Danes, we will invite the Swedes who are numerous in the States, and the Norwegians, to come around, and as there are Indians and half-breeds in British Columbia, and as British Columbia is a part of the immense North-West, we said not only for ourselves, but speaking of our children, we will make the proposition that if they help us to have our seventh on the two sides of the Rocky Mountains, they will each have a seventh.

If Judge Richardson got anything from this, it would have been that Riel had intended to invite ethnic immigrants from the United States to resettle in western Canada in order to help the Métis take political control. Strip away the bizarre ethno-religious aspects of the plan, and it is further evidence of Riel's treasonous intent.

CONCLUSION

Riel's rhetorical situation was virtually hopeless from the start. He had to defend himself before an English-speaking jury and a stipendiary magistrate in Regina, not, as he had imagined, before the Supreme Court of Canada in Ottawa or Montreal. Since he had no money to hire his own lawyers, he had to depend on counsel hired by others, who were determined to argue that he was not guilty by reason of insanity. He thus had to argue against prosecution and defence lawyers at the same time. But he made his predicament worse by trying to pack too much into his speeches. Instead of arguing that the government had neglected the Métis grievances and then attacked the community at Batoche – an argument that might conceivably have impressed the jury – he insisted on going over his whole career and spelling out his prophetic mission. As the old Greek proverb has it, "Whom the gods would destroy, they first make mad."

NOTES

1 Maggie Siggins, *Riel: A Life of Revolution* (Toronto: HarperCollins, 1994), 428.

2 George F.G. Stanley, *Louis Riel* (Toronto: McGraw-Hill Ryerson, 1963), 356; Desmond Morton, *The Last War Drum: The North West Campaign of 1885* (Toronto: Hakkert, 1972), 158.

3 See, for example, "Les Métis du Nord-Ouest," *Montreal Daily Star*, 28 November 1885, in George F.G. Stanley et al., eds, *The Collected Writings of Louis Riel/Les ecrits complets de Louis Riel* (hereafter CWLR), vol. 3, 278–94 (Edmonton: University of Alberta Press, 1985).

4 Louis Riel, "Address to the Jury," 31 July 1885, vol. 3, 536.

5 Louis Riel, "Address to the Court," 1 August 1885, *CWLR*, vol. 3, 547.

6 Tom Flanagan, Christopher Alcantara, and André Le Dressay, *Beyond the Indian Act: Restoring Aboriginal Property Rights* (Montreal and Kingston: McGill-Queen's University Press, 2010), 13.

7 Riel, "Jury," *CWLR*, vol. 3, 524.

8 Ibid.

9 Glen Campbell and Tom Flanagan, "Newly Discovered Writings of Louis Riel," in Christopher Adams, Gregg Dahl, and Ian Peach, eds, *Metis in Canada: History, Identity, Law and Politics* (Edmonton: University of Alberta Press, 2013), 259.

10 Riel, "Court," *CWLR*, vol. 3, 541.

11 See, as one example among many, the various drafts of his reply to the delegates who invited him to come to the Saskatchewan Valley, dated 5 June 1884, *CWLR*, vol. 3, 1–6.

12 Louis Riel, "Notes for the Address to the Jury, July 1885," *CWLR*, vol. 3, 155.

13 Riel, interview with Charles Bruce Pitblado, 19–20 May 1885, *CWLR*, vol. 3, 521.

14 Riel to Romuald Fiset, François-Xavier Lemieux, and Charles Fitzpatrick, 16 June 1885, *CWLR*, vol. 3, 104.

15 Riel to Richard Burton Deane and Edgar Dewdney, 24 June 1885, *CWLR*, vol. 3, 112. In the same vein is Riel to Richard Burton Deane and Edgar Dewdney, 27 June 1885, *CWLR*, vol. 3, 113–16.

16 Riel to Richard Burton Deane, Edgar Dewdney, and John A. Macdonald, 6 July 1885, *CWLR*, vol. 3, 125.

17 Riel to John A. Macdonald, 16 July 1885, *CWLR*, vol. 3, 130.

18 Riel to James W. Taylor, 21 July 1885, *CWLR*, vol. 3, 134–5.

19 See in general Thomas Flanagan, *Louis 'David' Riel: 'Prophet of the New World,'* 2nd ed. (Toronto: University of Toronto Press, 1996).

20 Riel to Alexandre-Antonin Taché, 24 July 1885, *CWLR*, vol. 3, 147.
21 Ibid., 149. "Vitale" is an untranslatable pun based on the village of St Vital, Riel's home in Manitoba. The usual French word in this phrase would be "vivante."
22 Riel, "Hymn," 9 June 1885, *CWLR*, vol. 4, 407. Note the pun, "Right in Montreal," which could mean "There in Montreal" or "Justice in Montreal." Although it is not a very effective pun, it illustrates Riel's knowledge of the various connotations of English words.
23 Riel to Romuald Fiset, 12 or 13 June 1885, *CWLR*, vol. 3, 99.
24 Thomas Flanagan, *Riel and the Rebellion: 1885 Reconsidered*, 2nd ed. (Toronto: University of Toronto Press, 2000), 138–40.
25 Desmond Morton, ed., *The Queen v Louis Riel: Canada's Greatest State Trial* (Toronto: University of Toronto Press, 1974), 214.
26 Ibid., 211–12.
27 Aristotle, *Rhetoric*, 1356a, in Richard McKeon, ed., *The Basic Works of Aristotle* (New York: Random House, 1941), 1329–30.
28 See note 3.
29 Flanagan, *Riel and the Rebellion*, 157, 171.
30 Riel, "Jury," *CWLR*, vol. 3, 526.
31 Ibid., 531.
32 Ibid., 534.
33 Ibid., 528.
34 Riel, "Court," *CWLR*, vol. 3, 541.
35 Riel, "Jury," *CWLR*, vol. 3, 524.
36 Ibid., 531.
37 Ibid., 524.
38 Ibid., 529.
39 Ibid., 535.
40 Manitoba Act of 1870, SC 1870, c. 3, s. 31. See at greater length Thomas Flanagan, *Metis Lands in Manitoba* (Calgary: University of Calgary Press, 1991). On the "seventh," see Flanagan, *Riel and the Rebellion*, 94–6.

4

The Use and Force of Rhetorical Strategies in Louis Riel's First Speech

CHRISTOPHER TINDALE

Riel's first speech to the jury can appear disconnected and unorganized. But viewed from the perspective of a rhetorical tradition with which we should expect him to be acquainted, it reflects the deliberate employment of rhetorical strategies, along with the kind of extemporaneous response to the exigencies of the trial that the rhetorical tradition taught. In this chapter, I explore some of the key rhetorical strategies apparent in Riel's first speech and consider the intentions behind them. The resultant picture reveals a man less excitable by the situation and clearer in his sense of himself and what his speech can accomplish.

THE TRIAL GENRE

Louis Riel's speech to the jury in 1885 fits into the historical genre of defence speeches, intentionally or otherwise. We might expect any such speech to include appeals to God and to duty, explanations of behaviour, and/or justifications that show how an individual has benefited those he or she served, yet it is difficult to show an intentional adherence to a pattern. On the other hand, such speeches, drawing from a range of rhetorical strategies available, can make more explicit allusion to their predecessors. And something of this will be suggested below.

RHETORICAL FEATURES OF THE FIRST SPEECH

I will look first at some of the shorter strategies employed before exploring a larger one in more detail. The first point I wish to draw attention to bears on the controversial question of Riel's sanity, about which there is some debate. The sensible thing might be for a defendant to avoid all such mention of his state of mind, unless there is reason to think that posing insanity as a possibility might support one's defence in some way. It is a clever device to draw attention to something by saying or indicating that you will *not* do so – that you will not say something since it is not worth mentioning. Socrates, in his trial speech, famously tells the jury that he will not bring his family before them in any sort of appeal. But in saying this, he does indeed bring forth his family – in the imaginations of the jury members, where they might have much greater effect than if they were physically brought into the court. There is something of this device – what rhetoricians call the *praeteritio* – in the way Riel handles the question of his sanity. He does not explicitly bring it to the jury's attention by saying he will not mention it, but there is an indirectness to the way he treats it that plays a similar role in allowing it to be considered on his terms and without inviting the criticism that he is placing attention on it. In one of the relevant parts of the speech, Riel says,

> Today, when I saw the glorious General Middleton bearing testimony that he thought I was not insane, and when Captain Young proved that I am not insane, I felt that God was blessing me, and blotting away from my name the blot resting upon my reputation on account of having been in the lunatic asylum of my good friend Dr Roy. I have been in an asylum, but I thank the lawyers for the Crown who destroyed the testimony of my good friend Dr Roy, because I have always believed that I was put in the asylum without reason.[1]

Ambiguity is compounded here by his reference to the asylum. Drawing attention to the favourable testimony should be sufficient to deal with the question of his sanity. But by proceeding to mention his time in the asylum, he draws the jury's attention back to the question, enhancing its plausibility. Still, he is using the testimony of others to point to a specific way of understanding his state of mind. This is not, of course, the same as not talking about something, but it has a similar effect.

More apparent, perhaps, is his strategy of turning the tables (or the *peritrope*), used in the later stages of the speech. The strategy involves turning back upon one's accusers charges that have been brought against you. Consider the following passages:

> British civilization which rules today the world, and the British constitution has defined such government as this is which rules the North-West Territories as irresponsible government, which plainly means that there is no responsibility, and by all the science which has been shown here yesterday you are compelled to admit if there is no responsibility, it is insane ...
>
> By the testimony laid before you during my trial witnesses on both sides made it certain that petition after petition had been sent to the Federal Government, and so irresponsible is that Government to the North-West that in the course of several years besides doing nothing to satisfy the people of this great land, it has even hardly been able to answer once or to give a single response. That fact would indicate an absolute lack of responsibility, and therefore insanity complicated with paralysis.[2]

In these remarks, Riel turns the charge of insanity back upon the government that accuses him, showing both that it is a contested concept and that, because of this, it can apply as much to an institution as to an individual. What matters, on this interpretation, is the quality of the actions of an agent. Irresponsible actions are not rational, and the absence of rationality indicates insanity. Thus Riel can proceed in the next paragraphs to refer to the actions of "an insane and irresponsible Government and its little one – the North-West Council" and to propose as part of his defence that he should be acquitted because he has been quarrelling with "an insane and irresponsible Government": "You are perfectly justified in declaring that having my reason and sound mind, I have acted reasonably and in self-defence, while the Government, my accuser, being irresponsible, and consequently insane, cannot but have acted wrong, and if high treason there is, it must be on its side and not on my part."[3] This last passage brings out the import of this device: he should hardly be accused, and therefore condemned, when his accusers (or the body they serve) are guilty of the same.

The matter of Riel's *ethos*, or character, how it is conveyed, and the role it plays in his defence deserves a full treatment of its own. But I address it here in less detail simply as a further rhetorical device used by the defendant. Aristotle lays considerable importance on *ethos* in his *Rhetoric*. Along with *logos* and *pathos*, it is one of three "proofs" by which an audience might be persuaded. Importantly, however, Aristotle stresses ways that *ethos* should be conveyed *through* the speech itself rather than by having a speaker draw on his or her prior reputation. "Character," writes Aristotle, "is almost, so to speak, the most authoritative form of persuasion," and there is persuasion through character when "the speech is spoken in such a way as to make the speaker worthy of credence; for we believe fair-minded people to a greater extent and more quickly on all subjects in general and especially where there is not exact knowledge but room for doubt."[4] In particular, Aristotle, in his ethical writings, is interested in how a speaker can convey qualities that are important, like practical wisdom, virtue, and goodwill. Of course, there are limitations to the account of *ethos* in the *Rhetoric*, and a situation like that faced by Riel brings these to the fore. He can hardly expect his jury to be unacquainted with his reputation (Aristotle seems to envisage situations where a speaker is largely unknown), so whatever he might attempt to accomplish in his speech, he faces this impediment, nor will it be easy for him to invest his character with aspects of practical wisdom or virtue and present himself as fair-minded. Yet at the same time we cannot ignore that judgments we make about character influence judgments we make about what people say. And to a certain degree, any speaker can strive to create a picture of character through speech that can challenge a view previously held. This Riel may be seen to do in the many paragraphs where he appeals to his duty, to the benefits he has provided, and to the sacrifices he has made in putting public interests before his own. He presents himself as an advocate of truth, a man of God, and a person who has respect for the police. In general, his speech is measured and not given to excited outbursts, suggesting a man of balance who is in control ("master of myself," as he says),[5] and he shows respect for the court, explicitly wishing not to give offence. In sum, the discourse he delivers exhibits some practical wisdom and a considerable measure of goodwill.

IMPLICIT AND EXPLICIT ALLUSION

The larger strategy active in the speech that I want to address is that of allusion, an important strategy of rhetorical argumentation. Allusion is one way that commonalities between speaker and audience can be invoked. It seemed a commonplace of ancient trials for defendants (or second-hand observers of the trials) to make allusion to other trials of similar circumstances. We see both implicit and explicit parallels, for example, between Plato's *Apology* of Socrates and Gorgias's *Palamedes* and between Isocrates's defence speech and Plato's *Apology*. Understanding that there are commonplaces within genres like the defence speech allows us to see the nature and legitimacy of such parallels rather than accusing authors like Plato and Isocrates of what in the present day would be termed plagiarism.

Allusion, Chaim Perelman and Lucie Olbrechts-Tyteca tell us in their *New Rhetoric*, "depends on a relationship with something that is not the immediate object of discourse."[6] A key variety of this strategy of rhetorical argumentation is textual allusion, where an arguer uses intertextual references and imitations to evoke ideas in the minds of audience members and thereby encourage them to draw a specific conclusion. Such allusions can convey an indirect reference in passing without making explicit mention of something. So those who employ this strategy must be confident that the reference alluded to is sufficiently present in what I would call the cognitive environment – that is, the beliefs, knowledge, and background information – of an audience in order for the association to be grasped and the conclusion drawn. On strictly logical terms, textual allusion and imitation of this nature would seem to have no argumentative force. But when audience considerations are highlighted in a rhetorical treatment of argumentation, the power of the strategy becomes evident.

If Riel invokes any predecessor through allusion, it is the figure of Socrates – another outspoken character who was ill-suited, it seems, for the times in which he found himself and was a victim of a justice system that conflicted with his own conception of what was just. Let us recall some of the salient features of Plato's *Apology* of Socrates. (As I mentioned above, we could also recover parallel features from Gorgias's trial speech of *Palamedes*, but we do not need to reach that far since it is the Platonic material with which Riel was more likely familiar.)

Socrates enters the unfamiliar environment of the law court stressing his inability to defend himself well with words. He addresses the charges and rumours that have been brought against him, speaking first to the longstanding negative reputation that he seems to have acquired and then to the specific current charges of believing in false gods and corrupting the youth. He argues that he would not have corrupted the youth in part because he would then have put himself in the company of those who might harm him. And he invites his accusers to bring forward witnesses to this corruption (even the relatives of the "corrupted" youth). He argues that Meletus (his chief accuser) is contradicting himself when he charges Socrates with believing in false gods because he also claims that Socrates does not believe in gods ("You cannot be believed, Meletus, even, I think, by yourself").[7] He then defends his own occupation and life in general, arguing that he has always done his duty and stood by what was just. He has shown this through his deeds, opposing the wrongful dictates of both the democracy and the oligarchy. He also speaks to his role (divinely appointed) as benefactor to the city. Then, having still been convicted by the jury in spite of all this, he proposes a penalty suitable to a benefactor such as himself. And after having been condemned to death, he addresses the jury, pointing to the negative reputation the city will now acquire for an unjust judgment.

The first set of parallels from Riel's speech is implicit. He begins his remarks with a similar warning that he cannot speak well. In his case, it is his alleged inability to speak English well and the excitement of the occasion that may put his mind "out of its ordinary condition."[8] As with Socrates, it is hard to see this as a real concern. Both speak well and are able to make their cases. But a point they have in common is that each is speaking in an unfamiliar mode: Socrates is not used to giving long speeches, having acquired his fame through the short dialectical exchanges of questions and answers; and Riel is not used to speaking in English. Each is out of his natural element and must adapt to a different manner of communication.[9]

The missions of each man can be considered divine. Socrates lays the origin of his practice at the feet of the oracle and judges he has the authority of the god to conduct himself as he does. Riel notes that it is the Crown's judgment that he is "naturally inclined to think of God at the beginning" of his actions.[10] This is a judgment he does not contest, but he worries that it will be seen as a symptom of his insanity. This is a point he will repeat: "I believed for years I had a mission, and when I speak of a mission you will understand me not as trying to play the role of insane before the grand jury so as to have a verdict of acquittal upon that ground."[11] And again: "I say that I have been blessed by God, and I hope that you will not take that as a presumptuous assertion."[12] These and his other appeals to God (at least six in this first speech) contribute, perhaps, to his *ethos*, but they also lay claim to an authority outside that of the court's authority. In this, he parallels Socrates's position. In each case, it is an ambivalent defence: for Riel it may support the concern for his sanity, and for Socrates it would seem to fuel the charge that he observes a new religious perspective. Riel, in fact, closes by placing his speech under God's protection, another statement that might be seen to undermine the authority of the court. There again, we should not find it unusual for speakers to make such appeals when addressing an audience who regard gods or God as important. But to invest themselves with an authority derived from that source might be at best unwise.

An appeal to duty is an important aspect of Socrates's defence and is connected with his claims that his actions have been beneficial to the state and have involved sacrifices on his part. We find all these points active in Riel's speech. Early on, he notes that he had directed his attention to help the Indians, the half-breeds, and the whites to the best of his ability. And this translates into the petitions of the next paragraph, in the pressing of which he concludes, "I have done my duty. I believe I have done my duty."[13] There is a compulsion underlying this belief, a compulsion that forces action in spite of the risks it involves. "[W]hat I have done," he says, "and risked, and to which I have exposed myself, rested certainly on the conviction, I had to do, was called upon to do something for my country."[14] That duty, clearly connected to the mission already noted, issues in what he calls "practical results,"[15] a series of benefits to specific groups and to society. He has been a big influence and, through the grace of God, the founder of Manitoba. He admits he has been outspoken and that sometimes his accusations have been forceful and caused offence. But the "one who has the courage to speak out in that way, instead of being an outrageous man, becomes in fact a benefactor to those men themselves, and to society."[16] As such a benefactor, he has been "acknowledged as a leader of good in this great country"[17] and even, more pretentiously, perhaps, as "the prophet of the new world."[18] An extreme claim, indeed, but I set it here against the similar claims to a prophetic voice made by Socrates and the predictions he makes about those who will follow him and act like gadflies on the state.

Such beneficial behaviour has not come without costs to Riel himself. He has put public interests ahead of his own, he claims, and his actions on the part of Manitoba and the people of the North-West involved fifteen years of suffering on his part. He reiterates this at the end of the speech: "For fifteen years I have been neglecting myself,"[19] citing as evidence his neglect of clothing and that his wife and children had to go without. And he concludes with the observation that although he worked to improve the condition of the people of the Saskatchewan Valley at the risk of his life, he "never had any pay."[20] Socrates also takes pains to insist that he took no fee for what he did and as a result went without many things. For both men, the duty and the force of the mission outweigh any thought of reward.

The claims of self-neglect for the benefit of those being served and the disclaimer regarding payment for such services emphasize that the actions performed were done foremost out of a sense of duty, grounded in the belief in a mission of divine origin. In these matters, Riel stands out as a Socratic figure. In one final respect, the sense of allusion is more striking still. At one point in his defence, Riel speaks of the guidance – and in this instance, consolation – he receives from a "spirit": "Last night while I was taking exercise the spirit who guides and assists me and consoles me, told me that to-morrow somebody will come ... if the spirit that directs me is the spirit of truth it is today that I expect help."[21] The nature of this guidance is unclear, but we might compare it to the famous inner voice associated with Socrates. This voice frequently spoke to him, even on small matters, if what he was going to do was wrong. "But now," he tells the jury, "this thing which might be thought, and is generally considered, the greatest of evils has come upon me; but the divine sign did not oppose me either when I left my home in the morning, or when I came here to the court, or at any point of my speech when I was going to say anything."[22] In the case of Socrates, this voice is essentially negative: it restrains him, telling him what *not* to do and never what to do. But it is no less obscure than Riel's "spirit" and continues to confound the commentators as to its nature and meaning. In both cases, the "spiritual manifestation," as we might call it, serves as a guide for action and, importantly, as a source of confidence. Socrates is confident that the course he is on is a correct one because if it were not, he believes, he would have been checked in that course. He calls this a "convincing proof" because otherwise "the accustomed sign would surely have opposed me if I had not been going to meet with something good."[23] Riel likewise believes that the events of the day after the spirit speaks to him (events involving the assistance promised to him) confirm the truthfulness of the "spirit" and thus, implicitly, the correctness of his mission and its divine origin. It is this confidence that sustains him in the courtroom.

There are, then, some fairly vivid parallels between the two cases, parallels that touch on the character and beliefs of the men involved, their justifications for the lives they have lived, and the specific ways that they make their respective defences and insist on a somewhat privileged position. Can we conclude from this that the allusions are deliberate on Riel's part? That is a more difficult case to make, and, of course, the power of this as a *strategy* of argumentation assumes that it is intentional. If it is the case, however, that an acquaintance with the *Apology* of Plato can be shown to be part of Riel's background or, less directly, of any educational program that he would have received, the case for deliberate allusion would appear plausible. In this regard, several historical matters are relevant, although none of them is definitive. We know that Louis Riel travelled to the East to receive advanced education, arriving in Montreal in July 1858. As a student at the Collège de Montréal, he would have received a classical education "based on the literature and philosophy of Greece, Rome and France."[24] The curriculum placed less emphasis on the sciences, concentrating instead on Latin, Greek, French, English, some mathematics, and, of course, philosophy. Beyond this, we learn, "Louis excelled in belles-lettres and rhetoric."[25] Thus any reservations we may have about Riel's familiarity with the relevant literature might be withdrawn. Plato's *Apology*, then as now, is central to any education in Greek philosophy, and Riel would have been acquainted with both the figure of Socrates and his arguments. Moreover, his study of rhetoric would have exposed him to the power of effective speaking and the devices that might be employed to this end. As historian Jack Bumsted notes, "Riel was a shrewd and capable orator."[26]

THE ARGUMENT FROM ACTION

In the trial speeches of the ancient world, the words of the defence were often supported by the actions of the defendant. Or, rather, the actions were advanced in lieu of words. We might call this the "argument from action," where the actions themselves serve as premises from which conclusions are to be drawn. In this way, Gorgias has Palamedes insist that words cannot bring "the truth of deeds clearly and certainly before their hearers"[27] and that his jury should thus take note of actions rather than words.[28] Socrates addresses his jury in a similar vein, offering as "powerful proofs" not "mere words" but what they "honour more – actions."[29] He then proceeds to describe two occasions when he opposed the wrongdoing of the state, once under the democrats and later under the oligarchs. Such a strategy might be the wise choice of any speaker who thinks he lacks a facility with words, as both the principal speakers considered here claim to do. But there is also vividness available to the descriptions of events that abstract proofs will lack. One can claim to be just in one's actions, but *showing* this makes the argument a concrete one. It creates an experiential link between speaker and hearer, who is invited to weigh the merits of the actions performed alongside the justifications given for those actions. Riel's speech constitutes such an argument from action. He believes, after all, that he has "acted reasonably and in self-defence." But more specifically, and in relation to the point being made here, he will use actions as premise material from which he expects conclusions to be drawn. Let's consider two instances of this. Charles Nolin, along with others, had spoken of Riel's ambition. But in response to this charge, Riel notes that he was approached to take a place in the council but declined. This action indicates he was not anxious for position. He speaks of it to defend his character; it is part of the defence that he is not egotistical. Again, witnesses have testified that he did not respect the police. He contests this, insisting that he has always respected the officers of the police. And as proof of this, he offers the fact that he wrote to Major Lief Crozier that he had respect for him. Such events are matters of immediate historical record, available to the court for confirmation. As such, although they might be subject to other interpretations, they have a clarity and straightforwardness about them that is often lacking in verbal arguments. With the latter, a speaker might always be accused of cleverness in the use of words. But the public facts of a man's life are open to all and to a certain degree speak for themselves.

COMMUNION

We might consider the choice of these rhetorical strategies. On one level, they may be simply what Riel judges appropriate for the circumstance in which he finds himself, given his training in rhetoric. The trial speech lends itself naturally to certain language and devices, as well as to specific moves of justification – both of character and of actions. But strategies like allusion and the argument from action, if they are deliberate, aim at a more personal goal: to effect communion with the audience. This term – "communion" – is used by Perelman and Olbrechts-Tyteca for the establishment of commonalities or identifications with the audience.[30] It involves meeting an audience on their terms, speaking in ways they understand, and drawing on experiences that they would be expected to have shared at some level. Riel is constrained by the nature of the event itself: he must address the charges brought against him, respond to the testimony of witnesses, and stay within the rules of the court. But he decides *how* he will do these things; the style of speech, the figures adopted, and the strategies used are under his inventive control. And the primary consideration in making these decisions is that of audience. Who is he addressing? What do they expect? How can he bring them to relate to him and thus understand the questions in his terms?

And yet, of course, on the face of things, the speech fails. In this, Riel keeps good company. For all his persuasiveness, Socrates fails to persuade his jury and suffers the consequences. If the allusion to Socrates is understood too strongly, then just as Riel would have his jury recall the similarities between these victims of injustice, so they might also recall the fate that Socrates met. Riel's fate is no different; he also fails to persuade.

Rhetorical speech is employed in situations of uncertainty, and any number of things can prevent its success. Riel's closing wish is for God to make it effective, and he believes this is possible because it "is proposed to good men, to good people, and to good ladies also."[31] But effectiveness is not the only measure for rhetorical speech because some audiences may be predisposed never to accept what is put to them and in that sense would be unmovable, no matter how "good" the arguments may be. And as W.M. Parish observes in his study of speeches, "Rhetoric, strictly speaking, is not concerned with the *effect* of a speech, but with its *quality*, and its quality can be determined quite apart from its effect."[32] The strategies Riel employs are appropriate for the occasion, and he employs them well in the sense that they are used correctly. They are designed to make the strongest case in the circumstances, and in that sense they serve him well and he does himself credit. He "speaks well."

There is a further feature of this rhetor's speech worth noting. I observed at the outset that Riel's speech could appear disconnected and unorganized. There is the sense among many that a trial speech should be well prepared in advance. But early orators attacked the idea of the prepared speech, and the rhetorical tradition also favours spontaneous, extempore speeches that address the exigencies of the occasion. That is, a speaker must not only have marshalled the arguments and strategies he will use in his defence; he must also be prepared to respond to what others say, to what happens in the courtroom, to the restlessness or attentiveness of the audience, and so forth. A good speaker is able to leave his prepared texts and work with the materials that the occasion presents. Riel had perhaps a 300-word text from which he extrapolated. For the rest, he spoke extemporaneously, after the best traditions of the past. He thus found a middle ground between the prepared and the spontaneous. If his speech appears disorganized, it is because we judge it from standards that it was never intended to meet. And in doing so, we fail to see its merits and power. In many respects, it was a makeshift speech in a makeshift courtroom, and the defendant did a masterful job of using the materials available to him – from the events of his life, the testimonies of others, and the exigencies of the courtroom – to construct a speech that rightfully stands among the best of the genre.

NOTES

I am grateful to audience members of the Riel's Defence conference at the University of Windsor, October 2010, whose discussion and comments led to a number of improvements of this chapter.

1 Desmond Morton, ed., *The Queen v Louis Riel: Canada's Greatest State Trial* (Toronto: University of Toronto Press, 1974), 316.

2 Ibid., 323–4.

3 Ibid., 324.

4 Aristotle, *On Rhetoric*, trans. G. Kennedy (Oxford: Oxford University Press, 2007), 1.2.4.

5 Morton, ed., *Queen v Louis Riel*, 322.
6 Chaim Perelman and Lucie Olbrechts-Tyteca, *The New Rhetoric: A Treatise on Argumentation*, ed. John Wilkinson and Purcell Weaver (Notre Dame, IN: University of Notre Dame Press, 1969), 170.
7 Plato, *The Apology*, trans. H.N. Fowler (Cambridge, MA: Harvard University Press, 1914), 26e.
8 Morton, ed., *Queen v Louis Riel*, 311.
9 Unlike the Socrates of the *Apology*, Riel reminds his audience at the end of the speech of his "imperfect way of speaking." See ibid., 325.
10 Ibid., 311.
11 Ibid., 314.
12 Ibid., 315.
13 Ibid., 312.
14 Ibid., 314.
15 Ibid., 313.
16 Ibid., 319.
17 Ibid., 311.
18 Ibid., 322.
19 Ibid., 324.
20 Ibid.
21 Ibid., 320.
22 Plato, *Apology*, 40a–b.
23 Ibid., 40c.
24 Maggie Siggins, *Riel: A Life of Revolution* (Toronto: HarperCollins, 1994), 50.
25 Ibid., 53.
26 J.M. Bumsted, *Louis Riel v. Canada: The Making of a Rebel* (Winnipeg: Great Plains, 2001), 292.
27 Hermann Diels and Walther Kranz, *Die Fragmente der Vorsokratiker* (Berlin: Weidmannsche Verlagsbuchhandlung, 1952), 11a, 35.
28 Ibid., 11a, 34.
29 Plato, *Apology*, 32a.
30 Perelman and Olbrechts-Tyteca, *New Rhetoric*, 51–6.
31 Morton, ed., *Queen v Louis Riel*, 325.
32 W.M. Parish, "The Study of Speeches," in *Readings in Rhetorical Criticism*, 2nd ed., ed. C.R. Burgchardt (State College, PA: Strata, 2000), 39.

5

Narrative and Logical Orders in Louis Riel's Address to the Jury

HANS V. HANSEN

My God! Through Jesus Christ, lend an immense effect to the speeches I made before the court on July 31 and August 1, 1885; and let them work for me in every way ... O Saint Joseph, intercede with God to set the great majority of people in the country on my side.

Riel, diary entry, August 1885[1]

INTRODUCTION

During his trial in Regina in July and August 1885, Louis Riel made two speeches. The first speech was made right after his lawyer had addressed the jury and before the Crown made its summation. Riel's second speech was made to the court after the jury announced its verdict and before the judge pronounced the sentence. It is the first speech that is best known, and it is the one to be considered here. About this speech, one of Riel's biographers has written that "Truly, he [Riel] had delivered the performance of his life!"[2] Jack Bumsted also thinks highly of the speech, saying that "Riel was a shrewd and capable orator, and he was nowhere more brilliant than in his speech[es] ... on 31 July and 1 August 1885."[3] Bob Beal and Rod Macleod say it was a remarkable performance and that "There can be few longer or more eloquent such statements in Canadian legal history."[4] Thomas Flanagan has found an eyewitness report that gives us a sense of Riel's speaking style as well as the effect his speech had on his immediate audience:

> At the outset he spoke in a quiet and low tone, many of his statements carrying home conviction to his hearers. "At any rate," was the subsequent comment, "Riel speaks with the belief that he is right." Gradually as he proceeded and got fairly launched into his subject, his eyes sparkled, his body swayed to and fro as if strongly agitated, and his hand accomplished a series of wonderful gestures as he warmed up and spoke with impassioned eloquence. His hearers were spell-bound, and well they might, as each concluding assertion with terrible earnestness was uttered with the effect and force of a trumpet blast. That every soul in the court was impressed is not untrue, and many ladies were moved to tears.[5]

These passages testify to Riel's intelligence and to his ability as an orator, but others, more circumspect, have reservations about the quality of the address to the jury. Joseph Boyden's observation that "Some say Louis's speech is brilliant, others that it is rambling and confused"[6] is indicative of the ambivalence surrounding the speech. George Stanley, for example, wrote that "At first [Riel] was nervous and hesitant in his speech. His sentences were short, jerky, incomplete. But none could question the sincerity of his words."[7] Stanley thought better of Riel's second speech than his first one: "Riel's second speech in the courtroom in Regina, was more coherent, less emotional, than his earlier one. It would almost seem as if his first efforts had been unprepared and delivered spontaneously in an effort to disprove the charge of insanity, while the second was more carefully prepared and thought out in advance. It may have been the speech he had intended to give in the first place."[8] Flanagan too comments that the first speech was rather poorly organized,[9] a judgment recorded twice earlier by Desmond Morton, first in his book about the North-West campaign and then in the introduction to his edition of the Riel trial transcript:

> When, to the despair of his lawyers, he was finally allowed to testify, he totally demolished the case they had sought to construct for him. It was an emotional speech, certainly disordered and rambling, but it contained the eloquent and desperate attempt of a man to prove his own sanity.[10]

> It was not an ordered speech. There were no notes and little evidence of a plan, but there was eloquence and pathos combined as

> the big, powerfully built man struggled simultaneously to plead his sanity and the wrongs of his people. For an hour he continued, sometimes quickly, sometimes slowly, struggling for the English words.[11]

That there is this variety of opinion about Riel's address to the jury – that it was, on the one hand, brilliant and eloquent and, on the other hand, confused and not well ordered – invites us to look at it anew and form our own opinion as to its merits.

In this chapter I want to begin a discussion of two aspects of Riel's speech. One aspect is the *narrative order* of the speech and the question of why the speech is, as Stanley, Morton, and Flanagan have observed, not well ordered as a presentation. The second aspect is the *logical order* of the argumentation in Riel's speech: what is the point of Riel's argumentation, and how do his various arguments fit together? Before proceeding to these two distinct inquiries, it will be helpful to have an overview of the contending positions at the trial, and this can be given by a stasis analysis of the perspectives taken by the Crown, the defence, and Riel himself.

A STASIS ANALYSIS OF THE CONTENDING POSITIONS

One of the many things that make Riel's trial fascinating is that there are three, rather than two, contending positions, each vying for the jury's favour. By the use of stasis theory, we can see how the three positions, that of the Crown, that of Riel's lawyers, and that of Riel himself, were different.

The three parties can be said to have agreed on the fact that the Métis, led by Riel, engaged in armed conflict with Canadian forces. But that is where the agreement ended. What the Crown classified as a rebellion (and the defence lawyers did not demur) Riel classified as self-defence. This leads to a further disagreement about the quality of the armed conflict that motivated Riel. The Crown held that Riel's actions were motivated by self-interest and amounted to high treason; his lawyers said Riel's actions were those of an insane man; Riel himself maintained they were reasonable actions intended to assure the rights of the inhabitants of the North-West. Finally, contrary to the Crown, both Riel and his lawyers felt that the trial was not legitimate, that there was something amiss with the jurisdiction, the judge, and

Table 5.1 Stasis analysis showing the three contending positions in Riel's trial

	Crown's position	*Defence's position*	*Riel's position*
Standpoint	Guilty	Not guilty because of insanity	Not guilty
Fact: *an sit* (is it)?	There was armed conflict between Riel's forces and the Canadian forces.	There was armed conflict between Riel's forces and the Canadian forces.	There was armed conflict between Riel's forces and the Canadian forces.
Definition/ classification: *quid sit* (what is it)?	Rebellion	Rebellion	Self-defence
Quality: *quale sit* (what kind is it)?	High treason, motivated by *self-interest*	Actions of an insane person, motivated by *megalomania*	Reasonable action of a sane person, motivated by a *desire for justice*
Procedure/ jurisdiction	Procedure was legitimate	Procedure was not legitimate	Procedure was not legitimate

the jury, which was unfair to the accused person. Altogether, this stasis analysis shows us the general shape that Riel's defence should take. To secure an acquittal from the jury, he would have to show either (a) that his actions on behalf of the Métis were in self-defence, or (b) that they were reasonable actions taken for a just cause, or (c) that the court that had him on trial was not the appropriate court to hear his case.

THE NARRATIVE ORDER OF THE SPEECH

Considering the narrative order of the speech brings us three questions: (1) What is the narrative order (i.e., how is it to be described)? (2) Why is the narrative order such as it is? (3) Could the same components of the speech have been arranged in a different narrative order so as to make for a better speech?

The speech as we have it from the court stenographers comes to us without headings or divisions, and with slight care to the formation of paragraphs. In attempting to determine the narrative order – the order of the ideas or arguments presented – I have divided the speech into the following eleven parts. (The numbers in parentheses refer to the paragraph numbering in the edition of the speech in this book; page references to the Morton edition of the transcript are in the notes.)

1 *Opening remarks* (I: 1–3).[12] Riel began with expressions of humility, helplessness, and piety. But he did not want his religiosity to be taken as a sign of insanity.
2 *Reasons for political actions* (I: 4, 5).[13] Riel gave his reasons for his political actions: that the conditions in the North-West were terrible and people were suffering. He referred to the fact that petitions sent to the Canadian government had gone unanswered and that he had tried to unite the people of the North-West in a political cause.
3 *Personal grievances* (I: 6–8).[14] Riel answered the charge of egotism and expressed reservations about the suitability of his lawyers; he also complained that the Crown had taken his personal papers from him.
4 *Riel's mission* (I: 9–14).[15] Riel asserted that what he did was for the sake of his country. He declared that he had a mission from God to help the people of the North-West. He also said that the clergy had blessed him and encouraged his work, which he took to be evidence that his mission was real.
5 *Questions about sanity and character* (I: 15–18).[16] Riel argued that he was not insane and not egotistical.
6 *Use of arms was in self-defence* (I: 19–20).[17] Riel maintained that the military action against Canadian troops was in self-defence.
7 *Defence of character* (I: 21–25).[18] Riel reminded the court of his past accomplishments in Manitoba and defended his criticisms of Archbishop Taché. He explained his desire to split from Rome and despaired of his lawyers' defence strategy.
8 *Prophetic abilities* (I: 26–28).[19] Riel admitted and discussed his prophetic abilities.
9 *Various remarks* (I: 29–36).[20] Riel commented on a number of issues, including his plan for immigrations to the North-West and his belief that he was the prophet of the New World.
10 *Criticisms of court and government* (I: 37–42).[21] Riel took the offensive and raised questions about the legitimacy of the court as well as the regional and national governments.

11 *Last remarks* (I: 43–6).[22] Riel made parting remarks to the jury, summing up his case in a dilemma, repeating the arguments of part 2, and appealing for pity.

Why does the speech have the order it does? Whereas Riel's second speech[23] exhibits some planning in that it traces the chronology of events in his political life and explains some of his political views, the speech to the jury does not exhibit any thought-out structure. I think this is best explained by the fact that in large part it is a reactive speech. Being upset by the course of the trial, Riel was distracted from conveying a systematic defence of his actions. As a consequence, his defence intermingled (1) responses to some of the Crown's remarks, (2) observations about the limitations of his own lawyers and the line of defence they were pursuing, and (3) corrections to misleading testimony made by witnesses.

We can see these distractions at work by first considering the last sentence of the Crown's opening statement: "The evidence will show that he [Riel] desired blood, that his only object was to obtain money, or gratify his desire for power and he was altogether reckless of the means he employed to further his ends."[24] This claim impugns Riel's motives as being neither moral nor political but entirely self-interested. Riel answers this allegation very near the beginning of his speech, not by denying that he had self-interested motives but by asserting that his self-interested actions took second place to his political activity (I: 6).[25]

It was also evident at several places throughout the trial that Riel was not happy with his team of defence lawyers. They were building the case that he was insane, and thereby undermining the defence he wanted to make, namely that he was justified in his actions. Although he sounded appreciative of their efforts, the result was that Riel and his lawyers were working at cross-purposes. Thus trying to unravel the case that his own defence team had constructed became a major preoccupation for Riel when his time came to address the jury (I: 15, 25).[26]

Not only did Riel think his lawyers were taking the wrong tack in his defence; he also thought that their examination of witnesses was inadequate because "they were absent of the situation, they did not know all the small circumstances as [he] did" (I: 7).[27] Riel thereby implied that his lawyers were not as effective on his behalf as he would have been if he had been doing the questioning. His frustration with his lawyers was especially evident in connection with Charles Nolin's testimony.[28] Riel wanted to put questions to Nolin himself, but his lawyers threatened to abandon him if he interfered with the way they were conducting the defence.[29] (The exchange between the lawyers, the judge, and Riel on this question occupies nine pages in the Morton edition of the trial transcript, so it was a significant moment during the proceedings.) The upshot was that Riel was forced to keep quiet during the examination of Nolin and therefore had to make room in his speech to the jury for his own reply to him (I: 17, 18, 28).[30] But at that point, his remarks and corrections to testimony seemed trivial and haphazardly inserted into the speech; hence they lost the impact that an immediate rebuttal might have had. Similarly, Riel remarked that he could have corrected the misleading effect of George Ness's testimony if he had been allowed to intervene in the proceedings (I: 22).[31] As with his reply to Nolin, Riel's defensive remarks with regard to Ness, removed from the immediacy of their provocation, seemed remote from the central concern of defending his actions in the North-West.

My suggestion is that because Riel had to counter both the Crown's and his own lawyers' cases, because he could not participate during his trial in the questioning of the witnesses, and because he felt compelled to dispel misleading evidence, his speech – which should have focused on the larger issue of defending his actions on behalf of the inhabitants of the North-West – became overloaded with digressions, some of which it might have been better to leave out. This lack of discernment, coupled with the urgency of the speech situation, goes a long way toward explaining the erratic nature of the speech. Riel was reacting to witnesses who had appeared before him in the preceding seventy-two hours (29–31 July) and to Charles Fitzpatrick, Riel's own lawyer, who, immediately before Riel was allowed to address the jury, had argued that Riel was insane (Friday, 31 July). Thus, given that "the circumstances [were] such as to excite any man" (I: 1)[32] and that he had only hours to arrange his responses to what he had heard from witnesses, it should not be surprising that Riel's speech lacks the finesse and narrative structure of a speech composed and rehearsed at leisure, well in advance.[33]

Given the material Riel chose to work with, could he have organized it so that it would have been a better, more effective speech?

Leaving the introduction and the conclusion aside, there seem to be three broad themes in the speech: (1) the justification for his political work, (2) a defence of his character, and (3) a challenge to the legitimacy of the court. If we reconfigure Riel's speech by reordering the eleven sections as follows, we have this revised (R) version of the original speech:

Introduction
- R1 *Opening remarks* (originally 1)

Personal factors: character, sanity, religion, and honour
- R2 *Personal grievances* (originally 3)
- R3 *Questions about sanity and character* (originally 5)
- R4 *Defence of character* (originally 7)
- R5 *Prophetic abilities* (originally 8)
- R6 *Riel's mission* (originally 4)
- R7 *Various remarks* (originally 9)

Complaints about the legal system
- R8 *Criticisms of court and government* (originally 10)

Political activity
- R9 *Reasons for political actions* (originally 2)
- R10 *Use of arms was in self-defence* (originally 6)

Conclusion
- R11 *Last remarks* (originally 11)

The advantage of this revised narrative order is that it imposes some thematic continuity on the content of Riel's speech, and it saves his best arguments for the end, where plausibly they would be more likely to stay with the jurors when they made their deliberations. (I do not mean that the sections can simply be moved around and the speech will suddenly be a smooth piece of oratory. Adjustments would have to be made to ease transitions.)

THE LOGICAL ORDER OF THE SPEECH

By "the logical order of the speech" is meant the ordering of the reasons and conclusions in the speech in a way that shows how they are evidentially connected to Riel's ultimate conclusion. We are concerned with the logical structure of the argumentation as we find it in Riel's speech, not with the complete and perhaps improved case that might be made for him in hindsight. We are thus engaged in a kind of reconstruction, the aim of which is to fit the discrete relevant arguments in the text into an overall structure that explicitly depicts the case Riel appears to have been making in his defence. Some of the arguments in the speech are immediately obvious; others come to the surface only as we try to understand Riel's intentions and the rhetorical situation.

Riel's ultimate conclusion is that the court should determine that he is not guilty and should acquit him. As noted in the last section, there are three broad themes in Riel's speech: one is about the justification for his political work, another is a defence of his character, and the third concerns the legitimacy of the court that has him on trial. With regard to each of these themes, Riel has a standpoint, and each standpoint is associated with a number of arguments. A fourth standpoint that will also be explored concerns Riel's appeals to sympathy. (I am using the term "standpoint" to indicate a position that a speaker feels strongly about and is prepared to support with arguments and to defend against objections.)

Standpoint 1: "My actions were justified"

Riel's plea that he was not guilty of high treason rested primarily on his standpoint that his actions were justified. The reasons he gave in support of this standpoint can be gathered together as his justification argument. It consists of two distinct arguments: one gives positive support for his actions by claiming they were based on justice; the other is intended to undermine the charge that his actions were treasonous. We may refer to these respectively as Riel's justice argument and his no-treason argument. Each of these arguments, by itself, gives a good reason against the charge; taken together, they make Riel's case stronger. The outline of the justification argument is as follows:[34]

1 Riel's actions were justified (*justification argument*):
  1.1 The people were suffering, the government was unresponsive to their needs, and there was no other way (*justice argument*).
  1.2 Riel's actions did not constitute treason (*no-treason argument*):
    1.2.1 Riel only wanted to extend rights of Canadians to the people of the North-West;
    1.2.2 Riel and his followers acted in self-defence (*self-defence argument*);

1.2.3 It was the Canadian government that was treasonous, not Riel.

(1.1) *Justice argument*. Riel said that when he returned to Canada in 1884 he found the Indians suffering, the Métis eating rotten pork, and the whites deprived of their public liberties (I: 4),[35] and accordingly, he was "called upon to do something for [his] country" (I: 9),[36] something that he took to be "his duty" (I: 5).[37] Although there is a paucity of factual information about the condition in the North-West in the speech, we may safely assume that these conditions were indeed bad and known to the jurors. Riel's reasoning thus takes the form of a moral argument: he had a duty – a moral reason – to do what he did, a duty to alleviate the suffering of the Indians and the Métis and to right the social injustice they and the whites suffered. His argument gains in significance when Riel depicts himself as a patriot acting in his mother country's interest (I: 3).[38]

Three separate factors underlie Riel's justice argument: (1) the bad conditions endured by the inhabitants of the North-West, (2) the fact that "petition after petition had been sent to the Federal Government" by the Métis (I: 41; see also I: 5, 6, 19, 42),[39] and (3) the perception that the government had been unresponsive to those petitions. Riel and other Métis came to believe that they were being treated unfairly and that something had to be done to end the injustice. Based on what may be called *two-wrongs* reasoning,[40] Riel sought to remedy the greater wrong (the negligence of the government) by a less egregious wrong (taking up arms against that same government). He appears to argue that his actions were justified because it was the only way to assuage the unjust conditions. In taking the course of action that they did, Riel and his associates were undoubtedly trying to duplicate the series of events that had been successful fifteen years before when the national government had been forced to negotiate with the Riel-led provisional government of the Red River settlement, leading to the creation of the Province of Manitoba. However, this instance of two-wrongs reasoning remains controversial. First, because it is difficult to weigh the two wrongs against each other and say which is the greater wrong. Second, because the question of whether the Métis really had exhausted all the possible legal avenues for bringing positive change to the North-West, and whether there was actually a need to take extraordinary means, is not clear. Flanagan believes that at the time the rebellion started, the government was taking "major step[s] towards settling the métis claims."[41] Whatever information might have been available to Riel, in this speech he depicts the situation in the North-West as no longer tolerable and himself as motivated by a call to end an injustice.

(1.2) *No-treason argument*. There are three related arguments that can be seen as supporting the conclusion that Riel's actions were not treasonous: one has to do with his intentions, another holds that the Métis acted in self-defence, and the third claims that it was the government that was the treasonous party.

(1.2.1) Riel made the claim seven times that his aim was to bring about *practical results* by his actions (I: 7, 24, 26, 35, 37),[42] by which he appears to have meant making improvements in the material and legal conditions of the people of the North-West. His repeated insistence that his aims were practical may have been meant to contrast with the perceptions that he wanted to do something more than what was practical – something grand, such as setting up a religious state or divorcing the North-West from Canada, goals that might well be seen as ideological. Against this perception, Riel's insistence that he wanted only practical results – better conditions and basic rights for the people of the North-West – becomes an argument that he meant no harm to Canada and that therefore his actions were not treasonous.

(1.2.2) Riel maintained that his military actions had been in self-defence (I: 43).[43] He argued that "The agitation in the North-West Territories would have been constitutional, and would certainly be constitutional to-day, if ... we had not been attacked" (I: 19)[44] and that "when they appeared and showed their teeth to devour [us], I was ready: that is what is called my crime of high treason, and to which they hold me today" (I: 42).[45] This argument cuts two ways: in implying that his actions were in self-defence, Riel adds weight to the justice argument, for it is just to defend oneself. But in these passages, Riel appears to be more concerned with showing that his actions should not be considered treasonous, which they might have been if the Métis forces had initiated the fighting with the Canadian soldiers. Since the Métis militia were not the aggressors, the charge of treason was a mistake. So the self-defence argument has the consequence of showing that Riel's actions did not fit the meaning of "treason" and therefore that he was wrongly charged.

(1.2.3) Riel also had an argument that "treason" should be redefined: "if high treason there is," he argued, "it must be on its [the government's] side and not on my part" (I: 43).[46] In this accusation, we may see Riel asking the question "What is treason, really?" Is it when neglected people unite and form a provisional government after being continuously ignored by a national government, or is it when a government uses a large professional army with superior weapons to attack a small group of citizens who are suffering and whose rights have been neglected?[47] Riel's position here has the ring of a persuasive definition of "treason": the negative connotation of the word remains, but the extension is changed to suit his defence. What was really treasonous, he claims, was what the Government of Canada did to the people of the North-West, not what Riel and his followers did to Canada.

Riel's arguments that his actions were justified are brief. They were given near the beginning of his speech and were mentioned again at the end of the speech. The justice argument is a strong argument and still speaks to us today. It gives positive support for the standpoint that his actions were justified. The function of the first two parts of the no-treason argument (1.2.1 and 1.2.2) is independent of the justice argument. The purpose of these parts of the argument is to show that the charge of high treason is inappropriate and that therefore Riel has been wrongly charged. These arguments stand on their own, and if they are right, they give considerable weight to the conclusion that Riel is not guilty of high treason. The exception is the last part of the no-treason argument (1.2.3), which alleges that the treason was on the part of the government. This part of the argument seems weak: whatever wrongs the government may have committed, they are not aptly classified as treasonous.

Standpoint 2: "I am not insane"

Riel did not want to be considered insane. At least two concerns motivated this: one was that if he was deemed insane, it would return him to the degraded life of an animal in an asylum;[48] the other was that if he was thought to be insane, no legitimacy could be given to either his political or religious aspirations. The collection of arguments in support of the standpoint that Riel was not insane can be called the sanity argument. It is directly opposed to the case his team of lawyers had been advancing as his defence.

Riel had three independent arguments in support of his standpoint that he was not insane: one of them stems from the testimony of witnesses, another from the validity of his mission, and a third can be construed as based on his behaviour in court. Together, they may be considered to constitute Riel's sanity argument, the outline of which is as follows:

2 Riel is not insane (*sanity argument*):
  2.1 The testimony of Crown witnesses is that Riel is not insane.
  2.2 Riel's religious mission is real:
    2.2.1 Church officers said Riel had a mission;
    2.2.2 Riel enjoyed divine protection;
    2.2.3 Riel really does have prophetic powers.
  2.3 Riel is comporting himself as a sane person at his trial (*ethotic argument*).

(2.1) One of Riel's arguments that he was not insane relied upon the testimony of two of the Crown's witnesses. It was General Frederick Middleton who commanded the Canadian forces during the uprising in the North-West, and it was to him that Riel was brought after he surrendered at Batoche. Middleton interviewed Riel and then passed charge of him to Captain George Holmes Young, who over a period of ten days brought Riel to Regina. Both these military leaders spoke at length with Riel and gave testimony that he was not insane (I: 15).[49] This evidence is interesting in two ways. First, when Riel made use of it, it had the flavour of an *ad hominem* argument, although it does not quite fit that mould: Riel is taking testimony from his accusers, the Crown, and turning it not against his accusers, as would be the case if it were a true *ad hominem* argument, but against his own defence team. A second way that this testimony is interesting is that it is pitted against the testimony of medical witnesses. The two military leaders were men of practical affairs who would be expected to be good judges of character and who had recent and extensive contact with Riel.[50] They were in an appropriate position to judge Riel's mental state at the time of the North-West Rebellion, which is what is relevant for the sake of the treason trial. But the medical experts, called by Riel's defence lawyers, testified as expected that Riel was insane,[51] and, *ceteris paribus*, their opinions should count for more than the opinions of the military men, who were not professionals of mental science. However, whereas the two military men based their judgment on Riel's behaviour immediately after the rebellion and were in complete agreement that he was sane, the doctors gave different accounts of insanity and could support their judgments only (1) by making reference to Riel's behaviour during his time in asylums (1876–78) seven years before the rebellion, (2) by drawing on brief interviews with Riel at the time of his trial (ten weeks after the end of the rebellion), and (3) by repeating what they heard other witnesses at the trial say about Riel's behaviour during the previous year. Therefore, although one would think that the question of Riel's sanity should rest with the medical experts, the uncertainty of their testimony – coupled with the less complicated testimony of the Crown's witnesses – lends uncertainty to the claim that Riel was mentally ill at the time of the rebellion.

(2.2) The primary reason that Riel was considered to be insane was that he had unconventional religious beliefs, among them that he had a mission from God. There is little trace of his radical theology in the speech, but Riel admitted to the court that he believed he had a mission, and since he believed his mission to be genuine, he denied that having a mission was a mark of insanity. There are three related but independent arguments we can identify in support of his claim that his mission was real: there is the apparent endorsement of his mission by church officers, there is alleged evidence that he enjoyed divine protection, and there are his claims to have prophetic powers.

(2.2.1) Riel argued that he was not delusional in thinking he had a mission. He interpreted the words of the bishop of Montreal, Ignace Bourget, and those of a priest at Worcester, New York, as supporting his view that he had a God-given mission to help the Métis (I: 11, 13).[52] The bishop was an important religious figure in Canada whose opinion a religious man would take seriously. Thus this can be seen as an instance of the *argument from authority*: religious authorities who had communicated with God said that Riel had a mission; hence Riel had a mission.

Riel was not the first person to claim to have a God-given mission, and the religious atmosphere of the nineteenth-century North-West might have been less skeptical of such a claim than we are inclined to be in our day. Crown attorneys when examining the sanity experts brought in by the defence tried to get them both to compare Riel to mid-nineteenth-century Mormons Joseph Smith and Brigham

Young[53] – divinely inspired leaders of the religious colony in the Utah Territory – on the assumption that they would say these men were not insane. The doctors did not cooperate. Nevertheless, arguments from authority are strong only when there is agreement among authorities about the evidence, and in religious matters this is seldom the case because religious arguments from authority are based on private experiences not shared by others; hence such arguments may not be accepted even by colleagues-in-faith, and they can be expected to have even less currency with audiences outside one's religion. Thus devout Protestant jurors might well be skeptical of the cogency of a Catholic's argument from authority. Therefore, Riel's argument from a religious authority, in addition to being logically weak, is also problematic in relation to the audience he was trying to persuade. It gave reason to think that Riel sincerely believed he had a mission but not reason to think that he really did.

(2.2.2) A priest had blessed Riel and his family before he returned to Canada in 1884 (I: 13).[54] This event Riel took as both a blessing of him and of his mission: he thought it had caused his captors to treat him kindly (I: 14)[55] and had made witnesses support him by saying that he was sane (I: 15)[56] – even that it had kept him safe by protecting him from bullets (I: 12).[57] All this Riel offered as evidence that he was blessed, that his mission was real, and that he was not insane. This argument – an *inference to the best explanation* – is weak given that less daring hypotheses than that he had divine protection could explain the same set of facts. Thus Riel can be criticized for failing to infer a much simpler and less ontologically loaded explanation: that his captors were basically fair and decent people, that the witnesses who testified that he was sane were supporting the Crown's case, and that he was too far out of range, or too safely positioned, to be struck by enemy fire.

(2.2.3) The third argument that Riel's mission was real was that he had prophetic powers. He gave examples of some of his successful prophecies: (1) that he would be allowed to make a speech (I: 27)[58] and (2) that there would be trouble in the North-West (I: 28).[59] He said, "it is not to be supposed that the half-breeds acknowledged me as a prophet if they had not seen that I could see something into the future" (I: 26).[60] So the Métis took Riel to be a prophet, he claims, at least in part because he had made successful predictions. This kind of argument is an *argument from sign*: having prophetic powers is a sign that someone really is a prophet. And if someone really is a prophet, this is a reason to think that his or her mission is real. Riel used this line of reasoning to bolster the claim that he was not insane in thinking he had a mission.[61]

Whether someone has prophetic powers is an empirical question to be determined over a period of time by testing specific precise and falsifiable predictions. Riel's two examples in this regard are not convincing: the first one because all defendants have a right to address the court in trials of this kind (so it was a risk-free prediction);[62] the second one, that there would be trouble in the North-West, is too vague a claim to be testable – what kind of trouble and when? Moreover, although he made an argument to the effect that he had prophetic powers, Riel also seemed to downplay their importance. On the one hand, he declared himself to be "through the grace of God ... the prophet of the New World" (I: 34),[63] implying that he really was a prophet because God had made him one; on the other hand, he said that he did not really want the name of "prophet" and that he allowed the half-breeds to think of him as a prophet only because it was practical to do so (given that he had practical aims) (I: 35).[64] Riel also attempted to defuse the taint of having prophetic powers by pointing out that the ability to foretell events is shared by Métis hunters, thereby implying that such powers were common among his people and therefore not necessarily a mark of being a prophet (I: 28).[65] One can't help but think that Riel was of two minds when speaking of his mission. Part of him wanted to tell what he took to be an important truth, but another part feared that if he gave too much attention to his belief that he had a mission, the jury would think he was insane. So this part of Riel's sanity argument was not only logically weak; it was also confusing to its audience, who could not tell how much stock Riel himself wanted to put in it.

(2.3) *Ethotic argument*. Riel admitted that he had been committed to an asylum earlier in his life, but he maintained that he had been put there "without reason" (I: 15).[66] This is not an argument but simply a denial that he had been mentally ill. However, he attempted to provide evidence that he was sane by the very way he comported himself in court; this could serve as indirect evidence that he was sane at the time of the conflicts earlier that year – the time for which his sanity legally mattered. The rhetorical situation in which he found himself when he made his speech – that of having to argue for his life – was such as would make any person anxious ("excited"), and Riel was afraid that his nervousness might be mistaken for mental illness (I: 1),[67] a worry that he returned to several times throughout his speech. "When I speak of a mission," he said, "you will understand me not as trying to play the role of [an] insane [person] before the grand jury so as to have a verdict of acquittal upon that ground" (I: 10).[68] These words indicate that Riel was aware that the way he presented himself would be taken as a mark of his mental health. He must therefore have determined to present himself as a sane, reasonable, and responsible person. This would be an ethotic argument (an argument based on character) and more generally an argument from sign.

Riel hoped, he said at the outset of his speech, that with God's help he "will maintain calmness and decorum" (I: 1),[69] allowing that he was excitable and that he could "pass from great passion to great calmness" (I: 16)[70] in a very short time. His objective must have been to remain calm throughout his speech and to be "master of [him]self" (I: 37).[71] For the most part, he is fairly successful in carrying this out. From his description of the deplorable conditions in the North-West at the beginning of his speech through his discussion of his mission and his prophetic abilities, Riel seems to be very much in control, defending his views with arguments – not always with strong arguments but for the most part not with arguments that called his sanity into question either. However, near the end of his speech (I: 37),[72] Riel appears to have sensed that the court thought he had gone on long enough. After saying he had just "a few more words to say" (I: 37),[73] his speech changed its rhythm. Ideas now came out more quickly, juxtaposing contrary thoughts: "you are only half a jury," Riel said and then quickly added, but that "does not prevent you from being just"; and the judge, Riel alleged, had acted "against the guarantees of liberty," but he was only doing his duty (I: 38); and even though his lawyers had referred to "the incompetency of the court," Riel said he had respect for it (I: 39).[74] These passages show that Riel's frustration with the course of his trial was starting to slip out. He was trying to remain calm and reasonable, so he immediately softened each accusation with accompanying words of respect. But this deference, as well as Riel's self-control, disappeared in the next few paragraphs when he hurriedly summarized the shortcomings of the territorial and national governments, climaxing in the charge that the government was "insane and irresponsible" (I: 42),[75] an argument considered below. This was an extravagant allegation, unwise in these circumstances, and not the pronouncement of a calm and reasonable person. I think it is here, in this part of the speech, more than in any other, that we see evidence that Riel could sometimes lose self-control. However, true to what Father Vital Fourmond said, Riel's anger waned as quickly as it waxed. Even though he repeated his claim that if there was any high treason, it must have been on the government's part (I: 43),[76] he went on to make his final remarks in a more dignified and reasonable manner. In itself, Riel's behaviour near the close of his speech is as easily interpretable as a momentary lapse in self-control under extreme pressure as it is interpretable as the conduct of an insane person. But having heard testimony that Riel had a medical history of insanity (and that mood swings were a symptom of the illness) and given that expert witnesses had said that he was insane, the jury might well have found some confirmation of the doctors' opinions in Riel's performance in the latter part of the speech. In conclusion, Riel's ethotic argument turns out not to be completely successful, logically or rhetorically.

Riel's standpoint that he was not insane was meant not only to validate his view that he had a mission and that his actions were justified; it also implied that he should not be acquitted on the basis of insanity, contrary to the wishes of his lawyers. Since the jury ultimately found Riel guilty, it must not have been persuaded by the defence's arguments that he was insane. Whether it was the Crown's case, or Riel's own speech, or some combination of the two that made the difference, we are not in a position to say. Still, we can't help but think that things might have ended differently for Riel if he had pursued the sanity argument with less fervour.

Standpoint 3: "These proceedings are not fair"

Riel argued that the proceedings against him in Regina were not fair. He had two main lines of argument for this: one was that the court in Regina did not have the jurisdiction to try his case; the other was that his accusers were insane and irresponsible. Together, these arguments can be called Riel's proceedings argument, and they fit together as follows:

3 The Regina trial of Riel is not fair (*proceedings argument*):
  3.1 The court in Regina is the wrong jurisdiction to try Riel for a capital offence (*jurisdiction argument*).
  3.2 The Government is insane and irresponsible (*insane-government argument*).

(3.1) *Jurisdiction argument*. Riel held that the proceedings against him were not legitimate because the court in Regina did not have jurisdiction over his case. His argument for this had to do with which of several statutes was in effect for someone charged with high treason in the North-West Territories. He did not state the legal arguments for this claim, but he alluded to the ones given by his lawyers at the outset of the trial.

The jurisdiction argument is not just an argument that concludes that the law is being wrongly applied, a technicality that might invalidate a verdict; it also had consequences that could materially affect the outcome of the trial. Riel thought he had a right to have his trial held in a court that would give proper consideration to the breadth and significance of his case (I: 39).[77] Thus he had hoped for some grander venue than the makeshift court room in Regina, where there was a jury of only six and a stipendiary judge. If Riel had been tried in Manitoba, as the government first directed, he would have been entitled to a twelve-person jury, half French and half English,[78] which may well have been sympathetic to his cause, and a regular magistrate who did not owe his position to the present government. So, when he said to the six jurors, "you are only half a jury" and remarked to the judge that his actions were "against the guarantees of liberty" (I: 38),[79] he was reminding the court that his trial had begun with two lengthy legal discussions: one was about the true nature of juries (whether a jury must be composed of twelve men); the other was about the proper venue for the trial (whether it should have been moved to Manitoba).[80] Riel recalled these issues for the jurors in an attempt to sway them in his favour by making them sense that there was something legally amiss, something of which they should not want to be a part. But the jury, influenced by the judge's instructions,[81] must have felt that the jurisdiction argument was not within its compass to decide, being a question for legal experts. That the trial should have been moved to one of the provinces would later be the subject for an (unsuccessful) appeal to the Court of Queen's Bench in Manitoba.[82]

(3.2) *Insane-government argument*. In his summation, delivered immediately before Riel's address to the jury, Charles Fitzpatrick, one of Riel's lawyers, relying on the testimony of witnesses for the defence, had said that Riel "is entirely insane and irresponsible for his acts."[83] In a way similar to his attempt to turn the charge of treason back on the government (see section 1.2.3 above), Riel tried to shift the diagnosis of insanity, with which he had been saddled, onto the government by relying on the connection between *sanity* and *responsibility*. If lack of responsibility implies insanity, and the government is not responsible, then, it may be alleged, the government is insane and, hence, unwarranted in pressing a case against Riel.

This argument deserves discussion. Flanagan and others, assuming it was meant to be comical, regard it as a pun, or a joke.[84] Indeed, there is a kinship between puns and fallacies, and the same factors that amuse us can also deceive us.[85] Because Riel repeats the allegation that the government is insane four times, we should try to understand on what ground he makes the claim, whether or not it was meant to be humorous, since it occurs in the context of very serious argumentation.

"British civilization," said Riel, "has defined such government as this is which rules the North-West Territories as irresponsible government" (I: 40).[86] A *responsible government* is one in which the ministers are responsible to the elected members of a legislature such as the House of Commons rather than to a governor general or a monarch.[87] Responsible government replaced the oligarchic form of government in British North America in the years following the Durham Report of 1839, and it also became the model for the federal government when Canada became a country in 1867 and for each of Canada's constituent provinces, including Manitoba in 1870. But responsible government had not been extended to the North-West Territories, as Riel observed near the beginning of his speech (I: 4).[88] The territories continued to be treated by Canada as a colony, just as Lower and Upper Canada had been ruled as colonies by Britain until the 1840s. The North-West Council, to which Riel referred (I: 40),[89] was created by the North-West Territories Act of 1875. This territorial government consisted of an appointed lieutenant governor and five appointed members of the council. In 1888, three years after the execution of Riel, "the Legislative Assembly came into being, and full representative – although not yet responsible – government became an accomplished fact."[90] Thus Riel was right to point out that – in 1885 and before – the government ruling the North-West Territories was not a responsible government; the ministers and the members of the council were responsible to those who appointed them, not to the elected members of the council, few as they were. Similarly, since the North-West had no elected members of Parliament, ministers of the government in Ottawa could not trace a line of responsibility to the people of the North-West Territories.

With this background, we can consider Riel's insane-government argument. These are the relevant passages:

> British civilization which rules today the world, and the British constitution has defined such government as this is which rules the North-West Territories as irresponsible government, which plainly means that there is no responsibility, and by all the science which has been shown here yesterday you are compelled to admit if there is no responsibility, it is insane. (I: 40)[91]

> That fact [that the government did nothing in response to many petitions] would indicate an absolute lack of responsibility, and therefore insanity complicated with paralysis. (I: 41)[92]

> The Ministers of an insane and irresponsible Government and its little one – the North-West Council – made up their minds to answer my petitions by surrounding me slyly. (I: 42)[93]

> If you take the plea of the defence that I am not responsible for my acts, acquit me completely since I have been quarrelling with an insane and irresponsible Government ... [T]he Government, my accuser, being irresponsible, and consequently insane, cannot but have acted wrong. (I: 43)[94]

There are different senses of "responsible" and "irresponsible" at large in these passages. One of them, already mentioned, is *government* responsibility, which we may call "G-responsibility": a government is a responsible government only if its ministers must answer to the elected members of a legislature rather than to a governor or a monarch. The other sense of "responsible" being used is that of *moral* responsibility, which we may call "M-responsibility." There is an important distinction between being *not M-responsible*, which is to be placed outside the scope of M-responsibility, and being *M-irresponsible*, which is to be within the scope of M-responsibility, but blameworthy. Thus we cannot infer that an agent is irresponsible from the proposition that he or she is not responsible. Moreover, being either praiseworthy or blameworthy presupposes M-responsibility, and consequently insane persons cannot be held morally or legally responsible for their actions and for the things they cause to happen. With these distinctions in mind, we can restate the insane-government argument and see how it might have appeared attractive at first glance:

1 Neither the North-West Council nor the Government of Canada is responsible (political fact);
2 Hence these are irresponsible governments (from 1);
3 To be insane is to be not responsible (a legal principle; also supported by testimony);
4 Hence the two governments are insane (from 1 and 3);
5 Hence the two governments are irresponsible and insane (from 2 and 4);
6 Charges made by an irresponsible and insane party should not be taken seriously (plausible claim);
7 The federal government has charged Riel with high treason (fact);
8 Hence the charges of high treason against Riel should not be taken seriously (from 5 to 7).

Let 1 and 3 be accepted. Line 2 does not follow from 1 because 2 is about M-responsibility and 1 is about G-responsibility; moreover, "*x* is not responsible" does not imply "*x* is irresponsible." Similarly, 4 does not follow from 1 and 3 because, again, 1 is about G-responsibility and 3 is about M-responsibility. Each of these inferences thus commits a fallacy of equivocation. Conclusion 5 does follow from lines 2 and 4 by the rule of conjunction, but since those premises have not been established, they give no support for 5, and hence no support for 8 either, even if we allow 6.

That Riel's argument commits fallacies does not show that he was mistaken in thinking that the government had failed to meet its obligations to the people of the North-West. We can identify a related argument that Riel may well have had in mind, namely that the government's not being G-responsible contributed to its failure to do what it should have done for the inhabitants of the North-West, and for that reason it should share in the responsibility for the rebellion and not charge Riel with high treason. This argument traces a route from the lack of responsible government to the reasonableness of Riel's actions, and would have strengthened the defence of his standpoint that the proceedings were not fair, at least to the extent that M-responsibility for the events in the spring of 1885 should be shared. However, it strains credibility to argue, as Riel did, that the government was mentally ill, and making that argument brought his judgment into question. Furthermore, if we continue Riel's turning of the tables, it has the consequence that the government should not be held accountable for its failings.

In summary, Riel defends his standpoint that the proceedings against him are not fair based on two lines of reasoning. There is first the jurisdiction argument, which has the consequence that Riel is put at a disadvantage in terms of judge and jury. It may weigh more with us today than it could with Riel's jury, whose duties were narrowly defined. Second, there is the insane-government argument, which holds that the charges against Riel should not have been made in the first place because his accuser is insane. It rests on a confusion of G-responsibility with M-responsibility. Hence this second part of the argument fails to give any support to the conclusion that the charge against Riel is not fair and that the proceedings against him should be dismissed.

Standpoint 4: "I deserve sympathy"

There is another strain of argument throughout Riel's speech that cannot be overlooked. In support of his position that he should not be found guilty of treason, he appeals to sympathy and gives reasons why he deserves it. The outline of the sympathy argument is as follows:

4 Riel deserves sympathy (*sympathy argument*):
- 4.1 Riel is at a disadvantage in having to address the court in English.
- 4.2 Riel is not able to defend himself in court, and his lawyers will not help him as he wants them to.
- 4.3 Riel is poor even though he has worked hard to serve the people of the North-West.

4.4 Riel has been humiliated by being banished and by being committed to an asylum.

(4.1) At the beginning of his speech, and twice afterward, Riel tells the court what it will easily be able to hear: that English is not his native language and that he does not have the same facility of expression in that language as he does in French. He thereby invites the jury to infer that he will not be able to defend himself as well as he might have done in his native tongue (I: 1, 10, 46).[95] Hence the jury should extend sympathy to Riel due to his cultural/linguistic situation.

(4.2) Riel furthermore depicts himself as being as helpless before the court as the day he was born (I: 3),[96] and he repeats this statement later in the speech (I: 25).[97] The idea is that he is as dependent on his lawyers as he was on his mother when he was an infant since he does not have the legal expertise to defend himself. Yet, as becomes clear throughout the speech, his lawyers and he disagree about how he should be "taken care of." The source of sympathy is thus not just that Riel lacks a skill that he needs but also that he is not being well served by those in whose care he has been placed. Perhaps the lawyers' refusal to assist Riel in making the case he thinks should be made can even be seen as a betrayal.

(4.3) Riel alludes to his poverty when he says that he has "not succeeded in wearing a fine coat" (I: 12)[98] and that he "never had much to buy any clothing" (I: 45).[99] He says that he "never had any pay," that "it has always been [his] hope to have a fair living one day" (I: 45),[100] and that his family had to accept charity from the church (I: 45).[101] Riel hopes for a just verdict that will console his friends and his family. Here the jury is invited to feel sympathy for Riel on the basis of an economic reason, a poverty ill-deserved, he thinks, because he has "worked to better the condition of the people of the Saskatchewan at the risk of [his] life" (I: 45).[102]

(4.4) Sympathy may also have been sought in Riel's remarks that he had been banished for five years (I: 11),[103] libelled for fifteen years (I: 21),[104] and "put in the asylum without reason" (I: 15).[105] Here the basis for sympathy is that Riel has been treated unjustly and humiliated by others. He also suffered by being in danger's way (I: 12, 42).[106]

That Riel deserves sympathy is plain enough. However, considerations of sympathy are normally thought to be relevant at the sentencing stage of a trial, where they might influence a judge to mitigate the severity of punishment – they are not relevant to whether or not a law has been broken. Accordingly, it could be said that Riel's appeals to sympathy should have been left out of his address to the jury. But high treason cases were special: there was but one punishment possible if found guilty – death by hanging[107] – so appeals to sympathy or pity could have had no effect at all on a judge after a guilty verdict had been entered. Perhaps Riel was aware of this, and it induced him to include his appeals to pity in the presentencing stage of the trial.

CONCLUSION

Riel defended four discernible standpoints in his speech to the jury: first, that he was justified in his actions; second, that he was not insane; third, that the proceedings against him were unfair; and fourth, that he deserved sympathy. How do these four standpoints all fit together as one extended piece of argumentation in support of the conclusion that Riel should not be found guilty of high treason? The first and third standpoints independently support the final conclusion. The second standpoint, that Riel is not insane, does not in itself lend any support to the conclusion, but it is needed as a presupposition for the first standpoint, that Riel was justified. Just as someone can be punished for their actions only if they are sane, so too is being sane a necessary condition for being justified in one's actions. Thus the components of the justification argument – the justice argument (1.1) and the no-treason argument (1.2) – presuppose the conclusion of the sanity argument. The fourth standpoint, that Riel deserves sympathy, is also part of the argumentation presented in support of the conclusion that Riel is not guilty or, at least, is intended to influence the jury to decide that way.

The logical order of the argumentation in Riel's speech to the jury, broadly described, is thus as outlined below in summary argument A. (Note that standpoint 2 is a presupposition of standpoint 1 rather than a reason for it; this is indicated by placing the sentence between asterisks.)

A. Riel should not be found guilty of high treason:
  A.1 (Standpoint 1) Riel's actions were morally justified (*justification argument*);
  (Standpoint 2) *Riel is not insane* (*sanity argument*).

A.2 (Standpoint 3) The Regina trial of Riel is not fair (*proceedings argument*).
A.3 (Standpoint 4) Riel deserves sympathy (*sympathy argument*).

By expanding the four component arguments in summary A to include all the subarguments that were identified as support for each of the four standpoints, we obtain the greater details of the logical order of the argumentation in this speech.

Riel was well aware of the sanity question as a central issue in his defence, as is evident from his own summary of his speech to the jury. "[If] I am not responsible for my acts, acquit me completely since I have been quarrelling with an insane and irresponsible Government ... [If] I am responsible, acquit me all the same ... [because] ... having my reason and sound mind, I have acted reasonably and in self-defence, while the Government, my accuser, being irresponsible, and consequently insane, cannot but have acted wrong, and if high treason there is, it must be on its side and not on my part" (I: 43).[108] Here, in a few words, Riel referred to many of the arguments he had made throughout his speech. The passage has the form of a dilemma, and we can reconstruct it as summary argument B.

B. Riel should be acquitted:
B.1 Either Riel is insane or he is sane;
B.2 If Riel is insane, then, because of the insane-government argument, Riel should be acquitted;
B.3 If Riel is sane, then, because of the justice argument, the self-defence argument, the insane-government argument, and the no-treason argument, Riel should be acquitted.

Argument B in effect says, "Acquit me whether or not you think I am sane." It poses the alternatives of "sanity" and "insanity" as a starting point and asserts as a consequence of each alternative that Riel should be acquitted. The premise B.3 summarizes what is packed into A.1, except that it includes the insane-government argument, which also appears in B.2 as a reason for acquitting Riel if he is insane. This is odd. If Riel is insane, that should be a reason to acquit him regardless of whether some other party is also insane. Furthermore, argument B seems out of character with the tenor of the speech that precedes it. Throughout the speech, until the passage quoted from paragraph 43,

Riel has insisted that he should not be considered insane. This insistence is captured in argument A but is weakened in argument B. That Riel should step back from his insistence that he is not insane is, of course, entirely understandable in the circumstances in which he finds himself: he will not be able to continue his struggle for the rights and welfare of the Métis people if he is found guilty of high treason. Finally, from the logical point of view, one has to notice that the options before the jury are not captured by the dilemma in argument B; rather, what is needed is a trilemma, the third option being that Riel is sane and should be found guilty – the option the jury in fact decided on. The jury could not do anything about the government's negligence in the North-West, nor did it have the temerity to question the legitimacy of the proceedings. But in the end, the jury did respond positively to two of Riel's standpoints: that he was sane and that he deserved sympathy. This is shown in the first instance by the fact that the jury found him guilty and in the second instance by its recommendation for mercy.[109]

This chapter has been an attempt to sketch the outlines of the narrative and logical orders of Riel's first speech, his address to the jury. It leaves many of the arguments and persuasive devices in the speech for future studies to identify and analyze. Nevertheless, I hope to have shown that beneath the troubled narrative surface of the speech there is a discernible logical order consisting of challenging and original arguments that are of durable interest to students of argumentation as well as to historians.

NOTES

Earlier versions of this chapter were presented at a meeting of the Rhetorical Society of America, Minneapolis, May 2010, and at the Riel's Defence conference, University of Windsor, October 2010. I am grateful to members of the Centre for Research in Reasoning, Argumentation and Rhetoric at the University of Windsor for their comments and suggestions.

1 Thomas Flanagan, ed., *The Diaries of Louis Riel* (Edmonton: Hurtig, 1976), 100.
2 E.B. Osler, *The Man Who Had to Hang: Louis Riel* (Toronto: Longmans Green and Company, 1961), 306.
3 J.M. Bumsted, *Louis Riel v. Canada: The Making of a Rebel* (Winnipeg: Great Plains, 2001), 292.

4 Bob Beal and Rod Macleod, *Prairie Fire: The North-West Rebellion* (Edmonton: Hurtig, 1984), 302.
5 *Montreal Daily Star*, 1 August 1885, quoted in Thomas Flanagan, ed., *The Collected Writings of Louis Riel/Les ecrits complets de Louis Riel*, vol. 3, *1884–1885* (Edmonton: University of Alberta Press, 1985), 537–8.
6 Joseph Boyden, *Louis Riel and Gabriel Dumont* (Toronto: Penguin, 2010), 155.
7 George F.G. Stanley, *Louis Riel* (Toronto: McGraw-Hill Ryerson, 1963), 354.
8 Ibid., 358.
9 Flanagan, ed., *Collected Writings*, 538.
10 Desmond Morton, *The Last War Drum: The North-West Campaign of 1885* (Toronto: Hakkert, 1972), 158.
11 Desmond Morton, "Introduction," in Desmond Morton, ed., *The Queen v Louis Riel: Canada's Greatest State Trial* (hereafter *Q v LR*) (Toronto: University of Toronto Press, 1974), xix–xx.
12 *Q v LR*, 311–12.
13 *Q v LR*, 312.
14 *Q v LR*, 313–14.
15 *Q v LR*, 314–16.
16 *Q v LR*, 316–17.
17 *Q v LR*, 317.
18 *Q v LR*, 318–19.
19 *Q v LR*, 319–21.
20 *Q v LR*, 321–2.
21 *Q v LR*, 322–4.
22 *Q v LR*, 324–5.
23 *Q v LR*, 350–71.
24 *Q v LR*, 75.
25 *Q v LR*, 313.
26 *Q v LR*, 316, 319.
27 *Q v LR*, 313.
28 *Q v LR*, 194–204.
29 *Q v LR*, 209.
30 *Q v LR*, 316–17, 320.
31 *Q v LR*, 318.
32 *Q v LR*, 311.
33 There is a draft of parts 10 and 11 (I: 38–43). See Flanagan, ed., *Collected Writings*, 155–6.
34 The way to understand the numbering here is that 1 is the standpoint, and

1.1 and 1.2 are reasons supporting 1; similarly, 1.2.1, 1.2.2, and 1.2.3 are different reasons supporting 1.2.

35 *Q v LR*, 312.

36 *Q v LR*, 314.

37 *Q v LR*, 312.

38 *Q v LR*, 312.

39 *Q v LR*, 323; see also 312, 313, 317, 324.

40 For an account of *two-wrongs reasoning*, see Leo A. Groarke and Christopher W. Tindale, *Good Reasoning Matters!* 4th ed. (Toronto: Oxford University Press, 2008), 352–60.

41 Thomas Flanagan, *Louis 'David' Riel: 'Prophet of the New World,'* 2nd ed. (Toronto: University of Toronto Press, 1996), 147.

42 *Q v LR*, 313, 319, 322.

43 Riel uses the term "self-defence" only once. See *Q v LR*, 324.

44 *Q v LR*, 317.

45 *Q v LR*, 324.

46 *Q v LR*, 324.

47 Paul Groarke sees a stronger version of this argument in Riel's favour. See chapter 7 in this volume.

48 "Here I have to defend myself against the accusation of high treason, or I have to consent to the animal life of an asylum. I don't care much about animal life if I am not allowed to carry with it the moral existence of an intellectual being." See *Q v LR*, 212.

49 *Q v LR*, 316.

50 "I could not give a medical opinion, but I consider that during the nine days I was living with him I would know if I was living with a lunatic." See testimony of Captain Young, who was in charge of Riel on his journey from Batoche to Regina, in *Q v LR*, 274. Also, General Middleton, who was first to interview Riel after his capture, testified, "I came to the conclusion that he was very far from being made [*sic*] or a fool." See *Q v LR*, 281.

51 *Q v LR*, 305.

52 *Q v LR*, 314, 315.

53 *Q v LR*, 256, 260.

54 *Q v LR*, 315.

55 *Q v LR*, 316.

56 *Q v LR*, 316.

57 *Q v LR*, 315.

58 *Q v LR*, 320.

59 *Q v LR*, 320.

60 *Q v LR*, 319.
61 Riel added more examples of his ability to see into the future at II: 44, *Q v LR*, 370.
62 Judge Hugh Richardson, on the first day of the trial, had told Riel that he would have "every opportunity" to speak later. See *Q v LR*, 205.
63 *Q v LR*, 322.
64 *Q v LR*, 322.
65 *Q v LR*, 320.
66 *Q v LR*, 316.
67 *Q v LR*, 311.
68 *Q v LR*, 314.
69 *Q v LR*, 311.
70 *Q v LR*, 316. See Fourmond's testimony in *Q v LR*, 240–2.
71 *Q v LR*, 322.
72 *Q v LR*, 322.
73 *Q v LR*, 322.
74 *Q v LR*, 323.
75 *Q v LR*, 324.
76 *Q v LR*, 324.
77 *Q v LR*, 323.
78 Flanagan, *Louis 'David' Riel*, 174.
79 *Q v LR*, 323.
80 *Q v LR*, 8–40.
81 In his charge to the jury, Judge Hugh Richardson said, "we have really nothing to do" with the question of jurisdiction. See *Q v LR*, 345.
82 Thomas Flanagan, *Riel and the Rebellion: 1885 Reconsidered*, 2nd ed. (Toronto: University of Toronto Press, 2000), 145.
83 *Q v LR*, 295.
84 Flanagan, *Louis 'David' Riel*, 175, says it is "a famous pun." Interestingly, although Flanagan's authoritative version of this speech notes that the speech provoked laughter on at least three occasions, there is not an indication that the present passage drew that kind of reaction. See Flanagan, ed., *Collected Writings*, 539.
85 The affinity of fallacies with jokes was noticed by Richard Whately at the very end of book 3, "Of Fallacies," in his *Elements of Logic*, 9th ed. (London: Longmans, 1875). He wrote, "*Jests* are mock-Fallacies, *i.e.*, Fallacies so palpable as not to be likely to deceive anyone, but yet bearing just that resemblance of Argument which is calculated to amuse."
86 *Q v LR*, 323.

87 See, for example, J.M.S. Careless, *Canada: A Story of Challenge*, 2nd ed. (Toronto: Macmillan, 1963), 188–9.
88 *Q V LR*, 312.
89 *Q V LR*, 323, passim.
90 George F.G. Stanley, *The Birth of Western Canada: A History of the Riel Rebellions* (1936; reprint, Toronto: University of Toronto Press, 1960), 192.
91 *Q V LR*, 323.
92 *Q V LR*, 324.
93 *Q V LR*, 324.
94 *Q V LR*, 324.
95 *Q V LR*, 311, 314, 325.
96 *Q V LR*, 312.
97 *Q V LR*, 319.
98 *Q V LR*, 315.
99 *Q V LR*, 324.
100 *Q V LR*, 324.
101 *Q V LR*, 325.
102 *Q V LR*, 324, 315.
103 *Q V LR*, 314.
104 *Q V LR*, 318.
105 *Q V LR*, 316.
106 *Q V LR*, 315, 324.
107 See Bumsted, *Louis Riel v. Canada*, 274.
108 *Q V LR*, 324.
109 *Q V LR*, 349–50.

6

"A New German-Indian World" in the North-West: A Métis Deconstruction of the Rhetoric of Immigration in Louis Riel's Trial Speeches

KERRY SLOAN

Deconstructive readings do not assert that texts have no meaning or that their meanings are undecipherable. Rather, deconstructionism argues that texts are always overflowing with complicated and often contradictory meanings ... Similarly, the point of deconstructing conceptual oppositions is not to show that concepts have no boundaries, but rather that their boundaries are fluid and appear differently as the opposition is placed into new interpretive contexts."[1]

The young woman (okay, perhaps now not so young) is lying on the prairie on a sunny summer morning, listening to the meadowlarks and the hissing sweetgrass, feeling the soft breath of the wind. She is trying to recall a dream she had as she dozed, a disordered dream that kept weaving in and out of high school history classes, Métis community meetings, and evenings with her family, full of laughter and argument. What was the argument about? Now she remembers – it was about Louis Riel. Who was he? Her second youngest uncle, a lawyer and wine connoisseur, thought Riel was a nut who distracted his followers with religious nonsense and prevented Gabriel Dumont, "the one with the common sense," from doing his job. It was Riel's fault the Métis had lost. Her youngest aunt, who had always been a bit of a radical, thought that Riel was a hero, a champion of Métis rights who gave up his work, family, and ultimately his life to defend the cause. Her grand-

father, a classic reactionary, thought Sir John A. Macdonald was right to "send in the gendarmes" – after all, Riel and his rabble had taken the guns and ammunition from the family's dry goods store and never paid for them. With a twinkle in her eye, her grandmother added that Louis was far too self-important for her liking – the romantic rejection he experienced served him right for trying to court her great-aunt when she was already spoken for.

The young(ish) woman smiles as she remembers these sallies and drifts back into her dream, but it has shifted. Instead of her family teasing and joking or her grade 9 history teacher droning on about the transcontinental railway and the fuss kicked up by those pesky half-breeds, she sees a figure, far off on the prairie. He has dark, curly hair and whiskers and an intelligent gaze, which is focused somewhere off in the distance. She looks down at her hands and is startled to see they have grown – both in size and in hairiness. Her light fleece jacket has been replaced by a large checked wool coat of old-fashioned cut. Her cheeks are unaccountably itchy. "Who have I become?" a man's voice asks in French, in sonorous but forceful tones. Could it be that she has spoken? "I am glad to be back on the prairie," the voice continues, "and would like to explain what I was trying to tell those people all those years ago. They didn't listen then, but now maybe people are starting to pay attention. At least I finally got all my writing published – it only took a hundred years! Maybe I got my spirit back too?" Her lips curve in a grin. Then she wakes herself laughing. "Am I crazy?" she asks herself. "Was I channelling Louis Riel?"

It is trite to say that Louis Riel was a complex person; the complexity of his personality, life, and political career have been the subject of much debate and speculation in both popular and scholarly discourse. Riel has been variously characterized as a traitor, a reprobate, a madman, and an egotist; and, conversely, as a hero, a prophet, a political genius, and a selfless proponent of his people's rights.[2] These oppositions are not surprising, considering that Riel was an intellectual but also a populist, a Canadian icon who was an American citizen, a devout Catholic who espoused the heresy of abandoning Rome, and a champion of Métis self-determination and land rights who invited multicultural immigration to the Métis homeland.

This latter paradox is illustrated in Riel's speeches at his 1885 Regina trial for high treason, particularly his second speech, addressed to the court after sentencing. For instance, although Riel's speeches evidence support for Métis cultural, territorial, and political autonomy in the face of Canada's expansionist policy of flooding the North-West with immigrants in order to outnumber the "Indians" and "half-breeds,"[3] in his second speech Riel expresses the hope that the Métis themselves will welcome thousands of immigrants of different cultures and religions to share the North-West with them and with the region's other Indigenous inhabitants.

Riel not only propounds his vision of cultural and religious harmony; he also advocates for more "mixing" between the new arrivals and the Métis and First Nations, with the hope of creating a "new nation." At the same time, he assumes that Métis distinctiveness will itself be preserved, and he upholds Métis rights and the "Indian" rights from which they flow. Riel does not see the ideas of distinctiveness and multiplicity as opposed but complementary, even mutually necessary. This necessity, for Riel, relates to his so-called "millenarian" vision[4] of a grand confederacy of nations – Métis, First Nation, and European – living together peacefully regardless of religion or culture. He sees the Métis as the spiritually appointed "guides" of these nations by virtue of their past experiences of integration. Riel envisions such relationships with these new nations as signalling the end of the spiritual and physical exile of the Métis.

Although Riel states in the second speech that he has "two sides to [his] head" (II: 19)[5] (perhaps in part a joking reference to those who had accused him of madness and/or treachery), his thinking cannot be explained as dualistic; neither can it be described as merely syncretic. Rather, his ideas intersected in fluid ways that changed over his lifetime, and they have continued to change with the flow of Métis history. Riel's views about multicultural immigration to the North-West encompass both his Indigenous and European heritage; they reflect Catholic teachings as well as Nehiyaw (Cree) and Anishinabek (Ojibway/Saulteaux)[6] religious philosophies. In fact, the political and the religious were not separable for Riel.[7] The complexities of Riel and his inclusive vision are reflections of the diversity of both his personal experiences and the experiences of the Métis nation over time. In his speeches to the court, and in other writings on the subject of multicultural immigration,[8] Riel presents a profoundly Métis vision of the fabric of the North-West.

However, in stating that Riel's multicultural, ecumenical vision is inherently Métis, I do not intend to promote vapid liberal pluralisms[9] or to essentialize either Riel or Métis philosophy. Riel himself, in his many manifestations, has eluded pat categorizations, which is perhaps why he still fascinates us.[10] Métis ideas span a range of views that are constantly being debated. Being Métis does not mean being only "part Native" and "part European"; nor does it mean being simply a "go-between."[11] Similarly, Riel's vision is not merely hybrid; it is complex and flows in and out of the boundaries between political, cultural, and religious domains.

In this chapter, I briefly outline some of the complexities of Riel's life and of Métis history. I then present Riel's views of Métis rights and his plan for multicultural immigration as expressed in his second trial speech, explaining how these ideas were connected for him. Following this, I explore the political and religious implications of these views, with a focus on the need for greater understanding of the Indigenous roots of Riel's religious philosophy. Finally, I explain why I think Riel's is a truly Métis vision that reflects the "boundary bashing" realities of his own life and of Métis history.

RIEL AND THE MÉTIS: MULTIPLICITY AND ABSENCE

The (maybe not so) young woman resigns herself. If Riel wants to talk to her, that is his business. The guy has had enough disrespect in his lifetime and beyond, so why should she deny him? Besides, she is curious. And sleepy. Yet, as she drifts back into slumber, she sees not a grown man but a small boy on the prairie, the great ocean of grass spread before him. He walks and walks, a small speck on the earth. Birds come and go overhead, disappearing into the clouds. The boy keeps walking, through a small wooden house, along a cobblestone street, past the fancy lattice-work of a church confessional, down the long halls of a dormitory. The images flicker quickly now, like an out-of-control film reel: the prairie, a fort, a trading post, more wooden houses, the Parliament buildings, a schoolhouse, a courthouse – a jailhouse. She sees the boy sitting at a desk in prison, writing, writing. His hands fly while birds fly above his hands, out through the window bars, and over the prairie grass.

As other authors in this volume argue, Riel's rhetorical task was a difficult one, and this difficulty may account to some extent for the complexities inherent in the text.[12] For instance, Riel was faced with having to argue against the prosecution – as well as his own lawyers –

while trying to persuade the court, jurors, and the public of the wrongs perpetrated against his people and others in the North-West. This was not an easy task and certainly one that might have led to structural convolutions.

Although I think this is an important insight, I also think that the complexity of Riel's trial speeches is a reflection of the complexity of his life and thought and of the complexity of Métis experience. In addition, the losses and exiles experienced by both Riel and the Métis influenced his speeches. However, these multiplicities and absences do not create merely fragmentary expressions, despite some of the difficulties of the text, but also interpenetrating and interdependent concepts that encompass Métisness.

Riel's personal and political complexities stem in part from the contrasts of his own experiences: as a prairie boy sent to the big city, equally adept in either milieu; as one of the few Métis at that time to have received a classical European education; as a social conservative who was also a political revolutionary; and as a trader and teacher who ultimately became a politician. Further, in his very being is a meeting of the oft-perceived solitudes of Canadian society: Indigenous and non-Indigenous, French and English, Catholic and Protestant, elite and grassroots, West and East.[13] Riel was aware of these tensions and sought to bring understanding to them.

A number of authors have pointed out ways that Riel transgresses dualistic expectations. For instance, Kevin Bruyneel states that "to Canada Riel is both sovereign body and sacred body, representing the experience of liberal inclusion and colonial oppression in such a way that they do not stand as contradictory forces";[14] Lauren L. Basson asserts that Riel confounded American racial oppositions.[15]

In terms of absence, Riel is notable for what he did not become: a priest, a lawyer, the husband of a well-to-do white woman from Quebec, a man who ultimately succeeded in his political aims. His life was a multitude of seeming failures, grief at the loss of family members, disappointed love, frustrated politics, poverty, institutionalization, exile, imprisonment – and ultimately an unjust death. Yet Riel in many ways was remarkably successful: he negotiated Manitoba's entry into Canadian Confederation; he brought French and English Métis together; he championed Métis, francophone, and Catholic rights, in many ways effectively; and he gave his people and the public a vision of greater unity that is still relevant today.

Multiplicity and absence met in Riel's encounters with the many ruptures of his time: colonialism, ethnic and religious tensions, destruction of the environment, economic change and uncertainty, political high-handedness, and ignorance of the elite about the experiences of ordinary people. These ruptures were and still are being negotiated by Métis people. Historically, the rupture of colonialism resulted in loss of Métis lands and governance structures, diminishment of cultural pride, and even erosion of culture itself to some extent.[16] Many Métis "went underground," disguising their identities where possible. Métis people were scattered across the country, exiles in their own lands. The multiplicities and diaspora of the Métis are in many ways a macrocosm of the complexities and exile of Riel.

Like most history, Métis history is messy. It has been characterized by high degrees and rapid rates of change, by adaptation and mobility, and by the proliferation of many political, religious, linguistic, and cultural differences.[17] Métis people may have French, Cree, Saulteaux, Irish, Scottish, and many other ethnic heritages; Métis people may speak French, English, Nehiyawewin (Cree), Anishinabemowin (Ojibway/Saulteaux), dialects of Michif (a Cree-Ojibway-French language) or (at one time) Bungee (a language derived from Ojibway, Cree, Scottish English, and Scottish Gaelic), or other languages;[18] Métis people may practise traditional Indigenous religions and/or Christianity (Catholic or Protestant), other religions, or no religion at all. Given these complexities, it is not surprising that scholarly and everyday debate surrounds questions of the formation of Métis identity. With due respect for postmodern and anticolonial critiques of cultural fluidity and hybridity as erasing real distinctiveness,[19] "boundary bashing" is a historical component of Métis individual and community identities, and Métis people have often viewed themselves as mutable, as living with multiple cultures and having multiple and overlapping identities.[20] Yet, as Métis people, we have not "unbounded" ourselves out of existence: we are distinctive, we have survived many years of oppression and ignorance, and we are now in a phase of resurgence. This was the crux of Riel's vision: that the Métis – even though they might "sleep for a hundred years"[21] – would get their spirits back and fulfil the promise of their flag's eternity symbol.[22]

I am suggesting that all of the above – absence/presence, mixing/distinctiveness, loss/continuity – may be valid and that these realities need not be dichotomized. I am also suggesting that there may be a "third space" that can be occupied by Métis individuals and communities, one that avoids duality, where these "elements encounter and transform each other."[23]

MÉTIS RIGHTS AND IMMIGRATION

The (somewhat) young woman looks hard at the prairie, hoping to see the boy, or even the wool coat. She listens closely for the manly self-deprecating voice but hears and sees only a few meadowlarks. Finally, one of these birds tells her, "You have to stand on the ground." So she does. When she stands up, she sees a dark line on the eastern horizon. As she continues to watch, the line becomes thicker and thicker. At first she thinks it must be a grasshopper plague, but finally she realizes this is a "plague" of people. People as far as the eye can see, and they are eating up the land, mowing down everything in their path. "Does it have to happen this way?" she cries.

The advancing column stops. The people form groups of large circles in the grass that is still standing. Then more people form an even larger circle around the groups of circles. The young(ish) woman hears the songs of birds, of sibilant grasses. Then human music rises, almost inaudible. One drum – a heartbeat – then many, slowly gaining volume. Fiddles, flutes, guitars, accordions, banjos, bouzoukis, ouds ... then singing in Cree, in French, in Michif, then in many languages she doesn't recognize. There is something about this music that makes her want to run and fly at the same time. After a while, the cacophony subsides, and individual instruments or melodies can be heard; then the musical babble returns again, but this is oddly calming. Rhythms change, keys change, cadences rise and fall. She wakes slowly and pleasantly, humming a tune.

In his trial speeches, Riel advocates for the Indigenous rights of the Métis and First Nations. He explains their mistreatment at the hands of the Canadian government and certain European-descended settlers from Ontario. He is adamant that the Métis are a distinct nation (e.g., II: 25) and have distinct claims. At the same time, he advocates for multicultural immigration to the North-West and for better understanding between Protestants and Catholics, hoping to prevent in North America the sectarian "evils which exist in Europe" (I: 24).[24]

In his second trial speech, Riel describes his advocacy for Métis concerns and the sufferings of the Métis themselves. He also argues that Métis Aboriginal rights come from "Indian" rights and expresses sympathy with First Nations experiences.[25] Intertwined with these statements on behalf of Indigenous peoples is Riel's expression of his vision of sharing the North-West with many groups of immigrants. These pro-Indigenous statements and pro-immigration statements are not separated but flow into each other.

For instance, Riel expresses the willingness of the Métis to assert their claims against any who would invade their territory, whether "Canadian" or "Indian."[26] With respect to the Canadians, he explains the connection between the "troubles" of Saskatchewan in 1885 and the "troubles" of Red River in 1869 and states that the events of 1885 "are the result of fifteen years' war" (II: 11). The root of this war was that "a certain number of individuals ... that can have passed for Canada ... wanted to take possession of the country without consulting the people" (II: 11). According to Riel, these people "came without notification, they came boldly. We said, who are they, they said, we are the possessors of the country. Well, knowing that it was not true, we done against those parties coming from the east, what we used to do against the Indians from the south and west, when they would invade us" (II: 11). The response to Canada's arrogant lack of consultation was to take up arms, to take prisoners, and from this position to negotiate. The Métis then negotiated a "treaty," which is how Riel viewed the Manitoba Act of 1870,[27] based on the list of rights demanded by the Métis (II: 13).[28]

In describing the Métis's protection of their territory and sovereignty,[29] Riel states that they were a nation with which the Canadian government treated, although he is likely doubtful whether his listeners accept the validity of the provisional government declared by the Métis in 1869 (II: 13). Notwithstanding that some of the Métis delegates were arrested on arriving in Ottawa, the treaty was concluded (II: 14). The "half-breeds" were thereby "granted" 1.4 million acres, or roughly one-seventh of the area of Manitoba in 1870, which was approximately 9.5 million acres (II: 16).[30] Riel says he wishes to inaugurate the principle of giving the Métis a seventh throughout the North-West; he stresses that he hopes to "bring to the half-breeds of the North-West at least as soon as possible the guarantee for the future that a seventh of the lands will also be given to them; and seeing and yourself understanding how it is difficult for a small population, as the half-breed population, to have their voices heard, I said what belongs to us ought to be ours. Our right to the North-West is acknowledged, our co-proprietorship with the Indians is acknowledged, since one-seventh of the land is given to us; but we have not the means to be heard" (II: 16).

For Riel, the problem of land entitlement was directly linked to his ideas of immigration. Riel relied on what had been agreed to in the Manitoba treaty, which in his view recognized Métis title and rights and effectively prevented American encroachment in the area. Riel also realized that immigration was coming, whether the Métis wanted it or not. Canada wanted to populate the North-West with Europeans, either from Canada itself or by encouraging European, mainly British, immigration. However, it feared a large influx of settlers from the United States. In his speech, Riel claims that if Canada will not fulfil its promises under the Manitoba Act, he will populate the North-West with people of his own choosing, who would be willing confederates supporting the Métis in the quest for their lands. He plays upon Canadian fears of losing the North-West to the United States. He cites his notoriety in the states and his support from American Fenians and politicians, as well as from the Métis south of the border – including Métis military leader Gabriel Dumont (II: 17, 18), who was then a resident of the United States.

Thus Riel's plan had two prongs: if Canada ignored Métis rights, he would invite immigrants from the United States, essentially amounting to American colonization; however, if Canada was just with respect to Métis claims, he would invite immigrants directly from Europe and would also ally with members of other Indigenous nations in the North-West. He states, "if we cannot have our seventh of the lands from Canada ... we will invite the Italians of the States, the Irish of the States, the Bavarians of the States, the Poles of the States, the Belgians of the States, and if they come and help us here to have the seventh, we will give them each a seventh" (II: 17). To show that the Métis are not Catholic "fanatics" or "partisans," since the groups he mentions are predominantly Catholic, Riel also invites the Protestant Danes, Swedes, and Norwegians from the states; he invites the Indians and "half-breeds" from British Columbia; he also invites the Jews (but on the condition that they convert to Christianity).[31] If all these groups come to the aid of the Métis, Riel states that they will each get a seventh.[32]

Riel reiterates his plan, stating that "if it is not possible that our views should be heard, we will meet as American citizens" (II: 17). He promises that, for example, the reward of the Germans for helping the Indians and the half-breeds would be "to take a part of the country and make a new German-Indian world somewhere in British North America" (II: 17). However, Riel states that recruiting immigrants from the United States is a last resort.

Riel clarifies that, by encouraging immigration, he is not abandoning the Métis cause. He declares he will

> invite the same parties from Europe as [enumerated in the American] emigration, but let it be well understood that as my right has been acknowledged as a co-proprietor of the soil with the Indians, I want to assert that right. It is constitutionally acknowledged in the Manitoba Act by the 31st clause of that Act, and it does not say to extinguish the Indian title. It says two words, extinguishing, and 1,400,000 acres of land, two words and as each child of the half-breeds got one-seventh, naturally I am at least entitled to the same. It is why I spoke of the seventh for the Indians, not of the lands but of the revenue as it increases (II: 20).[33]

Riel then goes on to explain why he thinks the Métis are entitled to one-seventh. One reason he gives is essentially that God has made it so (II: 20). He appeals to both the English and French that this is precisely why they have their own lands in Europe. He then chastises them for presuming that they are also entitled to Indigenous lands: "When they have crowded their country because they had no room to stay any more at home, it does not give them the right to come and take the share of all tribes besides them. When they come they ought to say, well, my little sister, the Cree tribe, you have a great territory, but that territory has been given to you as our own land, it has been given to our fathers in England or in France and of course you cannot exist without having that spot of land" (II: 21). Again, this highlights the position that the Cree, and the Métis, are nations in their own right, with rights to their own lands. Further, this is not merely by human political design but also by the will of God.

Having laid this authoritative foundation, Riel's next appeal is to his audience's belief in the "superior" civilizing influence of British Canadian culture.[34] He also suggests that, with increased settlement, the monetary value of First Nations and Métis lands will increase, probably sevenfold. Riel then mockingly quips, "I think it is a fair share to acknowledge the genius of civilization to such an extent as to give, when I have seven pair of socks, six, to keep one" (II: 22).[35] This comment, in addition to pointing out the reasonableness of the Métis position, reveals that Riel views all of Manitoba (and by extension all of the North-West) as being Métis territory. He does not mention shared use with other Indigenous groups, but this is suggested by actual practice and by the fact that the Métis made peace and resource-sharing treaties with various First Nations.[36]

Implicit in Riel's willingness to give up six-sevenths of the territory[37] he views as part of the Métis "homeland" is, likely, practical thinking: he realizes immigration is inevitable, and he also realizes that the value of the land and its profits will increase. Some authors take at face value Riel's statement that the potential increase would result from the Métis accepting European "civilization." This is true to the extent that Riel realized that the buffalo hunt, the fur trade, and the cartage trade were in decline and that the Métis would have to adapt to the coming "civilized" economy.[38] However, this same civilization that he praises – whether as an appeal to his listeners or to some degree sincerely – he also critiques as unjust, as a civilization that would allow its former benefactors, the First Nations and Métis, to starve while eating "rotten pork" instead of the buffalo and other game that once flourished in the land (I: 4).

Instead of railing against the inevitability of immigration, Riel welcomes it – but on Métis terms. In Riel's vision, immigrants would not displace or subsume the Métis; they would themselves contribute to creating a larger Métis nation, "a new German-Indian world" in the North-West. In this way, the territory that Riel had been willing to give up as part of the Manitoba treaty would once again become Métis; further, Métis territory would expand throughout the entire North-West, encompassing the Métis who left Manitoba after 1870.[39]

Riel continues to mock the injustices of "civilized" Canada in failing to grant him amnesty and in failing to live up to the written promises of the Manitoba Act.[40] He reiterates that the Métis were treated with as a nation with rights and that "the right of nations wanted that the treaty of Manitoba should be fulfilled towards the little community of Red River in the same condition that they were when they treated" (II: 25). Because Canada failed to honour its promises, many Métis sold their lands and went west, some only receiving half of what their lands were worth (II: 26).

After an interlude in which Riel talks about his career, he returns to the topic of immigration: "if Riel succeeded that he should one day, as a public man, invite emigration from the different parts of the different countries of the world, and because this North-West is acknowledged to be partly his own, as a half-breed of this population, to make bargains for this North-West here with the Canadian Government, in such a way so that when the English population has had a full and reasonable share of this land, other nationalities, with whom we are in sympathy, should have also their share of it" (II: 38). Despite Riel's praise of the English Constitution and "good government," the English are noticeably absent from Riel's list of immigrants;[41] in fact, he states, "When we gave the lands in Manitoba for one-seventh ... we did not give it only for the Anglo-Saxon race" (II: 38). He also has to consider the "Irish in the east and the French in the west" (II: 38).[42] These remarks indicate that, in addition to thinking about bringing immigrants from abroad, Riel is musing about how to incorporate Canadian settlers into his vision of the new Métis future.

The merging of existing groups with new immigrant groups is further stressed in Riel's hopes "of having a new Ireland in the North-West by constitutional means, inviting the Irish of the other side of the sea to come and have a share here; a new Poland in the North-West by the same way, a new Bavaria in the same way, a new Italy in the same way" (II: 40); at the same time, he asks for help from the French Canadians and from the "Indians" and "half-breeds" west of the Rocky Mountains (II: 40). Regarding settlement in the Far West, Riel suggests that the Belgians might be happy on Vancouver Island and that the Jews might be content "on the other side of the mountains, while the waves of the Pacific will chant sweet music for them to console their hearts for the mourning of 1,800 years" (II: 40). Returning to Canadian angst about American expansion, Riel again stresses he prefers "natural" immigration to immigration from the United States, as his thoughts "are for peace." He reminds his audience that en masse American immigration would be a disastrous "revolution." Riel states he can tolerate only those disasters "which I am bound to bring to defend my own life, and to avoid to take away from my country, disasters which threaten me and my friends and those who have confidence in me, and I don't abandon my ancestors either, the acknowledgment that I have from my ancestors" (II: 40). Here, Riel neatly parallels the potential of a North-West swamped by Americans and prevented from becoming part of Confederation with the impending Métis "disaster" of Canadian expansion;[43] his remarks encompass his concerns for his own people and his vision of immigration. Implied in his words are his belief both in the duty he owes to his ancestors and in their approval of his mission.

The ancestors whom Riel will not abandon he describes as being Irish by way of Scandinavia, French, and "Indian." Most biographers state that Riel's Indigenous ancestry was Chipewyan on his paternal grandmother's side, while describing both his mother's parents as French Canadian.[44] Riel refers again to Scandinavian and other European immigration:

> The Scandinavians, if possible, they will have a share, it is my plan, it is one of the illusions of my insanity, if I am insane, that they should have on the other side of these mountains, a new Norway, a new Denmark, and a new Sweden, so that those who spoke of the lands of the great North-West to be divided into seven, forgot that it was in ten. The French in Manitoba, the Bavarians, the Italians and ... the Poles and the Irish in the North-West, and then five on the other side too [this is probably a reference to the three Scandinavian groups, the Belgians, and the Jews living on the other side of the Rocky Mountains]. (II: 41)

Riel then returns to the threat of American immigration, this time not through his own influence but in the possible course of events. He tries to persuade his listeners that immigration "through the channel of constitutional means" would be preferable since, with him or without him, Americans have their eye on the land (II: 41). In the next sentence, he shifts from the idea of immigration to the idea of Métis entitlement to a "seventh": "that is about the seventh of lands. So you see that by the very nature of evidence that had been given here when the witnesses speak of a seventh of the land, that very same question originates from 1870, from the troubles of Red River which brought a treaty where the seventh of the lands took its existence" (II: 41).

Near the end of his speech to the court, Riel states, "I give myself as a prophet of the new world" (II: 43).[45] With this declaration, he sums up his purpose in acting as he has for the past fifteen years. As he had explained in his speech to the jury, he was "aiming all the time at practical results" (I: 7), such as relieving the sufferings of the Métis and ensuring respect for their rights; even after his death, he hoped his spirit would continue to bring these results. The Métis needed land: "God cannot create a tribe without locating it. We are not birds. We have to walk on the ground" (II: 21). They also needed to cope with the coming immigration. Riel's conceptual (if not immediately practical) response was to juxtapose territory and immigration in a way that would reposition both.

POLITICAL AND RELIGIOUS IMPLICATIONS

The (still somewhat) young woman is feeling happy, but also like she has overslept. Riel has really given her a run for her money. He has been pretty esoteric. Is he still trying to tell her something in her dreams? Maybe, but she can't figure out what and she is starting to get hungry. Not sharing with her ancestors the love of gopher pie, she walks until she finds a greasy spoon called "Blavatsky's." She leans her elbows on the Melmac bar counter and regards the bouffant-haired waitress. "Can I help you, dear?" asks the petite woman in Franco-Manitoban tones. She is of indeterminate age and – as her mother would have said (politically incorrectly, of course) – a "Heinz 57" like herself. The (youngish) woman replies that she would like a burger and a tea. The waitress, whose pin announces her name as "Julie," has green eyes flecked with brown, and these flicker with amusement as she says, "I'm afraid we only have fresh peas today. My son Lou is a bit of a health nut and talked me out of serving a lot of meat. Besides, it's Lent." Too hungry to debate the merits of either vegetarianism or self-abasement (or to remember it is already summer), the (relatively) young woman eats her peas and drinks her tea in silence. As her sugar low slowly wears off, she notices that the waitress's hairstyle has changed. Julie now sports long, dark braids, just like in the picture once taken of Riel's mother. This new coiffure attractively sets off her prominent cheekbones. "Yes, I know what you're thinking. All the history books say only my husband was Métis. But I was too. I mean, look at my pictures. I just thought it would be better for Louis if he could, you know, 'pass for white.' I mean, he was a really smart boy and we had big hopes for his education. We knew he could go far. But too much Indian blood could Scotch more than a French romance, if you'll pardon the mixed metaphor." Julie lets out a big belly laugh. "Come on, you know the story – your family is related to mine by marriage. All the Lagemodières are Métis." At this, the (vaguely) young woman begins to feel as if she is having some sort of allergic reaction to the peas. They are certainly creating a great deal of gas, but there seems to be more to it than that. She smells something odd and is feeling a little faint. "Ah, that's just the tea I made you, dear. You will see everything more clearly in a minute." True to Julie's word, when she looks out the window, she sees far off on the prairie that a series of grass fires is burning. The fires remind her of the story of the burning bush, because the grass is burning, and yet it is not being consumed. Moses speaks to her from the grass in the seventh fire: "You will wrestle with Riel, and ask why he has a new name. This question is not relevant. Riel has many names." Moses takes his leave as a straggly coyote runs from the fire, dragging a pair of socks.

Many observations have been made about Riel's political and religious thinking. Although some commentators acknowledge Riel's inclusiveness and ecumenism[46] and others, such as Thomas Flanagan and Gilles Martel (see below), attempt to trace the genealogy of these ideas, there has been almost no discussion of the Indigenous roots of Riel's philosophies.[47]

In terms of his political legacy, Jennifer Reid has described Riel as a man who envisioned a Canada that could accommodate the existence of immigrants from various countries without homogenizing them; she posits that Riel's encompassing of distinctiveness with *métissage* was foundational for Canada.[48] She suggests that, although "an overarching national discourse is not possible in the Canadian situation" owing to the multiplicities of Canada, "such a collectivity can develop a common identity in spite of its internal divergences."[49] Similarly, John Ralston Saul theorizes in *A Fair Country* that Canada is, essentially, a "Métis nation," owing much of its dialogic and confederalist nature to Indigenous political philosophy.[50] Of course, objections may be raised that such pronouncements, especially Saul's, gloss over the real legacy of colonialism and imply that we are all "one big happy family" or that they create a Riel whose raison d'être was to serve Canadian interests rather than the interests of the Métis. Nevertheless, these works acknowledge the Métis contributions to Canada and draw interesting parallels between Riel, the Métis, and the Canadian state, suggesting that distinctiveness can co-exist with the shifting of identities that mark the man, his people, and the country that he – and they – influenced.

Although the Métis were internally complex, exhibiting their own divisions as well as those that existed in the larger society, they had a great degree of political and cultural cohesion. They saw themselves both as an independent people (in Cree, the Otipemisiwak, "the people who own themselves") and as a people who were in interrelationship with Indigenous and European peoples, whose ancestry they shared. More than mere "go-betweens," the Métis negotiated political and trade relationships with all their "cousins," including treaties with other Indigenous nations. The idea of a confederacy of nations within the emerging state of Canada was therefore not farfetched to the Métis thinker.

For Riel, the idea of this confederacy was religious as well as political,[51] although I do not think he would have viewed these spheres as separable. He had come to believe that the Métis were the "chosen people" of the new world and that their mission was to lead by example the new multicultural, multireligious society he envisioned.[52] He saw his role as a prophet of this movement, which he would continue to influence after his death (e.g., I: 24).

Flanagan, Martel, and Manfred Mossman[53] describe Riel as being primarily influenced by Catholicism and to a lesser extent (and in agreement with J.M. Bumsted) by Protestant messianism, millennialism, and Theosophy.[54] Mossman compares Riel's religious ideas to those of the "cargo" cults of colonized Pacific countries, proposing that Riel's messianism and millennialism were a response to cultural disruption. Reid counters this in her article "'Faire Place à une Race Métisse'" as overly general, despite some unmistakable "resonances."[55] Flanagan takes up the issue of Riel's millennialism in his pioneering *Louis 'David' Riel: 'Prophet of the New World.'* Despite drawing parallels with "native resistance movements" such as the Ghost Dance of the "Plains Indians," Flanagan claims that it was the lack of syncretism with Native religion that set Riel's approach apart from other anticolonial religious movements.[56]

Flanagan, in his editor's notes to passages in Riel's diary that relate to immigration, suggests that Riel's ideas on this topic probably derived from the teachings of Bishop John Ireland of the Catholic Colonization Bureau.[57] Of course, the influence of Catholicism on Riel was considerable, but arguably Riel's Catholicism, like the Catholicism of many Métis, was indigenized.[58] That Indigenous religious beliefs and practices persisted among Catholic and Protestant Métis has been noted by Reid[59] and by the authors of "Métis Spirituality" in *Métis Legacy II*.[60]

What Chad S. Hamill says of the missionization of the Interior Salish people may equally be applied to the Métis:

> What the Jesuits perhaps failed to anticipate was the ingenuity and resilience of the cultures they encountered, comprising indigenous identities both fluid and "emergent." Despite [Father Nicholas] Point's assertion [that many medicine bundles had been destroyed], much remained of the old medicine ways during the process of Catholic indigenization ... Ultimately, this process of indigenization – of cultural integration, negotiation, and reinterpretation – could not be quelled. Nevertheless, the Jesuits found success.[61]

Perhaps the reluctance to assert that Riel's religious and political thought was influenced by Indigenous philosophies stems from the apparent lack of primary resources on the subject[62] and consequent difficulties in "proving" such claims. However, ignoring the common-sense clues that these connections must exist is rather like wilful blindness about the extent of European-Indigenous interrelationship in Canada.[63] For example, as Cole Harris points out in terms of sexual and family relationships in the early British Columbian context, there were many European men but almost no European women in the colony for many years.[64] In terms of Métis spirituality, most Métis of Riel's time practised both Christianity and Indigenous religions and viewed them as being syncretic rather than exclusive.[65] Thus, as a matter of common sense, it would be extremely unlikely that Riel was not aware of and influenced by Indigenous religion.[66] One example from the first trial speech that reflects a decidedly non-Catholic view is Riel's explanation of Métis foretelling: "Besides Nolin knows that among his nationality, which is mine, he knows that the half-breeds as hunters can foretell many things ... I have seen half-breeds who say, 'My hand is shaking, this part of my hand is shaking; you will see such a thing to-day,' and it happens. Others will say 'I feel the flesh on my leg move in such a way, it is a sign of such a thing,' and it happens. There are men who know that I speak right" (I: 28). However, Flanagan discounts this passage as providing any evidence of Indigenous religious influence, citing it merely as "local superstition" or "Indian lore."[67]

Riel's divergences from generally accepted Catholic doctrines were many and complex, changed over time, and have been discussed by a number of scholars and clerics. My intent is not to investigate these here but rather to illustrate that Riel's thinking transgressed the boundaries of Catholic colonialism. For instance, if Riel had been influenced only by such ideas, he would not have included the Protestant Scandinavians, non-Christian First Nations, or Jews (even as Christian converts). He would not have spoken about the "Episcopalian Catholics, the Methodist Catholics, the Lutheran Catholics, the Universalist Catholics, etc." as all belonging to a new "religious order ... to which all according to their own faith will belong."[68] Riel's thinking is also not completely congruent with Christian millienialism, which does not describe the thousand-year "peaceable kingdom" (or the subsequent eternal peace) as a confederacy.[69] The form of peace Riel describes is more reminiscent of the great confederacies of the First Nations, such as the Anishinabek Three Fires and the Haudenosaunee confederacies; these included confederacies created by treaties and political alliances, just as Riel often tried to persuade both his Indigenous and European relatives to join him in creating the North-West.

Anishinabek writer Leanne Simpson speaks about the spiritual importance of a treaty based on the laws of the Three Fires and Haudenosaunee confederacies in her work *Dancing on Our Turtle's Back*:[70]

> Gdoo-naaganinaa, meaning "Our Dish," is [a] relationship Nishinaabeg in the southeastern portion of the territory had with the Haudenosaunee Confederacy ... Gdoo-naaganinaa acknowledged that both the Nishinaabeg and the Haudenosaunee were eating out of the same dish through shared hunting territory and the ecological connections between those territories ... both parties were to be responsible for taking care of the Dish ... The Haudenosaunee refer to the treaty as the "Dish with One Spoon" treaty and there is an associated Wampum Belt. The concept behind the Dish with One Spoon Wampum reflects the principles that were given to the Haudenosaunee by the Peacemaker in the Kaienerekowa (Great Law of Peace).[71]

Simpson explains the boundaries of the treaty relationship:

> At no time did the Haudenosaunee assume that their participation in the Dish with One Spoon treaty meant that they could

> fully colonize Nishinaabeg territory or assimilate Nishinaabeg people into the Haudenosaunee culture. At no time did the Haudenosaunee assume that the Nishinaabeg intended to give up their sovereignty, independence or nationhood. Both political entities assumed that they would share the territory, that they would take care of their shared hunting grounds and that they would remain separate, sovereign, self-determining and independent nations.[72]

The idea of distinctive but interconnected political entities was part of the sacred philosophy of peace that was also enacted via the Great Law of Peace.[73] This covenant, thought to have been entered into between 1000 and 1450 CE,[74] created a confederacy of previously warring nations: the Mohawk, Onondaga, Seneca, Oneida, and Cayuga. The Tuscarora were admitted long afterward, in 1722. These nations kept their own laws and political structures, each preserving its own jurisdiction over its respective territory and laws. At the same time, each nation, through its female clan leaders, appointed a council of male civilian delegates to represent it at meetings of the larger Grand Council of the confederacy, where all the delegates discussed "issues of common concern."[75]

Similarly, Riel imagines each of the groups of immigrants he describes in the second trial speech as having its own territory in a different part of the North-West. These groups would mix with the Indigenous people of these territories, creating new Métis nations under the umbrella of the greater Métis nation.[76] Martel theorizes that Riel's intention was that the new Métis people would be governed by a form of republicanism similar to that of the United States but with certain modifications, including the creation of a number of separate councils that would periodically convene as a Grand Council. These modifications were designed "pour assurer l'harmonie dans la diversité."[77] Riel imagines in his second speech the confederacy he could never completely achieve in real life.

Riel's vision of a new Indigenous-settler confederacy of peace is strikingly similar to the Anishinabek Prophecy of the Seven Fires, which imagines "new people" who help to revitalize the Anishinabe and create an era of lasting peace throughout the world. This prophecy, which Indigenous elders state predates contact with Christianity by many millennia,[78] recounts that the era of peace will be brought on, in part, by settlers deciding to be confederates rather than colonialists. Simpson received permission from her elders to share a portion of the Prophecy of the Seven Fires in her essay "Oshkimaadiziig, the New People."[79] She points readers to a number of other works that explain the prophecy in more detail, but she gives us a brief interpretive summary that describes it as

> a series of sacred predictions that have foretold our history since the beginning of Creation. The later part of the prophecy relays that we are currently living in the Seventh Fire, a time when, after a long period of colonialism and cultural loss, a new people, the Oshkimaadiziig, emerge. It is the Oshkimaadiziig whose responsibilities involve reviving our language, philosophies, political and economic traditions, our ways of knowing and our culture. The foremost responsibility of the "new people" is to pick up those things previous generations have left behind by nurturing relationships with Elders that have not "fallen asleep." Oshkimaadiziig are responsible for decolonizing, for rebuilding our nation, and for forging new relationships with other nations by returning to the original Nishinaabeg visions of peace and justice. According to the prophecy, the work of the Oshkimaadiziig determines the outcome of the Eighth Fire, an eternal fire to be lit by all humans. It is an everlasting fire of peace, but its existence depends on our actions and choices today. In order for the Eighth Fire to be lit, settler society must also choose to change their ways, to decolonize their relationships with the land and Indigenous Nations, and to join with us in building a sustainable future based on mutual recognition, justice and respect.[80]

The prophetic traditions of the First Nations have influenced each other; and many prophets were also known to popular culture. Thus Riel was likely aware of at least some of the major figures, such as Dedanawiga, or the "Peacemaker," who brought peace to the warring Haudenosaunee; Tecumseh's brother, Tenskwatawa, a Shawnee political and religious leader; and the Paiute prophet Wovoka, who founded the Ghost Dance movement.[81]

Alfred A. Cave, in his *Prophets of the Great Spirit*,[82] asserts that it is impossible to generalize about the Indigenous American prophets active after the contact period. On the one hand, he argues that much of the rhetoric of these prophets was a syncretic mixture of Indigenous spirituality and Christian doctrines. On the other hand, he warns against ignoring the Indigenous antecedents of these prophets' teachings. In my view, Riel saw himself as being within Indigenous, Jewish, and Christian prophetic traditions.[83]

Although a thorough investigation of the similarities of Riel's prophecy to those of other Indigenous prophets is a subject for further research, a short comparison with some of the characteristics of Anishinabek prophets and prophecy as described by Anishinabek scholar Camille Bernier is revealing.[84] For instance, prophecy is not intended to be focused only on prediction, although prediction is important.[85] Riel says in his speech to the jury, "It is not to be supposed that the half-breeds acknowledged me as a prophet if they had not seen that I could see something into the future" (I: 26); however, he also admits that everyone can do this to some extent. According to Bernier, the ethos and usefulness of a prophet is more important than the mere ability of foretelling.[86] Riel assures his audience that he is a "decent" rather than an "evil" prophet (I: 26), that "the spirit that directs [him] is the spirit of truth" (I: 27), and that he is "a leader of good" (I: 30). He also claims, in concert with Bernier's idea of Anishinabek egalitarianism,[87] that he is "simply one of the flock, equal to the rest" (I: 34). Further, the prophet as "the receiver is merely a conveyer of the message";[88] Riel asserts he sees things only because he is "blessed without measure" (I: 26).

Rather than claiming that the connections I have made "prove" anything about Riel's understanding of the world, I wish to suggest, along with the authors of "Métis Spirituality," that "More research should be conducted to link Riel to the numerous spiritual beliefs of his Dene ancestors, the Plains First Nations and to the non-Aboriginal spiritual movements of the late nineteenth century."[89] I also do not wish to diminish the importance of connections between Riel's plan and Christian or other spiritual influences; indeed, I contend that these and Indigenous influences are mutually supportive. This is illustrated in the first trial speech by the multiple interpretations that may be given to Riel's beginning prayer;[90] by his appeal to the land as mother and his discussion of reciprocal obligations of people and the land;[91] by his defence of his own character, which is also a discourse on truth – one of the Seven Grandfather Teachings of the Anishinabe;[92] by his assertions that he has a guiding spirit; and by explanations of his prophetic role. Riel's second trial speech, as I have tried to show, also reveals that his politico-religious ideas have multiple sources – but have created something new and original, in the same way that the Métis are a "new" nation whose people have multiple, overlapping identities.

Métis blogger Claude Aubin, of Wabenaki heritage, writes in his article on the Great Peace of 1701 that this treaty was connected with the earlier Great Law of Peace, as well as with the Prophecy of the Seven Fires. The Métis, in his view, are central to the concerns of the spiritual laws:

> The Great Peace was to become the pattern, the way we would have to live and share together on this continent. In the same spirit of this treaty, the Great Peace of 1701 was for the Wabanaki people an alliance, a spiritual Confederacy of universal relatives, brothers and sisters. In the Native tradition, the goal of matrimonial exchanges was to join in spirit and in blood with all your allied communities.
>
> As it has always been in the past, the spiritual and physical hybridization process created by these marriages became the most irreversible link for any Peace and survival in North America.
>
> In other words, Kondiaronk the prophet with his knowledge of the Wampum belts, already knew about this process of lasting peace, from way before.
>
> The Métis are the creation of the Kitche Manitou, the spiritual writing, living parchemins of this Great Peace, carefully prepared by Kondiaronk to pave the road to fulfil the prophecy of the Seven Fires as told by Seven Prophets.
>
> The Métis are the children of that prophecy as told and seen by the Fourth Prophet. The Métis are living physical and spiritual treaties sealed in DNA ...
>
> The Prophecy of the Seven Fires speaks also of a mighty Confederacy of people, the creation of a mighty Nation where all the people will live in true Peace and harmony. The Confederacy is a concept that was provided to the Wabanaki people by the Kitche Manitou, the Life Force, as a political tool to keep peace and unity between people and the Nations, and to give an equal voice to all people, in order to avoid conflict and wars.[9]

This view reflects Riel's politico-spiritual vision of the purpose of the Métis.

MULTIPLICITY AND THE END OF EXILE

Riel's vision of "a new German-Indian world" in the North-West reveals a threshold[94] between apparently contradictory concepts: distinctiveness and multiplicity; exile and return. For me, Riel's description of himself as having "two sides" to his head does not indicate that his mind was divided but that he was Janusian. The Roman god Janus, whose head had two faces looking in opposite directions, presided over thresholds, doorways, and the passage from one year to another. Riel, like his people, lived an exile borne of liminality; but, in his vision, being in between is also a gateway to return.

In Riel's trial speeches, his fugal interweaving of his advocacy for Métis land rights – and the Métis right to exist –with his vision of multicultural, multireligious immigration "negotiates an initial incompatibility" across literal and metaphorical boundaries[95] and expands Métis consciousness and possibility.

Riel imagines his own people not as victims of Canadian colonialism but as authors of their own destiny. He reframes immigration in Métis terms: instead of being engulfed, the Métis will persist on an even grander scale. It is they who will link arms together[96] with their Indigenous and settler relatives, together lighting the eighth fire. Riel realizes his plan may take some time, especially if he is not there to implement it. But he has faith that his spirit can still accomplish "practical results" by helping the Métis nation to return from exile and reclaim its rightful place in the great confederacy of peace.

Well, so much for Blavatsky's diner, thinks the (so-called) young woman. Smoke inhalation, indigestion – not to mention the three-foot beehive hair-do – these are horrors she would rather forget. Besides, who wants to eat a whole plate of peas at one sitting? And what was in that tea? She doesn't know much about medicine but figures she needs an antacid. As she walks along a rut in the prairie grass back toward the centre of town, she decides to stop in at the local library. She might as well read up on Riel. Maybe his written words will speak more clearly to her than the bizarre metaphors he seems to be imparting in her dreams. She picks up a book that looks like a straight biography, but it is full of numbers that seem to be multiples of seven. Occasionally, she sees the infinity symbol, but math has always confounded her. She tries a basic history of the North-West "rebellions." Nothing. Far in the back reaches of the submarine-like stacks, she sees an old box, caked with dust. There are a few illegible scrawls on the box in something that looks vaguely like Cree syllabics. Or maybe it is ancient Hebrew. But when she looks inside, there are only a few scraps of paper with seemingly random words in different languages: "Janus," "Massinahican," "confederacy," "liberté." Disappointed, she decides she's had enough of Riel and his antics for one day. She leaves the library, walks out through the town, and is drawn back to the prairie. It is already dusk. The fires are still burning, and the circles of people are still standing. Slowly, people from every circle go and stand near the seventh fire. They take braided bundles of the long grass and set them close to the flames, releasing a sweet fragrance. The bundles ignite, coalescing into a new fire – the eighth fire. The tongues of flame leap like birds, the gathered sparks like thousands of tiny phoenixes. The people are not consumed, but their faces are illuminated and shine together like the moonlight.

NOTES

1 Jack Balkin, "Deconstruction's Legal Career," *Cardozo Law Review* 27 (2005): 719.

2 Various authors have pointed to the multitude of possible characterizations of Riel. See, for example, Maggie Siggins, *Riel: A Life of Revolution* (Toronto: HarperCollins, 1994); and George F.G. Stanley, "The Last Word on Louis Riel: The Man of Several Faces," in *Louis Riel: Selected Readings*, ed. Hartwell Bowsfield, 42–60 (Toronto: Copp Clark, 1988). Albert Braz tries to account for the divergence of these characterizations in *False Traitor: Louis Riel in Canadian Culture* (Toronto: University of Toronto Press, 2003).

3 I acknowledge that the terms "Indian" and "half-breed" have been (and are still) used derogatorily, although some Aboriginal people have "reclaimed" them; I use these terms here to reflect Riel's usage – common at the time – in his trial speeches. "Indian" in Riel's usage seems to denote a North American Indigenous person who is neither Inuit nor Métis (I use "First Nations"). In his trial speeches, Riel seems to include in the term "half-breed" both francophone Métis and English- or Gaelic-speaking Métis, whereas some people at the time used "half-breed" to mean nonfrancophone Métis only. In this chapter, I use the term "Métis" broadly to describe people of mixed Indigenous and non-Indigenous ancestry in the North-West, regardless of language, ancestry, or religion. I use "Indigenous," "Aboriginal," and "Native" interchangeably to describe the original inhabi-

tants of North America and their Métis relatives. A discussion of the reasons for imputing indigeneity to Métis people is beyond the scope of this chapter.

4 Authors who discuss Riel's religious ideas as being in the millenarian tradition include Thomas Flanagan, *Louis 'David' Riel: 'Prophet of the New World'* (Toronto: University of Toronto Press, 1979); Gilles Martel, *Le Messianisme de Louis Riel* (Waterloo, ON: Wilfrid Laurier University Press, 1984); Manfred Mossman, "The Charismatic Pattern: Canada's Riel Rebellion of 1885 as a Millenarian Protest Movement," *Prairie Forum* 10 (1985): 307–25, reprinted in *The Western Métis: Profile of a People*, ed. Patrick C. Douad, 185–202 (Regina: University of Regina Plains Research Centre, 2007); and J.M. Bumsted, "Louis Riel and the United States," *American Review of Canadian Studies* 29, no. 1 (1999): 17–41. For a review of Flanagan, Martel, and Mossmann's views on this topic, see Jennifer Reid, "'Faire Place à une Race Métisse': Colonial Crisis and the Visions of Louis Riel," in *Religion and Global Culture: New Terrain in the Study of Religion and the Work of Charles H. Long*, ed. Jennifer Reid, 51–66 (Lanham, MD: Lexington, 2003).

5 References to Riel's speeches are to paragraphs as they are numbered in the version of the texts printed in this volume. Here, "(II: 19)" indicates the second speech, paragraph 19.

6 Riel had Denesuline (Chipewyan) heritage and may have been influenced by the religious philosophy of this nation. He likely was also influenced by Cree and Ojibway/Saulteaux beliefs owing to the fact that this was the Indigenous heritage of many of the Métis people he knew. An example of the persistence of Dené concepts in Métis thought is a legend with Dunneza and Quebecois elements called "The Origin of the Grey Ducks," reproduced in Lawrence Barkwell, Leah M. Dorion, and Audreen Hourie, eds, *Metis Legacy II: Michif Culture, Heritage, and Folkways* (Saskatoon: Gabriel Dumont Institute and Pemmican, 2006), 46–54, which metaphorically explains the origin of the Métis. I found a Denesuline version of this story, called "The Crow," in Robert H. Lowie, "Chipewyan Tales," *Anthropological Papers of the American Museum of Natural History* 10, no. 3 (1912): 184–5.

7 See Reid, "'Faire Place,'" 56: "the conjuncture of this figure [Riel] and a revolutionary movement points to an exfoliation of meaning that transformed the uprisings from a purely political and economic problem to one that was religious."

8 These include entries in his diaries. See Thomas Flanagan, ed., *The Diaries of Louis Riel* (Edmonton: Hurtig, 1976), 52, 127–8, 169. I realize "multicultural" is a loaded term. I use it here rather than "multiethnic" because each of

the nations and regions Riel mentions as a source of immigrants is made up of people of many ethnicities. "Multinational," although in some ways more accurate, perhaps implies a larger group of source nations.

9 Rafael Pérez-Torres uses the phrase "a most vapid and liberal pluralism" in his article "Misegenation Now!" *American Literary History* 17, no. 2 (2005): 369.

10 See Kevin Bruyneel, "Exiled, Executed, Exalted: Louis Riel, *Homo Sacer*, and the Production of Canadian Sovereignty," *Canadian Journal of Political Science/Revue canadienne de science politique* 43, no. 3 (2010): 711–32. Bruyneel states, "It is not hyperbole to say that Canadians cannot get Riel out of their minds" (721). He cites Stanley, "Last Word on Louis Riel," 58, who refers to Riel's complexity as compelling; Stanley characterizes Riel as the Canadian Hamlet and states that "Riel's ghost" has "several faces." Bruyneel suggests that "standing symbolically for so many positions, Riel could not possibly represent any single one of them distinctly or adequately" (721).

11 This is a common metaphor used to describe Métis people, including Riel, who occasionally used it to describe himself. See, for example, Braz's chapter on this topic in *False Traitor*, entitled "The Go-Between: Riel as Cultural Mediator," 9. Métis poet Gregory Scofield illustrates the cultural divide Riel must negotiate in "Le Porte-parole/The Spokesman," in *Louis: The Heretic Poems* (Gibsons, BC: Nightwood, 2011), 45. Minelle Mahtani points out that the metaphors of "interpreter" and "translator" are often used by people of mixed ancestry to describe themselves; see Mahtani, "Mixed Metaphors: Positioning 'Mixed Race' Identity," in *Situating "Race" and Racisms in Space, Time, and Theory: Critical Essays for Activists and Scholars*, ed. Jo-Anne Lee and John Lutz (Montreal and Kingston: McGill-Queen's University Press, 2005), 77.

12 See the chapters by Thomas Flanagan and Hans V. Hansen in this volume.

13 Bruyneel, "Exiled, Executed, Exalted," 716, states, "Riel could never be a full member of the Canadian polity without upsetting its boundaries because his political figure served to demarcate those boundaries."

14 Ibid., 727.

15 Lauren L. Basson, "Savage Half-Breed, French Canadian or White US Citizen? Louis Riel and US Perceptions of Nation and Civilisation," *National Identities* 7, no. 4 (2005): 369–88; see also Lauren L. Basson, "Métis Americans: Louis Riel and the Northwest Territories," in *White*

Enough to Be American? Race Mixing, Indigenous People, and the Boundaries of State and Nation, 57–94 (Chapel Hill, NC: University of North Carolina Press, 2008). According to Basson, the United States also had difficulty with Riel's binationalism.

16 This is not to deny the many successes over the years of Métis people in revitalizing their political structures and culture(s).

17 See, for example, Sabrina Perissini, "From 'Great Lakes Metis' to 'Aboriginal People of Canada': The Changing Identity of Canadian Metis during the Eighteenth, Nineteenth and Twentieth Centuries," *Nexus* 14, no. 1 (2000): 87–107.

18 For instance, Riel spoke (at least) French, English, and Cree; see Siggins, *Riel*, 31. Gabriel Dumont spoke six Indigenous languages as well as French; see George Woodcock, *Gabriel Dumont: The Métis Chief and His Lost World* (Edmonton: Hurtig, 1975), 45.

19 See, for instance, Nikolas Kompridis, "Normativizing Hybridity/Neutralizing Culture," *Political Theory* 33, no. 3 (2005): 318–43. This paper is a critique of Seyla Benhabib, *The Claims of Culture: Equality and Diversity in the Global Era* (Princeton, NJ: Princeton University Press, 2002). Kompridis is concerned that overemphasis on "reinvention, reappropriation and subversion" (Benhabib, *Claims*, 68, cited at 334) will itself subvert the power of cultural traditions and practices and negate any rights that may flow from them. But he agrees that underemphasis on fluidity and multiplicity is problematic. Kompridis describes his own approach as one that re-envisions the relationship between identity and nonidentity, between old and new, and between hybridity and heterogeneity. In his view, neither essentialism nor anti-essentialism should have the last word (323). For a critique of hybridity from an Indigenous perspective, see, for example, Gordon Christie, "Indigenous Legal Theory: Some Initial Considerations," in *Indigenous People and the Law: Comparative and Critical Perspectives*, ed. Benjamin J. Richardson, Shin Imai, and Kent McNeil, 195–231 (Oxford: Hart, 2009).

20 For a similar Latino perspective, see Pérez-Torres, "Misegenation Now!" 369–70: "numerous ... critics have sought to demonstrate that concrete forms of hybridity – racial, social, cultural, and political – can promote a theoretical and critical apparatus that is not depoliticized, ahistorical, and aesthetically generalized."

21 According to Métis oral tradition, Riel prophesized, "My people will sleep for a hundred years. When they awake, it will be the artists who give them their spirits back."

22 The "lobes" of this symbol refer to the distinctiveness of the Indigenous and

European cultures from which the Métis were born; the continuity of the line represents the indivisibility of these cultures in being Métis.

23 See Nikos Papastergiadis, *The Turbulence of Migration: Globalization, Deterritorialization, and Hybridity* (Cambridge, UK: Polity, 2000), 170.

24 Riel's immigration plan involves only the Indigenous peoples of North America and Europeans, although in his writings he is sympathetic to North American blacks and the Indigenous peoples of South America. Similarly, his ecumenism does not extend to Islam, Hinduism, Buddhism, or other major religions, although he was apparently influenced by Theosophy, which borrowed from many Indian and Tibetan religious concepts. See note 54.

25 Flanagan contends that Riel's notion of Métis rights was not the same as the current conception; that is, Riel was willing to accept extinguishment of rights. See Thomas Flanagan, "Louis Riel and Aboriginal Rights," in *As Long as the Sun Shines and Water Flows: A Reader in Canadian Native Studies*, ed. Ian A.L. Getty and Antoine S. Lussier, 247–62 (Vancouver: UBC Press, 1983). This may be true insofar as Métis title to the 1.4 million acres granted to the "half-breeds" was concerned, although arguably this provision was invalid from inception or was voided by the Crown through repeal and/or nonperformance of treaty obligations. See the Manitoba Act of 1870, SC 1870, c. 3, s. 31; and *Manitoba Métis Federation v. Canada (Attorney General)*, 2007 MBQB 297, 223 Man. R. (2d) 42, aff'd 2010 MBCA 71, 255 Man. R. (2d) 167, rev'd in part 2013 SCC 14. See also Darren O'Toole, "Métis Claims to 'Indian' Title in Manitoba," *Canadian Journal of Native Studies* 28, no. 2 (2008): 241–71. O'Toole refutes Flanagan's assertion that the Métis never claimed Aboriginal title and suggests that Flanagan's view influenced the trial judge to rule against the Métis claimants in *Manitoba Métis Federation v. Canada.* Flanagan was an expert witness for the Crown in that case, in which the federation sought, among other things, a declaration that the Métis were deprived of their land rights promised in the Manitoba Act. Ultimately, the Supreme Court of Canada agreed that the federal government had failed to act in accordance with the "honour of the Crown" in administering Manitoba Act land grants for the Métis.

Another way of looking at Riel's "acceptance" of extinguishment is that he had few realistic long-term alternatives. Regardless, Riel says explicitly that extinguishment of "the Indian title" is not specifically prescribed in the Manitoba Act as concerns all of Manitoba; it applies only to the "seventh" allotted to the Métis (II: 20). Subsections 32(3) and (4) of the act, regarding "quieting of titles," refer merely to other portions of the province in which the "Indian title" was not extinguished. See also Riel's comments about unextinguished Métis title in the expanded Manitoba (II: 40) and about "co-proprietorship with the Indians" of the entire North-West (II: 16). For analyses of the nature of Métis land title and its potential for extinguishment, see Catherine Bell, "Métis Aboriginal Title" (MA thesis, University of British Columbia, 1989); and Joseph Eliot Magnet, *Métis Land Rights in Canada*, http://www.uottawa.ca/constitutional-law/metis.html (accessed 16 October 2013).

26 The reference to Métis territorial conflicts with First Nations relates in part to the Dakota/Métis territorial/land use conflict that was resolved by the 1859 treaty negotiated on behalf of the Métis by Jean-Baptiste Wilkie.

27 Riel viewed the oral promises, including the promise of amnesty, as being part of the Manitoba treaty.

28 See Siggins, *Riel*, addendum 1, 449.

29 A discussion of the philosophical and legal meanings of sovereignty is beyond the scope of this chapter. For a review of some of the debates around the Métis and sovereignty, see O'Toole, "Métis Claims."

30 Although Riel was speaking after the Manitoba borders were expanded in 1881, his calculations refer to lands of the "postage stamp" province in 1870, as promised in the Manitoba Act. See Riel's comments on the expansion of Manitoba at II: 40.

31 Although Riel saw himself as, in part, inheriting the tradition of the Hebrew prophets, he was not above commenting that the Jews would get a seventh "if they will help us with their money" (II: 17). Later, he refers to the Jews as being "rich and the lords of finance" (II: 40), although immediately afterward he sympathizes with their exile from their biblical homeland. The requirement of conversion for prospective Jewish immigrants would seem to negate Riel's ecumenism but was arguably progressive at the time, given prevailing anti-Jewish sentiments. Note Riel believed that the First Nations of North America were descended from the tribes of Israel; see Martel, *Le Messianisme*, 171.

32 He later acknowledges that there are really ten groups (II: 18).

33 Reading this passage together with Riel's letter from prison to James W. Taylor of the United States Consulate in Winnipeg suggests that only the Indian title from which Métis title flows had been extinguished in the "seventh." However, given the Métis's co-proprietorship with the "Indians," this may not imply complete extinguishment of the Indian title. In the letter, Riel writes,

> And as the half-breeds of Manitoba have had the constitutional guaranty of that seventh of the land on account of their Indian tribe, the Indian

> tribes of the North-west are naturally dissatisfied that the Indian title should carry less for them than for the half-breeds. Besides, the terms of the thirty-first clause of the Manitoba act state that the half-breed land grant is made toward extinguishing the Indian title, which give the Indians to understand that when the Federal Government will have given so much to the pure Indians as it gave the half-breeds, then only will the Indian title entirely be extinguished.

However, this could also refer to future First Nations treaty lands outside the "seventh." The passage at II: 20 implies that the First Nations will not have title to the lands but will be entitled to a share of the revenue. In his letter to Taylor, Riel explains that the Métis want that "their seventh be valued and understood to represent a capital in the hands of the government, and that an interest be paid to them every year on that capital." This undated letter, written after 1 August 1885, is reprinted in Ben J. Harrison, "Message to Congress from President Harrison," 11 March 1889, http://law2.umkc.edu/faculty/projects/ftrials/riel/message frompres.html (accessed 27 September 2013).

34 Martel, *Le Messianisme*, 373, has suggested and Braz, *False Traitor*, 92–3, has stated that Riel thought of European culture as superior. However, this does not imply that Riel thought disparagingly of Indigenous – including Métis – contributions to Canada and the North-West. In fact, as Riel expressed throughout his trial speeches and writings, it was Canadian society's deliberate spurning of Métis contributions that led to the "troubles" of 1869 and 1885.

35 Further to the comments in the above note, the irony that drips from the phrase "genius of civilization" suggests that Riel's praise for British culture is not entirely unfettered.

36 Métis understandings of territory are not exclusive. Although ideas of private property were part of Métis reality, Métis people also had more wide-ranging and communal ideas of property. See, for example, Jean Teillet, *Métis Law in Canada* (Vancouver: Pape Salter Teillet, 2013), xiii, 1–10 to 1–13, 1–36 to 1–37, http://www.pstlaw.ca/resources/Metis-Law-in-Canada-2013.pdf (accessed 15 December 2013). Riel's perspectives on First Nations are varied, and he is often critiqued for supposedly ignoring or even exploiting First Nations. On the latter point, see Blair Stonechild and Bill Waiser, *Loyal till Death: Indians and the North-West Rebellion* (Calgary: Fifth House, 1997). Riel does not acknowledge First Nations by nation in the "sevenths" (he does refer to the Crees tangentially), but this may reflect his assessment of his audience. Riel also says the territory is "partly his own" (II: 38). This

may refer to sharing territory with First Nations or to the division caused by the Manitoba treaty.

37 Note, however, that he earlier says that the Métis were not really willing to let the lands go (II: 15).

38 Although the centrality of these contributions to the North-West economy was being eroded, many Métis also engaged in farming, fishing, or small business. For an example of an alternative Métis economy, see Nicole J.M. St-Onge, "Variations in Red River: The Traders and Freemen Métis of Saint-Laurent, Manitoba," *Canadian Ethnic Studies/Études ethniques au Canada* 24, no. 2 (1992): 1–21.

39 This is not to imply Riel was planning for the colonization of First Nations. See further comments on this point at note 76.

40 Compare Riel's jibe at the "pioneers of civilization" (I: 21).

41 On this point, see Reid, "'Faire Place.'"

42 Riel may be referring to Nova Scotia and New Brunswick when he speaks of the "Irish in the east" and to Quebec when he speaks of the "French in the west," although there was at that time a sizable Irish population in Quebec and a significant francophone population in the Maritimes.

43 Although Riel's plan for immigration is a way of reclaiming immigration both as a positive force and on Métis terms, Riel wants the court to understand the impact of Canadian colonialism on the Métis.

44 Riel identified with King David, a Hebrew, whom the Bible says was one-eighth Moabite through his great-grandmother Ruth; Riel described himself as one-eighth "Indian." See Martel, *Le Messianisme*, 170–1, citing Riel's letter of 1 May 1876 to Father Ignace Bourget. However, Riel states in one of his essays, "The Métis hold to the name of their race. They don't preoccupy themselves with what proportions of Indian and European blood they have. For as little as there is of one or the other, it means, I am Métis." See Thomas Flanagan, ed., *The Collected Writings of Louis Riel/Les Ecrits complets de Louis Riel*, vol. 3, *1884–1885* (Edmonton: University of Alberta Press, 1985), 273, cited in and translated by Basson, "Métis Americans," 70. Note that some speculation exists among Métis as to whether Riel's mother was also Métis. Many Métis people claim to descend from Jean-Baptiste Lagemodière (Riel's maternal grandfather) and his first wife, a Cree woman known as "Josette." Maggie Siggins mentions Josette briefly in her biography of Marie-Anne Lagemodière, *Marie-Anne: The Extraordinary Life of Louis Riel's Grandmother* (Toronto: McClelland and Stewart, 2008), 90–3. A more sensationalized (and fictionalized) account can be found in Grant MacEwan,

Marie Anne: The Frontier Spirit of Marie Anne Lagemodière (Saskatoon: Western Producer Prairie Books, 1984).

45 Riel also states this in the first trial speech (I: 34).

46 In an interview, Métis author Joseph Boyden says of Riel, "I think he was one of the first proponents of something that was an absolutely foreign concept in the 1800s. If you read his diaries and read what he wanted to happen, it was exactly in many ways what Canada is now, where many different groups, many different countries and cultures can come together and live while maintaining their own culture, but live within the bigger sphere of Canada. Really his vision is very much what we are living today." See *Canadian Interviews*, 19 January 2011, copy on file with Kerry Sloan.

47 Siggins is an exception, although this topic is not a focus of her work. In *Riel*, 392, she states "Riel needed to establish a universe, no matter how metaphorically, where the oppressed minorities, the Métis, Indians, Irish, Slovaks, could live with dignity and equality. To suggest, as has been done endlessly for over a hundred years, that his vision was nothing more than the rantings of a madman is to say that a unique spirituality is beyond the pale, that conformity must smother all alternatives." See also the brief discussion of Riel's religious beliefs in Barkwell, Dorion, and Hourie, eds, *Metis Legacy II*, 185. Robert Choquette discusses Riel in his compendious *Canada's Religions: An Historical Introduction* (Ottawa: University of Ottawa Press, 2004), but he makes reference only to Riel's Catholic influences and goes so far as to state, "Louis Riel had no ties to the Amerindian people of Canada" (280).

48 Jennifer Reid, *Louis Riel and the Creation of Modern Canada: Mythic Discourse and the Postcolonial State* (Alberquerque: University of New Mexico Press, 2008). See also B.W. Powe, *Canada of Light* (Toronto: Somerville House, 1997).

49 Reid, *Louis Riel*, 181.

50 John Ralston Saul, *A Fair Country: Telling Truths about Canada* (Toronto: Penguin Canada, 2008).

51 Reid, "'Faire Place,'" 56.

52 This includes the confederacy of Métis and First Nations that Riel hoped to create. See, for example, Siggins, *Riel*, 285.

53 Flanagan, *Louis 'David' Riel*; Martel, *Le Messianisme*; Mossman, "Charismatic Pattern."

54 According to Bumsted, "Louis Riel," 26, "In any event, Riel's developing spiritual system was highly eclectic. Although fundamentally Catholic in origin, Riel's 'theology' appears to be much influenced by nineteenth-century

Protestant messianism and millennialism, to which he had undoubtedly become exposed on his American travels. His numerology was possibly indebted to the Millerite movement of the 1840s, although most nineteenth-century American Protestant millennialists, including Joseph Smith, liked to play with Biblical numbers. His doctrinal testament, the *Massinahican*, may well have been influenced by the theosophy of one Madame Blavatsky, who was active in New York State at the same time as Riel" (notes omitted). See also Thomas Flanagan, "On the Trail of the *Massinahican*: Riel's Encounter with Theosophy," *Journal of the Canadian Church Historical Society* 37, no. 2 (1995): 89–98. Note that Blavatsky travelled widely, acquainting herself with many cultures. She took the "five precepts" of Buddhist lay people while in Sri Lanka and was a supporter of Indian independence from Britain. See Isaac Lubelsky, *Celestial India: Madame Blavatsky and the Birth of Indian Nationalism* (London: Equinox, 2012).

55 Reid, "'Faire Place,'" 53.

56 Flanagan, *Louis 'David' Riel*, 179–80. Flanagan states, "This syncretism was absent from the teaching of Louis Riel, except insofar as he may have altered his ideas when presenting them to the illiterate métis" (180). He acknowledges that Riel makes a "brief allusion in his trial speech to métis traditions of prophecy" and may have been drawing on "local superstition or Indian lore" (180), but he does not acknowledge that Riel's audience at trial was not composed mainly of "illiterate métis." See below for further discussion of Flanagan's dismissal of Riel's reference to Métis prophecy in the second trial speech.

57 Flanagan, ed., *Diaries*, 127.

58 Flanagan, *Louis 'David' Riel*, 180, claims otherwise. An analysis of the mutual influences of Catholicism and Indigenous religions is beyond the scope of this chapter.

59 Reid, "'Faire Place.'"

60 Lawrence Barkwell, Darren Prefontaine, and Anne Carrière-Acco, "Metis Spirituality," in Barkwell, Dorion, and Hourie, eds, *Metis Legacy II*, 184. The authors cite Métis elder and author Maria Campbell as saying that Gabriel Dumont carried both Sioux and Cree medicine bundles (185).

61 Chad S. Hamill, *Songs of Power and Prayer in the Columbia Plateau: The Jesuit, the Medicine Man, and the Indian Hymn Singer* (Corvallis, OR: Oregon State University Press, 2012), 4.

62 Unfortunately, most of Riel's philosophical work *Massinahican* (Cree for "book" or "Bible") was lost. The fragments that remain have been reconstructed in Gilles Martel, ed., *The Collected Writings of Louis Riel/Les Ecrits*

complets de Louis Riel, vol. 2, *1875–1884* (Edmonton: University of Alberta Press, 1985). For an example, see note 76.

63 The historical and ongoing unwillingness to acknowledge such connections in British Columbia is examined in the work of Jean Barman, such as *Stanley Park's Secret: The Forgotten Families of Whoi Whoi, Kanaka Ranch and Brockton Point* (Madeira Park, BC: Harbour, 2005); "Lost Okanagan: In Search of the First Settler Families," *Okanagan History* 60 (1996): 8–20; "Invisible Women: Aboriginal Mothers and Mixed-Race Daughters in Rural British Columbia," in *Beyond the City Limits: Rural History in British Columbia*, ed. R.W. Sandwell, 159–79 (Vancouver: UBC Press, 1999); and "At the Edge of Law's Empire: Interraciality, Citizenship and the Law in British Columbia," *Windsor Yearbook of Access to Justice* 24, no. 1 (2006): 3–22.

64 See Cole Harris, *The Resettlement of British Columbia: Essays on Colonialism and Geographical Change* (Vancouver: UBC Press, 1997). On this point generally, see Saul, *Fair Country*.

65 Siggins, *Riel*, 31, gives an example of the syncretic religious views of Riël's mother, Julie, who, although a devout Catholic and apparently not Métis, "spent her formative years with Métis and Indian children and she must have been touched by Native myth and belief."

66 Métis elder Lottie Kozak, communication with author, 15 June 2012.

67 Flanagan, *Louis 'David'* Riel, 180, asserts that, whereas the founders of most "nativistic" religious movements borrow from both Christianity and Indigenous beliefs, Riel had not done this: "Riel's ideas were developed from Christian themes."

68 Martel, ed., *Collected Writings*, vol. 2, 77, quoted in Reid, "'Faire Place,'" 59.

69 Book of Revelation 20: 1–6; 21: 1–27; 22: 1–5.

70 Leanne Simpson, *Dancing on Our Turtle's Back: Stories of Nishinaabeg Re-Creation, Resurgence and a New Emergence* (Winnipeg: Arbeiter Ring, 2011).

71 Ibid., 112–13.

72 Ibid., 113.

73 For a description of this law from a Haudenosaunee (Mohawk) perspective, see Susan M. Hill, "'Travelling Down the River of Life Together Forever': Haudenosaunee Land Ethics and Treaty Agreements as the Basis for Restructuring the Relationship with the British Crown," in *Lighting the Eighth Fire: The Liberation, Resurgence, and Protection of Indigenous Nations*, ed. Leanne Simpson, 23–45 (Winnipeg: Arbeiter Ring, 2008).

74 Jack Weatherford, *Indian Givers: How the Indians of the Americas Transformed the World* (New York: Fawcett Columbine, 1988), 135. Some Haudenosaunee believe that this treaty is much more ancient.

75 Hill, "'Travelling Down the River,'" 43; Weatherford, *Indian Givers*, 137. This confederacy has lasted for hundreds, perhaps thousands, of years. The Great Law of Peace was the basis for treaties such as the 1613 treaty with the Dutch known as the Kaswentha, or Two Row Wampum, which was adopted and renewed by the British in 1664. The Dish with One Spoon treaty was also entered into in 1701 with France; the French referred to it as La Grande Paix. For a discussion about whether the Two Row Wampum implied Indigenous-European interdependence or noninterference, see John Borrows, *Recovering Canada: The Resurgence of Indigenous Law* (Toronto: University of Toronto Press, 2002), 148–50. James Tully discusses interdependence and the Two Row Wampum with respect to modern treaties in *Strange Multiplicity: Constitutionalism in an Age of Diversity* (Cambridge, UK: Cambridge University Press, 1995), 137.

76 In his trial speeches, Riel evidences support for First Nations' rights, title, and self-rule, which suggests that, in his vision, First Nations would have remained distinct, as well as contributing to *métissage*. However, Braz, *False Traitor*, 39, suggests that what Riel was proposing amounted to assimilation. This is an important issue that should be further explored. Some statements by Riel in a fragment from circa 1881 to 1884, thought to be part of his *Massinahican*, suggest the possibility of assimilation through *métissage*, although elsewhere in the fragment Riel is ambiguous. For example, he states that the Aboriginal peoples of North America "ferait place à une race nouvelle la race métisse qui varierait selon les pays." Here, "ferait place à" could mean either "would make way for" or "would make room for." Cited in Martel, ed., *Collected Writings*, vol. 2, 409. Regarding immigrants having their "own" territories in the North-West, Riel does not mention here sharing title or use with First Nations, although elsewhere he speaks of Métis "co-proprietorship with the Indians" (II: 16). For more on shared use of territories, see notes 26 and 36.

77 Riel to P.A.M., n.d., cited in Martel, *Le Messianisme*, 196–7.

78 This is not to necessarily state that the prophecy has not been influenced by Christianity. For a discussion of the origins and current Anishinabek understandings of this prophecy, including its adaptability, see Camille Bernier, "The Social Dynamics of Ojibwe Prophecy" (PhD diss., University of Wisconsin – Madison, 2011).

79 Leanne Simpson, "Oshkimaadiziig, the New People," in Simpson, ed., *Lighting the Eighth Fire*, 13–21. See also Edward Benton-Banai, *The Mishomis Book: The Voice of the Ojibway* (Hayward, WI: Indian Country Communications and Red School House Press, 1988).

80 Simpson, "Oshkimaadiziig," 14. Interestingly, according to Theosophy, there are also seven ages of humanity, although Blavatsky thought we were in the fifth age. See H.P. Blavatsky, *The Secret Doctrine*, vol. 2 (1888; reprint, Pasadena, CA: Theosophical University Press, n.d.), 249–50, http://www.theosociety.org/pasadena/sd/sd2-1-13.htm (accessed 28 September 2013).

81 Space does not permit description of the many Indigenous prophets and prophetic movements of the postcontact period. For an overview, see Alfred A. Cave, *Prophets of the Great Spirit: Native American Revitalization Movements in Eastern North America* (Lincoln, NB: University of Nebraska Press, 2006); Scott Peterson, *Native American Prophecies*, 2nd ed. (St Paul, MN: Paragon House, 1999); and Bernier, "Social Dynamics," ch. 2. See also John Tanner, *The Falcon: A Narrative of the Captivity and Adventures of John Tanner* (1830; reprint, New York: Penguin, 1994); and William W. Warren, *History of the Ojibway People*, 2nd ed. (1885; reprint, St Paul, MN: Minnesota Historical Society, 2009). Tanner, who was intimately familiar with the Métis and the events of the North-West, describes the behaviour and methods of many of the Algonquian prophets of his day.

82 Cave, *Prophets*.

83 David Day, author of *The Visions and Revelations of St. Louis the Métis* (Saskatoon: Thistledown, 1997), a book of "found poetry" based on Flanagan's edition of *The Diaries of Louis Riel*, states in his introduction, "As Riel himself recognized, the Métis were inheritors of the prophetic traditions of both the desert mystics of the ancient Hebrew tribes and the great plains shamans of the equally ancient Indian tribes ... The degree to which Louis Riel's language was influenced by old world prophets like Moses is obvious enough and has frequently been commented upon. What is less apparent is the influence of the visionary and ecstatic poetry of the Amerindian oral tradition. Yet his diary and notebooks are charged with that influence" (13). Day compares Riel's 1885 "Voice of the Indian" (Flanagan, ed., *Diaries*, 66) with a Sioux war song, which he says was recorded four years later in 1889. Riel's words (in translation): "I hear the voice of the Indian / He comes to join me / He is coming from the north / He is in the mood for war" (13). The Sioux text: "He comes from the north / He comes to fight / He comes from the north / I go to meet him" (13). Day does not provide a source for the Sioux war song and admits that he is not making any scholarly claims; his collection is "meant only as an observation and footnote on Louis Riel's life and times" (13).

84 The focus on Anishinabek prophets here is not meant to deny the possible influence on Riel of Cree or Chipewyan religious philosophy. There are

many shared elements in Ojibway and Cree religions. See, for example, Jennifer S.H. Brown and Robert Brightman, eds, *"The Orders of the Dreamed": George Nelson on Cree and Ojibwa Religion and Myth, 1823*, part 3 (Winnipeg: University of Manitoba Press, 1988). In an essay in *"The Orders of the Dreamed"* entitled "On the Ethics of Publishing Historical Documents," Emma LaRocque notes, "what amazes me about this manuscript [Nelson's accounts of Algonquin religion in Lac la Ronge, ca. 1823] is that my Plains-Cree Métis community in northeastern Alberta was still living and reciting in the 1950s and 1960s essentially the same religion, legends and myths discussed in the manuscript!" (202).

85 Bernier, "Social Dynamics," 86.

86 Ibid., 88.

87 Ibid., 86.

88 Ibid.

89 Barkwell, Prefontaine, and Carrière-Acco, "Métis Spirituality," 185.

90 In Riel's opening to his speech to the jury, he addresses a prayer to God, acknowledging that he is "naturally inclined to think of God at the beginning of [his] actions" (I: 2) and teasing those who would view his penchant for prayer as a mark of insanity. Although the text of the prayer is undeniably Christian, the ceremony of praying before any important undertaking or meeting is common in many First Nations traditions and is arguably an incident of procedural law, together with being a spiritual practice. Such protocols have been observed as part of treaty negotiations, as well as in modern court cases.

91 For a discussion of reciprocal mother-child and human-land obligations as referred to by Crown and First Nations negotiators of Treaty No. 1 of 1871, see Aimée Craft, *Breathing Life into the Stone Fort Treaty: An Anishinabe Understanding of Treaty One* (Saskatoon: Purich, 2013). Riel states, "The North-West is also my mother ... and although my mother country is sick and confined in a certain way, there are some from Lower Canada who came to help her take care of me ... I am sure that my mother country will not kill me more than my mother did forty years ago when I came into the world, because a mother is always a mother" (I: 3). Riel then describes his own care for his people, as well as for his relatives the "Indians" and "whites" (I: 4).

92 The Seven Grandfather Teachings are wisdom, love, respect, bravery, honesty, humility, and truth. These teachings are foundational to Anishinabek law and have parallels with Cree and Métis teachings. Riel in this passage is defending his criticism of a priest on the grounds that love required him to

speak the truth about what he saw as the priest's wrongdoing. John Borrows, *Drawing Out Law: A Spirit's Guide* (Toronto: University of Toronto Press, 2010), 104, states that Mishomis "learned ... that opposition was essential to understanding wisdom, love, respect, bravery, honesty, humility, and truth."

93 Claude Aubin, "1701: The Great Peace of Montreal & the Chest of Wampum Belts," 22 March 2012, paragraphs 3–8, http://www.claudeaubinmetis.com/viewtopic.php?f=32&t=124 (accessed 27 September 2013).

94 This idea derives from Bruyneel's application of Giorgio Agamben's concept of *homo sacer* to Louis Riel. See Bruyneel, "Exiled, Executed, Exalted."

95 Reid, "'Faire Place,'" 55.

96 Robert A. Williams Jr, *Linking Arms Together: American Indian Treaty Visions of Law and Peace, 1600–1800* (New York: Oxford University Press, 1997).

7

Reconstructing the Substantive Argument in Louis Riel's Address to the Jury

PAUL GROARKE

The purpose of the present chapter is to clarify the legal outlines of the defence presented by Louis Riel in his address to the jury.[1] I do not propose to canvass the different views in the literature as to the merits of the two speeches that comprise his address.[2] Nor do I propose to deal with the many procedural irregularities in the trial, which much of the recent literature on Riel has minimized.[3] Since the current book focuses specifically on Riel's speeches, I have restricted my comments to those aspects of the process that bear directly on the legal arguments that he presented in his address to the jury. Although Riel was a political rather than a legal figure, and his speeches serve as a kind of political testament, they also contain two substantive defences to the charge before the court. The remarkable fact is that this has largely been overlooked.

RIEL'S ADDRESS TO THE JURY IS A PRODUCT OF COMPETING FACTORS

The legal mechanics of Riel's address to the jury are a product of the crosscurrents in the trial. This accounts for the two factors that explain the legal composition of the address to the jury. The first factor is that Riel wanted to distance himself from the defence of insanity. The second factor is that he wanted to vindicate his actions legally. Although it is the second factor that is significant substantively, the literature on Riel has generally focused on his psychological state and the insanity defence.

THE DEFENCE LAWYERS REFUSED TO DEFEND RIEL ON THE MERITS OF THE CASE

The first factor in the formation of Riel's address to the jury had its origins in his lawyers' decision not to contest the merits of the case. The most defensible explanation for this decision, professionally, lies in the makeup of the jury. At the outset of the trial, the defence lawyers challenged the jury panel on the basis that the magistrate had personally selected the members of the panel.[4] When this challenge failed, and the jurors were chosen, the defence was left with a jury without francophones, Catholics, or Métis. The trial was accordingly in the hands of a magistrate and a jury, all of whom were anglophones and Protestants.[5] This was the sector of society that was the least sympathetic to any argument that the actions of the Métis could be justified.

This feature of the trial almost certainly played a significant part in determining why none of the defence lawyers dared to suggest that the so-called "rebellion" could be justified.[6] If this is the case, however, it seems to overlook the political facts on the ground, which were a good deal more complicated than the victory of the Canadian troops suggested. Many people in the North-West blamed the government for the uprising. It is nevertheless clear that the jury panel had been "sifted," since it appears to have been composed of those elements of society who had a personal stake in the Canadian government's assertion of sovereignty over the territories.

The issues raised by the composition of the jury are fundamental. The Declaration Which Offences Shall Be Adjudged Treason of 1351, under which Riel was charged, leaves it to the jury to decide the verdict on a charge of high treason.[7] This prerogative was given to the jury because the charge of treason had been used by the king to silence individuals who raised legitimate grievances against the state. The statute gave the people the final say in determining whether the actions of the accused constituted treason.

The plain fact of the matter is that the jury that tried Riel was not representative of the community and did not include his peers. Those elements of the community who were most likely to sympathize with his actions and his cause had been left out. This posed a major problem for the defence lawyers and raised ethical issues, since the participation of a lawyer in an unfair trial only lends it legitimacy. At this distance, one can only assume that the lawyers believed that any defence on the merits of the case would antagonize the magistrate and the jury – and ensure Riel's conviction. They accordingly looked for some other means of defence.

However, there are less attractive explanations for the lawyers' conduct lurking in the background. The leading lawyers for the defence were from Quebec and Upper Canada and did not share the regional sympathies of their client. Both Charles Fitzpatrick and François-Xavier Lemieux had judicial aspirations and probably had a personal interest in arguing insanity, since a plea of insanity meant that they did not have to attack the actions of those in political power in the courtroom.[8] The transcript of the trial makes it clear that the lawyers were quite unable to grasp the idea that the legal authority of Canada to enter the North-West Territories and use force against the inhabitants could be questioned in the courtroom.

THE DECISION OF THE DEFENCE LAWYERS TO ARGUE INSANITY WAS TACTICAL

It follows that the chief merit of the insanity defence from a tactical perspective was that it allowed the lawyers to plead for mercy without challenging the hard-held attitudes of the magistrate and the jury. The attractions of such a strategy from a political perspective were considerable. If Riel was insane, there was no need to justify the actions of the Métis or to determine whether they had a right to defend themselves. The nettlesome question of whether the Canadian government should have sent in troops could be ignored altogether.

The suggestion in the literature that the decision of the lawyers to advance the defence of insanity was based on a simple assessment of Riel's condition is naive. There was a significant degree of psychological manipulation in such a defence, which was attractive primarily because it would allow the jurors to express sympathy for Riel without calling their views of the conflict into question. However, there was also a patronizing element in such a tactic, which undermined the credibility of the Métis cause and catered to the racial prejudices of the time. Morally and historically, the defence lawyers took the victors' side, and refused to question the legitimacy of the government's actions.

The defence of insanity also came at a price. Indeed, the argument that Riel was insane was not a defence in the ordinary sense of the word, since the Crown took the position that the rules of pleading allowed the defence to either dispute the substance of the charges or admit the charges and argue insanity. The same line of reasoning appears in the magistrate's instructions to the jury.[9] As a result, the decision of the defence lawyers to pursue the defence of insanity was premised on the assumption that Riel had committed an act of treason and released the Crown of its usual burden in proving the charge.

RIEL REJECTED THE DEFENCE OF INSANITY

The immediate problem for the defence lawyers was that Riel rejected the strategy that they had adopted. Riel took the position that he was sane and felt that he had a defence on the merits of the case. This created tensions between Riel and his lawyers, which erupted into open conflict when Riel interrupted the proceedings and asked to cross-examine Charles Nolin. The transcript of the trial makes it clear that Riel was objecting to the failure of his lawyers to challenge the Crown's substantive case. "The witnesses are passing," Riel is recorded as saying, "and so are the opportunities."[10]

The defence lawyers responded scornfully. Charles Fitzpatrick stated on the record that Riel must not be allowed to interfere in their conduct of the case. When Riel continued to protest, the lawyers threatened to withdraw from the case. This was too much for Riel, who suffered from grandiosity, and the trial proceeded on the basis of the insanity defence. The insistence of the lawyers set up an impossible situation, however, since the defence that Riel wanted to advance and the defence advanced by his lawyers were incompatible. When Riel addressed the jurors, he accordingly asked them to reject the defence advanced by his own lawyers. He then advanced a defence on the merits of the case, which his lawyers had essentially admitted. The dynamics were hopeless.

THE CONDUCT OF THE DEFENCE LAWYERS RAISES ETHICAL AND PROFESSIONAL CONCERNS

The conduct of the defence lawyers raises a number of professional concerns. Their handling of the case was ill-considered and maladroit.[11] They also failed in their duty to their client, however, in at least two respects. The first failure was simply that they refused to take instructions from their client. This cannot be justified: nothing in the record suggests that Riel was incapable of instructing counsel. The second failure was that they refused to scrutinize the case put forward by the Crown.[12]

This is not a matter of applying contemporary standards to professionals who worked in different historical circumstances. John Langbein, the foremost authority in the area, has written that defence lawyers were originally prohibited in trials for treason. This ban was lifted in 1696 because the Crown had begun to use lawyers to prosecute such cases. "Thus, the main rationale for the 1696 Act was evening up for the Crown's perceived advantages in prosecuting treason."[13] The change in courtroom practice recognized that there was a "larger pattern of disadvantage for treason Defendants."[14] This disadvantage was a reflection of the political nature of such charges, which had often been used to suppress individuals who dared to challenge those in political power.

The same kind of rationale explains why English judges gave individuals charged with crimes a general right to counsel in the eighteenth century. The increasing use of lawyers to prosecute cases had led to many miscarriages of justice: Langbein opines that the adversarial nature of the common law process had brought out "the potential for distortion and fabrication" in the use of professional prosecutors.[15] Defence lawyers were allowed into the process in order to test the evidence presented by professional prosecutors, which "needed probing."[16]

This historical rationale is significant in the context of Riel's trial, since the lawyers who prosecuted Riel were highly partisan. Although Thomas Flanagan has argued that the fairness of Riel's trial can be judged only by comparing it to other trials of the time, comparisons are difficult, as J.M. Bumsted writes, because Riel was the only individual tried for high treason.[17] Bumsted nevertheless argues that the decision to try Riel for such an offence – which carried a mandatory sentence of death – shows that he was singled out for special treatment. It is not possible to settle the matter here, but Bob Beal and Barry Wright have provided ample evidence that the "justice system in all the 1885 rebellion cases demonstrated elements of harshness and carelessness unusual for the day."[18]

The transcript of Riel's trial makes it clear that the prosecutors were implacable. The competence of the defence lawyers in the face of the prosecution is also questionable. The principal issue, however, is that the defence lawyers neglected their responsibilities to their client. Riel's complaint that his lawyers were not cross-examining the witnesses was perfectly legitimate, and rather than probe the evidence led by the Crown, the defence lawyers insisted that the magistrate silence the accused. This essentially tells us that they were not willing to represent Riel if he insisted on defending himself on the merits of the case. The professional lapses are so severe that the trial process fits more easily into our idea of a mistrial.[19] This suggests that the case should not have gone to the jury.

The capitulation on the substantive charge before the court was almost total. It was collusion in practice, if not intention, and the defence lawyers must clearly bear the principal responsibility for depriving Riel of his substantive defence. There was a kind of betrayal in this, since they were obviously assigned with the responsibility to defend him. George Goulet has argued with a good deal of justification that the defence lawyers' reliance on the insanity defence simply diverted attention from the government's actions, which were highly disreputable. As a result, Riel was "deprived of his right to make full answer and defence" by his own lawyers.[20] Bumsted reaches a similar conclusion, observing that "Riel's own lawyers were the people who were most responsible for his failure to receive justice in 1885."[21]

It is true that the defence lawyers faced a difficult task, and the transcript of the trial makes it clear that the magistrate was determined to keep any suggestion that Riel might have a defence out of reach of the jury. However, this is no excuse: I am speaking as a trial lawyer when I say that a lawyer faced with such a situation should continue to press his client's case. The proper response would have been to force the magistrate's hand, so that the unfairness of the process appeared on the surface of the record. Riel's lawyers should have insisted that the magistrate rule openly, explicitly and repeatedly, and with reasons, in order to prepare the necessary grounds for an appeal.[22] Nor is this all; in a political case like Riel's, a trial serves a historical purpose, and it is wrong to sidestep the issues in the case. Bumsted has grasped the right dynamic and argues that the defence lawyers should have persisted in their demand for the documentary evidence that the Crown refused to produce. Instead, the defence gave in to the magistrate and the Crown and chose not to press the matter.[23]

The conduct of the defence lawyers is hard to understand. At a minimum, they should have put Riel's position before the court and the jury, whatever their views of its merits, and let the jury decide whether he had a defence. Thus the most striking feature of the trial from a professional perspective is not that the defence lawyers rejected their client's instructions; it is that they assumed the substantive elements of the offence had been made out. This assumption left only insanity, which Riel disputed, but in peril of his life, since that left the jury with nothing but the lawyer's assumption that he had committed the offence before the court. It was the jurors who understood that the lawyers had put them in an impossible situation and that Riel was arguing for his own conviction. Indeed, the jury foreman wept as he read the verdict.[24]

RIEL'S FIRST GOAL IN ADDRESSING THE JURY WAS TO REFUTE HIS LAWYERS' ARGUMENT THAT HE WAS INSANE

The legal predicament in which Riel found himself at the end of the trial can only be described as bizarre. I say this because Riel's major purpose in addressing the jury was to refute both of the positions adopted by his lawyers: that he was insane and that he had no substantive defence. His argument on the substantive question has been obscured by the endless discussion of insanity in the literature and by a tacit assumption that Canada's hold on the North-West Territories was beyond the reach of the trial. It is the second argument, however, that raises the more significant legal issues.

Riel's first goal in addressing the jury was accordingly to refute the argument of the defence that he was insane. At the outset of his remarks, he asks the jury not to interpret his "natural excitement," and his opening prayer, as any *indicia* that his mind is "out of its ordinary condition."[25] Riel declares that he has no wish to "play" insanity and implicitly accuses his lawyers of fabricating the defence in a misguided attempt to save him.

Riel acknowledged that he was taking the same position as the Crown. In fact, he thanked the prosecution for their work in "destroying" the testimony of Dr François Roy, who testified for the defence. "Even if I was going to be sentenced by you, gentlemen of the jury, I have this satisfaction if I die – that if I die I will not be reputed by all men as insane, as a lunatic."[26] It is not clear that Riel fully appreciated that insanity had become the only issue in the trial and that the jury's finding that he was sane would only secure his conviction for high treason.[27]

It is clear, however, that the jurors understood the dilemma in which they found themselves. In spite of this, they took Riel's side on the matter and rejected the insanity defence. The merits of their decision in this regard are better dealt with elsewhere. Since the purpose of this chapter is merely to sketch out the legal arguments in Riel's address to the jury, it is enough to say that the first goal of this address was accordingly met.[28] This left only his arguments on the merits, which were never seriously considered, and were never put to the jury.

RIEL'S SECOND GOAL WAS TO DEFEND HIMSELF ON THE MERITS OF THE CASE

Riel's second goal in his address to the jury was to defend himself on the substance of the charge. As a political figure, he clearly saw the trial as a historical opportunity to vindicate his people's cause. This carried a certain raw danger in the political climate of the time, and might have taken the trial directly into politics, but Riel seems to have recognized that he was in a legal forum and based his defence on legal principles. The fact that stands out professionally is that his lawyers refused to follow his instructions and seem to have been more concerned about the damage that this might do to their reputations.[29]

The information against Riel alleged that he had levied and made war against the Queen. This was high treason and a grave breach of a citizen's duty of allegiance to the Crown. Riel's legal response is muddied by his oratory, his many personal references, and his inevitable confusion, which reflects the fact that he was essentially representing himself. It is nevertheless possible to discern two substantive submissions in Riel's address to the jury.

The first substantive submission in the address to the jury is that the actions of the Government of Canada in the North-West were inherently illegitimate, since it did not enjoy sovereignty over the territories. It was therefore an invader and had no hold on the loyalties of the Métis. The second submission – which contains Riel's personal response to the charges – is that the Métis had a right to defend themselves from the force that attacked them. They had accordingly "acted reasonably and in self-defence." The clarity of the second submission is lost in Riel's final rhetorical flourish, when he inverts the argument against him and alleges that the Government of Canada is insane.

THE LAWYERS NEVERTHELESS REFUSED TO ACKNOWLEDGE THAT THE CASE RAISED ANY SUBSTANTIVE ISSUES

The initial difficulty in assessing Riel's defence on the merits of the case is that the lawyers refused to acknowledge that the trial raised any substantive issues. As a result, there are only incidental references to the substantive issues that Riel raises in the record of the trial and the appeal.[30] This is a telling defect, and in retrospect one has to wonder whether the lawyers knew that an analysis of Riel's substantive defence would take them into uncomfortable places. In any event, it was trial by assumption. The Crown simply took the position that the fact of the rebellion was plain and did not admit of a defence; the magistrate agreed. The defence lawyers refused to challenge the Crown's position and joined implicitly in the assumption.

The trial accordingly proceeded on the premise that there was an unlawful rebellion. It followed that anyone who participated in the rebellion had committed treason. Although Charles Fitzpatrick alluded to the historical grievances of the Métis in his jury address, these were put forward in mitigation rather than as a defence. The only person who challenged the rhetorical assumption that the Métis had rebelled against the lawfully constituted Government of the North-West Territories was Riel himself.

The magistrate put the substantive question very simply in his instructions to the jury and stated that, in his opinion, they "must" conclude that there was a rebellion. He then suggested that Riel had participated in the rebellion – and had therefore committed treason. The only material question on his instructions was whether Riel's mental state had relieved him of criminal responsibility.[31] If the jury found him sane, it was accordingly left only with a single verdict: guilty as charged.

The best that can be said of Riel's trial is accordingly that there was a trial on the issue of insanity. It follows that any evaluation of Riel's substantive defences is speculative. This is even more difficult without a re-enactment of the trial, since the evidence that should have been adduced by the defence was largely lacking. It is nevertheless possible to reconstruct the outline of Riel's substantive defence on the basis of his address to the jury. One of the things that stands out in such a reconstruction is that Riel saw himself as a statesman and recognized that the trial raised legal and constitutional issues that went far beyond the question of his guilt.

THE HISTORICAL LITERATURE HAS ALSO NEGLECTED THE SUBSTANTIVE DEFENCES

Riel had received some training as a lawyer in Montreal, and whatever its rhetorical merits, the most that can be said without further inquiry is that his address to the jury contains the elements of a viable defence on the merits of the case. In spite of this, there has been remarkably little debate in the historical literature as to the plausibility of such a defence. Some of this is probably a testament to the formidable influence of Thomas Flanagan, who has clearly taken the government's side in the trial. J.M. Bumsted has critiqued Flanagan's position in "Another Look at the Riel Trial for Treason." Bumsted's view deserves attention: he writes that Riel "wanted a defence based upon 'the merit of my actions,'" in effect justifying his rebellion in terms of the misconduct of the Canadian government in the North-West.[32]

Flanagan insists that Riel's choice of strategy in this regard was "legally hopeless because a government's mistakes can never be a sufficient defence against the charge of treason, at least as long as the same government continues to rule, and every lawyer would doubtless agree in principle."[33] This kind of comment rests on a number of assumptions as to the legitimacy of the actions of the Government of Canada in the North-West. It goes without saying that this is unsatisfactory: the main point in Riel's submissions was that these actions were illegitimate, and it is no answer to suggest that the fact of political power is sufficient to justify itself.

The questions raised by Riel's substantive arguments cannot be dismissed so easily. One question that Bumsted raises is whether Riel's substantive defence had any possibility of success. If nothing else, the prominence of jury nullification in the common law tradition lends support to Bumsted's suggestion that a jury might have acquitted Riel if his lawyers had tried to mount a proper defence. The injustice of Canada's conduct in the territories was well known, and Riel cut a sympathetic figure in the courtroom. Douglas Linder cites a letter that one of the jurors later wrote to a member of Parliament complaining that the trial was unfair: "Had the Government done their duty and redressed the grievances of the half-breeds of Saskatchewan ... there would never have been a second Riel Rebellion, and consequently no prisoner to try and condemn."[34]

RIEL'S FIRST DEFENCE ON THE MERITS WAS THAT HE HAD NO DUTY OF ALLEGIANCE TO THE QUEEN IN RIGHT OF CANADA

One of the deeper problems with the historical literature is that it suggests the real test of a substantive defence is whether it will succeed. However, this is a political rather than a legal test, and the question of whether Riel had a valid substantive defence exists independently of the question of whether the court was prepared to hear it. The possibility of an acquittal is accordingly a secondary issue.

Riel's underlying position was that the Government of Canada had no legal sovereignty over the North-West. Its actions were therefore illegitimate. This raises an important question regarding the terminology that has been used to describe the hostilities in the North-West. That question is whether there was a "rebellion"? The use of such a term is contentious, since it assumes that the Métis had risen up against the legitimate authority of the Canadian government. This was in fact the major substantive issue in the case. Riel's position was that Canada had annexed the North-West by force, without the consent of the people. The decision of the Canadian government to send in Canadian troops was accordingly a decision to invade another country. This helps to explain Riel's indignation in his address to the jury: it was positively "insane" to describe the actions of the Métis in response to the Canadian invasion as a "rebellion."

Riel's basic submission in his address to the jury is that Canada did not have *de facto* or *de jure* government of the North-West Territories. As a result, the Métis had no duty of allegiance to the Queen in right of Canada. It follows that there was no breach of duty and that the charge of high treason had not been proven. This argument deserves serious consideration, in spite of the fact that it was ignored by the lawyers, who simply assumed that Canada had government of the North-West Territories at the time that the violence broke out. The historical literature makes the same mistake and does not distinguish adequately between the different allegiances that came into play in the case. If Riel and the Métis owed a duty of allegiance to the Queen – a proposition that he ultimately rejected – it was an allegiance to the Queen of England, not the Queen in right of Canada.

The public position of the Government of Canada was that it had purchased the North-West Territories from the Hudson's Bay Company in 1869 and therefore enjoyed sovereignty in the North-West. Riel contested this position on two related fronts. His first line of argument was that the British Crown, under whom the Hudson's Bay Company had held the territories, had long since abandoned its title to government and could not convey it to the company. This was the position he took, after his conviction, when he petitioned the president of the United States: "The undersigned your humble petitioner, submits that ... the British Government have forfeited long ago all title and right of governing the North-West."[35] His second line of argument was that the Hudson's Bay Company had never received any title to government and accordingly could not convey it.

Riel is too excited to keep his arguments apart and merges his position with regard to the British Crown, the Hudson's Bay Company, and the Canadian government in a single argument. If Canada had acquired rights over the territories (a position he rejected), it too had forfeited those rights, like the British Crown and the Hudson's Bay Company before it. "When I came into the North-West in July, the first of July 1884," Riel complains to the jury, "I found the Indians suffering. I found the half-breeds eating the rotten pork of the Hudson Bay Company and getting sick and weak every day. Although a half-breed, and having no pretension to help the whites, I also paid attention to them. I saw they were deprived of responsible government, I saw that they were deprived of their public liberties."[36] A government that fails in its duties to the people has no legal hold on their loyalties.

Riel's address to the jury reflects the politics of the time. There had been an undisguised competition for government of the North-West Territories. Tensions had arisen between the traditional inhabitants of the territories and the settlers who had come in under the auspices of the Canadian government. There was no clear dividing line, however, and people on both sides of the conflict shared a general feeling of resentment toward the Canadian government. Riel accordingly makes a broader argument, which takes its force from the fact that the territories were under the administration of the lieutenant governor and were governed basically by decree. In his submission, which has a distinctly American flavour, this was government without representation. The North-West Council, which ostensibly governed the territories, he argues, was "only a sham representative legislature and no representative government at all."

RIEL'S SECOND DEFENCE ON THE MERITS WAS THAT THE MÉTIS WERE DEFENDING THEMSELVES

Riel's second argument on the merits was that the Métis were defending themselves. He was therefore entitled to an acquittal. Riel's argument in this regard was based on the inherent right of a people to preserve itself, which was historically seen more as a duty than a right. This idea is present in Riel's submissions to the jury, since he took the position that the Métis had no choice but to defend themselves. Although there are differences between the collective right of a people to defend itself and a personal right of self-defence, it seems clear that each of these defences entails the other.[37] In either case, the fundamental feature of self-defence, like the legal doctrines of necessity and compulsion, is that it supersedes our ordinary legal obligations.

If Riel was right in arguing that the Government of Canada was an invader, his submission that the Métis had the right to use justifiable force in defending themselves seems to provide a complete answer to the charge. Even if the Government of Canada had some claim on the loyalties of the North-Westerners, however, there are other doctrines that apply. Patrick J. Charles has written that the doctrine of lawful rebellion reached its high point in England during the "Glorious Revolution" of 1688-89 and was taken up again in the American Revolution.[38] The doctrine was at least arguably a tenet of the common law and was relied upon to justify the use of force by Parliament to "restore" the Constitution if a king abused the laws, liberties, religion, and estates of the people. Charles argues that it is the doctrine of lawful rebellion that provides the real source of the right to bear arms that is found in the American Constitution.

Riel accordingly had a respectable argument that the rebellion against the Queen in right of Canada – if the Crown was right in characterizing it as a "rebellion" – was lawful and necessary to preserve the people. The political genesis of such an argument is clearly significant. In a pamphlet written in 1643, Stephen Marshall argues that "a people, especially the representative body of a State, may (after all humble Remonstrances) defend themselves against the unlawfull violence of the Sup[reme] Magistrate ... Endeavoring ... to deprive them of their lawfull Liberties."[39] William Prynne actually describes the king who "offers violence" to the people as rebellious, since the king has his own duty of allegiance to the people.[40]

The exact formulation of these arguments seems less important in a courtroom than their appeal to a jury's natural sense of justice. In his address to the jury, Riel submits that the Métis had brought their grievances to the attention of the Government of Canada. Rather than redress these grievances, the government had invaded the country. "By the testimony laid before you during my trial witnesses on both sides made it certain that petition after petition had been sent to the Federal Government, and so irresponsible is that Government to the North-West that in the course of several years besides doing nothing to satisfy the people of this great land, it has even hardly been able to answer once or to give a single response."[41] It was the Canadian government that had accordingly failed to honour the loyalties that it owed to the people.

There is no mistaking Riel's position. The government has deprived the people of their liberties. It has oppressed and attacked them.[42]

> The agitation in the North-West Territories would have been constitutional, and would certainly be constitutional to-day if, in my opinion, we had not been attacked. Perhaps the Crown has not been able to find out the particulars, that we were attacked, but as we were on the scene it was easy to understand. When we sent petitions to the Government, they used to answer us by sending police, and when the rumors were increasing every day that Riel had been shot here or there, or that Riel was going to be shot by such and such a man, the police would not pay any attention to it.[43]

The ethical plea in this appeal is unmistakeable. How could the Métis be liable in law for defending themselves, as defend themselves they must?[44] There is an echo of William Prynne at the end of Riel's address to the jury. Like earlier revolutionaries, and with considerable justification, Riel reverses the charge against him and argues that it is the government and the Queen in right of Canada that have committed treason against the people.[45]

This is not the place to pursue these arguments any further. The discussion in this chapter is nevertheless enough to set out the substantive defences that Riel raised in his address to the jury. As we have seen, Riel was alone in raising these defences. Instead of following his instructions, the defence lawyers attacked him on another front and argued that he was insane. This was a tacit surrender: Riel's oratory had its impact on the jury, who recommended mercy, but the point was moot and the merits of the case were never seriously examined.[46] There may be a good historical argument that this was inevitable in the overheated and rather uneasy political climate of the time. Those excuses are long past, however, and it is evident that a reconsideration of the merits of the case against Louis Riel is long overdue.

NOTES

1 References to the trial transcript, including Riel's address to the jury, are to Desmond Morton, ed., *The Queen v Louis Riel: Canada's Greatest State Trial* (Toronto: University of Toronto Press, 1974), subsequently referred to as *The Queen v LR*. The same material is available in Douglas Linder, *Louis Riel Trial, 1885*, Famous Trials website (2004), http://law2.umkc.edu/faculty/projects/ftrials/riel/riel.html (accessed 12 September 2013), along with the decision of the Court of Queen's Bench of Manitoba on appeal and the decision of the Privy Council refusing leave to appeal.

2 See the discussion at the beginning of Hans V. Hansen's chapter in this volume.

3 These irregularities have received considerable attention on the political as well as the historical front. See, for example, the website of David Kilgour, the member of Parliament for Edmonton-Beaumont, at http://www.david-kilgour.com (accessed 30 September 2013): "Should the presiding magistrate, Hugh Richardson, as a member of the anti-Catholic Orange Order and a part-time magistrate serving only at the pleasure of the federal government, not have disqualified himself from the case? Why was the trial held in Regina and not in Winnipeg? Why were the six jurors selected all English-speaking Protestants who were thus obliged to depend on interpreters for much of the testimony? Why did Judge Richardson select the names of the 36 prospective jurors?"

4 *The Queen v LR*, 14: "Mr. Fitzpatrick: ... the number [of jurors] is immaterial, whether 12, 20 or 25, if the summoning of the jury is not of such a nature as to guarantee a proper and fit trial. That is the point Lord Coke refers to when he says the appointment of the sheriff is taken out of the hands of the Crown, because, in a case in which the Crown might be a party, he might

be suspected of having returned a corrupt jury, and surely this is one of these cases."

5 See George F.G. Stanley, *Louis Riel* (Toronto: McGraw-Hill Ryerson, 1963), 347.

6 See *The Queen v LR*, 229, where one of Riel's lawyers states, "I do not want to justify the rebellion."

7 See A Declaration Which Offences Shall Be Adjudged Treason of 1351, 25 Edw. 3, stat. 5, which is available in the official repository of UK statutes at http://www.legislation.gov.uk (accessed 29 September 2013). The relevant provision makes it an offence to "levy War against our Lord the King in his Realm ... and thereof be ... attainted of open Deed by [the People] of their Condition." The word "attainted" refers to a legal finding that "taints" an accused in the eyes of the law and therefore deprives him of his civil rights.

8 See J.M. Bumsted, "Another Look at the Riel Trial for Treason," in *Canadian State Trials*, vol. 3, *Political Trials and Security Measures, 1840–1914*, ed. Barry Wright and Susan Binnie, 411–50 (Toronto: Published for the Osgoode Society for Canadian Legal History by University of Toronto Press, 2009), 444. Both men later became chief justices – one of Canada and the other of Quebec. It is notable that Fitzpatrick's criticisms of the Canadian government in his address to the jury were incidental and did not challenge the legitimacy of its actions in the territories.

9 The Crown took full advantage of the fact that Riel was tried at an awkward time in the history of the insanity defence, when the law was still in a state of flux.

10 *The Queen v LR*, 208.

11 Bumsted writes in "Another Look," 430–2, that the conduct of the trial by the defence lawyers left a good deal to be desired. It is evident, for example, that the defence did not realize that Riel was being tried under the Statute of Treasons of 1352 rather than the Dominion's Lawless Agressions Act, also known as the Fenian Act of 1867, 31 Vict., c. 14, until the end of a "bungled" first day. Bumsted is apparently referring to A Declaration Which Offences Shall Be Adjudged Treason of 1351, which is cited in note 7.

12 See Bumsted, "Another Look," 440: "In his two speeches to the jury, Riel suggested the outlines of a defence based on his actions. Three observations can be made about Riel's arguments. First, as far as was humanly possible, his own lawyers attempted to discredit Riel's perfectly rational discussion of the politics of the rebellion and his reasons for the rebellion by presenting him as insane. Secondly, Riel was forced to spend far too much time and

energy defending his sanity. Thirdly, the defence did not call many witnesses or introduce documentation attempting to demonstrate the truth of Riel's criticisms of government policy in the west. Indeed, it allowed the Crown to evade the production of evidence that Riel regarded as essential to his case."

13 John H. Langbein, "The Prosecutorial Origins of Defence Counsel in the Eighteenth Century: The Appearance of Solicitors," *Cambridge Law Journal* 58, no. 2 (July 1999): 314–65, at 317. There is a relatively new literature in the field of history that deals with the origins of the lawyer's role in common law criminal trials. This literature includes Langbein's *The Origins of Adversary Criminal Trial* (Oxford: Oxford University Press, 2003); David J.A. Cairns, *Advocacy and the Making of the Adversarial Criminal Trial, 1800–1865* (Oxford: Clarendon, 1998); Stephan Landsman, "The Rise of the Contentious Spirit: Adversary Procedure in Eighteenth-Century England," *Cornell Law Review* 75 (1990): 498–609; David Lemmings, "Criminal Trial Procedure in Eighteenth-Century England: The Impact of Lawyers," *Journal of Legal History* 26 (2005): 63–70; and Allyson May, "Advocates and Truth-Seeking in the Old Bailey Courtroom," *Journal of Legal History* 26 (2005): 71–7.

14 Ibid., 341.

15 Ibid., 356.

16 Ibid., 321.

17 See Bumsted, "Another Look," 420: "There are really no standards of comparison for high treason trials conducted in Canada in 1885, since Riel was the only insurrectionist of that year tried under the Statute of Treasons of 1352. (This is in fact A Declaration Which Offences Shall Be Adjudged Treason, which is cited in note 7, under which treason was a capital offence.) The other 1885 convictions were for the lesser offence of treason-felony [which was not a capital offence], as were the 1848 Irish treason trials and the 1883 English cases involving the Fenian 'dynamiters.'" Bumsted is responding to the position developed by Thomas Flanagan in *Riel and the Rebellion: 1885 Reconsidered* (Saskatchewan: Western Producer Prairie Books, 1983) and *Louis 'David' Riel: 'Prophet of the New World,'* 2nd ed. (Toronto: University of Toronto Press, 1996), where he argues that the fairness of Riel's trial can be judged only in accordance with the standards of the time. From a moral and a legal perspective, this idea seems fallacious, since it is entirely possible that the legal process of the time was inherently unfair. This is certainly true of the trials of the natives who were charged for their participation in the conflict.

18 Bob Beal and Barry Wright, "Summary and Incompetent Justice: Legal Responses to the 1885 Crisis," in *Canadian State Trials*, vol. 3, *Political Trials and Security Measures, 1840–1914*, ed. Barry Wright and Susan Binnie, 353–410 (Toronto: Osgoode Society for Canadian Legal History and University of Toronto Press, 2009), 353.

19 The general legal test of whether an accused was deprived of the right to counsel is whether the assistance of counsel was "ineffective." See *R. v. X.X.S.*, for example, at 2006 Can LII 20 (ONSC). The situation in the case of Riel goes far beyond such issues, however, since the conduct of the defence lawyers aided the prosecution.

20 George R.D. Goulet, *The Trial of Louis Riel: Justice and Mercy Denied* (Calgary: Tellwell, 1999), 139.

21 Bumsted, "Another Look," 445.

22 A review of the transcript in *The Queen v LR* makes it clear that the magistrate was avoiding such rulings, which are easily appealed.

23 See Bumsted, "Another Look," 434: "More and continued pressure on all the points in this defence initiative was what was required, not least to force Justice Richardson into rulings that could be challenged on appeal. Instead, the defence focused on the insanity defence."

24 See Douglas Linder, "The Trial of Louis Riel" (2007), http://papers.ssrn.com (accessed 30 September 2013): "The jury of six men deliberated Riel's fate for an hour. They filed back into the courtroom. The foreman, Francis Cosgrove, 'crying like a baby' announced the verdict. 'Guilty,' he said, and then added, 'Your Honor, I have been asked by my brother jurors to recommend the prisoner to the mercy of the Crown.'"

25 See the first paragraph of Riel's address to the jury in *The Queen v LR*, 311: "Your Honors, gentlemen of the jury: It would be easy for me today to play insanity, because the circumstances are such as to excite any man, and under the natural excitement of what is taking place today (I cannot speak English very well, but am trying to do so, because most of those here speak English), under the excitement which my trial causes me would justify me not to appear as usual, but with my mind out of its ordinary condition. I hope with the help of God I will maintain calmness and decorum as suits this honorable court, this honorable jury."

26 *The Queen v LR*, 316.

27 *The Queen v LR*, 324: "If you pronounce in favor of the Crown, which contends that I am responsible, acquit me all the same." Riel may have been asking the jury to refuse to follow the instructions of the magistrate and nullify the case, but the conduct of his own lawyers was against him. The

groundwork for nullification had not been done. The magistrate and the Crown had made sure moreover that the jury was made up of individuals who were unlikely to nullify the process.

28 Legally it needs to be said that the decision of the jury deserves some deference. The jurors had a natural grasp of the habits of the time and were in a better position than we are to judge the normality of Riel's conduct.

29 The response of François-Xavier Lemieux, Riel's lead, was merely to hold Riel accountable for the consequences. See *The Queen v LR*, 311: "MR. LEMIEUX: May it please your Honors. At a former stage of the trial you will remember that the prisoner wished to cross-examine the witnesses, we objected at the time, thinking that it was better for the interest of the prisoner that we should do so. The prisoner at this stage is entitled to make any statement he likes to the jury and he has been so warned by your Honor, but I must declare before the court that we must not be considered responsible for any declaration he may make."

30 There are hints in the written decisions from the Manitoba Court of Queen's Bench that the defence may have tried to raise the substantive issues in the oral arguments before that court. However, the written decisions from the court do not address these issues seriously.

31 See *The Queen v LR*, 346–7: "I will be very brief in making what remarks I have to make to you. The questions really for you to determine are, first, are you satisfied that there was a rebellion? If you are satisfied that there was a rebellion, as I think you must be, the first question I will ask then is it brought home conclusively to you that the prisoner at the bar was implicated? In charges of this sort there are no classes, no accessories, all are principals. If you are conclusively convinced that the prisoner was implicated, then has anything been shown here to relieve him from responsibility?"

32 Bumsted, "Another Look," 131.

33 The quotation from Flanagan is from *Riel and the Rebellion*, which is cited in ibid., 130n17.

34 Linder, "Trial of Louis Riel."

35 See "Message to Congress from President Harrison," 11 March 1889, in Douglas Linder, *Louis Riel Trial, 1885*, Famous Trials website (2004), as cited in note 1.

36 *The Queen v LR*, 312.

37 Indeed, the collective manifestation of the right was apparently described, simply, as "self-defence" during the course of the American Revolution.

38 Patrick J. Charles, "The Right of Self-Preservation and Resistance: A True Legal and Historical Understanding of the Anglo-American Right to Arms," *Cardozo Law Review de novo* 18 (2010): 18–60.

39 Quoted in ibid., 28n56.
40 Quoted in ibid., 27n51.
41 *The Queen v LR*, 324.
42 The word "liberty" casts a larger net in this context than it does in its current usage and includes the perquisites and benefits that come with our membership in society.
43 *The Queen v LR*, 317.
44 Many of the facts in the case bear out Riel's argument that the actions of the Métis were defensive in nature. Riel argued, for example, that he chose not to request the assistance of "the nationalities" across the border (*The Queen v LR*, 321). This was a conscious decision not to wage war. Riel had also ordered Gabriel Dumont not to attack the Canadian troops as they advanced.
45 It is plain that Riel was a skilled orator and enjoyed sudden reversals and other dramatic devices, which heightened the emotional impact of his speeches. Whether this served the purposes of legal analysis is another matter.
46 There were many ways that the substantive issues could have been addressed by the defence lawyers, if only in protest, and it is the completeness of their surrender in this regard that merits attention.

8

"An Insane and Irresponsible Government": Louis Riel and the Representation of Responsibility

BENJAMIN AUTHERS

Louis Riel's defence speech was his first sustained opportunity to address the court at his 1885 trial for high treason.[1] During the speech, Riel argued for the legitimacy of the North-West Resistance by pointing to Ottawa's mismanagement of the North-West Territories, highlighting its failure to respond to the suffering of Indigenous and Métis inhabitants or to grant European settlers "responsible government" and "public liberties."[2] Riel argued that his actions in the resistance, far from being treasonous, were a reasoned and legitimate political response to government inaction, one that would have remained "constitutional,"[3] and thus legal, if the Dominion had not reacted with aggression.[4]

Riel's eagerness to refute his lawyers' use of the insanity defence in the trial, although at first glance more explicitly personal, also carried with it an important political meaning. Riel strenuously refused the "blot resting upon my reputation"[5] that he saw symbolized by the defence. Instead, he argued that his actions were justified given the Dominion government's treatment of the territories. Indeed, in resisting the Dominion, Riel argued that it was he who acted "reasonably and in self-defence."[6] Ottawa, on the other hand, "being irresponsible, and consequently insane, cannot but have acted wrong."[7]

This chapter examines the language of responsibility and insanity deployed in Louis Riel's "case" and considers how representations of Riel's mental state came to illustrate disparate forms of "responsibility knowledge."[8] Ideas about responsibility circulate in markedly different ways in texts by and about Riel. In the characterization put forward by his defence lawyers, Riel is legally insane. He is incapable of knowing the quality of his otherwise treasonous acts and thus cannot be held responsible for them. This position, rebutted by the Crown and ultimately rejected by the jury, constitutes one of the most persistent representations of Riel, a legalized formulation whose connotations permeate (as many such legally inflected concepts do) the broader culture. Contemporaneous literary works like Ralph Connor's *Patrol of the Sundance Trail*[9] and Joseph Edmund Collins's *The Story of Louis Riel the Rebel Chief*[10] speak both to this legal understanding of culpability and to the complex ideological framework that it evokes and helps to constitute. Although such novels do not engage directly with the legal question of whether Riel should have been declared not guilty by reason of insanity in his trial, their discussions of Riel's mental state and culpability owe much to the law's conceptions of responsibility, even as they nuance and challenge those conceptions. Connor's and Collins's novels demonize Riel as both crazed and criminal, a characterization of responsibility made even more emphatically in John Mackie's representation of Riel in *The Rising of the Red Man: A Romance of the Louis Riel Rebellion*.[11] In Mackie's text, it becomes apparent that ideas of deviance and criminality define the Métis leader, framing him as an impostor who is at once irrational and culpable. Legalized terms such as "criminality" and "imposture" shape Mackie's representation of Riel, but they are only one of many discourses to do so.

Riel's self-representation in his defence speech also extrapolates the legal meaning of "insanity" and "responsibility." For Riel, his actions during the North-West Resistance were those of a rational man acting in response to an "irresponsible, and consequently insane" government. Creating an analogy between the legal defence of insanity and an idea of unresponsive, irresponsible government, Riel not only exerts some control over his own depiction but also offers a political critique of the Dominion's administration of the North-West Territories. Each of these texts relies in different ways on ideas of insanity and responsibility, and each demonstrates how potent and overdetermined the concept of responsibility is in discussions of Louis Riel and the 1885 resistance. Their respective representations also indicate how the figure of Riel himself speaks to legalized, but also more broadly cultural, manifestations of responsibility circulating in anglophone culture in the final decades of the nineteenth century.

Much interesting work has been done since the late nineteenth century on the question of Riel's mental state and whether he should have been held legally culpable for his actions in the resistance. Although I am cognizant of the importance of this issue to the historical record and to the questions of law and justice that the defence's failure raises, this chapter supplements, rather than directly contributing to, these discussions, looking to depictions of Riel's responsibility rather than asking whether the man himself should have been found not guilty by reason of insanity at trial.

RESPONSIBILITY AND RIEL

In his 2010 biography *Louis Riel and Gabriel Dumont*, Métis writer Joseph Boyden notes that Riel's lawyers had a small number of possible options available to them in formulating his defence. Faced with the capital offence of high treason, the lawyers argued that the charge should be reduced to the less serious crime of treason-felony, while also asserting that the "rebellion was justified and a product of acute and continued federal mishandling and even criminal ignorance of Métis petitions for land claim review."[12] "Most useful" and "obvious" to them, however, was the defence of insanity.[13] Riel's confinement in asylums in the 1870s, the apparently bizarre nature of his writings, and the religious mania that appeared to drive his actions – including his claim to be a "prophet of the new world"[14] – all seemed to support the defence. At the very least, being found insane would spare Riel's life.

Following Riel's defence speech and the closing statements of the defence and prosecution, Lieutenant Colonel Hugh Richardson, the stipendiary magistrate presiding over Riel's trial, set out the following test for the jury: "To establish a defence on the ground of insanity, it must be clearly proved that at the time he committed the act, the party accused was labouring under such defective reasoning from a diseased mind as not to know the nature and quality of the act he was committing, or that if he did know it, that he did not know that he was doing wrong. That I propound to you as the law."[15] Richardson's directions restate for the jury the *M'Naghten* Rules, a series of criteria for determining legal responsibility established in the aftermath of the 1843 English trial of Daniel M'Naghten. Having believed for a number of years that he was being persecuted by the British Tories, M'Naghten killed Edmund Drummond, private secretary to then prime minister Sir Robert Peel, while believing Drummond to be Peel. The verdict of not guilty by reason of insanity that the jury subsequently passed down generated considerable public disquiet, including critique from Queen Victoria. In a private letter to Peel, the queen articulated what has proven to be a persistent anxiety about legal insanity: that an offender might escape punishment when "everybody is morally convinced that ... [he or she was] perfectly conscious and aware of what they did."[16]

Appearing before the House of Lords to clarify the state of the law following *M'Naghten*, the judges of the Central Criminal Court responded to a series of questions posed by the Lords. In a majority "decision," they asserted the general rule that an accused suffering from a mental defect is to be found not guilty unless "he knew at the time of committing such crime that he was acting contrary to law."[17] This knowledge did not, they argued, need to be a specific knowledge of the formal law. Rather, it must be proved by the defence that "the party accused was laboring under such a defect of reason, from disease of the mind, as not to know the nature and quality of the act he was doing; or, if he did know it, that he did not know he was doing what was wrong."[18] In this formulation, being responsible for one's acts is premised on an idealized knowledge of what the act signifies within a discourse that presumes law to be socially normative and that meaningfully equates knowledge of the law with an apparently innate and moral sense of what is "wrong."[19] Legal responsibility depends on the accused being able to recognize the "nature and quality" of their actions as contrary to law. Knowledge about the right and wrong of specific acts thus becomes the central question of legal responsibility; this knowledge determines whether a defect in reason, caused by a mental illness, means that an accused is not competent to be considered liable for the consequences (punishment) of an otherwise criminal deed.

As framed in *Daniel M'Naghten's Case* and restated by Justice Richardson in *The Queen v. Louis Riel*, the insanity defence served as the primary legal formulation for negotiating an individual's responsibility in cases of mental impairment.[20] In Riel's trial, his lawyers offered the jury the following choice: "either this man is the lunatic that we his counsel have tried to make him, or he is an entirely sane man in the full possession of all his mental faculties, and was responsible in the eyes of God and man for everything that he has done."[21] Arguing the former as a defence against Riel's culpability for his actions in the North-West Resistance, his lawyers rooted their argument in racialized assumptions about Riel's Métis heritage that implicitly suggested that the "mysticism" of his "untutored" Indigenous forebears predisposed him to a simplicity that could veer into instability.[22] More overtly, the defence argued that because of his involvement in the Red River Uprising of 1869-70, Riel had developed a "disease of the mind known as megalomania ... an extraordinary love of power and extraordinary development of ambition," where the sufferer may act "under the insane delusion that he is either a great poet or a god or a king or that he is in direct communication with the Holy Ghost."[23] Riel's actions were not, the defence asserted, rational: he would not, for example, have surrendered himself to General Frederick Middleton if he had not been acting in the unreasonable belief that God was protecting him from harm.[24] Offering an alternative characterization to the "deep, designing, cunning rascal that he is represented to be"[25] by the prosecution, the defence's interpretation asserted that he was "entirely insane and irresponsible for his acts"[26] and thus incapable of assuming legal responsibility for his role in the North-West Resistance. Riel was unable to rationally understand the quality and consequence of his actions and had acted subject to the terms of his delusions, including his megalomanic belief that there could be "no opposition of his objects."[27] Given the evidence proffered in support of this by the defence, Charles Fitzpatrick, one of Riel's lawyers, appealed to the jury in his closing statement to uphold the "principles of English liberty [that] have always found a safe resting place in the hearts of English jurors."[28] Riel, "a poor confirmed lunatic,"[29] should be protected by the fairness of English justice, exemplified in the insanity defence, even though the man himself, "an alien in race and an alien in religion"[30] – and an alien because of his "foreign," unbalanced mental state – might otherwise be impossible to sympathize with.

The legalized conception of personal responsibility articulated in the *M'Naghten* Rules, as a hermeneutics that formalizes a potential divorce between the commission of an act and culpability for that act, forms a significant aspect of how responsibility is known and understood by cultures shaped by the common law tradition. Yet, although legal concepts like abrogated responsibility due to mental impairment are deeply influential of the broader social context, law does not define such ideas exclusively. I want now to move from the specifics of Riel's trial to a consideration of how these legalized conceptions of his sanity and responsibility were understood outside of the courtroom. Juridical determinations frequently have connotations beyond those of individual cases, with judicial findings and legislative acts, as well as legal tests, rules, and procedures, impacting the society of which they are a part. Law's aesthetics influence how the culture at large apprehends a situation, defining it "in a certain way and encourag[ing] us all to look at it likewise."[31]

As interest in Riel's mental state[32] (and, indeed, the outcry over the *M'Naghten* decision) indicates, the legal terms of insanity were the subject of considerable popular scrutiny in the nineteenth century, and a public discussion existed in which law's privileged understandings of responsibility and culpability were interrogated by other sociocultural discourses. Ideas about responsibility, as Kimberley White has argued in relation to early-twentieth-century murder trials in Canada, are constantly being "produced, ordered, and contested,"[33] even within individual cases. This complex production of meaning is, White argues, "an intensely cultural process" that brings "together a cross-section of ideas, institutions, and individuals with the common goal of trying to make sense of, and determine just responses to, acts of murder."[34] Looking beyond "the official verdict in each case," White finds that "the law was anything but fixed on the subject of responsibility" and that "legal outcomes were not simply the result of a measured application of legal rules."[35] Rather, seemingly extralegal ideas of responsibility and insanity shape legal reasoning in a reciprocal manner. Although law constitutes a particularly influential conception of responsibility, it is but one aspect of the multiple and often conflicting body of "responsibility knowledge" that White argues is at work in determinations of the nature of culpable behaviour.

White's conception of responsibility knowledge as a cumulative and contested discourse presents a useful way of thinking about how ideas of treason and national belonging, mental health, politics, and law all cohere in representations of Louis Riel. One such genre of representation, Riel's depiction in literature, presents a multitude of distinct portrayals that span the past century and a half. As it is beyond the scope of this chapter to effectively survey all of these representations,[36] my primary focus will be on John Mackie's *Rising of the Red Man*, one example taken from a corpus of novels published between 1885 and 1914 that engage, like the trial, in a discussion of Riel's mental state, political action, and responsibility.

The description of Louis Riel as a manipulative "fanatic and rebellion maker"[37] illustrates how the legal characterization of Riel's responsibility belies an intersecting cultural investment in his capacity. As Albert Braz and Wolfgang Klooss have both noted, the representation of Louis Riel in Canadian literature has been marked by a "tremendous fluidity,"[38] a shifting series of depictions that frequently say more about how Riel (and the Métis more generally) has been "exploited ideologically by the antagonistic forces that determined the fate of the young Dominion" than about Riel's "individual, psychological or social dimensions."[39] Understanding the cultural work to which the figure of Riel has been put in literary texts is consequently often "less about him than about their authors and their specific social activity."[40]

In the decades immediately following the North-West Resistance, a flurry of novels by Canadian and British writers sought to represent its causes and consequences. Joseph Edmund Collins's *The Story of Louis Riel the Rebel Chief* (1885), a hurriedly written quasi-fictional account of Riel's life authored concurrently with the Saskatchewan campaign,[41] acknowledges Métis grievances, deems the Dominion "criminally remiss"[42] in its treatment of the Métis, and finds that their hatred of the government is "not altogether undeserved."[43] Ralph Connor's later *Patrol of the Sundance Trail* (1914), a sequel to his *Corporal Cameron of the North West Mounted Police* (1912),[44] also evinces sympathy for Indigenous and Métis people in the North-West, depicting the hero explaining to his wife that much of the land they stood on was ceded "under persuasion"[45] and noting through an authoritative North-West Mounted Police superintendent that the Métis "have real grievances."[46] Yet, as with many authors of the period, compassion for the Métis "rarely extended to their leader."[47] For Connor, Riel is a "hair-brained four-flusher" and blood-thirsty agitator,[48] a "traitor ... liar and a coward"[49] who, despite his apparent ineptitude, egoism, and untrustworthiness, "possesses a wonderful power over the half-breeds."[50] His great transgression is inciting the Plains Indians against the government, an act that will not only disrupt the progress of settlement but also, ultimately, bring ruin upon the Indians themselves when they are confronted with British might.[51] Collins's Riel is still more despicable. He is bestial, even monstrous, a "heartless Rebel ruffian" "insensible to every human impulse"[52] who seeks to seduce the Métis girl Marie before casting "her off among the worthless and degraded ones of her sex."[53] The Métis's choice of leader ultimately mediates any claims their plight might have to the sympathies of the novels' (presumably white, anglophone) readers.

However, although Collins in particular suggests that Riel is immoral and irrational, he does not also argue that the Métis leader's irrationality is sufficient to deny his responsibility for his actions. Acknowledging that his representation of Riel might appear "absolutely illogical, and unreasonable,"[54] the narrator of Collins's text describes the novel as a completely faithful depiction of "the unstaid, unreasoning character of Riel, and [of] how far passion and impulse will carry him away from sound understanding."[55] For Collins, this capacity to be carried "away from sound understanding" does not serve as a denial of responsibility. Riel's irrationality never exculpates his guilt. Instead, Collins's ultimate judgment is that Riel and his allies must be captured and executed as "menace[s] to public peace," lest they "be given the opportunity again of covering the land with blood."[56] Such representations of Riel depict the Métis leader both as responsible in the sense that he should be punished for his actions in the North-West Resistance (and for the other crimes and misdeeds assigned to him) and as a menace whose ideas and actions put him outside the purview of appropriate behaviour. This conception of responsibility knowledge, framed differently from that envisioned in the insanity defence, conceives of irrationality and criminality as potentially concurrent and of mental incapacity, in Riel's case at least, as being of an insufficient degree to constitute a moral defence.

In a similar fashion, British novelist John Mackie's *Rising of the Red Man: A Romance of the Louis Riel Rebellion* (1902) makes much of Riel's purported barbarism and irrationality, while also stopping short of suggesting that this might release Riel from responsibility for his actions. Mackie, author of a number of adventure novels written between the nineteenth and twentieth centuries and published in Britain, Canada, and the United States, was a former officer of the North-West Mounted Police who also worked in Australia and served in the Boer War.[57] His novels often stress their own veracity, with

Mackie positing on one occasion that he had "gleaned from real life and nature a knowledge of men and things which stands me in better stead than any flight of the imagination."[58] This aura of presumed truth pervades Mackie's writing and would have been a familiar promise to readers of his earlier works. Such a promise might also have imbued Mackie's representation of Riel with particular authority for his readers and spoken to still-fresh memories and prejudices about the resistance.[59] Riel drifts in and out of *The Rising of the Red Man*, ostensibly replaced in the reader's interest by the spirited, beautiful heroine, Dorothy, and by the brave North-West Mounted Police hero, Sergeant Pasmore. Despite Riel's frequent absences from the text, however, Mackie's novel remains a depiction of "the Louis Riel Rebellion," an attribution that underscores who, for the author, is culpable for the 1885 insurrection and the bloodshed it brings.

The Rising of the Red Man opens with a prologue depicting "Louis David Riel, fanatic and rebellion-maker ... addressing a great general meeting of the half-breeds and Indians near Batoche."[60] In what Braz describes as an "exposé"[61] of Riel, Mackie represents the Métis leader both as an irrational fanatic – the reader never gets a clear sense of why Riel would seek to rebel other than because of his own wickedness and religious unorthodoxy – and as a duplicitous villain who is manipulating the superstition of his audience. To gain support, the "knavish fanatic"[62] has planned for this meeting to coincide with a solar eclipse that he will interpret as evidence of the Manitou's support for his schemes. Having known about the eclipse because "had not the almanac told him and all the world – with the exception of the ignorant half-breeds and Indians whom he was addressing,"[63] Riel persuades his audience to join him through both this show of supernatural approval and more mundane "prognostications concerning that happy coming era in which unlimited food, tobacco and firewater would make merry the hearts of all."[64] As Mackie's narrator wryly asserts, "being an educated man there was a good deal of method in his madness,"[65] a representation that emphasizes Riel's capacity to know his actions (and therefore, in concordance with legal discourse, to be responsible for them), while asserting the simultaneous existence of some form of "madness."

Mackie's depiction of Riel here is of a charlatan who succeeds in fomenting rebellion largely because his audience is unsophisticated, superstitious, and gullible. There is no suggestion either that his Aboriginal listeners might have legitimate complaints (indeed, the only reason given for any of them being there is a reference to "Poundmaker and his Stonies, who were always spoiling for trouble")[66] or that Riel himself has any policy beyond vague and "pleasant prognostications"[67] and "hanky-panky."[68] Indeed, Mackie closes his prologue by asserting that "never perhaps in the history of impostors from Mahomet to the Mahdi had an almanac proved so useful,"[69] dismissing political grievances by equating Riel with those whom Mackie deems false prophets. In doing so, Mackie also gestures to the suspicion, articulated by Queen Victoria in her letter to Peel, that culpability might be masked by the appearance of insanity. Such an anxiety was a persistent presence in the British context in which Mackie wrote: as Roger Smith notes, although some viewed the insanity plea as progress, there was also a sense that "the success of the plea correlated with a spread of unpunished violence."[70] Riel, characterized by his defence counsel as insane, is here represented by Mackie both as "mad" to an ill-defined degree and as an impostor, an "arrant fraud,"[71] who is the latest in a line of foreign (i.e., not British) pretenders to both religious and political authority. Mackie determines Riel to be both illegitimate because of madness (and thus incapable of valid political action) and an impostor, sufficiently cognizant of his actions to be acting with intent and knowledge. Casting Riel as a fraud delegitimizes his politics as effectively as a finding of insanity.

Riel reappears in the novel a few chapters after the prologue, capturing Dorothy as she tries to make her way to the fort at Battleford. When Dorothy is first brought to him, Riel fails to notice her. Instead, "he was so carried away by the exuberance of his own eloquence ... that he ... kept on in a state of rapt ecstasy. His semi-mystical oration was a weird jumble of religion and lawlessness, devout exhortation, riot, plunder, prayer, and pillage."[72] Mackie's alliterative account of Riel's ramblings reaffirms Riel's incoherence for the reader, his religious extremism signifying the illegitimacy of his, and thus the resistance's, transgression. Despite the buffoonish quality of his speech, Mackie's Riel constitutes a real threat to the British social order in the Canadian North-West. Not appearing to be "the dangerous, religious fanatic that he was in reality,"[73] Riel has a physical presence that belies his actual, dangerous meaning, although here, more than in the prologue, his apparently unconscious "oration" suggests irrationality rather than cunning imposture. Duplicity, whether intentional or not, thus becomes central to Mackie's contingent and inconsistent characterization of Riel and to his shifting articulation of Riel's responsibility. Because Riel is never legitimate in Mackie's formulation – he perpetually appears to be something that he is not, always an impostor of some form – he is deserving of punishment for his actions. Thus Mackie finds no reason to reconcile Riel's apparently sincere ramblings with his earlier duplicity or to characterize him as other than a villain. The author is uninterested in either explaining or exculpating the state of mind of a man he describes as plotting to have white settlers "utterly exterminated" so that his Aboriginal "elect might possess the land undisturbed."[74] Rather, Mackie bluntly condemns Riel, preempting his inevitable punishment in a vision of "the great beams of the gallows-tree looming up blackly."[75] Avoiding the *M'Naghten* binary, Mackie instead develops a conception of responsibility in his novel wherein madness and culpability coexist in such a way that the legitimacy of Riel's punishment *for both* is unquestioned. *The Rising of the Red Man* consequently elides any questioning of guilt through a discourse of insanity – or, indeed, of political legitimacy – in favour of the self-evident assertion of culpable villainy by the novelist.

SELF AND STATE IN RIEL'S DEFENCE SPEECH

Mackie's representation of Riel resists the binarized conception of mental states privileged in the *M'Naghten* defence to assert a differently articulated form of responsibility knowledge. But even as *The Rising of the Red Man* elides the law's privileged test in favour of an alternative, nonexculpatory conception of irrationality, the novel still gestures to juridical discourse in its reiteration of cultural anxieties about the misuse of the insanity defence and a belief in the "gallows-tree" as an appropriate, legalized punishment for Riel. The law provides an important, if implicit, discursive frame in the novel, inviting readers to engage with Mackie's notion of responsibility alongside its legal forms.

Riel himself resists the denial of responsibility signified by legalized insanity in his defence speech, although to very different ends than Mackie does. From the beginning of the speech, Riel acknowledges that insanity offers a ready explanation for his actions and that "it would be easy for me to-day to play insanity, because the circumstances are such as to excite any man."[76] However, he then immediately rejects this characterization, calling on God to help him "maintain calmness and decorum as suits this honorable court, this honorable jury."[77] Fitting the norms expected by the court allows Riel to represent himself and his actions as rational. And it is necessary that he be seen as rational if he is then to have the North-West Resistance, and his involvement in it, judged as a legitimate and meaningful way of responding to the Dominion's treatment of the North-West Territories. Mackie's novel emphasizes Riel's responsibility regardless of his mental state; in his defence speech, Riel rejects the insanity plea because it cannot coexist with legitimate political action.

What does an assertion of insanity, with its subsequent refutation of responsibility for one's actions, mean for an individual claiming political agency? A determination of legal insanity constitutes a denial of personal capacity, a paternal act that is also a delegitimizing one. As Smith notes, "to describe an intentional act assigns responsibility to the individual. It also reinforces a society in which values are located within an individual's mental attributes. The guilty verdict concerns an occasion on which these values should have received expression but did not."[78] Alternatively, legal determinations of insanity speak to the limits of individual responsibility and self-governance as legal and political categories. In "the absence of a state of mind capable of 'knowing,'" a defendant is "no longer definable as a person" in the eyes of both the law and the society whose norms law enforces precisely because that defendant is no longer capable of having responsibility assigned to him or her.[79] In asserting that their client was insane, and thus not culpable for his acts in the North-West Resistance, Riel's lawyers were certainly aware of what the defence meant for public perceptions of their client. As Fitzpatrick stated in his summation of the defence case, "if he is a lunatic, we, in the exercise of a sound discretion, have done right to endeavour to prove it. If he is a sane man, what humiliation have we passed upon that man, we his counsel endeavoring, despite his orders, despite his desire, despite his instructions, to make him out a fool."[80] Such an articulation acknowledges the degrading social meaning given to mental illness, the "humiliation" presupposed by rendering the defendant "no longer definable as a person" in important ways.[81]

John Gardner asserts similarly that as self-respecting individuals we want to be able to give rational explanations for our behaviour, to take responsibility for it. "Being responsible for one's own wrongs is a distinctly human capacity and one that is central to all distinctively human lives," Gardner argues. "It is the capacity that makes self-respect possible, for self-respect is the attitude of one who can sincerely say, of anything she did wrong, that she was justified in doing it, or, failing that, excused."[82] In making a defence of insanity, then, "one does not explain oneself as a rational being. One explains oneself *away* as a rational being. One casts oneself in the role of object rather than subject; one explains oneself (including one's reasons) in terms of facts that are not reasons, or are not figuring in one's explanation as reasons."[83] Such a defence does nothing to authorize one's actions. Rather, it delegitimates them. As Annalise Acorn notes, "by claiming non-responsibility on grounds of mental illness, one excludes oneself from full membership in the moral community of rational and accountable human beings. Conceding mental illness concedes an inability fully to participate in human life."[84]

Riel attempted to assert control over how his mental state was represented throughout his trial in order to counter this discursive removal from "full membership in the moral community of rational and accountable human beings." Seeking to put questions to Charles Nolin in order to refute Nolin's claims of Riel's erratic behaviour, Riel stated, "I cannot abandon my dignity. Here I have to defend myself against the accusation of high treason, or I have to consent to the animal life of an asylum. I don't care much about animal life if I am not allowed to carry with it the moral existence of an intellectual being."[85] This assertion underscores Riel's awareness of the consequences of the insanity plea, with its intertwined personal and political connotations. Insanity means that Riel loses his "dignity," his capacity to represent himself as having a "moral existence" as an "intellectual being." An "animal life," for Riel, is precisely one without claim to human capacity and intellectual and moral being. Moreover, a finding of insanity would also cast the legitimacy of his role in the North-West Resistance in doubt. Because the legal test is concerned with the accused's mental state at the time of committing the act, a finding of insanity would mean Riel was incompetent at the time of the uprising. Thus his justifications for resistance come under doubt and potentially lose their legitimacy. As Boyden's Riel recognizes, if the insanity defence had been successful, "the rest of Canada and the world may think," as a consequence, "that the Métis cause is just as insane."[86] Defending himself against the insanity defence was, for Riel, a necessary political move, as "he knew he had to be judged lucid if he were to have political legitimacy."[87]

Asserting that if he is found guilty and sentenced to death, he will have the "satisfaction," as he terms it, not to be "reputed by all men as insane, as a lunatic," Riel acknowledges that the nature of his responsibility is in this case a very public question.[88] To be considered insane, he argues, is to be a political impostor, a "blot" on his reputation that renders him politically illegitimate.[89] After the court's finding of guilt, Riel acknowledges that at least he is reattributed with agency: "up to this moment I have been considered by a certain party as insane, by another party as a criminal ... I suppose that after having been condemned, I will cease to be called a fool, and for me, it is a great advantage ... If I have a mission ... I cannot fulfil my mission as long as I am looked upon as an insane being."[90] Here, like Gardner and Acorn, Riel notes the disparity between being represented as insane and being considered fully, competently human. If he is considered an "insane being," Riel argues, he cannot legitimately pursue a mission on behalf of the North-West Territories. Countering this characterization necessitates that Riel represent himself as knowing the "nature and quality" of his actions and that he reassert them (and so himself) as politically legitimate and rational. As Daniel Clarke notes, "Riel's great aim, even at the trial, was to falsify the charge of insanity, and to show by his words his mental capacity to be a leader of men."[91] Clearing his personal reputation and reinscribing his responsibility for his actions were necessary aims for Riel to achieve if his politics were to be vindicated. They also formed the basis for his subsequent argument that it was Ottawa, not him, that lacked responsibility in its actions in the North-West.

Arguing that the Dominion failed to provide the people of the North-West Territories with adequate political representation[92] and that the North-West Council had reiterated "the great defect of its parent," Riel notes that the people of the North-West sent "petition after petition" to the federal government about the state of their affairs.[93] "So irresponsible is that Government to the North-West," however, that "besides doing nothing to satisfy the people of this great land, it has even hardly been able to answer once or to give a single response. That fact would indicate an absolute lack of responsibility, and therefore insanity coupled with paralysis."[94] Unlike his own responsible responsiveness to the needs of the territories – Riel later notes, "although I am simply a guest of this country ... I worked to better the condition of the people of the Saskatchewan at the risk of my life"[95] – the "Ministers of an insane and irresponsible Government"[96] replied to these petitions first with silence and then with aggression. Such an idea of irresponsibility, based on a model of politics that foregrounds responsiveness to constituents' needs as the criterion for good governance, analogizes the legal terms of the insanity defence to emphasize the incapacity of Canadian administration of the territories.

Riel then turns to imperial British political and legal norms in order to further assert the illegitimacy of institutional governance in the North-West Territories: "British civilization which rules to-day the world, and the British constitution has defined such government as this is which rules the North-West Territories as irresponsible government, which plainly means that there is no responsibility, and by all the science which has been shown here yesterday you are compelled to admit if there is no responsibility, it is insane."[97] Characterizing the government's actions as "insane," Riel refers back to his lawyers' representation of him, rearticulating and redeploying the term as a critique of what he determines to be irresponsible government in the North-West. Riel's understanding of responsibility is here informed by the politico-legal norms of British civilization and constitutional standards, a concept informed by "fair-play" in the courts[98] and by the self-evident values of "good sense,"[99] as well as by the ostensible objectivity of science. Appealing to these ideals, Riel is able not only to point to the Dominion government's failure to adhere to "British civilization" but also to cast himself, because he has been responsible, in the government's place. Collapsing the responsibility denied by the insanity defence with the idea of responsible government – government that is responsible to the people, as Riel has argued himself to be[100] – Riel analogizes and reinterprets the concept of responsibility so as to exculpate himself of the "blot" of insanity and to argue that this blot should instead mark the Dominion. Rather than align himself with Canadian governance, Riel appeals to a superior, British conception of civilization and constitutionalism given virtue by its imperial pre-eminence (it "rules today the world"). Such an ideal is set at odds with Canadian administration in the territories ("irresponsible government") and is instead aligned with Riel's self-representation of his own actions, through which he has attempted "to help the Indians, to help the half-breeds and to help the whites to the best of my ability."[101] By pathologizing the actions of an "insane and irresponsible Government,"[102] Riel can represent his own actions as appropriate, responsible, and sane, whereas the North-West is administered by a government that has shirked its responsibilities.

In this analogy lies the complex heart of Riel's invocation and reinterpretation of responsibility as the key concept of both self-governance and political governance, a responsibility knowledge that slips discursively both between responsiveness to constituents and those forms of responsible behaviour that the insanity defence denies and between a lack of responsibility and irresponsibility. For Riel, such slippage can be read through the idea of legal insanity, but the form of responsible behaviour by which he is seeking to justify his actions cannot be delimited by this idea:

> If you take the plea of the defence that I am not responsible for my acts, acquit me completely since I have been quarrelling with an insane and irresponsible Government. If you pronounce in favour of the Crown, which contends that I am responsible, acquit me all the same. You are perfectly justified in declaring that having my reason and sound mind, I have acted reasonably and in self-defence, while the Government, my accuser, being irresponsible, and consequently insane, cannot but have acted wrong, and if high treason there is, it must be on its side and not on my part.[103]

Having already established that the Dominion government is irresponsible because in failing to represent its constituents properly, it has not met British standards of civilization – and thus has failed to be responsible to and for the North-West Territories – Riel can then assert a dual defence to the charges against him, both of which present the North-West Resistance as in some fashion legitimate. On the one hand, if the jury finds Riel to be without legal capacity, and thus not responsible for his actions, he should be acquitted. Responding to the insanity defence's connection between mental illness and irresponsibility, Riel asserts that any insanity on his part occurs in the face of a similar "disease" in the figurative mind of the government. Riel consequently finds that even if he is determined to be legally insane, since the government remains similarly incompetent, such a finding does not necessarily undermine the aims of the resistance.

In the alternative, if he is found to be sane, Riel posits that he should, nevertheless, still be acquitted. Because his actions in defiance of a government that "cannot but have acted wrong" are necessarily justifiable, Riel has acted in self-defence of both the people of the North-West and himself. Riel argues that he responded reasonably to an institution that had so failed to perform its function that it had, seemingly paradoxically, itself committed an act of insanity and high treason. Structured around a slippage between ideas of the state's responsibility *for* its actions and its responsibility *to* its constituents, Riel's concept of government locates irresponsibility in a failure to act in the interests of a people. To carry Riel's analogy further, such a failure warrants an intervention, a supplanting of the government's inability to act for its citizens with a new, responsible institution, a shift of value and power that would then reinforce norms of British civilization and constitutional law. So, although Riel does not fully account for the knowledge-based terms of the *M'Naghten* defence, the language of insanity that he evokes is a powerful tool in his articulation of the political legitimacy of the North-West Resistance. It enables Riel to depict his actions not only as sane but also as a responsible answer to neglect, as well as to suggest that in his self-representation, we might find a new, and sane, model for governance in the North-West Territories.

Louis Riel's rearticulation of the insanity defence, then, carries with it an analysis of the interrelations between personal and political responsibility, a conception of responsibility knowledge that not only underscores the pervasive cultural presence of legalized conceptions of insanity and culpability but also refigures them in a more politically effective manner. Referencing conceptions of idealized British civilization and constitutional law as standards that the Dominion has failed to meet in its treatment of the North-West Territories, Riel invokes the empire in his own actions, justifying the resistance and his role in it as upholding an imperial standard that Canada has otherwise failed to maintain. Thus the North-West Resistance, in symbolizing the territories' aspirations to exemplify the values of the empire, comes to stand for an idealized form of responsibility and for the possibility of "a single postimperial global formation."[104] Arguing through a conception of insanity that mirrors the multifaceted public understanding of the concept as at once medical, legal, moral, and political, Riel articulates the government's actions as insane and thus requiring a response; analogizing them to personal actions, he finds that they cannot be the acts of a valid political actor. He also, simultaneously, argues that the government failed to promulgate appropriate British conceptions of law and civilization. The Dominion would seem to be both without responsibility and irresponsible: whether by negligence or misdeed, it has failed by its actions and through its inactions to fulfil the very task that it exists to fulfil, that of caring for and representing the inhabitants of the North-West Territories. Riel, in seeking to act where the government has not, comes to embody responsible governance in both of these senses. He has undertaken a mission to help (to be responsible to and for) the Métis, Indigenous, and European settlers in the territories, and in pursuing his mission as a responsible political actor, he has proven himself sane and legitimate, whereas the government has not.

In his self-representation in his defence speech, Louis Riel articulates a form of responsibility knowledge that invokes the terms of the legal defence of insanity, and he seeks to rearticulate these terms through a linking of constitutional politics and personal governance. Consequently, the speech engages with some of the formulations of responsibility circulating around Riel in both legal and literary texts, including the speaking for Riel that his lawyers must necessarily do if they are to employ insanity in his defence and the complicated, even legally contradictory, conception of Riel's personal, moral, and political responsibility that we see in *The Rising of the Red Man*. The capacity for responsibility is at the heart of all these texts, and that they serve such significantly different ideological functions is indicative of the work that the disparate forms of responsibility knowledge performed in late-nineteenth- and early-twentieth-century Canadian and British law and literature. Moreover, such texts reiterate Braz and Klooss's assertions that depictions (including, I would argue, self-depictions) of Louis Riel have always been overdetermined with meaning. Texts like *The Rising of the Red Man* and Louis Riel's defence speech serve to emphasize the potent interdisciplinary role that ideas of responsibility play, and continue to play, in representations of Riel as a political symbol and as an individual with a significant cultural afterlife.

NOTES

1 Riel gave two substantial speeches at the trial: one between his lawyers' closing arguments and before the jury's deliberations and one after he was found guilty. I refer to the former speech as his "defence speech" throughout this chapter.

2 Desmond Morton, ed., *The Queen v Louis Riel: Canada's Greatest State Trial* (hereafter *Q v LR*) (Toronto: University of Toronto Press, 1974), 312.

3 Ibid., 317.

4 Ibid., 324.

5 Ibid., 316.

6 Ibid., 324.

7 Ibid.

8 Kimberley White, *Negotiating Responsibility: Law, Murder, and States of Mind* (Vancouver: UBC Press, 2008), 10.

9 Ralph Connor, *The Patrol of the Sun Dance Trail* (Toronto: McClelland and Stewart, 1914).

10 Joseph Edmund Collins, *The Story of Louis Riel the Rebel Chief* (1885; reprint, Toronto: Coles, 1970).

11 John Mackie, *The Rising of the Red Man: A Romance of the Louis Riel Rebellion* (London: Jarrold, 1902).

12 Joseph Boyden, *Louis Riel and Gabriel Dumont* (Toronto: Penguin, 2010), 140.

13 Ibid., 141.

14 Ibid. See Riel's self-assertion of this in his defence speech (*Q v LR*, 322).

15 *Q v LR*, 348.

16 Cited in Donald West and Alexander Walk, *Daniel McNaughton: His Trial and the Aftermath* (Ashford: Gaskell Books, 1977), 10. Victoria had already been the subject of three assassination attempts at this point in her reign, including one by Edward Oxford, who was subsequently declared insane. See "Queen Victoria," in *Orlando: Women's Writing in the British Isles from the Beginnings to the Present*, ed. Susan Brown, Patricia Clements, and Isobel Grundy, database (Cambridge, UK: Cambridge University Press, 2006), http://orlando.cambridge.org (accessed 19 October 2013).

17 *Daniel M'Naghten's Case*, 10 *Clark and Finnelly* 200 (1843), 209.

18 Ibid., 210.

19 Ibid.

20 The *M'Naghten* test has been amended over time in Canada. For example, the 1892 Canadian Criminal Code modified the law from a test of knowl-

edge to one of "appreciating" the nature of one's actions, "a change generally thought to make it easier to invoke the defence of insanity." See Thomas Flanagan, *Louis 'David' Riel: 'Prophet of the New World,'* 2nd ed. (Toronto: University of Toronto Press, 1996), 175.

21 *QVLR*, 294.

22 Ibid., 295.

23 Ibid., 296.

24 Ibid., 309.

25 Ibid., 299.

26 Ibid., 295.

27 Ibid., 301.

28 Ibid., 311.

29 Ibid.

30 Ibid., 310.

31 Desmond Manderson, *Songs without Music: Aesthetic Dimensions of Law and Justice* (Berkeley: University of California Press, 2000), 27. Consider, for example, that common law motifs of the trial, with its staging of witnessing, examining, cross-examining, and a final declaration of truth, are accepted ideals of justice and fairness. In a different way, "artistic merit" is conceptually relevant to both literary studies and law, while also giving meaning to the political concept of freedom of expression. For more on this idea, see Manderson, *Songs without Music*; and Benjamin Authers, "Truth in the Telling: Procedure, Testimony, and the Work of Improvisation in Legal Narrative," *Critical Studies in Improvisation* 6, no. 1 (2010): n.p.

32 See, among other examples, Daniel Clarke, *A Psycho-Medical History of Louis Riel* (1887; reprint, Ottawa: Canadian Institute for Historical Microreproductions, 1982).

33 White, *Negotiating*, 9.

34 Ibid., 11.

35 Ibid., 9.

36 For a comprehensive examination of literary representations of Riel, see Albert Braz, *The False Traitor: Louis Riel in Canadian Culture* (Toronto: University of Toronto Press, 2003).

37 Mackie, *Rising*, 9.

38 Braz, *False Traitor*, 3.

39 Wolfgang Klooss, "Stereotyping in Canadian Literature: The Métis in Anglo- and Francophone Writing," in *Images of Louis Riel in Canadian Culture*, ed. Ramon Hathorn and Patrick Holland, 131–74 (Lewiston, NY: Edwin Mellen, 1992), 171.

40 Braz, *False Traitor*, 3.
41 Ibid., 45.
42 Collins, *Story*, 136; see also 176.
43 Ibid., 8.
44 Ralph Connor, *Corporal Cameron of the North West Mounted Police* (Toronto: Westminster, 1912).
45 Connor, *Patrol*, 33.
46 Ibid., 23.
47 Braz, *False Traitor*, 67. Anne Mercier and Violet Watt's *The Red House by the Rockies: A Tale of Riel's Rebellion* (London: Society for Promoting Christian Knowledge, 1896) stands as a notable exception to this general rule.
48 Connor, *Patrol*, 180, 192–3.
49 Ibid., 205.
50 Ibid., 24. Connor is similarly dismissive of "the ignoble part played by the vain and empty-headed Riel" in the Red River Uprising, which he terms "the fantastic and futile rebellion of 1870" (206).
51 Ibid., 206.
52 Collins, *Story*, 96.
53 Ibid., 68.
54 Ibid., 86.
55 Ibid.
56 Ibid., 175.
57 Mackie's service records, along with an obituary detailing his life, have been digitized as a part of the North-West Mounted Police Collection at Library and Archives Canada. Search John Mackie, in "North West Mounted Police (NWMP) – Personnel Records, 1873–1904," http://www.collectionscanada.gc.ca/databases/nwmp-pcno/index-e.html (accessed 2 October 2013).
58 John Mackie, "Preface," in *The Heart of the Prairie* (London: Nisbet, 1899), n.p.
59 Collins's *Story of Louis Riel* was presumed by many to be factual. In response to this misconception, Collins included a note in his subsequent novel, *Annette the Métis Spy: A Heroine of the N.W. Rebellion* (Toronto: Rose, 1886), at 142–3, confirming that despite the "story [being] printed again and again as truth," "there is no historic truth in" it.
60 Mackie, *Rising*, 9.
61 Braz, *False Traitor*, 56.
62 Ibid., 11.
63 Mackie, *Rising*, 10.
64 Ibid., 12.

65 Ibid., 10.
66 Ibid., 9.
67 Ibid., 10.
68 Ibid., 12.
69 Ibid., 13. Collins, *Story*, 49, makes a similar comparison: "M. Riel, like Mohammed, El Mahdi, and other great patrons of race and religion, is strong of will; but he is weaker than a shorn Samson when a lovely woman chooses to essay a conquest."
70 Roger Smith, *Trial by Medicine: Insanity and Responsibility in Victorian Trials* (Edinburgh, UK: Edinburgh University Press, 1981), 6.
71 Mackie, *Rising*, 13.
72 Ibid., 61.
73 Ibid., 60.
74 Ibid., 61.
75 Ibid., 149.
76 *Q V LR*, 311.
77 Ibid.
78 Smith, *Medicine*, 172.
79 Ibid., 11.
80 *Q V LR*, 294.
81 Notably, this includes the denial of liberty that psychiatric incarceration constitutes, as well as other legal and administrative incapacities.
82 John Gardner, *Offences and Defences: Selected Essays in the Philosophy of Criminal Law* (Oxford: Oxford University Press, 2007), 276. Gardner underscores the importance of this by arguing that the legal form of the trial is predicated on allowing us to give rational account of ourselves (188–9).
83 John Gardner, "Hart and Feinberg on Responsibility," in *The Legacy of H.L.A. Hart: Legal, Political, and Moral Philosophy*, ed. Matthew H. Kramer et al., 121–40 (Oxford: Oxford University Press, 2008), 123.
84 Annalise Acorn, "Responsibility, Self-respect and the Ethics of Self-pathologization," in *Rethinking Criminal Law Theory*, ed. François Tanguay-Renaud and James Stripopolous, 141–60 (Oxford: Hart, 2011), 141–2.
85 *Q V LR*, 211–12.
86 Boyden, *Louis Riel*, 147.
87 Braz, *False Traitor*, 152. The representation of Riel as insane, and the political delegitimizing that it entails, continues to be a source of contention. As Desmond Morton notes, "Later generations of Indians and Métis have found it insulting to be told that their ancestors accepted the leadership of a lunatic." See Desmond Morton, "Introduction," in *Q V LR*, vii–xxxv, at xvi.

88 *QVLR*, 316.

89 Compare Mackie's description of Riel as an impostor in *The Rising of the Red Man*. Riel would also invoke the term at the end of his second trial speech, where he calls for a commission to address his sanity so that he might not be seen as an "impostor" but rather as a legitimate political actor.

90 *QVLR*, 350–1.

91 Clarke, *Psycho-Medical History*, 13. Clarke was medical superintendent at the Toronto Insane Asylum and provided expert testimony about Riel's insanity for the defence at the trial. He went on to publish *A Psycho-Medical History of Louis Riel*, in which he also argued that Riel was insane.

92 *QVLR*, 312.

93 Ibid., 323.

94 Ibid., 323–4.

95 Ibid., 324.

96 Ibid.

97 Ibid., 323.

98 Ibid., 315.

99 Ibid., 323.

100 See, for example, ibid., 324.

101 Ibid., 312.

102 Ibid., 324.

103 Ibid.

104 Duncan Bell, *The Idea of Greater Britain: Empire and the Future of World Order, 1860–1900* (Princeton, NJ: Princeton Univerity Press, 2007), 10. See also Douglas Owram, *Promise of Eden: The Canadian Expansionist Movement and the Idea of the West, 1856–1900* (Toronto: University of Toronto Press, 1980), ch. 6. A similar gesture toward the perceived values of the British Empire can be seen in Fitzpatrick's invocation of the "principles of English liberty," referenced above.

9

"Who Starts the Nations?" Louis Riel and the Question of Geopolitical Legitimacy

JENNIFER REID

Louis Riel was a political reformer who was driven by religious visions. In this sense he was not unique in the modern period. The story of the Americas is rife with figures like him, leaders who materialized out of oppressed nations and cultures, carrying visions of other kinds of modernities that were grounded in their profound experiences of the sacred. Here we might think, for instance, of Wovoka, whose visionary experiences were the catalyst for the Ghost Dance, a movement of reform that began first among the Paiute in 1889. Wovoka had been instructed by God to introduce a dance among his people that would recreate the colonial world into which they had been thrust by the westward expansion of the United States. Or again, we might think of the Brazilian reformer Antônio Conselheiro, who spent twenty years preaching against slavery and the newly formed republic in peasant communities in northeastern Brazil. He and thousands of disenfranchised followers created a community in the 1890s where together they waited for a millennium that would bring prosperity. Like the Métis resisters in the Canadian North-West, they were crushed by military forces.

In one sense, Riel, Wovoka, Conselheiro, and countless others experienced nineteenth-century colonialism as a religious problem that required a religious resolution. Pre-existing frameworks of the sacred – mainstream Catholicism, for instance, or traditional tribal knowledge – were challenged significantly as these figures struggled to account for a changed and chaotic world; and in each instance visionary religious experiences re-established a kind of order in which the lives of their communities, albeit briefly, promised to acquire new kinds of broad cultural significance. What has come to interest me in recent years is the political content of these visions and the social movements they brought into being. More specifically, with respect to Riel, I find myself with an abiding interest in the way that Riel's experience was undeniably linked to state creation. I explored this issue from the perspective of the creation of the Canadian state in a book I published a few years ago, tracing Riel's impact on both the geopolitical development of Confederation-era Canada and the more ontological development of Canadian mythologies of place and collective identity.[1] Since that time, I have found it difficult to turn my attention away from Riel, who continues to fascinate me. Although it is obvious that colonialism was experienced as a local and religious problem for those who were forced to undergo it, in Riel's discourses we can also hear an analysis of the rise of modern nation-states more broadly – a critique of modernity with profoundly religious implications.

Of course, contemporary scholars have linked the emergence of modern states with an Enlightenment secularism that rejected earlier constructions of geopolitical identity grounded in religious frameworks. Benedict Anderson's now famous argument concerning "imagined" political identities is a case in point. The European idea of the nation-state, he has argued, became viable only when medieval assumptions about language, time, and political authority – and their relationships with the transcendent authority of God – were rejected.[2] In Riel's discourses, however – and especially in the two speeches he delivered in the Regina courtroom as his trial for high treason was coming to a close – we can discern the outlines of a very different understanding of the rise of the nation-state, an understanding of modernity that is decidedly ontological.

A few years ago, a writer for *Maclean's Magazine* described the 1885 rebellion in the Saskatchewan Valley in the following way: "[It] began as a western protest movement, it grew into an act of political defiance and soon became a matter of cultural survival. And as Riel grew increasingly delusional, it took on the characteristics of a religious movement. Riel was going to create a New Jerusalem on the plains: the Métis were the chosen people and he would be their 'infallible pontiff.' With God on their side, how could they lose?"[3] The cynicism we hear in this passage is not unique. Although most Canadians have generally come to regard Riel sympathetically, it remains that his religious life and visions still seem enigmatic to most of us. As Joseph Boyden has put it, Riel's mind was situated "somewhere between" sanity and insanity.[4] Generally, Riel's political actions have been viewed as expressions of the former, whereas his religious life has been regarded to varying degrees as a byproduct of the latter. This is not to say that his religious life has been thoroughly stigmatized. Among scholars, for instance, there has been some notable work that has treated it with sensitivity, interpreting Riel's religious life within the context of modern nativistic movements. They have thus delivered Riel's religious inspiration from the denigrating constraints of psychosis and repositioned it as part of a broader movement among oppressed peoples to maintain a sense of cultural integrity in the face of colonial domination. Notable in this respect has been the work of Thomas Flanagan.[5] This has resulted in recent years in a less derogatory interpretation of Riel's religious life and has created an opening, I think, for new considerations of his visions and experiences. That is where my interest primarily lies, particularly as regards the religious significance of Riel's understanding of the limits and possibilities of the modern nation-state as he presented it in his final speeches to the jury and the court in Regina.

The colonial period initiated an unprecedented global shift in the nature of human community in our world. That change was experienced in the first instance by non-Europeans as a problem of sheer survival. The slave trade and westward and southern expansion resulted in large-scale displacements of peoples, destruction of cultures, and marginalization of communities from centres of political and economic power – a chronicle of injustice that haunts the West to this day. The story of obvious resistance in this context is well known: slave revolts and Aboriginal resistance movements marked the period from its outset and have reverberated into the present in various forms. At one level, these movements have been about land, wealth, and cultural self-determination. When we think of the Métis resistances of 1869–70 and 1885, these were obviously central issues. As the boundary line between Canada and the Métis moved westward, land and economic survival became crucial concerns. In this respect, we see a community that was forced to retreat as much as possible from a protracting physical boundary. This is a phenomenon that we have witnessed repeatedly throughout the history of the Americas.

But this is only one perspective on the problem of boundary lines. Although marginalization was an unmistakable result – and the equally inescapable legacy – of colonial expansion, the theoretical frameworks on which it was enacted have not been as inviolable as dominant geopolitical forces have perceived them to be. In the midst of tangible pressure from expanding states, oppressed societies have also consistently engaged in critical analyses of the historical, cultural, and ethical foundations of those states. The ostensibly stable assumptions that underlie dominant visions of modernity have constituted another kind of boundary that voices like Riel's have called into question. They are, from the perspective of those who have found themselves on the underbelly of Western modernity, contested sites of legitimate cultural meaning.

Critiques fostered by this vantage point have rarely, if ever, posed a serious threat to reigning ideologies. As Frederick Cooper has noted in *Colonialism in Question*, the ability to constructively advance interpretations "and to alter definitions of what is a debatable issue and what is not" is always imbalanced in this context.[6] Consequently, there has generally been no necessity for dominant sectors of society to pay attention to these definitions. Yet as we confront an emerging twenty-first-century world that is at once increasingly global and increasingly fragmented, it may well be a good time to stop and begin to listen more closely to some of these alternate perceptions of modernity. We have the colonial period to thank for many of the global economies and ethnic nationalisms that are coming to define our world. They point to what Dipesh Chakrabarty has called the "fragmentary histories of human belonging that never constitute a one or a whole";[7] they point to the way that the history of the modern West has been "heterotemporal." I have borrowed the concept of "heterotemporality" from David Schoenbrun because I think it is a particularly apt way of speaking about the simultaneous and divergent understandings of time and space that emerged from European and non-European cultures during the colonial and postcolonial periods.[8] In some sense, we have not all been living in precisely the same temporal and spatial frame. Moreover, voices emerging from other times and spaces of modernity may speak to us more profoundly than we have imagined. I believe that Riel's speeches at the end of his trial present us with a heterotemporal Canada within which certain prevailing assumptions about the modern period are challenged in significant ways.

So what are the particular "debatable issues" that came to the fore in Riel's trial? At a basic level, of course, they revolved around questions related to the trial itself. The question of Riel's innocence or guilt was primary among these as the foundation for the trial. But also at stake was the question of Riel's sanity, something on which the jury and Riel himself would ultimately agree. And then there was the question of an appropriate sentence for a man found guilty of high treason in a case presided over by a magistrate committed to meting out the maximum penalty despite the jury's recommendation for leniency. These are issues that are, to varying degrees, adjudicated in any criminal trial: culpability, fitness to stand trial, and the parameters for handing down a sentence. There was, however, a second order of debatable issues related directly to the trial that were also more specific to the situation involving Riel. Was it appropriate for the Canadian government, for instance, to move the location of the trial from Winnipeg (where a jury could well have included persons who were French Catholics) to Regina (where the jury was comprised solely of English Protestants)? Could these persons have been construed in any way to constitute a jury of Riel's peers? Or again, given the magnitude of the trial and the events that led to it, was it appropriate for that jury to have been comprised of six, rather than twelve, jurors? Should a part-time magistrate who was an Orangeman have been permitted to preside over the proceedings that transpired in a makeshift courtroom? Should a naturalized United States citizen even have been convicted by a Canadian court of a crime established in a 500-year-old obsolete statute of British law? These are questions related directly to the authority of the court, and they point to a third order of debatable issues that Riel confronted in his two speeches and that ideologically underlay the authority of the expanding Canadian state: issues that related to the legitimacy of the state and to its political and legal apparatus. I believe that because of a general unease with their religious character, his reflections in these discourses on the broad issue of geopolitical legitimacy in the modern period have not been fully appreciated. This chapter is an attempt, in a preliminary way, to do that.

"Who starts the nations?"[9] To my mind this is the religious question that undergirds the two speeches on which this volume is focused. Religion is an integral part of these speeches, mapping out both personal and public spaces that rest at the heart of Riel's understanding of himself, his later life's work, and the evolving Canadian state. The personal aspects of Riel's religious life are conspicuous in these two discourses, and I want to talk a little about these. But my principal interest in this chapter revolves around the way that his very individual experience of the sacred spurred him on to a more collective vision that was a constructive – and religious – critique of nineteenth-century modes of state creation as they were playing themselves out in the Canadian context.

Riel's religious life was rich. When we read these speeches, we find a portrait of a man with an intense visionary life, a profound sense of the reality and power of God, and a belief in his own divinely ordained mission. During the last years of his life, he regularly found himself visited by the Holy Spirit, who provided him with direction and encouragement. His diaries from his period of incarceration, for instance, are riddled with references to these kinds of visionary experiences.[10] On 31 July 1885, he also described one of these encounters to the jury, an experience he had the night before in his jail cell. "The spirit who guides and assists me and consoles me," he said, had indicated that someone would come to his assistance in the courtroom the following day.[11] In his subsequent address to the court on 1 August, he spoke of the spirit having provided him with a kind of precognition of events that had transpired during the final confrontation of the Métis with Canadian forces.[12] These visionary experiences reinforced his distinct sense of having been blessed by God[13] and set apart as "the prophet of the new world," in the image of King David.[14]

Although this comparatively self-referential mode of religious life is significant in itself, I find it all the more intriguing when it is placed in the context of the more public facets of Riel's vision. Here we find religion functioning as the foundation for a critique of the colonial enterprise as it was being played out in the context of Canadian expansion. We should make no mistake: Riel had no problem with the basic idea of a Canadian state. He had hoped, he reminded the court, that the "agitation in the North-West Territories would have been constitutional," and he felt certain that it would have been so if he and his compatriots had not been attacked by military forces. Even as he faced conviction for high treason, he spoke well of Canada, saying that he still hoped that he would one day "have an opportunity of being acknowledged as a leader of good in this great country."[15] Canada per se was not the problem. What he had a distinct issue with, however, was the formation of a state that systematized a disequilibrium of power, privileging one ethnic group within an obviously multi-ethnic New World territory. In other words, he did not believe that the geopolitical space that was to be the Canadian state should be politically and culturally dominated by an Anglo-Canadian elite.

This was not simply a Métis problem but also one that affected everyone living in the North-West, and Riel had spent much of his public life advocating for the rights of a variety of communities whom he believed were being variously marginalized by central Canadian expansionist forces. He told the court,

> When I came into the North-West in July, the first of July, 1884, I found the Indians suffering. I found the half-breeds eating the rotten pork of the Hudson Bay Company and getting sick and weak every day. Although a half-breed, and having no pretension to help the whites, I also paid attention to them. I saw they were deprived of responsible government, I saw that they were deprived of their public liberties ... I have directed my attention to help the Indians, to help the half-breeds and to help the whites to the best of my ability.[16]

Although Riel was unmistakably moved to action by a consciousness of these inequalities, it was the idea of ethnic dominance that most troubled him. He saw the New World as a metamorphic ethnic space in which the concept of discreet nationalities typical of the Old World was no longer operative or appropriate. He regarded himself as a prime example of this process of transformation, and he explained in court, "My ancestors were among those that came from Scandinavia and the British Isles, 1,000 years ago. Some of them went to Limerick and were called Reilson, and then they crossed into Canada and they were called Riel; so in me there is Scandinavian, and well rooted; there is the Irish, and there is the French, and there is some Indian blood."[17]

Riel believed that in a world defined by this kind of ethnic transformation, a single ethnic group could not claim ascendance. Ontario, he told the court in Regina, had no special status: it was merely "our sister colony in the east."[18] His vision of the emerging Canadian state was one of a confederation of new nations based on transformed ethnicities. The number of such communities varied, but what Riel ultimately foresaw were new nationalities that would all in some fashion be indigenized and become part of a new hyphenated (i.e., "German-Indian") world – whether Italians, Irish, Bavarians, and Poles; Belgians, Scandinavians, Germans, and Jews; or First Nations, Métis, English, and French.[19] Aboriginality figured prominently in these new configurations as an underlying structure of identity, the only authentically North American foundation, perhaps, on which Riel could imagine a legitimate modern society in Canada. Regarding himself as a "co-proprietor of the soil with the Indians," he envisioned a geopolitical community that was to be a "whole Cree world"[20] divided evenly among its constitutive ethnic parts. The question of fairness was at the forefront of his thinking. Reflecting back on the agreement that was reached with the Canadian government in the wake of the 1869–70 resistance at Red River, for instance, he emphasized during his trial this necessity for an equitable division of territory: "when the English population has had a full and reasonable share of this land, other nationalities, with whom we are in sympathy, should have also their share of it. When we gave the lands in Manitoba for one-seventh, we did not explain, we gave it to the Canadian Government, but in giving it to the Canadian Government ... we did not give it only for the Anglo-Saxon race."[21]

The question of how the incipient Canadian state was to be structured was not simply a political and ethnic issue for Riel. It was also fundamentally religious. As he told the court a number of times, he had a mission.[22] In the first instance, that mission undoubtedly involved a reformation of the Catholic Church, and he held fast to the words of a spiritual director who told him that God had singled him out for special work: "Father Jean Baptiste [Primeau], the priest of Worcester, who was my director of conscience, said to me: 'Riel, God has put an object into your hands, the cause of the triumph of religion in the world, take care, you will succeed when most believe you have lost.'" Riel added, "I have got those words in my heart."[23] What this meant was that he bore the responsibility for creating a New World Catholic Church, the nature of which he described in detail elsewhere in his writing,[24] a responsibility that he believed to be based on the fact that God had revealed more "to the New World than He had judged appropriate to reveal to the Old."[25] This new reformed church was also enmeshed in his understanding of the radical difference of the New World, which in this case required diffusing Old World antipathies between Catholicism and Protestantism: "If I have any influence in the new world it is to help in that way ... then my children's children will shake hands with the protestants of the new world in a friendly manner. I do not wish these evils which exist in Europe to be continued."[26] It is worth noting here that during his trial he freely employed the concept of Protestantism in a very new way, pointing again to a break with the Old World as well as a politicization of the idea in the Canadian context. Reflecting on a witness's statement that Riel had called a Catholic priest a Protestant, he explained "that we were protesting against the Canadian Government, and that he was protesting against us, and that we were two protestants in our different ways."[27] This statement provides as good an entrée as any into the geopolitical significance of Riel's religious mission. It was not simply about the church. The cause of religion in the world involved a broader mandate for reforming the developing Canadian polity, and it was to this mandate that he referred when he said in court that his mission was aimed at "practical results."[28] He was called "to do something which, at least in the North-West, nobody has done yet."[29] His understanding of himself as the prophet of the New World involved a mission to unseat central Canada's expansionist plans.

Although Riel's mission was undoubtedly religious, I suggest that he understood Canadian expansion to be an issue that was also fundamentally religious. His mission was obviously based on his own sense of a divine call to struggle for change. But equally, the nature of the issues that he believed were at stake in the conflict marked out another kind of religious meaning. Implicit in his struggle was a critique of the universalizing structure of authority undergirding Anglo-Canadian colonialism. The ideology of modern state creation, which Canadian expansionist interests shared, sprang from the new political communities that emerged from the European Enlightenment. These states rejected earlier modes of claiming territory and imagining geopolitical identity on religious grounds. In place of the universal authority of religion, modern states relied on the notion of the sovereignty of the state itself for legitimacy. Nineteenth-century imperialist interests were well served by the ideology of state sovereignty. Indigenous and creolized societies, however, were not. Colonial expansion based on the universalizing principle of state sovereignty entailed the marginalization of nations and cultures occupying territories earmarked for absorption by the new colonial states. Riel recognized that these states had simply replaced one universalizing structure (God) with another (state sovereignty), and he regarded the new ideology as inherently flawed. The state's assumption that it could extend itself with impunity was having devastating consequences for other communities like the Métis.[30]

Riel's concern was with a state apparatus that appeared to be answerable only to itself with respect to its formation. As a good Catholic, Riel knew that an entity cannot create itself. For him, that would have been a power reserved for God. His mission was consequently to ground the emerging Canadian state in a structure of authority that was not subject to the comparatively transitory whims of an elite and powerful class of nineteenth-century colonials. In this respect, he approached the problem from two different angles that were apparent in his trial speeches. The first was an appeal to the authority of the international community, particularly Great Britain. With respect to the international community, he appealed three times in his second speech to the Law of Nations (to which he referred as "the right of nations").[31] The Law of Nations refers to a tradition of international law involving rights and obligations presumed to exist among distinct nations or nation-states. Riel clearly regarded this as a mode of authority that exceeded that of an individual state. When speaking of the treatment in Ottawa of Red River emissaries in 1870, for instance, he claimed that it was counter to the Law of Nations: "Our delegates had been invited three times, how were they received in Canada? They were arrested – to show exactly what is the right of nations."[32] More incisively perhaps, later on in this address he made it clear that he regarded the Red River community as a distinct nation deserving of rights articulated by international law:

> there were two societies who treated together; one was small, but in its smallness it had its rights. The other was great, but by its greatness it had no greater rights than the rights of the small, because the rights is the same for everyone, and when they began by treating the leaders of the small community as bandits, as outlaws, leaving them without protection, they disorganized that community. The right of nations wanted that the treaty of Manitoba should be fulfilled towards the little community of Red River in the same condition that they were when they were treated. That is the right of nations.[33]

Turning to the authority of Britain, Riel focused specifically on the British concepts of liberty and impartial justice, the British Constitution, and Britain's role in world history. Speaking to the presiding magistrate, Hugh Richardson, Riel called the trial to account for its failure to uphold the tradition of British justice: "you have acted according to your duty, and while it is, in our view, against the guarantees of liberty, I trust the Providence of God will bring out good of what you have done conscientiously." He similarly told the court, "When I see British people sitting in the court to try me, remembering that the English people are proud of that word 'fair-play,' I am confident that I will be blessed by God and by man also." He spoke also of the British Constitution and the idea of civilization, suggesting that on the basis of these, the existing government in the North-West was unacceptable: "British civilization ... rules to-day the world, and the British constitution has defined such government as this is which rules the North-West Territories as irresponsible government."[34] For Riel, both the Constitution and British civilization were destined for broader historical purpose than merely the benefit of England. The Constitution was "an institution which has been perfected for the nations of the world," and although it had failed to do so, civilization had the capacity to improve the lives of Aboriginal peoples (including the Métis). On this point, Riel chastised the "gentlemen" seated in the courtroom: "If I give offence, I do not speak to insult. Yes, you are the pioneers of civilization, the whites are the pioneers of civilization, but they bring among the Indians demoralization."[35]

The second perspective on the problem of the origin of nations that Riel expressed in the court was noticeably more transcendent in character and provided a context within which perhaps to situate conventions of international law and to definitely locate the tradition of British justice, the British Constitution, and the concept of British civilization. This framework was the foundation of both his vision and his critique, and it revolved around the reality of God as a defining and world-creating power. With respect to the Law of Nations (*jus gentium*), we might recall that its European foundations stem back to the Roman Empire and were picked up by later Christian theologians and jurists. The first European treatises on the subject were written by Francisco de Vitoria (who held a chair in theology at the University of Salamanca in the mid-sixteenth century). Vitoria, greatly influenced by Aquinas, argued that issues in international law had to be adjudicated in terms of natural law based on Christian anthropology.[36] Having attended seminary, Riel may well have been acquainted with this tradition, and his appeal to the Law of Nations could have been a means of calling for not only a universal, but also a transcendent, mode of ordering human relationships in modern states, particularly the Canadian state.

Although I am admittedly speculating with respect to the Law of Nations, we can be more certain of Riel's belief in the transcendent nature of the British legal and political traditions. Regardless of what the British Empire had achieved, Riel believed that it was ultimately subject to God and a part of God's plan for humanity. He told the court,

> As among the nations of Europe 2,000 years ago, the Roman people were the leading race, and were teaching the other nations good government; that is my opinion of the Anglo-Saxon race. I am not insane enough to [regret] the great glory of the Anglo-Saxon race God has given to that race, and when God gives something to somebody it is for a good purpose, and because God gave glory to England, it is because He wanted the Anglo-Saxon race to work for His own glory, and I suppose it is not finished yet.[37]

Riel had no doubt that God was also at work in the Canadian situation. He had a God-given mission, after all, to resist colonial expansion and to work for the creation of a different kind of state defined by an alliance of a group of transformed ethnic communities. He had no doubt that God was deeply implicated in his struggle. He told the jury that God had been present in the struggle for the recognition of rights in the North-West: "there was a Providence in the battles of the Saskatchewan." Likewise, God was present with him in the courtroom where he was working to assure his acquittal: God, he said, was "in this box with me, and He is on the side of my lawyers, even with the honorable court, the Crown and the jury, to help me, and to prove by the extraordinary help that there is a Providence today in my trial." Furthermore, he believed that God would reward the members of the jury who were determining his fate if they returned a verdict of not guilty, and that their reward would be given to them not only in this life but also in the next: "What you will do in justice to me, in justice to my family, in justice to my friends, in justice to the North-West, will be rendered a hundred times to you in this world, and to use a sacred expression, life everlasting in the other."[38]

Riel took the idea of a Canadian confederation very seriously. In some ways, perhaps he took it more seriously than did the political apparatus in Ottawa against which he struggled. A confederation is a configuration in which independent states, or nations, are united for limited purposes (e.g., defence and foreign relations) and in which the federal government does not hold the balance of power. Settlers in the North-West were not part of such an arrangement. Lacking responsible government, their territory was, as the *Regina Leader*'s editor, Nicholas Flood Davin, put it early in 1885, "disestablished and disendowed, and outside the pall of the Constitution."[39] The Métis were not part of such a configuration as they were forced progressively westward after 1870 in search of land, rapidly disappearing buffalo, and employment to sustain themselves. And the region's First Nations were certainly not part of such an arrangement, as they found themselves coerced into treaties, removed from their lands, and sinking into destitution and starvation. Riel's advocacy for the rights of all these communities betrayed a belief not only in the need for responsible government but also in the legitimacy of the Métis and First Nations *as* nations. His appeal to the Law of Nations during his remarks to the court underscored this belief and, together with his support for an expanding Canadian state, indicated that his conception of a confederation may well have been closer to the spirit of such a geopolitical arrangement than that of its architects and promoters in the Canadian context.

What was lacking in the expansion of the Canadian state was, first, a sense of regard for the integrity of the nations that Canada intended to incorporate into itself. "When they have crowded their country," he told the court after he had been found guilty of treason, "because they had no room to stay any more at home, it does not give them the right to come and take the share of all tribes besides them ... We are not birds. We have to walk on the ground."[40] Riel's belief in the need for a basic level of regard for the rights of other peoples was perhaps the reason why he stressed again and again at his trial his own respect for those who opposed him: the police, the "honorable" court and jury, the presiding magistrate, and the Crown attorneys.[41] But also absent from Canada's expansionist aspirations was a mitigating presence of another authority aside from the state itself. He acknowledged the value of a

Canadian state, but he wanted it to be founded on something other than the vagaries of elite ambitions:

> In England, in France, the French and the English have lands, the first was in England, they were the owners of the soil and they transmitted to generations. Now, by the soil they have had their start as a nation. Who starts the nations? The very one who creates them, God. God is the master of the universe, our planet is his land, and the nation and the tribes are members of His family, and as a good father, he gives a portion of his lands to that nation, to that tribe, to everyone, that is his heritage, that is his share of the inheritance, of the people, or nation or tribe. Now, here is a nation strong as it may be, it has its inheritance from God.[42]

Riel's confrontation with Canada was not simply a political issue. His religiously inspired vision of the state confronted an opposing force that had assumed for itself the power to create its world – an undertaking that in any place or time outside the modern West would be reserved for the gods.

Riel did not reside in the same modernity as those who opposed him. In the time and space he inhabited, nation-states owed their existence to the will of something other than themselves, and their unilateral assumption of god-like powers of world creation was illogical and humanly destructive. By the time he delivered his two trial speeches in July 1885, he had, to quote Marshal Sahlins, integrated his "experience of the world system in[to] something that is logically and ontologically more inclusive: [his] own system of the world."[43] Inherent in Riel's religious mission was a vision of the modern state that was fundamentally critical of the notion of state sovereignty as the source of political legitimacy. He knew first hand that a state that regarded itself as its own ultimate authority was a potentially destructive geopolitical entity.

This chapter began with the suggestion that the notion of heterotemporality opens a space in which to reconsider fundamental assumptions about the word we inhabit. Riel's trial speeches provide us with one such temporal frame. Emerging out of this frame was a critique of the sacrosanct ideology of state sovereignty. Fundamental to the notion of sovereignty is the principle of autonomy from external authority; another state's intervention in the affairs of one's own state, for instance, is unacceptable. Yet we know that this principle has been largely rhetorical in our time, not only with respect to the rights of Indigenous and other ethnic nations but also at the level of the state: it has too often been claimed by states with the power to deny it to those with less political or military leverage. And again and again we have seen it subverted – for example, by the Soviet Union in Eastern Europe and by the United States in Nicaragua and Granada, as well as more recently in Iraq.[44] Riel knew that the ideology of state sovereignty was a precarious foundation for geopolitical identity in a global community. Ultimate authority borne of an entity itself is no authority at all.

NOTES

1 Jennifer Reid, *Louis Riel and the Creation of Modern Canada: Mythic Discourse and the Postcolonial State* (2008; reprint, Winnipeg: University of Manitoba Press, 2011).

2 Benedict Anderson, *Imagined Communities: Reflections on the Origin and Spread of Nationalism* (London and New York: Verso, 1991), 36. See also Brian C.J. Singer, *Society, Theory, and the French Revolution: Studies in the Revolutionary Imaginary* (London: Macmillan, 1986), 103–4.

3 Will Ferguson, "Ghosts of a Nation: The Louis Riel Rebellion Still Haunts us Today," *Maclean's*, 17 November 2003, 128, 133.

4 Boyden is cited in "Was Louis Riel Insane?" *Maclean's*, 4 October 2010, 76.

5 See, for instance, Thomas Flanagan, *Louis 'David' Riel: 'Prophet of the New World,'* 2nd ed. (Toronto: University of Toronto Press, 1996).

6 Frederick Cooper, *Colonialism in Question: Theory, Knowledge, History* (Berkeley: University of California Press, 2005), 149.

7 Dipesh Chakrabarty, *Provincializing Europe: Postcolonial Thought and Historical Difference* (Princeton, NJ: Princeton University Press, 2000), 255.

8 David L. Schoenbrun, "Conjuring the Modern in Africa: Durability and Rupture in Histories of Public Healing between the Great Lakes of East Africa," *American Historical Review* 111 (December 2006), 1403–39, at 1410.

9 Louis Riel, "Address to the Court," 1 August 1885 (hereafter "Court"), in *The Collected Writings of Louis Riel/Les ecrits complets de Louis Riel*, vol. 3, *1884–1885*, ed. Thomas Flanagan (Edmonton: University of Alberta Press, 1985), 547.

10 Thomas Flanagan, ed., *The Diaries of Louis Riel* (Edmonton: Hurtig, 1976).

11 Louis Riel, "Address to the Jury," 31 July 1885 (hereafter "Jury"), in *The Collected Writings of Louis Riel/Les ecrits complets de Louis Riel*, vol. 3, *1884–1885*, ed. Thomas Flanagan (Edmonton: University of Albert Press, 1985), 532.
12 Riel, "Court," 559.
13 "Bishop Bourget wrote again and said: 'Be ye blessed by God and man and take patience in your evils.' Am I not taking patience? Will I be blessed by man as I have been by God?" See Riel, "Jury," 527.
14 "I say humbly, through the grace of God, I believe I am the prophet of the new world." See Riel, "Jury," 534; see also Riel, "Court," 541, 559.
15 Riel, "Jury," 529, 533.
16 Ibid., 524–5.
17 Riel, "Court," 557.
18 Ibid., 548.
19 Ibid., 545, 555–7.
20 Ibid., 547, 556.
21 Ibid., 555.
22 See for instance, Riel, "Jury," 526, 531.
23 Ibid., 527.
24 See Flanagan, *Louis 'David' Riel*; and Flanagan, ed., *Diaries*, 80.
25 Flanagan *Louis 'David' Riel*, 27.
26 Riel, "Jury," 531.
27 Ibid.
28 Ibid.
29 Riel, "Court," 541.
30 Elsewhere, Riel spoke of his view of new revolutionary states as undermining the authority of the Catholic Church: "Revolutionary ideas are trying to make Rome into a city of tradesmen and merchants. A majority of governments has contributed to despoiling the Papacy of its possessions. Their machinations have gone to the extreme of daring to rob the successor of St. Peter of his royal patrimony and give it to Umberto, who purposely remains in Rome to mock Catholicism by making the scepter he has stolen from us shine before our eyes." Cited in Thomas Flanagan, "Louis Riel's Religious Beliefs: A Letter to Bishop Taché," *Saskatchewan History* 27, no. 1 (1974): 21.
31 See Riel, "Court," 543, 550.
32 Ibid., 543.
33 Ibid., 549–50. On the treatment of the delegation from Red River, see 543.
34 Riel, "Jury," 534–5, 527.
35 Riel, "Court," 555, 548; Riel, "Jury," 530, 539n27.

36 See Thomas D. Williams, "Francisco de Vitoria and the Pre-Hobbesian Roots of Natural Rights Theory," *Alpha Omega* 7, no. 1 (2004): 47–59.

37 Riel, "Court," 555.

38 Riel, "Jury," 526, 536.

39 Cited in J.M. Bumsted, *Louis Riel v. Canada: The Making of a Rebel* (Winnipeg: Great Plains, 2001), 246–7.

40 Riel, "Court," 547.

41 Riel, "Jury," 529, 524, 526, 534–5.

42 Riel, "Court," 547.

43 Marshal Sahlins, *Culture in Practice: Selected Essays* (New York: Zone Books, 2000), 417.

44 Norihisa Yamashita, "Empire or Post-Empire: The Concept of 'Long Century' and the Consequences of Globalization," in *Emerging Meso-Areas in the Former Socialist Countries: Histories Revived or Improvised?*, ed. Kimitaka Matsuzato, 335–48 (Sapporo, Japan: Hokkaido University, 2005), 343.

10

Louis Riel's *Ethos* and the *Différend*

MAURICE CHARLAND

Riel suffered. Riel had claims. Riel again and again sought a remedy for his suffering and the sufferings of the North-West. His story can be understood as a search for a proper hearing, for an instance that could hear his case and provide justice. This is what animated his creation of a provisional government in Manitoba in 1869; this is what led him to write the president, while an American citizen living in Montana, proposing American annexation of the North-West; this is what led him to battle Canadian troops at Duck Lake. It is also what led him to surrender to General Frederick Middleton rather than flee across the border. Riel had hoped that his claims would finally be heard, that he could put into words his story of the North-West's and his suffering, so that he and his cause could be judged against the horizon of justice.

Riel, of course, would not find that hearing. Riel's fantasy was that justice, on his terms no less, could be phrased outside of a legal framework. Riel sought political justice, a form of justice that has not settled into a concept and is beyond actualization.[1] Riel had given himself up to the Crown in the hope that his story would be heard by the Supreme Court in an extraordinary session. Riel thought that such a court could offer relief in the name of justice. Tragically, although Riel was trained in law, he did not take the law into account. Nothing in the law provided for the Supreme Court to assume such a role, nor can legal principles vindicate extralegal actions. Riel had failed to recognize both the pragmatics of the jurisdiction to which he had submitted and his place within it. Under the law, the aggrieved party was the Crown. Riel was charged with high treason because he "most wickedly, maliciously and traitorously did levy and make war against our said Lady the Queen."[2] Although Riel did enter a court charged with treating him justly, it was not constituted to proceed on his terms. He had entered a realm of difficult and strained speech. As on other occasions, Riel found himself within what Jean-François Lyotard terms a *différend*.

Lyotard develops the category of the *différend* to account for cases where a wrong cannot be phrased.[3] Influenced by the work of Ludwig Wittgenstein, Lyotard asserts against the dominant traditions of political thought from Plato to Marx that one should not consider justice to be a particular form, concept, or state of affairs.[4] Rather, justice operates as a regulative idea within certain types of speech acts. Justice is the structuring principle of forensic discourse, where claims are made, contested, and judged. In other words, justice is both the raison d'être and immanent principle of a certain kind of language game. By following the linguistic turn, Lyotard is able to consider the nature of forensic speech and the limits that language places on the "presencing" of justice. Specifically, when there is a *différend*, justice cannot appear because the game cannot be played.[5]

Language games consist of vocabularies, speaking positions, and procedures that enable, regulate, and empower speech acts as "moves." Not all speech is the same, and there is no one set of procedures that enables and gives traction to all utterances. Rather, there are many forms of language games. Some, such as philosophy, law, and science, are learned, but most are not. Some games are prescriptive, whereas others are descriptive; and if we follow Thomas Kuhn's discussion of "paradigms," there are radically distinct descriptive games, such as alchemy and chemistry.[6] Lyotard, more metaphysical than Michel Foucault, maintains that distinct categories of thought are produced in and through distinct language games and that language games are bounded, proceeding through "phrase regimens" that are often incommensurable with each other. Furthermore, there is no such thing as a game-free language, nor can any one language game enable or give meaning to all possible phrases.

From this perspective, Lyotard understands that justice cannot be fixed into a concept. Rather, as Aristotle's *Rhetoric* makes clear, justice is rendered in situated judgments by a recognized adjudicator following the *stasis* of conflicting claims. For Aristotle, such judgment is prudential.[7] Lyotard uses the term *différend* to name those instances where such judgment cannot be realized adequately. As Lyotard puts it, a *différend* is a "case where the plaintiff is divested of the means to argue and for that reason becomes a victim" and "when the 'regulation' of the conflict that opposes them is done in the idiom of one of the parties while the wrong suffered by the other is not signified in that idiom."[8] Lyotard opposes the *différend* to the *litige*, or litigation, where parties to a dispute (1) phrase their claims in a common language and (2) both recognize and are recognized by an adjudicator. In a *litige*, there is some form of law, some form of court, and both parties have standing. A *différend* occurs when communication fails.

Being "divested of the means to argue" can mean many things, and although Lyotard does not treat specifically the various types of divestment, he offers some examples. Victims of the Holocaust obviously cannot speak, but even those who survived the camps might be unable to phrase the wrong that they have suffered: trauma can inhibit speech, and suffering can be beyond words. Furthermore, in a dispute, the Holocaust denier will set an impossible evidentiary standard. At quite another level, Lyotard points out that no court has jurisdiction to hear a Martinican's claim to be wronged by the very fact of his French citizenship. Thus the *différend* arises from a number of lacks: the lack of a faculty of speech; the lack of an idiom in which the wrong can be phrased; the lack of an idiom intelligible to others; and to use legal terms, the lack of a jurisdiction in which a claim is actionable and the victim's lack of standing. In all of these cases, there can be no forensic dispute. Justice cannot appear. One cannot speak of injustice, only of victimage, and even that is occulted because a proper claim cannot be brought to presence.

Although Lyotard's approach to the *différend* is forensic, it can be extended beyond the courts to the political domain if one considers that for Lyotard, politics is not a language game but occurs between language games and indeed is the "threat of the differend [sic]."[9] Political claims are also normative, call for justice, and are addressed to judging instances. When politics is normalized, it proceeds as if it were a *litige* and thus takes the form of a language game. The parties of normalized politics have the means to argue, share an idiom in which to make claims, and both recognize and are recognized by those who judge, such as legislators, members of the executive, and an electorate. Lyotard describes such politics as being of the *polis*, extending the idea of a city constituted under law to mean a rationalized order regulated through institutions of speech. Athens, of course, is the prototype. Members of the *polis*, the *viri* of Athens, speak as citizens under the law. They have standing.

In law, standing is the right to initiate a suit in court. To have standing, one must have a legally protected interest; one must be at risk or have suffered a wrong; and the law must be able to provide redress. Without standing, one does not appear and consequently one cannot even be heard, as the United States Supreme Court ruled in *Dred Scott v. Sanford*,[10] denying citizenship and the right to sue in federal court to both slaves and their nonslave descendants. Consequently, standing is a principle of appearance that is relational because one must be seen in order to appear. It is also perspectival: one stands before a court and is seen in a particular way. Standing is fundamental to the exercise of law and is also fundamental to all language games. One always speaks as someone from somewhere to someone, and this fact defines a pragmatic relation. When the *viri* of Athens spoke, they did so as citizens: there was no *différend.* In contrast, noncitizens, women, slaves, and other types of *others* had no standing. They were not recognized by authority but misrecognized as incapable of proper speech. They could not expect to be heard on their own terms, nor did they have rights to which to appeal. They could speak politically, but they might not be heard: for them, politics was a struggle between language games. They must speak like *pagans*; that is, they must negotiate the terrain between normative orders and their language games, exploit *aporia*, and try to establish new linkages. They must know the risks of merely reacting but learn how to reply artfully.[11] Extending this metaphor, pagans belong to what Lyotard calls the *pagus*, the unincorporated regions of the life-world that have not been rationalized. For Lyotard, the *pagus* is not governed by law or disciplinary reason. At least metaphorically, it is the wild country that Odysseus traversed as he sought to return home. It is populated by minor deities, barbarians, and marginal figures who rely upon ruses and strange powers to seek advantage. As Odysseus and the *coureurs des bois* discovered, the *pagus* is both dangerous and exhilarating. It is not compatible, however, with the idea of law.

What has this to do with Riel? As a Métis in the North-West, Riel was not fully of the *polis*, even while viewing himself as being within the institutionalized order. The difficulties that Riel faced, both in the courtroom and throughout his public life, can be understood as arising through a *différend.* Riel often spoke from a position where his standing was problematic; he made claims for which no proper tribunal was constituted, and indeed, his claims were often in a language that appeared fantastic or incoherent. In 1869 he had made claims on behalf of the North-West, but no institutions existed to act upon them. He had an idiom, but there was no proper jurisdiction. The North-West was a *pagus.* The territory was not yet part of Canada, and the Crown's proxy, the Hudson's Bay Company, was no longer asserting its authority. Thus there was a *différend.* Riel was divested of the means to argue on behalf of the North-West. Riel's response, however, was not to adopt the *persona* of the pagan or the metic. He did not play the "outlaw." He did not use wildly sophistic tactics but implemented a strategy. Trained in law, he saw himself as a member of some *polis* and within some kind of law and thus declared a provisional government under "a fundamental principle," "admitted by all men," that people have the sovereign right to choose their own form of government.[12] Although a rhetoric of constitution is in a sense *pagan* because it proceeds outside the law, its ruse is to act as though it is within the law, and in doing so, it refigures its own pragmatics. It evokes a higher law and proclaims a new jurisdiction. It seeks to transform the *pagus* into a *polis* and to bring a new sovereignty and law into the world. This genre is well suited to those on the boundary of empire, including both colonized and alienated colonists who do not want to be pagan but rather seek a *polis* of their own. However, its power requires more than illocutionary force. Revolutionary constitution requires an initial "coup," a treasonous moment of founding that must be sustained through a new monopoly of violence. In 1870 Riel and his compatriots materialized this new order, enacting sovereignty, by executing Thomas Scott, an Ontario-born Orangeman accused of plotting against Riel. This strategy was effective, for although Ottawa did not formally recognize the provisional government, it entered into communication with some of its representatives, incorporated many of their demands into the Manitoba Act of 1870, and did not seek to prosecute Scott's execution.

In 1884 the situation had changed radically. Riel, then in Montana, was invited by Métis to come to what is now Saskatchewan to assist them in receiving a settlement similar to the one granted to Manitoba fifteen years previous. As in 1869, Riel lacked proper standing to press claims. However, although the North-West was not part of Canada in 1869, Ottawa now had jurisdiction over the territory, which was, except for Manitoba, an interior colony lacking parliamentary representation. As Riel correctly observed in his trial speeches, the government was not responsible. Riel and the Métis sent Ottawa petitions, but these had little traction. There may not have been a *différend* at the level of intelligibility, but the institutions that in principle could provide relief were far away and not compelled to do so. This is why Riel again resorted to a rhetoric of constitution, appealing to popular sovereignty. In 1870 this coup was initially successful, permitting Manitoba's entry into Confederation in part under its own terms. However, Riel lacked the means to sustain such a rhetoric in 1885. After an initial victory at Duck Lake, the rebellion was defeated, and Riel was forced to make a fateful choice.

Lyotard tells the story of two Greeks who ran afoul of the law. Although their crime was not treason but one of its lesser forms, impiety, they faced the ultimate penalty. The Greek best known to us is Socrates; the other is Protagoras, a Sophist and teacher of rhetoric. Socrates could imagine himself only within the law. He belonged to the *polis* and lived by its rules. He stood trial, refused to recant or beg for mercy, was sentenced to death, and drank the hemlock. Protagoras knew better: not an Athenian, he saw no reason to submit to a court of hypocritical moralists. He skipped town. He lived to speak another day.[13] Riel had the option of exercising pagan wisdom and following Gabriel Dumont across the border and beyond General Middleton's reach. After all, he had not lived his life as a member of the *polis* under its laws. Indeed, by then he was an American citizen. But no! Riel saw himself as belonging to the North-West and as residing within truth and righteousness. Perhaps, like Socrates, he valued his cause and his truth over his life. Refusing pagan strategies, he surrendered and ran the risk of martyrdom.

The Crown claimed injury: Riel had waged war against her majesty, and Canadian blood had stained the white snow. Riel was charged with the capital crime of high treason and delivered for trial to stipendiary magistrate Hugh Richardson's court. Riel was before a tribunal in which he would have an opportunity to speak but not on his own terms. Riel had been hoping for another forum. He had proposed in a letter to John A. Macdonald that his case be heard before the Supreme Court sitting as a board of inquiry.[14] Riel thought that he could exonerate himself and his cause by telling his story on his own terms to a disinterested tribunal. Of course, no such hearing was forthcoming. It is difficult not to conclude that Riel was naive or profoundly misunderstood jurisdiction and standing. Perhaps his faith had led him to believe that others would recognize him under a higher law. General Middleton's promise of a fair hearing was in a sense kept. Riel was delivered to civilian authorities rather than summarily executed, but Riel would not find a tribunal to hear his case, his suffering, and his claim for relief as he wanted to phrase them. Only the Crown had standing to seek relief. Riel stood as defendant. There was a *différend.*

Riel's lawyers were well versed in the pragmatics of criminal trials and were certain that his speech could not save him, that no narrative of rational action could counter the evidence against him, and they knew better than to argue a right to treason. For this reason, they sought to exploit the *différend.* As in 1869, Riel had called for responsible government and a settlement of Métis land claims, but since then he had also grown convinced that he was a prophet with a divine mission. He heard God's voice and called for a new church in the North-West or for the papacy's relocation there. He envisioned partitioning the North-West into seven parts, each to be colonized by a different ethnic group. In addition, he had been treated for over a year in two Quebec asylums for "megalomania" and was reported to become agitated and irrational when discussing political and religious subjects. Consequently, Riel's lawyers sought to demonstrate that he was insane and thus not responsible for his actions. They offered the court a narrative of noble motives but irrational actions in order to free Riel from the burden of agency and *mens rea*. They sought to both garner sympathy and shift his standing under the law, which reserves a special place for the insane. The law's conception of justice requires free moral agents who are responsible for their actions. The insane are considered to act under compulsion or to lack a moral faculty. They thus raise a different kind of *différend*: they cannot "own" either their words or their acts, so they cannot be punished, punishment being a moral concept, but only sequestered for treatment or public safety. The defence strategy was thus to gain Riel's life at the cost of his voice, his reason, his *ethos*, and his cause. This was as much as his trial could allow.

Riel's lawyers contested few of the facts of the prosecution's case. They staked their defence on Riel himself, on his *ethos* and motives. They praised the Métis as neglected civilizers who tempered Indian savagery and ensured the relative tranquillity of the North-West. They further argued that Riel's cause was noble. Charles Fitzpatrick, Riel's junior counsel and a future chief justice of Canada, went so far as to say that although one cannot defend treason, "the plant of liberty requires the nourishment of blood occasionally."[15] At the same time, they sought to insulate Riel from his own speech. They tried their best to stop him from speaking at trial. They did not permit him to cross-examine witnesses and made it clear to the court that they preferred that he not speak at the end of their case. Their insanity plea was intended to invalidate his speech, for they feared it would be damning if uttered in a sane-sounding manner. Thus, against the prosecution's argument that Riel's treason was motivated by self-interest, they portrayed a noble but mad Riel who was in over his head and lacked the capacity to appreciate the moral quality of his acts.

Lyotard develops the concept of the *différend* through speech act theory, focusing on the pragmatics of demonstration, obligation, and judgment. Fundamental to his analysis is that "injustice" is a meaningful category only through judgment. Suffering in itself is neither just nor unjust, but becomes so only through a moral compass. Furthermore, the pragmatics of forensic judgment includes argument and deliberation. There is a *différend* when one is divested of the means to *argue*, which means to make and support claims. The *différend* thus renders forensic judgment impossible. What then can happen? Lyotard's answer is that third parties must bear witness to the heterogeneity of language games. This means that they must be sensitive to conspicuous silences and to improper utterances and "moves," including those that Lyotard calls pagan. In addition, if we return to Lyotard's examples of divestment, we can imagine *praxes* that would foster a movement toward forensic inquiry: for trauma, there could be therapy; for the lack of an idiom, there could be poetry; rhetoric can serve to extend intelligibility; and judicial interpretation and revolution can refigure jurisdiction. These praxes might begin by bearing witness to unrepresentability, but all of them also attend, albeit in an unfamiliar way, to what is uttered. That is to say, these forms of praxis promote *aletheia*, or uncovering, by listening for what exceeds the bounds of given language games, by placing games in conflict, or by rendering them noisier and less hermetic than they first appear.

Even if Riel's case was practically hopeless, his fate rested nevertheless with the jury, which could have seen fit to free him. Indeed, no one doubted that Riel might be acquitted if the trial were held in Winnipeg before a jury equally composed of anglophone Protestants and francophone Catholics. Although juries are mandated to be judges only of fact, they may deliberate broadly and render judgments informed by an intuition regarding the appropriateness (we might say *la justesse*) of applying the law. In a sense, juries may bear witness to, and thus respond to, a *différend.* Even in Regina, in a magistrate's court with a jury of six English-speaking Protestants, the improbable possibility remained that Riel could transform his standing rhetorically, appear to them as a plaintiff, and win an acquittal; and so, against his lawyer's wishes and considered judgment, he exercised his right to address the jury. This was Riel's "rhetorical situation."

Rhetoric was defined by Aristotle as the art of determining the available means of persuasion in a given case. Rhetoric is manifest in situations where speech crafted in the face of constraints addresses an audience and calls upon it to resolve an exigence, "an imperfection marked by urgency."[16] Riel's situation was highly constrained, formed by the pragmatics of the trial. Riel was not at a political rally, with great leeway to select an audience and define the stakes. He was in a criminal court where the primary exigence was established by the charge against him, and his audience was chosen by the court and mandated only to judge the facts of the case, including Riel's intentions, actions, sanity, and guilt under the law. Furthermore, Riel was constrained by the court's purpose, rules, and procedures, the pertinent statute, the prosecution's case against him, and his counsel's defence. Most significantly, he could address the jury only once. He would speak at the very end of the trial, after his counsel's summation and before the prosecutor's final statement. He could neither produce nor cross-examine witnesses nor introduce evidence. All that he could produce was himself.

Riel's ultimate objective was to vindicate his cause and secure justice for the North-West. Riel, like both the prosecution and defence, recognized the importance to the case of his character, his *ethos*. The defence had argued that his motives were noble but that his actions arose from mental illness. The prosecution did their best to present him otherwise. Consider how Britton Bath Osler, the first prosecutor to address the court, described the crime: "this was no constructive treason ... [n]o treason such as may be made out from meetings, treasonable acts or letters ... but treason, involving the shedding of brave men's blood; treason which roused the whole country, treason sounding from the dead bodies lying on the blood-stained snow."[17] Riel was not on trial for his ideas but for waging war and bringing death. Furthermore, the prosecution challenged the very idea of a noble Riel: as lead Crown prosecutor Christopher Robinson asserted in his summation, "[The defence] cannot claim for the client what is called a niche in the temple of fame and at the same time assert that he is entitled to a place in a lunatic asylum."[18]

Riel may have feared the asylum, but he also recognized that his name and cause depended on the jury judging him sane. Furthermore, sanity was not enough. Riel also did not wish to die or enter history a traitor. To secure freedom and vindication, the jury would have to judge him favourably, excusing his actions. In order to fulfil his objective, Riel would have to undermine the pragmatics of his trial and stand as a plaintiff against the government. In essence, this would require a pagan rhetorical moment that brought the jury into the *pagus*, the place outside law, and that invoked the idea of justice in accordance with a law that was to come. This would require Riel figuring himself as worthy of their respect and trust, as having *ethos*.

Aristotle understood *ethos* as having three components: good will, virtue, and prudence. *Ethos* requires (1) appearing to care for the good of the audience, (2) appearing to hold and act in accordance with the community's values, and (3) appearing to act effectively to advance the common good. In Riel's statement to the court, one can observe his mastery of the genre of petition, which is most suited to the powerless. To the modern critic, this genre seems obsequious, but it is the dominant genre under monarchy. It proceeds by praising a potential benefactor in order to establish a familial relation of mutual obligation.[19] So Riel opens his speech by expressing his respect and admiration for the court, the jury, and even the prosecution: "Oh, my God, bless me, bless this honourable court, bless this honourable jury, bless my good lawyers who have come 700 leagues to try to save my life, bless also the lawyers for the Crown, because they have done, I am sure, what they thought their duty. They have shown me fairness which at first I did not expect from them."[20] Through this speech act, Riel performs good will. His narrative seeks to highlight his virtue and the purity of his motives by describing his concern for the Métis, Indians, and whites of the North-West who suffered under government neglect:

> When I came into the North-West in July, the first of July 1884, I found the Indians suffering. I found the half-breeds eating the rotten pork of the Hudson Bay Company and getting sick and weak every day. Although a half-breed, and having no pretension to help the whites, I also paid attention to them. I saw they were deprived of responsible government, I saw that they were deprived of their public liberties. I remembered that half-breed meant white and Indian, and while I paid attention to the suffering Indians and the half-breeds I remembered that the greatest part of my heart and blood was white and I have directed my attention to help the Indians, to help the half-breeds and to help the whites to the best of my ability. We have made petitions, I have made petitions with others to the Canadian Government asking to relieve the conditions of this country.[21]

Through this statement, Riel attempted, with considerable success, to create sufficient identification to garner sympathy. He cares for all, advocates for all, and even shares a white heart and blood with the jurors. He is not an alien but of the North-West. Paradoxically, however, this performance places him within the law and the community, which makes it more difficult to effect a radical restructuring of his context in order to give rise to new law.

Ethos requires prudence. The prudent individual, Aristotle's *phronimos*, acts successfully to realize the contingent good. Clearly, prudence requires sanity, but there is more to it. Prudence is an intellectual virtue based on good practical judgment and effective action. Riel could perform sanity. Not even the defence had asserted that his malady was immediately obvious, only that it was manifest on occasion, particularly when his authority was challenged on political and religious matters. Riel could thus undermine his own lawyers through a rhetoric of enactment – where the rhetorical performance materializes and stands as proof of what is asserted – by calling attention to his apparent calmness and lucidity under stressful circumstances:

> It would be easy for me to-day to play insanity, because the circumstances are such as to excite any man, and under the natural excitement of what is taking place to-day ... under the excitement which my trial causes me would justify me not to appear as usual, but with my mind out of its ordinary condition. I hope with the

> help of God I will maintain calmness and decorum as suits this honourable court, this honourable jury.[22]

> I wish you to believe that I am not trying to play insanity, there is in the manner, in the standing of a man, the proof that he is sincere, not playing.[23]

Riel, then, figured himself as unorthodox rather than insane: "What is my insanity about [religion]? My insanity, your Honours, gentlemen of the jury, is that I wish to leave Rome aside, inasmuch as it is the cause of division between Catholics and Protestants."[24] With God on his side, he even called attention to the prosecution's own case for his sanity: "To-day when I saw the glorious General Middleton bearing testimony that he thought I was not insane ... I felt that God was blessing me and blotting away from my name the blot resting upon my reputation."[25]

Despite the defence narrative of an underlying mental illness, Riel managed to appear responsible and in control of his faculties. However, appearing to have good practical judgment would be another matter. Riel had to bring the jury to believe that he followed the best course of action under the circumstances. In other words, he needed to make them feel as he had, to the point that they could imagine themselves joining him in arms, as Wilfrid Laurier said he would have in his stirring speech on the Champ-de-Mars the day after Riel's death.[26] Persuasion proceeds through identification.[27] Riel claimed that his was a divine mission – that God had spoken to him. This might not be considered evidence of madness in the nineteenth century, but it certainly would undermine identification by a Protestant jury. That is to say, although Riel sought to act like a member of the community against an illegitimate state, his religious visions placed him again in the *pagus*, outside of the political community's rational order. How could the jury attribute good judgment when he claimed to hear voices that they could not? How could they place themselves in his shoes when he marked himself as other in this way? Although there may not be madness, there is a *différend.*

Riel asked the jury to find him both sane and innocent, even though both prosecution and defence had treated these as antithetical. Rhetorical discourse is successful when it offers a "fitting response" to a given exigence. As Lloyd Bitzer has observed, multiple exigences undermine the possibility of crafting such responses.[28] Riel's paramount objective was to redeem his sanity, but he also sought to save his life by claiming that the rebellion was an act of self-defence. Riel argued that he and the Métis ambushed Middleton's men because they feared for their lives, but his account is awkward tropologically. To be sane, Riel must figure himself as responsible, yet to be innocent of treason, he must figure himself as powerless to have acted otherwise. Tropologically, being powerless is a form of irresponsibility, of not being responsible. Indeed, this is implicit even in Riel's grand sophistic pun: "British civilization which rules to-day the world, and the British constitution has defined such government as this is which rules the North-West Territories as irresponsible government, which plainly means that there is no responsibility, and by all the science which has been shown here yesterday you are compelled to admit if there is no responsibility, it is insane."[29] Riel claimed to be sane, and thus responsible, even as he asserted lacking control. In keeping with his situation and the rhetoric of petition, Riel emphasized his powerlessness. Before the jury, he likened himself to an infant, figured the North-West as his mother, and expressed his faith in the North-West's mercy and kindness, asking only for British fair play.[30] Continuing, he amplified his powerlessness, figuring himself as a noble victim. In his account, the petitions for redress that he and the Métis sent to Ottawa were answered by police. Consequently, and despite his claim to great influence, Riel presented himself as having little agency: he was ineffective. Although he had noble intentions, he could not master the flow of events. Furthermore, in order to secure this claim of irresponsibility, he would have to counter the prosecution's narrative, which was supported by witnesses and material evidence. Riel's attempt to do so did not result in a clear and compelling story. He alluded to some "foreign policy,"[31] to his planned "partition of lands into sevenths,"[32] and to the mystical beliefs of his people.[33] Continuing, he fretted about the status of his private papers.[34] Through all of this, his structure of motives became opaque. As a result, despite his virtue and good will, he *lacked prudence.* His *ethos* was deficient. Why would a reasonable person approve his course of action? Although they might have pitied him, the jury could not stand beside him. Imprudent, in over his head, and caught up in events, Riel could earn sympathy without gaining acquittal. He could be rational but lack practical reason. Poor judgment does not excuse treason.

Riel could have proceeded differently. He could have been more pagan: Lyotard counsels today's marginal figures to eschew grand principles and aim for minor victories through sophistry and opportunistic rhetoric.[35] The strength of the weak, for Lyotard, consists not in playing power's game but in exploiting its inconsistencies and *aporia*.[36] Riel could have followed Protagoras, fled the jurisdiction, and evaded prosecution. He could have collaborated with his counsel and figured himself as insane in order to live another day and plot an escape from his asylum. He could also have boldly admitted treason as the only politically expedient course of action, echoed Fitzpatrick's allusion to the Glorious Revolution, and dramatically asserted that treason against an irresponsible and illegitimate government was justified in the name of reason, liberty, and history. He would then, of course, be a different Riel: neither a victim nor a religious visionary but a bold revolutionary. Even if still convicted, he would have entered history with a much different *ethos.* With a clearer focus, not mired in details, his speech would have gained in eloquence. Had such a speech succeeded, it would have constituted a new law and thus displaced the *différend.*

Riel's speech did not secure his freedom but only brought to presence his inability to address the law. The jury foreman cried while delivering the guilty verdict,[37] bearing witness to the *différend* even as he recommended Riel to the mercy of the Crown. Riel's call for justice would have to wait to be properly heard. Riel entered history as one that others bore witness to, felt compelled to speak for, or invoked in the name of their own causes.

NOTES

1 Jean-François Lyotard and Jean-Loup Thébaud, *Just Gaming* (Minneapolis: University of Minnesota Press, 1985), 87–8.

2 Desmond Morton, ed., *The Queen v Louis Riel: Canada's Greatest State Trial* (hereafter *Q v LR*) (Toronto: University of Toronto Press, 1974), 6.

3 Jean-François Lyotard, *The Differend: Phrases in Dispute* (Minneapolis: University of Minnesota Press, 1988), 208.

4 Lyotard and Thébaud, *Just Gaming*, 22–3.

5 Ultimately, in Lyotard's account, a *différend* lurks in every case. Thus justice never appears fully. Even the best judgments are only approximations of what justice could be, and that is always receding.

6 T.S. Kuhn, *The Structure of Scientific Revolutions* (Chicago, IL: University of Chicago Press, 1962), 172.
7 Lyotard and Thébaud, *Just Gaming*, 26.
8 Lyotard, *Differend*, 9.
9 Lyotard and Thébaud, *Just Gaming*, 138.
10 *Dred Scott v. Sanford*, 60 US 393, 1857.
11 Jean-François Lyotard, "Lessons in Paganism," in *The Lyotard Reader*, ed. A. Benjamin, 122–54 (Cambridge, UK: Basil Blackwell, 1989), 137.
12 "Declaration of the People of Rupert's Land and the North West," Archives of Manitoba, MG 3, A1–6, http://manitobia.ca/cocoon/launch/en/records/RRS/RRS_1869_1207 (accessed 5 October 2013).
13 Jean-François Lyotard, "On the Strength of the Weak," in *Toward the Postmodern*, ed. M.S. Roberts and R. Harvey, 62–72 (Atlantic Highlands, NJ: Humanities Press International, 1993), 71–2.
14 Desmond Morton, "Introduction," in *Q V LR*, vii–xxxv, at ix.
15 *Q V LR*, 288.
16 Lloyd Bitzer, "The Rhetorical Situation," *Philosophy and Rhetoric* 1, no. 1 (1968): 1–14.
17 *Q V LR*, 71–2.
18 Ibid., 326–7.
19 Michael Dorland and Maurice Charland, *Law, Rhetoric and Irony in the Formation of Canadian Civil Culture* (Toronto: University of Toronto Press, 2002), 295–301.
20 *Q V LR*, 311–12.
21 Ibid., 312.
22 Ibid., 311.
23 Ibid., 322.
24 Ibid., 319.
25 Ibid., 316.
26 "Denouncing Riel's Execution," *New York Times*, 23 November 1885, 1.
27 Kenneth Burke, *A Rhetoric of Motives* (Berkeley: University of California Press, 1969), 340.
28 Bitzer, "Rhetorical Situation," 12.
29 *Q V LR*, 323.
30 Ibid., 312.
31 Ibid., 313.
32 Ibid., 321.
33 Ibid., 320.

34 Ibid., xx.
35 Lyotard, "Lessons in Paganism," 152.
36 Lyotard, "Strength of the Weak."
37 *QVLR*, 321, 71.

11

The Case of Two Trials, Louis Riel and Socrates: Intersections, Overlaps, Divergences

LOUIS GROARKE

INTRODUCTION

The purpose of this chapter is not to revisit the details of an episode in Canadian history that specialists have, by this time, picked clean but, through a study of Riel's trial speeches, to come to some informed appreciation of his political and philosophical ideas and his proper place in this nation's history. Expressed in more technical terms, this chapter is partly an exercise in argumentation theory and partly an exercise in applied ethics. Combining a careful analysis of his reasoning strategy with a scrupulously fair account of his ethical, religious, and political ideals, I hope to move beyond the peculiarly Canadian debates that swirl around Riel's legacy and consider what he represents in light of the larger history of ideas. Although I will focus on the strategy Riel uses to defend himself at his trial, I will also try to arrive at a more balanced appreciation of the successes and failures of Riel's personal life and political career. To anyone versed in philosophy, an inevitable comparison with Socrates recommends itself. Both men are depicted as martyrs for a cause. Both were put on trial for what amounts to treason, both gave long speeches in their own defence, both were spectacularly unsuccessful when it came to convincing jurors of their innocence, both were executed by the state, both enjoyed a reputation for moral heroism after their deaths, and so on. At the same time, the two men are conspicuously different in important ways.

Is Louis Riel a Canadian Socrates? Almost? Not quite? Not by any stretch of the imagination? Although the resemblances between the two narratives are striking, there are key differences. Comparing the two men – their public careers, their personal lives, and, in particular, the legal and rhetorical strategies they adopt in their own defence at their respective trials – provides a unique perspective on the life and times of Louis Riel. This leads to larger questions about the nature of legal argument, about court procedures, about religious appeals to higher powers, about the defining traits of sanity and insanity, about the connect (or disconnect) between morality and political justice, and even about the basic nature of historical investigation. And it leads invariably to a fairer evaluation of Louis Riel and his place in history.

Comparing a modern individual to a great historical figure over an immense gap of time and across a wide cultural and religious gap is no longer familiar methodology in academic research. Some would argue that the inevitable heterogeneity of such historical circumstances presents an insurmountable barrier to this kind of approach. But as I intend to demonstrate, this is a useful way into attendant issues. As we shall see, there are reasonably objective standards that can be used to evaluate figures across timespans and cultures. By comparing these two men, we can come to a better understanding of the Riel episode in Canadian history, one that escapes invidious disputes mired in important but parochial debates about the role of this or that ethnicity, or language group, or religion in Canadian history.

RIEL AND SOCRATES

Textual Sources

Before embarking on a comparison of the trials, ideas, and careers of Riel and Socrates, it is important to note some conspicuous differences in the available textual sources. Our knowledge of Socrates mostly comes to us through Plato, his devoted student, who did his best to portray Socrates in a complimentary light.[1] (Authors such as Xenophon and Diogenes Laertius take similarly laudatory tones when discussing Socrates, as did ancient philosophers generally.) We must assume, then, that our picture of Socrates may be inaccurate or exaggerated, at least when it comes to details. Even an early, journalistic work like the *Apology*, Plato's account of Socrates's trial, probably presents us with a cleaned-up Socrates, a figure who has been adeptly buffed and polished by Plato's prose and turned into the hero of philosophy. Socrates gives three speeches to the audience: his defence, his recommendation for sentencing, and a sort of afterword to his friends after he has been condemned to death.[2]

The situation with Riel is quite different. To begin with, there is a wealth of journalistic material about his life and trial, from diverse sources, some of it neutral, some of it sympathetic, and some of it decidedly hostile to Riel and his cause. Then there are Riel's private writings: his poetry, his diaries, his devotional reflections, and his letters.[3] These provide an immediate conduit into his inner life, but they include the kinds of passages a vulnerable, excitable man commits to paper in the dark hours of the night, which look absurd or weak or pompous in the cold, stern light of day. This is not to say that Riel would have been embarrassed by someone else reading what he wrote. He seems to have been a compulsive writer, an Augustine-like figure, who felt an almost exhibitionist need to chronicle his inner state of mind and record his spiritual sojourn for himself and for posterity. But we must keep in mind that a poem or a diary entry is not a political speech or an exercise in logic. Finally, in the case of Riel's trial, we have a verbatim transcript of two rambling statements: one that Riel makes in his own defence and one that he makes to the courtroom after being found guilty. In both cases, we have a nervous man, a francophone expressing himself in English, who is speaking with some premeditation but also extemporaneously. He did not read from notes, and there is a tone of sincerity and directness that marks spontaneous discourse. It is generally accepted that his second speech, which is more political, is the better of the two, but it comes after he has already been found guilty. I will focus, for the most part, on the first speech, which contains his defence before the jury. (There is some suggestion that Riel had originally intended to make use of the second speech in his defence but changed his mind.)[4]

When comparing Riel with Socrates, then, we cannot overlook these differences in the textual sources. We do not have Riel as he might have looked, tidied up and artfully embellished by a master stylist for posterity. We have the real flesh and blood man. We should not be surprised that Riel comes across as an intensely human and fallible character in contrast to the aloof, imperturbable, and even godlike Socrates.

Utter, Unshakeable Conviction

Although the circumstances differ in ways I will not explore here, Riel and Socrates were both tried and executed for what amounted to treason and insurrection. In both cases, their detractors viewed them as troublemakers, as dangerous men, as schemers and demagogues who threatened the public peace and questioned rightful authority. In both cases, their followers were mesmerized by these charismatic men and interpreted their behaviour as moral heroism, as loyalty to a higher ideal. What these two contrasting figures share most, however, is the sense of epistemological certainty about their mission and their role in history.

Both men were wholly convinced that their cause was just; both were utterly unimpressed by the arguments of their opponents, and both displayed an unwavering loyalty to what they saw as the right course of action. In the *Apology*, Plato depicts a slightly irascible Socrates, uncowed by hostile jurors. Riel, to be sure, may have had doubts about individual decisions he had made, but he was just as certain that he would be posthumously exonerated and, indeed, remembered and honoured for his moral righteousness. It is this sense of utter conviction, based partly on religious and partly on moral motives, that marks these men out as different from most of their fellows.

In a courtroom, a good defence presupposes a kind of cleverness that requires more than bare moral conviction. Neither Riel nor Socrates was well suited to argue his case before a hostile court and win concessions. For all Socrates's professions of ignorance, his speeches in the *Apology* are unbending in their sense of righteous self-affirmation; they are not intended to placate or reconcile an unsympathetic audience (who had just lived through one of the most traumatic political episodes in Athens's troubled history due, in part, to one of Socrates's associates).[5] Socrates ignores the formal decorum of the court, insults many of his listeners, does not admit any wrongdoing, does not ask for mercy, and approaches the sentencing phase of the trial flippantly. Although he displays some force of logical argument – he is a philosopher, after all – he ignores eloquent persuasion in favour of something more akin to the blunt, plain-spoken truth. He begins his speech by saying that he will speak the way he normally expresses himself in the marketplace:

> You shall hear from me the whole truth: not, however, delivered after their manner, in a set oration duly ornamented with words and phrases. No indeed! but I shall use the words and arguments which occur to me at the moment ... And I must beg of you to grant me one favour, which is this – If you hear me using the same words in my defence which I have been in the habit of using, and which most of you may have heard in the agora [the marketplace], and at the tables of the money-changers, or anywhere else, I would ask you not to be surprised at this, and not to interrupt me.[6]

Riel is more reconciliatory than Socrates, but his speech before the jurors is equally ineffective. Partly rambling and disjointed but also eloquent at times, it effectively destroys the insanity plea his own lawyers have constructed in his defence. (It would have been better for him not to speak at all, as his lawyers would have preferred.) For whatever reasons, Riel calmly downplays his history of mental illness in a way that completely ignores the feverish pitch of his megalomania and the wild scope of his private revelations. He is mostly concerned about clearing his name and discussing his political, moral, and religious motives and concerns. All in all, he loses sight of the narrow purpose and the limits of court proceedings. He remarks, "I will not be reputed by all men as insane, as a lunatic ... You are perfectly justified in declaring that having my reason and sound mind, I have acted reasonably."[7]

As George Stanley reports and as scholarly consensus has it, Riel's statement to the court was, in a legal sense anyway, a fatal miscalculation: "Riel had done more [by his speech] than any Crown witness to destroy the whole edifice of insanity that Fitzpatrick [his lawyer] had so carefully erected."[8] Riel's own words, in effect, sealed his fate. The jury could not, in good conscience, accept this talk about insanity on the basis of what they saw played out in front of them. As juror Edwin Brooks explained, "We were in a dilemma ... We had to give our finding on Riel according to the evidence. We refused to find him insane. The only thing we could do was to add the clause to our verdict, recommending mercy."[9]

It seems strange that an accomplished, persuasive man with a charismatic personality should find himself floundering in a courtroom setting. The case of Riel is complicated, as he was labouring under the mighty influence of his religious delusions. Still, one has the sense that it was not his religious delusions but ordinary moral righteousness and steadfast character that lay behind his determination to act as he did. Riel is more fearful about death, but one observes in both men the same attitude of moral courage. Or is it moral obstinacy? This brings up a larger issue. What is the point of legal argument? Surely, to win the case. Is it possible, then, that an excess of moral conviction – however justified, however sincere – is a legal disadvantage?

These men told the truth in court as they saw it without adapting themselves to the rhetorical context. They ignored the specific requirements of the legal arena and used the courtroom as a platform to launch into a general defence of their reputations. Both of them were responding to what they saw as a negative stereotype fabricated over the years. When Socrates undertakes to address "the accusation ... which has given rise to this slander of me," he claims that he is mostly worried about older, unnamed accusers who have been spreading false rumours about him for many years.[10] Their unfounded charges have taken a powerful grip on the collective imagination and have the ring of truth, which is difficult to counter. Socrates complains, "I have had many accusers, who accused me of old, and their false charges have continued during many years."[11] Riel likewise asserts, "I have been libelled for fifteen years."[12] But defending oneself against a generalized bad reputation is a difficult (and, perhaps, impossible) feat. And it distracts from the main issue: meeting specific legal criteria for acquittal.

Both Socrates and Riel try to articulate what they believe to be true in the face of hostile prejudice and, arguably, lose sight of what a courtroom "competition" is about. One can depict their uncompromising posture as purity of purpose or as culpable negligence. (All things being equal, one is morally obliged, after all, to take reasonable measures to preserve one's life.) But neither man adequately adapts himself to the rules of the legal game. Perhaps these were hopeless cases; perhaps there was nothing more that could have been done. But focusing on the general task of salvaging one's good reputation in the face of malicious rumour and settled habit is a risky legal strategy. For whatever reason, both Socrates and Riel were unable to do much to effectively counter the ill will of their opponents, nor were they able to escape, respectively, the cup of hemlock or the noose.

A Calling from on High

One might think that the main difference between Riel and Socrates arises from Riel's fervent religious convictions. The average modern reader might assume that because Riel was a deeply religious individual, he must be different from Socrates, who was a first-rate philosopher and, therefore, a master of logical argument. We tend, in the present epoch, to posit implicit dichotomies between logic and faith, science and religion, empirical knowledge and theological speculation. But things are not so simple. In fact, both men make an appeal to supernatural phenomena and point to religious motivations as the source of their "divine" missions.

The religious motivations of Socrates and Riel are not as different as we might first suppose. Socrates is a more subtle and rigorous thinker, he displays a healthy moderation in line with the Greek virtue of *sophrosune*, he strives to be humble, and he is capable of a healthy skepticism and an ironic detachment. Nonetheless, a central feature of Socrates's story is the philosopher's adamant insistence that he was personally chosen by God to serve the Athenian people in a philosophical way, a vocation that was, in turn, confirmed by invisible signs and dreams and prophecies.

We tend to overlook this aspect of the Socratic narrative because of the positivist tendencies of contemporary philosophy. Socrates was obviously a religious man. We cannot address this issue in depth here, but I will consider briefly the death scene in the *Phaedo*, where there are a series of references to religion and the supernatural. To begin with, the central theme of the reported discussion is the immortality of the soul, hardly a secular topic. Socrates claims, at one point, that we human beings do not belong to ourselves. We should be considered "property" that belongs to the gods: "the gods are our guardians."[13] (This is why suicide is wrong. We should leave it to the gods to decide when we should die.) He also discusses the nonphysical nature of the soul, reincarnation, life after death, and a myth involving a kind of last judgment.[14] But Socrates does not only pay serious attention to religious subject matter; his actions also follow suit. The dialogue begins with Socrates explaining that he has been writing poems as well as a hymn to Apollo because he has had a recurrent dream that he must not disobey.[15] He reminds Crito to sacrifice a rooster to Aesculapius, the demigod of medicine, after he has passed away.[16] And before he drinks the hemlock, he asks whether he can make a libation to a god, and when that is impossible, he offers a prayer instead.[17]

At his trial, religious themes also make an appearance. In the *Apology*, Socrates upbraids members of the audience who claim he is an atheist: "I do believe that there are gods, and in a far higher sense than that in which any of my accusers believe in them."[18] But even more important, in the context of the present comparison with Riel, is his firm insistence that he has a divine mission he must carry out, whatever a human court decrees. This mission, he insists, has been communicated to him in paranormal ways.

First, Socrates tells the audience in the courtroom that he acts in accordance with an invisible, supernatural sign: "You have often heard me speak of an oracle or sign which comes to me. ... This sign I have had ever since I was a child. [It] is a voice which comes to me and always forbids me to do something which I am [wrongly] going to do, but never commands me to do anything."[19] Socrates is claiming, in plain words, that he has an immediate connection to the supernatural world through a divine sign, that this sign personally speaks to him, and that this private sign is a reliable guide for his own conduct! This is not so different from the beliefs Riel subscribes to.

Indeed, Riel describes, with disarming artlessness, an incident of spiritual communication at his trial. "Last night," he says,

> While I was taking exercise the spirit who guides and assists me and consoles me, told me that "to-morrow somebody will come *t'aider*," five English and one French word, *t'aider*, that is, "to help you." I am consoled by that. While I was recurring to my God, to our God ... these words came to me ... in the morning someone will come *t'aider*, that is today. I said that to my two guards and you can go for the two guards. I told them that if the spirit that directs me is the spirit of truth it is today that I expect help. This morning the good doctor ... said you will speak to-day before the court. I thought I would not be allowed to speak; those words were given to me to tell me that I would have liberty to speak. There was one French word in it, it meant I believe that there was to be some French influence in it, but the most part English.[20]

Riel hears a voice speak to him, a voice that reassures him that something positive is about to happen. It is perhaps a more detailed message than Socrates claims to hear, but the same mechanism of paranormal communication applies in both cases.

Furthermore, Socrates, in a manner not unlike Riel, is consoled by his familiar sign at the end of his trial when he has been sentenced to death. He tells his supporters,

> You are my friends, and I should like to show you the meaning of this event which has happened to me. O my judges ... I should like to tell you of a wonderful circumstance. Hitherto the familiar oracle within me has constantly been in the habit of opposing me even about trifles, if I was going to make a slip or error about anything; and now as you see there has come upon me that which ... is generally believed to be, the last and worst evil. But the oracle made no sign of opposition, either as I was leaving my house and going out in the morning, or when I was going up into this court, or while I was speaking, at anything which I was going to say; and yet I have often been stopped in the middle of a speech; but now in nothing I either said or did touching this matter has the oracle opposed me. What do I take to be the explanation of this? I will tell you. I regard this as a proof that what has happened to me is a good, and that those of us who think that death is an evil are in error. This is a great proof ... for the customary sign would surely have opposed me had I been going to evil and not to good.[21]

(Remarkably, Riel expresses similar sentiments in his statement to the court just after he has been found guilty: "I think it is for good and not for my loss ... I consider it as a great advantage.")[22]

But there is still more. Socrates defends his allegedly treasonous actions by appealing to – of all things! – an enigmatic and prophetic pronouncement by the Oracle at Delphi. Plato depicts the unruly scene in the courtroom as Socrates proceeds:

> O men of Athens, I must beg you not to interrupt me, even if I seem to say something extravagant ... I will refer you to a witness who is worthy of credit ... and that witness shall be the god of Delphi. You must have known Chaerephon ... Well, Chaerephon ... was very impetuous in all his doings, and he went to Delphi and boldly asked the oracle to tell him whether – as I was saying, I must beg you not to interrupt – he asked the oracle to tell him whether

> there was anyone wiser than I was, and the Pythian prophetess answered that there was no man wiser. Chaerephon is dead himself, but his brother ... will confirm the truth of this story.[23]

What is going on here? Socrates claims that he has received a concrete sign of his philosophical vocation from the god's representative: the priestess. This is very similar to Riel's repeated insistence that he received confirmation of his holy mission at the hands of God's representatives on earth, namely various priests and bishops. Riel tells the courtroom,

> It is true, gentlemen, I believed for years I had a mission ... I have not assumed to myself that I had a mission ... the late Archbishop Bourget said: "Riel ... is a man to accomplish great things" ... And in another letter ... Bishop Bourget wrote again and said: "Be ye blessed by God and man" ... I say that I have been blessed by God, and I hope that you will not take that as a presumptuous assertion ... I am blessed by God ... Not only Bishop Bourget spoke to me in that way, but Father Jean Baptiste [Primeau], the priest of Worcester, who was my director of conscience, said to me: "Riel, God has put an object into your hands, the cause of the triumph of religion in the world" ... I have got those words in my heart, those words of J.B. [Primeau] and the late Archbishop Bourget. But last year, while I was yet in Montana ... the Reverend Father Frederick Ebeville, curate of the church of the Immaculate Conception, at Benton, said to me: "Go and bring [your family] to the altar, I want to bless you" [And] the priest put on his surplice and he took holy water and [blessed] us ... When the glorious General Middleton fired on us during three days ... I said that nothing but the blessing without measure of Father Frederick Ebeville could save me, and that can save me to-day from these charges. The benediction promised to me surrounded me all the time in the Saskatchewan.[24]

One may complain, doubtless, that Riel is reading an extravagant message into commonplace pieties, a matter considered below. But whatever one thinks about Riel's interpretations of these sacred events, his understanding of the benevolent spiritual economy that controls the world is akin to Socrates's interpretation: there is a God who gives commands to individuals through wondrous signs on earth. Socrates is as adamant as Riel. He tells his listeners, "This duty of cross-examining other men [i.e., of practicing philosophy,] has been imposed upon me by God; and has been signified to me by oracles, visions, and in every way in which the will of divine power was ever intimated to any one."[25]

If both Riel and Socrates claim an intimate connection with the divine, there is a heavy moral responsibility that naturally accompanies this kind of epistemological claim. An individual must dutifully carry out any divine commission bestowed upon him by God through certain signs, for God's authority is higher than human authority. Socrates, like Riel, is unyielding in this particular. He tells the court that he must carry out the god's commands even if a (merely) human court goes so far as to forbid such action. He tells the Athenians in no uncertain terms,

> Strange, indeed, would be my conduct, O men of Athens, if I who, when I was ordered by the generals ... at Potidaea and Amphipolis and Delium, remained where they placed me ... facing death; if ... when, as I conceive and imagine, God orders me to fulfil the philosopher's mission ... I were to desert my post through fear of death, or any other fear; that would indeed be strange, and I might justly be arraigned in court for denying the existence of the gods, if I disobeyed the oracle because I was afraid of death: ... [And therefore] if you say to me, Socrates, this time we ... will let you off, but upon one condition, that you are not to inquire and speculate in this way any more, and that if you are caught doing this again you shall die; – if this was the condition on which you let me go, I should reply: Men of Athens, I honor and love you; but I shall obey God rather than you, and ... I shall never cease from the practice and teaching of philosophy ... For this is the command of God ... and I believe that to this day no greater good has ever happened in the state than my service to the God ... Wherefore, O men of Athens, I say to you ... either acquit me or not; but whatever you do, know that I shall never alter my ways, not even if I have to die many times.[26]

It is this unyielding sense of a divine imperative that is the source of the moral "intransigence" of both Socrates and Riel. One may complain that Riel was insane or that he always teetered precariously on the edge of mental illness. But choosing between Socrates and Riel, on this point at least, is not an easy matter. It makes no sense to point to Socrates as a hero for the ages (a common enough trope) while denigrating Riel's behaviour. Both were honourable men; both were following a privately understood revelation. If such revelations have epistemological authority, as Socrates believes they do, one cannot so easily dispose of Riel's scruples and his dogged persistence at following his conscience.

Riel's religious delusions are discussed in more detail below. But one cannot arrive at any adequate sense of who Riel was and what motivated him without some insight into the role of religion in his life. Thomas Flanagan has discussed Louis Riel's self-description as a "prophet of the new world."[27] As Flanagan explains, Riel adopted the middle name "David" because, in his religious enthusiasm, he saw himself as a modern King David, divinely appointed to lead the new Hebrew people, the Métis, in the establishment of a new millenarian religion that would fuse Judaism, Protestantism, and Catholicism.[28] Flanagan writes, "Riel's identity as a new David meant that he aspired to gather all power, both temporal and spiritual, into his hands, and wield it under divine inspiration vouchsafed to prophets."[29]

In one diary entry at the end of his life, an overwrought Riel writes, "The spouse of Christ, the Church, my mother, will shower me with her own blessings. Her solicitude will honour me because I, little David in the service of the great King, had the courage to go outside the camp of Israel for a moment, to try and hold off the giant that was marching against all of us with his redoubtable strength and reputation."[30] Riel saw his own role in the North-West Rebellion as that of the divinely appointed Métis king fending off the Canadian Goliath in a political battle for natural rights and justice. This grandiose but sincere description reveals his perception of himself. James Willard Schultz, a trader who had business dealings with Riel, recalled that the Métis leader had confided to him (sometime between 1880 and 1882), "Do you know, these people of mine are just as were children of Israel, a persecuted race deprived of their heritage. But I will wrest justice for them from the tyrant. I will be unto them a second David."[31]

In Riel's mind, the secular aims of the North-West campaign were always second to religious aims he often kept to himself. We must not underestimate the scope of Riel's religious ambitions. In 1876, at the asylum of St Michel-Archange at Beauport, where he had been committed for insanity, Riel went so far as to sign an angry note to the chaplain, "Louis 'David' Riel. By the grace of Jesus Christ. Prophet, Infallible Pontiff, and Priest-King."[32] Not a modest title! Rather, in a single, histrionic phrase, it weds the idea of political and religious authority to the idea of epistemic certainty derived from direct revelation.

Of course, Riel was in the full throes of madness when he wrote those words, but, although his religious neurosis ebbed and flowed, it never properly relinquished its hold on his mind (as discussed below). Throughout his later life, Riel firmly believed he was in close communication with the supernatural realm and that he had been personally given a divine mission to lead the Métis people to salvation, a vocation that was confirmed, innumerable times, by private revelations, mystical ecstasies, inner volitions, thoughts, and dreams. But, as Socrates's example illustrates, one cannot ask a man, even a mentally ill man, who believes that he is commanded by God to act otherwise. A just society can find ways of limiting such behaviour. It can fairly restrain such a person, commit him to an asylum, force-feed him medicine, visit him with therapy, or hand him over to sympathetic religious authorities to deal with him. But, as I suggest below, it makes a mockery of the law and morality to maintain that we should hang him from a gibbet. Although there are ways of distinguishing between the religious attitudes of Socrates and Riel, the latter's behaviour is not, in the final analysis, so different from that of Socrates. It depends on a similar account of religious or spiritual knowledge.

Practice versus Theory, Outsider versus Insider: A Divide

Riel was, in the traditional sense of the term, a man of action; Socrates, conversely, aspired to the contemplative life. In this sense at least, the careers of the two men diverged. Socrates dedicated his life to the ethical improvement of individual citizens through logical argument, whereas Riel dedicated his life to social justice and political reform. It is also the case that Socrates always saw himself as an insider, a loyal citizen of Athens, whereas Riel inevitably saw himself as an outsider. He was, after all, someone who had already fought with the federal government, someone who had been exiled, an American citizen in Canada, a half-breed in white society, a francophone outside Quebec, and a Catholic in largely Protestant English Canada.

Turn back to Socrates. In the modern world, we tend to associate philosophy with the production of books, papers, and written arguments. Socrates was not a writer. He had, as far as we can tell, a great mind; he possessed a wonderful clarity and rigorous sense of argument. He aspired to a life of meaningful conversation, believing that virtue required knowledge and that rigorous dialogue and debate were the best means to virtuous knowledge. Although he fulfilled his duties to the state through military service and a one-time stint as a senator, he studiously avoided any prolonged involvement in politics. At the same time, Socrates believed that his practice of philosophy improved listeners and, indirectly, Athens, for a city-state composed of knowledgeable, virtuous citizens is better than one composed of ignorant, vicious people. He set about improving Athens one individual at a time. (This is, ironically, the opposite of what was said in the charge brought against him: that he was *corrupting* the youth of Athens.)

If, however, Socrates saw himself as an "apolitical" Athenian, he lived and died (in his own mind at least) a loyal citizen. He is an Athenian through and through. In the *Crito*, Socrates imagines the personified "Laws of Athens" saying to him:

> Of all Athenians you [Socrates] have been the most constant resident in the city, which, as you never leave, you may be supposed to love. For you never went out of the city either to see the games, except once ... or to any other place unless when you were on military service; nor did you travel as other men do. Nor had you any curiosity to know other states or their laws: your affections did not go beyond us and our state; we were your especial favourites, and you acquiesced in our government of you; and here in this city you begat your children, which is a proof of your satisfaction. Moreover ... in the course of the trial, ... you [claimed] that you preferred death to exile.[33]

Riel had some talent and a penchant for writing, but he lacked the rigorous mind and the retiring temperament needed for a scholarly life and embraced instead the life of political involvement and social progress. He aimed not at theoretical understanding but at social transformation understood in terms of some change in the basic structures that mediate interactions between groups in society. In his trial statement, Riel reminds the jurors that he has practical goals: "I think I could safely tell you that what Captain Young said that I am aiming all the time at practical results was true, and I could have proved it. During my life I have aimed at practical results. I have writings, and after my death I hope that my spirit will bring practical results."[34] Again, speaking about reconciling Catholics and Protestants, he insists that his work will bear *practical* fruit in the long term: "If I have any influence in the new world ... even if it takes 200 years to become practical, then after my death that will bring out practical results, and then my children's children will shake hands with the Protestants of the new world in a friendly manner."[35] He reiterates the same point three more times in his speech: (1) "What is the kind of mission that I have? Practical results,"[36] (2) "I have to attend to practical results,"[37] and (3) "I am aiming all the time at practical results."[38]

One of the reasons that Riel chooses politics over philosophy is, of course, natural aptitude. He was a charismatic man and a born organizer. He had talents as an orator, a writer, and a leader, but he was not, by temperament, a philosopher. Riel is also pushed into his active role by a sense of grievous injustice he feels called upon to address. He sees himself as an outsider in Canada. He had deliberately exiled himself from his home – something that Socrates could not countenance – and had become an American citizen. He takes the position of someone who represents a group that has been unfairly excluded from Canadian society – hence the need for the threat of secession, by brute force if necessary, in order to achieve proper representation within the Canadian system. As Riel tells the court, somewhat flamboyantly,

> The only things I would like to call your attention to before you retire to deliberate are: 1st That the House of Commons, Senate and Ministers of the Dominion, and [those] who make laws for this land and govern it, are no representation whatever of the people of the North-West. 2nd That the North-West Council generated by the Federal Government has the great defect of its parent. 3rd The number of members elected for the Council by the people make it only a sham representative legislature and no representative government at all.[39]

There is a divide, then, that separates Riel, the discontented outsider, from Socrates, the contented insider. Riel is the outsider who wants to forcibly enlarge white, Canadian, British society to include his own kind; Socrates is an insider who tries to change individuals through philosophical discussion while accepting the political and social structure of Athenian democracy.[40] Still, there is a curious parallel here. In explaining why he has refused a political career, Socrates utters a warning about politics. As it turns out, Socrates avoids politics because of his inner voice, just as Riel engages in politics because of his inner voice. In the *Apology*, Socrates tells the crowd,

> Someone may wonder why I go about in private, giving advice and busying myself with the concerns of others, but do not venture to come forward in public and advise the state. I will tell you the reason of this. You have often heard me speak of an oracle or sign which comes to me ... This is what stands in the way of my being a politician. And rightly, as I think. For I am certain ... that if I had engaged in politics, I should have perished long ago and done no good either to you or to myself. And don't be offended at my telling you the truth: for the truth is that no man who goes to war with you or any other multitude, honestly struggling against the commission of unrighteousness and wrong in the state, will save his life; he who will really fight for the right, if he would live even for a little while, must have a private station and not a public one.[41]

Riel and his followers might have seen these as prophetic words.

Virtuous Poverty

Both Socrates and Riel point to their own poverty as evidence of their purity of purpose. In both cases, we meet with the same trope: the indigent labourer who has been busy toiling away at some moral function, neglecting his own interests for the betterment of others. Socrates tells the Athenians,

> And that I am given to you by God is proved by this: – that if I had been like other men, I should not have neglected all my own concerns, or patiently seen the neglect of them during all these

> years ... coming to you individually, like a father or elder brother, exhorting you to regard virtue; this I say, would not be like human nature. And had I gained anything, or if my exhortations had been paid, there would have been some sense in that: but now, as you will perceive, not even the impudence of my accusers dares to say that I have ever exacted or sought pay of anyone; they have no witness of that. And I have a witness of the truth of what I say; my poverty is a sufficient witness.[42]

Elsewhere, Socrates repeats himself: "I am in utter poverty by reason of my devotion to the god."[43]

Riel, for his part, was responding to allegations that he was conniving to extort money from the government. (An overblown accusation, it seems; no doubt, he wanted money for his family, which he saw as his just due, but there were much larger religious and political reasons for the insurrection.) Riel tells the courtroom,

> For fifteen years I have been neglecting myself. Even one of the most hard witnesses on me said that with all my vanity, I never was particular to my clothing; yes, because I never had much to buy any clothing. The Rev. Father André has often had the kindness to feed my family with a sack of flour, and [so has] Father Fourmand. My wife and children are without means, while I am working more than any representative in the North-West. Although ... I worked to better the condition of the people of the Saskatchewan at the risk of my life, to better the condition of the people of the North-West, I have never had any pay. It has always been my hope to have a fair living one day.[44]

In both cases, the accused points to his impecunious state as poof of noble purpose and as sure evidence of his selfless devotion to the divine cause.

A Defence of Necessity: A Métis Robespierre?

We must turn now to an examination of the specific argument Riel uses in his own defence at his trial. Although Riel, like Socrates, brushes off the charges against him, he has a considerably more challenging task when it comes to convincing a jury of his innocence.

As we have seen, Socrates was not a rebel; whatever his students may have done, he saw himself as an insider, as someone who believed in the Constitution and the laws of Athens. At the very beginning of his speech in the *Apology*, Socrates declares, "in obedience to the law I make my defence."[45] In the *Crito*, Plato depicts his teacher as someone who accepts cheerfully his execution because he does not want to destroy the laws of Athens.[46] Riel, in plain contrast, set out, by his own admission, through a very public act of insurrection, to *destroy* the laws of Canada. That is, after all, what rebellion is about. Socrates was an insider negotiating with other insiders, but Riel finds himself in the unenviable position of an outsider, trying to justify his rebellious acts to insiders. He is unable to convince them, in part, because of his earlier role in the Red River Uprising of 1869–70 and the execution of Thomas Scott, an act for which Riel had already been granted amnesty (and five years banishment) but a score that his opponents still aimed to settle. One of the jurors at his trial reflected, fifty years later: "We tried Riel for treason, and he was hanged for the murder of Thomas Scott."[47]

The contrast between Riel and Socrates, in this connection, could not be greater. It is not only that Socrates goes out of his way *not* to rebel against established authority, at least not in any overtly political way, but also that he is able to point to a concrete instance where, unlike Riel, he *refuses* to play any role in a political execution. Socrates tells his audience,

> When the oligarchy of the Thirty was in power, they sent for me and four others into the rotunda, and bade us bring Leon the Salaminian from Salamis, as they wanted to execute him. This was a specimen of the sort of commands which they were always giving with the view of implicating as many as possible in their crimes; ... My only fear was the fear of doing an unrighteous or unholy thing. For the strong arm of that oppressive power did not frighten me into doing wrong; and when we came out of the rotunda the other four went to Salamis and fetched Leon, but I went quietly home. For which I might have lost my life, had not the power of the Thirty shortly afterwards come to an end.[48]

Whereas Socrates refused to cooperate with the execution of Leon of Salamis, Riel oversaw the execution of Thomas Scott. In the minds of his opponents, he had assumed the role of the Thirty Tyrants, *not* the role of Socrates. (Note that Riel was fully aware of the ill will caused by the execution. In his second statement to the court, he asks for a general investigation into the question "Was Riel a murderer of Thomas Scott, when Thomas Scott was executed?")[49]

With time (and with a bit of luck, perhaps!) the acrimony that attends historical debates usually dies away to less vehement discourse. Contemporary readers may look askance at the passionate rhetoric that swirled around the Métis leader when he was alive, but it is hard to exaggerate the pitch of his opponents' antagonistic feelings. In their eyes, he was a diabolical figure. He was a *rebel*, a demagogue, a French Métis Robespierre, who had, on two occasions, tried to institute his own "reign of terror" and subvert the proper rule of law and established authority through arms and through his wild and treacherous scheming.

An anonymously published work entitled *The Story of Louis Riel: The Rebel Chief* exemplifies this utterly hostile attitude.[50] Written, in fact, by Canadian author Joseph Edmund Collins (1855–92), the book is a strange mixture of scurrilous fairy tale and popular history. Collins describes Riel in the following terms: "this apostle of insurrection and unrest,"[51] "the rebel and murderer,"[52] "that dangerous person,"[53] "your malignant persecutor,"[54] "this monster,"[55] "the Arch-Agitator,"[56] "the outwitted tyrant-libertine,"[57] "an ambitious, short-sighted demagogue, who palmed off his low cunning for brilliant leadership, upon the credulous half-breeds,"[58] "the vengeful Riel,"[59] "the miscreant,"[60] "the ruffian"[61], "the miscreant-fiend,"[62] "the greasy murderer,"[63] "this beastly, murderous tyrant,"[64] "the man whose hands were red with the blood of his fellow creature,"[65] "the worthless vagabond rascal,"[66] "the untried felon,"[67] "the criminal,"[68] "the plausible disturber,"[69] and "the murderous apostle of tumult."[70]

But the calumny does not end there. Collins's descriptions of Riel are little more than opportunistic attempts at bombastic and hopelessly tendentious caricature. Riel is "swarthy"[71] with "wolfish eyes,"[72] he has "[a] murderous eye"[73] and "[a] voice resembling that of some foul beast,"[74] he keeps a "greasy pocket book,"[75] he practises "wallowing and filthy habits,"[76] he possesses "[an] unstaid, unreasoning character,"[77] and he is "absolutely illogical, and unreasonable."[78] Collins informs the reader that when Riel was defeated at Fort Garry, he "scurried out," "with hysterical precipitation," "mounted a horse, and rode away in mortal terror."[79] He has a terrible temper: when a wicked stratagem fails, he begins "to storm and blaspheme, and would [have] remind[ed] anyone who saw him of some wild beast foiled of his prey."[80] He is capable of "diabolic vengeance."[81] He drinks to excess: liquor raises "the savage part of his blood to the highest pitch of unreasoning and confident passion."[82] He is, of course, a womanizer, too: "sometimes working, sometimes hunting, always wooing."[83] Collins reports matter-of-factly, in a salaciously suggestive footnote, that "it is stated upon certain authority, how good I don't know, that the brave Mr. Riel rejoices in the possession of three wives."[84]

Although the author does express some (qualified) sympathy for "half-breeds," he has no patience for their political aims and organized militancy. Collins describes Riel's political aspirations as self-interested sham: "The plain object of this plausible disturber was cash. The lazy rascal had failed to earn a livelihood among the half-breeds of Montana and now was resolved to get some help from the Dominion Treasury."[85] We are told: "[Riel] did not care a fig for the condition of 'his people'! but like the long-winged petrel, he [was] a bad weather bird, and here was his opportunity. He went abroad among the people, fomenting the discord, and assuring them that if all other means failed they would obtain their rights by rising against the authorities."[86] Although he is distrusted by the English and Scottish settlers, the author adds, "I need hardly say that most of the Irish settlers were heart and soul with Riel. It was not that they had any particular grievance to resent, or any grievance at all for that matter. It was as natural for them to rise in revolt, since the rising meant resistance to lawful authority, as it is for the little duck first cast into the pond, to swim."[87]

We will not try to disentangle all the ethnic, political, and moral stereotyping inherent in this kind of popular treatment laced with half-truths. One may quibble over details, but Riel's Robespierre-like image as rabble-rouser and troublemaker, despite some basis in fact, was largely fuelled by party prejudice and age-old ethnic and religious divisions. With the passing of time, these sociological factors come more easily into focus, although even later debate over Riel is not spared the echoes of such partisan thinking. Clément Chartier, president of the Métis National Council, writes,

> What [the usual account of Canadian] history does not adequately portray is the political manipulation that lay behind the trial:

> Canada's government ensured the proceedings would take place in Regina, where it was guaranteed six English Protestant jurors and a part-time magistrate (who was also a political advisor to the government) – as opposed to Winnipeg, where there would be a regular jury of 12 that most likely would have included some Métis or Frenchmen, presided over by a qualified Queen's Bench judge.[88]

We can assume that the jury at the trial did their duty as best they could. Still, to have Riel's case tried before an all-white, Protestant, anglophone jury made his attempt to clear his name and win acquittal well-nigh impossible.

Socrates may have been considered a minor irritation and even a nefarious influence by some Athenians, but he was still seen as an insider; he did not claim to be a rebel, a self-proclaimed destroyer of Athens. Riel had a bad reputation, at least among a large segment of the population, but he was also an outsider, a self-proclaimed rebel, the leader of an insurrection. And he could not bring himself to accept the insanity plea constructed by his own lawyers. But he had to say something. So he elaborated a just war argument.

Riel argues that his cause was just, that the Canadian government had mistreated all the people of Saskatchewan, and that he had acted in self-defence. He explains his motives: "When I came into the North-West in July, the first of July 1884, I found the Indians suffering. I found the half-breeds eating the rotten pork of the Hudson Bay Company and getting sick and weak every day. Although a half-breed, and having no pretension to help the whites, I also paid attention to them. I saw they were deprived of responsible government, I saw that they were deprived of their public liberties."[89] According to Riel, then, no one in Saskatchewan was being adequately governed or represented: not the Indians, not the Métis, not the whites. So he claims that the insurrection was a just war of sorts, regrettably forced on him and his men by drastic circumstances.

One could see Riel's strategy in terms of the standard defence of necessity. He was obliged to act, he had no choice, so he defended himself and his people. Riel uses language of biblical proportions: "The Ministers of an insane and irresponsible Government and its little one – the North-West Council – made up their minds to answer my petitions by surrounding me slyly and by attempting to jump upon me suddenly and upon my people in the Saskatchewan. Happily when they appeared and showed their teeth to devour, I was ready: that is what is called my crime of high treason ... Oh, my good jurors, in the name of Jesus Christ ... they have tried to tear me to pieces."[90] Compare Riel's phrasing to what we read, for example, in different psalms.[91] Psalm 7: "Lest they tear my heart to pieces like a lion."[92] Psalm 35: "Withdraw my soul ... from young lions ... They have opened wide their jaws against me."[93] Psalm 22: "The powerful ... have surrounded me; they open their jaws wide against me like a roaring lion tearing something to pieces."[94] And so on.

Riel's own defence against the charge of treason is surprisingly brief. He is mostly interested in rehabilitating his reputation and spends considerable time discussing personal and religious matters. But there is another factor that comes into play. He seems to believe that everyone knows that the Métis were mistreated and that his cause was just. At the end of his own defence to the jury, Riel suggests a cumbersome chiasmus: "I have acted reasonably and in self-defence, while the Government, my accuser, being irresponsible, and consequently insane, cannot but have acted wrong, and if high treason there is, it must be on its side and not on my part."[95] To paraphrase his idea more simply: *You charge me with high treason; I charge you with high treason.* Hardly reconciliatory words from someone facing the death penalty!

Unlike Socrates, Riel had to prove that he was justified in overthrowing the government by direct, violent means. This was an extra burden he had to bear, but it was too much. As his lawyers correctly believed, his best hope of acquittal, given the circumstances, was through an insanity plea. But Riel was ill-disposed to such a plea for reasons discussed next. One might say that Riel's insanity prevented him from realizing how mentally ill he really was. As the accused himself quipped during the trial: "If I am insane, of course I don't know it, it is a property of insanity to be unable to know it."[96] These are poignant, prophetic words and sadly true.

Riel's Mental State: Socrates Gone Mad?

According to Diogenes Laertius, Plato described the Cynic philosopher Diogenes (whom later generations pictured as living in a barrel) as "a Socrates gone mad."[97] We might well wonder whether Riel, to the extent that he bore any resemblance to Socrates, could be described in similar terms.

I will take it as a given that the level-headed Socrates, whatever his religious proclivities, was not mentally ill. The same cannot be said for Riel. Indeed, the most plausible view is that Riel suffered from mental illness from the time he was first admitted to the St Jean de Dieu asylum at Longue Pointe in 1876 right up until the end of his life. Psychiatrist E.R. Markson summarizes his condition during the years just before the North-West Rebellion: "[Riel] was a victim of his own prophetic and megalomaniac zeal, and progressive deterioration of his mental state ensued. His grandiosity and religious fervour increased. His heterodoxy became extreme and he reviled and abused the local clergy, claiming his own supremacy and his appointment by God to re-organize the spiritual and temporal affairs of the territory."[98]

In Riel's case, religion and madness intertwine in intricate ways. This is not to say that Riel was insincere in his beliefs – he was absolutely sincere. Nor is this to say that all religious belief is deluded. But perhaps, because religion, in moving toward the transcendental, lends itself to grandiose-sounding claims, Riel found in religion a way to disguise and nurse a hysterical, self-absorbed mania that drove him to the kind of public grandstanding and radical political intervention that eventually led to the scaffold.

In the preface to his excellent book on Riel's religion, Flanagan writes: "Louis Riel's religion was a form of millenarianism and can be understood through the use of concepts which have been developed in the study of that field. The major competing theory ... is the psychiatric thesis that he was mentally ill and that his religious ideas were symptoms of his disease."[99] But the insanity thesis and the millenarianism thesis are not mutually exclusive. Riel's politico-religious worldview seems to have been most clearly influenced by three different sources: traditional Roman Catholic piety, a form of millenarianism, and his mental illness. Millenarianism plays a role but not to the exclusion of the two other factors.

When Riel took over the parish church at Batoche in the North-West Rebellion, he shouted to onlookers: "Rome is fallen."[100] In this kind of moment, Riel does not resemble the modest, self-deprecating Socrates as much as he does the German philosopher Frederick Nietzsche, who was coincidentally going mad at just about this time. (Nietzsche suffered a complete mental breakdown in 1888, never to recover.) Despite obvious religious differences, both men nursed a tremendously exaggerated sense of their own role in world history, a propensity for eclectic, oracular-sounding pronouncements, and a gift for fiery, war-like rhetoric that seems to capture something of the millenarian mood.

Riel does not hesitate to compare himself to the founders of Islam, Judaism, and Christianity. At one point, he confides to himself in a notebook: "God has given me a genius greater than Mahomet and I can found a religion and empires more famous than his, even if I wanted to work like him, without putting myself under the discipline of divine authorization. But that glory does not tempt me."[101] Riel also saw himself, in Flanagan's words, as "the earthly messiah whom the Jews expected."[102] His tendency to identify himself with important Judaic figures extended not only to King David, as already mentioned, but also to other equally important figures such as Moses, Elias, and Daniel. Of course, Riel describes his prophetic mission in traditional Roman Catholic terms as well. At Batoche, he has the provisional government (called the Exovedate, "from out of the flock") pass a motion:

> That the French-Canadian Métis Exovedate recognize Louis 'David' Riel as a prophet of Jesus Christ the Son of God and the only Redeemer of the world: as a prophet at the feet of Mary Immaculate under the powerful and most favourable protection of the Virgin Mother of Christ; as a prophet under the visible and most comforting protection of St. Joseph, the chosen patron of the Métis, the patron of the universal church; as a prophet humbly imitating in many ways St. John the Baptist, the glorious patron of the French Canadians and the French-Canadian Métis.[103]

Riel's individual religious beliefs, which encompassed bizarre extremes, were not per se insane. Most of his "heresies" are rather pedestrian, theological commonplaces and do not derive from any original or systematic theological study. Riel is a theological dabbler. He dabbles – a little bit of this, a little bit of that – in line with ideas gathered from an eclectic range of sources. At different moments in his career, he decides that the Sabbath should be changed from Sunday to Saturday, that all are to be saved without exception (universalism), that hell cannot endure forever, that the pope is not infallible, that the papacy is to be moved to St Boniface, Manitoba, that everyone is a priest, that the Virgin Mary must be addressed as "the Mother of the

Son of God," not (as in the traditional formula) as "the Mother of God," that circumcision, polygamy, and even incest are to be tolerated because the Jews had practised them, that secular government must be placed in the hands of the episcopate, and so on. Riel's fluid state of belief betrays the source of his speculations in his own religious enthusiasm. At times, his theological efforts seem likely motivated by political and social considerations – for example, by the desire to make a political alliance with Protestants and Jews.

Flanagan vividly describes one scene that fairly captures the essence of Riel's excited state of mind:

> Riel came to Prince Albert on 3 December [1884] and had another quarrel with [Father Alexis] André. Soon after, he went to St. Laurent, where the missionaries of the district were gathered for their monthly retreat. When Riel vigorously attacked them, André declared that the Oblates would treat him as an enemy from now on and openly oppose the Métis. At that Riel burst into tears, fell to his knees, and begged forgiveness. The priests led him to the chapel, where he promised before the tabernacle never to lead a revolt against the civil authorities. After this melodramatic scene, the fathers seriously debated whether Riel should be barred from the sacraments as a heretic. They decided against such drastic action, on the grounds, as [Father Vital] Fourmand put it, that his extravagant words and wrathful outbursts were due "to the spiritual ordeals and misfortunes of his part, which were enough ... to put his unstable mind into a state akin to madness, where he was not really responsible for what he said."[104]

Riel's religious hysteria, his megalomania, or in the language of the time, his "ambituous mania,"[105] must be understood within the context of the Roman Catholic faith that he grew up in and returned to eventually. (Remember that he had been a student in a seminary.) Seen from this Catholic perspective, which emphasizes submission to legitimate spiritual authority, his later claims are nothing short of preposterous. Stanley describes a revealing incident: "While he was awaiting trial, Riel received a visit from his old friend and associate of 1869/70, the Abbé Georges Dugast, who had been sent by Archbishop [Alexandre-Antonin] Taché to see him. The two men talked together amicably in the presence of several policemen, and Dugast returned to St. Boniface convinced that the former Métis leader was as 'mad as a hatter.'"[106] Although Riel recanted his heresies (formally, in writing) and rejoined the Roman Church before he died, it was an uncomfortable reconciliation in that Riel, however sincere and submissive to the Holy Mother Church, could not give up the idea of his central role in salvation history. A creative, exuberant, even feverish mysticism continued to express itself in all sorts of extravagant ways. Still, his last written words were, "I die Catholic and in the only true faith."[107]

Even in these matters, the comparison to Socrates is instructive. Socrates is a religious man but does not try to found his own religion; he does not set himself up as leader of a new cult. He claims supernatural guidance, but he operates within conventional limits. Most important, he respects tradition and established religious authority. In the *Euthyphro* dialogue (which takes place just before Socrates's trial), Plato suggests an implicit comparison between the young Euthyphro, who is a religious fanatic, and Socrates, the reasonable believer. The dialogue is not, as many suppose, a comment on the irrationality of religious belief but a more narrow criticism of religious fanaticism. When Euthyphro tells Socrates about his plan to bring his father to trial for murder on flimsy charges, Socrates replies, "Good heavens, Euthyphro! and is your knowledge of religion and of things pious and impious so very exact, that ... you are not afraid lest you too may be doing an impious thing?"[108]

Riel, of course, resembles Euthyphro, not Socrates. Euthyphro is brimming over with spiritual self-confidence and religious innovations. Socrates, in flagrant opposition, is very cautious on religious matters: he recoils from the extraordinary and the untraditional; he recognizes his limits. When Euthyphro assures Socrates that he knows what he is talking about in matters related to the gods, Socrates (ironically) exclaims, "Rare friend! I think that I cannot do better than be your disciple. Then before the trial with Meletus comes on I shall challenge him, and say that I have always had a great interest in religious questions, and now, as he charges me with rash imaginations and innovations in religion, [I will say that] I have become your disciple."[109] Socrates does not make any claim to special knowledge about esoteric religious matters; indeed, he is willing to submit to anyone who has such knowledge. (He discovers, of course, that Euthyphro is just an enthusiastic young man; he has no genuine knowledge at all.)

It is odd that the very thing that makes Riel an ideal liberal icon – his fierce repudiation of established authority (the Che Guevara of Canada?) – is, on a religious level, the most conspicuous expression of his delusions. It is not so strange that mystically minded religious people experience odd, incongruous visions and ecstasies. This is rather commonplace. What is telling about Riel's mental mania is his claim to a special, privileged, detailed knowledge of occult things, his usurpation of religious authority, and his unbounded, Euthyphro-like confidence in his own inerrant judgment. This clashes with the humility required in mainstream religious practice, which requires submission to a higher authority, one that is larger than one individual, that has lasted through the ages, and that operates within widely accepted limits. It is this faithful acceptance of the judgments, statements, and ritual usages of a tradition that, in more ordinary cases, prevents the excesses that clouded Riel's judgment.

Socrates presents a conspicuous foil to Riel on these issues. Socrates does not display any false modesty about his own virtue, but he is smart enough to develop a philosophical doctrine that can explain away his claims to special inspiration as a form of humility. After the oracle at Delphi proclaimed that there was no one wiser than himself, Socrates came to view this assertion as an oblique comment on his own self-evident ignorance. He describes how he pondered the oracle's utterance:

> I said to myself, What can the god mean? ... for I know that I have no wisdom, small or great. What can he mean when he says that I am the wisest of men? And yet he is a god and cannot lie; that would be against his nature ... Accordingly I went to one who had the reputation of wisdom, and observed to him ... and the result was as follows: When I began to talk with him, I could not help thinking that he was not really wise, although he was thought wise by many, and wiser still by himself ... The truth is, O men of Athens, that God only is wise; and in this oracle he means to say that the wisdom of men is little or nothing; he is not speaking of Socrates, he is only using my name as an illustration, as if he said, He, O men, is the wisest, who, like Socrates, knows that his wisdom is in truth worth nothing.[110]

To be sure, some scholars dispute Socrates's sincerity. But Socrates is able to construct a defence that logically revolves around a kind of self-deprecation: he was given his divine mission because he was the only Athenian who understood the full extent of human ignorance. Using this clever device, he is able to secure his epistemologically privileged status in a way that simultaneously precludes any extravagant claims about the extent of his own wisdom and spiritual insight. This approach to his philosophical vocation ties in with his persistent claim that he never taught anyone.[111]

Riel was simply incapable of invoking, in any effective way, this kind of stratagem. Doubtless, he strenuously tries to adopt a pose of spiritual humility. As he explains at his trial, "It is probably for that reason I have found the word '*exovede*' [used to describe members of his provisional government at Batoche], I prefer to be called one of the flock; I am no more than you are, I am simply one of the flock, equal to the rest."[112] But this suggestion of spiritual egalitarianism is disingenuous. Riel was driven by his mania to ascribe to himself powers of epistemological certainty in religious matters and a superhuman authority. Self-glorification permeates all of his behaviour. This was a well-known feature of his personality and contributed to his bad reputation. When General Frederick Middleton described him as a man "imbued with a strong, morbid, religious feeling mingled with intense personal vanity," he was inaccurately describing, in moral terms, a narcissistic disorder that had reached crisis proportions.[113]

Although Riel is mentally ill, he does everything he can at his trial to prove that he is not insane. This adds another wrinkle to his speech to the jury that sets it apart from Socrates's more straightforward affirmation of his own ignorance. Flanagan writes, "Riel's presentation at his trial of his religious views and of his utopian political schemes cannot be taken as an accurate expression of his thinking. He did not directly falsify his ideas, but he selected and emphasized according to what he thought an English and Protestant audience would want to hear."[114]

Riel's words to the jury were obviously counterproductive – indeed, almost suicidal – from a legal point of view. They further prove his illogical state of mind. Briefly consider three aspects of his defence. First, Riel individually thanks Crown witnesses because they have undermined the insanity defence his own lawyers have constructed! He informs the jury,

> To-day when I saw the glorious General Middleton bearing testimony that he thought I was not insane, and when Captain Young

> proved that I am not insane, I felt that God was blessing me, and blotting away from my name the blot resting upon my reputation on account of having been in the lunatic asylum of my good friend Dr Roy. I have been in an asylum, but I thank the lawyers for the Crown who destroyed the testimony of ... Dr Roy, because I have always believed that I was put in the asylum without reason ... that is a blessing too in that way.[115]

Viewed from a legal perspective, thanking the Crown, in gushing terms, for destroying the basis of one's own insanity defence is nothing less than high absurdity. Consider an apt analogy. It is as if those on a sport's team were to effusively thank those on the opposing team for scoring points against them so they could lose the game! Riel wants, at all costs, to be remembered as a seminal religious figure, not as a madman. The possibility that he might be returned to an insane asylum is disturbing enough, but he is desperate, above all, to preserve what he regards as his rightful place in Canadian political and religious history. His miscalculation of his precarious legal situation, his befuddled idea that he could win his case through some sincere appeal to social justice, and his misunderstanding of his own mental health point to his profoundly unbalanced state of mind.

Second, Riel thanks the court for not putting his private thoughts and spiritual reflections on display. He declares, "The learned lawyers for the Crown have produced all the papers and scribbling that was under their hands. I thank them for not having brought out those papers which are so particular to myself, though as soon as they saw what they were they should not have looked at them ... I told Parenteau, one of the prisoners, to put all my books under ground ... I say I thank the learned lawyers for the Crown for having reserved so many things."[116] Again, he tells the jury, "Captain Deane has seen my papers ... I have written these documents and they are in the hands of those whom I trust. I do not want to make them public during my trial."[117] Riel wanted to hide his private revelations, the very ones that would have demonstrated the sweeping magnitude of his neurosis, from the jury. One can understand why, but this was tantamount to withholding evidence that could have produced an acquittal. So, here again, Riel is expressing gratitude for a development that arguably led to his conviction.

Third, Riel continually downplays the extent of his religious beliefs in his descriptions to the jury. For example, he assures them that he didn't want to be the pope – "To set myself up as Pope, no, no."[118] That he "did not wish to force [his] views" on others.[119] That he wished to leave Rome only "inasmuch as it is the cause of division between Catholics and Protestants."[120] That he did not have extraordinary prophetic powers. That his claims to see into the future could best be understood in terms of a familiar Métis tradition of reading spiritual meanings into ordinary things. Indeed, he assures them, "we all see into the future more or less."[121]

Riel had to explode the defence his lawyers had constructed in order to save his misplaced sense of spiritual mission. If he was shown to be insane, his spiritual mission would be worthless. Not surprisingly, he remained faithful to what he saw as his religious vocation. This was, paradoxically, both admirable and deluded. Socrates, in sharp contrast, did not have to labour under such constraints. He was clear-thinking enough to deprecate his own importance and to use it as a way to defend a much more modest claim to a divinely inspired mission.

EPILOGUE

This comparison of the trials of Riel and Socrates pushes two issues into the foreground for further consideration: one has to do with Louis Riel; the other involves a more general point about historical methodology.

In Canadian political circles, there is perennial talk of granting Riel some kind of posthumous pardon. Flanagan, in a carefully worded and even courageous statement, has argued firmly against any such decision: "There is no reason to believe that his trial was unfair."[122] Given the composition of the jury, however, a guilty verdict seems to have been a foregone conclusion. It is highly improbable that a jury composed of Irish Catholics, French Canadians, or Métis would have come to a similar conclusion.

But even if we accept Flanagan's claim that Riel received as fair a trial as could have been hoped for given the historical circumstances, can the guilty verdict be justified, in hindsight, in light of what we now know about the case? Imagine that we could use newly discovered DNA techniques to show that someone who was hanged for murder fifty years ago was, in fact, innocent. That the jury at the time did its level best does not mean that the innocent person does not *now* deserve a pardon. There is then an argument to be made that it would be unfair and irresponsible not to formally acknowledge that a mistake was made in Riel's trial.

I will not take up the issue here of whether or not the rebellious Riel was engaged in a just war. Whether he was or was not, Riel was a mentally ill man who was obliged, by his own conscience, to do what he did. He felt himself pushed forward to a grandiose fate by a divine will that he could not, in good conscience, resist. As he himself is reported to have said, "I have a task to accomplish by reason of a divine vocation."[123] It is, admittedly, a complicated case. In some sense, Riel knew the difference between right and wrong and understood the nature of his actions (the M'Naghten Rules), but in another sense, he clearly did not.[124] In his own understanding, he was obeying God – a right (and reasonable) thing to do. He was, in effect, emulating Socrates. Modern secular approaches fail us here. We cannot enter into Riel's understanding of the nature of his own actions without taking into account their religious dimension, but it was precisely this religious dimension that was the source and expression of his insanity.

If Riel was, in some sense at least, a "Socrates gone mad," he cannot be properly blamed for his own mental illness. Craziness may be dangerous but it is not a crime. A civilized society does not visit capital punishment on the mentally ill. It does not execute people because they have gone mad. In the case of Riel, the stigma attached to mental illness has, no doubt, precluded proper discussion of these issues. Those who view Riel as a hero are naturally reluctant to stress his mental instability insomuch as it seems to detract from the cause he supported. But this is a mistake. One might, just as easily, admire how much Riel did accomplish in the political arena in spite of his delusions and mental instability.

I have compared the trial appearances and the careers of Louis Riel and Socrates. Riel is not a Canadian Socrates, but there are enough resemblances and differences to make the comparison a useful tool for appreciating the man and his actions. Ethical investigation does not have to end in simplistic judgments. Riel was a gifted and charismatic individual. Although he was not a profound thinker or rigorous theologian, he had a creative mind and was capable of a sometimes extravagant eloquence. Mostly, he was a man of action. Despite his foibles and flaws, he was a proud and noble individual who set out to do good. But he was also mentally ill. His religious megalomania turned his good impulses in the direction of inevitable disaster.

Despite the usual protestations about the impossibility of transcultural comparisons across distant historical epochs, human nature remains more or less the same throughout the ages. Historical individuals who find themselves in similar situations often respond in predictably similar ways. There are familiar enough tropes that can be usefully applied when we attempt to understand the precise motivations and behaviour of figures from widely divergent eras and backgrounds. History does not happen in any old way. Human nature is not infinitely malleable, despite what a certain fashionable relativism would have us believe. If we hope to make objective sense of history, we need to take a certain distance from localized concerns that sometimes cloud our view of a figure like Louis Riel. One way of doing this is to view him in the light of a classical figure like Socrates.

NOTES

1 There are some differences among ancient sources as to what actually happened at Socrates's trial. In this chapter, I generally rely on Plato's account while, of course, keeping in mind that what posterity has been left with is largely Socrates *as he was viewed through Plato's eyes*.

2 Classical scholars are generally agreed that Plato's *Apology* dates from an early stage in his career and is therefore more reliable on matters of fact than later works, which present more elaborate philosophical content.

3 See George F.G. Stanley, ed., *The Collected Writings of Louis Riel/Les écrits complets de Louis Riel*, 5 vols (Edmonton: University of Alberta Press, 1985); Thomas Flanagan, ed., *The Diaries of Louis Riel* (Edmonton: Hurtig, 1976); Gilles Martel, Glen Campbell, and Thomas Flanagan, eds, *Poésies de jeunesse* (St Boniface, MB: Éditions du Blé, 1977); Louis Riel, *Poésies religieuses et politiques* (Montreal: Imprimerie de l'Étendard, 1886).

4 George F.G. Stanley, *Louis Riel* (Toronto: McGraw-Hill Ryerson, 1963), 358.

5 Socrates's associate Critias, an uncle of Plato, was a leading member of the violent regime of the Thirty Tyrants.

6 Plato, *Apology*, trans. Benjamin Jowett, 17b7–d. (Abundant versions of Jowett's still-masterful translations are available in print and online. Nothing said here hinges on precise matters of philological dispute.)
7 Louis Riel, "Address to the Jury," in Desmond Morton, ed., *The Queen v Louis Riel: Canada's Greatest State Trial* (hereafter Q v LR) (Toronto: University of Toronto Press, 1974), 316, 324.
8 Stanley, *Louis Riel*, 356.
9 Cited in William McCartney Davidson, *The Life and Times of Louis Riel* (Calgary: Albertan Printers, 1952), 105.
10 Plato, *Apology*, 19a8–b.
11 Ibid., 18b–b3.
12 Riel, "Jury," Q v LR, 318.
13 Plato, *Phaedo*, trans. Benjamin Jowett, 62b.
14 See, for example, ibid., 80b–c ff., 81a ff., 84b–c, 107d ff.
15 Ibid., 60d.
16 Ibid., 118a.
17 Ibid., 117b.
18 Plato, *Apology*, 35d7–e.
19 Ibid., 31d–5.
20 Riel, "Jury," Q v LR, 320.
21 Plato, *Apology*, 40a–c5.
22 Louis Riel, "Address to the Court," Q v LR, 350–1.
23 Plato, *Apology*, 21a–b.
24 Riel, "Jury," Q v LR, 314–16.
25 Plato, *Apology*, 33c5–d.
26 Ibid., 28d5–30c.
27 Riel, "Jury," Q v LR, 322; Riel, "Court," Q v LR, 370. See Thomas Flanagan, *Louis 'David' Riel: 'Prophet of the New World,'* 1st ed. (Toronto: University of Toronto Press, 1979).
28 Thomas Flanagan, "Louis Riel's Name 'David,'" in *Riel and the Métis*, ed. A.S. Lussier, 48–65 (Winnipeg: Manitoba Métis Federation Press, 1979).
29 Flanagan, "Louis Riel's Name," 57.
30 Flanagan, ed., *Diaries*, 352.
31 Cited in Flanagan, *Louis 'David' Riel*, 116.
32 Cited in ibid., 68.
33 Plato, *Crito*, trans. Benjamin Jowett, 52b–c8.
34 Riel, "Jury," Q v LR, 313. One wonders whether Riel, knowing who was in the jury box, could have thought that this kind of emphasis would have appealed to a work-oriented Protestant mindset.

35 Riel, "Jury," *Q V LR*, 319.
36 Ibid.
37 Ibid., 322. He continues, "Is it practical that you be acknowledged as a prophet? It is practical to say it."
38 Ibid.
39 Ibid., 323.
40 At least, this is how Plato depicts things; we might suspect that Socrates was really in favour of aristocracy.
41 Plato, *Apology*, 31c5–32a3.
42 Ibid., 31a7–c3.
43 Ibid., 23c.
44 Riel, "Jury," *Q V LR*, 324.
45 Plato, *Apology*, 19a7.
46 See Plato, *Crito*, 52d, 54d. This text may have been part of a larger strategy by Plato to rehabilitate the posthumous reputation of Socrates.
47 Cited in George F.G. Stanley, *Louis Riel: Patriot or Rebel?* (Ottawa: Canadian Historical Association, 1979), 23.
48 Plato, *Apology*, 32c5–e.
49 Riel, "Court," *Q V LR*, 369.
50 Anonymous (Joseph Edmund Collins), *The Story of Louis Riel: The Rebel Chief* (1885; facsimile ed., Toronto: Coles Canadianna Collection, 1970).
51 Ibid., 5.
52 Ibid., 55.
53 Ibid., 76.
54 Ibid., 76.
55 Ibid., 77.
56 Ibid., 87.
57 Ibid., 88.
58 Ibid., 90.
59 Ibid., 106.
60 Ibid., 114.
61 Ibid., 117.
62 Ibid., 125.
63 Ibid., 126.
64 Ibid.
65 Ibid., 129.
66 Ibid.
67 Ibid., 130.
68 Ibid.

69 Ibid., 136.
70 Ibid., 176.
71 Ibid., 59.
72 Ibid., 48.
73 Ibid.
74 Ibid., 115.
75 Ibid., 63.
76 Ibid., 126.
77 Ibid., 86.
78 Ibid.
79 Ibid., 126.
80 Ibid., 87.
81 Ibid., 117.
82 Ibid., 87.
83 Ibid., 132.
84 Ibid., 131.
85 Ibid., 136.
86 Ibid., 135–6.
87 Ibid., 90.
88 Clément Chartier, "Louis Riel Needs No Pardon," Métis Nation of Alberta website, 3 May 2004, http://www.albertametis.ca/MNAHome/News-Archive/Louis-Riel-Needs-No-Pardon.aspx (accessed 7 October 2013).
89 Riel, "Jury," *Q v LR*, 312.
90 Ibid., 324.
91 These are all taken from the French edition of the Darby Bible, http://darbybible.com (accessed 7 October 2013). Translations into English are mine.
92 Psalm 7: 1–2: "Éternel, mon Dieu! ... sauve-moi ... de peur qu'il ne déchire mon âme comme un lion qui met en pièces."
93 Psalm 35: 17, 21: "Seigneur! ... Retire mon âme ... des jeunes lions ... Ils ont élargi leur bouche contre moi."
94 Psalm 22: 11–13: "Des puissants ... m'ont entouré; ils ouvrent leur gueule contre moi, comme un lion déchirant et rugissant."
95 Riel, "Jury," *Q v LR*, 324.
96 Ibid., 319
97 Diogenes Laertius, *Lives of the Eminent Philosophers*, trans. R.D. Hicks (1925; reprint, Cambridge, MA: Harvard University Press, 1972), bk 6, ch. 54.
98 E.R. Markson, "The Life and Death of Louis Riel: A Study in Forensic Psychiatry," part 1—A, "Psychoanalytic Commentary," *Canadian Psychiatric Asso-*

ciation Journal 10, no. 4 (August 1965), reprinted in *Louis Riel: Rebel of the Western Frontier or Victim of Politics and Prejudice?* ed. H. Bowsfield, 177–81 (Toronto: Copp Clark, 1969), 178.

99 Flanagan, *Louis 'David' Riel*, vii–viii.

100 Stanley, *Louis Riel*, 305.

101 Cited in Flanagan, *Louis 'David' Riel*, 115.

102 Ibid., 185

103 Ibid., 139.

104 Ibid., 129.

105 A term used by Dr François Roy, of the Beauport Asylum.

106 Stanley, *Louis Riel*, 346; the French is "un fou à double platine."

107 Ibid., 370.

108 Plato, *Euthyphro*, trans. Benjamin Jowett, 4e2–5.

109 Ibid., 5a. There is some pleasant irony and witty inversion expressed in these lines.

110 Plato, *Apology*, 21b2–23b3.

111 See ibid., 21b–23b.

112 Riel, "Jury," *Q V LR*, 322.

113 Cited in Stanley, *Louis Riel*, 339.

114 Flanagan, *Louis 'David' Riel*, 161.

115 Riel, "Jury," *Q V LR*, 316.

116 Ibid., 313.

117 Ibid., 321.

118 Ibid., 322.

119 Ibid., 319.

120 Ibid.

121 Ibid., 320.

122 Thomas Flanagan, *Riel and the Rebellion: 1885 Reconsidered*, 2nd ed. (Toronto: University of Toronto Press, 2000), 173. See also the final chapter, "Exoneration?" 169–89.

123 Cited in Flanagan, *Louis 'David' Riel*, 129.

124 A lot more could be said on this score.

Contributors

BENJAMIN AUTHERS is an Australian Research Council Laureate Postdoctoral Fellow at the Centre for International Governance and Justice, Australian National University. Focusing on the interrelations between law and literature, his research has examined ideas of responsibility, law enforcement, and nation building in nineteenth-century Canada and the legal and cultural work of rights in Canadian and international contexts. His writing has been published in journals such as the *University of Toronto Quarterly*, *Law Text Culture*, and the *Journal of the Association for the Study of Australian Literature*.

MAURICE CHARLAND is a professor of communication studies at Concordia University in Montreal. He is the author of two foundational essays in rhetoric and communication studies that address the question of identity: "Technological Nationalism" (*Canadian Journal of Political and Social Theory*, 1986) and "Constitutive Rhetoric: The Case of the *Peuple Québécois*" (*Quarterly Journal of Speech*, 1987), the latter of which received the Charles H. Woolbert Research Award in 2000. His concern with Louis Riel stems from the book that he co-authored with Michael Dorland of Carleton University entitled *Law, Rhetoric, and Irony in the Formation of Canadian Civil Culture* (2002), which was awarded the 2002 G.J. Robinson Prize for best Canadian book of the year in communication studies.

THOMAS FLANAGAN is a retired professor of political science at the University of Calgary. He is a co-editor of *The Collected Writings of Louis Riel* (5 vols, 1985) and the author of *Louis 'David' Riel: 'Prophet of the New World'* (1979), *Riel and Rebellion: 1885 Reconsidered* (1983), and *Metis Lands in Manitoba* (1991). He is a fellow of the Royal Society of Canada.

LOUIS GROARKE grew up in Edmonton and Calgary and has retained an interest in Western history and issues. He has an undergraduate degree from Colorado State University and a doctorate in philosophy from the University of Waterloo. He now teaches philosophy at St Francis Xavier University in Nova Scotia. His publications include a history of Western ethics, *Moral Reasoning: Rediscovering The Ethical Tradition* (2011); a historical study, *An Aristotelian Account of Induction* (2009); and a work on the concepts of individual and political freedom, *The Good Rebel* (2002). He has published in various philosophy journals. One previous paper deals with popular perceptions of the French revolutionary martyr Jean-Paul Marat.

PAUL GROARKE is a lawyer and an academic who has taught the legal side of the criminology curriculum. He has practised law for almost thirty years, initially as a defence lawyer in Calgary and more recently in an adjudicative capacity as a member of the Canadian Human Rights Tribunal. He also earned a doctorate in philosophy from the University of Waterloo in 2000 and has taught philosophy of law. Dr Groarke has an interest in historical cases and has also written on the trial of Socrates. His forthcoming book, *Legal Theories: An Historical Introduction to Philosophy of Law and Jurisprudence*, provides a historical account of the common law tradition.

HANS V. HANSEN's interest in Louis Riel stems from his days as a master's student at the University of Manitoba. He went on to earn a doctorate in philosophy from Wayne State University and now teaches at the University of Windsor. His research interests include the study of argumentation, both its history and normative models, and some of his papers have appeared in *Synthese*, *Logique et Analyse*, *Philosophy and Rhetoric*, and *Argumentation*, among other journals. He is a co-editor of the journal *Informal Logic* and the editor of the present volume.

DESMOND MORTON was born in Calgary to a military family and spent his adolescent years in schools in Regina and Winnipeg. His great grandfather, General Sir William Otter, played a significant role in the North-West campaign of 1885, and among Morton's earliest books are *The Last War Drum* (1972), *A Canadian General* (1974), and *The Queen v Louis Riel* (1974). He has published more than forty books on Canada's social, political, and military history. He was principal of the University of Toronto's Mississauga campus and founding director of the McGill Institute for the Study of Canada. He retired as Hiram Mills Emeritus Professor of History at McGill University in 2007 and has been working on *Une histoire militaire des canadiens-français*.

NICOLE C. O'BYRNE first became interested in Louis Riel during her master's research at McGill University on the history of the British North America Act of 1930. She continues to explore historical Métis claims and the evolution of Métis policy in Alberta and Saskatchewan through her doctoral work in the Law and Society Program at the University of Victoria. She is currently an assistant professor in the Faculty of Law, University of New Brunswick, where she teaches Aboriginal law.

JENNIFER REID has written extensively on Louis Riel, most recently in her book *Louis Riel and the Creation of Modern Canada* (2008). Her other books include *In Search of Kluskap: A Journey into Mi'kmaw Myth* (2013), *Religion, Writing, and Colonial Resistance: Mathias Carvalho's Louis Riel* (2011), *'Worse Than Beasts': An Anatomy of Melancholy and the Literature of Travel in 17th and 18th Century England* (2005), and *Myth, Symbol, and Colonial Encounter: British and Mi'kmaq in Acadia, 1700–1867* (1995). Reid is a historian of religions who teaches at the University of Maine, Farmington.

KERRY SLOAN is a doctoral candidate in the Law and Society Program at the University of Victoria. Her research focuses on how the perception of Métis history and communities in British Columbia affects Métis rights in the province. A former practitioner of Aboriginal rights law and general litigation in Alberta, she received a master's degree in English from the University of British Columbia, where she studied rhetorical analysis and comparative literature. Louis Riel unsuccessfully courted her great-great-great aunt Louise-Anne Marion.

CHRISTOPHER TINDALE is a professor of philosophy and the director of the Centre for Research in Reasoning, Argumentation, and

Rhetoric (CRRAR) at the University of Windsor. He received his doctorate from the University of Waterloo. Among his key publications are *Good Reasoning Matters*, with Leo Groarke (1989), *Acts of Arguing* (1999), *Rhetorical Argumentation* (2004), *Fallacies and Argument Appraisal* (2007), *Reason's Dark Champions* (2010), and *Einführung in die Informelle Logik* (2011). His interest in Riel's speeches stems from a general interest in the nature and uses of rhetoric.

Index